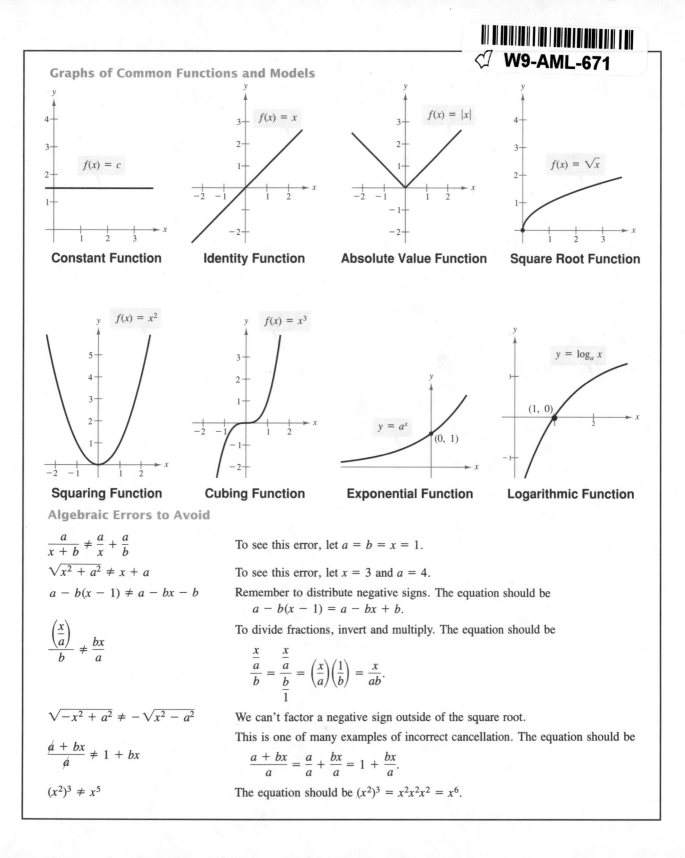

Graphs of Common Functions and Models

Constant Function
$f(x) = c$

Identity Function
$f(x) = x$

Absolute Value Function
$f(x) = |x|$

Square Root Function
$f(x) = \sqrt{x}$

Squaring Function
$f(x) = x^2$

Cubing Function
$f(x) = x^3$

Exponential Function
$y = a^x$
$(0, 1)$

Logarithmic Function
$y = \log_a x$
$(1, 0)$

Algebraic Errors to Avoid

$\dfrac{a}{x + b} \neq \dfrac{a}{x} + \dfrac{a}{b}$

To see this error, let $a = b = x = 1$.

$\sqrt{x^2 + a^2} \neq x + a$

To see this error, let $x = 3$ and $a = 4$.

$a - b(x - 1) \neq a - bx - b$

Remember to distribute negative signs. The equation should be
$$a - b(x - 1) = a - bx + b.$$

$\dfrac{\left(\dfrac{x}{a}\right)}{b} \neq \dfrac{bx}{a}$

To divide fractions, invert and multiply. The equation should be
$$\frac{\dfrac{x}{a}}{b} = \frac{\dfrac{x}{a}}{\dfrac{b}{1}} = \left(\frac{x}{a}\right)\left(\frac{1}{b}\right) = \frac{x}{ab}.$$

$\sqrt{-x^2 + a^2} \neq -\sqrt{x^2 - a^2}$

We can't factor a negative sign outside of the square root.

$\dfrac{a + bx}{a} \neq 1 + bx$

This is one of many examples of incorrect cancellation. The equation should be
$$\frac{a + bx}{a} = \frac{a}{a} + \frac{bx}{a} = 1 + \frac{bx}{a}.$$

$(x^2)^3 \neq x^5$

The equation should be $(x^2)^3 = x^2 x^2 x^2 = x^6$.

College Algebra

Concepts and Models

Instructor's Annotated Edition

College Algebra

Concepts and Models

Second Edition

Roland E. Larson   Robert P. Hostetler

The Pennsylvania State University
The Behrend College

Anne V. Hodgkins

Phoenix College

D. C. Heath and Company
Lexington, Massachusetts   Toronto

Address editorial correspondence to:
D. C. Heath and Company
125 Spring Street
Lexington, MA 02173

Acquisitions Editors: Ann Marie Jones, Charles Hartford
Managing Editor: Catherine B. Cantin
Development Editor: Emily Keaton
Production Editor (art): Rachel D'Angelo Wimberly
Marketing Manager: Christine Hoag
Designer: Henry Rachlin
Photo Researchers: Derek Wing and Billie Porter
Production Coordinator: Lisa Merrill
Composition: Meridian Creative Group
Art: Folium, Inc.; Meridian Creative Group; Patrice Rossi; Illustrious, Inc.
Cover Image: "Untitled (1940–1945)" by Rolph Scarlett. From the collection of Mr. and Mrs. Barney A. Ebsworth
Cover Photo Reseracher: Linda Finigan

Trademark Acknowledgments: TI and CBL are registered trademarks of Texas Instruments, Inc. Casio is a registered trademark of Casio, Inc. Sharp is a registered trademark of Sharp Electronics Corp.

Published simultaneously in Canada.

Printed in the United States of America.

International Standard Book Number: 0-669-39618-4

Library of Congress Catalog Number: 95-75644

10 9 8 7 6 5 4 3 2

Preface

The primary goals of *College Algebra: Concepts and Models,* Second Edition, are to encourage students to develop their understanding of algebra and to show how algebra is a modern modeling language for real-life problems.

New to the Second Edition

In the Second Edition, all text elements were considered for revision, and many new examples, exercises, and applications were added to the text. Following are the major changes in the Second Edition.

Improved Coverage   The Second Edition begins with a Prerequisites chapter, which reviews the fundamental concepts of algebra. All or part of this material may be covered or it can be omitted, offering greater flexibility in designing the course syllabus. Graphing is now introduced in Chapter 2, earlier than in the previous edition. Throughout the text, greater emphasis is given to geometry, collecting and interpreting data and statistics, data analysis, and creating models, as well as to the NCTM Standards and Addenda and the AMATYC Guidelines.

Technology   Recognizing that graphing technology is becoming increasingly available, the Second Edition offers the opportunity to use graphing utilities throughout, without requiring their use. This opportunity is presented through a combination of features, including—at point of use—discovery opportunities that require scientific or graphing calculators (see pages 202 and 435), graphing utility instructions in the text margin (see pages 214 and 295), and clearly labeled exercises that require the use of a graphing utility (see pages 411 and 455).

Group Activities   Each section ends with a Group Activity. This exercise reinforces students' understanding by exploring mathematical concepts in a variety of ways: You Be the Instructor, Extending the Concept, Problem Solving, Exploring with Technology, and Communicating Mathematically. Some Group Activities encourage interpretation of mathematical concepts and results (see pages 50, 183, and 204); some provide opportunities for problem posing and error analysis (see pages 85, 355, 509, and 610); and others reinforce methods of constructing mathematical models, tables, and graphs (see pages 192, 297, 307, 327, 392, and 409). Designed to be completed in class or as homework assignments, all Group Activities give students the opportunity to work cooperatively as they think, talk, and write about mathematics.

Data Analysis/Modeling   Throughout the Second Edition, students are offered many more opportunities to collect and interpret data, make conjectures, and construct mathematical models. Students are exposed to modeling problems with experimental and theoretical probabilities (see the Group Activity on page 69); encouraged to use mathematical models to make predictions from real data (see Exercises 71 and 72 on page 146, the Chapter Project on page 158, and Exercise 51 on page 467); and asked to use curve-fitting techniques to write their own models from data (see Exercises 45–48 on page 220, the Chapter Project on page 312, and Exercises 47–54 on page 455). This edition encourages greater use of charts, tables, scatter plots, and graphs to summarize, analyze, and interpret data.

Applications   To emphasize for students the connection between mathematical concepts and real-world situations, numerous up-to-date, real-life applications are integrated throughout the text. Appearing as examples (see Example 7 on page 215), exercises (see Exercises 42–46 on page 423), group activities (see page 440), and projects (see page 425), these applications offer students frequent opportunities to use and review their problem-solving skills. A wide range of disciplines are represented by the applications, including physics, chemistry, the social sciences, biology, and business. And the career interviews cover areas such as real estate, ecology, construction, and biochemistry.

Connections   In addition to highlighting the connections between algebra and areas outside mathematics through real-world applications, this text also emphasizes the connections between algebra and other branches of mathematics, such as probability (see Exercises 71 and 72 on page 88 and Section 8.6), geometry (see pages 111 and 466), logic (see Math Matters on page 61), and statistics (see Appendix B). Too, many examples and exercises throughout the text reinforce the connections among graphical, numerical, and algebraic representations of important algebraic concepts.

The Second Edition contains many other new features as well, including Discovery, Study Tips, Historical Notes, Mid-Chapter Quizzes, Chapter Summaries, and Chapter Projects. These and other features of the Second Edition are described in greater detail in the following pages.

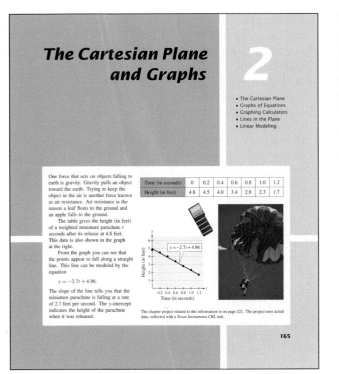

Chapter Opener

Each chapter opens with a look at a real-life application that is explored in depth in the Chapter Project at the end of the chapter. Real data is manipulated using graphical, numerical, and algebraic techniques. In addition, a list of the section titles shows students how the topics fit into the overall development of algebra.

Section Outline

Each section begins with a list of the major topics covered in that section. These topics are also the subsection titles and can be used for easy reference and review by students.

Graphics

Visualization is a critical problem-solving skill. To encourage the development of this ability, the text has over 1000 figures in examples, exercises, and answers to odd-numbered exercises. Included are graphs of equations and functions, geometric figures, displays of statistical information, scatter plots, and numerous screen outputs from graphing technology. All graphs of equations and functions, computer- or calculator-generated for accuracy, are designed to resemble students' actual screen outputs as closely as possible. Graphics are also used to emphasize graphical interpretation, comparison, and estimation.

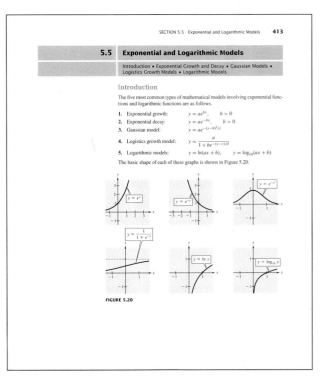

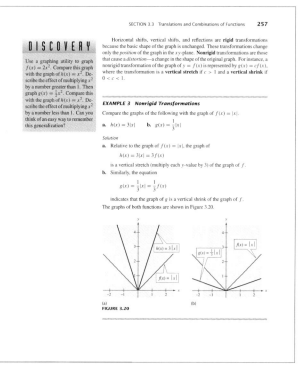

Discovery

Throughout the text, Discovery notes encourage active participation by students, taking advantage of the power of technology (graphing calculators or scientific calculators) to explore mathematical concepts and discover mathematical relationships. Using a variety of approaches, including visualization, verification, pattern recognition, and modeling, students develop an intuitive understanding of theoretical topics.

Examples

Each of the more than 400 text examples was carefully chosen to illustrate a particular mathematical concept, problem-solving approach, or computational technique, and to enhance students' understanding. The examples in the text cover a wide variety of problem types, including theoretical, real-life applications (many with real data), and those requiring the use of graphing technology. Each example is titled for easy reference, and real-life applications are labeled. Many examples include side comments in color, which clarify the steps of the solution.

Graphing Utilities

Instructions for using graphing utilities offer convenient reference for students using graphing technology. Appearing in the margins, the instructions can be easily omitted if desired. Additionally, problems in the Exercise Sets that require a graphing utility have been identified with a graphing calculator icon.

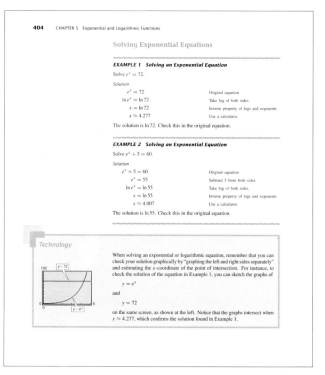

Definitions and Rules

All rules, formulas, theorems, guidelines, properties, definitions, and summaries are highlighted for emphasis. Each is also titled for easy reference.

Study Tips

Study Tips appear in the margin at point of use. They offer students specific, helpful, and insightful suggestions for studying algebra. "How to Study Algebra" in the front of the text outlines a general plan designed to improve student study skills.

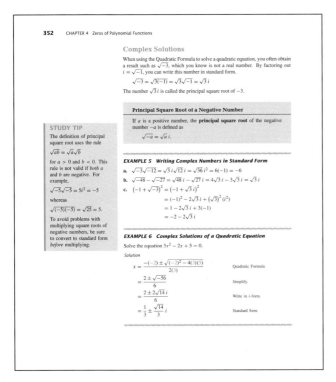

Historical Notes

To help students understand that algebra has a past, historical notes featuring mathematical artifacts or mathematicians and their work are included in each chapter.

---

394 CHAPTER 5 Exponential and Logarithmic Functions

5.3 Properties of Logarithms

Change of Base • Properties of Logarithms •
Rewriting Logarithmic Expressions • Application

Change of Base

Most calculators have only two types of log keys, one for common logarithms (base 10) and one for natural logarithms (base e). Although common logs and natural logs are the most frequently used, you may occasionally need to evaluate logarithms to other bases. To do this, you can use the following *change-of-base formula*.

Change-of-Base Formula

Let a, b, and x be positive real numbers such that $a \neq 1$ and $b \neq 1$. Then $\log_a x$ is given by

$$\log_a x = \frac{\log_b x}{\log_b a}$$

One way to look at the change-of-base formula is that logarithms to base a are simply *constant multiples* of logarithms to base b. The constant multiplier is

$$\frac{1}{\log_b a}.$$

John Napier, a Scottish mathematician, developed logarithms as a way to simplify some of the tedious calculations of his day. Beginning in 1594, Napier worked about 20 years on the invention of logarithms. Napier was only partially successful in his quest to simplify tedious calculations. Nonetheless, the development of logarithms was a step forward and received immediate recognition.

EXAMPLE 1 *Changing Bases Using Common Logarithms*

a. $\log_4 30 = \dfrac{\log_{10} 30}{\log_{10} 4} \approx \dfrac{1.47712}{0.60206} \approx 2.4534$

b. $\log_2 14 = \dfrac{\log_{10} 14}{\log_{10} 2} \approx \dfrac{1.14613}{0.30103} \approx 3.8074$

EXAMPLE 2 *Changing Bases Using Natural Logarithms*

a. $\log_4 30 = \dfrac{\ln 30}{\ln 4} \approx \dfrac{3.40120}{1.38629} \approx 2.4534$

b. $\log_2 14 = \dfrac{\ln 14}{\ln 2} \approx \dfrac{2.63906}{0.693147} \approx 3.8074$

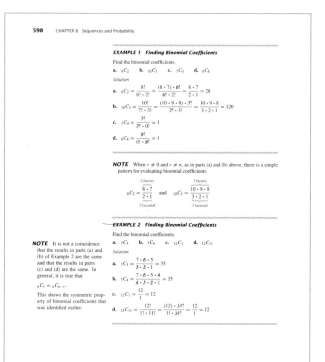

Notes

Notes anticipate students' needs by offering additional insight, pointing out common errors, and describing generalizations.

Applications

Real-life applications are integrated throughout the text in examples and exercises. These applications offer students constant review of problem-solving skills and emphasize the relevance of the mathematics. Many of the applications use recent, real data, and all are titled for reference. Photographs with captions throughout the text also encourage students to see the link between mathematics and real life.

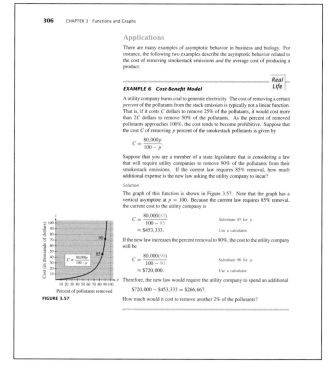

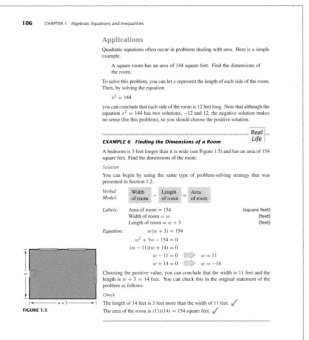

106   CHAPTER 1   Algebraic Equations and Inequalities

Applications

Quadratic equations often occur in problems dealing with area. Here is a simple example.

A square room has an area of 144 square feet. Find the dimensions of the room.

To solve this problem, you can let x represent the length of each side of the room. Then, by solving the equation

$$x^2 = 144$$

you can conclude that each side of the room is 12 feet long. Note that although the equation $x^2 = 144$ has two solutions, -12 and 12, the negative solution makes no sense (for this problem), so you should choose the positive solution.

Real Life

EXAMPLE 6   Finding the Dimensions of a Room

A bedroom is 3 feet longer than it is wide (see Figure 1.5) and has an area of 154 square feet. Find the dimensions of the room.

Solution

You can begin by using the same type of problem-solving strategy that was presented in Section 1.2.

| Verbal Model: | Width of room | · | Length of room | = | Area of room |

Labels: Area of room = 154   (square feet)
           Width of room = w (feet)
           Length of room = $w + 3$ (feet)

Equation: $w(w + 3) = 154$
                 $w^2 + 3w - 154 = 0$
                 $(w - 11)(w + 14) = 0$
                 $w - 11 = 0$ ⟹ $w = 11$
                 $w + 14 = 0$ ⟹ $w = -14$

Choosing the positive value, you can conclude that the width is 11 feet and the length is $w + 3 = 14$ feet. You can check this in the original statement of the problem as follows.

Check

The length of 14 feet is 3 feet more than the width of 11 feet. ✓
The area of the room is $(11)(14) = 154$ square feet. ✓

FIGURE 1.5

Problem Solving

The text provides ample opportunity for students to hone their problem-solving skills. In both the exercises and the examples in the Second Edition, students are asked to apply verbal, analytical, graphical, and numerical approaches to solving problems. Students are also encouraged to use the graphing utility as a tool in problem solving. They are taught the following approach to solving applied problems: (1) Construct a verbal model; (2) Label variable and constant terms; (3) Construct an algebraic model; (4) Using the model, solve the problem; and (5) Check the answer in the original statement of the problem. In the Second Edition, there is increased emphasis on identifying units of measure, and many solutions were rewritten with explanations and additional help in the form of comments adjacent to the computation. There is also increased use of color to emphasize and clarify the solution steps.

Group Activities

The Group Activities that appear at the end of sections reinforce students' understanding by approaching mathematical concepts in a variety of ways: Communicating Mathematically, You Be the Instructor, Extending the Concept, Problem Solving, and Exploring with Technology. Designed to be completed as group projects in class or as homework assignments, the Group Activities give students opportunities for interactive learning and to think, talk, and write about mathematics.

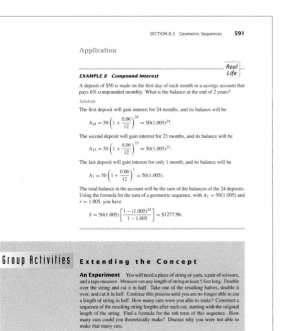

SECTION 8.3   Geometric Sequences   **591**

Application

Real Life

EXAMPLE 8   Compound Interest

A deposit of $50 is made on the first day of each month in a savings account that pays 6% compounded monthly. What is the balance at the end of 2 years?

Solution

The first deposit will gain interest for 24 months, and its balance will be

$$A_{24} = 50\left(1 + \frac{0.06}{12}\right)^{24} = 50(1.005)^{24}.$$

The second deposit will gain interest for 23 months, and its balance will be

$$A_{23} = 50\left(1 + \frac{0.06}{12}\right)^{23} = 50(1.005)^{23}.$$

The last deposit will gain interest for only 1 month, and its balance will be

$$A_1 = 50\left(1 + \frac{0.06}{12}\right)^{1} = 50(1.005).$$

The total balance in the account will be the sum of the balances of the 24 deposits. Using the formula for the sum of a geometric sequence, with $A_1 = 50(1.005)$ and $r = 1.005$, you have

$$S = 50(1.005)\left[\frac{1 - (1.005)^{24}}{1 - 1.005}\right] = \$1277.96.$$

Group Activities   Extending the Concept

An Experiment   You will need a piece of string or yarn, a pair of scissors, and a tape measure. Measure out any length of string at least 5 feet long. Double over the string and cut it in half. Take one of the resulting halves, double it over, and cut it in half. Continue this process until you are no longer able to cut a length of string in half. How many cuts were you able to make? Construct a sequence of the resulting string lengths after each cut, starting with the original length of the string. Find a formula for the nth term of this sequence. How many cuts could you theoretically make? Discuss why you were not able to make that many cuts.

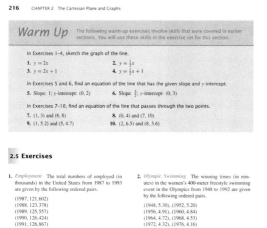

Warm Up The following warm-up exercises involve skills that were covered in earlier sections. You will use these skills in the exercise set for this section.

In Exercises 1–4, sketch the graph of the line.

1. $y = 2x$
2. $y = \frac{1}{2}x$
3. $y = 2x + 1$
4. $y = \frac{1}{2}x + 1$

In Exercises 5 and 6, find an equation of the line that has the given slope and y-intercept.

5. Slope: 1; y-intercept: $(0, 2)$
6. Slope: $\frac{2}{3}$; y-intercept: $(0, 3)$

In Exercises 7–10, find an equation of the line that passes through the two points.

7. $(1, 3)$ and $(6, 8)$
8. $(0, 4)$ and $(7, 10)$
9. $(1, 5.2)$ and $(5, 4.7)$
10. $(2, 6.5)$ and $(8, 3.6)$

2.5 Exercises

1. *Employment* The total numbers of employed (in thousands) in the United States from 1987 to 1993 are given by the following ordered pairs.

(1987, 121,602)
(1988, 123,378)
(1989, 125,557)
(1990, 126,424)
(1991, 126,867)
(1992, 128,548)
(1993, 129,525)

A linear model that approximates this data is

$y = 113,336.2 + 1265.0t$, $7 \le t \le 13$

where y represents the number of employed (in thousands) and $t = 0$ represents 1980. Plot the actual data *and* the model on the same graph. How closely does the model represent the data? (Source: U. S. Bureau of Labor Statistics)

2. *Olympic Swimming* The winning times (in minutes) in the women's 400-meter freestyle swimming event in the Olympics from 1948 to 1992 are given by the following ordered pairs.

(1948, 5.30), (1952, 5.20)
(1956, 4.91), (1960, 4.84)
(1964, 4.72), (1968, 4.53)
(1972, 4.32), (1976, 4.16)
(1980, 4.15), (1984, 4.12)
(1988, 4.06), (1992, 4.12)

A linear model that approximates this data is

$y = 5.4 - 0.03t$, $8 \le t \le 52$

where y represents the winning time in minutes and $t = 0$ represents 1940. Plot the actual data *and* the model on the same graph. How closely does the model represent the data? (Source: Olympic Committee)

Warm-Up Exercises

The exercise set in each section is preceded by a set of ten warm-up exercises to help students review skills that are needed in the main exercise set. Answers to all warm-up exercises are given in the back of the text.

Exercises

The nearly 5000 numbered exercises include a broad range of conceptual, computational, and applied problems to accommodate a variety of teaching and learning styles, including multi-part, exploration and discovery, real-life applications, mathematical modeling, writing, estimation, geometry, challenging, data interpretation and analysis, and exercises that require graphing technology. Designed to build competence, skill, and understanding, each exercise set is graded in difficulty to allow students to gain confidence as they progress. Detailed solutions to selected odd-numbered exercises are given in the Student Solutions Guide, with answers appearing in the back of the text.

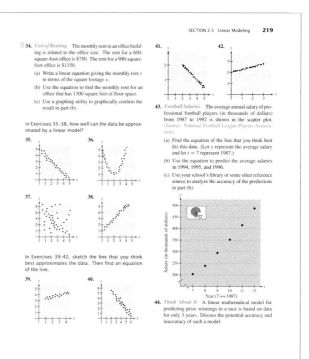

34. *Cost of Renting* The monthly rent in an office building is related to the office size. The rent for a 600-square-foot office is $750. The rent for a 900-square-foot office is $1150.

(a) Write a linear equation giving the monthly rent r in terms of the square footage x.

(b) Use the equation to find the monthly rent for an office that has 1300 square feet of floor space.

(c) Use a graphing utility to graphically confirm the result in part (b).

In Exercises 35–38, how well can the data be approximated by a linear model?

35.
36.
37.
38.

In Exercises 39–42, sketch the line that you think best approximates the data. Then find an equation of the line.

39.
40.
41.
42.

43. *Football Salaries* The average annual salary of professional football players (in thousands of dollars) from 1987 to 1992 is shown in the scatter plot. (Source: National Football League Players Association)

(a) Find the equation of the line that you think best fits this data. (Let y represent the average salary and let $t = 7$ represent 1987.)

(b) Use the equation to predict the average salaries in 1994, 1995, and 1996.

(c) Use your school's library or some other reference source to analyze the accuracy of the predictions in part (b).

44. *Think About It* A linear mathematical model for predicting prize winnings in a race is based on data for only 3 years. Discuss the potential accuracy and inaccuracy of such a model.

SECTION 1.4 The Quadratic Formula **121**

39. $4x^2 - 15 = 25$ **40.** $4x^2 + 2x + 4 = 2x - 8$

41. $x^2 + 3x + 1 = 0$ **42.** $x^2 + 3x - 4 = 0$

43. $(x - 1)^2 = 9$ **44.** $2x^2 - 4x - 6 = 0$

45. $100x^2 - 400 = 0$

46. $2x^2 + 4x - 9 = 2(x - 1)^2$

Writing Real-Life Problems In Exercises 47–50, solve the number problem *and* write a real-life problem that could be represented by this verbal model. For instance, an applied problem that could be represented by Exercise 47 is as follows.

The sum of the length and width of a one-story house is 100 feet. The house has 2500 square feet of floor space. What are the length and width of the house?

47. Find two numbers whose sum is 100 and whose product is 2500.

48. One number is 1 more than another number. The product of the two numbers is 72. Find the numbers.

49. One number is 1 more than another number. The sum of their squares is 113. Find the numbers.

50. One number is 2 more than another number. The product of the two numbers is 440. Find the numbers.

Cost Equation In Exercises 51–54, use the cost equation to find the number of units x that a manufacturer can produce for the cost C. (Round your answer to the nearest positive integer.)

51. $C = 0.125x^2 + 20x + 5000$ $C = \$14,000$

52. $C = 0.5x^2 + 15x + 5000$ $C = \$11,500$

53. $C = 800 + 0.04x + 0.002x^2$ $C = \$1680$

54. $C = 800 - 10x + \dfrac{x^2}{4}$ $C = \$896$

55. *Seating Capacity* A rectangular classroom seats 72 students. If the seats were rearranged with three more seats in each row, the classroom would have two fewer rows. Find the original number of seats in each row.

56. *Dimensions of a Corral* A rancher has 200 feet of fencing to enclose two adjacent rectangular corrals (see figure). Find the dimensions such that the enclosed area will be 1400 square feet.

$4x + 3y = 200$

57. *Geometry* An open box is to be made from a square piece of material by cutting 2-inch squares from the corners and turning up the sides (see figure). The volume of the finished box is to be 200 cubic inches. Find the size of the original piece of material.

58. *Geometry* An open box (see figure) is to be constructed from 108 square inches of material. Find the dimensions of the square base.

59. *On the Moon* An astronaut on the moon throws a rock straight up into space. The initial velocity is 40 feet per second and the initial height is 5 feet. How long will it take the rock to hit the surface? If the rock had been thrown with the same initial velocity and height on earth, how long would it remain in the air?

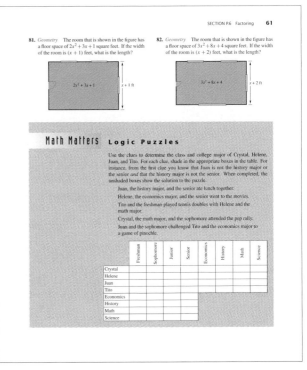

Geometry

Geometric formulas and concepts are reviewed throughout the text. For reference, common formulas are listed inside the back cover of this text.

Career Interviews

Career Interviews with people who use algebra in their jobs help students understand that algebra is a modern problem-solving language.

Math Matters

Each chapter contains a Math Matters feature that engages student interest by discussing an historical note or mathematical problem. For those features that pose a question, the answers appear in the back of the text.

SECTION P.6 Factoring **61**

81. *Geometry* The room that is shown in the figure has a floor space of $2x^2 + 3x + 1$ square feet. If the width of the room is $(x + 1)$ feet, what is the length?

$2x^2 + 3x + 1$ $x + 1$ ft

82. *Geometry* The room that is shown in the figure has a floor space of $3x^2 + 8x + 4$ square feet. If the width of the room is $(x + 2)$ feet, what is the length?

$3x^2 + 8x + 4$ $x + 2$ ft

Math Matters Logic Puzzles

Use the clues to determine the class and college major of Crystal, Helene, Juan, and Tito. For each clue, shade in the appropriate boxes in the table. For instance, from the first clue you know that Juan is not the history major or the senior *and* that the history major is not the senior. When completed, the unshaded boxes show the solution to the puzzle.

Juan, the history major, and the senior ate lunch together.

Helene, the economics major, and the senior went to the movies.

Tito and the freshman played tennis doubles with Helene and the math major.

Crystal, the math major, and the sophomore attended the pep rally.

Juan and the sophomore challenged Tito and the economics major to a game of pinochle.

	Freshman	Sophomore	Junior	Senior	Economics	History	Math	Science
Crystal								
Helene								
Juan								
Tito								
Economics								
History								
Math								
Science								

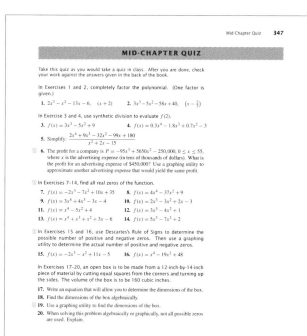

MID-CHAPTER QUIZ

Take this quiz as you would take a quiz in class. After you are done, check your work against the answers given in the back of the book.

In Exercises 1 and 2, completely factor the polynomial. (One factor is given.)

1. $2x^3 - x^2 - 13x - 6$, $(x + 2)$
2. $3x^3 - 5x^2 - 58x + 40$, $(x - \frac{2}{3})$

In Exercise 3 and 4, use synthetic division to evaluate $f(2)$.

3. $f(x) = 3x^3 - 5x^2 + 9$
4. $f(x) = 0.3x^4 - 1.8x^3 + 0.7x^2 - 3$

5. Simplify: $\dfrac{2x^4 + 9x^3 - 32x^2 - 99x + 180}{x^2 + 2x - 15}$

6. The profit for a company is $P = -95x^3 + 5650x^2 - 250,000$, $0 \le x \le 55$, where x is the advertising expense (in tens of thousands of dollars). What is the profit for an advertising expense of $450,000? Use a graphing utility to approximate another advertising expense that would yield the same profit.

In Exercises 7–14, find all real zeros of the function.

7. $f(x) = -2x^3 - 7x^2 + 10x + 35$
8. $f(x) = 4x^4 - 37x^2 + 9$
9. $f(x) = 3x^4 + 4x^3 - 3x - 4$
10. $f(x) = 2x^3 - 3x^2 + 2x - 3$
11. $f(x) = x^4 - 5x^2 + 4$
12. $f(x) = 3x^3 - 4x^2 + 1$
13. $f(x) = x^4 + x^3 + x^2 + 3x - 6$
14. $f(x) = 5x^3 - 7x^2 + 2$

In Exercises 15 and 16, use Descartes's Rule of Signs to determine the possible number of positive and negative zeros. Then use a graphing utility to determine the actual number of positive and negative zeros.

15. $f(x) = -2x^3 - x^2 + 11x - 5$
16. $f(x) = x^4 - 19x^2 + 48$

In Exercises 17–20, an open box is to be made from a 12-inch-by-14-inch piece of material by cutting equal squares from the corners and turning up the sides. The volume of the box is to be 160 cubic inches.

17. Write an equation that will allow you to determine the dimensions of the box.
18. Find the dimensions of the box algebraically.
19. Use a graphing utility to find the dimensions of the box.
20. When solving this problem algebraically or graphically, not all possible zeros are used. Explain.

Mid-Chapter Quizzes

Each chapter contains a Mid-Chapter Quiz. This feature allows the student to perform a self-assessment midway through the chapter.

Chapter Project

Chapter Projects, referenced in the chapter opener, are extended applications that use real data, graphs, and modeling to enhance students' understanding of mathematical concepts. Designed as individual or group projects, they offer additional opportunities to discuss and write about mathematics. Real data (collected with a *CBL* in half of the projects) is used throughout. Many projects include research assignments that give students the opportunity to collect, analyze, and interpret their own data.

CHAPTER PROJECT: Gravity and Air Resistance

This chapter project (as well as the projects in Chapters 3, 5, and 8) utilizes data that was collected by the Texas Instruments Calculator-Based Laboratory *(CBL)* System. The *CBL* is used in conjunction with either the *TI-82* or *TI-85* calculator to collect real data in real time. This data may then be stored in the *TI-82* or *TI-85* for further analysis. It is not necessary to have access to a *CBL* to complete most of this project. However, if you do have access to a *CBL*, the last question of the project (indicated by a *CBL* icon) gives the opportunity to explore the topic further with a *CBL*. Instruction on how to use a *CBL* for your own data collection is given in the Appendix.

There are two forces that act on any falling object—gravity and air resistance. Gravity pulls an object toward the earth, and air resistance tries to keep an object from falling. However, the force of gravity is much stronger than the force of air resistance; thus, air resistance only slows the fall of an object.

In this project, you will consider the fall of a miniature parachute. The parachute is released approximately 5 feet above the ground. A motion detector attached to a *CBL* unit tracks the fall of a weight attached to the parachute.

To determine the effects of gravity and air resistance on a weighted parachute as it falls, one records the height of the falling object at timed intervals. The table shows the data collected with a *Texas Instruments CBL* unit.[*]

Use this information to investigate the following questions.

Time	Height
0	4.83262
0.10002	4.70658
0.199997	4.47971
0.300005	4.23124
0.400013	3.97196
0.499992	3.68748
0.599972	3.39579
0.699962	3.10771
.0799942	2.81962
0.899922	2.52434
0.999902	2.25066
1.099872	1.95897
1.199872	1.67809

1. *Predicting the Outcome*   Which of the following predictions describes how the parachute will fall immediately after its release? Give a reason for your prediction.

   (a) Air resistance will cause the parachute to fall slowly at first, then more rapidly as gravity takes over.

   (b) The parachute will fall rapidly at first, then more slowly as air resistance affects it.

   (c) Gravity and air resistance will work together to cause the parachute to fall at a constant rate.

2. *Graphing the Data*   Input the data from the table at the left into a graphing utility, and use it to make a scatter plot of the data. How does the scatter plot compare to your prediction in Question 1?

3. *Fitting a Model to Data*   Use the statistical features of your graphing utility to find a linear model for the data.

   (a) Write the linear model.

   (b) What is the slope of the line? Explain what the slope of the line tells you about the fall of the parachute.

   (c) What is the y-intercept of the line? Explain what the y-intercept tells you about the fall of the parachute.

4. *Comparing Actual Data with the Model*   Enter the linear model from Question 3(a) in your graphing utility. Using the TRACE feature, find the height of the parachute at $t = 0.2$ seconds and at $t = 0.6$ seconds. How do the model values compare to the values from the table?

5. *Extended CBL Exploration*   Use the CBL instructions in Appendix C to collect and analyze data about falling objects. Attach binder clips or paperweights to a parachute to vary the weight. How does the amount of weight affect a falling object? What happens when the size of the parachute decreases?

*Instructions for using a CBL to collect your own data are given in Appendix C.

Chapter Summary

The Chapter Summary reviews the skills covered in the chapter. Section references for the major topics make this an effective study tool, and correlation to the review exercises offers guided practice.

Review Exercises

The Review Exercises at the end of each chapter offer the student an opportunity for additional practice. Answers to odd-numbered review exercises are given in the back of the text.

Chapter Tests

Each chapter ends with a Chapter Test, an effective tool for student self-assessment.

Cumulative Tests

The Cumulative Tests that appear after Chapters 2, 5, and 8 help students judge their mastery of previously covered material, as well as reinforce the knowledge students have been accumulating throughout the text—preparing them for other exams and for future courses.

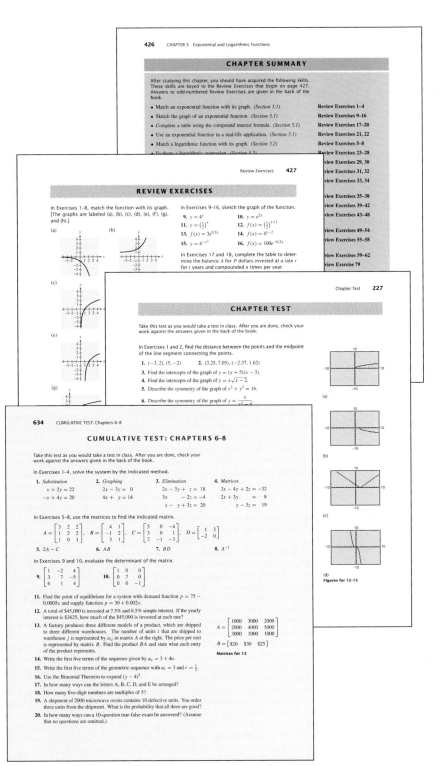

Supplements

College Algebra: Concepts and Models, Second Edition, by Larson, Hostetler, and Hodgkins, is accompanied by a comprehensive supplements package. Most items are keyed to the text.

Printed Resources

For the student

Study and Solutions Guide by Dianna L. Zook, Indiana University and Purdue University at Fort Wayne
- Section summaries of key concepts
- Detailed, step-by-step solutions to selected odd-numbered section exercises and review exercises
- Detailed, step-by-step solutions to all Mid-Chapter Quiz, Chapter Test, and Cumulative Test questions

Graphing Technology Keystroke Guide by Benjamin N. Levy
- Keystroke instructions for six different models of graphing calculators from Texas Instruments, Sharp, Casio, and Hewlett-Packard
- BestGrapher instructions for both IBM and Macintosh
- Examples with step-by-step solutions
- Extensive graphics screen output
- Technology tips

Problem Solving, Modeling, and Data Analysis Labs by Wendy Metzger, Palomar College
- Objectives, a brief discussion, and multi-part, guided discovery activities or applications in each assignment
- Keystrokes instructions for *Derive* and *TI-82*
- Keyed to the text by topic
- Funded in part by NSF (National Science Foundation, Instrumentation and Laboratory Improvement) and California Community College Fund for Instructional Improvement

For the instructor

Instructor's Annotated Edition
- Entire student edition of the text, with the student answers section: Answers to all odd-numbered exercises, and answers to all Mid-Chapter Quizzes, Chapter Tests, Cumulative Tests, and Math Matters
- Instructor's answers section: Answers to all even-numbered exercises, and answers to all Discovery Boxes, Technology Boxes, and Group Activities
- Annotations at point of use, offering specific teaching strategies and suggestions for implementing Group Activities, pointing out common student

errors, and giving additional examples, exercises, class activities, and group activities

Complete Solutions Guide by Dianna L. Zook, Indiana University and Purdue University at Fort Wayne
- Detailed, step-by-step solutions to all section and review exercises; Mid-Chapter Quiz, Chapter Test, and Cumulative Test questions; and Chapter Project questions.

Test Item File and Instructor's Resource Guide
- Printed test bank with approximately 1700 test items (multiple-choice, open-ended, and writing) coded by level of difficulty
- Technology-required test items coded for easy reference
- Bank of premade chapter test forms with answer keys
- Two premade final exams
- Transparency masters
- Lab activities contributed by Phyllis Shaw, Paradise Valley Community College

Media Resources

For the student

Tutor (IBM, Macintosh)
- Interactive tutorial software keyed to the text by section
- Diagnostic feedback
- Additional practice
- Chapter warm-ups and self-tests
- Glossary

Videotapes by Dana Mosely
- Comprehensive coverage keyed to the text by section
- For media/resource centers
- Additional explanation of important concepts, sample problems, and applications

For the instructor

Computerized Testing
- Test-generating software for both IBM and Macintosh computers
- Approximately 1700 test items
- Also available as a printed test bank

Acknowledgments

We would like to thank the many people who have helped us at various stages of this project to prepare the text and supplements package. Their encouragement, criticisms, and suggestions have been invaluable to us.

Second Edition Advisory Panel: Carol Edwards, St. Louis Community College at Florissant Valley; Carl Hughes, Fayetteville State University; and David Surowski, Kansas State University.

Second Edition Survey Respondents: Over 60 professors took time to respond to the *College Algebra: Concepts and Models* survey. We appreciate their comments.

Career Interviews: Our thanks to Laura Balaoro, Mary Kay Brown, Mark P. DesMeules, and Stephen T. King for their help in creating the career interviews. We appreciate their time and effort.

Thanks to Darrel Thoman, William Jewell College, for contributing to the *Test Item File and Resource Guide*.

Thanks to all of the people at D. C. Heath and Company who worked with us in the development and production of the text, especially Charles Hartford and Ann Marie Jones, Mathematics Acquisitions Editors; Cathy Cantin, Managing Editor; Emily Keaton, Developmental Editor; Carolyn Johnson, Editorial Associate; Rachel Wimberly, Production Editor; Henry Rachlin, Designer; Gary Crespo, Art Editor; Lisa Merrill, Production Coordinator; and Billie L. Porter, Photo Researcher.

We would also like to thank the staff at Larson Texts, Inc., who assisted with proofreading the manuscript; preparing and proofreading the art package; and checking and typesetting the supplements.

On a personal level, we are grateful to our wives, Deanna Gilbert Larson and Eloise Hostetler, for their love, patience, and support. Also, a special thanks goes to R. Scott O'Neil.

If you have suggestions for improving the text, please feel free to write to us. Over the past two decades, we have received many useful comments from both instructors and students, and we value these very much.

Roland E. Larson
Robert P. Hostetler
Anne V. Hodgkins

Contents

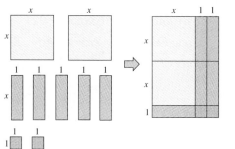
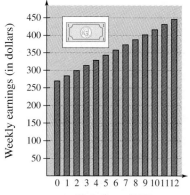

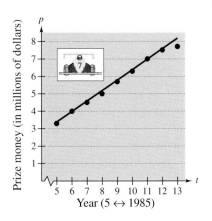

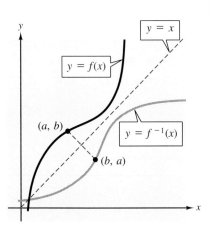

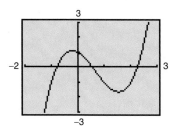

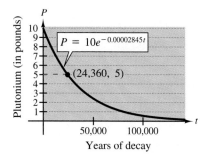

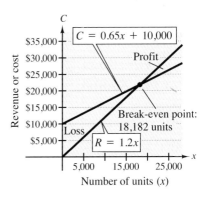

Sign Pattern for Cofactors

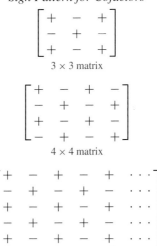

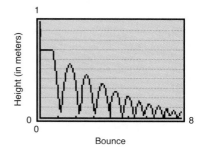

Bounce

How to Study Algebra

After years of teaching and guiding students through algebra courses, we have compiled the following list of suggestions for studying algebra. These study tips may take some time and effort—but they work!

Making a Plan   Make your own course plan right now! Determine the number of hours you need to spend on algebra each week. Write your plans on your calendar or some other schedule planner, and then *stick to your plan.*

Preparing for Class   Before attending class, read the portion of the text that is to be covered. This takes a lot of self-discipline, but it pays off. By going to class prepared, you will be able to benefit much more from your instructor's presentation. Algebra, like most other technical subjects, is easier to understand the second or third time you hear it.

Attending Class   Attend every class. Arrive on time with your text, a pen or pencil, paper for notes, and your calendar.

Participating in Class   As you are reading the text before class, write down any questions that you have about the material. Then, ask your instructor during class.

Taking Notes   Take notes in class, especially on definitions, examples, concepts, and rules. Then, as soon after class as possible, read through your notes, adding any explanations that are necessary to make your notes understandable *to you.*

Doing the Homework   Learning algebra is like learning to play the piano or learning to play basketball. You cannot become skilled by just watching someone else do it. You must also do it yourself. A general guideline is to spend two to four hours of study outside of class for each hour in class. When working exercises, your ultimate goal is to be able to solve the problems accurately and quickly. When you start a new exercise set, however, understanding is much more important than speed.

Finding a Study Partner   When you get stuck on a problem, it may help to try to work with someone else. Even if you feel you are giving more help than you are getting, you will find that an excellent way to learn is by teaching others.

Working Cooperatively   Begin by agreeing on what you have to do. Make a plan. Check and proofread all work. Listen carefully, ask for help when you need it, and offer help when asked.

Building a Math Library   Start building a library of books that can help you with this and future math courses. You might consider using the *Study and Solutions Guide* that accompanies the text. Also, since you will probably be taking other math courses after you finish this course, we suggest that you keep the text. It will be a valuable reference book. Tutorial software and videos available with this text will also be valuable additions to your mathematics library.

Keeping Up with the Work   Don't let yourself fall behind in the course. If you think that you are having trouble, seek help immediately. Ask your instructor, attend your school's tutoring services, talk with your study partner, use additional study aids such as videos or software tutorials—but do something. If you are having trouble with the material in one chapter of your algebra text, there is a good chance that you will also have trouble in later chapters.

Getting Stuck   *Everyone* who has ever taken a math course has had this experience: You are working on a problem and cannot see how to solve it, or you have solved it but your answer does not agree with the answer given in the back of the book. People have different approaches to this sort of problem. You might ask for help, take a break to clear your thoughts, sleep on it, rework the problem, or reread the section in the text. The point is, try not to get frustrated or spend too much time on a single problem.

Keeping Your Skills Sharp   Before each exercise set in the text we have included a short set of *Warm-Up Exercises.* These exercises will help you review skills that you learned in previous exercises. These sets are designed to take only a few minutes to solve. We suggest working the entire set before you start each new exercise set. (All of the Warm-Up Exercises are answered in the back of the text.)

Assessing Your Progress   In the middle of each chapter is a *Mid-Chapter Quiz.* Take the quiz as you would if you were in class, then check your answers in the back of the text.

Checking Your Work   One of the nice things about algebra is that you don't have to wonder whether your solution is correct. You can tell whether it is correct by checking it in the original statement of the problem. If, in addition to your "solving skills," you work on your "checking skills," you should find your test scores improving.

Preparing for Exams   Cramming for algebra exams seldom works. If you have kept up with the work and followed the suggestions given here, you should be almost ready for the exam. At the end of each chapter, we have included three features that should help as a final preparation. Read the *Chapter Summary,* work the *Review Exercises,* and set aside an hour to take the sample *Chapter Test.*

Taking Exams   Most instructors suggest that you do *not* study right up to the minute you are taking a test. This tends to make people anxious. The best cure for anxiousness during tests is to prepare well before taking the test. Once the test has begun, read the directions carefully, and try to work at a reasonable pace. (You might want to read the entire test first, then work the problems in the order with which you feel most comfortable.) Hurrying tends to cause people to make careless errors. If you finish early, take a few moments to clear your thoughts and then take time to go over your work.

Learning from Mistakes   When you get an exam back, be sure to go over any errors that you might have made. Don't be too quick to pass off an error as just a "dumb mistake." Take advantage of any mistakes by hunting for ways to continually improve your test-taking abilities.

What Is Algebra?

To some, algebra is manipulating symbols or performing mathematical operations with letters instead of numbers. To others, it is factoring, solving equations, or solving word problems. And to still others, algebra is a mathematical language that can be used to model real-world problems. In fact, algebra is all of these!

As you study this text, it is helpful to view algebra from the "big picture"—to see how the various rules, operations, and strategies fit together.

The rules of arithmetic form the foundation of algebra. These rules are generalized through the use of symbols and letters to form the basic rules of algebra, which are used to *rewrite* algebraic expressions and equations in new, more useful forms. The ability to rewrite algebraic expressions and equations is the common skill involved in the three major components of algebra—*simplifying* algebraic expressions, *solving* algebraic equations, and *graphing* algebraic functions. The following chart shows how this college algebra text fits into the "big picture" of algebra.

Construct
↓
Rewrite   *Construct*
↓   ↓
Graph   *Rewrite*   *Construct*
             ↓   ↓
             Solve   *Rewrite*   **Rules of Arithmetic**
                          ↓   **1** Concepts of Elementary
                          Simplify   Algebra
                                       Rules of Algebra
                                       1 Concepts of Elementary Algebra

Algebraic Expressions
- **1** Concepts of Elementary Algebra
- **5** Rational Expressions and Rational Functions
- **6** Radicals and Complex Numbers
- **9.1** Exponential and Logarithmic Expressions
- **10** Additional Topics in Algebra

Algebraic Equations
- **2** Introduction to Equations, Graphs and Functions
- **3** Linear Functions, Equations and Inequalities
- **4** Systems of Linear Equations and Inequalities
- **5.6** Solving Equations Involving Rational Expressions
- **6.5** Solving Equations Involving Radicals
- **7** Quadratic Functions, Equations and Inequalities
- **8.3** Variation and Mathematical Models
- **9.5** Solving Exponential and Logarithmic Equations

Functions and Graphs
- **2** Introduction to Equations, Graphs, and Functions
- **3** Linear Functions, Equations and Inequalities
- **4** Systems of Linear Equations and Inequalities
- **5.5** Graphing Rational Functions
- **7.5** Graphing Quadratic Functions
- **8** Additional Functions and Relations
- **9** Exponential and Logarithmic Functions

Review of Fundamental Concepts of Algebra

P

- Real Numbers: Order and Absolute Value
- The Basic Rules of Algebra
- Integer Exponents
- Radicals and Rational Exponents
- Polynomials and Special Products
- Factoring
- Fractional Expressions and Probability

Musical notes are characterized by their frequencies and amplitudes. High and low frequencies correspond to high and low notes.

The musical note A-440 (the first A above middle C) has a frequency of 440 vibrations per second. This note is often the first to be tuned on a musical instrument. The frequency F of any note can be found by the model

$$F = 440 \cdot \sqrt[12]{2^n}$$

where n represents the number of notes (black and white keys on a piano) above or below A-440.

Two notes are said to harmonize (sound pleasing to the ear) if the ratio of their frequencies is an integer or simple rational number. For instance, any two notes shown in the table and graph at the right harmonize.

n	−48	−36	−24	−12	0	12	24	36
F	27.5	55	110	220	440	880	1760	3520

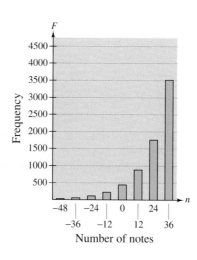

The chapter project related to this information is on page 73.

1

P.1   Real Numbers: Order and Absolute Value

Real Numbers ▪ The Real Number Line and Ordering ▪
Absolute Value and Distance ▪ Application

Real Numbers

The formal term that is used in mathematics to refer to a collection of objects is
the word **set.** For instance, the set

{1, 2, 3}

contains the three numbers 1, 2, and 3. Note that a pair of braces { } is used to
enclose the members of the set. In this text, a *pair* of braces will always indicate
the members of a set. Parentheses ( ) and brackets [ ] are used to represent
other ideas.

The set of numbers that is used in arithmetic is the **set of real numbers.** The
term *real* distinguishes real numbers from *imaginary* or *complex* numbers—a type
of number you will study later in this text.

A set *A* is called a **subset** of a set *B* if every member of *A* is also a member
of *B*. Here are some examples

- {1, 2, 3} is a subset of {1, 2, 3, 4}.
- {0, 4} is a subset of {0, 1, 2, 3, 4}.

One of the most commonly used subsets of real numbers is the set of **natural
numbers** or **positive integers**

{1, 2, 3, 4, . . .}.   Set of positive integers

Note that the three dots indicate that the pattern continues. For instance, the set
also contains the numbers 5, 6, 7, and so on.

Positive integers can be used to describe many quantities that you encounter
in everyday life—for instance, you might be taking four classes this term, or you
might be paying $240 dollars a month for rent. But even in everyday life, positive
integers cannot describe some concepts accurately. For instance, you could have
a zero balance in your checking account, or the temperature could be $-10°$ (10
degrees below zero). To describe such quantities you need to expand the set of
positive integers to include **zero** and the **negative integers.** The expanded set is
called the set of **integers,** which can be written as follows.

The set of integers is a **subset** of the set of real numbers—this means that every
integer is a real number.

STUDY TIP

Whenever a mathematical
term is formally introduced
in this text, the word will
appear in boldface type.
Be sure you understand
the meaning of each new
word—it is important that
each word becomes part of
your mathematical vocabulary.

Even with the set of integers, there are still many quantities in everyday life that you cannot describe accurately. The costs of many items are not in whole-dollar amounts, but in parts of dollars, such as $1.19 or $39.98. You might work $8\frac{1}{2}$ hours, or you might miss the first *half* of a movie. To describe such quantities, the set of integers is expanded to include **fractions.** The expanded set is called the set of **rational numbers.** Formally, a real number is called **rational** if it can be written as the ratio p/q of two integers, where $q \neq 0$. (The symbol $\neq$ means **not equal to.**) For instance,

$$2 = \frac{2}{1}, \quad 0.333\ldots = \frac{1}{3}, \quad 0.125 = \frac{1}{8}, \quad \text{and} \quad 1.126126\ldots = \frac{125}{111}$$

are rational numbers. Real numbers that cannot be written as ratios of two integers are called **irrational.** For instance, the numbers

$$\sqrt{2} = 1.4142135\ldots \quad \text{and} \quad \pi = 3.1415926\ldots$$

are irrational. The decimal representation of a rational number is either *terminating* or *repeating*. For instance, the decimal representation of $\frac{1}{4} = 0.25$ is terminating, and the decimal representation of $\frac{4}{11} = 0.363636\ldots = 0.36\overline{36}$ is repeating. (The line over "36" indicates which digits repeat.)

The decimal representation of an irrational number neither terminates nor repeats. When you perform calculations using decimal representations of non-terminating decimals, you usually use a decimal approximation that has been **rounded** to a certain number of decimal places. For instance, rounded to four decimal places, the decimal approximations of $\frac{2}{3}$ and π are

$$\frac{2}{3} \approx 0.6667 \quad \text{and} \quad \pi \approx 3.1416.$$

The symbol $\approx$ means **approximately equal to.**

Figure P.1 shows several commonly used subsets of real numbers and their relationships to each other.

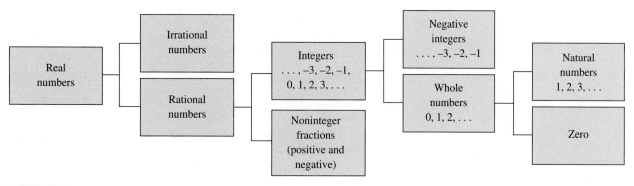

FIGURE P.1

The Real Number Line and Ordering

The picture that is used to represent the real numbers is the **real number line.** It consists of a horizontal line with a point (the **origin**) labeled as 0. Points to the left of 0 are associated with **negative numbers,** and points to the right of 0 are associated with **positive numbers,** as shown in Figure P.2.

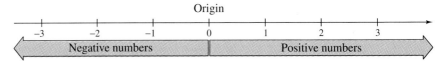

FIGURE P.2   The Real Number Line

NOTE   The real number zero is neither positive nor negative. Thus, when you want to talk about real numbers that might be positive *or* zero, you can use the term **nonnegative real numbers.**

Each point on the real number line corresponds to exactly one real number, and each real number corresponds to exactly one point on the real number line, as shown in Figure P.3. The number associated with a point on the real number line is the **coordinate** of the point.

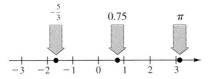

FIGURE P.3   Every real number corresponds to a point on the real number line.

The real number line provides you with a way of comparing any two real numbers. For instance, if you choose any two (different) numbers on the real number line, one of the numbers must be to the left of the other number. The number to the left is **less than** the number to the right, and the number to the right is **greater than** the number to the left.

Definition of Order on the Real Number Line

If the real number a lies to the left of the real number b on the real number line, a is **less than** b, which is denoted by

$$a < b$$

as shown in Figure P.4. This relationship can also be described by saying that b is **greater than** a and writing $b > a$. The symbol $a \leq b$ means that a is **less than or equal to** b, and the symbol $b \geq a$ means that b is **greater than or equal to** a.

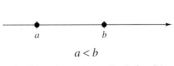

$$a < b$$

FIGURE P.4 a is to the left of b.

The symbols $<$, $>$, $\leq$, and $\geq$ are called **inequality symbols.** Inequalities are useful in denoting subsets of real numbers, as shown in Examples 1 and 2.

EXAMPLE 1   *Interpreting Inequalities*

a. The inequality $x \leq 2$ denotes all real numbers that are less than or equal to 2, as shown in Figure P.5(a).

b. The inequality $-2 \leq x < 3$ means that $x \geq -2$ *and* $x < 3$. This **double inequality** denotes all real numbers between -2 and 3, including -2 but *not* including 3, as shown in Figure P.5(b).

c. The inequality $x > -5$ denotes all real numbers that are greater than -5, as shown in Figure P.5(c).

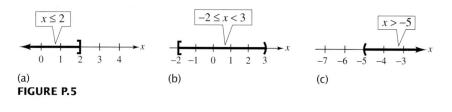

(a)   (b)   (c)

FIGURE P.5

In Figure P.5, notice that a bracket is used to *include* the endpoint of an interval and a parenthesis is used to *exclude* the endpoint.

EXAMPLE 2   *Inequalities and Sets of Real Numbers*

a. "c is nonnegative" means that c is greater than or equal to zero, which you can write as $c \geq 0$.

b. "b is at most 5" can be written as $b \leq 5$.

c. "d is negative" can be written as $d < 0$, and "d is greater than -3" can be written as $-3 < d$. Combining these two inequalities produces $-3 < d < 0$.

d. "x is positive" can be written as $0 < x$, and "x is not more than 6" can be written as $x \leq 6$. Combining these two inequalities produces $0 < x \leq 6$.

The following property of real numbers is the **Law of Trichotomy.** As the "tri" in its name suggests, this law tells you that for any two real numbers a and b, precisely one of *three* relationships is possible.

$$a < b, \qquad a = b, \qquad \text{or} \qquad a > b \qquad \text{Law of Trichotomy}$$

Absolute Value and Distance

The **absolute value** of a real number is its *magnitude*, or its value disregarding its sign. For instance, the absolute value of -3 is $|-3|$, which has the value of 3.

NOTE The absolute value of any real number is either positive or zero. Moreover, 0 is the only real number whose absolute value is zero. That is, $|0| = 0$.

Definition of Absolute Value

Let a be a real number. The **absolute value** of a, denoted by $|a|$, is

$$|a| = \begin{cases} a, & \text{if } a \geq 0 \\ -a, & \text{if } a < 0. \end{cases}$$

Be sure you see from this definition that the absolute value of a real number is never negative. For instance, if $a = -5$, then $|-5| = -(-5) = 5$.

EXAMPLE 3   Finding Absolute Value

a. $|-7| = 7$

b. $\left|\frac{1}{2}\right| = \frac{1}{2}$

c. $|-4.8| = 4.8$

d. $-|-9| = -(9) = -9$

EXAMPLE 4   Comparing Real Numbers

Place the correct symbol ($<$, $>$, or $=$) between the real numbers.

a. $|-4|$ $|4|$ **b.** $|-5|$ 3   **c.** $-|-1|$ $|-1|$

Solution

a. $|-4| = |4|$, because both are equal to 4.

b. $|-5| > 3$, because $|-5| = 5$ and 5 is greater than 3.

c. $-|-1| < |-1|$, because $-|-1| = -1$ and $|-1| = 1$.

The list at the top of the next page gives four useful properties of absolute value. When you see a list such as this, try to formulate verbal descriptions of the properties—they are easier to remember that way. For instance, the third property tells you that the absolute value of a product of two numbers is equal to the product of the absolute values of the two numbers.

Properties of Absolute Value

Let a and b be real numbers. Then the following properties are true.

1. $|a| \geq 0$ **2.** $|-a| = |a|$

3. $|ab| = |a|\,|b|$ **4.** $\left|\dfrac{a}{b}\right| = \dfrac{|a|}{|b|}$

Absolute value can be used to define the distance between two numbers on the real number line. To see how this is done, consider the numbers -3 and 4, as shown in Figure P.6. To find the distance between these two numbers, subtract *either* number from the other and then take the absolute value of the difference. For instance,

(Distance between -3 and 4) $= |-3 - 4| = |-7| = 7$.

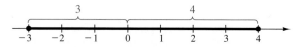

FIGURE P.6   The distance between -3 and 4 is 7.

Distance Between Two Numbers

Let a and b be real numbers. The **distance between a and b** is given by

Distance $= |b - a| = |a - b|$.

EXAMPLE 5   *Finding the Distance Between Two Numbers*

a.   The distance between 2 and 7 is

Distance $= |2 - 7| = |-5| = 5$.

b.   The distance between 0 and -4 is

Distance $= |0 - (-4)| = |4| = 4$.

c.   The statement "the distance between x and 2 is at least 3" can be written as

$|x - 2| \geq 3$.

Application

Real Life

EXAMPLE 6 Budget Variance

You work in the accounting department of a company. One of your jobs is to check whether the monthly expenses from the other departments vary too much from their budgets. Your company's definition of "too much" is that the difference between actual and budgeted expenses must be less than or equal to $500 *and* less than or equal to 5% of the budgeted expenses. By letting *a* represent the actual expenses and *b* represent the budgeted expenses, you can translate these requirements as follows.

$$|a - b| \leq 500 \quad \text{and} \quad |a - b| \leq 0.05b.$$

For travel, the budgeted expense was $12,500 and the actual expense was $12,872.56. For office supplies, the budgeted expense was $750 and the actual expense was $704.15. For wages, the budgeted expense was $84,600 and the actual expense was $85,143.95. Are these amounts within budget restrictions?

Solution

One way to determine whether these three expenses are within budget restrictions is to create a table, as follows.

	Budgeted Expense, b	Actual Expense, a	$|a - b|$	0.05b
Travel	$12,500.00	$12,872.56	$372.56	$625.00
Office Supplies	$750.00	$704.15	$45.85	$37.50
Wages	$84,600.00	$85,143.95	$543.95	$4230.00

From this table, you can see that the expense for travel passes both tests, but the expenses for office supplies and wages each fails one of the tests. Can you see why?

NOTE This book has two basic goals. The first is to present the rules of algebra, and the second is to show how algebra is used to describe real-life situations and solve real-life problems.

The algebra that you will study in this book has applications in the real world (business, physical sciences, life and behavioral sciences, and engineering). We cannot, of course, include every type of application. But from those that are included in the book, we hope that you will gain a new appreciation for the usefulness of algebra as a "modeling language" for the real world. For instance, the example at the right shows how the concept of absolute value can be used to describe a common type of problem in accounting.

Group Activities   Extending the Concept

Understanding Absolute Value Discuss the differences among $|-a|$, $-|a|$, and $-|-a|$. Illustrate these differences with several examples using both negative and positive values of *a*. Which of the three expressions give equivalent results?

P.1 Exercises

In Exercises 1–6, determine which numbers in the set are (a) natural numbers, (b) integers, (c) rational numbers, and (d) irrational numbers.

1. $\left\{-9, -\frac{7}{2}, 5, \frac{2}{3}, \sqrt{2}, 0.1\right\}$

2. $\left\{\sqrt{5}, -7, -\frac{7}{3}, 0, 3.12, \frac{5}{4}\right\}$

3. $\left\{12, -13, 1, \sqrt{4}, \sqrt{6}, \frac{3}{2}\right\}$

4. $\left\{\frac{8}{2}, -\frac{8}{3}, \sqrt{10}, -4, 9, 14.2\right\}$

5. $\left\{3, -1, \frac{1}{3}, \frac{6}{3}, -\frac{1}{2}\sqrt{2}, -7.5\right\}$

6. $\left\{25, -17, \frac{12}{5}, \sqrt{9}, \sqrt{8}, -\sqrt{8}\right\}$

In Exercises 7–12, plot the two real numbers on the real number line and place the appropriate inequality sign ($<$ or $>$) between them.

7. $\frac{3}{2}, 7$

8. $-3.5, 1$

9. $-4, -8$

10. $1, \frac{16}{3}$

11. $\frac{5}{6}, \frac{2}{3}$

12. $-\frac{8}{7}, -\frac{3}{7}$

In Exercises 13–22, describe the subset of real numbers that is represented by the inequality, and sketch the subset on the real number line.

13. $x \le 5$

14. $x \ge -2$

15. $x < 0$

16. $x > 3$

17. $x \ge 4$

18. $x < 2$

19. $-2 < x < 2$

20. $0 \le x \le 5$

21. $-1 \le x < 0$

22. $0 < x \le 6$

In Exercises 23–28, use inequality notation to describe the set of real numbers.

23. x is negative.

24. y is greater than 5 and less than or equal to 12.

25. The person's age A is at least 30.

26. The yield Y is no more than 45 bushels per acre.

27. The annual rate of inflation r is expected to be at least 3.5%, but no more than 6%.

28. The price p of unleaded gasoline is not expected to go above $1.45 per gallon during the coming year.

In Exercises 29–36, write the expression without using absolute value signs.

29. $|-10|$

30. $|0|$

31. $|3 - \pi|$

32. $|4 - \pi|$

33. $\dfrac{-5}{|-5|}$

34. $-3 - |-3|$

35. $-3|-3|$

36. $|-1| - |-2|$

In Exercises 37–42, place the correct symbol ($<$, $>$, or $=$) between the two real numbers.

37. $|-4| \quad\rule{1cm}{0.4pt}\quad |4|$

38. $|-3| \quad\rule{1cm}{0.4pt}\quad -|-3|$

39. $-5 \quad\rule{1cm}{0.4pt}\quad -|5|$

40. $-|-6| \quad\rule{1cm}{0.4pt}\quad |-6|$

41. $-|-2| \quad\rule{1cm}{0.4pt}\quad -|2|$

42. $-(-2) \quad\rule{1cm}{0.4pt}\quad -2$

In Exercises 43–52, find the distance between a and b.

43.

44.

45.

46.

47. $a = -\frac{7}{2}, b = 0$

48. $a = \frac{3}{4}, b = \frac{9}{4}$

49. $a = 126, b = 75$

50. $a = -126, b = -75$

51. $a = \frac{16}{5}, b = \frac{112}{75}$

52. $a = 9.34, b = -5.65$

In Exercises 53–58, use absolute value notation to describe the expression.

53. The distance between x and 5 is no more than 3.

54. The distance between x and -10 is at least 6.

55. The distance between z and $\frac{3}{2}$ is greater than 1.

56. The distance between z and 0 is less than 8.

57. y is at least six units from 0.

58. y is at most two units from a.

In Exercises 59 and 60, use a calculator to order the numbers from smallest to largest.

59. $\frac{7071}{5000}$, $\frac{584}{413}$, $\sqrt{2}$, $\frac{47}{33}$, $\frac{127}{90}$

60. $\frac{26}{15}$, $\sqrt{3}$, $1.73\overline{20}$, $\frac{381}{220}$, $\sqrt{10} - \sqrt{2}$

In Exercises 61–64, use a calculator to find the decimal form of the rational number. If it is a nonterminating decimal, write the repeating pattern.

61. $\frac{5}{8}$

62. $\frac{1}{3}$

63. $\frac{41}{333}$

64. $\frac{6}{11}$

Budget Variance In Exercises 65–70, the accounting department of a company is checking to see whether the actual expenses of a department differ from the budgeted expenses by more than $500 or 5%. Complete the missing parts of the table and determine whether each actual expense passes the "budget variance test."

	Budgeted Expense, b	Actual Expense, a	$\lvert a - b \rvert$	$0.05b$
65.	$25,000.00	$24,872.12		
66.	$112,700.00	$113,356.52		
67.	$9400.00	$9972.59		
68.	$7500.00	$7104.68		
69.	$37,640.00	$36,968.25		
70.	$2575.00	$2613.15		

Median Incomes In Exercises 71–76, use the bar graph, which shows the median income for American households that purchase equipment for skiing, tennis, golf, fishing, camping, and hunting. In each exercise you are given a household income. Find the amount by which the income differs from the median income for the particular sport. (Source: National Sporting Goods Association)

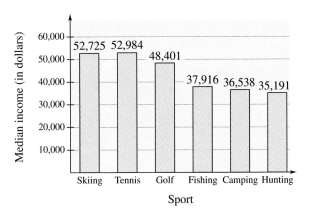

	Median Income, y	Household Income, x	$\lvert y - x \rvert$
71. Skiing		$65,400	
72. Hunting		$75,400	
73. Tennis		$37,300	
74. Golf		$18,760	
75. Camping		$21,300	
76. Fishing		$26,370	

P.2	**The Basic Rules of Algebra**

Algebraic Expressions ▪ Basic Rules of Algebra ▪ Equations ▪ Calculators and Rounding

Algebraic Expressions

The French mathematician Nicolas Chuquet (ca 1500) wrote *Triparty en la science des nombres,* in which a form of exponent notation was used. Our expressions $6x^3$ and $10x^2$ were written as $.6.^3$ and $.10.^2$. Zero and negative exponents were also represented, so x^0 would be written as $.1.^0$ and $3x^{-2}$ as $.3.^{2.m.}$. Chuquet wrote that $.72.^1$ divided by $.8.^3$ is $.9.^{2.m.}$. That is, $72x \div 8x^3 = 9x^{-2}$.

One of the basic characteristics of algebra is the use of letters (or combinations of letters) to represent numbers. The letters used to represent numbers are called **variables,** and combinations of letters and numbers are called **algebraic expressions.** Here are a few examples of algebraic expressions.

$$5x, \qquad 2x - 3, \qquad \frac{4}{x^2 + 2}, \qquad 7x + y$$

Algebraic Expression

A collection of letters (called **variables**) and real numbers (called **constants**) that are combined using the operations of addition, subtraction, multiplication, and division is an **algebraic expression.** (Other operations, which you will study later, can also be used to form an algebraic expression.)

The **terms** of an algebraic expression are those parts that are separated by addition. For example, the algebraic expression $x^2 - 5x + 8$ has three terms: x^2, $-5x$, and 8. Note that $-5x$, rather than $5x$, is a term, because

$$x^2 - 5x + 8 = x^2 + (-5x) + 8.$$

The terms x^2 and $-5x$ are the **variable terms** of the expression, and 8 is the **constant term** of the expression. The numerical factor of a variable term is the **coefficient** of the variable term. For instance, the coefficient of the variable term $-5x$ is -5, and the coefficient of the variable term x^2 is 1.

EXAMPLE 1   Identifying the Terms of an Algebraic Expression

Algebraic Expression	*Terms*
a. $4x - 3$	$4x, -3$
b. $2x + 4y - 5$	$2x, 4y, -5$

To **evaluate** an algebraic expression, substitute numerical values for each of the variables in the expression. Here are some examples.

Expression	Value of Variable	Substitute	Value of Expression
$-3x + 5$	$x = 3$	$-3(3) + 5$	$-9 + 5 = -4$
$3x^2 + 2x - 1$	$x = -1$	$3(-1)^2 + 2(-1) - 1$	$3 - 2 - 1 = 0$
$-2x(x + 4)$	$x = -2$	$-2(-2)(-2 + 4)$	$-2(-2)(2) = 8$
$\dfrac{1}{x - 2}$	$x = 2$	$\dfrac{1}{2 - 2}$	Undefined

EXAMPLE 2 Evaluating Algebraic Expressions

Evaluate the following algebraic expressions when $x = -2$ and $y = 3$.

a. $4y - 2x$ **b.** $5 + x^2$ **c.** $5 - x^2$

Solution

a. When $x = -2$ and $y = 3$, the expression $4y - 2x$ has a value of

$$4(3) - 2(-2) = 12 + 4 = 16.$$

b. When $x = -2$, the expression $5 + x^2$ has a value of

$$5 + (-2)^2 = 5 + 4 = 9.$$

c. When $x = -2$, the expression $5 - x^2$ has a value of

$$5 - (-2)^2 = 5 - 4 = 1.$$

Basic Rules of Algebra

The four basic arithmetic operations are **addition, multiplication, subtraction, and division,** denoted by the symbols $+$, $\times$ or $\cdot$, $-$, and $\div$. Of these, addition and multiplication are considered to be the two primary arithmetic operations. Subtraction and division are defined as the inverse operations of addition and multiplication, as follows.

Subtraction	*Division*
$a - b = a + (-b)$	If $b \neq 0$, then $a \div b = a\left(\dfrac{1}{b}\right) = \dfrac{a}{b}.$

In these definitions, $-b$ is called the **opposite** (or additive inverse) of b, and $1/b$ is called the **reciprocal** (or multiplicative inverse) of b. In place of $a \div b$, you can use the fraction symbol a/b. In this fractional form, a is called the **numerator** of the fraction and b is called the **denominator.**

Be sure you see that the **basic rules of algebra,** listed below, are true for variables and algebraic expressions as well as for real numbers.

Basic Rules of Algebra

Let a, b, and c be real numbers, variables, or algebraic expressions.

Property	*Example*
Commutative Property of Addition	
$a + b = b + a$	$4x + x^2 = x^2 + 4x$
Commutative Property of Multiplication	
$ab = ba$	$(4 - x)x^2 = x^2(4 - x)$
Associative Property of Addition	
$(a + b) + c = a + (b + c)$	$(-x + 5) + 2x^2 =$
	$-x + (5 + 2x^2)$
Associative Property of Multiplication	
$(ab)c = a(bc)$	$(2x \cdot 3y)(8) = (2x)(3y \cdot 8)$
Distributive Property	
$a(b + c) = ab + ac$	$3x(5 + 2x) = 3x \cdot 5 + 3x \cdot 2x$
$(a + b)c = ac + bc$	$(y + 8)y = y \cdot y + 8 \cdot y$
Additive Identity Property	
$a + 0 = a$	$5y^2 + 0 = 5y^2$
Multiplicative Identity Property	
$a \cdot 1 = 1 \cdot a = a$	$(4x^2)(1) = (1)(4x^2) = 4x^2$
Additive Inverse Property	
$a + (-a) = 0$	$5x^3 + (-5x^3) = 0$
Multiplicative Inverse Property	
$a \cdot \dfrac{1}{a} = 1, \ a \neq 0$	$(x^2 + 4)\left(\dfrac{1}{x^2 + 4}\right) = 1$

NOTE   Because subtraction is defined as "adding the opposite," the Distributive Properties are also true for subtraction. For instance, the "subtraction form" of $a(b + c) = ab + ac$ is

$$a(b - c) = a[b + (-c)] = ab + a(-c) = ab - ac.$$

EXAMPLE 3   Identifying the Basic Rules of Algebra

Identify the rule of algebra illustrated in each of the following.

a. $(4x^2)5 = 5(4x^2)$

b. $(2y^3 + y) - (2y^3 + y) = 0$

c. $(4 + x^2) + 3x^2 = 4 + (x^2 + 3x^2)$

d. $(x - 5)7 + (x - 5)x = (x - 5)(7 + x)$

e. $2x \cdot \dfrac{1}{2x} = 1, \quad x \neq 0$

Solution

a. This equation illustrates the Commutative Property of Multiplication.

b. This equation illustrates the Additive Inverse Property.

c. This equation illustrates the Associative Property of Addition. In other words, to form the sum $4 + x^2 + 3x^2$, it doesn't matter whether 4 and x^2 are added first or x^2 and $3x^2$ are added first.

d. This equation illustrates the Distributive Property in reverse order.

$$ab + ac = a(b + c) \qquad \text{Distributive Property}$$
$$(x - 5)7 + (x - 5)x = (x - 5)(7 + x)$$

e. This equation illustrates the Multiplicative Inverse Property. Note that it is important that x be a nonzero number. If x were allowed to be zero, you would be in trouble because the reciprocal of zero is undefined.

The following three lists summarize the basic properties of negation, zero, and fractions. When you encounter such lists, we suggest that you not only *memorize* a verbal description of each property, but that you also try to gain an *intuitive feeling* for the validity of each.

NOTE   Be sure you see the difference between the *negative* (or *opposite*) *of a number* and a *negative number*. If a is already negative, then its additive inverse, $-a$, is positive. For instance, if $a = -5$, then $-a = -(-5) = 5$.

Properties of Negation

Let a and b be real numbers, variables, or algebraic expressions.

Property	Example
1. $(-1)a = -a$	$(-1)7 = -7$
2. $-(-a) = a$	$-(-6) = 6$
3. $(-a)b = -(ab) = a(-b)$	$(-5)3 = -(5 \cdot 3) = 5(-3)$
4. $(-a)(-b) = ab$	$(-2)(-6) = 12$
5. $-(a + b) = (-a) + (-b)$	$-(3 + 8) = (-3) + (-8)$

Properties of Zero

Let a and b be real numbers, variables, or algebraic expressions. Then the following properties are true.

1. $a + 0 = a$ and $a - 0 = a$

2. $a \cdot 0 = 0$

3. $\dfrac{0}{a} = 0, \quad a \neq 0$

4. $\dfrac{a}{0}$ is undefined.

5. Zero-Factor Property: If $ab = 0$, then $a = 0$ or $b = 0$.

The "or" in the Zero-Factor Property includes the possibility that both factors are zero. This is called an **inclusive or,** and it is the way the word "or" is always used in mathematics.

Properties of Fractions

Let a, b, c, and d be real numbers, variables, or algebraic expressions such that $b \neq 0$ and $d \neq 0$. Then the following properties are true.

1. *Equivalent Fractions:* $\dfrac{a}{b} = \dfrac{c}{d}$ if and only if $ad = bc$.

2. *Rules of Signs:* $-\dfrac{a}{b} = \dfrac{-a}{b} = \dfrac{a}{-b}$ and $\dfrac{-a}{-b} = \dfrac{a}{b}$

3. *Generate Equivalent Fractions:* $\dfrac{a}{b} = \dfrac{ac}{bc}, \quad c \neq 0$

4. *Add or Subtract with Like Denominators:* $\dfrac{a}{b} \pm \dfrac{c}{b} = \dfrac{a \pm c}{b}$

5. *Add or Subtract with Unlike Denominators:* $\dfrac{a}{b} \pm \dfrac{c}{d} = \dfrac{ad \pm bc}{bd}$

6. *Multiply Fractions:* $\dfrac{a}{b} \cdot \dfrac{c}{d} = \dfrac{ac}{bd}$

7. *Divide Fractions:* $\dfrac{a}{b} \div \dfrac{c}{d} = \dfrac{a}{b} \cdot \dfrac{d}{c} = \dfrac{ad}{bc}, \quad c \neq 0$

In Property 1 (equivalent fractions) the phrase "if and only if" implies two statements. One statement is: If $a/b = c/d$, then $ad = bc$. The other statement is: If $ad = bc$, where $b \neq 0$ and $d \neq 0$, then $a/b = c/d$.

EXAMPLE 4 *Properties of Zero and Properties of Fractions*

a. $x - \dfrac{0}{5} = x - 0 = x$ — Properties 3 and 1 of zero

b. $\dfrac{x}{5} = \dfrac{3 \cdot x}{3 \cdot 5} = \dfrac{3x}{15}$ — Generate equivalent fractions.

c. $\dfrac{x}{3} + \dfrac{2x}{5} = \dfrac{x \cdot 5 + 3 \cdot 2x}{15}$ — Add fractions with unlike denominators.

d. $\dfrac{7}{x} \div \dfrac{3}{2} = \dfrac{7}{x} \cdot \dfrac{2}{3} = \dfrac{14}{3x}$ — Divide fractions.

If a, b, and c are integers such that $ab = c$, then a and b are **factors** or **divisors** of c. For example, 2 and 3 are factors of 6 because $2 \cdot 3 = 6$. A **prime number** is a positive integer that has exactly two factors: itself and 1. For example, 2, 3, 5, 7, and 11 are prime numbers, whereas 1, 4, 6, 8, 9, and 10 are not. The numbers 4, 6, 8, 9, and 10 are **composite** because they can be written as the products of two or more prime numbers. The number 1 is neither prime nor composite. The **Fundamental Theorem of Arithmetic** states that every positive integer greater than 1 can be written as the product of prime numbers in precisely one way (disregarding order). For instance, the prime factorization of 24 is

$$24 = 2 \cdot 2 \cdot 2 \cdot 3.$$

When adding or subtracting fractions with unlike denominators, you can use Property 4 of fractions by rewriting both fractions so that they have the same denominator. This is called the **least common denominator** method.

EXAMPLE 5 *Adding and Subtracting Fractions*

Evaluate $\dfrac{2}{15} - \dfrac{5}{9} + \dfrac{4}{5}$.

Solution

By prime factoring the denominators ($15 = 3 \cdot 5$, $9 = 3 \cdot 3$, and $5 = 5$) you can see that the least common denominator is $3 \cdot 3 \cdot 5 = 45$. Thus, it follows that

$$\frac{2}{15} - \frac{5}{9} + \frac{4}{5} = \frac{2 \cdot 3}{15 \cdot 3} - \frac{5 \cdot 5}{9 \cdot 5} + \frac{4 \cdot 9}{5 \cdot 9}$$
$$= \frac{6 - 25 + 36}{45}$$
$$= \frac{17}{45}.$$

Equations

An **equation** is a statement of equality between two expressions. Thus, the statement

$$a + b = c + d$$

means that the expressions $a + b$ and $c + d$ represent the same number. For instance, because $1 + 4$ and $3 + 2$ both represent the number 5, you can write $1 + 4 = 3 + 2$. Three important properties of equality are as follows.

Properties of Equality

Let a, b, and c be real numbers, variables, or algebraic expressions.

1. Reflexive: $a = a$

2. Symmetric: If $a = b$, then $b = a$.

3. Transitive: If $a = b$ and $b = c$, then $a = c$.

In algebra, you often rewrite expressions by making substitutions that are permitted under the **Substitution Principle.** "If $a = b$, then a can be replaced by b in any expression involving a." Two important consequences of the Substitution Principle are the following rules.

1. If $a = b$, then $a + c = b + c$.   Add c to both sides.

2. If $a = b$, then $ac = bc$.   Multiply both sides by c.

The first rule allows us to add the same number to both sides of an equation. The second allows us to multiply both sides of an equation by the same number. The converses of these two rules are called the **Cancellation Laws** for addition and multiplication.

1. If $a + c = b + c$, then $a = b$.   Subtract c from both sides.

2. If $ac = bc$ and $c \neq 0$, then $a = b$.   Divide both sides by c.

When adding, subtracting, multiplying, or dividing more than two numbers, it is important to use symbols of grouping (such as parentheses) to indicate the order of operations.

EXAMPLE 6   Symbols of Grouping

a. $7 - 3(4 - 2) = 7 - 3(2) = 7 - 6 = 1$

b. $(4 - 5) - (3 - 6) = (-1) - (-3) = -1 + 3 = 2$

Calculators and Rounding

This text includes two types of calculator problems. The first type can be done with a scientific calculator, whereas the second type requires a graphing calculator. The specific keystrokes that are listed in this text are those that correspond to a standard scientific calculator and to the *Texas Instruments TI-82*, which is a graphing calculator. Here are some comparisons between graphing calculator keys and scientific calculator keys.

1. The key marked ENTER is similar to = .

2. The key marked (−) is similar to +/− .

3. The key marked ∧ is similar to y^x .

4. The key marked x^{-1} is similar to $1/x$.

For example, you can evaluate 13^3 on a graphing calculator or a scientific calculator, as follows.

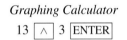

Graphing Calculator	*Scientific Calculator*
13 ∧ 3 ENTER	13 y^x 3 =

EXAMPLE 7 Using a Calculator

Scientific Calculator

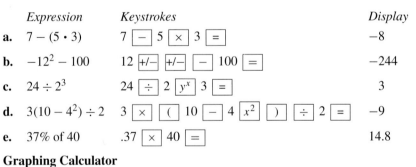

	Expression	*Keystrokes*	*Display*
a.	$7 - (5 \cdot 3)$	7 − 5 × 3 =	−8
b.	$-12^2 - 100$	12 +/− +/− − 100 =	−244
c.	$24 \div 2^3$	24 ÷ 2 y^x 3 =	3
d.	$3(10 - 4^2) \div 2$	3 × (10 − 4 x^2) ÷ 2 =	−9
e.	37% of 40	.37 × 40 =	14.8

Graphing Calculator

NOTE Be sure you see the difference between the change sign keys +/− or (−) and the subtraction key − , as used in Example 7(b).

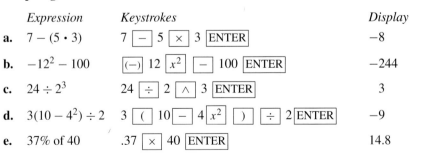

	Expression	*Keystrokes*	*Display*
a.	$7 - (5 \cdot 3)$	7 − 5 × 3 ENTER	−8
b.	$-12^2 - 100$	(−) 12 x^2 − 100 ENTER	−244
c.	$24 \div 2^3$	24 ÷ 2 ∧ 3 ENTER	3
d.	$3(10 - 4^2) \div 2$	3 (10 − 4 x^2) ÷ 2 ENTER	−9
e.	37% of 40	.37 × 40 ENTER	14.8

For all their usefulness, calculators do have a problem representing some numbers because they are limited to a finite number of digits. For instance, what does your calculator display when you compute $2 \div 3$? Some calculators simply truncate (drop) the digits that exceed their display range and display .66666666. Others will round the number and display .66666667. Although the second display is more accurate, *both* of these decimal representations of $\frac{2}{3}$ contain rounding errors. When rounding decimals in this text, use the following rule: *round up on 5 or greater, round down on 4 or less.*

EXAMPLE 8   Rounding Decimal Numbers

	Number	Rounded to Three Decimal Places	
a.	$\sqrt{2} = 1.4142135\ldots$	1.414	Round down.
b.	$\pi = 3.1415926\ldots$	3.142	Round up.
c.	$\dfrac{7}{9} = 0.7777777\ldots$	0.778	Round up.

One of the best ways to minimize error due to rounding is to leave numbers in your calculator until your calculations are complete. If you want to save a number for future use, store it in your calculator's memory.

Group Activities   Extending the Concept

Roundoff Error   The following sets of fractions add up to 1.

$$\textbf{a.}\ \ \frac{1}{6} + \frac{4}{6} + \frac{1}{6} = 1 \qquad \textbf{b.}\ \ \frac{1}{6} + \frac{2}{6} + \frac{2}{6} + \frac{1}{6} = 1$$

Convert the fractions in each set to decimals *rounded* to three places, and add again. What happens? For set (a), the rounded decimals will not add up to 1, *no matter how many decimal places of accuracy you keep.* Show this by trying again with the fractions rounded to several other numbers of decimal places. However, the rounded decimals for set (b) will always add up to 1 when all the fractions are rounded to the same number of decimal places. Round the fractions in this set to four decimal places and re-add; try again by rounding to two decimal places and then to one decimal place. Can you explain why there seems to be no rounding error when these decimals are added?

Warm Up

The following warm-up exercises involve skills that were covered in earlier sections. You will use these skills in the exercise set for this section.

In Exercises 1–4, place the correct inequality symbol ($<$ or $>$) between the two numbers.

1. -4 ___ -2 **2.** 0 ___ -3

3. $\sqrt{3}$ ___ 1.73 **4.** $-\pi$ ___ -3

In Exercises 5–8, find the distance between the two numbers.

5. $4, 6$ **6.** $-2, 2$

7. $0, -5$ **8.** $-1, 3$

In Exercises 9 and 10, evaluate the expression.

9. $|-7| + |7|$ **10.** $-|8 - 10|$

P.2 Exercises

In Exercises 1–6, identify the terms of the algebraic expression.

1. $7x + 4$ **2.** $-5 + 3x$

3. $x^2 - 4x + 8$ **4.** $3x^2 - 8x - 11$

5. $4x^3 + x - 5$ **6.** $3x^4 + 3x^3$

In Exercises 7–14, evaluate the expression for the values of x. (If not possible, state the reason.)

Expression	Values	
7. $4x - 6$	(a) $x = -1$	(b) $x = 0$
8. $9 - 7x$	(a) $x = -3$	(b) $x = 3$
9. $x^2 - 3x + 4$	(a) $x = -2$	(b) $x = 2$
10. $-x^2 + 5x - 4$	(a) $x = -1$	(b) $x = 1$
11. $-x^3 + 4x$	(a) $x = 0$	(b) $x = 2$
12. $2x^3 - x - 1$	(a) $x = 1$	(b) $x = 0$
13. $\dfrac{x + 1}{x - 1}$	(a) $x = 1$	(b) $x = -1$

Expression	Values	
14. $\dfrac{x}{x + 2}$	(a) $x = 2$	(b) $x = -2$

In Exercises 15–30, identify the rule(s) of algebra illustrated by the equation.

15. $3 + 4 = 4 + 3$ **16.** $x + 9 = 9 + x$

17. $-15 + 15 = 0$ **18.** $(x + 3) - (x + 3) = 0$

19. $2(x + 3) = 2x + 6$ **20.** $2\left(\frac{1}{2}\right) = 1$

21. $(5 + 11) \cdot 6 = 5 \cdot 6 + 11 \cdot 6$

22. $\dfrac{1}{h + 6}(h + 6) = 1, \ h \neq -6$

23. $h + 0 = h$

24. $57 \cdot 1 = 57$

25. $(z - 2) + 0 = z - 2$

26. $1 \cdot (1 + x) = 1 + x$

27. $6 + (7 + 8) = (6 + 7) + 8$

28. $x + (y + 10) = (x + y) + 10$

29. $x(3y) = (x \cdot 3)y = (3x)y$

30. $\frac{1}{7}(7 \cdot 12) = \left(\frac{1}{7} \cdot 7\right)12 = 1 \cdot 12 = 12$

In Exercises 31–34, write the prime factorization of the integer.

31. 36

32. 30

33. 14

34. 56

In Exercises 35–54, perform the operations. (Write fractional answers in reduced form.)

35. $(8 - 17) + 3$

36. $3 - (-6)$

37. $10 - 6 - 2$

38. $-3(5 - 2)$

39. $(4 - 7)(-2)$

40. $(-5)(-8)$

41. $2\left(\dfrac{77}{-11}\right)$

42. $\dfrac{27 - 35}{4}$

43. $8(-6)(-2)$

44. $\dfrac{3}{16} + \dfrac{5}{16}$

45. $\dfrac{6}{7} - \dfrac{4}{7}$

46. $\dfrac{10}{11} + \dfrac{6}{33} - \dfrac{13}{66}$

47. $\dfrac{2}{5} \cdot \dfrac{7}{8}$

48. $4 \cdot \dfrac{3}{4}$

49. $\dfrac{4}{5} \cdot \dfrac{1}{2} \cdot \dfrac{3}{4}$

50. $\dfrac{2}{3} \cdot \dfrac{5}{8} \cdot \dfrac{3}{4}$

51. $\dfrac{2}{3} \div 8$

52. $\dfrac{11}{16} \div \dfrac{3}{4}$

53. $12 \div \dfrac{1}{4}$

54. $\left(\dfrac{3}{5} \div 3\right) - \left(6 \cdot \dfrac{4}{8}\right)$

In Exercises 55–64, use a calculator to evaluate the expression. (Round to two decimal places.)

55. $4 + \dfrac{5}{11}$

56. $-3 + \dfrac{3}{7}$

57. $\dfrac{1}{8} + \dfrac{1}{7}$

58. $\dfrac{3}{4} + \dfrac{1}{3}$

59. $3\left(-\dfrac{5}{12} + \dfrac{3}{8}\right)$

60. $2\left(-7 + \dfrac{1}{6}\right)$

61. $\dfrac{11.46 - 5.37}{3.91}$

62. $\dfrac{-8.31 + 4.83}{7.65}$

63. $\dfrac{\frac{1}{3}(-4 + 8)}{6}$

64. $\dfrac{\frac{1}{5}(-8 - 9)}{-\frac{1}{3}}$

In Exercises 65–70, use a calculator to solve the percent problem.

65. 33% of 57

66. 17% of 29

67. 29% of 685

68. 87% of 1578

69. 149% of 12

70. 121% of 34

In Exercises 71 and 72, find the percent that corresponds to the unlabeled portion of the circle graph.

71.

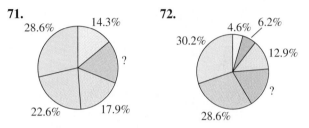

28.6% 14.3% ? 22.6% 17.9%

72.

30.2% 4.6% 6.2% 12.9% ? 28.6%

73. *Vehicle Sales*   The circle graph shows the market shares for American vehicle sales (cars and trucks) for 1990. What percent of the market did General Motors have? If 15 million vehicles were sold during 1990, find the number of vehicles sold by General Motors, Ford, Chrysler, Toyota, Honda, and Nissan. (Round your answers to the nearest tenth of a million vehicles.)   (Source: *USA Today*)

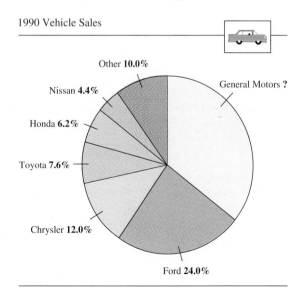

1990 Vehicle Sales

Other **10.0%**
General Motors **?**
Nissan **4.4%**
Honda **6.2%**
Toyota **7.6%**
Chrysler **12.0%**
Ford **24.0%**

74. *Federal Government Expenses* The circle graph shows the types of expenses for the federal government in 1992. What percent of the total budget was spent by the Treasury Department? If the total expenses were 1,381,800,000,000 dollars, find the expense for each of the indicated categories. (Round your answers to the nearest billion dollars.) (Source: Office of Management and Budget)

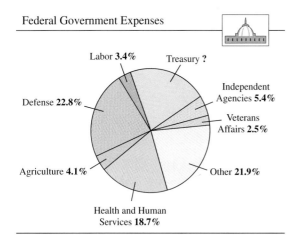

Federal Government Expenses

Labor **3.4%**
Treasury **?**
Independent Agencies **5.4%**
Defense **22.8%**
Veterans Affairs **2.5%**
Agriculture **4.1%**
Other **21.9%**
Health and Human Services **18.7%**

75. *Calculator Keystrokes* Write the algebraic expression that corresponds to the following keystrokes.

5 × (2.7 − 9.4) =   Scientific

5 (2.7 − 9.4) ENTER   Graphing

76. *Calculator Keystrokes* Write the algebraic expression that corresponds to the following keystrokes.

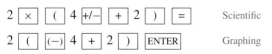

2 × (4 +/− + 2) =   Scientific

2 ((−) 4 + 2) ENTER   Graphing

77. *Calculator Keystrokes* Write the keystrokes used to evaluate the following algebraic expression. (Base your description on either a scientific or a graphing calculator.)

$5(18 - 2^3) \div 10$

78. *Calculator Keystrokes* Write the keystrokes used to evaluate the following algebraic expression. (Base your description on either a scientific or a graphing calculator.)

$-6^2 - \left[7 + (-2)^3 \right]$

79. *College Enrollment* The percent of students 24 years old or older at a college is 44.7%. If the college enrollment for the 1992-1993 academic year was 13,385 students, how many students were under 24 years old?

80. *College Enrollment* The percent of students who attend evening classes at a college is 44.1%. If the total enrollment for the college is 11,875 students, how many students attend evening classes?

C A R E E R   I N T E R V I E W

Laura Balaoro

Real Estate Broker

Laura Balaoro and Associates Real Estate Services San Jose, CA 95110

I own and operate an independent real estate brokerage. I deal primarily in residential sales.

One of the more important decisions to be made in the actual purchase of a house is the size of the buyer's down payment. The larger the down payment a buyer can make, the less mortgage interest he or she will pay in the long run. A low down payment can also result in higher closing costs as a result of additional mortgage insurance. However, it is often difficult for a first-time buyer to make a substantial down payment. We normally figure the size of the down payment as a percent of the price of the house, so we use the formula down payment = house price × percent. It is standard in my area for first-time buyers to use 10% as a minimum down payment. With an average home price of $210,000 in my region, a first-time buyer might expect to need a down payment of at least $210,000 × 10%, or $21,000.

P.3 Integer Exponents

Properties of Exponents ▪ Exponents and Calculators ▪ Scientific Notation

Properties of Exponents

Repeated multiplication of a real number by itself can be written in **exponential form.** Here are some examples.

Repeated Multiplication	*Exponential Form*
$7 \cdot 7$	7^2
$a \cdot a \cdot a \cdot a \cdot a$	a^5
$(-4)(-4)(-4)$	$(-4)^3$
$(2x)(2x)(2x)(2x)$	$(2x)^4$

Exponential Notation

Let a be a real number, a variable, or an algebraic expression, and let n be a positive integer. Then

$$a^n = \underbrace{a \cdot a \cdot a \cdots a}_{n \text{ factors}}$$

where n is the **exponent** and a is the **base.** The expression a^n is read as "a to the nth **power**" or simply "a to the nth."

When multiplying exponential expressions with the same base, *add* exponents.

$$a^m \cdot a^n = a^{m+n} \qquad \text{Add exponents when multiplying.}$$

For instance, to multiply 2^2 and 2^3, you can write

$$2^2 \cdot 2^3 = \overbrace{(2 \cdot 2)}^{\substack{\text{Two} \\ \text{factors}}} \cdot \overbrace{(2 \cdot 2 \cdot 2)}^{\substack{\text{Three} \\ \text{factors}}} = \overbrace{2 \cdot 2 \cdot 2 \cdot 2 \cdot 2)}^{\substack{\text{Five} \\ \text{factors}}}$$
$$= 2^{2+3} = 2^5.$$

On the other hand, when dividing exponential expressions, *subtract* exponents. That is,

$$\frac{a^m}{a^n} = a^{m-n}, \qquad a \neq 0. \qquad \text{Subtract exponents when dividing.}$$

These and other properties of exponents are summarized in the list on the following page.

Properties of Exponents

Let a and b be real numbers, variables, or algebraic expressions, and let m and n be integers. (Assume all denominators and bases are nonzero.)

Property	*Example*
1. $a^m a^n = a^{m+n}$	$3^2 \cdot 3^4 = 3^{2+4} = 3^6$
2. $\dfrac{a^m}{a^n} = a^{m-n}$	$\dfrac{x^7}{x^4} = x^{7-4} = x^3$
3. $a^{-n} = \dfrac{1}{a^n}$	$y^{-4} = \dfrac{1}{y^4}$
4. $a^0 = 1, \quad a \neq 0$	$(x^2 + 1)^0 = 1$
5. $(ab)^m = a^m b^m$	$(5x)^3 = 5^3 x^3 = 125x^3$
6. $(a^m)^n = a^{mn}$	$(y^3)^{-4} = y^{3(-4)} = y^{-12} = \dfrac{1}{y^{12}}$
7. $\left(\dfrac{a}{b}\right)^m = \dfrac{a^m}{b^m}$	$\left(\dfrac{2}{x}\right)^3 = \dfrac{2^3}{x^3} = \dfrac{8}{x^3}$
8. $\left(\dfrac{a}{b}\right)^{-n} = \left(\dfrac{b}{a}\right)^n, a \neq 0, b \neq 0$	$\left(\dfrac{3}{2}\right)^{-3} = \left(\dfrac{2}{3}\right)^3$
9. $|a^2| = |a|^2 = a^2$	$|(-2)^2| = |-2|^2 = (-2)^2 = 4$

NOTE Be sure you see that these properties of exponents apply for *all* integers m and n, not just positive ones. For instance, by Property 2, you have

$$\frac{3^4}{3^{-5}} = 3^{4-(-5)}$$
$$= 3^{4+5}$$
$$= 3^9.$$

DISCOVERY

Using your calculator, find the values of $10^3, 10^2, 10^1, 10^0, 10^{-1}$, and 10^{-2}. What do you observe? Explain how to use the fact that $10^2 = 100$ and $10^1 = 10$ to find 10^0 and 10^{-1}.

EXAMPLE 1 Using Properties of Exponents

a. $3^4 \cdot 3^{-1} = 3^{4-1} = 3^3 = 27$

b. $\dfrac{5^6}{5^4} = 5^{6-4} = 5^2 = 25$

c. $5\left(\dfrac{2}{5}\right)^3 = 5 \cdot \dfrac{2^3}{5^3} = 5 \cdot 5^{-3} \cdot 2^3 = 5^{-2}2^3 = \dfrac{2^3}{5^2} = \dfrac{8}{25}$

d. $(-5 \cdot 2^3)^2 = (-5)^2 \cdot (2^3)^2 = 25 \cdot 2^6 = 25 \cdot 64 = 1600$

EXAMPLE 2 Using Properties of Exponents

a. $(-3ab^4)(4ab^{-3}) = -12a \cdot a(b^4)(b^{-3}) = -12a^2 b$

b. $3a(-4a^2)^0 = 3a(1) = 3a, \ a \neq 0$

c. $\left(\dfrac{5x^3}{y}\right)^2 = \dfrac{5^2(x^3)^2}{y^2} = \dfrac{25x^6}{y^2}$

The next two examples show how expressions involving negative exponents can be rewritten using positive exponents.

EXAMPLE 3   Rewriting with Positive Exponents

a. $x^{-1} = \dfrac{1}{x}$

Property 3: $a^{-n} = \dfrac{1}{a^n}$

b. $\dfrac{1}{3x^{-2}} = \dfrac{1(x^2)}{3} = \dfrac{x^2}{3}$

-2 exponent does not apply to 3.

c. $\dfrac{12a^3 b^{-4}}{4a^{-2} b} = \dfrac{12a^3 \cdot a^2}{4b \cdot b^4} = \dfrac{3a^5}{b^5}$

EXAMPLE 4   Quotients Raised to Negative Powers

Rewrite $\left(\dfrac{3x^2}{y}\right)^{-2}$ and simplify.

Solution

$$\left(\frac{3x^2}{y}\right)^{-2} = \frac{3^{-2}(x^2)^{-2}}{y^{-2}} = \frac{3^{-2}x^{-4}}{y^{-2}} = \frac{y^2}{3^2 x^4} = \frac{y^2}{9x^4}$$

Real Life

EXAMPLE 5   Ratio of Volume to Surface Area

The volume V and surface area S of a sphere are given by

$$V = \frac{4}{3}\pi r^3 \qquad \text{and} \qquad S = 4\pi r^2$$

where r is the radius of the sphere. A spherical weather balloon has a radius of 2 feet, as shown in Figure P.7. Find the quotient of the volume and the surface area.

Solution

Form the quotient V/S, and simplify, as follows.

$$\begin{aligned}
\frac{V}{S} &= \frac{\left(\frac{4}{3}\right)\pi r^3}{4\pi r^2} \\
&= \frac{\left(\frac{4}{3}\right)\pi 2^3 \text{ ft}^3}{4\pi 2^2 \text{ ft}^2} \\
&= \frac{1}{3}(2) \text{ feet} \\
&= \frac{2}{3} \text{ feet}
\end{aligned}$$

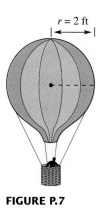

$r = 2$ ft

FIGURE P.7

Exponents and Calculators

One of the most useful features of a calculator is its ability to evaluate exponential expressions. Note how this is done in the next example.

EXAMPLE 6 Using a Calculator to Raise a Number to a Power

Scientific Calculator

	Expression	*Keystrokes*	*Display*
a.	$13^4 + 5$	13 $\boxed{y^x}$ 4 $\boxed{+}$ 5 $\boxed{=}$	28566
b.	$3^{-2} + 4^{-1}$	3 $\boxed{y^x}$ 2 $\boxed{+/-}$ $\boxed{+}$ 4 $\boxed{y^x}$ 1 $\boxed{+/-}$ $\boxed{=}$	.3611111111
c.	$\dfrac{3^5 + 1}{3^5 - 1}$	$\boxed{(}$ 3 $\boxed{y^x}$ 5 $\boxed{+}$ 1 $\boxed{)}$ $\boxed{\div}$ $\boxed{(}$ 3 $\boxed{y^x}$ 5 $\boxed{-}$ 1 $\boxed{)}$ $\boxed{=}$	1.008264463

Graphing Calculator

	Expression	*Keystrokes*	*Display*
a.	$13^4 + 5$	13 $\boxed{\wedge}$ 4 $\boxed{+}$ 5 $\boxed{\text{ENTER}}$	28566
b.	$3^{-2} + 4^{-1}$	3 $\boxed{\wedge}$ $\boxed{(-)}$ 2 $\boxed{+}$ 4 $\boxed{\wedge}$ $\boxed{(-)}$ 1 $\boxed{\text{ENTER}}$	.3611111111
c.	$\dfrac{3^5 + 1}{3^5 - 1}$	$\boxed{(}$ 3 $\boxed{\wedge}$ 5 $\boxed{+}$ 1 $\boxed{)}$ $\boxed{\div}$ $\boxed{(}$ 3 $\boxed{\wedge}$ 5 $\boxed{-}$ 1 $\boxed{)}$ $\boxed{\text{ENTER}}$	1.008264463

The following box gives a formula for finding the balance in a savings account. (You will study this formula in detail in Section 5.1.)

Balance in an Account

The balance A in an account that earns an annual interest rate of r (in decimal form) for t years is given by one of the following.

$$A = P(1 + rt) \qquad \text{Simple interest}$$

$$A = P\left(1 + \frac{r}{n}\right)^{nt} \qquad \text{Compound interest}$$

In both formulas, P is the principal (or the initial deposit). In the formula for compound interest, n is the number of compoundings *per year*.

Real Life

EXAMPLE 7   *Finding the Balance in an Account*

a. If a friend borrows $250 at 8% annual interest for 6 months, the amount due after 6 months (at simple interest) is

$$A = P(1 + rt) = 250\left[1 + 0.08\left(\tfrac{1}{2}\right)\right] = \$260.00.$$

b. If you deposit $1000 at 6% annual interest compounded quarterly for 5 years, the balance after 5 years is

$$A = P\left(1 + \frac{r}{n}\right)^{nt} = 1000\left(1 + \frac{0.06}{4}\right)^{(4)(5)} = \$1346.86.$$

In addition to finding the balance in an account, the compound interest formula can also be used to determine the rate of inflation. To apply the formula, you must know the cost of an item for two different years, as demonstrated in Example 8.

Real Life

EXAMPLE 8   *Finding the Rate of Inflation*

In 1974, the cost of a first-class postage stamp was $0.10. By 1995, the cost had risen to $0.32, as shown in Figure P.8. Find the average annual rate of inflation for first-class postage over this 21-year period.   (Source: U.S. Postal Service)

Solution

To find the average annual rate of inflation, use the formula for compound interest with *annual* compounding. Thus, you need to find the value of r that will make the following equation true.

$$A = P\left(1 + \frac{r}{n}\right)^{nt}$$

$$0.32 = 0.10(1 + r)^{21}$$

You can begin by guessing that the average annual rate of inflation was 5%. Entering $r = 0.05$ in the formula, you find that $0.10(1 + 0.05)^{21} \approx 0.2786$. Because this result is less than 0.32, try some larger values of r. Finally, you can discover that

$$0.10(1 + 0.057)^{21} \approx 0.32.$$

Therefore, the average annual rate of inflation for first-class postage from 1974 to 1995 was about 5.7%.

There have been many periods in U.S. history that have had little inflation. For instance, the cost of a first-class letter was $0.03 from 1851 to 1958.

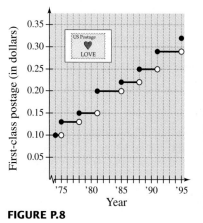

FIGURE P.8

Scientific Notation

Exponents provide an efficient way of writing and computing with very large (or very small) numbers. For instance, a drop of water contains more than 33 billion billion molecules—that is, 33 followed by 18 zeros.

$$33,000,000,000,000,000,000$$

It is convenient to write such numbers in **scientific notation.** This notation has the form $c \times 10^n$, where $1 \leq c < 10$ and n is an integer. Thus, the number of molecules in a drop of water can be written in scientific notation as

$$3.3 \times 10,000,000,000,000,000,000 = 3.3 \times 10^{19}.$$

The *positive* exponent 19 indicates that the number is large (10 or more) and that the decimal point has been moved 19 places. A *negative* exponent in scientific notation indicates that the number is *small* (less than 1). For instance, the mass (in grams) of one electron is approximately

$$9.0 \times 10^{-28} = 0.00000000000000000000000000009.$$

28 decimal places

EXAMPLE 9   Converting to Scientific Notation

a. $0.0000572 = 5.72 \times 10^{-5}$ Number is less than 1.

b. $149,400,000 = 1.494 \times 10^8$ Number is greater than 10.

EXAMPLE 10   Converting to Decimal Notation

a. $3.125 \times 10^2 = 312.5$ Number is greater than 10.

b. $3.73 \times 10^{-6} = 0.00000373$ Number is less than 1.

Activities

1. Rewrite using the Distributive Property: $3(x - y)$.

   Answer: $3x - 3y$

2. Simplify: $\left(\dfrac{6x^{-1}y}{y^3} \right)^{-2}$.

   Answer: $\dfrac{x^2 y^4}{36}$

3. Write in scientific notation: 39,000,000.

   Answer: 3.9×10^7

Most calculators automatically switch to scientific notation when they are showing large (or small) numbers that exceed the display range. Try multiplying $86,500,000 \times 6000$. If your calculator follows standard conventions, its display should be

5.19   11   or   5.19 E 11 .

This means that $c = 5.19$ and the exponent of 10 is $n = 11$, which implies that the number is 5.19×10^{11}. To *enter* numbers in scientific notation, your calculator should have an exponential entry key labeled EE .

EXAMPLE 11   The Speed of Light

The distance between earth and the sun is approximately 93 million miles, as shown in Figure P.9. How long does it take for light to travel from the sun to earth? Use the fact that light travels at a rate of approximately 186,000 miles per second.

Solution

Using the formula Distance = (rate)(time), you find the time as follows.

$$\text{Time} = \frac{\text{distance}}{\text{rate}} = \frac{93 \text{ million miles}}{186,000 \text{ miles per second}}$$

$$= \frac{9.3 \times 10^7 \text{ miles}}{1.86 \times 10^5 \text{ miles/second}}$$

$$= 5 \times 10^2 \text{ seconds}$$

$$\approx 8.33 \text{ minutes}$$

Note that to convert 500 seconds to 8.33 minutes, you divided by 60 because there are 60 seconds in a minute.

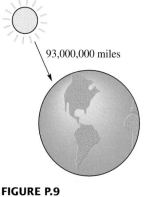

93,000,000 miles

FIGURE P.9

Group Activities   Problem Solving

Compound Interest   Suppose that you have $5000 to invest for 5 years in an account that pays compound interest according to the formula

$$A = 5000 \left(1 + \frac{r}{n}\right)^{5n}.$$

Which of the following combinations would produce the highest balance?

	Annual Interest Rate	Type of Compounding
a.	5.60%	Daily ($n = 365$)
b.	5.65%	Weekly
c.	5.70%	Monthly
d.	5.75%	Quarterly

Choose one annual interest rate and construct a bar chart or table showing the balance that could be obtained with each type of compounding. Discuss how the balance is affected by the type of compounding.

Warm Up The following warm-up exercises involve skills that were covered in earlier sections. You will use these skills in the exercise set for this section.

In Exercises 1–10, perform the indicated operation(s) and simplify.

1. $\left(\frac{2}{3}\right)\left(\frac{3}{2}\right)$

2. $\left(\frac{1}{4}\right)(5)(4)$

3. $3\left(\frac{2}{7}\right) + 11\left(\frac{2}{7}\right)$

4. $\frac{1}{2} \div 2$

5. $\frac{1}{3} + \frac{1}{2} - \frac{5}{6}$

6. $\frac{1}{3} \div \frac{1}{3}$

7. $\frac{1}{7} + \frac{1}{3} - \frac{1}{21}$

8. $11\left(\frac{1}{4}\right) + \frac{5}{4}$

9. $\frac{1}{12} - \frac{1}{3} + \frac{1}{8}$

10. $\left(\frac{1}{2} - \frac{1}{3}\right) \div \frac{1}{6}$

P.3 **Exercises**

In Exercises 1–22, evaluate the expression. Write fractional answers in simplified form.

1. $2^2 \cdot 2^4$

2. $3 \cdot 3^3$

3. $\frac{5^5}{5^2}$

4. $\frac{2^6}{2^3}$

5. $(3^3)^2$

6. $(2^4)^2$

7. -3^4

8. $(-3)^4$

9. $\left(\frac{1}{2}\right)^{-3}$

10. $-(-4)^2$

11. $5^{-1} + 2^{-1}$

12. $5^{-1} - 2^{-1}$

13. $\left(\frac{3}{4}\right)^2$

14. $\left(\frac{3}{4}\right)^{-2}$

15. $(2^3 \cdot 3^2)^2$

16. $(-5 \cdot 4^2)^3$

17. $\left(-\frac{3}{5}\right)^3 \left(\frac{5}{3}\right)^2$

18. $4\left(\frac{1}{2}\right)^5$

19. $6^5 \cdot 6^{-3}$

20. $6 \cdot 2^{-3} \cdot 3^{-1}$

21. 3^0

22. $(-2)^0$

In Exercises 23–28, evaluate the expression for the indicated value of x.

	Expression	Value
23.	$-3x^3$	$x = 2$
24.	$\dfrac{x^2}{2}$	$x = 6$
25.	$4x^{-3}$	$x = 2$
26.	$7x^{-2}$	$x = 4$
27.	$6x^0 - (6x)^0$	$x = 10$
28.	$5(-x)^3$	$x = 3$

In Exercises 29–56, simplify the expression.

29. $(3x)^2$

30. $(-5z)^3$

31. $5x^4(x^2)$

32. $(8x^4)(2x^3)$

33. $10(x^2)^2$

34. $(4x^3)^2$

35. $6y^2(2y^4)^2$

36. $(-z)^3(3z^4)$

37. $\dfrac{3x^5}{x^3}$

38. $\dfrac{25y^8}{10y^4}$

39. $\dfrac{7x^2}{x^3}$

40. $\dfrac{r^4}{r^6}$

41. $\dfrac{12(x+y)^3}{9(x+y)}$

42. $\left(\dfrac{4}{y}\right)^3\left(\dfrac{3}{y}\right)^4$

43. $(x+5)^0, \quad x \neq -5$

44. $(2x^5)^0, \quad x \neq 0$

45. $(2x^2)^{-2}$

46. $(z+2)^{-3}(z+2)^{-1}$

47. $(-2x^2)^3(4x^3)^{-1}$

48. $(4y^{-2})(8y^4)$

49. $\left(\dfrac{x}{10}\right)^{-1}$

50. $\left(\dfrac{4}{z}\right)^{-2}$

51. $\left(\dfrac{3z^2}{x}\right)^{-2}$

52. $\left(\dfrac{x^{-3}y^4}{5}\right)^{-3}$

53. $3^n \cdot 3^{2n}$

54. $2^m \cdot 2^{3m}$

55. $\dfrac{x^n \cdot x^{2n}}{x^{3n}}$

56. $\dfrac{x^2 \cdot x^n}{x^3 \cdot x^n}$

In Exercises 57–64, write the number in scientific notation.

57. *Land Area of Earth:*   57,500,000 square miles

58. *Ocean Area of Earth:*   139,400,000 square miles

59. *Light Year:*   9,461,000,000,000,000 kilometers

60. *Approximate World Population:*   5,000,000,000

61. *Relative Density of Hydrogen:*   0.0000899

62. *One Point (Printer's Measure):*   0.013837 inch

63. *One Micron:*   0.00003937 inch

64. *Thickness of Soap Bubble:*   0.0000001 meter

In Exercises 65–72, write the number in decimal form.

65. *Daily Coca-Cola Consumption:* 5.24×10^8 servings

66. *Number of Air Sacs in Lungs:* 3.5×10^8

67. *Temperature of Sun:* $1.3 \times 10^{7\circ}$ Celsius

68. *Number of Insect Species:* 9.5×10^5

69. *Electron Charge:* 4.8×10^{-10} electrostatic unit

70. *Width of Human Hair:* 9.0×10^{-4} meter

71. *Width of Air Molecule:* 1.0×10^{-9} meter

72. *Gravity Constant:* 6.673×10^{-8} c^3/gram-sec^2

In Exercises 73–78, use a calculator to evaluate the expression. (Round to three decimal places.)

73. (a) $2400(1 + 0.06)^{20}$

   (b) $750\left(1 + \dfrac{0.11}{365}\right)^{800}$

74. (a) $5000(1 + 0.02)^{60}$

   (b) $2500\left(1 + \dfrac{0.085}{12}\right)^{60}$

75. (a) $0.000345(8,900,000,000)$

   (b) $\dfrac{67,000,000 + 93,000,000}{0.0052}$

76. (a) $0.000045(9,200,000)$

   (b) $\dfrac{0.0000928 - 0.0000021}{0.0061}$

77. (a) $(9.3 \times 10^6)^3(6.1 \times 10^{-4})$

   (b) $\dfrac{(2.414 \times 10^4)^6}{(1.68 \times 10^5)^5}$

78. (a) $(1.2 \times 10^2)^2(5.3 \times 10^{-5})$

   (b) $\dfrac{(3.28 \times 10^{-6})^{10}}{(5.34 \times 10^{-3})^{22}}$

In Exercises 79 and 80, complete the table by finding the balances.

Number of Years	5	10	15	20	25
Balance					

79. *Balance in an Account*   One thousand dollars is deposited in an account with an annual percentage rate of 8%, compounded monthly.

80. *Balance in an Account*   Five thousand dollars is deposited in an account with an annual percentage rate of 6%, compounded quarterly.

81. *Balance in an Account* Ten thousand dollars is deposited in an account with an annual percentage rate of 6.5% for 10 years. What is the balance in the account if the interest is compounded (a) quarterly and (b) monthly?

82. *Balance in an Account* Two thousand dollars is deposited in an account with an annual percentage rate of 7.75% for 15 years. What is the balance in the account if the interest is compounded (a) quarterly and (b) monthly?

83. *Reasoning* A student calculates the amount in an account after 5 years to be $10,616.78 when $2000 was invested at 6% interest compounded monthly. Determine the error the student made in the calculation if the problem was set up correctly. Explain to the student how to correct the error.

84. *Reasoning* A student tells you that $1000 invested at 8% for 10 years compounded quarterly will yield $2208.04. The student then concludes that $2000 invested at 8% for 10 years compounded quarterly will yield $4416.08. Do you agree? Explain.

85. *College Tuition* The bar chart shows the average tuition at private universities in the United States from 1982 to 1992. Estimate the average annual rate of inflation for tuition over this 10-year period. (Source: College Board)

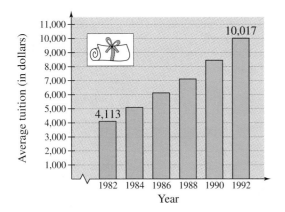

86. *Minimum Wage* The bar chart shows the minimum wage in the United States from 1973 to 1993. Estimate the average annual rate of inflation for minimum wages over this 20-year period. (Source: U.S. Department of Labor)

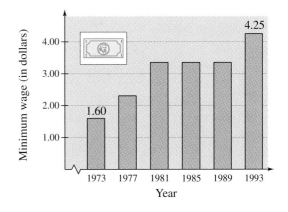

87. *Calculator Keystrokes* Write the algebraic expression that corresponds to the following keystrokes.

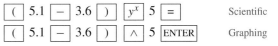

88. *Calculator Keystrokes* Write the algebraic expression that corresponds to the following keystrokes.

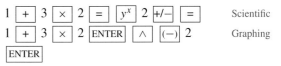

89. *Think about It* Because $-2^3 = -8$ and $(-2)^3 = -8$, a student concludes that $-a^n = (-a)^n$ where n is an integer. Do you agree? Can you find an example where $-a^n \neq (-a)^n$?

90. *Think about It* Because $2^0 = 1$ and $3^0 = 1$, a student concludes that $0^0 = 1$. Do you agree? What argument would you give to show that $0^0 \neq 1$?

P.4	**Radicals and Rational Exponents**

Radicals and Properties of Radicals ▪ Simplifying Radicals ▪
Rational Exponents ▪ Radicals and Calculators

Radicals and Properties of Radicals

A **square root** of a number is defined as one of its equal factors. For example, 5 is a square root of 25 because 5 is one of the two equal factors of 25. In a similar way, a **cube root** of a number is one of its three equal factors. Here are some examples.

Number	*Equal factors*	*Root*	
$25 = (-5)^2$	$(-5)(-5)$	-5	(square root)
$-64 = (-4)^3$	$(-4)(-4)(-4)$	-4	(cube root)
$81 = 3^4$	$3 \cdot 3 \cdot 3 \cdot 3$	3	(fourth root)

In general, an ***n*th root** of a number is defined as follows.

Definition of *n*th Root of a Number

Let a and b be real numbers and let n be a positive integer. If

$$a = b^n$$

then b is an ***n*th root of *a*.** If $n = 2$, the root is a **square root,** and if $n = 3$, the root is a **cube root.**

From this definition, you can see that some numbers have more than one nth root. For example, both 5 and -5 are square roots of 25. The following definition distinguishes between these two roots.

Principal *n*th Root of a Number

Let a be a real number that has at least one real nth root. The **principal *n*th root of *a*** is the nth root that has the same sign as a, and it is denoted by the **radical symbol**

$$\sqrt[n]{a}.$$ Principal nth root

The positive integer n is the **index** of the radical, and the number a is the **radicand.** If $n = 2$, omit the index and write $\sqrt{a}$ rather than $\sqrt[2]{a}$.

NOTE   The plural of index is *indexes* or *indices*.

EXAMPLE 1   Evaluating Expressions Involving Radicals

a. The principal square root of 121 is $\sqrt{121} = 11$ because $11^2 = 121$.

b. The principal cube root of $\frac{125}{64}$ is $\sqrt[3]{\frac{125}{64}} = \frac{5}{4}$ because $\left(\frac{5}{4}\right)^3 = \frac{5^3}{4^3} = \frac{125}{64}$.

c. The principal fifth root of -32 is $\sqrt[5]{-32} = -2$ because $(-2)^5 = -32$.

d. $-\sqrt{49} = -7$ because $7^2 = 49$.

e. $\sqrt[4]{-81}$ is not a real number because there is no real number that can be raised to the fourth power to produce -81.

From Example 1, you can make the following generalizations about nth roots of a real number.

1. If a is a positive real number and n is a positive *even* integer, then a has exactly two real nth roots, which are denoted by $\sqrt[n]{a}$ and $-\sqrt[n]{a}$.

2. If a is any real number and n is an *odd* integer, then a has only one (real) nth root. It is the principal nth root and is denoted by $\sqrt[n]{a}$.

3. If a is negative and n is an *even* integer, then a has no (real) nth root.

Integers such as 1, 4, 9, 16, 49, and 81 are called **perfect squares** because they have integer square roots. Similarly, integers such as 1, 8, 27, 64, and 125 are called **perfect cubes** because they have integer cube roots.

Properties of Radicals

Let a and b be real numbers, variables, or algebraic expressions such that the indicated roots are real numbers, and let m and n be positive integers. Then the following properties are true.

	Property	*Example*
1.	$\sqrt[n]{a^m} = \left(\sqrt[n]{a}\right)^m$	$\sqrt[3]{8^2} = \left(\sqrt[3]{8}\right)^2 = (2)^2 = 4$
2.	$\sqrt[n]{a} \cdot \sqrt[n]{b} = \sqrt[n]{ab}$	$\sqrt{5} \cdot \sqrt{7} = \sqrt{5 \cdot 7} = \sqrt{35}$
3.	$\dfrac{\sqrt[n]{a}}{\sqrt[n]{b}} = \sqrt[n]{\dfrac{a}{b}}, \quad b \neq 0$	$\dfrac{\sqrt[4]{27}}{\sqrt[4]{9}} = \sqrt[4]{\dfrac{27}{9}} = \sqrt[4]{3}$
4.	$\sqrt[m]{\sqrt[n]{a}} = \sqrt[mn]{a}$	$\sqrt[3]{\sqrt{10}} = \sqrt[6]{10}$
5.	$\left(\sqrt[n]{a}\right)^n = a$	$\left(\sqrt{3}\right)^2 = 3$
6.	For n even, $\sqrt[n]{a^n} = |a|$.	$\sqrt{(-12)^2} = |-12| = 12$
	For n odd, $\sqrt[n]{a^n} = a$.	$\sqrt[3]{(-12)^3} = -12$

NOTE   A common special case of Property 6 is $\sqrt{a^2} = |a|$.

Simplifying Radicals

An expression involving radicals is in **simplest form** when the following conditions are satisfied.

1. All possible factors have been removed from the radical.
2. All fractions have radical-free denominators (accomplished by a process called *rationalizing the denominator*).
3. The index of the radical has been reduced as far as possible.

To simplify a radical, factor the radicand into factors whose powers are multiples of the index. The roots of these factors are written outside the radical, and the "leftover" factors make up the new radicand.

NOTE   In Example 2(c), note that absolute value is included in the answer because $\sqrt[4]{x^4} = |x|$.

EXAMPLE 2   Simplifying Even Roots

a. $\sqrt[4]{48} = \sqrt[4]{16 \cdot 3}$ Find largest fourth-power factor.

$\quad\quad = \sqrt[4]{2^4 \cdot 3}$ Rewrite.

$\quad\quad = 2\sqrt[4]{3}$ Find fourth root.

b. $\sqrt{75x^3} = \sqrt{25x^2 \cdot 3x}$ Find largest square factor.

$\quad\quad = \sqrt{(5x)^2 \cdot 3x}$ Rewrite.

$\quad\quad = 5x\sqrt{3x}, \quad x \geq 0$ Find root of perfect square.

c. $\sqrt[4]{(5x)^4} = |5x| = 5|x|.$

EXAMPLE 3   Simplifying Odd Roots

a. $\sqrt[3]{24} = \sqrt[3]{8 \cdot 3}$ Find largest cube factor.

$\quad\quad = \sqrt[3]{2^3 \cdot 3}$ Rewrite.

$\quad\quad = 2\sqrt[3]{3}$ Find root of perfect cube.

b. $\sqrt[3]{24a^4} = \sqrt[3]{8a^3 \cdot 3a}$ Find largest cube factor.

$\quad\quad = \sqrt[3]{(2a)^3 \cdot 3a}$ Rewrite.

$\quad\quad = 2a\sqrt[3]{3a}$ Find root of perfect cube.

c. $\sqrt[3]{-40x^6} = \sqrt[3]{(-8x^6) \cdot 5}$ Find largest cube factor.

$\quad\quad = \sqrt[3]{(-2x^2)^3 \cdot 5}$ Rewrite.

$\quad\quad = -2x^2\sqrt[3]{5}$ Find root of perfect cube.

Use pattern recognition to help students identify perfect squares, cubes, etc., of both positive and negative integers when simplifying radicals. Have students construct a table of powers for several integers. For example:

n	n^2	n^3	n^4
-3	9	-27	81
-2	4	-8	16
-1	1	-1	1
0	0	0	0
1	1	1	1
2	4	8	16
3	9	27	81

To **rationalize the denominator** of a fraction, multiply the numerator and denominator by an appropriate factor. To find the "appropriate factor," make use of the form $a + b\sqrt{m}$ and its **conjugate** $a - b\sqrt{m}$. The product of this conjugate pair has no radical. For instance,

$$\left(4 + \sqrt{3}\right)\left(4 - \sqrt{3}\right) = 4^2 - \left(\sqrt{3}\right)^2 = 16 - 3 = 13.$$

Therefore, to rationalize a denominator of the form $a - b\sqrt{m}$ (or $a + b\sqrt{m}$), you can multiply the numerator and denominator by the conjugate factor. If $a = 0$, the rationalizing factor of $\sqrt{m}$ is itself, $\sqrt{m}$.

EXAMPLE 4   Rationalizing Single-Term Denominators

a. To rationalize the following denominator, multiply *both* the numerator and the denominator by $\sqrt{3}$ to obtain

$$\frac{5}{2\sqrt{3}} = \frac{5}{2\sqrt{3}} \cdot \frac{\sqrt{3}}{\sqrt{3}} = \frac{5\sqrt{3}}{2\sqrt{3^2}} = \frac{5\sqrt{3}}{2(3)} = \frac{5\sqrt{3}}{6}.$$

b. To rationalize the following denominator, multiply *both* the numerator and the denominator by $\sqrt[3]{5^2}$. Note how this eliminates the radical from the denominator by producing a perfect *cube* in the radicand.

$$\frac{2}{\sqrt[3]{5}} = \frac{2}{\sqrt[3]{5}} \cdot \frac{\sqrt[3]{5^2}}{\sqrt[3]{5^2}} = \frac{2\sqrt[3]{5^2}}{\sqrt[3]{5^3}} = \frac{2\sqrt[3]{25}}{5}$$

EXAMPLE 5   Rationalizing a Denominator with Two Terms

$$\frac{2}{3 + \sqrt{7}} = \frac{2}{3 + \sqrt{7}} \cdot \frac{3 - \sqrt{7}}{3 - \sqrt{7}} \qquad \text{Multiply numerator and denominator by conjugate.}$$

$$= \frac{2\left(3 - \sqrt{7}\right)}{3^2 - \left(\sqrt{7}\right)^2} \qquad \text{Multiply fractions.}$$

$$= \frac{2\left(3 - \sqrt{7}\right)}{9 - 7} \qquad \text{Simplify.}$$

$$= \frac{2\left(3 - \sqrt{7}\right)}{2} \qquad \text{Cancel like factors.}$$

$$= 3 - \sqrt{7} \qquad \text{Simplify.}$$

Activities

1. Simplify: $\sqrt[3]{250x^6y^4}$.

   Answer: $5x^2y\sqrt[3]{2y}$

2. Evaluate: $\sqrt[3]{64^2}$.

   Answer: 16

3. Are the two expressions equivalent?

   $$\frac{2}{\sqrt{3} - 1} \qquad \sqrt{3} + 1$$

   Answer: Yes

Don't confuse an expression such as $\sqrt{2} + \sqrt{7}$ with $\sqrt{2 + 7}$. In general,

$$\sqrt{x + y} \neq \sqrt{x} + \sqrt{y}.$$

Rational Exponents

The following definition shows how radicals are used to define **rational exponents**. Until now, work with exponents has been restructured to integer exponents.

Definition of Rational Exponents

If a is a real number and n is a positive integer such that the principal nth root of a exists, then $a^{1/n}$ is defined to be

$$a^{1/n} = \sqrt[n]{a}.$$

If m is a positive integer that has no common factor with n, then

$$a^{m/n} = \left(a^{1/n}\right)^{m} = \left(\sqrt[n]{a}\right)^{m}$$

and

$$a^{m/n} = (a^{m})^{1/n} = \sqrt[n]{a^{m}}.$$

NOTE   The numerator of a rational exponent denotes the *power* to which the base is raised, and the denominator denotes the *index* or the *root* to be taken. Moreover, it doesn't matter in which order the two operations are performed, provided the nth root exists. Here is an example.

$$8^{2/3} = \left(\sqrt[3]{8}\right)^{2} = 2^{2} = 4$$
$$8^{2/3} = \sqrt[3]{8^{2}} = \sqrt[3]{64} = 4$$

The properties of exponents that were listed in Section P.3 also apply to rational exponents (provided the roots indicated by the denominators exist). Those properties are relisted here, with different examples.

Properties of Exponents

Let r and s be rational numbers, and let a and b be real numbers, variables, or algebraic expressions. If the roots indicated by the rational exponents exist, then the following properties are true.

Property	Example
1. $a^{r}a^{s} = a^{r+s}$	$4^{1/2}(4^{1/3}) = 4^{5/6}$
2. $\dfrac{a^{r}}{a^{s}} = a^{r-s}, \quad a \neq 0$	$\dfrac{x^{2}}{x^{1/2}} = x^{2-(1/2)} = x^{3/2}$
3. $a^{-r} = \dfrac{1}{a^{r}}, \quad a \neq 0$	$4^{-1/2} = \dfrac{1}{4^{1/2}} = \dfrac{1}{2}$
4. $\left(\dfrac{a}{b}\right)^{-r} = \left(\dfrac{b}{a}\right)^{r}, \quad a \neq 0, \; b \neq 0$	$\left(\dfrac{x}{4}\right)^{-1/2} = \left(\dfrac{4}{x}\right)^{1/2} = \dfrac{2}{x^{1/2}}$
5. $(ab)^{r} = a^{r}b^{r}$	$(2x)^{1/2} = 2^{1/2}(x^{1/2})$
6. $(a^{r})^{s} = a^{rs}$	$(x^{3})^{1/3} = x$
7. $\left(\dfrac{a}{b}\right)^{r} = \dfrac{a^{r}}{b^{r}}, \quad b \neq 0$	$\left(\dfrac{x}{3}\right)^{1/3} = \dfrac{x^{1/3}}{3^{1/3}}$

Rational exponents are particularly useful for evaluating roots of numbers on a calculator, for reducing the index of a radical, and for simplifying (and factoring) algebraic expressions. Examples 6 and 7 demonstrate some of these uses.

EXAMPLE 6   Simplifying with Rational Exponents

a. $(27)^{1/3} = \sqrt[3]{27} = 3$

b. $(-32)^{-4/5} = \left(\sqrt[5]{-32}\right)^{-4} = (-2)^{-4} = \dfrac{1}{(-2)^4} = \dfrac{1}{16}$

c. $\left(-5x^{2/3}\right)\left(3x^{-1/3}\right) = -15x^{(2/3)-(1/3)} = -15x^{1/3}, \; x \neq 0$

EXAMPLE 7   Reducing the Index of a Radical

a. $\sqrt[6]{a^4} = a^{4/6} = a^{2/3} = \sqrt[3]{a^2}$

b. $\sqrt[3]{\sqrt{125}} = \left(125^{1/2}\right)^{1/3}$ Rewrite with rational exponents.

$= (125)^{1/6}$ Multiply exponents.

$= (5^3)^{1/6}$ Rewrite base as perfect cube.

$= 5^{3/6}$ Multiply exponents.

$= 5^{1/2}$ Reduce exponent.

$= \sqrt{5}$ Rewrite as radical.

Radical expressions can be combined (added or subtracted) if they are **like radicals**—that is, if they have the same index and radicand. For instance, $2\sqrt{3x}$, and $\frac{1}{2}\sqrt{3x}$ are like radicals, but $\sqrt[3]{3x}$ and $2\sqrt{3x}$ are not like radicals. To determine whether two expressions involve like radicals, you should first simplify each radical.

EXAMPLE 8   Simplifying and Combining Like Radicals

a. $2\sqrt{48} + 3\sqrt{27} = 2\sqrt{16 \cdot 3} + 3\sqrt{9 \cdot 3}$ Find square factors.

$= 8\sqrt{3} + 9\sqrt{3}$ Find square roots.

$= 17\sqrt{3}$ Combine like terms.

b. $\sqrt[3]{16x} - \sqrt[3]{54x} = \sqrt[3]{8 \cdot 2x} - \sqrt[3]{27 \cdot 2x}$ Find cube factors.

$= 2\sqrt[3]{2x} - 3\sqrt[3]{2x}$ Find cube roots.

$= -\sqrt[3]{2x}$ Combine like terms.

Radicals and Calculators

There are two methods of evaluating radicals on most calculators. For square roots, you can use the *square root key*. For other roots, you should first convert the radical to exponential form and then use the *exponential key*.

EXAMPLE 9   *Evaluating a Cube Root with a Calculator*

Use a calculator to evaluate $\sqrt[3]{25}$. Round to three decimal places.

Solution

First write $\sqrt[3]{25}$ as $25^{1/3}$. Then use the following keystrokes.

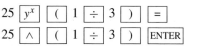

| 25 | y^x | (| 1 | ÷ | 3 |) | = | | Scientific |

25 $\wedge$ (1 ÷ 3) ENTER   Graphing

For either of these keystroke sequences, the calculator display should read 2.9240177. Thus, you have

$$\sqrt[3]{25} \approx 2.924.$$

EXAMPLE 10   *Evaluating Radicals with a Calculator*

Evaluate the following. Round to three decimal places.

a. $\sqrt[3]{-4}$ **b.** $(1.4)^{-2/5}$

Solution

a. Because

$$\sqrt[3]{-4} = \sqrt[3]{(-1)(4)} = \sqrt[3]{-1} \cdot \sqrt[3]{4} = -\sqrt[3]{4}$$

you can attach the negative sign of the radicand as follows.

4 y^x (1 ÷ 3) = +/−   Scientific

(−) (4 $\wedge$ (1 ÷ 3)) ENTER   Graphing

The calculator display is -1.5874011, which implies that

$$\sqrt[3]{-4} \approx -1.587.$$

b. Using the following keystroke sequence

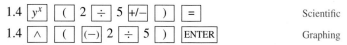

1.4 y^x (2 ÷ 5 +/−) =   Scientific

1.4 $\wedge$ ((−) 2 ÷ 5) ENTER   Graphing

you obtain a calculator display of 0.8740751. Thus, you have

$$(1.4)^{-2/5} \approx 0.874.$$

Group Activities

Problem Solving

Statistical Explorations When one is dealing with a set of data, there are several different ways to describe or summarize the data. One way is to show a picture of it by graphing the data in one or several of a variety of ways. Another way is to compute an *average*, the value around which the data tends to gather. One of the most commonly used averages is the **arithmetic mean.** If your data set contains n values, the arithmetic mean, denoted by $\bar{x}$, is computed as $\bar{x} = \dfrac{x_1 + x_2 + x_3 + \ldots + x_n}{n}$. A verbal description of the arithmetic mean is "add up all the values in the data set and divide by the number of values you added."

Most statisticians, however, will say that the mean alone does not describe a set of data fully. It often helps to know how much variation there is among the values. Are all the data values clustered closely around the value of the mean, or are the values spread out through a wide range around the mean? One way of measuring variation in a set of data is known as **standard deviation.** You may have seen this term in journal, magazine, or newspaper articles, or you may have heard an instructor use it when summarizing exam results. A formula for computing the standard deviation σ of a data set containing n values is

$$\sigma = \sqrt{\frac{(x_1 - \bar{x})^2 + (x_2 - \bar{x})^2 + (x_3 - \bar{x})^2 + \ldots + (x_n - \bar{x})^2}{n}}.$$

Suppose you have a set of four quiz scores: $\{20, 18, 17, 13\}$. The mean of this data is 17. The following shows the computation of the standard deviation of these scores.

$$\sigma = \sqrt{\frac{(20 - 17)^2 + (18 - 17)^2 + (17 - 17)^2 + (13 - 17)^2}{4}}$$

$$= \sqrt{\frac{3^2 + 1^2 + 0^2 + (-4)^2}{4}} \approx 2.55$$

Compute the means and standard deviations of the following sets of data.

a. $\{46, 49, 50, 51, 54\}$ **b.** $\{30, 35, 50, 65, 70\}$

c. $\{1, 1, 10, 10, 19, 19\}$ **d.** $\{8, 9, 10, 10, 11, 12\}$

Compare the mean and standard deviation in part (a) with those in part (b), and compare those in part (c) with those in part (d). What can you conclude about the relationship between the standard deviation and the structure of the data set (in terms of how spread out the data is) in each case?

Warm Up

The following warm-up exercises involve skills that were covered in earlier sections. You will use these skills in the exercise set for this section.

In Exercises 1–10, simplify the expression.

1. $\left(\frac{1}{3}\right)\left(\frac{2}{3}\right)^2$

2. $3(-4)^2$

3. $(-2x)^3$

4. $(-2x^3)(-3x^4)$

5. $(7x^5)(4x)$

6. $(5x^4)(25x^2)^{-1}, \quad x \neq 0$

7. $\dfrac{12z^6}{4z^2}, \quad z \neq 0$

8. $\left(\dfrac{2x}{5}\right)^2 \left(\dfrac{2x}{5}\right)^{-4}, \quad x \neq 0$

9. $\left(\dfrac{3y^2}{x}\right)^0, \quad x \neq 0, \ y \neq 0$

10. $\left[(x+2)^2(x+2)^3\right]^2$

P.4 Exercises

In Exercises 1–12, fill in the missing form.

Radical Form	Rational Exponent Form
1. $\sqrt{9} = 3$	
2. $\sqrt[3]{64} = 4$	
3.	$32^{1/5} = 2$
4.	$-\left(144^{1/2}\right) = -12$
5.	$196^{1/2} = 14$
6. $\sqrt[3]{614.125} = 8.5$	
7. $\sqrt[3]{-216} = -6$	
8.	$(-243)^{1/5} = -3$
9.	$27^{2/3} = 9$
10. $\left(\sqrt[4]{81}\right)^3 = 27$	
11. $\sqrt[4]{81^3} = 27$	
12.	$16^{5/4} = 32$

In Exercises 13–34, evaluate the expression.

13. $\sqrt{9}$

14. $\sqrt{49}$

15. $\sqrt[3]{8}$

16. $\sqrt[3]{64}$

17. $\sqrt{36}$

18. $\sqrt[3]{\frac{27}{8}}$

19. $-\sqrt[3]{-27}$

20. $\sqrt[3]{0}$

21. $\dfrac{4}{\sqrt{64}}$

22. $\dfrac{\sqrt[4]{81}}{3}$

23. $\left(\sqrt[3]{-125}\right)^3$

24. $\sqrt[4]{562^4}$

25. $16^{1/2}$

26. $27^{1/3}$

27. $36^{3/2}$

28. $16^{3/2}$

29. $32^{-3/5}$

30. $100^{-3/2}$

31. $\left(\frac{16}{81}\right)^{-3/4}$

32. $\left(\frac{9}{4}\right)^{-1/2}$

33. $\left(-\frac{1}{64}\right)^{-1/3}$

34. $\left(-\frac{125}{27}\right)^{-1/3}$

In Exercises 35–40, simplify the expression.

35. $\sqrt{8}$

36. $\sqrt[3]{\frac{16}{27}}$

37. $\sqrt[3]{16x^5}$

38. $\sqrt[4]{(3x^2)^4}$

39. $\sqrt{75x^2y^{-4}}$ **40.** $\sqrt[5]{96x^5}$

In Exercises 41–48, rewrite the expression by rationalizing the denominator. Simplify the result.

41. $\dfrac{1}{\sqrt{3}}$ **42.** $\dfrac{5}{\sqrt{10}}$

43. $\dfrac{8}{\sqrt[3]{2}}$ **44.** $\dfrac{5}{\sqrt[3]{(5x)^2}}$

45. $\dfrac{2x}{5 - \sqrt{3}}$ **46.** $\dfrac{5}{\sqrt{14} - 2}$

47. $\dfrac{3}{\sqrt{5} + \sqrt{6}}$ **48.** $\dfrac{5}{2\sqrt{10} - 5}$

In Exercises 49–54, simplify the expression.

49. $5\sqrt{x} - 3\sqrt{x}$ **50.** $3\sqrt{x + 1} + 10\sqrt{x + 1}$

51. $2\sqrt{50} + 12\sqrt{8}$ **52.** $4\sqrt{27} - \sqrt{75}$

53. $2\sqrt{4y} - 2\sqrt{9y}$ **54.** $2\sqrt{80} + \sqrt{125}$

In Exercises 55–64, simplify the expression.

55. $\sqrt{2}\sqrt{3}$ **56.** $\sqrt{2}\sqrt{5}$

57. $5^{1/2} \cdot 5^{3/2}$ **58.** $3^{1/3} \cdot 3^{4/3}$

59. $\dfrac{6^{1/2}}{6^{1/3}}$ **60.** $\dfrac{2^{3/2}}{2}$

61. $\dfrac{x^2}{x^{1/2}}$ **62.** $\dfrac{x \cdot x^{1/2}}{x^{3/2}}$

63. $\sqrt{\sqrt{32}}$ **64.** $\sqrt{\sqrt{x^4}}$

In Exercises 65–68, use rational exponents to reduce the index of the radical.

65. $\sqrt[4]{3^2}$ **66.** $\sqrt[6]{x^3}$

67. $\sqrt[6]{(x + 1)^4}$ **68.** $\sqrt[4]{(3x^2)^4}$

In Exercises 69–76, use a calculator to approximate the given number. (Round to three decimal places.)

69. $\sqrt{57}$ **70.** $\sqrt[3]{45}$

71. $5.7^{2/5}$ **72.** $36.8^{1.2}$

73. $0.26^{-0.8}$ **74.** $3.75^{-1/2}$

75. $\dfrac{3 - \sqrt{5}}{2}$ **76.** $\dfrac{-4 + \sqrt{12}}{4}$

In Exercises 77–82, fill in the blank with $<$, $=$, or $>$.

77. $\sqrt{5} + \sqrt{3}$ _____ $\sqrt{5 + 3}$

78. $\sqrt{3} - \sqrt{2}$ _____ $\sqrt{3 - 2}$

79. 5 _____ $\sqrt{3^2 + 2^2}$

80. 5 _____ $\sqrt{3^2 + 4^2}$

81. $\sqrt{3} \cdot \sqrt[4]{3}$ _____ $\sqrt[8]{3}$

82. $\sqrt{\dfrac{3}{11}}$ _____ $\dfrac{\sqrt{3}}{\sqrt{11}}$

83. *Geometry*   Find the dimensions of a cube that has a volume of 13,824 cubic inches (see figure).

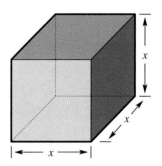

84. *Geometry*   Find the dimensions of a square classroom that has 800 square feet of floor space (see figure).

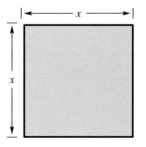

Declining Balances Depreciation In Exercises 85 and 86, find the annual depreciation rate r. To find the annual depreciation rate by the declining balances method, use the formula

$$r = 1 - \left(\frac{S}{C}\right)^{1/n}$$

where n is the useful life of the item (in years), S is the salvage value (in dollars), and C is the original cost (in dollars).

85. A truck whose original cost is $75,000 is depreciated over an 8-year period, as shown in the figure.

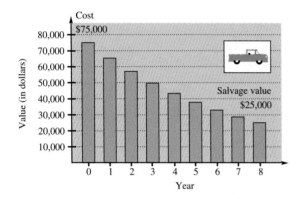

86. A printing press whose original cost is $125,000 is depreciated over a 10-year period, as shown in the figure.

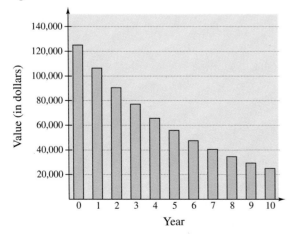

87. *Period of a Pendulum* The period T in seconds of a pendulum is given by

$$T = 2\pi\sqrt{\frac{L}{32}}$$

where L is the length of the pendulum in feet. Find the period of a pendulum whose length is 2 feet.

88. *Length of a Pendulum* Use the formula given in Exercise 87 to find the period of a pendulum whose length is 1.5 feet.

Notes on a Musical Scale In Exercises 89 and 90, find the frequency of the indicated note on a piano (see figure). The musical note A above middle C has a frequency of 440 vibrations per second. If we denote this frequency by F_1, then the frequency of the next higher note is given by $F_2 = F_1 \cdot 2^{1/12}$. Similarly, the frequency of the next note is given by $F_3 = F_2 \cdot 2^{1/12}$.

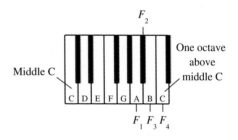

89. Find the frequency of the musical note B above middle C.

90. Find the frequency of the musical note C that is one octave above middle C.

91. *Calculator Experiment* Enter any positive real number in your calculator and repeatedly take the square root. What real number does the display appear to be approaching?

92. *Calculator Experiment* Square the real number $2/\sqrt{5}$ and note that the radical is eliminated from the denominator. Is this equivalent to rationalizing the denominator? Why or why not?

MID-CHAPTER QUIZ

Take this quiz as you would take a quiz in class. After you are done, check your work against the answers given in the back of the book.

In Exercises 1 and 2, place the correct symbol ($<$, $>$, or $=$) between the two real numbers.

1. $-|-7|$ ▓▓▓ $|-7|$ **2.** $-(-3)$ ▓▓▓ $|-3|$

In Exercises 3 and 4, use inequality notation to describe the set of real numbers.

3. x is positive or x is equal to zero.

4. The apartment occupancy rate r will be at least 96.5% during the coming year.

5. Describe the subset of real numbers that is represented by the inequality, and sketch the subset on the real number line.

$0 \leq x < 3$

6. Identify the terms of the algebraic expression $3x^2 - 7x + 2$.

In Exercises 7–10, perform the indicated operation(s). (Write fractional answers in reduced form.)

7. $-5 - (-7)$ **8.** $\dfrac{28 - 4}{(-6)}$ **9.** $\dfrac{2}{3} \cdot \dfrac{5}{4} \cdot \dfrac{3}{7}$ **10.** $\dfrac{11}{15} \div \dfrac{3}{5}$

In Exercises 11–14, simplify the expression.

11. $(-x)^3(2x^4)$ **12.** $\dfrac{5y^7}{15y^3}$ **13.** $(3t^{-2})(6t^5)$ **14.** $\left(\dfrac{x^{-2}y^2}{3}\right)^{-3}$

15. One thousand dollars is deposited in an account with an annual rate of 8%, compounded monthly. Find the balance in the account after 10 years.

In Exercises 16 and 17, evaluate the expression.

16. $\dfrac{-\sqrt[4]{81}}{3}$ **17.** $\left(\sqrt[3]{-64}\right)^3$

In Exercises 18 and 19, simplify the expression.

18. $3^{1/2} \cdot 3^{3/2}$ **19.** $\sqrt[3]{81} - 4\sqrt[3]{3}$

20. Find the dimensions of a cube that has a volume of 15,625 cubic centimeters (see figure).

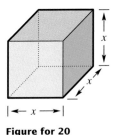

Figure for 20

P.5	**Polynomials and Special Products**

Polynomials ▪ Operations with Polynomials ▪
Special Products ▪ Applications

Polynomials

One of the simplest and most common types of algebraic expression is a
polynomial. Here are some examples.

$$2x + 5, \qquad 3x^4 - 7x^2 + 2x + 4, \qquad 5x^2y^2 - xy + 3$$

The first two are *polynomials in x* and the third is a *polynomial in x and y.* The
terms of a polynomial in x have the form ax^k, where a is the **coefficient** and k is
the **degree** of the term. Because a polynomial is defined as an algebraic sum, the
coefficients take on the signs between the terms. For instance, the polynomial

$$2x^3 - 5x^2 + 1 = 2x^3 + (-5)x^2 + (0)x + 1$$

has coefficients 2, −5, 0, and 1.

NOTE   Polynomials with one,
two, and three terms are called
monomials, binomials, and
trinomials, respectively.

Definition of a Polynomial in x

Let $a_n, \ldots, a_2, a_1, a_0$ be real numbers and let n be a *nonnegative
integer.* A **polynomial in x** is an expression of the form

$$a_n x^n + \cdots + a_2 x^2 + a_1 x + a_0$$

where $a_n \neq 0$. The polynomial is of **degree n**, and the number a_n is the
leading coefficient. The number a_0 is the **constant term.** The constant
term is considered to have a degree of zero.

Note in the definition of a polynomial that the polynomial is written with
descending powers of x. This is called the **standard form** of a polynomial.

EXAMPLE 1   *Rewriting a Polynomial in Standard Form*

	Polynomial	*Standard Form*	*Degree*
a.	$4x^2 - 5x^3 - 2 + 3x$	$-5x^3 + 4x^2 + 3x - 2$	3
b.	$4 - 9x^2$	$-9x^2 + 4$	2
c.	8	$8 \; (8 = 8x^0)$	0

A polynomial that has all zero coefficients is called the **zero polynomial,** denoted by 0. This particular polynomial is not considered to have a degree.

EXAMPLE 2 Identifying a Polynomial and Its Degree

a. $-2x^3 + x^2 + 3x - 2$ is a polynomial of degree 3.

b. $\sqrt{x^2 - 3x}$ is not a polynomial because the radical sign indicates a noninteger power of x.

c. $x^2 + 5x^{-1}$ is not a polynomial because of the negative exponent.

For polynomials in more than one variable, the *degree of a term* is the sum of the powers of the variables in the term. The *degree of the polynomial* is the highest degree of all its terms. For instance, the polynomial

$$5x^3y - x^2y^2 + 2xy - 5$$

has two terms of degree 4, one term of degree 2, and one term of degree 0. The degree of the polynomial is 4.

Operations with Polynomials

You can **add** and **subtract** polynomials in much the same way that you add and subtract real numbers—you simply add or subtract the *like terms* (terms having the same variables to the same powers) by adding their coefficients. For instance, $-3x^2$ and $5x^2$ are like terms and their sum is given by

$$-3x^2 + 5x^2 = (-3 + 5)x^2$$
$$= 2x^2.$$

EXAMPLE 3 Sums and Differences of Polynomials

a. $(5x^3 - 7x^2 - 3) + (x^3 + 2x^2 - x + 8)$

$\qquad = (5x^3 + x^3) + (2x^2 - 7x^2) - x + (8 - 3)$ Group like terms.

$\qquad = 6x^3 - 5x^2 - x + 5$ Combine like terms.

b. $(7x^4 - x^2 - 4x + 2) - (3x^4 - 4x^2 + 3x)$

$\qquad = 7x^4 - x^2 - 4x + 2 - 3x^4 + 4x^2 - 3x$ Distribute sign.

$\qquad = (7x^4 - 3x^4) + (4x^2 - x^2) + (-3x - 4x) + 2$ Group like terms.

$\qquad = 4x^4 + 3x^2 - 7x + 2$ Combine like terms.

NOTE   A common mistake is to fail to change the sign of *each* term inside parentheses preceded by a minus sign. For instance, note the following.

$$-(3x^4 - 4x^2 + 3x) = -3x^4 + 4x^2 - 3x \qquad \text{Correct}$$

$$\overline{-(3x^4 - 4x^2 + 3x)} = -3x^4 - 4x^2 + 3x \qquad \text{Common mistake}$$

To find the **product** of two polynomials, you can use the left and right Distributive Properties. For example, if you treat $(5x + 7)$ as a single quantity, you can multiply $(3x - 2)$ by $(5x + 7)$ as follows.

$$(3x - 2)(5x + 7) = 3x(5x + 7) - 2(5x + 7)$$
$$= (3x)(5x) + (3x)(7) - (2)(5x) - (2)(7)$$
$$= 15x^2 + 21x - 10x - 14$$

Product of **First terms**	Product of **Outer terms**	Product of **Inner terms**	Product of **Last terms**

$$= 15x^2 + 11x - 14$$

With practice you should be able to multiply two binomials without writing all of the above steps. In fact, the four products in the boxes above suggest that you can put the product of two binomials in the FOIL form in just one step. This is called the **FOIL Method.**

When multiplying two polynomials, be sure to multiply *each* term of one polynomial by *each* term of the other. The following vertical pattern is a convenient way to multiply two polynomials.

EXAMPLE 4   *Using a Vertical Format to Multiply Polynomials*

Multiply $(x^2 - 2x + 2)$ by $(x^2 + 2x + 2)$.

Solution

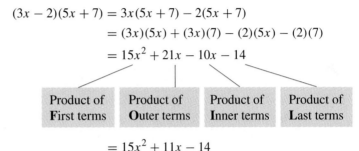

$$
\begin{array}{l}
\quad x^2 - 2x + 2 \qquad\qquad\qquad\qquad \text{Standard form} \\
\quad x^2 + 2x + 2 \qquad\qquad\qquad\qquad \text{Standard form} \\
\hline
x^4 - 2x^3 + 2x^2 \qquad\qquad\qquad x^2(x^2 - 2x + 2) \\
\qquad\quad 2x^3 - 4x^2 + 4x \qquad\qquad 2x(x^2 - 2x + 2) \\
\qquad\qquad\qquad 2x^2 - 4x + 4 \qquad 2(x^2 - 2x + 2) \\
\hline
x^4 + 0x^3 + 0x^2 + 0x + 4 = x^4 + 4
\end{array}
$$

Thus,

$$(x^2 - 2x + 2)(x^2 + 2x + 2) = x^4 + 4.$$

Special Products

Special Products

Let u and v be real numbers, variables, or algebraic expressions.

Special Product	*Example*

Sum and Difference of Two Terms

$$(u + v)(u - v) = u^2 - v^2 \qquad\qquad (x + 4)(x - 4) = x^2 - 16$$

Square of a Binomial

$$(u + v)^2 = u^2 + 2uv + v^2 \qquad\qquad (x + 3)^2 = x^2 + 6x + 9$$

$$(u - v)^2 = u^2 - 2uv + v^2 \qquad\qquad (3x - 2)^2 = 9x^2 - 12x + 4$$

Cube of a Binomial

$$(u + v)^3 = u^3 + 3u^2v + 3uv^2 + v^3 \qquad (x + 2)^3 = x^3 + 6x^2 + 12x + 8$$

$$(u - v)^3 = u^3 - 3u^2v + 3uv^2 - v^3 \qquad (x - 1)^3 = x^3 - 3x^2 + 3x - 1$$

EXAMPLE 5   Sum and Difference of Two Terms

$$(5x + 9)(5x - 9) = (5x)^2 - 9^2 = 25x^2 - 81$$

EXAMPLE 6   Square of a Binomial

$$(6x - 5)^2 = (6x)^2 - 2(6x)(5) + 5^2$$
$$= 36x^2 - 60x + 25$$

EXAMPLE 7   Cube of a Binomial

$$(3x + 2)^3 = (3x)^3 + 3(3x)^2(2) + 3(3x)(2)^2 + 2^3$$
$$= 27x^3 + 54x^2 + 36x + 8$$

EXAMPLE 8   The Product of Two Trinomials

$$(x + y - 2)(x + y + 2) = [(x + y) - 2][(x + y) + 2]$$
$$= (x + y)^2 - 2^2$$
$$= x^2 + 2xy + y^2 - 4$$

Applications

EXAMPLE 9   A Savings Plan

In 1993, 12% of all college students were on work-study programs. Many more worked part-time or full-time in private enterprise.

During the summers prior to your four college years you are able to save $1500, $1800, $2400, and $2600. During the summer after graduation you save $3000. This money is deposited in an account that pays 7% interest, compounded annually. Find the balance in the account at the end of the summer after graduation.

Solution

Using the formula for compound interest, for *each* deposit you have

$$\text{Balance} = P\left(1 + \frac{r}{n}\right)^{nt} = P(1 + 0.07)^t = P(1.07)^t.$$

For the first deposit, $P = 1500$ and $t = 4$. For the second deposit, $P = 1800$ and $t = 3$, and so on. The balances for the five deposits are as follows.

Date	Deposit	Time in Account	Balance in Account
First Summer	$1500	4 years	$1500(1.07)^4$
Second Summer	$1800	3 years	$1800(1.07)^3$
Third Summer	$2400	2 years	$2400(1.07)^2$
Fourth Summer	$2600	1 year	$2600(1.07)$
Fifth Summer	$3000	0 years	3000

By adding these five balances, you can find the total balance in the account to be

$$1500(1.07)^4 + 1800(1.07)^3 + 2400(1.07)^2 + 2600(1.07) + 3000.$$

Note that this expression is in polynomial form. By evaluating the expression, you can find the balance to be $12,701.03, as shown in Figure P.10.

Activities

1. Explain what happens when the parentheses are removed from the algebraic statement $-(3x^2 - 4x + 2)$.

2. Multiply using special products: $\left(\sqrt{2} + \sqrt{3}\right)^2$.

   Answer: $5 + 2\sqrt{6}$

3. Multiply: $(2x + 1)(x - 5)$.

   Answer: $2x^2 - 9x - 5$

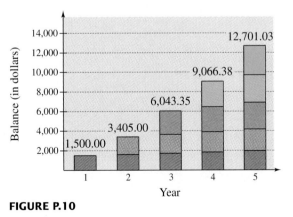

FIGURE P.10

Real Life

EXAMPLE 10   *Geometry: Volume of a Box*

An open box is made by cutting squares from the corners of a piece of metal that is 16 inches by 20 inches, and turning up the sides, as shown in Figure P.11. If the edge of each cut-out square is x inches, what is the volume of the box? Find the volume when $x = 1$, $x = 2$, and $x = 3$.

Solution

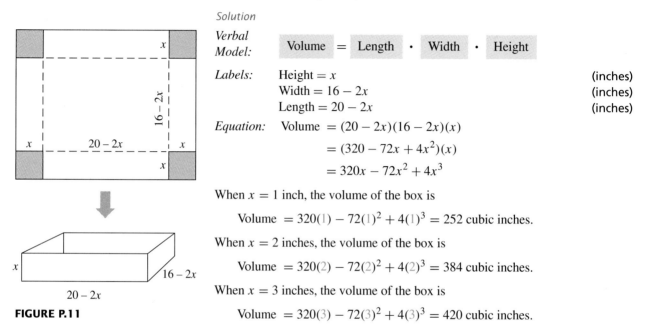

Verbal Model:

| Volume | = | Length | · | Width | · | Height |

Labels:

Height $= x$ (inches)
Width $= 16 - 2x$ (inches)
Length $= 20 - 2x$ (inches)

Equation:

$$\text{Volume} = (20 - 2x)(16 - 2x)(x)$$
$$= (320 - 72x + 4x^2)(x)$$
$$= 320x - 72x^2 + 4x^3$$

When $x = 1$ inch, the volume of the box is

$$\text{Volume} = 320(1) - 72(1)^2 + 4(1)^3 = 252 \text{ cubic inches.}$$

When $x = 2$ inches, the volume of the box is

$$\text{Volume} = 320(2) - 72(2)^2 + 4(2)^3 = 384 \text{ cubic inches.}$$

When $x = 3$ inches, the volume of the box is

$$\text{Volume} = 320(3) - 72(3)^2 + 4(3)^3 = 420 \text{ cubic inches.}$$

FIGURE P.11

Group Activities   Communicating Mathematically

Understanding Special Products   For each of the special products listed on page 48, show the equivalence in each statement—that is, show that $(u + v)(u - v)$ does in fact equal $(u^2 - v^2)$—by multiplying out each product and simplifying. Discuss how recognizing these special product patterns can save time. Write a verbal description for each special product.

Warm Up The following warm-up exercises involve skills that were covered in earlier sections. You will use these skills in the exercise set for this section.

In Exercises 1–10, perform the indicated operation(s).

1. $(7x^2)(6x)$
2. $(10z^3)(-2z^{-1})$

3. $(-3x^2)^3$
4. $-3(x^2)^3$

5. $\dfrac{27z^5}{12z^2}$
6. $\sqrt{24} \cdot \sqrt{2}$

7. $\left(\dfrac{2x}{3}\right)^{-2}$
8. $16^{3/4}$

9. $\dfrac{4}{\sqrt{8}}$
10. $\sqrt[3]{-27x^3}$

P.5 Exercises

In Exercises 1–6, find the degree and leading coefficient of the polynomial.

1. $2x^2 - x + 1$
2. $-3x^4 + 2x^2 - 5$

3. $x^5 - 1$
4. 3

5. $4x^5 + 6x^4 - x - 1$
6. $2x$

In Exercises 7–12, determine whether the algebraic expression is a polynomial. If it is, write the polynomial in standard form and state its degree.

7. $2x - 3x^3 + 8$
8. $2x^3 + x - 3x^{-1}$

9. $\dfrac{3x + 4}{x}$
10. $\dfrac{x^2 + 2x - 3}{2}$

11. $y^2 - y^4 + y^3$
12. $\sqrt{y^2 - y^4}$

In Exercises 13–18, evaluate the polynomial for the indicated values of x.

13. $4x + 3$ 　(a) $x = -1$ 　(b) $x = 0$
　　　　　　　(c) $x = 1$ 　(d) $x = 2$

14. $-5x + 7$ 　(a) $x = -1$ 　(b) $x = 0$
　　　　　　　　(c) $x = 1$ 　(d) $x = 2$

15. $x^2 - 2x + 3$ 　(a) $x = -2$ 　(b) $x = -1$
　　　　　　　　　(c) $x = 0$ 　(d) $x = 1$

16. $-2x^2 + 3x + 4$ 　(a) $x = -2$ 　(b) $x = -1$
　　　　　　　　　　(c) $x = 0$ 　(d) $x = 1$

17. $x^3 - 4x^2 + x$ 　(a) $x = -1$ 　(b) $x = 0$
　　　　　　　　　(c) $x = 1$ 　(d) $x = 2$

18. $-x^3 + x$ 　(a) $x = -1$ 　(b) $x = 0$
　　　　　　　(c) $x = 1$ 　(d) $x = 2$

In Exercises 19–32, perform the indicated operations and write the resulting polynomial in standard form.

19. $(6x + 5) - (8x + 15)$

20. $(2x^2 + 1) - (x^2 - 2x + 1)$

21. $-(x^3 - 2) + (4x^3 - 2x)$

22. $-(5x^2 - 1) - (-3x^2 + 5)$

23. $(15x^2 - 6) - (-8x^3 - 14x^2 - 17)$

24. $(15x^4 - 18x - 19) - (13x^4 - 5x + 15)$

25. $5z - [3z - (10z + 8)]$

26. $(y^3 + 1) - [(y^2 + 1) + (3y - 7)]$

27. $3x(x^2 - 2x + 1)$ **28.** $y^2(4y^2 + 2y - 3)$

29. $-5z(3z - 1)$ **30.** $-4x(3 - x^3)$

31. $(-2x)(-3x)(5x + 2)$ **32.** $(1 - x^3)(4x)$

In Exercises 33–64, find the product.

33. $(x + 3)(x + 4)$ **34.** $(x - 5)(x + 10)$

35. $(3x - 5)(2x + 1)$ **36.** $(7x - 2)(4x - 3)$

37. $(x + 6)^2$ **38.** $(3x - 2)^2$

39. $(2x - 5y)^2$ **40.** $(5 - 8x)^2$

41. $[(x - 3) + y]^2$ **42.** $[(x + 1) - y]^2$

43. $(x + 10)(x - 10)$ **44.** $(2x + 3)(2x - 3)$

45. $(x + 2y)(x - 2y)$ **46.** $(2x + 3y)(2x - 3y)$

47. $(m - 3 + n)(m - 3 - n)$ **48.** $(x + y + 1)(x + y - 1)$

49. $(2r^2 - 5)(2r^2 + 5)$ **50.** $(3a^3 - 4b^2)(3a^3 + 4b^2)$

51. $(x + 1)^3$ **52.** $(x - 2)^3$

53. $(2x - y)^3$ **54.** $(3x + 2y)^3$

55. $\left(\sqrt{x} + \sqrt{y}\right)\left(\sqrt{x} - \sqrt{y}\right)$

56. $\left(5 + \sqrt{x}\right)\left(5 - \sqrt{x}\right)$

57. $(4x^3 - 3)^2$ **58.** $(x^2 + 9)(x^2 - x - 4)$

59. $(8x + 3)^2$ **60.** $(x - 2)(x^2 + 2x + 4)$

61. $(x^2 - x + 1)(x^2 + x + 1)$

62. $(x^2 + 3x - 2)(x^2 - 3x - 2)$

63. $5x(x + 1) - 3x(x + 1)$

64. $(2x - 1)(x + 3) + 3(x + 3)$

65. *Savings Plan* A person who is attending 4 years of college has a summer job. During the summers prior to each of the four college years, the person is able to save $1200, $1400, $800, and $2300. During the summer after graduation the person saves $2900. This money is deposited in an account that pays 6% interest, compounded annually. Find the balance in the account at the end of the summer after graduation.

66. *Savings Plan* Suppose you have an investment that pays an annual dividend in July. Each year for 6 years, you reinvest this dividend in an account that pays 6.5% interest, compounded annually. For the dividends shown in the table, find the balance in the account at the end of 6 years. (*Note:* The first deposit will earn 5 years of interest, and the sixth deposit will earn no interest.)

Year	1	2	3
Dividend	$840	$930	$760

Year	4	5	6
Dividend	$1020	$980	$1130

67. *Compound Interest* After 2 years, an investment of $500 compounded annually at an interest rate of r will yield an amount of

$$500(1 + r)^2.$$

Write this polynomial in standard form.

68. *Compound Interest* After 3 years, an investment of $1200 compounded annually at an interest rate of r will yield an amount of

$$1200(1 + r)^3.$$

Write this polynomial in standard form.

69. *Geometry* An open box is made by cutting squares from the corners of a piece of metal that is 18 inches by 26 inches, and folding up the sides (see figure). If the edge of each cut-out square is x inches, what is the volume of the box? Find the volume when $x = 1$, $x = 2$, and $x = 3$.

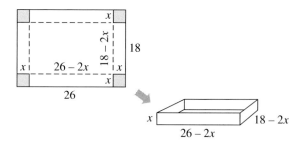

70. *Geometry* An open box is made by cutting squares from the corners of a piece of cardboard that is 20 inches by 28 inches, and turning up the sides (see figure). If the edge of each cut-out square is x inches, what is the volume of the box? Find the volume when $x = 1$, $x = 2$, and $x = 3$.

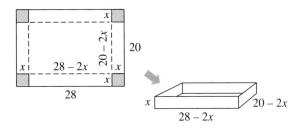

71. *Geometry* Find the area of the shaded region in the figure. Write your answer as a polynomial in standard form.

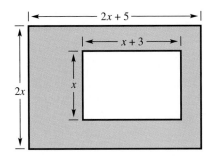

72. *Geometry* Find the area of the shaded region in the figure. Write your answer as a polynomial in standard form.

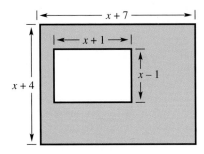

73. *Geometry* Find a polynomial that represents the total number of square feet for the following floor plan.

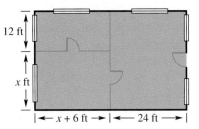

74. *Geometry* Find a polynomial that represents the total number of square feet for the following floor plan.

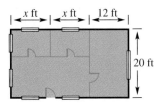

75. *Graphing Calculator* You may evaluate an expression on a *TI-82* graphing calculator by using $\boxed{Y=}$ and notation. For example, if you want to evaluate $x^2 + 6x + 9$ for $x = 1$, use the following keystrokes to enter the expression into the calculator.

$\boxed{Y=}$ $\boxed{X, T, \theta}$ $\boxed{x^2}$ $\boxed{+}$ 6 $\boxed{X, T, \theta}$ $\boxed{+}$ 9 $\boxed{QUIT}$

To evaluate the expression for $x = 1$:

$\boxed{Y\text{-VARS}}$ $\boxed{ENTER}$ $\boxed{ENTER}$ $\boxed{(}$ 1 $\boxed{)}$ $\boxed{ENTER}$

Your display should be 16. Evaluate the expression for $x = -2, 6$, and -9.

76. *Graphing Calculator* Use the example given in Exercise 75 to evaluate the polynomial

$$Y_1 = 3x^3 - 2x^2 + 5x + 2$$

when $x = -1, 0, 1$, and 2.

P.6	**Factoring**
	Common Factors ▪ Factoring Special Polynomial Forms ▪ Trinomials with Binomial Factors ▪ Factoring by Grouping

Common Factors

The process of writing a polynomial as a product is called **factoring.** It is an important tool for solving equations and reducing fractional expressions.

A polynomial that cannot be factored using integer coefficients is called **prime** or **irreducible over the integers.** For instance, the polynomial $x^2 - 3$ is irreducible over the integers. [Over the *real numbers*, this polynomial can be factored as $x^2 - 3 = \left(x + \sqrt{3}\right)\left(x - \sqrt{3}\right)$.]

A polynomial is **completely factored** when each of its factors is prime. For instance,

$$x^3 - x^2 + 4x - 4 = (x - 1)(x^2 + 4) \qquad \text{Completely factored}$$

is completely factored, but

$$x^3 - x^2 - 4x + 4 = (x - 1)(x^2 - 4) \qquad \text{Not completely factored}$$

is not completely factored. Its complete factorization would be $x^3 - x^2 - 4x + 4 = (x - 1)(x + 2)(x - 2)$.

The simplest type of factoring involves a polynomial that is the product of a monomial and another polynomial. To factor such a polynomial, you can use the Distributive Property in the *reverse* direction.

$$ab + ac = a(b + c) \qquad \text{a is a common factor.}$$

EXAMPLE 1 Removing Common Factors

Factor each of the following.

a. $6x^3 - 4x$ **b.** $(x - 2)(2x) + (x - 2)(3)$

Solution

a. Each term of this polynomial has $2x$ as a common factor.

$$6x^3 - 4x = 2x(3x^2) - 2x(2)$$
$$= 2x(3x^2 - 2)$$

b. The binomial factor $(x - 2)$ is common to both terms.

$$(x - 2)(2x) + (x - 2)(3) = (x - 2)(2x + 3)$$

Factoring Special Polynomial Forms

Factoring Special Polynomial Forms

Factored Form	*Example*

Difference of Two Squares

$$u^2 - v^2 = (u + v)(u - v) \qquad\qquad 9x^2 - 4 = (3x + 2)(3x - 2)$$

Perfect Square Trinomial

$$u^2 + 2uv + v^2 = (u + v)^2 \qquad\qquad x^2 + 6x + 9 = (x + 3)^2$$
$$u^2 - 2uv + v^2 = (u - v)^2 \qquad\qquad x^2 - 6x + 9 = (x - 3)^2$$

Sum or Difference of Two Cubes

$$u^3 + v^3 = (u + v)(u^2 - uv + v^2) \qquad x^3 + 8 = (x + 2)(x^2 - 2x + 4)$$
$$u^3 - v^3 = (u - v)(u^2 + uv + v^2) \qquad 27x^3 - 1 = (3x - 1)(9x^2 + 3x + 1)$$

STUDY TIP

In Example 2, note that the first step in factoring a polynomial is to check for common factors. Once the common factor is removed, it is often possible to recognize patterns that were not obvious at first glance.

EXAMPLE 2   Removing a Common Factor First

Factor $3 - 12x^2$.

Solution

$$
\begin{aligned}
3 - 12x^2 &= 3(1 - 4x^2) & \text{Common factor} \\
&= 3[1^2 - (2x)^2] & \text{Difference of squares} \\
&= 3(1 + 2x)(1 - 2x) & \text{Completely factored.}
\end{aligned}
$$

EXAMPLE 3   Factoring the Difference of Two Squares

a.
$$
\begin{aligned}
(x + 2)^2 - y^2 &= [(x + 2) + y][(x + 2) - y] \\
&= (x + 2 + y)(x + 2 - y) \\
&= (x + y + 2)(x - y + 2)
\end{aligned}
$$

b. Applying the difference of two squares formula twice, you have

$$
\begin{aligned}
16x^4 - 81 &= (4x^2)^2 - 9^2 \\
&= (4x^2 + 9)(4x^2 - 9) & \text{First application} \\
&= (4x^2 + 9)[(2x)^2 - 3^2] \\
&= (4x^2 + 9)(2x + 3)(2x - 3). & \text{Second application}
\end{aligned}
$$

A perfect square trinomial is the square of a binomial, and it has the following form.

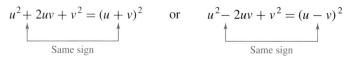

$$u^2 + 2uv + v^2 = (u + v)^2 \qquad \text{or} \qquad u^2 - 2uv + v^2 = (u - v)^2$$

Same sign   Same sign

Note that the first and last terms are squares and the middle term is twice the product of u and v.

EXAMPLE 4 Factoring Perfect Square Trinomials

a. $16x^2 + 8x + 1 = (4x)^2 + 2(4x)(1) + 1^2$

$$= (4x + 1)^2$$

b. $x^2 - 10x + 25 = x^2 - 2(x)(5) + 5^2$

$$= (x - 5)^2$$

The next two formulas show that sums and differences of cubes factor easily. Pay special attention to the signs of the terms.

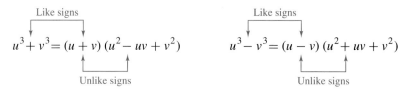

Like signs   Like signs

$$u^3 + v^3 = (u + v)(u^2 - uv + v^2) \qquad\qquad u^3 - v^3 = (u - v)(u^2 + uv + v^2)$$

Unlike signs   Unlike signs

EXAMPLE 5 Factoring the Sum and Difference of Cubes

Factor the following.

a. $x^3 - 27$ b. $3x^3 + 192$

Solution

a. $x^3 - 27 = x^3 - 3^3$

$$= (x - 3)(x^2 + 3x + 9)$$

b. $3x^3 + 192 = 3(x^3 + 64)$ Common factor

$$= 3(x^3 + 4^3)$$

$$= 3(x + 4)(x^2 - 4x + 16)$$

Trinomials with Binomial Factors

To factor a trinomial of the form $ax^2 + bx + c$, use the following pattern.

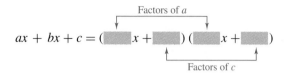

Factors of a

$$ax + bx + c = (\boxed{}x + \boxed{}) (\boxed{}x + \boxed{})$$

Factors of c

The goal is to find a combination of factors of a and c so that the outer and inner products add up to the middle term bx. For instance, in the trinomial $6x^2 + 17x + 5$, you have $a = 6$, $c = 5$, and $b = 17$. After some experimentation, you can find that the factorization is $(2x + 5)(3x + 1)$.

EXAMPLE 6   Factoring a Trinomial: Leading Coefficient Is 1

Factor the trinomial $x^2 - 7x + 12$.

Solution

For this trinomial, you have $a = 1$, $b = -7$, and $c = 12$. Because b is negative and c is positive, both factors of 12 must be negative. That is, $12 = (-2)(-6)$, $12 = (-1)(-12)$, or $12 = (-3)(-4)$. Therefore, the possible factorizations of $x^2 - 7x + 12$ are

$$(x - 2)(x - 6), \quad (x - 1)(x - 12), \quad \text{and} \quad (x - 3)(x - 4).$$

Testing the middle term, you can find the correct factorization to be

$$x^2 - 7x + 12 = (x - 3)(x - 4).$$

EXAMPLE 7   Factoring a Trinomial: Leading Coefficient Is Not 1

Factor the trinomial $2x^2 + x - 15$.

Solution

For this trinomial, you have $a = 2$ and $c = -15$, which means that the factors of -15 must have unlike signs. The eight possible factorizations are as follows.

$$(2x - 1)(x + 15) \qquad (2x + 1)(x - 15)$$
$$(2x - 3)(x + 5) \qquad (2x + 3)(x - 5)$$
$$(2x - 5)(x + 3) \qquad (2x + 5)(x - 3)$$
$$(2x - 15)(x + 1) \qquad (2x + 15)(x - 1)$$

Testing the middle term, you can find the correct factorization to be

$$2x^2 + x - 15 = (2x - 5)(x + 3).$$

Factoring by Grouping

Sometimes polynomials with more than three terms can be **factored by grouping.**

EXAMPLE 8 Factoring by Grouping

$$x^3 - 2x^2 - 3x + 6 = (x^3 - 2x^2) - (3x - 6)$$ Group terms.

$$= x^2(x - 2) - 3(x - 2)$$ Factor groups.

$$= (x - 2)(x^2 - 3)$$ Common factor

Group Activities   Extending the Concept

A 3-D View of a Special Product   The figure below shows two cubes: a large cube whose volume is a^3 and a smaller cube whose volume is b^3. If the smaller cube is removed from the larger, the remaining solid has a volume of $a^3 - b^3$ and is composed of three rectangular boxes, labeled Box 1, Box 2, and Box 3. Find the volume of each box and describe how these results are related to the following special product formula.

$$a^3 - b^3 = (a - b)(a^2 + ab + b^2)$$

$$= (a - b)a^2 + (a - b)ab + (a - b)b^2$$

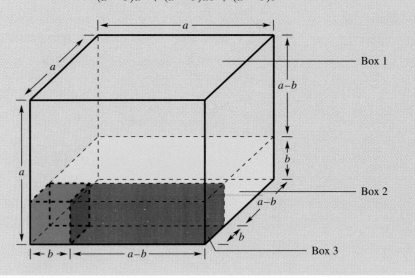

Warm Up The following warm-up exercises involve skills that were covered in earlier sections. You will use these skills in the exercise set for this section.

In Exercises 1–10, find the product.

1. $3x(5x - 2)$ **2.** $-2y(y + 1)$

3. $(2x + 3)^2$ **4.** $(3x - 8)^2$

5. $(2x - 3)(x + 8)$ **6.** $(4 - 5z)(1 + z)$

7. $(2y + 1)(2y - 1)$ **8.** $(x + a)(x - a)$

9. $(x + 4)^3$ **10.** $(2x - 3)^3$

P.6 **Exercises**

In Exercises 1–6, factor out the common factor.

1. $3x + 6$ **2.** $5y - 30$

3. $2x^3 - 6x$ **4.** $4x^3 - 6x^2 + 12x$

5. $(x - 1)^2 + 6(x - 1)$ **6.** $3x(x + 2) - 4(x + 2)$

In Exercises 7–12, factor the difference of two squares.

7. $x^2 - 36$ **8.** $x^2 - \frac{1}{4}$

9. $16y^2 - 9$ **10.** $49 - 9y^2$

11. $(x - 1)^2 - 4$ **12.** $25 - (z + 5)^2$

In Exercises 13–18, factor the perfect square trinomial.

13. $x^2 - 4x + 4$ **14.** $x^2 + 10x + 25$

15. $4t^2 + 4t + 1$ **16.** $9x^2 - 12x + 4$

17. $25y^2 - 10y + 1$ **18.** $z^2 + z + \frac{1}{4}$

In Exercises 19–32, factor the trinomial.

19. $x^2 + x - 2$ **20.** $x^2 + 5x + 6$

21. $s^2 - 5s + 6$ **22.** $t^2 - t - 6$

23. $y^2 + y - 20$ **24.** $z^2 - 5z - 24$

25. $x^2 - 30x + 200$ **26.** $x^2 - 13x + 42$

27. $3x^2 - 5x + 2$ **28.** $2x^2 - x - 1$

29. $9z^2 - 3z - 2$ **30.** $12x^2 + 7x + 1$

31. $5x^2 + 26x + 5$ **32.** $5u^2 + 13u - 6$

In Exercises 33–38, factor the sum or difference of cubes.

33. $x^3 - 8$ **34.** $x^3 - 27$

35. $y^3 + 64$ **36.** $z^3 + 125$

37. $8t^3 - 1$ **38.** $27x^3 + 8$

In Exercises 39–44, factor by grouping.

39. $x^3 - x^2 + 2x - 2$ **40.** $x^3 + 5x^2 - 5x - 25$

41. $2x^3 - x^2 - 6x + 3$ **42.** $5x^3 - 10x^2 + 3x - 6$

43. $6 + 2x - 3x^3 - x^4$ **44.** $x^5 + 2x^3 + x^2 + 2$

In Exercises 45–70, completely factor the expression.

45. $4x^2 - 8x$ **46.** $12x^3 - 6x^2$

47. $x^3 - 9x$ **48.** $12x^2 - 48$

49. $x^3 - 4x^2$

50. $6x^2 - 54$

51. $x^2 - 2x + 1$

52. $16 + 6x - x^2$

53. $1 - 4x + 4x^2$

54. $9x^2 - 6x + 1$

55. $-2x^2 - 4x + 2x^3$

56. $2y^3 - 7y^2 - 15y$

57. $9x^2 + 10x + 1$

58. $13x + 6 + 5x^2$

59. $3x^3 + x^2 + 15x + 5$

60. $5 - x + 5x^2 - x^3$

61. $x^4 - 4x^3 + x^2 - 4x$

62. $3u - 2u^2 + 6 - u^3$

63. $25 - (x + 5)^2$

64. $(t - 1)^2 - 49$

65. $(x^2 + 1)^2 - 4x^2$

66. $(x^2 + 8)^2 - 36x^2$

67. $2t^3 - 16$

68. $5x^3 + 40$

69. $4x(2x - 1) + (2x - 1)^2$

70. $5(3 - 4x)^2 - 8(3 - 4x)(5x - 1)$

In Exercises 71 and 72, factor the expression by using the following formula.

$(x + a)^3 = x^3 + 3x^2a + 3xa^2 + a^3$

71. $x^3 + 6x^2 + 12x + 8$ **72.** $27x^3 + 27x^2 + 9x + 1$

Geometric Modeling   In Exercises 73–76, make a "geometric factoring model" to represent the given factorization. For instance, a factoring model for $2x^2 + 5x + 2 = (2x + 1)(x + 2)$ is shown below.

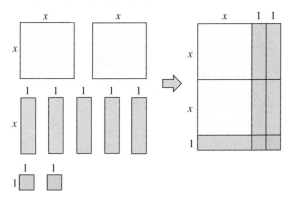

73. $3x^2 + 7x + 2 = (3x + 1)(x + 2)$

74. $x^2 + 4x + 3 = (x + 3)(x + 1)$

75. $2x^2 + 7x + 3 = (2x + 1)(x + 3)$

76. $x^2 + 3x + 2 = (x + 2)(x + 1)$

Geometric Modeling   In Exercises 77–80, match the "geometric factoring model" with its formula. [The models are labeled (a), (b), (c), and (d).]

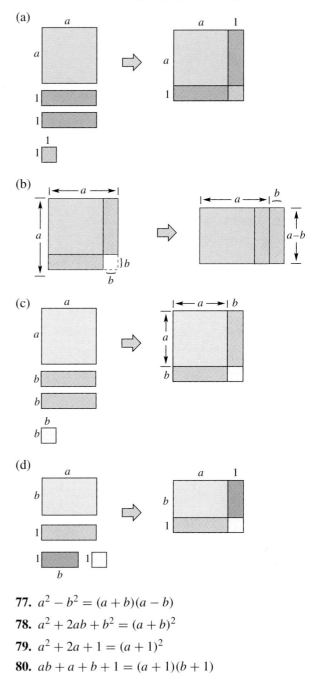

77. $a^2 - b^2 = (a + b)(a - b)$

78. $a^2 + 2ab + b^2 = (a + b)^2$

79. $a^2 + 2a + 1 = (a + 1)^2$

80. $ab + a + b + 1 = (a + 1)(b + 1)$

81. *Geometry*   The room that is shown in the figure has a floor space of $2x^2 + 3x + 1$ square feet. If the width of the room is $(x + 1)$ feet, what is the length?

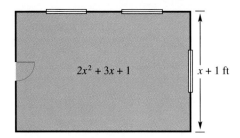

$2x^2 + 3x + 1$

$x + 1$ ft

82. *Geometry*   The room that is shown in the figure has a floor space of $3x^2 + 8x + 4$ square feet. If the width of the room is $(x + 2)$ feet, what is the length?

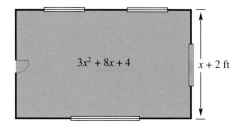

$3x^2 + 8x + 4$

$x + 2$ ft

Math Matters   Logic Puzzles

Use the clues to determine the class and college major of Crystal, Helene, Juan, and Tito. For each clue, shade in the appropriate boxes in the table. For instance, from the first clue you know that Juan is not the history major or the senior *and* that the history major is not the senior. When completed, the unshaded boxes show the solution to the puzzle.

Juan, the history major, and the senior ate lunch together.

Helene, the economics major, and the senior went to the movies.

Tito and the freshman played tennis doubles with Helene and the math major.

Crystal, the math major, and the sophomore attended the pep rally.

Juan and the sophomore challenged Tito and the economics major to a game of pinochle.

	Freshman	Sophomore	Junior	Senior	Economics	History	Math	Science
Crystal								
Helene								
Juan								
Tito								
Economics								
History								
Math								
Science								

P.7	**Fractional Expressions and Probability**
	Domain of an Expression ▪ Rational Expressions ▪ Complex Fractions ▪ Introduction to Probability

Domain of an Expression

The end of this section includes a brief discussion of probability, covering basic concepts such as outcomes, sample spaces, and events. You may wish to cover this material either as part of Chapter P or when appropriate in later chapters.

The set of all real numbers for which an algebraic expression is defined is called the **domain** of the expression. For instance, the domain of

$$\frac{1}{x}$$

is all real numbers other than $x = 0$. Two algebraic expressions are said to be **equivalent** if they yield the same value for all numbers in their domain. For instance, the expressions

$$[(x + 1) + (x + 2)] \quad \text{and} \quad 2x + 3$$

are equivalent.

STUDY TIP

For all algebraic expressions, exclude from the domain all values that could create *division by zero* or *the square root of a negative number*.

EXAMPLE 1   Finding the Domain of an Algebraic Expression

a. The domain of the polynomial

$$2x^3 + 3x + 4$$

is the set of all real numbers. In fact, the domain of any polynomial is the set of all real numbers (unless the domain is specifically restricted).

b. The domain of the polynomial

$$x^2 + 5x + 2, \qquad x > 0$$

is the set of positive real numbers, because the polynomial is specifically restricted to that set.

c. The domain of the radical expression

$$\sqrt{x}$$

is the set of nonnegative real numbers, because the square root of a negative number is not a real number.

d. The domain of the expression

$$\frac{x + 2}{x - 3}$$

is the set of all real numbers except $x = 3$, because the value of $x = 3$ would produce an undefined division by zero.

Activities

1. Find an expression equivalent to $9x - 4$.

2. The implied domain excludes which values of x from the domain of $\dfrac{x^2 + 6x + 9}{x^2 - 9}$?

   Answer: $x = 3, x = -3$

3. What is the domain of $\sqrt{3 - x}$?

   Answer: the set of all reals less than or equal to 3

Rational Expressions

The quotient of two algebraic expressions is a **fractional expression.** Moreover, the quotient of two *polynomials* such as

$$\frac{1}{x}, \qquad \frac{2x - 1}{x + 1}, \qquad \text{or} \qquad \frac{x^2 - 1}{x^2 + 1}$$

is a **rational expression.** Recall that a fraction is in reduced form if its numerator and denominator have no factors in common aside from ± 1. To write a fraction in reduced form, you can apply the following **Cancellation Law.**

$$\frac{a \cdot \cancel{c}}{b \cdot \cancel{c}} = \frac{a}{b}, \qquad b \neq 0, \quad c \neq 0$$

The key to success in simplifying rational expressions lies in your ability to *factor* polynomials. For example,

$$\frac{18x^2 - 18}{6x - 6} = \frac{3(6)(x + 1)(\cancel{x - 1})}{6(\cancel{x - 1})} = 3(x + 1), \qquad x \neq 1.$$

Note that the original expression is undefined when $x = 1$ (because division by zero is undefined). Thus, in order to make sure that the reduced expression is *equivalent* to the original expression, we must restrict the domain of the reduced expression by excluding the value $x = 1$.

EXAMPLE 2   Reducing a Rational Expression

$$\frac{x^2 + 4x - 12}{3x - 6} = \frac{(x + 6)(\cancel{x - 2})}{3(\cancel{x - 2})} \qquad \text{Factor completely.}$$

$$= \frac{x + 6}{3}, \qquad x \neq 2 \qquad \text{Cancel common factors.}$$

NOTE   In Example 2, do not make the mistake of trying to further reduce by canceling *terms*.

$$\frac{\cancel{x + 6}}{\cancel{3}} \cdot \frac{x + \overset{2}{\cancel{6}}}{\cancel{3}} = x + 2$$

Remember that to reduce fractions, you cancel *factors*, not terms.

When simplifying rational expressions, be sure to factor each polynomial completely before concluding that the numerator and denominator have no factors in common. Moreover, changing the sign of a factor may allow further reduction, as demonstrated in part (b) of the next example.

EXAMPLE 3   Reducing Rational Expressions

a. $\dfrac{x^3 - 4x}{x^2 + x - 2} = \dfrac{x(x+2)(x-2)}{(x+2)(x-1)}$ Factor completely.

$\qquad = \dfrac{x(x-2)}{x-1}, \qquad x \neq -2$ Cancel common factors.

b. $\dfrac{12 + x - x^2}{2x^2 - 9x + 4} = \dfrac{(4-x)(3+x)}{(2x-1)(x-4)}$ Factor completely.

$\qquad = \dfrac{-(x-4)(3+x)}{(2x-1)(x-4)}$ $4 - x = -(x-4)$

$\qquad = -\dfrac{3+x}{2x-1}, \qquad x \neq 4$ Cancel common factors.

To multiply or divide rational expressions, use the properties of fractions (see Section P.2). Recall that to divide fractions you invert the divisor and multiply.

EXAMPLE 4   Multiplying Rational Expressions

$\dfrac{6x^2 - 6x}{x^2 + 2x - 3} \cdot \dfrac{x^2 + x - 6}{2x}$

$\qquad = \dfrac{6x(x-1)(x+3)(x-2)}{(x-1)(x+3)(2x)}$ Factor and multiply.

$\qquad = \dfrac{3(2x)(x-1)(x+3)(x-2)}{(x-1)(x+3)(2x)}$ Cancel common factors.

$\qquad = 3(x-2), \quad x \neq 0, \ x \neq 1, \ x \neq -3$ Simplify.

EXAMPLE 5   Dividing Rational Expressions

$\dfrac{2x}{3x-12} \div \dfrac{x^2 - 2x}{x^2 - 6x + 8} = \dfrac{2x}{3x-12} \cdot \dfrac{x^2 - 6x + 8}{x^2 - 2x}$ Invert and multiply.

$\qquad = \dfrac{(2x)(x-2)(x-4)}{(3)(x-4)(x)(x-2)}$ Factor and multiply.

$\qquad = \dfrac{(2x)(x-2)(x-4)}{(3)(x-4)(x)(x-2)}$ Cancel common factors.

$\qquad = \dfrac{2}{3}, \quad x \neq 0, \ x \neq 2, \ x \neq 4$ Simplify.

To add or subtract rational expressions, use the least common denominator method or the following basic definition for adding two fractions.

$$\frac{a}{b} \pm \frac{c}{d} = \frac{ad \pm bc}{bd}, \qquad b \neq 0, \ d \neq 0$$

This definition is efficient for adding or subtracting *two* fractions that have no common factors in their denominators.

EXAMPLE 6   *Adding Rational Expressions*

$$\frac{x}{x-3} + \frac{2}{3x+4} = \frac{x(3x+4) + 2(x-3)}{(x-3)(3x+4)} \qquad \frac{a}{b} + \frac{c}{d} = \frac{ad+bc}{bd}$$

$$= \frac{3x^2 + 4x + 2x - 6}{(x-3)(3x+4)} \qquad \text{Remove parentheses.}$$

$$= \frac{3x^2 + 6x - 6}{(x-3)(3x+4)} \qquad \text{Collect like terms.}$$

EXAMPLE 7   *Combining Rational Expressions*

Perform the given operations and simplify.

$$\frac{3}{x-1} - \frac{2}{x} + \frac{x+3}{x^2-1}$$

Solution

Using the factored denominators $(x-1)$, x, and $(x+1)(x-1)$, you can see that the least common denominator is $x(x+1)(x-1)$.

$$\frac{3}{x-1} - \frac{2}{x} + \frac{x+3}{x^2-1}$$

$$= \frac{3(x)(x+1)}{x(x+1)(x-1)} - \frac{2(x+1)(x-1)}{x(x+1)(x-1)} + \frac{(x+3)(x)}{x(x+1)(x-1)}$$

$$= \frac{3(x)(x+1) - 2(x+1)(x-1) + (x+3)(x)}{x(x+1)(x-1)}$$

$$= \frac{3x^2 + 3x - 2x^2 + 2 + x^2 + 3x}{x(x+1)(x-1)}$$

$$= \frac{2x^2 + 6x + 2}{x(x+1)(x-1)}$$

$$= \frac{2(x^2 + 3x + 1)}{x(x+1)(x-1)}$$

Complex Fractions

Problems involving the division of two rational expressions are sometimes written as **complex fractions.** The rules for dividing fractions still apply in such cases: invert the denominator and multiply, as follows.

$$\frac{\left(\dfrac{x-1}{5}\right)}{\left(\dfrac{x+3}{x}\right)} = \frac{x-1}{5} \cdot \frac{x}{x+3}$$

$$= \frac{x(x-1)}{5(x+3)}$$

EXAMPLE 8   Simplifying a Complex Fraction

$$\frac{\left(\dfrac{x^2+2x-3}{x-3}\right)}{(4x+12)} = \frac{\left(\dfrac{x^2+2x-3}{x-3}\right)}{\left(\dfrac{4x+12}{1}\right)} \qquad \text{Rewrite denominator.}$$

$$= \frac{x^2+2x-3}{x-3} \cdot \frac{1}{4x+12} \qquad \text{Invert and multiply.}$$

$$= \frac{(x-1)(x+3)}{(x-3)(4)(x+3)} \qquad \text{Factor and multiply.}$$

$$= \frac{(x-1)(\cancel{x+3})}{(x-3)(4)(\cancel{x+3})} \qquad \text{Cancel common factors.}$$

$$= \frac{x-1}{4(x-3)}, \quad x \neq -3 \qquad \text{Simplify.}$$

EXAMPLE 9   Simplifying a Complex Fraction

$$\frac{\left(\dfrac{x}{x+1}\right)}{\left(\dfrac{2x}{x^2-1}\right)} = \frac{x}{x+1} \cdot \frac{x^2-1}{2x} \qquad \text{Invert and multiply.}$$

$$= \frac{\cancel{x}(x-1)(\cancel{x+1})}{(\cancel{x+1})(2)(\cancel{x})} \qquad \text{Multiply and factor.}$$

$$= \frac{x-1}{2}, \quad x \neq 0, \ x \neq -1 \qquad \text{Cancel common factors.}$$

Introduction to Probability

As a member of a complex society, you are used to living with varying amounts of uncertainty. For example, you may be questioning the likelihood of getting a good job after graduation, or winning a state lottery, or the probability of rain tomorrow.

In assigning measurements to uncertainties in everyday life, we often use ambiguous terminology, such as "fairly certain," "probable," or "highly unlikely." In mathematics, we attempt to remove this ambiguity by assigning a number to the likelihood of the occurrence of an event. We call this measurement the **probability** that the event will occur. For example, if we toss a fair coin, we say that the probability that it will land heads up is one-half, or 0.5.

In the study of probability, we call any happening whose result is uncertain an **experiment.** The possible results of the experiment are called **outcomes,** the set of all possible outcomes of an experiment is called the **sample space** of the experiment, and any subcollection of a sample space is called an **event.**

To calculate the probability of an event, we count the number of outcomes in the event and in the sample space. The *number of outcomes* in event E is denoted by $n(E)$ and the number of outcomes in the sample space S is denoted by $n(S)$.

The Probability of an Event

If an event E has $n(E)$ equally likely outcomes and its sample space has $n(S)$ equally likely outcomes, the **probability** of event E is

$$P(E) = \frac{n(E)}{n(S)}.$$

Because the number of outcomes in an event must be *less than or equal to* the number of outcomes in the sample space, we can see that the probability of an event must be a number between 0 and 1. That is, for any event E, it must be true that $0 \le P(E) \le 1$, as indicated in Figure P.12.

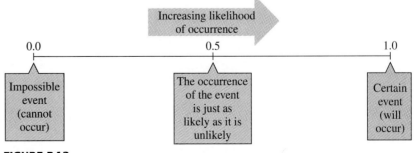

FIGURE P.12

Properties of the Probability of an Event

Let E be an event that is a subset of a finite sample space S.

1. $0 \le P(E) \le 1$

2. If $P(E) = 0$, E *cannot occur* and is called an **impossible event.**

3. If $P(E) = 1$, E *must occur* and is called a **certain event.**

Real Life

EXAMPLE 10   Finding the Probability of an Event

A card is drawn at random from a standard deck of 52 playing cards.

a. What is the probability that the card is the ace of hearts?

b. What is the probability that the card is an ace?

Solution

a. Because there are 52 cards in the deck, the number of possible outcomes in the sample space is 52. Moreover, because only one of these cards is the ace of hearts, the number of outcomes in the event "drawing the ace of hearts" is 1. Thus, the probability of drawing the ace of hearts is

$$P(\text{drawing the ace of hearts}) = \frac{1}{52}.$$

b. Because there are 52 cards in the deck, the number of possible outcomes in the sample space is 52. Moreover, because four of these cards are aces, the number of outcomes in the event "drawing an ace" is 4. Thus, the probability of drawing an ace is

$$P(\text{drawing an ace}) = \frac{4}{52} = \frac{1}{13}.$$

This probability is shown graphically in Figure P.13.

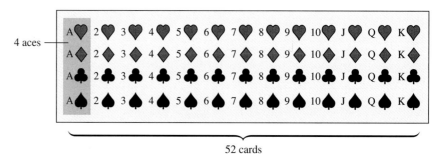

FIGURE P.13

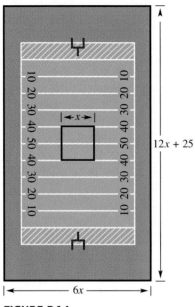

FIGURE P.14

EXAMPLE 11 *Finding the Probability of an Event*

A parachutist plans to land in the middle of a football stadium during a half-time show. The parachutist is certain of landing in the stadium field, but cannot guarantee a landing in a square area in the center of the field, as shown in Figure P.14. Assuming that the probability of landing at any given point in the field is the same, what is the probability that the parachutist will land in the center square?

Solution

The area of the entire field is given by

$$\text{Area of field} = (\text{length})(\text{width}) = (6x)(12x + 25).$$

The area of the center square is given by

$$\text{Area of center square} = (\text{length})(\text{width}) = x^2.$$

Therefore, the probability of landing in the center square is

$$P(\text{landing in center square}) = \frac{x^2}{(6x)(12x + 25)} = \frac{x}{6(12x + 25)}.$$

Group Activities   Extending the Concept

Group Activity Suggestion

This group activity lends itself to groups of two or three students (one or two as data recorders and one as pencil roller). You may wish to pool the data reported by each group and discuss it as an entire class. You may want to point out that, by pooling the data, the proportion of experiments resulting in the engraved side up is closer to the theorized value.

Investigating Probability   Locate a six-sided wooden pencil with an engraving (such as "No. 2") on one of its flat faces. As an experiment, roll the pencil on a flat surface, and observe the side facing up when it stops rolling. What is the probability that, when the pencil rolls to a stop, the engraved side is facing straight up? What is the probability that the side facing up is blank?

Now roll the pencil 50 times, keep track of the outcome of each roll, and find the proportion of experiments that resulted in the engraved side facing up. To what number do you think the proportion should be close? Do you think the probabilities you calculated are reasonable for the pencil you used? Why or why not?

Warm Up The following warm-up exercises involve skills that were covered in earlier sections. You will use these skills in the exercise set for this section.

In Exercises 1–10, completely factor the polynomials.

1. $5x^2 - 15x^3$ **2.** $16x^2 - 9$

3. $9x^2 - 6x + 1$ **4.** $9 + 12y + 4y^2$

5. $z^2 + 4z + 3$ **6.** $x^2 - 15x + 50$

7. $3 + 8x - 3x^2$ **8.** $3x^2 - 46x + 15$

9. $s^3 + s^2 - 4s - 4$ **10.** $y^3 + 64$

P.7 Exercises

In Exercises 1–10, find the domain of the expression.

1. $3x^2 - 4x + 7$ **2.** $2x^2 + 5x - 2$

3. $4x^3 + 5x + 3, \ x \geq 0$ **4.** $6x^2 + 7x - 9, \ x > 0$

5. $\dfrac{1}{x-2}$ **6.** $\dfrac{x+1}{2x+1}$

7. $\dfrac{x-1}{x^2-4x}$ **8.** $\dfrac{2x+1}{x^2-9}$

9. $\sqrt{x+1}$ **10.** $\dfrac{1}{\sqrt{x+1}}$

In Exercises 11–16, find the missing factor in the numerator so that the two fractions will be equivalent.

11. $\dfrac{5}{2x} = \dfrac{5()}{6x^2}$

12. $\dfrac{3}{4} = \dfrac{3()}{4(x+1)}$

13. $\dfrac{x+1}{x} = \dfrac{(x+1)()}{x(x-2)}$

14. $\dfrac{3y-4}{y+1} = \dfrac{(3y-4)()}{y^2-1}$

15. $\dfrac{3x}{x-3} = \dfrac{3x()}{x^2-x-6}$ **16.** $\dfrac{1-z}{z^2} = \dfrac{(1-z)()}{z^3+z^2}$

In Exercises 17–30, write the rational expression in reduced form.

17. $\dfrac{15x^2}{10x}$ **18.** $\dfrac{18y^2}{60y^5}$

19. $\dfrac{2x}{4x+4}$ **20.** $\dfrac{9x^2+9x}{2x+2}$

21. $\dfrac{x-5}{10-2x}$ **22.** $\dfrac{x^2-25}{5-x}$

23. $\dfrac{x^3+5x^2+6x}{x^2-4}$ **24.** $\dfrac{x^2+8x-20}{x^2+11x+10}$

25. $\dfrac{y^2-7y+12}{y^2+3y-18}$ **26.** $\dfrac{x+1}{x^2+11x+10}$

27. $\dfrac{2-x+2x^2-x^3}{x-2}$ **28.** $\dfrac{x^2-9}{x^3+x^2-9x-9}$

29. $\dfrac{z^3-8}{z^2+2z+4}$ **30.** $\dfrac{y^3-2y^2-3y}{y^3+1}$

In Exercises 31–54, perform the indicated operations and simplify.

31. $\dfrac{5}{x-1} \cdot \dfrac{x-1}{25(x-2)}$

32. $\dfrac{x+13}{x^3(3-x)} \cdot \dfrac{x(x-3)}{5}$

33. $\dfrac{(x-9)(x+7)}{x+1} \cdot \dfrac{x}{9-x}$

34. $\dfrac{(x+5)(x-3)}{x+2} \cdot \dfrac{1}{(x+5)(x+2)}$

35. $\dfrac{r}{r-1} \cdot \dfrac{r^2-1}{r^2}$

36. $\dfrac{4y-16}{5y+15} \cdot \dfrac{2y+6}{4-y}$

37. $\dfrac{t^2-t-6}{t^2+6t+9} \cdot \dfrac{t+3}{t^2-4}$

38. $\dfrac{y^3-8}{2y^3} \cdot \dfrac{4y}{y^2-5y+6}$

39. $\dfrac{x^2+x-2}{x^3+x^2} \cdot \dfrac{x}{x^2+3x+2}$

40. $\dfrac{x^3-1}{x+1} \cdot \dfrac{x^2+1}{x^2-1}$

41. $\dfrac{3(x+y)}{4} \div \dfrac{x+y}{2}$

42. $\dfrac{x+2}{5(x-3)} \div \dfrac{x-2}{5(x-3)}$

43. $\dfrac{\left[\dfrac{x^2}{(x+1)^2}\right]}{\left[\dfrac{x}{(x+1)^3}\right]}$

44. $\dfrac{\left(\dfrac{x^2-1}{x}\right)}{\left[\dfrac{(x-1)^2}{x}\right]}$

45. $\dfrac{5}{x-1} + \dfrac{x}{x-1}$

46. $\dfrac{2x-1}{x+3} + \dfrac{1-x}{x+3}$

47. $6 - \dfrac{5}{x+3}$

48. $\dfrac{3}{x-1} - 5$

49. $\dfrac{3}{x-2} + \dfrac{5}{2-x}$

50. $\dfrac{2x}{x-5} - \dfrac{5}{5-x}$

51. $\dfrac{2}{x^2-4} - \dfrac{1}{x^2-3x+2}$

52. $\dfrac{x}{x^2+x-2} - \dfrac{1}{x+2}$

53. $-\dfrac{1}{x} + \dfrac{2}{x^2+1} + \dfrac{1}{x^3+x}$

54. $\dfrac{2}{x+1} + \dfrac{2}{x-1} + \dfrac{1}{x^2-1}$

In Exercises 55–62, simplify the complex fraction.

55. $\dfrac{\left(\dfrac{x}{2}-1\right)}{(x-2)}$

56. $\dfrac{(x-3)}{\left(\dfrac{x}{4}-\dfrac{4}{x}\right)}$

57. $\dfrac{\left(\dfrac{1}{x}-\dfrac{1}{x+1}\right)}{\left(\dfrac{1}{x+1}\right)}$

58. $\dfrac{\left(\dfrac{5}{y}-\dfrac{6}{2y+1}\right)}{\left(\dfrac{5}{y}+4\right)}$

59. $\dfrac{\left(\dfrac{1}{[x+2]^2}-\dfrac{1}{x^2}\right)}{2}$

60. $\dfrac{\left(\dfrac{x+4}{x+5}-\dfrac{x}{x+1}\right)}{4}$

61. $\dfrac{\left(\sqrt{x}-\dfrac{1}{2\sqrt{x}}\right)}{\sqrt{x}}$

62. $\dfrac{3x^{1/3}-x^{-2/3}}{3x^{-2/3}}$

63. *Random Selection*   One of the 35 students in your class is chosen at random. What is the probability that it will be you?

64. *Misplaced Test*   Thirty-eight of the students in your class took a test. Your instructor lost two of the tests. What is the probability that yours got lost?

65. *Income Tax Audit*   In 1990, the Internal Revenue Service audited 388,000 of the 43,694,000 income tax returns that were filed with Form 1040A. If you filed a Form 1040A return during that year, what is the probability that your return was audited?   (Source: Internal Revenue Service)

66. *Lottery*   The "Daily Number" lottery in many states involves choosing an integer from 000 to 999. Suppose that you have purchased a lottery ticket that has one number. What is the probability that your number will be selected?

67. *Defective Product*   Suppose that you just bought a new camera. The manufacturer of your camera made 250,000 cameras of your model, and 5000 of them had defective lenses. What is the probability that your new camera has a defective lens?

68. *Playing Cards*   Suppose that you draw a single card from a standard deck of playing cards. What is the probability that the card is a heart?

Monthly Payment In Exercises 69 and 70, use the following formula, which gives the approximate annual percentage rate r of a monthly installment loan:

$$r = \frac{\left[\dfrac{24(NM - P)}{N}\right]}{\left(P + \dfrac{NM}{12}\right)}$$

where N is the total number of payments, M is the monthly payment, and P is the amount financed.

69. (a) Approximate the rate r for a 4-year car loan of $15,000 that has monthly payments of $400.

(b) Simplify the expression for the annual percentage rate r, and then rework part (a).

70. (a) Approximate the rate r for a 5-year car loan of $18,000 that has monthly payments of $400.

(b) Simplify the expression for the annual percentage rate r, and then rework part (a).

Probability In Exercises 71–74, consider an experiment in which a marble is tossed into a box whose dimensions are $2x + 1$ inches by x inches. Find the probability that the marble will come to rest in the shaded portion of the box.

71.

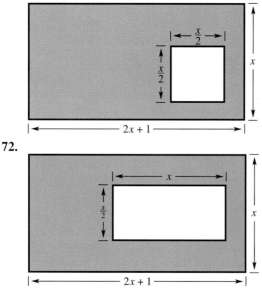

72.

73.

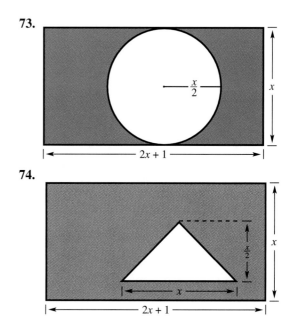

74.

75. *Refrigeration* When food (at room temperature) is placed in a refrigerator, the time required for the food to cool depends on the amount of food, the air circulation in the refrigerator, the original temperature of the food, and the temperature of the refrigerator. Consider the following model, which gives the temperature of food that is at $75°$F and is placed in a $40°$F refrigerator:

$$T = 10\left(\frac{4t^2 + 16t + 75}{t^2 + 4t + 10}\right), \quad 0 \le t$$

where T is the temperature in degrees Fahrenheit and t is the time in hours. Sketch a bar graph showing the temperature of the food when $t = 0, 1, 2, 3, 4,$ and 5 hours.

76. *Oxygen Level* The mathematical model

$$O = \frac{t^2 - t + 1}{t^2 + 1}, \quad 0 \le t$$

measures the percentage of the normal level of oxygen in a pond, where t is the time in weeks after organic waste is dumped into the pond. Sketch a bar graph showing the oxygen level of the pond when $t = 0, 1, 2, 3, 4,$ and 5 weeks.

CHAPTER PROJECT: Musical Notes and Frequencies

The musical note A-440 has a frequency of 440 vibrations per second. The frequency of any note can be found by the equation $F = 440 \cdot \sqrt[12]{2^n}$, where n represents the number of black and white keys below or above A-440. The figure below shows a piano keyboard with its corresponding written notes. It also shows the approximate ranges of the human voice.

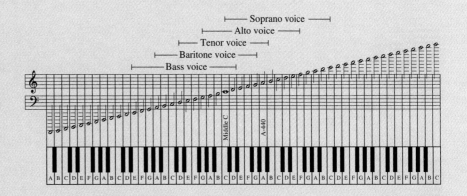

Use this information to investigate the following questions.

1. Approximate the lowest and highest frequencies of a soprano voice.

2. Approximate the lowest and highest frequencies of a tenor voice.

3. Approximate the lowest and highest frequencies of a bass voice.

4. From the results of Questions 1–3, approximate the range of the human voice.

5. Notes that are an octave apart differ by 12 notes. Thus, you can represent the two notes by n and $n + 12$. What can you say about the frequencies of notes that are one octave apart? Two octaves apart?

6. There are 88 keys on a piano—39 keys above A-440 and 48 keys below A-440. Find the highest and lowest frequencies on a piano.

7. Two notes harmonize if the ratio of their frequencies is an integer or simple rational number. Find a key such that the frequency ratio of A-440 to that key is about 3 to 4. Find a key such that the frequency ratio of A-440 to that key is about 3 to 2.

8. Make a table listing at least 20 notes and their frequencies. Enter the data into a calculator or computer program with power regression capabilities. Find a power regression equation to fit the data that you entered. Does the equation agree with the model given? Explain.

CHAPTER SUMMARY

After studying this chapter, you should have acquired the following skills. These skills are keyed to the Review Exercises that begin on page 75. Answers to odd-numbered Review Exercises are given in the back of the book.

- Classify real numbers as natural numbers, integers, rational numbers, or irrational numbers. *(Section P.1)* — **Review Exercises 1, 2**

- Plot real numbers on a real number line. *(Section P.1)* — **Review Exercises 3, 4**

- Give a verbal description of numbers represented by an inequality. *(Section P.1)* — **Review Exercises 5, 6**

- Use inequality notation to describe sets of real numbers. *(Section P.1)* — **Review Exercises 7, 8**

- Interpret absolute value notation. *(Section P.1)* — **Review Exercises 9–12**

- Find the distance between two numbers on the real number line. *(Section P.1)* — **Review Exercises 13–16**

- Identify the terms of an algebraic expression. *(Section P.2)* — **Review Exercises 17, 18**

- Evaluate an expression. *(Section P.2)* — **Review Exercises 19, 20, 31, 32**

- Identify rules of algebra. *(Section P.2)* — **Review Exercises 21, 22**

- Perform operations on real numbers. *(Section P.2)* — **Review Exercises 23–30**

- Simplify an expression. *(Section P.3)* — **Review Exercises 33, 34**

- Use scientific notation. *(Section P.3)* — **Review Exercises 35–40**

- Use tables to solve problems. *(Section P.3)* — **Review Exercises 41, 42**

- Use rational exponent notation. *(Section P.4)* — **Review Exercises 43, 44**

- Evaluate radicals. *(Section P.4)* — **Review Exercises 45, 46, 57, 58**

- Simplify radicals. *(Section P.4)* — **Review Exercises 47–56**

- Perform operations with polynomials. *(Section P.5)* — **Review Exercises 59–66**

- Factor polynomials. *(Section P.6)* — **Review Exercises 67–72, 75, 76**

- Find the domain of an expression. *(Section P.7)* — **Review Exercises 73, 74**

- Simplify a rational expression. *(Section P.7)* — **Review Exercises 77–80**

- Perform operations with rational expressions. *(Section P.7)* — **Review Exercises 81–86**

- Find the probability of an event. *(Section P.7)* — **Review Exercises 87–90**

REVIEW EXERCISES

In Exercises 1 and 2, determine which numbers in the set are (a) natural numbers, (b) integers, (c) rational numbers, and (d) irrational numbers.

1. $\{11, -14, -\frac{8}{9}, \frac{5}{2}, \sqrt{6}, 0.4\}$

2. $\{\sqrt{15}, -22, -\frac{10}{3}, 0, 5.2, \frac{3}{7}\}$

In Exercises 3 and 4, plot the two real numbers on the real number line and place the appropriate inequality sign ($<$ or $>$) between them.

3. $-4, -3$ **4.** $\frac{1}{2}, \frac{1}{3}$

In Exercises 5 and 6, give a verbal description of the subset of real numbers that is represented by the inequality, and sketch the subset on the real number line.

5. $x \le 7$ **6.** $x \ge 1$

In Exercises 7 and 8, use inequality notation to describe the set of real numbers.

7. x is nonnegative.

8. x is greater than 2 and less than or equal to 5.

In Exercises 9 and 10, write the expression without using absolute value signs.

9. $-|-14|$ **10.** $|-4 - 2|$

In Exercises 11 and 12, place the correct symbol ($<$, $>$, or $=$) between the two real numbers.

11. $|-12|$ ⬚ $-|12|$

12. $|9|$ ⬚ $|-9|$

In Exercises 13 and 14, find the distance between a and b.

13. $a = 48, \quad b = 45$

14. $a = 2, \quad b = -8$

In Exercises 15 and 16, use absolute value notation to describe the expression.

15. The distance between x and 7 is at least 4.

16. The distance between x and 25 is no more than 10.

In Exercises 17 and 18, identify the terms of the algebraic expression.

17. $2x^2 - 3x + 4$ **18.** $3x^3 - 9x$

In Exercises 19 and 20, evaluate the expression for the values of x.

19. $-4x^2 - 6x$ (a) $x = -1$ (b) $x = 0$

20. $12 - 5x^2$ (a) $x = -2$ (b) $x = 3$

In Exercises 21 and 22, identify the rule(s) of algebra illustrated by the equation.

21. $5(x^2 + x) = 5x^2 + 5x$

22. $x + (2x + 3) = (x + 2x) + 3$

In Exercises 23–28, perform the indicated operations. (Write fractional answers in reduced form.)

23. $-3 - 2(4 - 5)$ **24.** $12(-3 + 5) - 20$

25. $\frac{1}{2} + \frac{1}{3} - \frac{1}{6}$ **26.** $\frac{5}{12} + \frac{3}{4}$

27. $5^2 \cdot 5^{-1}$ **28.** $(4^2)^2$

In Exercises 29 and 30, use a calculator to evaluate the expression. (Round your answer to two decimal places.)

29. $4(\frac{1}{6} - \frac{1}{7})$ **30.** $-2 + 3(\frac{1}{2} - \frac{1}{3})$

In Exercises 31 and 32, evaluate the expression for the value of x.

31. $-2x^2, \quad x = -3$

32. $-\dfrac{(-x)^2}{3}, \quad x = -1$

In Exercises 33 and 34, simplify the expression.

33. $\dfrac{(4x)^2}{2x}$ **34.** $4(-3x)^3$

In Exercises 35 and 36, write the number in scientific notation.

35. *Daily U.S. Consumption of Dunkin' Donuts:*

2,740,000 donuts

36. *Number of Meters in One Foot:*

0.3048 meters per foot

In Exercises 37 and 38, write the number in decimal form.

37. *Distance Between Sun and Jupiter:*

4.833×10^8 miles

38. *Ratio of Day to Year:*

2.74×10^{-3}

In Exercises 39 and 40, use a calculator to evaluate the expression. (Round your answer to three decimal places.)

39. (a) $1800(1 + 0.08)^{24}$

(b) $0.0024(7,658,400)$

40. (a) $50,000\left(1 + \dfrac{0.075}{12}\right)^{48}$

(b) $\dfrac{28,000,000 + 34,000,000}{87,000,000}$

In Exercises 41 and 42, complete the table by finding the balances.

41. *Balance in an Account* Two thousand dollars is deposited in an account with an annual percentage rate of 6%, compounded monthly.

Years	5	10	15	20	25
Balance					

42. *Balance in an Account* Ten thousand dollars is deposited in an account with an annual percentage rate of 5.5%, compounded quarterly.

Years	5	10	15	20	25
Balance					

In Exercises 43 and 44, fill in the missing form.

Radical Form	*Rational Exponent Form*
43. $\sqrt{16} = 4$	
44.	$16^{1/4} = 2$

In Exercises 45 and 46, evaluate the expression.

45. $\sqrt{144}$ **46.** $\sqrt[3]{125}$

In Exercises 47 and 48, simplify by removing all possible factors from the radical.

47. $\sqrt{4x^4}$ **48.** $\sqrt[3]{\dfrac{2x^3}{27}}$

In Exercises 49 and 50, rewrite the expression by rationalizing the denominator. Simplify your answer.

49. $\dfrac{1}{2 - \sqrt{3}}$ **50.** $\dfrac{2}{3 - \sqrt{10}}$

In Exercises 51–54, simplify the expression.

51. $2\sqrt{x} - 5\sqrt{x}$ **52.** $\sqrt{72} + \sqrt{128}$

53. $\sqrt{5}\sqrt{2}$ **54.** $4^{1/3} \cdot 4^{5/3}$

In Exercises 55 and 56, use rational exponents to reduce the index of the radical.

55. $\sqrt[4]{5^2}$ **56.** $\sqrt[8]{x^4}$

In Exercises 57 and 58, use a calculator to approximate the number. (Round your answer to three decimal places.)

57. $\sqrt{127}$ **58.** $\sqrt[3]{52}$

In Exercises 59–66, perform the operations and write the resulting polynomial in standard form.

59. $2(x - 3) - 4(2x - 8)$

60. $3(x^2 - 5x + 2) + 3x(2 - 4x)$

61. $x(x - 2) - 2(3x + 7)$

62. $2x(x + 1) + 3(x^2 - x)$

63. $(x + 1)(x - 2)$

64. $(x - 5)(x + 5)$

65. $(x - 1)(x^2 + 2)$

66. $(2x + 1)^2$

In Exercises 67–72, completely factor the expression.

67. $4x^2 - 36$

68. $x^2 - 4x - 5$

69. $-3x^2 - 6x + 3x^3$

70. $x^3 - 16x$

71. $x^3 - 4x^2 - 2x + 8$

72. $x^3 - 125$

In Exercises 73 and 74, find the domain of the expression.

73. $\dfrac{2x + 1}{x - 3}$

74. $4\sqrt{2x}$

In Exercises 75 and 76, find the missing factor of the numerator so that the two fractions will be equivalent.

75. $\dfrac{4}{3x} = \dfrac{4()}{9x^2}$

76. $\dfrac{5}{7} = \dfrac{5()}{7(x + 2)}$

In Exercises 77–80, write the rational expression in reduced form.

77. $\dfrac{x^2 - 4}{2x + 4}$

78. $\dfrac{x^2 - 2x - 15}{x + 3}$

79. $\dfrac{2x^2 + 4x}{2x}$

80. $\dfrac{x^3 + 2x^2 - 3x}{x - 1}$

In Exercises 81–84, perform the operations and simplify.

81. $\dfrac{2x - 1}{x + 1} \cdot \dfrac{x^2 - 1}{2x^2 - 7x + 3}$

82. $\dfrac{x + 2}{x - 4} \div \dfrac{2x + 4}{8x}$

83. $\dfrac{x}{x - 1} + \dfrac{2x}{x - 2}$

84. $\dfrac{2}{x + 2} - \dfrac{3}{x - 2}$

In Exercises 85 and 86, simplify the complex fraction.

85. $\dfrac{\left(\dfrac{1}{x} - \dfrac{1}{y}\right)}{(x^2 - y^2)}$

86. $\dfrac{\left(\dfrac{1}{x} - \dfrac{1}{y}\right)}{\left(\dfrac{1}{x} + \dfrac{1}{y}\right)}$

87. *Dialing a Phone* Suppose that you are dialing a phone number. After dialing the first six digits correctly, your finger slips and you are not sure which number (from 0 to 9) you dialed as the seventh digit. What is the probability that you dialed the correct phone number?

88. *Raffle Ticket* A volunteer fire department sells 2450 raffle tickets. Ten of these tickets are chosen at random for ten different prizes. If you purchased only one ticket, what is the probability that you will win one of the prizes?

Probability In Exercises 89 and 90, consider an experiment in which a marble is tossed into a box whose dimensions are $x + 2$ inches by $2x$ inches. Find the probability that the marble will come to rest in the blue portion of the box.

89.

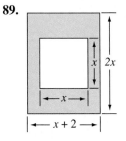

90.

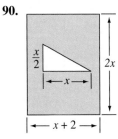

CHAPTER TEST

Take this test as you would take a test in class. After you are done, check your work against the answers given in the back of the book.

In Exercises 1 and 2, evaluate the expression.

1. $-3x^2 - 5x$, $x = -3$

2. $-\dfrac{(-x)^3}{4}$, $x = 2$

In Exercises 3–8, simplify the expression.

3. $8(-2x^2)^3$

4. $3\sqrt{x} - 7\sqrt{x}$

5. $5^{1/4} \cdot 5^{7/4}$

6. $\sqrt{128} - \sqrt{72}$

7. $\sqrt{12x^3}$

8. $\dfrac{2}{5 - \sqrt{7}}$

In Exercises 9 and 10, write the polynomial in standard form.

9. $(3x + 5)^2$

10. $3x(x + 5) - 2x(4x - 7)$

11. Completely factor $5x^2 - 80$.

12. Completely factor $x^3 - 6x^2 - 3x + 18$.

13. Simplify $\dfrac{x^2 - 16}{3x + 12}$.

14. Multiply and simplify $\dfrac{3x - 5}{x + 3} \cdot \dfrac{x^2 + 7x + 12}{9x^2 - 25}$.

15. Add and simplify $\dfrac{x}{x - 3} + \dfrac{3x}{x - 4}$.

16. Subtract and simplify $\dfrac{3}{x + 5} - \dfrac{4}{x - 2}$.

17. Complete the table at the right, given that $3000 is deposited in an account with an annual percentage rate of 8%, compounded monthly. What can you conclude from the table?

Years	Balance
5	
10	
15	
20	
25	

Table for 17

18. Simplify the complex fraction $\dfrac{\left(\dfrac{1}{x} + \dfrac{1}{y}\right)}{\left(\dfrac{1}{x} - \dfrac{1}{y}\right)}$.

19. You have just bought a new microwave oven. The manufacturer made 500,000 microwaves of your model, and 10,000 of them had defective timers. What is the probability that your new microwave oven has a defective timer?

20. Forty-two of the students in your class took a test. Your instructor lost two of the tests. What is the probability that your test got lost?

Algebraic Equations and Inequalities

1

- Linear Equations
- Linear Equations and Modeling
- Quadratic Equations
- The Quadratic Formula
- Other Types of Equations
- Linear Inequalities
- Other Types of Inequalities

During each year in recent history, the federal government has had less income than expenses. The difference between expenses and income for a year is called the annual federal deficit. By 1994, the national debt (or sum of annual federal deficits) had risen to over 5 trillion dollars.

The table at the right shows the national debt (in billions) at 10-year intervals from 1940 through 1990. This same data is shown graphically in the bar graph at the right.

From the graph, it is clear that the national debt is increasing. This statement alone, however, doesn't tell the whole story. The more distressing observation is that the annual federal deficit is accelerating. That is, the annual deficit is getting larger each year. Many economists believe that a rising national debt such as this is harmful to the economy.

Year	1940	1950	1960	1970	1980	1990
National Debt	$43.0	$256.1	$284.1	$370.1	$907.7	$3233.3

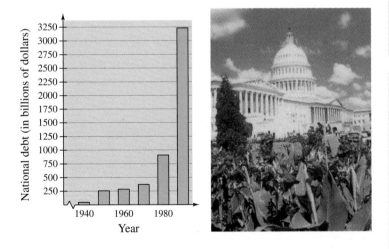

The chapter project related to this information is on page 158.

1.1	**Linear Equations**
	Equations and Solutions ▪ Linear Equations in One Variable ▪ Equations Involving Fractional Expressions ▪ Application

Equations and Solutions

An **equation** is a statement that two algebraic expressions are equal. Some examples of equations in x are

$$3x - 5 = 7, \quad x^2 - x - 6 = 0, \quad \text{and} \quad \sqrt{2x} = 4.$$

To **solve** an equation in x means that you find all values of x for which the equation is **true.** Such values are called **solutions.** For instance, $x = 4$ is a solution of the equation $3x - 5 = 7$, because $3(4) - 5 = 7$ is a true statement.

An equation that is true for *every* real number in the domain of the variable is an **identity.** Two examples of identities are

$$x^2 - 9 = (x + 3)(x - 3) \quad \text{and} \quad \frac{x}{3x^2} = \frac{1}{3x}, \quad x \neq 0.$$

The first equation is an identity because it is a true statement for all real values of x. The second is an identity because it is true for all nonzero real values of x.

An equation that is true for just *some* (or even none) of the real numbers in the domain of the variable is called a **conditional equation.** For example, the equation $x^2 - 9 = 0$ is conditional because $x = 3$ and $x = -3$ are the only values in the domain that satisfy the equation.

EXAMPLE 1   Classifying Equations

Classify each of the following as an identity or a conditional equation.

a. $2(x + 3) = 2x + 6$ **b.** $2(x + 3) = x + 6$ **c.** $2(x + 3) = 2x + 3$

Solution

a. This equation is an identity because it is true for every real value of x.

b. This equation is a conditional equation because there are real values of x (such as $x = 1$) for which the equation is not true.

c. This equation is a conditional equation because there are no real number values of x for which the equation is true.

Be sure you understand that equations are used in algebra for two distinct purposes: (1) *identities* are used to state mathematical properties and (2) *conditional equations* are used to model and solve problems that occur in real life.

Linear Equations in One Variable

The most common type of conditional equation is a **linear equation.**

Definition of a Linear Equation

A **linear equation** in one variable x is an equation that can be written in the standard form

$$ax + b = 0$$

where a and b are real numbers with $a \neq 0$.

A linear equation has exactly one solution. To see this, consider the following steps. (Remember that $a \neq 0$.)

$$ax + b = 0 \qquad \text{Original equation}$$
$$ax = -b \qquad \text{Subtract } b \text{ from both sides.}$$
$$x = -\frac{b}{a} \qquad \text{Divide both sides by } a.$$

Thus, the equation $ax + b = 0$ has exactly one solution, $x = -b/a$.

To solve a linear equation in x, you should isolate x by a sequence of **equivalent** (and usually simpler) equations, each having the same solution as the original equation. The operations that yield equivalent equations come from the basic rules of algebra reviewed in Chapter P.

Forming Equivalent Equations

A given equation can be transformed into an equivalent equation by one or more of the following steps.

	Given Equation	*Equivalent Equation*
1. Remove symbols of grouping, combine like terms, or simplify one or both sides of the equation.	$2x - x = 4$ $3(x - 2) = 5$	$x = 4$ $3x - 6 = 5$
2. Add (or subtract) the same quantity to (from) *both* sides of the equation.	$x + 1 = 6$	$x = 5$
3. Multiply (or divide) *both* sides of the equation by the same *nonzero* quantity.	$2x = 6$	$x = 3$
4. Interchange sides of the equation.	$2 = x$	$x = 2$

EXAMPLE 2   Solving a Linear Equation

Solve $3x - 6 = 0$.

Solution

$$3x - 6 = 0 \qquad \text{Original equation}$$
$$3x = 6 \qquad \text{Add 6 to both sides.}$$
$$x = 2 \qquad \text{Divide both sides by 3.}$$

After solving an equation, you should **check each solution** in the *original* equation. For instance, in Example 2, you can check that 2 is a solution by substituting 2 for x in the original equation $3x - 6 = 0$, as follows.

Check

$$3x - 6 = 0 \qquad \text{Original equation}$$
$$3(2) - 6 \stackrel{?}{=} 0 \qquad \text{Substitute 2 for } x.$$
$$6 - 6 = 0 \qquad \text{Solution checks.} \checkmark$$

EXAMPLE 3   Solving a Linear Equation

$$6(x - 1) + 4 = 3(7x + 1) \qquad \text{Original equation}$$
$$6x - 6 + 4 = 21x + 3 \qquad \text{Distributive Property}$$
$$6x - 2 = 21x + 3 \qquad \text{Simplify.}$$
$$-15x = 5 \qquad \text{Add 2 and subtract } 21x.$$
$$x = -\frac{1}{3} \qquad \text{Divide both sides by } -15.$$

The solution is $-\frac{1}{3}$. You can check this as follows.

Check

$$6(x - 1) + 4 = 3(7x + 1) \qquad \text{Original equation}$$
$$6\left(-\frac{1}{3} - 1\right) + 4 \stackrel{?}{=} 3\left[7\left(-\frac{1}{3}\right) + 1\right] \qquad \text{Substitute } -\frac{1}{3} \text{ for } x.$$
$$6\left(-\frac{4}{3}\right) + 4 \stackrel{?}{=} 3\left(-\frac{7}{3} + 1\right) \qquad \text{Add fractions.}$$
$$-\frac{24}{3} + 4 \stackrel{?}{=} -\frac{21}{3} + 3 \qquad \text{Multiply.}$$
$$-8 + 4 \stackrel{?}{=} -7 + 3 \qquad \text{Simplify.}$$
$$-4 = -4 \qquad \text{Solution checks.} \checkmark$$

Equations Involving Fractional Expressions

To solve an equation involving fractional expressions, you can multiply every term in the equation by the least common denominator (LCD) of the terms.

EXAMPLE 4   An Equation Involving Fractional Expressions

$$\frac{x}{3} + \frac{3x}{4} = 2 \qquad\qquad \text{Original equation}$$

$$(12)\frac{x}{3} + (12)\frac{3x}{4} = (12)2 \qquad\qquad \text{Multiply by least common denominator.}$$

$$4x + 9x = 24 \qquad\qquad \text{Reduce and multiply.}$$

$$13x = 24 \qquad\qquad \text{Combine like terms.}$$

$$x = \frac{24}{13} \qquad\qquad \text{Divide both sides by 13.}$$

The solution is $\frac{24}{13}$. Check this in the original equation.

When an equation is multiplied or divided by a *variable* expression, it is possible to introduce an **extraneous** solution—one that does not satisfy the original equation. In such cases a check is especially important.

EXAMPLE 5   An Equation with an Extraneous Solution

Solve $\dfrac{1}{x - 2} = \dfrac{3}{x + 2} - \dfrac{6x}{x^2 - 4}$.

Solution

The least common denominator is $x^2 - 4 = (x + 2)(x - 2)$. Multiplying each term by this LCD and reducing produces the following.

$$\frac{1}{x - 2} = \frac{3}{x + 2} - \frac{6x}{x^2 - 4}$$

$$\frac{1}{x - 2}(x + 2)(x - 2) = \frac{3}{x + 2}(x + 2)(x - 2) - \frac{6x}{x^2 - 4}(x + 2)(x - 2)$$

$$x + 2 = 3(x - 2) - 6x, \qquad x \neq \pm 2$$

$$x + 2 = 3x - 6 - 6x$$

$$4x = -8$$

$$x = -2$$

After checking $x = -2$, you can see that it yields a denominator of zero. Therefore, $x = -2$ is extraneous, and the equation has *no solution*.

An equation with a *single fraction* on each side can be cleared of denominators by **cross-multiplying,** which is equivalent to multiplying by the least common denominator and then reducing.

EXAMPLE 6   Cross-Multiplying to Solve an Equation

Solve $\dfrac{3y - 2}{2y + 1} = \dfrac{6y - 9}{4y + 3}$.

Solution

$$\dfrac{3y - 2}{2y + 1} = \dfrac{6y - 9}{4y + 3} \qquad \text{Original equation}$$

$$(3y - 2)(4y + 3) = (6y - 9)(2y + 1) \qquad \text{Cross-multiply.}$$

$$12y^2 + y - 6 = 12y^2 - 12y - 9 \qquad \text{Multiply.}$$

$$13y = -3 \qquad \text{Isolate } y\text{-term on left.}$$

$$y = -\dfrac{3}{13} \qquad \text{Divide both sides by 13.}$$

The solution is $-\frac{3}{13}$. Check this in the original equation.

EXAMPLE 7   Using a Calculator to Solve an Equation

Solve $\dfrac{1}{9.38} - \dfrac{3}{x} = \dfrac{5}{0.3714}$.

Solution

Roundoff error will be minimized if you solve for x before performing any calculations. The least common denominator is $(9.38)(0.3714)(x)$.

$$\dfrac{1}{9.38} - \dfrac{3}{x} = \dfrac{5}{0.3714}$$

$$(9.38)(0.3714)(x) \left(\dfrac{1}{9.38} - \dfrac{3}{x} \right) = (9.38)(0.3714)(x) \left(\dfrac{5}{0.3714} \right)$$

$$0.3714x - 3(9.38)(0.3714) = (9.38)(5)(x), \qquad x \neq 0$$

$$[0.3714 - 5(9.38)]x = 3(9.38)(0.3714)$$

$$x = \dfrac{3(9.38)(0.3714)}{0.3714 - 5(9.38)}$$

$$x \approx -0.225 \qquad \text{Round to three places.}$$

NOTE   Because of roundoff error, a check of a decimal solution may not yield exactly the same values for both sides of the original equation. The difference, however, should be quite small.

Application

EXAMPLE 8   Weekly Earnings

The average weekly earnings y (in dollars) for workers in the United States from 1980 to 1992 can be modeled by the linear equation

$$y = 14.6t + 270, \qquad 0 \le t$$

where $t = 0$ represents 1980, as shown in Figure 1.1. Use the model to estimate when the average weekly earnings will reach $500. (Source: U.S. Bureau of Labor Statistics)

Solution

To answer this question, let $y = 500$ and solve the resulting equation for t.

$$y = 14.6t + 270 \qquad \text{Given model}$$

$$500 = 14.6t + 270 \qquad \text{Substitute 500 for } y.$$

$$230 = 14.6t \qquad \text{Subtract 270 from both sides.}$$

$$t = \frac{230}{14.6} \approx 16 \qquad \text{Divide both sides by 14.6.}$$

Because $t = 0$ corresponds to 1980, it follows that $t = 16$ corresponds to 1996. Thus, you can expect the average weekly earnings to reach $500 by 1996.

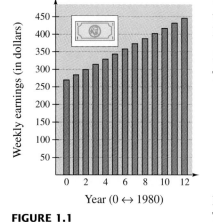

Year (0 ↔ 1980)

FIGURE 1.1

Group Activities   You Be the Instructor

Creating Quiz Questions   Suppose you are making a quiz for your algebra class that covers an introduction to equations. Make a list of eight equations that your students must identify as either identities or conditional equations (make sure there are some of each). Include at least one linear equation. Discuss how an equation in one variable may be "disguised" so that it is a linear equation and yet is not written in standard form.

Warm Up The following warm-up exercises involve skills that were covered in earlier sections. You will use these skills in the exercise set for this section.

In Exercises 1–10, perform the indicated operations and simplify your answer.

1. $(2x - 4) - (5x + 6)$ **2.** $(3x - 5) + (2x - 7)$

3. $2(x + 1) - (x + 2)$ **4.** $-3(2x - 4) + 7(x + 2)$

5. $\dfrac{x}{3} + \dfrac{x}{5}$ **6.** $x - \dfrac{x}{4}$

7. $\dfrac{1}{x + 1} - \dfrac{1}{x}$ **8.** $\dfrac{2}{x} + \dfrac{3}{x}$

9. $\dfrac{4}{x} + \dfrac{3}{x - 2}$ **10.** $\dfrac{1}{x + 1} - \dfrac{1}{x - 1}$

1.1 Exercises

In Exercises 1–6, determine whether the equation is an identity or a conditional equation.

1. $2(x - 1) = 2x - 2$ **2.** $3(x + 2) = 3x + 6$

3. $2(x - 1) = 3x + 4$ **4.** $3(x + 2) = 2x + 4$

5. $2(x + 1) = 2x + 1$ **6.** $3(x + 4) = 3x + 4$

In Exercises 7–14, determine whether the values of x are solutions of the equation.

7. $5x - 3 = 3x + 5$
   (a) $x = 0$ (b) $x = -5$ (c) $x = 4$ (d) $x = 10$

8. $7 - 3x = 5x - 17$
   (a) $x = -3$ (b) $x = 0$ (c) $x = 8$ (d) $x = 3$

9. $3x^2 + 2x - 5 = 2x^2 - 2$
   (a) $x = -3$ (b) $x = 1$ (c) $x = 4$ (d) $x = -5$

10. $5x^3 + 2x - 3 = 4x^3 + 2x - 11$
   (a) $x = 2$ (b) $x = -2$ (c) $x = 0$ (d) $x = 10$

11. $\dfrac{5}{2x} - \dfrac{4}{x} = 3$
   (a) $x = -\frac{1}{2}$ (b) $x = 4$ (c) $x = 0$ (d) $x = \frac{1}{4}$

12. $3 + \dfrac{1}{x + 2} = 4$
   (a) $x = -1$ (b) $x = -2$ (c) $x = 0$ (d) $x = 5$

13. $(x + 5)(x - 3) = 20$
   (a) $x = 3$ (b) $x = -2$ (c) $x = 0$ (d) $x = -7$

14. $\sqrt[3]{x - 8} = 3$
   (a) $x = 2$ (b) $x = -5$ (c) $x = 35$ (d) $x = 8$

In Exercises 15–52, solve the equation and check your answer. (Some equations have no solution.)

15. $x + 10 = 15$ **16.** $7 - x = 18$

17. $7 - 2x = 15$ **18.** $7x + 2 = 16$

19. $8x - 5 = 3x + 10$ **20.** $7x + 3 = 3x - 13$

21. $2(x + 5) - 7 = 3(x - 2)$

22. $2(13t - 15) + 3(t - 19) = 0$

23. $6[x - (2x + 3)] = 8 - 5x$

24. $8(x + 2) - 3(2x + 1) = 2(x + 5)$

25. $\dfrac{5x}{4} + \dfrac{1}{2} = x - \dfrac{1}{2}$ **26.** $\dfrac{x}{5} - \dfrac{x}{2} = 3$

27. $\frac{3}{2}(z + 5) - \frac{1}{4}(z + 24) = 0$

28. $\dfrac{3x}{2} + \dfrac{1}{4}(x - 2) = 10$

29. $0.25x + 0.75(10 - x) = 3$

30. $0.60x + 0.40(100 - x) = 50$

31. $x + 8 = 2(x - 2) - x$

32. $3(x + 3) = 5(1 - x) - 1$

33. $\dfrac{100 - 4u}{3} = \dfrac{5u + 6}{4} + 6$

34. $\dfrac{17 + y}{y} + \dfrac{32 + y}{y} = 100$

35. $\dfrac{5x - 4}{5x + 4} = \dfrac{2}{3}$

36. $\dfrac{10x + 3}{5x + 6} = \dfrac{1}{2}$

37. $10 - \dfrac{13}{x} = 4 + \dfrac{5}{x}$

38. $\dfrac{15}{x} - 4 = \dfrac{6}{x} + 3$

39. $\dfrac{1}{x - 3} + \dfrac{1}{x + 3} = \dfrac{10}{x^2 - 9}$

40. $\dfrac{1}{x - 2} + \dfrac{3}{x + 3} = \dfrac{4}{x^2 + x - 6}$

41. $\dfrac{x}{x + 4} + \dfrac{4}{x + 4} + 2 = 0$

42. $\dfrac{2}{(x - 4)(x - 2)} = \dfrac{1}{x - 4} + \dfrac{2}{x - 2}$

43. $\dfrac{7}{2x + 1} - \dfrac{8x}{2x - 1} = -4$

44. $\dfrac{4}{u - 1} + \dfrac{6}{3u + 1} = \dfrac{15}{3u + 1}$

45. $\dfrac{3}{x(x - 3)} + \dfrac{4}{x} = \dfrac{1}{x - 3}$

46. $3 = 2 + \dfrac{2}{z + 2}$

47. $(x + 2)^2 + 5 = (x + 3)^2$

48. $(x + 1)^2 + 2(x - 2) = (x + 1)(x - 2)$

49. $(x + 2)^2 - x^2 = 4(x + 1)$

50. $4(x + 1) - 3x = x + 5$

51. $(2x + 1)^2 = 4(x^2 + x + 1)$

52. $(2x - 1)^2 = 4(x^2 - x + 6)$

In Exercises 53–58, use a calculator to solve the equation. (Round your answer to three decimal places.)

53. $0.275x + 0.725(500 - x) = 300$

54. $2.763 - 4.5(2.1x - 5.1432) = 6.32x + 5$

55. $\dfrac{x}{0.6321} + \dfrac{x}{0.0692} = 1000$

56. $\dfrac{2}{7.398} - \dfrac{4.405}{x} = \dfrac{1}{x}$

57. $(x + 5.62)^2 + 10.83 = (x + 7)^2$

58. $\dfrac{x}{2.625} + \dfrac{x}{4.875} = 1$

In Exercises 59–64, evaluate each expression in two ways. (a) Calculate entirely on your calculator using appropriate parentheses, and then round the answer to two decimal places. (b) Round both the numerator and the denominator to two decimal places before dividing, and then round the final answer to two decimal places. Does the second method introduce an additional roundoff error?

59. $\dfrac{1 + 0.73205}{1 - 0.73205}$

60. $\dfrac{1 + 0.86603}{1 - 0.86603}$

61. $\dfrac{2 - 1.63254}{(2.58)(0.135)}$

62. $\dfrac{2 + 0.57735}{1 + 0.57735}$

63. $\dfrac{333 + \dfrac{1.98}{0.74}}{4 + \dfrac{6.25}{3.15}}$

64. $\dfrac{1.73205 - 1.19195}{3 - (1.73205)(1.19195)}$

65. *Legal Services Advertising* The amount of money spent on advertising for legal services on television from 1980 to 1991 can be approximated by the linear equation

$$y = 9t - 0.78, \qquad 0 \le t$$

where y is the amount in millions of dollars per year and t is the calendar year, with $t = 0$ corresponding to 1980. If this linear pattern continues, when will the amount of advertising for legal services on television reach \$150 million per year? (Source: Television Bureau of Advertising)

66. *Croquet Clubs*   The number of croquet clubs in the United States from 1985 to 1990 can be approximated by the linear equation

$$y = 47.2t - 135, \qquad 5 \le t$$

where y is the number of clubs and t is the calendar year, with $t = 5$ corresponding to 1985. If this linear pattern continues, when will there be 500 croquet clubs in the United States?   (Source: U.S. Croquet Association)

Human Height   In Exercises 67 and 68, use the following information. The relationship between the length of an adult's thigh bone and the height of the adult can be approximated by the linear equations

$$y = 0.432x - 10.44 \qquad \text{Female}$$
$$y = 0.449x - 12.15 \qquad \text{Male}$$

where y is the length of the thigh bone (femur) in inches and x is the height of the adult in inches (see figure).

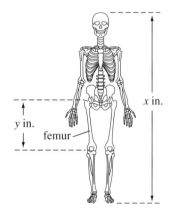

67. An anthropologist discovers a thigh bone belonging to an adult human female. The bone is 16 inches long. How tall would you estimate the female to have been?

68. From the foot bone of an adult human male, an anthropologist estimates that the male was 69 inches tall. A few feet away from the sight where the foot bone was discovered, the anthropologist discovers an adult male thigh bone that is 19 inches long. Is it possible that both bones came from the same person?

Social Security Benefits   In Exercises 69 and 70, use the following information. Under the Social Security laws (as existing in 1990) your expected maximum annual Social Security benefit at retirement depends on your age. (The younger you are, the greater your annual maximum benefit will be, as shown in the figure.) The relationship between present age and maximum annual benefit is approximately

$$y = 23{,}851 - 185x, \quad 20 \le x \le 65$$

where y is the maximum benefit in dollars and x is the present age in years.   (Source: *Money Magazine*)

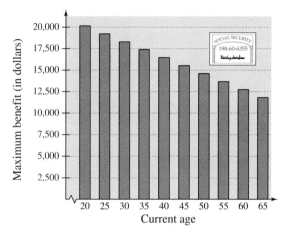

69. Another person in the class told you that she used this formula and discovered that her maximum benefit is $19,596. How old is she?

70. You receive a brochure describing a retirement plan. The brochure claims that your maximum retirement benefit from Social Security is $17,376. How old would you have to be for this to be true?

Probability   In Exercises 71 and 72, use the equation $p + p' = 1$, which states that the sum of the probability that an event *will* occur p and the probability that it *will not* occur p' is 1.

71. If the probability that an event will occur is 0.54, what is the probability that the event will not occur?

72. If the probability that an event will not occur is 0.87, what is the probability that the event will occur?

1.2   **Linear Equations and Modeling**

Introduction to Problem Solving ▪ Using Mathematical Models ▪
Mixture Problems ▪ Common Formulas

Introduction to Problem Solving

This section presents an introduction to problem solving, along with guidelines for mathematical modeling. Key words and phrases for translating words into algebra and common formulas are given to help students develop their own mathematical models. Students should be urged to organize the solutions to equations in a vertical manner. This also applies to inequalities in Sections 1.6 and 1.7.

In this section, you will study ways to use algebra to solve real-life problems. To do this, you will construct one or more equations that represent the real-life problem. This procedure is called **mathematical modeling.**

A good approach to mathematical modeling is to use two stages. In the first stage, the verbal description of the problem is used to form a *verbal model.* Then, after assigning labels to each of the quantities in the verbal model, you form a *mathematical model* or *algebraic equation.*

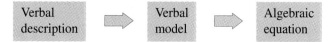

When you are trying to construct a verbal model, it is helpful to look for a *hidden equality*—a statement that two algebraic expressions are equal. For instance, in the following example the hidden equality equates your annual income to 24 paychecks and one bonus check.

*Real
Life*

EXAMPLE 1   *Using a Verbal Model*

You have accepted a job for which your annual salary will be $24,740. This salary includes a year-end bonus of $500. If you are paid twice a month, what will your gross pay be for each paycheck?

Solution

Because there are 12 months in a year and you will be paid twice a month, it follows that you will receive 24 paychecks during the year.

*Verbal
Model:*

Labels:	Income for year = 24,740	(dollars)
	Amount of each paycheck = x	(dollars)
	Bonus = 500	(dollars)

Equation: $24{,}740 = 24x + 500$

The algebraic equation for this problem is a *linear equation* in the variable x. Using the techniques discussed in Section 1.1, you can solve this equation for x. If you do this, you will find that the solution is $x = \$1010$.

Encourage students to write both the verbal model and the algebraic equation for homework problems and testing.

Translating Key Words and Phrases

Key Words and Phrases	Verbal Description	Algebraic Statement
Consecutive		
Next, subsequent	Consecutive integers	$n, n+1$
Addition		
Sum, plus, greater, increased by,	The sum of 5 and x	$5 + x$
more than, exceeds, total of	Seven more than y	$y + 7$
Subtraction		
Difference, minus, less than,	Four decreased by b	$4 - b$
decreased by, subtracted from,	Three less than z	$z - 3$
reduced by, the remainder		
Multiplication		
Product, multiplied by, twice,	Two times x	$2x$
times, percent of		
Division		
Quotient, divided by, per	The quotient of x and 8	$\dfrac{x}{8}$

Real Life

EXAMPLE 2   Constructing Mathematical Models

a. A salary of $22,400 is increased by 9%. What is the new salary?

Verbal Model: New salary $=$ 9%(salary) $+$ Salary

Labels:
Salary $= 22,400$ (dollars)
New salary $= S$ (dollars)
Percent $= 0.09$ (percent in decimal form)

Equation: $S = 0.09(22,400) + 22,400$

STUDY TIP

In Example 2, notice that part of the labeling process is to list the unit of measure for each labeled quantity. If you develop this habit, you will find it is a big help in checking the validity of your verbal model.

b. All items of clothing in a store are reduced by 20%. Find the original price of a suit selling for $140.

Verbal Model: Original price $-$ 20%(original price) $=$ Sale price

Labels:
Original price $= p$ (dollars)
Sale price $= 140$ (dollars)
Percent $= 0.2$ (percent in decimal form)

Equation: $p - 0.2p = 140$

Using Mathematical Models

Study the next several examples carefully. Your goal should be to develop a *general problem-solving strategy.*

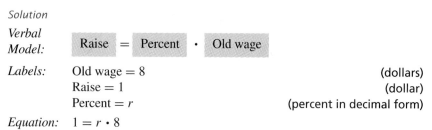

EXAMPLE 3 *Finding the Percent of a Raise*

You accept a job that pays $8 an hour. You are told that after a 2-month probationary period, your hourly wage will be increased to $9 an hour. What percent raise will you receive after the 2-month period?

Solution

Verbal Model: | Raise | = | Percent | · | Old wage |

Labels: Old wage = 8   (dollars)
Raise = 1   (dollar)
Percent = r   (percent in decimal form)

Equation: $1 = r \cdot 8$

By solving this equation, you will find that you will receive a raise of $\frac{1}{8} = 0.125$ or 12.5%.

EXAMPLE 4 *Finding the Percent of a Benefit Package*

Your annual salary is $22,000. In addition to your salary, your employer also provides the following benefits. The total of this benefit package represents what percent of your annual salary?

Social Security (Employer's Portion):	7.65% of salary	$1683
Workman's Compensation:	0.5% of salary	$110
Unemployment Compensation:	0.75% of salary	$165
Medical Insurance:	$2240 per year	$2240
Retirement Contribution:	6.5% of salary	$1430

Solution

Verbal Model: | Benefit package | = | Percent | · | Salary |

Labels: Salary = 22,000   (dollars)
Benefit package = 5628   (dollars)
Percent = r   (percent in decimal form)

Equation: $5628 = r \cdot 22{,}000$

By solving this equation, you will find that your benefit package is $r = \frac{5628}{22,000}$ or about 25.6% of your salary.

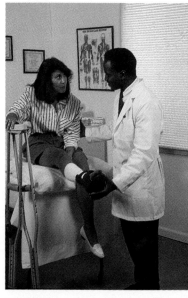

In 1992, 14.7% of the population of the United States had no health insurance.

Point out to your students that words such as *is, are, was, will, will be,* and *represents* indicate equality and often represent the equals sign (=) in the verbal model.

EXAMPLE 5 Finding the Dimensions of a Room

A rectangular family room is twice as long as it is wide, and its perimeter is 84 feet. Find the dimensions of the family room.

Solution

For this problem, it helps to sketch a picture (see Figure 1.2).

*Verbal
Model:* $2 \cdot$ [Length] $+ 2 \cdot$ [Width] $=$ [Perimeter]

Labels: Perimeter $= 84$ (feet)
Width $= w$ (feet)
Length $= l = 2w$ (feet)

Equation: $2(2w) + 2w = 84$
$4w + 2w = 84$
$6w = 84$
$w = 14$ feet
$l = 2w = 28$ feet

FIGURE 1.2

The dimensions of the room are 14 feet by 28 feet.

EXAMPLE 6 A Distance Problem

A plane is flying nonstop from New York to San Francisco, a distance of about 2700 miles, as shown in Figure 1.3. After 1.5 hours in the air, the plane flies over Chicago (a distance of 800 miles from New York). How long will it take the plane to fly from New York to San Francisco? (Assume that the plane flies at a constant speed during the entire flight.)

Solution

To solve this problem, use the formula that relates distance, rate, and time. That is, (distance) = (rate)(time). Because it took the plane 1.5 hours to travel a distance of 800 miles, you can conclude that its rate (or speed) must have been

$$\text{Rate} = \frac{\text{distance}}{\text{time}} = \frac{800 \text{ miles}}{1.5 \text{ hours}} \approx 533.33 \text{ miles per hour.}$$

Because the entire trip is 2700 miles, the time for the entire trip is

$$\text{Time} = \frac{\text{distance}}{\text{rate}} = \frac{2700 \text{ miles}}{533.33 \text{ miles per hour}} \approx 5.06 \text{ hours.}$$

FIGURE 1.3

Because 0.06 hours represents about 4 minutes, you can conclude that the trip will take about 5 hours and 4 minutes.

Another way to solve the distance problem in Example 6 is to use the concept of **ratio and proportion.** To do this, let x represent the time required to fly from New York to San Francisco, and set up the following proportion.

$$\frac{\text{Time to San Francisco}}{\text{Time to Chicago}} = \frac{\text{Distance to San Francisco}}{\text{Distance to Chicago}}$$

$$\frac{x}{1.5} = \frac{2700}{800}$$

$$x = 1.5 \cdot \frac{2700}{800}$$

$$x \approx 5.06$$

Notice how a ratio and proportion are used with a result from geometry to solve the problem in the following example.

Real Life

EXAMPLE 7   An Application Involving Similar Triangles

To measure the height of the twin towers of the World Trade Center (in New York City), you measure the shadow cast by one of the towers to be 170.25 feet long, as shown in Figure 1.4. Then you measure the shadow cast by a 4-foot post and find that its shadow is 6 inches long. How can this information be used to determine the height of the tower?

Solution

To find the height of the tower, you can use a result from geometry that states that the ratios of corresponding sides of similar triangles are equal.

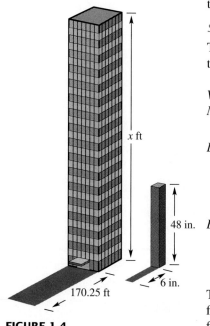

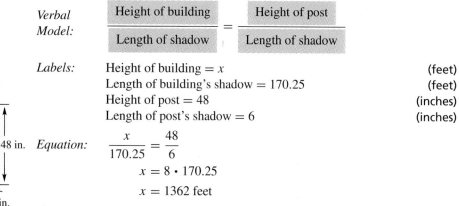

Verbal Model: $$\frac{\text{Height of building}}{\text{Length of shadow}} = \frac{\text{Height of post}}{\text{Length of shadow}}$$

Labels:
Height of building $= x$ (feet)
Length of building's shadow $= 170.25$ (feet)
Height of post $= 48$ (inches)
Length of post's shadow $= 6$ (inches)

Equation: $$\frac{x}{170.25} = \frac{48}{6}$$

$$x = 8 \cdot 170.25$$

$$x = 1362 \text{ feet}$$

The tower is about 1362 feet high. (This is the building from which King Kong fell in the remake of the movie *King Kong*. In the original movie, King Kong fell from the Empire State Building, which has a height of 1250 feet.)

x ft

48 in.

6 in.

170.25 ft

FIGURE 1.4

Mixture Problems

The next example is called a **mixture problem** because it involves two different unknown quantities that are *mixed* in a specific way. Watch for a *hidden product* in the verbal model.

Real Life

EXAMPLE 8 A Simple Interest Problem

You have received an inheritance of $10,000 that has been invested in two ways. Part of the money was invested at $9\frac{1}{2}\%$ simple interest and the remainder was invested at 11%. After 1 year, the two investments have returned a combined interest of $1038.50. How much was invested in each type of account?

Solution

Simple interest problems are based on the formula $I = Prt$, where I is the interest, P is the principal, r is the annual percentage rate (in decimal form), and t is the time in years.

Verbal Model: $\boxed{\text{Interest from } 9\frac{1}{2}\%} + \boxed{\text{Interest from } 11\%} = \boxed{\text{Total interest}}$

Labels:
Amount invested at $9\frac{1}{2}\% = x$ (dollars)
Amount invested at $11\% = 10,000 - x$ (dollars)
Interest from $9\frac{1}{2}\% = Prt = (x)(0.095)(1)$ (dollars)
Interest from $11\% = Prt = (10,000 - x)(0.11)(1)$ (dollars)
Total interest $= 1038.50$ (dollars)

Equation:
$$0.095x + 0.11(10,000 - x) = 1038.5$$
$$0.095x + 1100 - 0.11x = 1038.5$$
$$-0.015x = -61.5$$
$$x = \$4100 \text{ at } 9\frac{1}{2}\%$$
$$10,000 - x = \$5900 \text{ at } 11\%$$

The amount invested at $9\frac{1}{2}\%$ is $4100 and the amount invested at 11% is $5900. Check these results in the original statement of the problem.

NOTE In Example 8, did you recognize the hidden products in the two terms on the left side of the equation? Both hidden products come from the common formula

$$\boxed{\text{Interest}} = \boxed{\text{Principal}} \cdot \boxed{\text{Rate}} \cdot \boxed{\text{Time}}$$
$$I = Prt.$$

Programming

You can use a programmable calculator to solve simple interest problems using the program below. The program may be entered into a *TI-82* calculator. Programs for other calculator models may be found in the Appendix.

```
PROGRAM: SIMPINT
: Fix2
: Disp "PRINCIPAL"
: Input P
: Disp "INTEREST RATE IN"
: Disp "DECIMAL FORM"
: Input R
: Disp "NO. OF YEARS"
: Input T
: PRT → I
: Disp "THE INTEREST IS"
: Disp I
: Float
```

Use the program with Example 8 to find how much interest was earned on just the portion of the inheritance invested at 11%.

Common Formulas

Many common types of geometric, scientific, and investment problems use ready-made equations, called **formulas.** Knowing formulas such as those in the following lists will help you translate and solve a wide variety of real-life problems involving perimeter, area, volume, temperature, interest, and principal.

Common Formulas for Area, Perimeter, and Volume

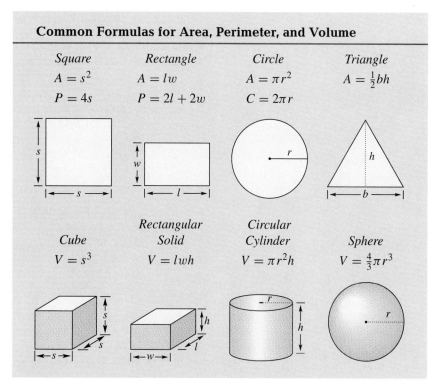

Square	Rectangle	Circle	Triangle
$A = s^2$	$A = lw$	$A = \pi r^2$	$A = \frac{1}{2}bh$
$P = 4s$	$P = 2l + 2w$	$C = 2\pi r$	

Cube	Rectangular Solid	Circular Cylinder	Sphere
$V = s^3$	$V = lwh$	$V = \pi r^2 h$	$V = \frac{4}{3}\pi r^3$

Miscellaneous Common Formulas

Temperature: F = degrees Fahrenheit, C = degrees Celsius

$$F = \frac{9}{5}C + 32$$

Simple interest: I = interest, P = principal, r = interest rate, t = time

$$I = Prt$$

Distance: d = distance traveled, r = rate, t = time

$$d = rt$$

When working with applied problems you often need to rewrite one of the common formulas. For instance, the formula $P = 2l + 2w$ for the perimeter of a rectangle can be rewritten or solved for w to produce $w = \frac{1}{2}(P - 2l)$.

Strategy for Solving Word Problems

1. *Search* for the hidden equality—two expressions said to be equal or known to be equal. A sketch may be helpful.

2. *Write* a verbal model that equates these two expressions.

3. *Assign* numbers to the known quantities and letters (or algebraic expressions) to the variable quantities.

4. *Rewrite* the verbal model as an algebraic equation using the assigned labels.

5. *Solve* the resulting algebraic equation.

6. *Check* to see that the answer satisfies the word problem as stated. (Remember that "solving for x" or some other variable may not completely answer the question.)

Group Activities Problem Solving

Red Herrings Applied problems in textbooks usually give precisely the right amount of information that is necessary to solve a given problem. In real life, however, you often must sort through the given information and discard information that is irrelevant to the problem. Such irrelevant information is called a **red herring.** Find the red herrings in the following problems.

a. From 100 to 200 feet beneath the surface of the ocean, pressure changes at a rate of approximately 4.4 pounds per square inch for every 10-foot change in depth. A diver takes 30 minutes to ascend 25 feet from a depth of 150 feet. What change in pressure does the diver experience?

b. A store manager marks up the price of a bread machine by 35% to $200, which is $50 less than a competitor's price. If sales tax is 5.5%, what will be the final price of the bread machine to the consumer?

Warm Up The following warm-up exercises involve skills that were covered in earlier sections. You will use these skills in the exercise set for this section.

In Exercises 1–10, solve the equation (if possible) and check your answer.

1. $3x - 42 = 0$

2. $64 - 16x = 0$

3. $2 - 3x = 14 + x$

4. $7 + 5x = 7x - 1$

5. $5[1 + 2(x + 3)] = 6 - 3(x - 1)$

6. $2 - 5(x - 1) = 2[x + 10(x - 1)]$

7. $\dfrac{x}{3} + \dfrac{x}{2} = \dfrac{1}{3}$

8. $\dfrac{2}{x} + \dfrac{2}{5} = 1$

9. $1 - \dfrac{2}{z} = \dfrac{z}{z + 3}$

10. $\dfrac{x}{x + 1} - \dfrac{1}{2} = \dfrac{4}{3}$

1.2 Exercises

Creating a Mathematical Model In Exercises 1–10, write an algebraic expression for the verbal expression.

1. The sum of two consecutive natural numbers

2. The product of two natural numbers whose sum is 25

3. *Distance Traveled* The distance traveled in t hours by a car traveling at 50 miles per hour

4. *Travel Time* The travel time for a plane that is traveling at a rate of r miles per hour for 200 miles

5. *Acid Solution* The amount of acid in x gallons of a 20% acid solution

6. *Discount* The sale price for an item that is discounted by 20% of its list price L

7. *Geometry* The perimeter of a rectangle whose width is x and whose length is twice the width

8. *Geometry* The area of a triangle whose base is 20 inches and whose height is h inches

9. *Total Cost* The total cost of producing x units for which the fixed costs are $1200 and the cost per unit is $25

10. *Total Revenue* The total revenue obtained by selling x units at $3.59 per unit

Using a Mathematical Model In Exercises 11–16, write a mathematical model for the number problem, and solve the problem.

11. The sum of two consecutive natural numbers is 525. Find the two numbers.

12. Find three consecutive natural numbers whose sum is 804.

13. One positive number is five times another positive number. The difference between the two numbers is 148. Find the numbers.

14. One number is one-fifth of another number. The difference between the two numbers is 76. Find the numbers.

15. Find two consecutive integers whose product is 5 less than the square of the smaller number.

16. Find two consecutive natural numbers such that the difference of their reciprocals is one-fourth the reciprocal of the smaller number.

17. *Weekly Paycheck* Your weekly paycheck is 15% *more* than your coworker's. Your two paychecks total $645. Find the amount of each paycheck.

18. *Monthly Profit* The total profit for a company in February was 20% higher than it was in January. The total profit for the two months was $157,498. Find the profit for each month.

19. *Weekly Paycheck* Your weekly paycheck is 15% *less* than your coworker's. Your two paychecks total $645. Find the amount of each paycheck.

20. *Monthly Profit* The total profit for a company in February was 20% less than it was in January. The total profit for the two months was $157,498. Find the profit for each month.

Price Increase In Exercises 21–24, the prices of different items are given for 1940 and 1990. Find the percent increase for each item. (Source: *USA Today*)

Item	1940	1990
21. Worker's annual earnings	$1500	$27,000
22. A ncw car	$800	$16,400
23. A pound of ground beef	$0.15	$1.44
24. $2\frac{1}{2}$ acres on the Potomac	$7500	$2,500,000

25. *Average Age of a Car* In 1970 the average age of a passenger car in the United States was 5.6 years. By 1980 the average age had increased by 17.86%, and by 1990 the average age had increased (over the 1980 age) by an additional 18.18%. (Source: American Automobile Manufacturers Association)

(a) Find the average age in 1980.

(b) Find the average age in 1990.

26. *Highest Football Salaries* In 1970 the highest paid professional football player (Joe Namath) had a salary of $150,000 per year. By 1980 the highest paid player (Walter Payton) had a salary that was 216.67% higher. By 1990 the highest paid player (Joe Montana) had a salary that was 742.11% higher than the 1980 high. Find the salaries of Walter Payton in 1980 and Joe Montana in 1990. (Source: *Sport Magazine*)

27. *Retirement Nest Eggs* Based on a survey of 740 companies, the most common types of investments by corporate retirement plans are those shown in the bar graph. How many respondents in the survey used each type of investment? (Most used more than one type of investment.) (Source: Wyatt Company)

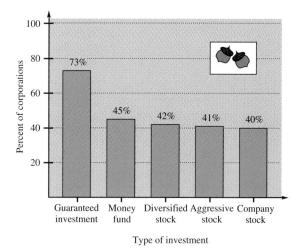

28. *School Lunch Snacks* Based on a survey of 426 mothers, the most common types of snacks put in children's school lunches are those shown in the bar graph. How many respondents put each type of snack in their children's lunches? (Most put in more than one type of snack.) (Source: Nabisco Brands)

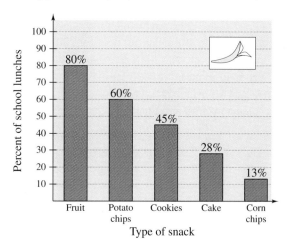

29. *Enjoying Nature* The Defenders of Wildlife organization conducted a survey in which 300 adults were asked how they enjoyed nature. (Respondents were allowed to answer in more than one category.) Of the 300 people, 189 answered "feeding birds," 186 answered "watching wildlife in neighborhood," 129 answered "driving to wildlife site," 108 answered "camping, biking, photography," and 105 answered "hiking." Suppose that one of the 300 people was chosen at random. What is the probability that the person said that feeding birds was a way that he or she enjoyed nature? (Source: Defenders of Wildlife)

30. *Business Headaches* In a survey, 414 small-business owners were asked what their biggest business headache was. (Respondents were allowed to answer in only one category.) The responses were as follows.

Government regulations	128
Employee relations	83
Cash flow	46
Insurance	37
Paperwork	25
Taxes	25
Other	70

Suppose that one of the 414 people was chosen at random. What is the probability that the person said that government regulations was his or her biggest headache? (Source: MasterCard)

31. *Geometry* A room is 1.5 times as long as it is wide, and its perimeter is 75 feet (see figure). Find the dimensions of the room.

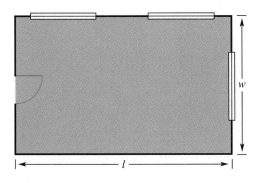

32. *Geometry* A picture frame has a total perimeter of 3 feet (see figure). The width of the frame is 0.62 times its length. Find the dimensions of the frame.

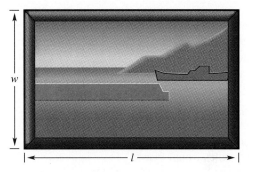

33. *Course Grade* To get an A in a course you must have an average of at least 90 percent on four tests that have 100 points each. If your scores on the first three tests were 87, 92, and 84, what must you score on the fourth test to get an A for the course?

34. *Course Grade* Suppose you are taking a course that has four tests. The first three tests have 100 points each and the fourth test has 200 points. To get an A in the course you must have an average of at least 90 percent on the four tests. Your scores on the first three tests were 87, 92, and 84. What must you score on the fourth test to get an A for the course?

35. *Loan Payments* A family has annual loan payments totaling $13,077.75, or 58.6% of its annual income. What is the family's income?

36. *Food Budget* Suppose your annual budget for food (eating at home and eating at restaurants) is 23.2% of your annual income. During the year, you spent $4832.12 on food. What is your annual income?

37. *List Price* The price of a swimming pool has been discounted 16.5%. The sale price is $849. Find the original list price of the pool.

38. *List Price* The price of a compact disc player has been discounted 27.5%. The sale price is $628. Find the original list price of the player.

39. *Discount Rate* The price of a television set has been discounted by $150. The sale price is $245. What percent is the discount of the original list price?

40. *Discount Rate* The price of a shirt has been discounted by \$20. The sale price is \$29.95. What percent is the discount of the original list price?

41. *Travel Time* Suppose you are driving on a freeway to another town that is 150 miles from your home. After 30 minutes, you pass a freeway exit that you know is 25 miles from your home. Assuming that you continue at the same constant speed, how long will it take for the entire trip?

42. *Travel Time* A plane is flying from Orlando to Denver, a distance of about 1950 miles. After 1 hour and 15 minutes, the plane flies over a town that is 600 miles from Orlando. How long will it take the plane to fly from Orlando to Denver? (Assume that the plane flies at a constant speed during the entire flight.)

43. *Travel Time* Two cars start at a given point and travel in the same direction at average speeds of 40 miles per hour and 55 miles per hour. How much time must elapse before the cars are five miles apart?

44. *Catch-Up Time* Students are traveling in two cars to a football game 135 miles away. The first car leaves on time and travels at an average speed of 45 miles per hour. The second car starts $\frac{1}{2}$ hour later and travels at an average speed of 55 miles per hour. At these speeds, how long will it take the second car of students to catch up to the first car?

45. *Travel Time* Two families meet at a park for a picnic. At the end of the day one family travels east at an average speed of 42 miles per hour and the other family travels west at an average speed of 50 miles per hour. Both families have approximately 160 miles to travel. Find the time it takes each family to get home.

46. *Average Speed* A truck traveled at an average speed of 55 miles per hour on a 200-mile trip to pick up a load of freight. On the return trip (with the truck fully loaded), the average speed was 40 miles per hour. Find the average speed for the round trip.

47. *Radio Waves* Radio waves travel at the same speed as light, 3.0×10^8 meters per second. Find the time required for a radio wave to travel from mission control in Houston to NASA astronauts on the surface of the moon 3.86×10^8 meters away.

48. *Distance to a Star* Find the distance to a star that is 50 light years (distance traveled by light in 1 year) away. (Light travels at 186,000 miles per second.)

49. *Height of a Tree* Suppose that you want to measure the height of a tree that is in your yard. To do this, you measure the tree's shadow and find that it is 25 feet long. You also measure the shadow of a 5-foot lamppost and find its shadow to be 2 feet long (see figure). How tall is the tree?

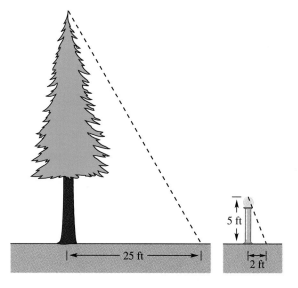

50. *Height of a Building* Suppose that you want to measure the height of a building. To do this, you measure the building's shadow and find that it is 50 feet long. You also measure the shadow of a 4-foot stake and find its shadow to be $3\frac{1}{2}$ feet long. How tall is the building?

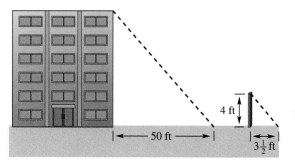

51. *Projected Expenses*   From January through May, a company's expenses have totaled $234,980. If the monthly expenses continue at this rate, what will the total expenses for the year be?

52. *Projected Revenue*   From January through August, a company's revenues have totaled $345,950. If the monthly revenue continues at this rate, what will the total revenue for the year be?

53. *Investment Mix*   Suppose you invest $12,000 in two funds paying $10\frac{1}{2}\%$ and 13% simple interest. The total annual interest is $1447.50. How much is invested in each fund?

54. *Investment Mix*   Suppose you invest $25,000 in two funds paying 11% and $12\frac{1}{2}\%$ simple interest. The total annual interest is $2975.00. How much is invested in each fund?

55. *Comparing Investment Returns*   You invest $12,000 in a fund paying $9\frac{1}{2}\%$ simple interest and $8000 in a fund for which the interest rate is variable. At the end of the year you receive notification that the total interest for both funds is $2054.40. Find the equivalent simple interest rate on the variable rate fund.

56. *Comparing Investment Returns*   You have $10,000 on deposit earning simple interest that is linked to the *prime rate*. When the prime rate dropped, your rate dropped by $1\frac{1}{2}\%$ for the last quarter of the year. Your total annual interest was $1112.50. What was your interest rate for the first three quarters and for the last quarter?

57. *Production Limit*   A company has fixed costs of $10,000 per month and variable costs of $8.50 per unit manufactured. The company has $85,000 available to cover the monthly costs. How many units can the company manufacture? (*Fixed costs* are those that occur regardless of the level of production. *Variable costs* depend on the level of production.)

58. *Production Limit*   A company has fixed costs of $10,000 per month and variable costs of $9.30 per unit manufactured. The company has $85,000 available to cover the monthly costs. How many units can the company manufacture? (*Fixed costs* are those that occur regardless of the level of production. *Variable costs* depend on the level of production.)

59. *Length of a Tank*   The diameter of a cylindrical propane gas tank is 4 feet (see figure). The total volume of the tank is 603.2 cubic feet. Find the length of the tank.

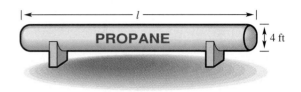

60. *Water Depth*   A trough is 12 feet long, 3 feet deep, and 3 feet wide (see figure). Find the depth of the water when the trough contains 70 gallons. (1 gallon ≈ 0.13368 cubic feet.)

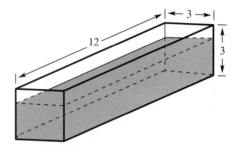

61. *Mixture*   A 55-gallon barrel contains a mixture with a concentration of 40%. How much of this mixture must be withdrawn and replaced by 100% concentrate to bring the mixture up to 75% concentration? (See figure.)

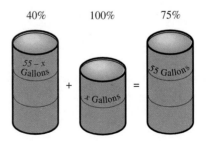

62. *Mixture*   A farmer mixed gasoline and oil to make 2 gallons of mixture for his two-cycle chain saw engine. This mixture was 32 parts gasoline and 1 part two-cycle oil. How much gasoline must be added to bring the mixture to 40 parts gasoline and 1 part oil?

New York City Marathon   In Exercises 63 and 64, find the average speed of the record-holding runners in the New York City Marathon. The length of the course is 26 miles, 385 yards. (Note that 1 mile = 5280 feet = 1760 yards.)

63. Men's record time: 2 hours, 8 minutes

64. Women's record time: 2 hours, $25\frac{1}{2}$ minutes

In Exercises 65–78, solve for the variable.

65. *Area*   Solve for h in $A = \frac{1}{2}bh$.

66. *Perimeter*   Solve for l in $P = 2l + 2w$.

67. *Volume*   Solve for l in $V = lwh$.

68. *Volume*   Solve for h in $V = \pi r^2 h$.

69. *Markup*   Solve for C in $S = C + RC$.

70. *Discount*   Solve for L in $S = L - RL$.

71. *Interest*   Solve for r in $A = P + Prt$.

72. *Interest*   Solve for P in $A = P\left(1 + \dfrac{r}{n}\right)^{nt}$.

73. *Area*   Solve for b in $A = \frac{1}{2}(a + b)h$.

74. *Area*   Solve for θ in $A = \dfrac{\pi r^2 \theta}{360}$.

75. *Sequence*   Solve for n in $L = a + (n - 1)d$.

76. *Series*   Solve for r in $S = \dfrac{rL - a}{r - 1}$.

77. *Lateral Surface Area*   Solve for h in $A = 2\pi rh$.

78. *Surface Area*   Solve for r in $A = 4\pi r^2$.

1.3	**Quadratic Equations**

Solving Quadratic Equations by Factoring ▪
Extracting Square Roots ▪ Applications

Solving Quadratic Equations by Factoring

In the first two sections of this chapter, you studied linear equations in one variable. In this and the next section, you will study quadratic equations.

Definition of a Quadratic Equation

A **quadratic equation** in x is an equation that can be written in the standard form

$$ax^2 + bx + c = 0$$

where a, b, and c are real numbers with $a \neq 0$. Another name for a quadratic equation in x is a **second-degree polynomial equation in x.**

There are three basic techniques for solving quadratic equations: factoring, extracting square roots, and the Quadratic Formula. (The Quadratic Formula is discussed in the next section.) The first technique is based on the Zero-Factor Property given in Section P.2.

If $ab = 0$, then $a = 0$ or $b = 0$.   Zero-Factor Property

To use this property, rewrite the left side of the standard form of a quadratic equation as the product of two linear factors. Then find the solutions of the quadratic equation by setting each factor equal to zero.

STUDY TIP

Be sure you see that the Zero-Factor Property applies *only* to equations written in standard form (in which one side of the equation is zero). Therefore, all terms must be collected on one side *before* factoring. For instance, in the equation

$$(x - 5)(x + 2) = 8$$

it is *incorrect* to set each factor equal to 8. Can you solve this equation correctly?

EXAMPLE 1   Solving a Quadratic Equation

Solve $x^2 - 3x - 10 = 0$.

Solution

$x^2 - 3x - 10 = 0$		Original equation
$(x - 5)(x + 2) = 0$		Factor.
$x - 5 = 0$ $\Longrightarrow$ $x = 5$		Set 1st factor equal to 0.
$x + 2 = 0$ $\Longrightarrow$ $x = -2$		Set 2nd factor equal to 0.

The solutions are 5 and -2. Check these in the original equation.

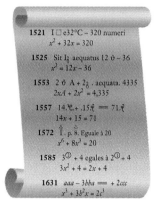

In ancient and medieval times, quadratic equations were classified in three categories.

$$x^2 + px = q$$
$$x^2 = px + q$$
$$x^2 + q = px$$

Solutions of these types of quadratic equations have been found in Babylonian texts over 4000 years old. Quadratic equations of the form $x^2 + px + q = 0$, where p and q are positive, were not solved because such equations have no positive solution.

A common error occurs when there are only two terms in an equation (as in Example 2).

$$6x^2 - 3x = 0$$
$$6x^2 = 3x$$
$$2x = 1$$
$$x = \tfrac{1}{2}$$

The root $x = 0$ is lost when dividing by $3x$.

EXAMPLE 2   Solving a Quadratic Equation by Factoring

Solve $6x^2 - 3x = 0$.

Solution

$6x^2 - 3x = 0$	Original equation
$3x(2x - 1) = 0$	Factor.
$3x = 0 \implies x = 0$	Set 1st factor equal to 0.
$2x - 1 = 0 \implies x = \dfrac{1}{2}$	Set 2nd factor equal to 0.

The solutions are 0 and $\frac{1}{2}$. Check these by substituting in the original equation as follows.

Check

$6x^2 - 3x = 0$	Original equation
$6(0)^2 - 3(0) \overset{?}{=} 0$	Substitute 0 for x.
$0 - 0 = 0$ ✓	First solution checks.
$6\left(\frac{1}{2}\right)^2 - 3\left(\frac{1}{2}\right) \overset{?}{=} 0$	Substitute $\frac{1}{2}$ for x.
$\frac{6}{4} - \frac{3}{2} = 0$ ✓	Second solution checks.

If the two factors of a quadratic expression are the same, the corresponding solution is a **double** or **repeated** solution.

EXAMPLE 3   A Quadratic Equation with a Repeated Solution

Solve $9x^2 - 6x = -1$.

Solution

$9x^2 - 6x = -1$	Original equation
$9x^2 - 6x + 1 = 0$	Write in standard form.
$(3x - 1)^2 = 0$	Factor.
$3x - 1 = 0$	Set repeated factor equal to 0.
$x = \dfrac{1}{3}$	Solution

The only solution is $\frac{1}{3}$. Check this in the original equation.

Extracting Square Roots

There is a nice shortcut for solving equations of the form $u^2 = d$, where $d > 0$. By factoring, you can see that this equation has two solutions.

$u^2 = d$	Original equation
$u^2 - d = 0$	Standard form
$(u + \sqrt{d})(u - \sqrt{d}) = 0$	Factor.
$u + \sqrt{d} = 0 \implies u = -\sqrt{d}$	Set 1st factor equal to 0.
$u - \sqrt{d} = 0 \implies u = \sqrt{d}$	Set 2nd factor equal to 0.

Solving an equation of the form $u^2 = d$ without going through the steps of factoring is called **extracting square roots.**

Extracting Square Roots

The equation $u^2 = d$, where $d > 0$, has exactly two solutions:

$$u = \sqrt{d} \quad \text{and} \quad u = -\sqrt{d}.$$

These solutions can also be written as $u = \pm\sqrt{d}$.

EXAMPLE 4   Extracting Square Roots

Solve $4x^2 = 12$.

Solution

$4x^2 = 12$	Original equation
$x^2 = 3$	Divide both sides by 4.
$x = \pm\sqrt{3}$	Extract square roots.

The solutions are $\sqrt{3}$ and $-\sqrt{3}$. Check these in the original equation.

EXAMPLE 5   Extracting Square Roots

$(x - 3)^2 = 7$	Original equation
$x - 3 = \pm\sqrt{7}$	Extract square roots.
$x = 3 \pm \sqrt{7}$	Add 3 to both sides.

The solutions are $3 \pm \sqrt{7}$. Check these in the original equation.

Applications

Quadratic equations often occur in problems dealing with area. Here is a simple example.

> A square room has an area of 144 square feet. Find the dimensions of the room.

To solve this problem, you can let x represent the length of each side of the room. Then, by solving the equation

$$x^2 = 144$$

you can conclude that each side of the room is 12 feet long. Note that although the equation $x^2 = 144$ has two solutions, -12 and 12, the negative solution makes no sense (for this problem), so you should choose the positive solution.

Real Life

EXAMPLE 6   Finding the Dimensions of a Room

A bedroom is 3 feet longer than it is wide (see Figure 1.5) and has an area of 154 square feet. Find the dimensions of the room.

Solution

You can begin by using the same type of problem-solving strategy that was presented in Section 1.2.

Verbal Model:
$$\boxed{\text{Width of room}} \cdot \boxed{\text{Length of room}} = \boxed{\text{Area of room}}$$

Labels: Area of room $= 154$ (square feet)
            Width of room $= w$ (feet)
            Length of room $= w + 3$ (feet)

Equation: $w(w + 3) = 154$

$$w^2 + 3w - 154 = 0$$

$$(w - 11)(w + 14) = 0$$

$$w - 11 = 0 \implies w = 11$$

$$w + 14 = 0 \implies w = -14$$

Choosing the positive value, you can conclude that the width is 11 feet and the length is $w + 3 = 14$ feet. You can check this in the original statement of the problem as follows.

Check

The length of 14 feet is 3 feet more than the width of 11 feet. ✓

The area of the room is $(11)(14) = 154$ square feet. ✓

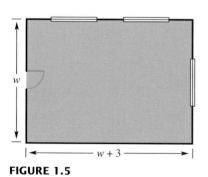

w

$w + 3$

FIGURE 1.5

Another application of quadratic equations involves an object that is falling (or vertically projected into the air). The equation that gives the height of such an object is called a **position equation,** and on earth's surface it has the form

$$s = -16t^2 + v_0t + s_0.$$

In this equation, s represents the height of the object (in feet), v_0 represents the original velocity of the object (in feet per second), s_0 represents the original height of the object (in feet), and t represents the time (in seconds).

Real Life

EXAMPLE 7 *Falling Object*

A construction worker is working on the 24th floor of a building project, as shown in Figure 1.6. The worker accidentally drops a wrench and immediately yells "Look out below!" Could a person at ground level hear this warning in time to get out of the way of the falling wrench?

Solution

Let's assume that each floor of the building is 10 feet high, so that the wrench was dropped from a height of 240 feet. Because sound travels at about 1100 feet per second, you can assume that a person at ground level hears the warning within 1 second of the time the wrench is dropped. To set up a mathematical model for the height of the wrench, you can use the position equation

$$s = -16t^2 + v_0t + s_0.$$

Because the object is dropped (rather than thrown), you can conclude that the initial velocity is $v_0 = 0$. Moreover, because the initial height is $s_0 = 240$ feet, you have the following model.

$$s = -16t^2 + 240$$

After falling for 1 second, the wrench is at a height of $-16(1^2) + 240 = 224$ feet. After falling for 2 seconds, the wrench is at a height of $-16(2^2) + 240 = 176$ feet. To find the number of seconds it takes the wrench to hit the ground, let the height s be zero and solve the resulting equation for t.

$s = -16t^2 + 240$	Position equation
$0 = -16t^2 + 240$	Set height equal to 0.
$16t^2 = 240$	Add $16t^2$ to both sides.
$t^2 = 15$	Divide both sides by 16.
$t = \sqrt{15} \approx 3.87$	Extract positive square root.

The wrench will take about 3.87 seconds to hit the ground. If the person heard the warning 1 second after the wrench was dropped, the person would still have almost 3 seconds to get out of the way.

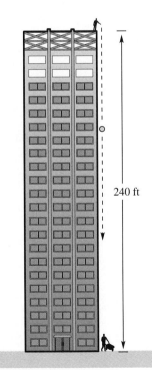

240 ft

FIGURE 1.6

A third type of application that often involves a quadratic equation is one dealing with the hypotenuse of a right triangle. Recall from geometry that the sides of a right triangle are related by a formula called the **Pythagorean Theorem.** This theorem states that if a and b are the lengths of the sides of the triangle and c is the length of the hypotenuse (see Figure 1.7),

$$a^2 + b^2 = c^2. \qquad \text{Pythagorean Theorem}$$

Notice how this formula is used in the next example.

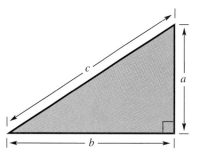

FIGURE 1.7

Real Life

EXAMPLE 8   *Cutting Across the Lawn*

Your house is on a large corner lot. Several children in the neighborhood cut across your lawn in such a way that the distance across the lawn is 32 feet, as shown in Figure 1.8. Assuming that Figure 1.8 is drawn to scale, how many feet does a person save by walking across the lawn instead of walking on the sidewalk?

Solution

In Figure 1.8, let x represent the length of the shorter part of the sidewalk. Using a ruler, it appears that the length of the longer part of the sidewalk is twice the shorter, so you can represent its length by $2x$.

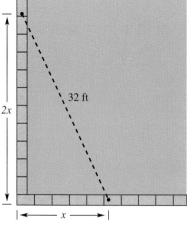

32 ft

$2x$

x

FIGURE 1.8

$$
\begin{aligned}
x^2 + (2x)^2 &= 32^2 && \text{Pythagorean Theorem} \\
5x^2 &= 1024 && \text{Add like terms.} \\
x^2 &= 204.8 && \text{Divide both sides by 5.} \\
x &= \sqrt{204.8} && \text{Extract positive square root.}
\end{aligned}
$$

The total distance on the sidewalk is

$$x + 2x = 3x = 3\sqrt{204.8} \approx 42.9 \text{ feet.}$$

Thus, cutting across the lawn saves a person about 10.9 feet of walking.

A fourth type of common application of a quadratic equation is one in which a quantity y is changing over time t according to a quadratic model.

EXAMPLE 9   The Cost of a New Car

From 1970 to 1990, the average cost of a new car can be modeled by

$$\text{Cost} = 30.5t^2 + 4192, \qquad 0 \le t$$

where $t = 0$ represents 1970. If the average cost of a new car continued to increase according to this model, when would the average cost reach $20,000? (Source: Commerce Department, American Autodatum)

Solution

To solve this problem, let the cost be $20,000 and solve the equation for t.

$30.5t^2 + 4192 = 20,000$	Set cost equal to 20,000.
$30.5t^2 = 15,808$	Subtract 4192 from both sides.
$t^2 \approx 518.295$	Divide both sides by 30.5.
$t \approx \sqrt{518.295}$	Extract positive square root.

The solution is $t \approx 23$. Because $t = 0$ represents 1970, you can conclude that the average cost of a new car reached $20,000 in 1993.

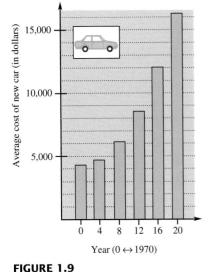

FIGURE 1.9

Average cost of new car (in dollars)

15,000

10,000

5,000

0   4   8   12   16   20

Year (0 ↔ 1970)

Group Activities   Extending the Concept

Geometry   A **Pythagorean triple** is a set of three positive integers (a, b, c) that satisfy the Pythagorean Theorem for right triangles, $a^2 + b^2 = c^2$. One such Pythagorean triple is (3, 4, 5), because $3^2 + 4^2 = 5^2$. How many other Pythagorean triples can you find? (Remember that each member of the triple must be an integer.) Make a list of these triples, and discuss the method you used to find them.

Warm Up

The following warm-up exercises involve skills that were covered in earlier sections. You will use these skills in the exercise set for this section.

In Exercises 1–4, simplify the expression.

1. $\sqrt{\frac{7}{50}}$

2. $\sqrt{32}$

3. $\sqrt{7^2 + 3 \cdot 7^2}$

4. $\sqrt{\frac{1}{4} + \frac{3}{8}}$

In Exercises 5–10, factor the expression.

5. $3x^2 + 7x$

6. $4x^2 - 25$

7. $16 - (x - 11)^2$

8. $x^2 + 7x - 18$

9. $10x^2 + 13x - 3$

10. $6x^2 - 73x + 12$

1.3 Exercises

In Exercises 1–10, write the quadratic equation in standard form.

1. $2x^2 = 3 - 5x$

2. $4x^2 - 2x = 9$

3. $x^2 = 25x$

4. $10x^2 = 90$

5. $(x - 3)^2 = 2$

6. $12 - 3(x + 7)^2 = 0$

7. $x(x + 2) = 3x^2 + 1$

8. $x^2 + 1 = \dfrac{x - 3}{2}$

9. $\dfrac{3x^2 - 10}{5} = 12x$

10. $x(x + 5) = 2(x + 5)$

In Exercises 11–22, solve the quadratic equation by factoring.

11. $6x^2 + 3x = 0$

12. $9x^2 - 1 = 0$

13. $x^2 - 2x - 8 = 0$

14. $x^2 - 10x + 9 = 0$

15. $x^2 + 10x + 25 = 0$

16. $16x^2 + 56x + 49 = 0$

17. $3 + 5x - 2x^2 = 0$

18. $2x^2 = 19x + 33$

19. $x^2 + 4x = 12$

20. $x^2 + 4x = 21$

21. $-x^2 - 7x = 10$

22. $-x^2 + 8x = 12$

In Exercises 23–36, solve the equation by extracting square roots. List both the exact answer *and* a decimal answer that has been rounded to two decimal places.

23. $x^2 = 16$

24. $x^2 = 144$

25. $x^2 = 7$

26. $x^2 = 27$

27. $3x^2 = 36$

28. $9x^2 = 25$

29. $(x - 12)^2 = 18$

30. $(x + 13)^2 = 21$

31. $(x + 2)^2 = 12$

32. $(x + 5)^2 = 20$

33. $12x^2 = 300$

34. $6x^2 = 250$

35. $3x^2 + 2(x^2 - 4) = 15$

36. $x^2 + 3(x^2 - 5) = 10$

In Exercises 37–58, solve the equation by any convenient method.

37. $x^2 = 64$

38. $7x^2 = 32$

39. $x^2 - 2x + 1 = 0$

40. $x^2 - 6x + 5 = 0$

41. $16x^2 - 9 = 0$

42. $11x^2 + 33x = 0$

43. $4x^2 - 12x + 9 = 0$

44. $x^2 - 14x + 49 = 0$

45. $(x + 3)^2 = 81$

46. $(x - 5)^2 = 8$

47. $4x = 4x^2 - 3$

48. $80 + 6x = 9x^2$

49. $50 + 5x = 3x^2$

50. $144 - 73x + 4x^2 = 0$

51. $12x = x^2 + 27$

52. $26x = 8x^2 + 15$

53. $50x^2 - 60x + 10 = 0$

54. $9x^2 + 12x + 3 = 0$

55. $(x + 3)^2 - 4 = 0$

56. $(x - 2)^2 - 9 = 0$

57. $(x + 1)^2 = x^2$

58. $(x + 1)^2 = 4x^2$

59. *Geometry*   A one-story building is 14 feet longer than it is wide (see figure). The building has 1632 square feet of floor space. What are the dimensions of the building?

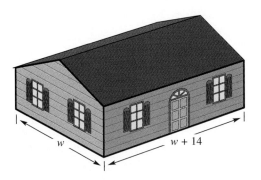

60. *Geometry*   A billboard is 10 feet longer than it is high (see figure). The billboard has 336 square feet of advertising space. What are the dimensions of the billboard?

61. *Geometry*   A triangular sign has a height that is equal to its base. The area of the sign is 4 square feet. Find the base and height of the sign.

62. *Geometry*   The building lot shown in the figure has an area of 8000 square feet. What are the dimensions of the lot?

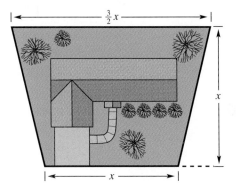

In Exercises 63–66, assume that air resistance is negligible, which implies that the position equation discussed in this section is a reasonable model.

63. *Royal Gorge Bridge*   The Royal Gorge Bridge near Canon City, Colorado, is the highest suspension bridge in the world. The bridge is 1053 feet above the Arkansas river. If a rock were dropped from the bridge, how long would it take to hit the water?

64. *Falling Object*   A rock is dropped from the top of a 200-foot cliff that overlooks the ocean. How long will it take for the rock to hit the water?

65. *Olympic Diver*   The high-dive platform in the Olympics is 10 meters above the water. A diver wants to perform a double flip into the water. How long will the diver be in the air?

66. *The Owl and the Mouse*   An owl is circling a field and sees a mouse. The owl folds its wings and begins to dive. If the owl starts its dive from a height of 100 feet, how long does the mouse have to escape?

67. *Geometry*   The hypotenuse of an isosceles right triangle is 5 centimeters long. How long are the sides? (An isosceles triangle is one that has two sides of equal length.)

68. *Geometry* An equilateral triangle has a height of 1 foot. How long are each of its sides? (*Hint:* Use the height of the triangle to partition the triangle into two right triangles of the same size.)

69. *Flying Distance* The cities of Chicago, Atlanta, and Buffalo approximately form the vertices of a right triangle. The distance from Atlanta to Buffalo is about 650 miles and the distance from Atlanta to Chicago is about 565 miles. Use the map in the figure to approximate the flying distance from Atlanta to Buffalo *by way of* Chicago.

70. *Depth of a Submarine* The sonar of a navy cruiser detects a submarine that is 3000 feet from the cruiser. The angle formed by the ocean surface and a line from the cruiser to the submarine is 45° (see figure). How deep is the submarine?

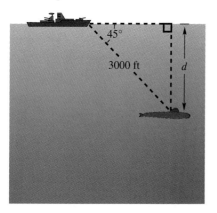

71. *Salvage Depth* A marine salvage ship detects a wreck 2000 feet off the bow. If the angle between the ocean surface and a line from the ship to the wreck is 45°, how deep is the wreck?

72. *Golf Winnings* The leading golf winnings (in dollars) from 1983 to 1988 in the United States can be approximated by the model

Winnings $= 29{,}035t^2 + 429{,}200, \quad 0 \le t$

where $t = 0$ represents 1983 (see figure). Assuming that the leading winnings continue to follow the pattern given by this quadratic model, when would the leading winnings exceed \$3,500,000? (Source: Professional Golfers' Association)

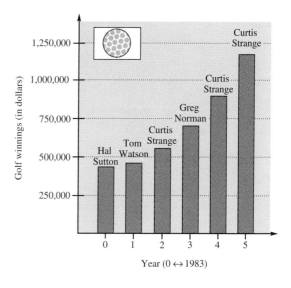

73. *Total Revenue* The demand equation for a certain product is $p = 20 - 0.0002x$, where p is the price per unit and x is the number of units sold. The total revenue for selling x units is given by

Revenue $= xp = x(20 - 0.0002x)$.

How many units must be sold to produce a revenue of \$500,000?

74. *Total Revenue*   The demand equation for a certain product is $p = 30 - 0.0005x$, where p is the price per unit and x is the number of units sold. The total revenue for selling x units is given by

Revenue $= xp = x(30 - 0.0005x)$.

How many units must be sold to produce a revenue of $450,000?

75. *Production Cost*   A company determines that the average monthly cost of raw materials for manufacturing a product line can be modeled by

Cost $= 20.75t^2 + 5104, \quad 0 \leq t$

where $t = 0$ represents 1990. If the average monthly cost of raw materials continues to increase according to this model, when will the average monthly cost reach $10,000?

76. *Monthly Cost*   A company determines that the average monthly cost for staffing temporary positions can be modeled by

Cost $= 115.35t^2 + 12,072, \quad 0 \leq t$

where $t = 0$ represents 1990. If the average monthly cost for staffing temporary positions continues to increase according to this model, when will the average monthly cost reach $20,000?

77. *U.S. Population*   The population of the United States from 1800 to 1890 can be approximated by the model

Population $= 694.2t^2 + 6183.3, \quad 0 \leq t \leq 9$

where the population is given in thousands of people and the time t represents the calendar year with $t = 0$ corresponding to 1800, $t = 1$ corresponding to 1810, and so on (see figure). If this model had continued to be valid up through the present time, when would the population of the United States have reached 250,000,000? Judging from the given figure, would you say this model was a good representation of the population through 1890? How about through 1990?   (Source: U.S. Bureau of Census)

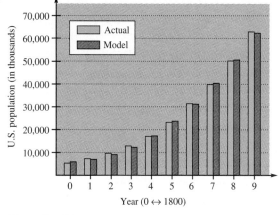

Figure for 77

78. *A Short Deck*   A standard deck of cards is missing several cards. One of the cards in the deck, however, is known to be the ace of hearts. A card is drawn from the deck at random, and then replaced, and then a card is drawn again. The probability that both drawings produced the ace of hearts is given by

$p^2 \approx 0.000567.$

How many cards are in the deck? (*Hint:* Write p in the form $1/n$, where n is the number of cards in the deck.)

79. *Airline Flight*   Suppose that you are going on a *round-trip* airline flight. You are told that the probability that both flights will be late is

$p^2 = 0.2016.$

What is the probability p that you will be late on the return trip?

1.4	**The Quadratic Formula**

Development of the Quadratic Formula ▪ The Discriminant ▪
Using the Quadratic Formula ▪ Applications

Development of the Quadratic Formula

In Section 1.3 you studied two methods for solving quadratic equations. These two methods are efficient for special quadratic equations that are factorable or that can be solved by extracting square roots.

There are, however, many quadratic equations that cannot be solved efficiently by either of these two techniques. Fortunately, there is a general formula that can be used to solve *any* quadratic equation. It is called the **Quadratic Formula.** You can develop this important formula by a process called **completing the square,** as follows.

$$ax^2 + bx + c = 0 \qquad \text{Standard form, } a \neq 0$$

$$ax^2 + bx = -c \qquad \text{Subtract } c \text{ from both sides.}$$

$$x^2 + \frac{b}{a}x = -\frac{c}{a} \qquad \text{Divide both sides by } a.$$

$$x^2 + \frac{b}{a}x + \left(\frac{b}{2a}\right)^2 = -\frac{c}{a} + \left(\frac{b}{2a}\right)^2 \qquad \text{Add } \left(\frac{b}{2a}\right)^2 \text{ to both sides.}$$

$$\left(x + \frac{b}{2a}\right)^2 = \frac{b^2 - 4ac}{4a^2} \qquad \text{Simplify.}$$

$$x + \frac{b}{2a} = \pm\sqrt{\frac{b^2 - 4ac}{4a^2}} \qquad \text{Extract square roots.}$$

$$x = \frac{-b \pm \sqrt{b^2 - 4ac}}{2a} \qquad \text{Solutions}$$

This result is summarized below.

STUDY TIP

The Quadratic Formula is one of the most important formulas in algebra, and you should memorize it. We have found that it helps to try to memorize a verbal statement of the rule. For instance, you might try to remember the following verbal statement of the Quadratic Formula: "The opposite of b, plus or minus the square root of b squared minus $4ac$, all divided by $2a$."

The Quadratic Formula

The solutions of

$$ax^2 + bx + c = 0, \qquad a \neq 0$$

are given by the **Quadratic Formula,**

$$x = \frac{-b \pm \sqrt{b^2 - 4ac}}{2a}.$$

The Discriminant

In the Quadratic Formula, the quantity under the radical sign, $b^2 - 4ac$, is called the **discriminant** of the quadratic expression $ax^2 + bx + c$.

$$b^2 - 4ac \qquad\qquad\qquad \text{Discriminant}$$

It can be used to determine the nature of the solutions of a quadratic equation.

Because this course stresses the applications of algebra to real-life situations, a discussion of complex numbers is presented in Section 4.4 rather than in Chapter 1. However, if instead you would like to discuss complex numbers with quadratic equations, you may teach Section 4.4 now with no difficulty.

NOTE   In case 3 at the right, the equation has no real solutions.

Solutions of a Quadratic Equation

The solutions of a quadratic equation

$$ax^2 + bx + c = 0, \quad a \neq 0$$

can be classified by the discriminant, $b^2 - 4ac$, as follows.

1. If $b^2 - 4ac > 0$, the equation has *two* distinct real solutions.

2. If $b^2 - 4ac = 0$, the equation has *one* repeated real solution.

3. If $b^2 - 4ac < 0$, the equation has *two* distinct imaginary solutions. (In Section 4.4, you will study imaginary numbers.)

EXAMPLE 1   *Using the Discriminant*

Use the discriminant to determine the number of real solutions of each of the following quadratic equations.

a. $4x^2 - 20x + 25 = 0$ **b.** $13x^2 + 7x + 1 = 0$ **c.** $5x^2 = 8x$

Solution

a. Using $a = 4$, $b = -20$, and $c = 25$, the discriminant is

$$b^2 - 4ac = (-20)^2 - 4(4)(25) = 400 - 400 = 0.$$

Thus, there is *one* repeated real solution.

b. Using $a = 13$, $b = 7$, and $c = 1$, the discriminant is

$$b^2 - 4ac = (7)^2 - 4(13)(1) = 49 - 52 = -3.$$

Thus, there are *no* real solutions.

c. In standard form, this equation is $5x^2 - 8x = 0$, with $a = 5$, $b = -8$, and $c = 0$, which implies that the discriminant is

$$b^2 - 4ac = (-8)^2 - 4(5)(0) = 64.$$

Thus, there are *two* distinct real solutions.

Activities

1. Use the discriminant to find the number of solutions to $3x^2 - 3x - 1 = 0$.

   Answer: two

2. Write $6x - 15 = 7x^2$ in standard form.

   Answer: $7x^2 - 6x + 15 = 0$

3. Use the discriminant to find the number of solutions to $\frac{1}{4}x^2 - x + 1 = 0$

   Answer: one

Using the Quadratic Formula

When using the Quadratic Formula, remember that *before* the formula can be applied, you must first write the quadratic equation in standard form.

Use a graphing utility to graph $y = x^2 + 3x - 9$. What do you observe about the number of times the graph crosses the x-axis and the number of solutions found in Example 2? Using the TRACE feature, compare the approximate values for x where the graph crosses the x-axis and the decimal approximations for the solutions in Example 2. What is the relationship between the solutions of a quadratic equation and the behavior of its graph? Do your conclusions apply in Example 3?

EXAMPLE 2   Two Distinct Solutions

Solve $x^2 + 3x = 9$.

Solution

$$x^2 + 3x = 9 \qquad\qquad \text{Original equation}$$

$$x^2 + 3x - 9 = 0 \qquad\qquad \text{Write in standard form.}$$

$$x = \frac{-3 \pm \sqrt{(3)^2 - 4(1)(-9)}}{2(1)} \qquad\qquad \text{Quadratic Formula}$$

$$x = \frac{-3 \pm \sqrt{45}}{2} \qquad\qquad \text{Simplify.}$$

$$x = \frac{-3 \pm 3\sqrt{5}}{2} \qquad\qquad \text{Simplify.}$$

The solutions are

$$x = \frac{-3 + 3\sqrt{5}}{2} \qquad \text{and} \qquad x = \frac{-3 - 3\sqrt{5}}{2}.$$

Check these in the original equation.

EXAMPLE 3   One Repeated Solution

Solve $8x^2 - 24x + 18 = 0$.

Solution

Begin by dividing both sides by the common factor of 2.

$$8x^2 - 24x + 18 = 0 \qquad\qquad \text{Original equation}$$

$$4x^2 - 12x + 9 = 0 \qquad\qquad \text{Divide both sides by 2.}$$

$$x = \frac{-(-12) \pm \sqrt{(-12)^2 - 4(4)(9)}}{2(4)} \qquad\qquad \text{Quadratic Formula}$$

$$x = \frac{12 \pm \sqrt{0}}{8} \qquad\qquad \text{Simplify.}$$

$$x = \frac{3}{2} \qquad\qquad \text{Repeated solution}$$

The only solution is $\frac{3}{2}$. Check this in the original equation.

The discriminant in Example 3 is a perfect square (zero in this case), and you could have factored the quadratic as

$$4x^2 - 12x + 9 = 0$$
$$(2x - 3)^2 = 0$$

to conclude that the solution is $x = \frac{3}{2}$. Because factoring is easier than applying the Quadratic Formula, try factoring first when solving a quadratic equation. If, however, factors cannot be easily found, then use the Quadratic Formula. For instance, try solving the quadratic equation

$$x^2 - x - 12 = 0$$

in two ways—by factoring and by the Quadratic Formula—to see that you get the same solutions either way.

When using a calculator with the Quadratic Formula, you should get in the habit of using the memory key to store intermediate steps. This will save steps and minimize roundoff error.

Technology

Try to calculate the value of x in Example 4 by using additional parentheses instead of storing the intermediate result, 196.43478, in your calculator.

EXAMPLE 4   Using a Calculator with the Quadratic Formula

Solve $16.3x^2 - 197.6x + 7.042 = 0$.

Solution

In this case, $a = 16.3$, $b = -197.6$, $c = 7.042$, and you have

$$x = \frac{-(-197.6) \pm \sqrt{(-197.6)^2 - 4(16.3)(7.042)}}{2(16.3)}.$$

To evaluate these solutions, begin by calculating the square root of the discriminant, as follows.

Scientific Calculator Steps

Graphing Calculator Steps

In either case, the result is 196.43478. Storing this result and using the recall key, you can find the following two solutions.

$$x \approx \frac{197.6 + 196.43478}{2(16.3)} \approx 12.087 \qquad \text{Add stored value.}$$

$$x \approx \frac{197.6 - 196.43478}{2(16.3)} \approx 0.036 \qquad \text{Subtract stored value.}$$

Programming

The *TI-82* program below can be used to solve equations with the Quadratic Formula. Programs for other calculator models may be found in the Appendix. Enter a, b, and c when prompted.

```
PROGRAM: QUADRAT
: Disp "AX² + BX + C = 0"
: Prompt A
: Prompt B
: Prompt C
: B² - 4AC → D
: If D ≥ 0
: Then
: (-B + √D) / (2A) → M
: Disp M
: (-B - √D) / (2A) → N
: Disp N
: Else
: Disp "NO REAL SOLUTION"
: End
```

Applications

In Section 1.3, you studied four basic types of applications involving quadratic equations: area, falling bodies, Pythagorean Theorem, and quadratic models. The solution of each of these types of problems can involve the Quadratic Formula. For instance, Example 5 shows how the Quadratic Formula can be used to analyze a quadratic model for school expenditures.

Real Life

EXAMPLE 5 School Expenditures

From 1975 to 1989, the total school expenditures (elementary through university) in the United States closely followed the quadratic model

$$\text{Expenditures} = 459.5t^2 + 9170.3t + 107{,}298, \qquad 0 \le t$$

where the expenditures are measured in millions of dollars and the time $t = 0$ represents 1975. These expenses are shown graphically in Figure 1.10. Assuming that this model continues to be valid, when will the total school expenditures reach \$400,000,000,000 per year?   (Source: U.S. National Center for Education Statistics)

Solution

Because the expenditures are measured in millions of dollars, you need to solve the equation

$$459.5t^2 + 9170.3t + 107{,}298 = 400{,}000.$$

To begin, write the equation in standard form as

$$459.5t^2 + 9170.3t - 292{,}702 = 0.$$

Then apply the Quadratic Formula with $a = 459.5$, $b = 9170.3$, and $c = -292{,}702$ to obtain

$$t = \frac{-9170.3 \pm \sqrt{(9170.3)^2 - 4(459.5)(-292{,}702)}}{2(459.5)}.$$

Choosing the positive solution, you will find that

$$t = \frac{-9170.3 + \sqrt{(9170.3)^2 - 4(459.5)(-292{,}702)}}{2(459.5)}$$

$$\approx 17.$$

Because $t = 0$ corresponds to 1975, it follows that $t = 17$ must correspond to 1992. Thus, from the model you can conclude that the total school expenditures reached \$400,000,000,000 in 1992. A table that shows the expenditures for several years can help confirm this result.

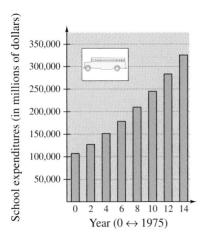

FIGURE 1.10

Technology

Use the recall expression feature of the graphing calculator to find the time in the air for the rock on earth. Simply replace -2.7 with -16 in the expression for t.

In Section 1.3, you learned that the position equation for a falling object is of the form

$$s = -16t^2 + v_0 t + s_0$$

where s is the height (in feet), v_0 is the initial velocity (in feet per second), t is the time (in seconds), and s_0 is the initial height (in feet). This equation is valid only for free-falling objects near earth's surface. Because of differences in gravitational force, position equations are different on other planets or moons. For instance, the next example looks at a position equation on our moon.

Real Life

EXAMPLE 6 *Throwing an Object on the Moon*

An astronaut standing on the surface of the moon throws a rock into space, as shown in Figure 1.11. The height of the rock is given by

$$s = -2.7t^2 + 27t + 6.$$

How much time will elapse before the rock strikes the lunar surface?

Solution

Because s gives the height of the rock at any time t, you can find the time that the rock hits the surface of the moon by setting s equal to zero and solving for t.

$$-2.7t^2 + 27t + 6 = 0$$

$$t = \frac{-27 \pm \sqrt{(27)^2 - 4(-2.7)(6)}}{2(-2.7)}$$

$$\approx 10.2 \text{ seconds} \qquad \text{Choose positive solution.}$$

Thus, about 10.2 seconds will elapse before the rock hits the surface.

FIGURE 1.11

Group Activities   Extending the Concept

Comparing Gravitational Forces   In Example 6, suppose the rock were thrown on earth's surface, rather than on the moon.

a. If the rock were thrown with the same initial velocity, would it remain in the air for a longer time or a shorter time? Explain your reasoning.

b. Use the Quadratic Formula to find the actual time that the rock would remain in the air.

Warm Up The following warm-up exercises involve skills that were covered in earlier sections. You will use these skills in the exercise set for this section.

In Exercises 1–4, simplify the expression.

1. $\sqrt{9 - 4(3)(-12)}$ **2.** $\sqrt{36 - 4(2)(3)}$

3. $\sqrt{12^2 - 4(3)(4)}$ **4.** $\sqrt{15^2 + 4(9)(12)}$

In Exercises 5–10, solve the quadratic equation by factoring.

5. $x^2 - x - 2 = 0$ **6.** $2x^2 + 3x - 9 = 0$

7. $x^2 - 4x = 5$ **8.** $2x^2 + 13x = 7$

9. $x^2 = 5x - 6$ **10.** $x(x - 3) = 4$

1.4 Exercises

In Exercises 1–8, use the discriminant to determine the number of real solutions of the quadratic equation.

1. $4x^2 - 4x + 1 = 0$ **2.** $2x^2 - x - 1 = 0$

3. $3x^2 + 4x + 1 = 0$ **4.** $x^2 + 2x + 4 = 0$

5. $2x^2 - 5x + 5 = 0$ **6.** $3x^2 - 6x + 3 = 0$

7. $\frac{1}{5}x^2 + \frac{6}{5}x - 8 = 0$ **8.** $\frac{1}{3}x^2 - 5x + 25 = 0$

In Exercises 9–30, use the Quadratic Formula to solve the equation.

9. $2x^2 + x - 1 = 0$ **10.** $2x^2 - x - 1 = 0$

11. $16x^2 + 8x - 3 = 0$ **12.** $25x^2 - 20x + 3 = 0$

13. $2 + 2x - x^2 = 0$ **14.** $x^2 - 10x + 22 = 0$

15. $x^2 + 14x + 44 = 0$ **16.** $6x = 4 - x^2$

17. $x^2 + 8x - 4 = 0$ **18.** $4x^2 - 4x - 4 = 0$

19. $12x - 9x^2 = -3$ **20.** $16x^2 + 22 = 40x$

21. $36x^2 + 24x = 7$ **22.** $3x + x^2 - 1 = 0$

23. $4x^2 + 4x = 7$ **24.** $16x^2 - 40x + 5 = 0$

25. $28x - 49x^2 = 4$ **26.** $9x^2 + 24x + 16 = 0$

27. $8t = 5 + 2t^2$ **28.** $25h^2 + 80h + 61 = 0$

29. $(y - 5)^2 = 2y$ **30.** $(x + 6)^2 = -2x$

In Exercises 31–36, use a calculator to solve the equation. (Round your answer to three decimal places.)

31. $5.1x^2 - 1.7x - 3.2 = 0$

32. $10.4x^2 + 8.6x + 1.2 = 0$

33. $7.06x^2 - 4.85x + 0.50 = 0$

34. $-0.005x^2 + 0.101x - 0.193 = 0$

35. $422x^2 - 506x - 347 = 0$

36. $2x^2 - 2.50x - 0.42 = 0$

In Exercises 37–46, solve the equation by any convenient method.

37. $3x + 4 = 2x - 7$ **38.** $x^2 - 2x + 5 = x^2 - 5$

39. $4x^2 - 15 = 25$ **40.** $4x^2 + 2x + 4 = 2x - 8$

41. $x^2 + 3x + 1 = 0$ **42.** $x^2 + 3x - 4 = 0$

43. $(x - 1)^2 = 9$ **44.** $2x^2 - 4x - 6 = 0$

45. $100x^2 - 400 = 0$

46. $2x^2 + 4x - 9 = 2(x - 1)^2$

Writing Real-Life Problems In Exercises 47–50, solve the number problem *and* write a real-life problem that could be represented by this verbal model. For instance, an applied problem that could be represented by Exercise 47 is as follows.

The sum of the length and width of a one-story house is 100 feet. The house has 2500 square feet of floor space. What are the length and width of the house?

47. Find two numbers whose sum is 100 and whose product is 2500.

48. One number is 1 more than another number. The product of the two numbers is 72. Find the numbers.

49. One number is 1 more than another number. The sum of their squares is 113. Find the numbers.

50. One number is 2 more than another number. The product of the two numbers is 440. Find the numbers.

Cost Equation In Exercises 51–54, use the cost equation to find the number of units x that a manufacturer can produce for the cost C. (Round your answer to the nearest positive integer.)

51. $C = 0.125x^2 + 20x + 5000$ $C = \$14,000$

52. $C = 0.5x^2 + 15x + 5000$ $C = \$11,500$

53. $C = 800 + 0.04x + 0.002x^2$ $C = \$1680$

54. $C = 800 - 10x + \dfrac{x^2}{4}$ $C = \$896$

55. *Seating Capacity* A rectangular classroom seats 72 students. If the seats were rearranged with three more seats in each row, the classroom would have two fewer rows. Find the original number of seats in each row.

56. *Dimensions of a Corral* A rancher has 200 feet of fencing to enclose two adjacent rectangular corrals (see figure). Find the dimensions such that the enclosed area will be 1400 square feet.

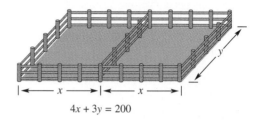

$4x + 3y = 200$

57. *Geometry* An open box is to be made from a square piece of material by cutting 2-inch squares from the corners and turning up the sides (see figure). The volume of the finished box is to be 200 cubic inches. Find the size of the original piece of material.

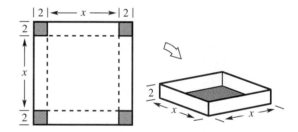

58. *Geometry* An open box (see figure) is to be constructed from 108 square inches of material. Find the dimensions of the square base.

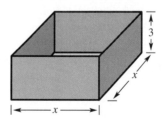

59. *On the Moon* An astronaut on the moon throws a rock straight up into space. The initial velocity is 40 feet per second and the initial height is 5 feet. How long will it take the rock to hit the surface? If the rock had been thrown with the same initial velocity and height on earth, how long would it remain in the air?

60. *Hot Air Balloon* Two people are riding in a hot air balloon that is 200 feet above the ground. One person drops a sack of sand and the balloon starts to rise. One second later, the balloon is 20 feet higher and the other person throws a sack of sand toward the ground (see figure). The position equation for the first sack is $s = -16t^2 + 200$, and (using an initial velocity of -20 feet per second) the position equation for the second sack is

$$s = -16t^2 - 20t + 220.$$

Which sack of sand will hit the ground first? (*Hint:* Remember that the first sack was dropped one second before the second sack.)

61. *Flying Distance* A small commuter airline flies to three cities whose locations form the vertices of a right triangle (see figure). The total flight distance (from City A to City B to City C and back to City A) is 1400 miles. It is 600 miles between the two cities that are farthest apart. Find the other two distances between cities.

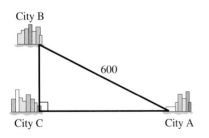

62. *Distance from the Dock* A windlass is used to tow a boat to a dock. The rope is attached to the boat at a point 15 feet below the level of the windlass (see figure). Find the distance from the boat to the dock when there is 75 feet of rope out.

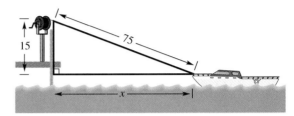

63. *Recreational Spending* The total number of dollars (in millions) spent on recreation in the United States from 1980 to 1991 can be approximated by the model

$$\text{Spending} = 114{,}599 + 13{,}748.4t + 261.11t^2$$

where $t = 0$ represents 1980. The figure shows the actual spending and the spending represented by the model. (Source: U.S. Bureau of Economic Analysis)

(a) During which year was 200 billion dollars spent on recreation?

(b) From the model, when would you say the total annual recreational spending will reach 400 billion dollars?

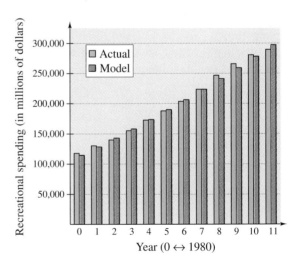

64. *U.S. Sales and Inventories*   The total amount of average monthly sales (in millions of dollars) from manufacturing and trade in the United States from 1980 to 1992 can be approximated by the model

Sales $= 28t^2 + 20{,}247t + 322{,}121$

where $t = 0$ represents 1980. The figure shows the actual spending and the spending represented by the model. From this model, when would you say the total sales will reach \$650,000,000,000?   (Source: U.S. Bureau of the Census)

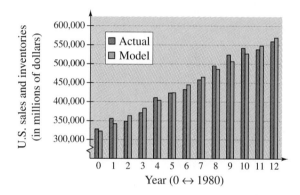

65. *Computers in the Classroom*   The number of personal computers (in millions) used in grades K through 12 in the United States from 1980 to 1988 can be modeled by

Number $= 0.016t^2 + 0.1t - 0.04$

where $t = 0$ represents 1980. From this model, when will 3 million computers be in use in grades K through 12?   (Source: Future Computing/Data Inc.)

66. *Physicians' Services*   The average amount of money spent annually (per person, including insurance payments) for physicians' services in the United States from 1970 to 1991 is approximated by the model

Amount $= 1.054t^2 + 21.168t + 175.15$

where the amount spent is in dollars and t represents the year, with $t = 0$ corresponding to 1980. From this model, when will the average amount reach \$800? (Source: Social Security Administration)

67. *Flying Speed*   Two planes leave simultaneously from the same airport, one flying due east and the other due south (see figure). The eastbound plane is flying 50 miles per hour faster than the southbound plane. After 3 hours the planes are 2440 miles apart. Find the speed of each plane.

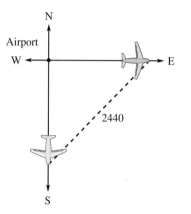

68. *Flying Speed*   Two planes leave simultaneously from the same airport, one flying due east and the other due south. The eastbound plane is flying 100 miles per hour faster than the southbound plane. After 2 hours the planes are 1500 miles apart. Find the speed of each plane.

69. *Total Revenue*   The demand equation for a certain product is

$p = 60 - 0.0004x$

where p is the price per unit and x is the number of units sold. The total revenue for selling x units is given by $R = xp$. How many units must be sold to produce a revenue of \$220,000?

70. *Total Revenue*   The demand equation for a certain product is

$p = 50 - 0.0005x$

where p is the price per unit and x is the number of units sold. The total revenue for selling x units is given by $R = xp$. How many units must be sold to produce a revenue of \$250,000?

MID-CHAPTER QUIZ

Take this quiz as you would take a quiz in class. After you are done, check your work against the answers given in the back of the book.

In Exercises 1–4, solve the equation and check the solution.

1. $2(x + 3) - 5(2x - 3) = 5$

2. $\dfrac{3x + 3}{5x - 2} = \dfrac{3}{4}$

3. $\dfrac{2}{x(x - 1)} + \dfrac{1}{x} = \dfrac{1}{x - 4}$

4. $(x + 3)^2 - x^2 = 6(x + 2)$

In Exercises 5 and 6, solve the equation. (Round to three decimal places.)

5. $\dfrac{x}{3.057} + \dfrac{x}{4.392} = 100$

6. $0.378x + 0.757(500 - x) = 215$

In Exercises 7 and 8, write an algebraic expression for the verbal description.

7. The total cost of producing x units for which the fixed costs are \$1400 and the cost per unit is \$36

8. The amount of acid in x gallons of a 25% acid solution

9. *Production Limit*   A company has fixed costs of \$20,000 per month and variable costs of \$7.50 per unit manufactured. The company has \$80,000 available to cover monthly costs. How many units can the company manufacture?

In Exercises 10–15, solve the equation.

10. *Factoring:* $3x^2 + 13x = 10$

11. *Extracting Roots:* $3x^2 = 15$

12. *Extracting Roots:* $(x + 3)^2 = 17$

13. *Quadratic Formula:* $3x^2 + 7x - 2 = 0$

14. *Quadratic Formula:* $2x + x^2 = 5$

15. *Quadratic Formula:* $3x^2 - 4.50x - 0.32 = 0$

In Exercises 16–18, use the discriminant to determine the number of real solutions.

16. $x^2 + 3x - 5 = 0$ **17.** $2x^2 - 4x + 9 = 0$ **18.** $x^2 + 3 = 0$

19. A fire retardant is dropped from a height of 250 feet. With no initial velocity, what is the minimum time it will take to hit the ground? Explain your reasoning.

20. An open box 5 inches deep and 180 cubic inches in volume is to be constructed. Find the dimensions of the square base.

1.5   Other Types of Equations

Polynomial Equations ▪ Solving Equations Involving Radicals ▪ Equations with Fractions or Absolute Values ▪ Applications

Polynomial Equations

In this section you will extend the techniques for solving equations to nonlinear and nonquadratic equations. At this point in the text, you have only three basic methods for solving nonlinear equations—*factoring, extracting roots,* and the *Quadratic Formula.* So the main goal of this section is to learn to *rewrite* nonlinear equations in a form to which you can apply one of these methods.

EXAMPLE 1   Solving a Polynomial Equation by Factoring

Solve $3x^4 = 48x^2$.

Solution

The basic approach here is first to write the polynomial equation in standard form with zero on one side, to factor the other side, and then to set each factor equal to zero.

$$3x^4 = 48x^2 \qquad \text{Original equation}$$
$$3x^4 - 48x^2 = 0 \qquad \text{Collect terms on left side.}$$
$$3x^2(x^2 - 16) = 0 \qquad \text{Common monomial factor}$$
$$3x^2(x + 4)(x - 4) = 0 \qquad \text{Difference of two squares}$$
$$3x^2 = 0 \quad \Longrightarrow \quad x = 0 \qquad \text{Set 1st factor equal to 0.}$$
$$x + 4 = 0 \quad \Longrightarrow \quad x = -4 \qquad \text{Set 2nd factor equal to 0.}$$
$$x - 4 = 0 \quad \Longrightarrow \quad x = 4 \qquad \text{Set 3rd factor equal to 0.}$$

You can check these solutions by substituting in the original equation, as follows.

Check

$$3x^4 = 48x^2 \qquad \text{Original equation}$$
$$3(0)^4 = 48(0)^2 \qquad \text{0 checks. } \checkmark$$
$$3(-4)^4 = 48(-4)^2 \qquad \text{−4 checks. } \checkmark$$
$$3(4)^4 = 48(4)^2 \qquad \text{4 checks. } \checkmark$$

After checking, you can conclude that the solutions are 0, −4, and 4.

STUDY TIP

A common mistake that is made in solving an equation such as that in Example 1 is dividing both sides of the equation by the variable factor x^2. This loses the solution $x = 0$. When using factoring to solve an equation, be sure to set each factor equal to zero. Don't divide both sides of an equation by a variable factor in an attempt to simplify the equation.

EXAMPLE 2   Solving a Polynomial Equation by Factoring

Solve $x^3 - 3x^2 - 3x + 9 = 0$.

Solution

$$x^3 - 3x^2 - 3x + 9 = 0 \qquad \text{Original equation}$$

$$x^2(x - 3) - 3(x - 3) = 0 \qquad \text{Group terms.}$$

$$(x - 3)(x^2 - 3) = 0 \qquad \text{Factor by grouping.}$$

$$x - 3 = 0 \implies x = 3 \qquad \text{Set 1st factor equal to 0.}$$

$$x^2 - 3 = 0 \implies x = \pm\sqrt{3} \qquad \text{Set 2nd factor equal to 0.}$$

The solutions are 3, $\sqrt{3}$, and $-\sqrt{3}$. Check these in the original equation.

Occasionally, mathematical models involve equations that are of **quadratic type.** In general, an equation is of quadratic type if it can be written in the form

$$au^2 + bu + c = 0$$

where $a \neq 0$ and u is an algebraic expression.

EXAMPLE 3   Solving an Equation of Quadratic Type

Solve $x^4 - 3x^2 + 2 = 0$.

Solution

This equation is of quadratic type with $u = x^2$.

$$(x^2)^2 - 3(x^2) + 2 = 0$$

To solve this equation, you can factor the left side of the equation as the product of two second-degree polynomials.

$$x^4 - 3x^2 + 2 = 0 \qquad \text{Original equation}$$

$$(x^2 - 1)(x^2 - 2) = 0 \qquad \text{Partially factor.}$$

$$(x + 1)(x - 1)(x^2 - 2) = 0 \qquad \text{Completely factor.}$$

$$x + 1 = 0 \implies x = -1 \qquad \text{Set 1st factor equal to 0.}$$

$$x - 1 = 0 \implies x = 1 \qquad \text{Set 2nd factor equal to 0.}$$

$$x^2 - 2 = 0 \implies x = \pm\sqrt{2} \qquad \text{Set 3rd factor equal to 0.}$$

The solutions are -1, 1, $\sqrt{2}$, and $-\sqrt{2}$. Check these in the original equation.

Solving Equations Involving Radicals

The steps involved in solving the remaining equations in this section will often introduce *extraneous solutions*, as discussed in Section 1.1. Operations such as squaring both sides of an equation, raising both sides of an equation to a rational power, or multiplying both sides by a variable quantity all have this potential danger. Thus, when you use any of these operations, a check is crucial.

EXAMPLE 4   An Equation Involving a Rational Exponent

Solve $4x^{3/2} - 8 = 0$.

Solution

$4x^{3/2} - 8 = 0$	Original equation
$4x^{3/2} = 8$	Add 8 to both sides.
$x^{3/2} = 2$	Isolate $x^{3/2}$.
$x = 2^{2/3}$	Raise both sides to $\frac{2}{3}$ power.
$x \approx 1.578$	Round to three decimal places.

NOTE   The essential technique used in Example 4 is to isolate the factor with the rational exponent, and raise both sides to the *reciprocal power*. In Example 5, this is equivalent to isolating the square root and squaring both sides.

Check

$4x^{3/2} - 8 = 0$	Original equation
$4(2^{2/3})^{3/2} \overset{?}{=} 8$	Substitute $2^{2/3}$ for x.
$4(2) \overset{?}{=} 8$	Property of exponents
$8 = 8$	Solution checks. ✓

EXAMPLE 5   An Equation Involving a Radical

$\sqrt{2x + 7} - x = 2$	Original equation
$\sqrt{2x + 7} = x + 2$	Isolate the square root.
$2x + 7 = x^2 + 4x + 4$	Square both sides.
$0 = x^2 + 2x - 3$	Standard form
$0 = (x + 3)(x - 1)$	Factor.
$x + 3 = 0 \implies x = -3$	Set 1st factor equal to 0.
$x - 1 = 0 \implies x = 1$	Set 2nd factor equal to 0.

By checking these values, you can determine that the only solution is 1.

Equations with Fractions or Absolute Values

To solve an equation involving fractions, multiply both sides of the equation by the least common denominator of each term in the equation. This procedure will "clear the equation of fractions." For instance, in the equation

$$\frac{2}{x^2 + 1} + \frac{1}{x} = \frac{2}{x}$$

you can multiply both sides of the equation by $x(x^2 + 1)$. Try doing this and solving the resulting equation. You should obtain one solution: $x = 1$.

EXAMPLE 6   An Equation Involving Fractions

Solve $\dfrac{2}{x} = \dfrac{3}{x - 2} - 1$.

Solution

For this equation, the least common denominator of the three terms is $x(x - 2)$, so we begin by multiplying each term in the equation by this expression.

$$\frac{2}{x} = \frac{3}{x - 2} - 1$$

$$x(x - 2)\frac{2}{x} = x(x - 2)\frac{3}{x - 2} - x(x - 2)(1)$$

$$2(x - 2) = 3x - x(x - 2)$$

$$2x - 4 = -x^2 + 5x$$

$$x^2 - 3x - 4 = 0$$

$$(x - 4)(x + 1) = 0$$

$$x - 4 = 0 \implies x = 4$$

$$x + 1 = 0 \implies x = -1$$

Check

$$\frac{2}{x} = \frac{3}{x - 2} - 1 \qquad\qquad \text{Original equation}$$

$$\frac{2}{4} \overset{?}{=} \frac{3}{4 - 2} - 1 \qquad\qquad \text{Substitute 4 for } x.$$

$$\frac{1}{2} = \frac{3}{2} - 1 \qquad\qquad \text{4 checks. } ✓$$

$$\frac{2}{-1} \overset{?}{=} \frac{3}{-1 - 2} - 1 \qquad\qquad \text{Substitute } -1 \text{ for } x.$$

$$-2 = -1 - 1 \qquad\qquad -1 \text{ checks. } ✓$$

The solutions are 4 and -1.

To solve an equation involving an absolute value, remember that the expression inside the absolute value signs can be positive or negative. This results in *two* separate equations, each of which must be solved. For instance, the equation

$$|x - 2| = 3$$

results in the two equations

$$x - 2 = 3 \qquad \text{and} \qquad -(x - 2) = 3$$

which implies that the equation has two solutions: 5 and -1.

EXAMPLE 7   An Equation Involving Absolute Value

Solve $|x^2 - 3x| = -4x + 6$.

Solution

Because the variable expression inside the absolute value signs can be positive or negative, you must solve the following two equations.

First Equation

$x^2 - 3x = -4x + 6$	Use positive expression.
$x^2 + x - 6 = 0$	Standard form
$(x + 3)(x - 2) = 0$	Factor.
$x + 3 = 0 \implies x = -3$	Set 1st factor equal to 0.
$x - 2 = 0 \implies x = 2$	Set 2nd factor equal to 0.

Second Equation

$-(x^2 - 3x) = -4x + 6$	Use negative expression.
$x^2 - 7x + 6 = 0$	Standard form
$(x - 1)(x - 6) = 0$	Factor.
$x - 1 = 0 \implies x = 1$	Set 1st factor equal to 0.
$x - 6 = 0 \implies x = 6$	Set 2nd factor equal to 0.

Check

$|(-3)^2 - 3(-3)| = -4(-3) + 6$	-3 checks. ✓
$|(2)^2 - 3(2)| \neq -4(2) + 6$	2 does not check. ✗
$|(1)^2 - 3(1)| = -4(1) + 6$	1 checks. ✓
$|(6)^2 - 3(6)| \neq -4(6) + 6$	6 does not check. ✗

The solutions are -3 and 1.

Applications

It would be impossible to categorize the many different types of applications that involve nonlinear and nonquadratic models. However, from the few examples and exercises that are given, we hope that you will gain some appreciation for the variety of applications that can occur.

Real Life

EXAMPLE 8   Reduced Rates

In 1993, downhill skiing and cross country skiing were ranked as the 22nd and 26th most popular sports in the United States. The three most popular were exercise walking, swimming, and bicycling.

A ski club chartered a bus for a ski trip at a cost of $480. In an attempt to lower the bus fare per skier, the club invited nonmembers to go along. After five nonmembers joined the trip, the fare per skier decreased by $4.80. How many club members are going on the trip?

Solution

Begin the solution by creating a verbal model and assigning labels, as follows.

Verbal Model: | Cost per skier | · | Number of skiers | = | Cost of trip |

Labels:
Cost of trip $= 480$ (dollars)
Number of ski club members $= x$ (people)
Number of skiers $= x + 5$ (people)
Original cost per member $= \dfrac{480}{x}$ (dollars per person)
Cost per skier $= \dfrac{480}{x} - 4.80$ (dollars per person)

Equation:

$$\left(\frac{480}{x} - 4.80\right)(x + 5) = 480$$

$$\left(\frac{480 - 4.8x}{x}\right)(x + 5) = 480$$

$$(480 - 4.8x)(x + 5) = 480x$$

$$480x - 4.8x^2 - 24x + 2400 = 480x$$

$$-4.8x^2 - 24x + 2400 = 0$$

$$x^2 + 5x - 500 = 0$$

$$(x + 25)(x - 20) = 0$$

$$x + 25 = 0 \quad \Longrightarrow \quad x = -25$$

$$x - 20 = 0 \quad \Longrightarrow \quad x = 20$$

Choosing the positive value of x, you can conclude that 20 ski club members are going on the trip. Check this in the original statement of the problem.

Activities

1. Solve for x: $9x^4 - 9x^2 + 2 = 0$.

   Answer: $x = \pm\dfrac{\sqrt{3}}{3}, \pm\dfrac{\sqrt{6}}{3}$

2. Solve for x: $\sqrt{3x + 10} - x = 4$.

   Answer: $x = -2, -3$

3. Solve for x: $|5x + 2| = 22$.

   Answer: $x = -\dfrac{24}{5}, 4$

Interest in a savings account is calculated by one of three basic methods: simple interest, interest compounded n times per year, and interest compounded continuously. The next example uses the formula for interest that is compounded n times per year.

$$A = P\left(1 + \frac{r}{n}\right)^{nt}$$

In this formula, A is the balance in the account, P is the principal (or original deposit), r is the annual percentage rate (in decimal form), n is the number of compoundings per year, and t is the time in years. Later, in Chapter 5, you will study the derivation of this formula for compound interest.

Real Life

EXAMPLE 9   Compound Interest

Suppose that when you were born, your grandparents deposited $5000 in a long-term investment in which the interest was compounded quarterly. On your 25th birthday the balance in the account is $25,062.59. What was the annual percentage rate for this investment?

Solution

Formula: $A = P\left(1 + \dfrac{r}{n}\right)^{nt}$

Labels: Amount $= A = 25{,}062.59$ (dollars)
Principal $= P = 5000.00$ (dollars)
Time $= t = 25$ (years)
Compoundings per year $= n = 4$ (compoundings)
Annual rate $= r$ (percent in decimal form)

Equation: $25{,}062.59 = 5000\left(1 + \dfrac{r}{4}\right)^{4(25)}$

$$\frac{25{,}062.59}{5000} = \left(1 + \frac{r}{4}\right)^{100}$$

$$5.0125 \approx \left(1 + \frac{r}{4}\right)^{100}$$

$$(5.0125)^{1/100} \approx 1 + \frac{r}{4}$$

$$1.01625 \approx 1 + \frac{r}{4}$$

$$0.01625 \approx \frac{r}{4}$$

$$0.065 \approx r$$

The annual percentage rate is $0.065 = 6.5\%$. Check this in the original statement of the problem.

EXAMPLE 10   *Market Research*

The marketing department at a publisher is asked to determine the price of a book. The department determines that the demand for the book depends on the price of the book according to the formula

$$p = 40 - \sqrt{0.0001x + 1}, \qquad 0 \le x$$

where p is the price per book in dollars and x is the number of books sold at the given price. For instance, in Figure 1.12 note that if the price were \$39 then (according to the model) no one would be willing to buy the book. On the other hand, if the price were \$17.60, 5 million copies could be sold. If the publisher set the price at \$12.95, how many copies would be sold?

Solution

$$p = 40 - \sqrt{0.0001x + 1} \qquad \text{Given model}$$

$$12.95 = 40 - \sqrt{0.0001x + 1} \qquad \text{Set price at 12.95.}$$

$$\sqrt{0.0001x + 1} = 27.05 \qquad \text{Isolate radical.}$$

$$0.0001x + 1 = 731.7025 \qquad \text{Square both sides.}$$

$$0.0001x = 730.7025 \qquad \text{Subtract 1 from both sides.}$$

$$x = 7{,}307{,}025 \qquad \text{Divide both sides by 0.0001.}$$

Thus, by setting the book's price at \$12.95, the publisher can expect to sell about 7.3 million copies.

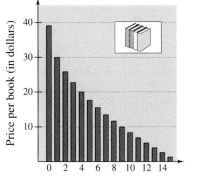

Price per book (in dollars)

Number of books sold (in millions)

FIGURE 1.12

Group Activities   You Be the Instructor

Error Analysis   Find the error(s) in the following solution of $\sqrt{3}x = \sqrt{7x + 4}$.

$$\sqrt{3}x = \sqrt{7x + 4}$$

$$3x^2 = 7x + 4$$

$$x = \frac{-7 \pm \sqrt{7^2 - (4)(3)(4)}}{2(3)}$$

$$x = -1 \quad \text{and} \quad x = -\frac{4}{3}$$

Warm Up

The following warm-up exercises involve skills that were covered in earlier sections. You will use these skills in the exercise set for this section.

In Exercises 1–10, find the real solutions of the equation.

1. $x^2 - 22x + 121 = 0$

2. $x(x - 20) + 3(x - 20) = 0$

3. $(x + 20)^2 = 625$

4. $5x^2 + x = 0$

5. $3x^2 + 4x - 4 = 0$

6. $12x^2 + 8x - 55 = 0$

7. $x^2 + 4x - 5 = 0$

8. $4x^2 + 4x - 15 = 0$

9. $x^2 - 3x + 1 = 0$

10. $x^2 - 4x + 2 = 0$

1.5 Exercises

In Exercises 1–58, find the real solutions of the equation. Check your solutions.

1. $4x^4 - 18x^2 = 0$

2. $20x^3 - 125x = 0$

3. $x^3 - 2x^2 - 3x = 0$

4. $2x^4 - 15x^3 + 18x^2 = 0$

5. $x^4 - 81 = 0$

6. $x^6 - 64 = 0$

7. $5x^3 + 30x^2 + 45x = 0$

8. $9x^4 - 24x^3 + 16x^2 = 0$

9. $x^3 - 3x^2 - x + 3 = 0$

10. $x^3 + 2x^2 + 3x + 6 = 0$

11. $x^4 - x^3 + x - 1 = 0$

12. $x^4 + 2x^3 - 8x - 16 = 0$

13. $x^4 - 10x^2 + 9 = 0$

14. $x^4 - 29x^2 + 100 = 0$

15. $x^4 + 5x^2 - 36 = 0$

16. $x^4 - 4x^2 + 3 = 0$

17. $4x^4 - 65x^2 + 16 = 0$

18. $36t^4 + 29t^2 - 7 = 0$

19. $x^6 + 7x^3 - 8 = 0$

20. $x^6 + 3x^3 + 2 = 0$

21. $\sqrt{2x} - 10 = 0$

22. $4\sqrt{x} - 3 = 0$

23. $\sqrt{x - 10} - 4 = 0$

24. $\sqrt{5 - x} - 3 = 0$

25. $\sqrt[3]{2x + 5} + 3 = 0$

26. $\sqrt[3]{3x + 1} - 5 = 0$

In Exercises 35–40, remind students to use $\pm$ when taking the square root of both sides of an equation. This is not immediately obvious to most students when the solution requires raising both sides of an equation to the 3/2 power.

27. $2x + 9\sqrt{x} - 5 = 0$

28. $6x - 7\sqrt{x} - 3 = 0$

29. $x = \sqrt{11x - 30}$

30. $2x - \sqrt{15 - 4x} = 0$

31. $-\sqrt{26 - 11x} + 4 = x$

32. $x + \sqrt{31 - 9x} = 5$

33. $\sqrt{x + 1} - 3x = 1$

34. $\sqrt{2x + 1} + x = 7$

35. $(x - 5)^{2/3} = 16$

36. $(x + 3)^{4/3} = 16$

37. $(x + 3)^{3/2} = 8$

38. $(x^2 + 2)^{2/3} = 9$

39. $(x^2 - 5)^{2/3} = 16$

40. $(x^2 - x - 22)^{4/3} = 16$

41. $\dfrac{1}{x} - \dfrac{1}{x + 1} = 3$

42. $\dfrac{x}{x^2 - 4} + \dfrac{1}{x + 2} = 3$

43. $\dfrac{20 - x}{x} = x$

44. $\dfrac{4}{x} - \dfrac{5}{3} = \dfrac{x}{6}$

45. $x = \dfrac{3}{x} + \dfrac{1}{2}$

46. $4x + 1 = \dfrac{3}{x}$

47. $\dfrac{1}{x} = \dfrac{4}{x - 1} + 1$

48. $x + \dfrac{9}{x + 1} = 5$

49. $\dfrac{4}{x + 1} - \dfrac{3}{x + 2} = 1$

50. $\dfrac{x + 1}{3} - \dfrac{x + 1}{x + 2} = 0$

51. $|x + 1| = 2$

52. $|x - 2| = 3$

53. $|2x - 1| = 5$

54. $|3x + 2| = 7$

55. $|x| = x^2 + x - 3$

56. $|x^2 + 6x| = 3x + 18$

57. $|x - 10| = x^2 - 10x$

58. $|x + 1| = x^2 - 5$

In Exercises 59–62, use a calculator to find the real solutions of the equation. (Round your answers to three decimal places.)

59. $3.2x^4 - 1.5x^2 - 2.1 = 0$

60. $7.08x^6 + 4.15x^3 - 9.6 = 0$

61. $1.8x - 6\sqrt{x} - 5.6 = 0$

62. $4x + 8\sqrt{x} + 3.6 = 0$

63. *Sharing the Cost*   A college charters a bus for $1700 to take a group of students to the World's Fair. When six more students join the trip, the cost per student drops by $7.50. How many students were in the original group?

64. *Sharing the Cost*   Three students are planning to rent an apartment for a year and share equally on the monthly rent. By adding a fourth person to the group, each person could save $75 a month. How much is the monthly rent?

65. *Compound Interest*   A deposit of $2500 reaches a balance of $3544.06 after 5 years. The interest in the account is compounded monthly. What is the annual percentage rate for this investment?

66. *Compound Interest*   A sales representative for a mutual fund company describes a "guaranteed investment fund" that the company is offering to new investors. You are told that if you deposit $10,000 in the fund you will be guaranteed a return of at least $25,000 after 20 years.   (a) If after 20 years you received the minimum guarantee, what annual percentage rate did you receive?   (b) If after 20 years you received $35,000, what annual percentage rate did you receive?   (Assume that the interest in the fund is compounded quarterly.)

67. *Borrowing Money*   Suppose you borrow $100 from a friend and agree to pay the money back, plus $10 in interest, after 6 months. Assuming that the interest is compounded monthly, what annual percentage rate are you paying?

68. *Cash Advance*   You take out a cash advance on a credit card for $500. After 2 months, you owe $515.75. The interest is compounded monthly. What is the annual percentage rate for this loan?

69. *Airline Passengers*   An airline offers daily flights between Chicago and Denver. The total monthly cost of these flights is modeled by

$$C = \sqrt{0.2x + 1}$$

where C is measured in millions of dollars and x is measured in thousands of passengers (see figure). The total cost of the flights for a month is 2.5 million dollars. How many passengers flew that month?

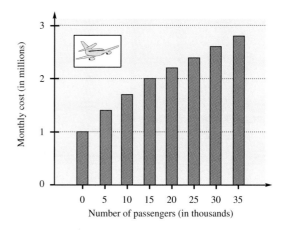

70. *Farmers Markets*   The number of farmers markets F in the United States for the period from 1970 to 1990 can be approximated by the model

$$F = \sqrt{4672t^2 + 94{,}841t + 112{,}148}$$

where $t = 0$ represents 1970.   (Source: U.S. Department of Agriculture)

(a) Complete the table.

t	0	5	10	15	20
F					

(b) From this model, when does it appear that there will be 2000 farmers markets in the United States?

71. *Life Expectancy* The life expectancy table (for ages 58–75) used by the U.S. National Center for Health Statistics is modeled by

$$y = \sqrt{0.6632x^2 - 110.55x + 4680.24}$$

where x represents a person's current age and y represents the average number of years the person is expected to live (see figure). If a person's life expectancy is estimated to be 20 years, how old is the person? (Source: U.S. National Center for Health Statistics)

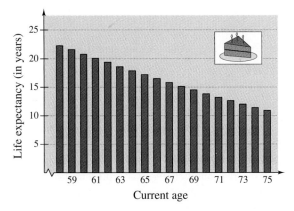

72. *Federal Deficit* The deficit (in millions of dollars) for the federal government from 1985 to 1992 can be modeled by

$$\text{Deficit} = 8{,}372{,}681.1 - 152.19t^{5/2}$$

where $t = 85$ represents 1985 (see figure). Predict the year that the deficit will reach 6 trillion dollars. (Source: U.S. Department of the Treasury)

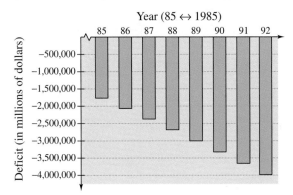

73. *Market Research* The demand equation for a certain product is modeled by

$$p = 30 - \sqrt{0.0001x + 1}$$

where x is the number of units demanded per day and p is the price per unit. Find the demand if the price is set at $13.95.

74. *Power Line* A power station is on one side of a river that is $\frac{1}{2}$ mile wide. A factory is 6 miles downstream on the other side of the river. It costs $18 per foot to run power lines over land and $24 per foot to run them under water. The cost of the project is $616,877.27. Find the length x as labeled in the figure.

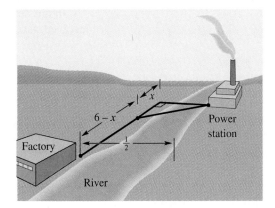

75. *Sailboat Stays* Two stays for the mast on a sailboat are attached to the boat at two points, as shown in the figure. One point is 10 feet from the mast and the other point is 15 feet from the mast. The total length of the two stays is 35 feet. How high on the mast are the stays attached?

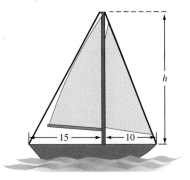

1.6 Linear Inequalities

Introduction ▪ Properties of Inequalities ▪ Solving a Linear Inequality ▪ Inequalities Involving Absolute Value ▪ Applications

Introduction

Simple inequalities were reviewed in Section P.1 to *order* the real numbers. There, you used inequality symbols $<$, $\leq$, $>$, and $\geq$ to compare two numbers and to denote subsets of real numbers. For instance, the simple inequality

$$x \geq 3$$

denotes all real numbers x that are greater than or equal to 3.

In this section you will expand your work with inequalities to include more involved statements such as

$$5x - 7 < 3x + 9 \qquad \text{and} \qquad -3 \leq 6x - 1 < 3.$$

As with an equation, you **solve an inequality** in the variable x by finding all values of x for which the inequality is true. Such values are **solutions** and are said to **satisfy** the inequality. The set of all real numbers that are solutions of an inequality is the **solution set** of the inequality.

The set of all points on the real number line that represent the solution set is the **graph** of the inequality. Graphs of many types of inequalities consist of intervals on the real number line. The four different types of **bounded** intervals are summarized below.

Bounded Intervals on the Real Number Line

Let a and b be real numbers such that $a < b$. The following intervals on the real number line are **bounded.** The numbers a and b are the **endpoints** of each interval.

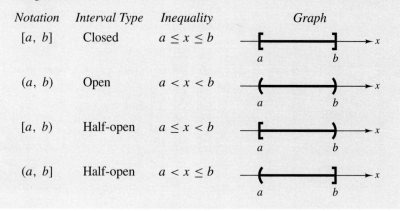

Notation	Interval Type	Inequality	Graph
$[a, b]$	Closed	$a \leq x \leq b$	
(a, b)	Open	$a < x < b$	
$[a, b)$	Half-open	$a \leq x < b$	
$(a, b]$	Half-open	$a < x \leq b$	

Note that a closed interval contains both of its endpoints, a half-open interval contains only one of its endpoints, and an open interval does not contain either of its endpoints. Often, the solution of an inequality is an interval on the real line that is **unbounded.** For instance, the interval consisting of all positive numbers is unbounded.

Unbounded Intervals on the Real Number Line

Let a and b be real numbers. The following intervals on the real number line are **unbounded.**

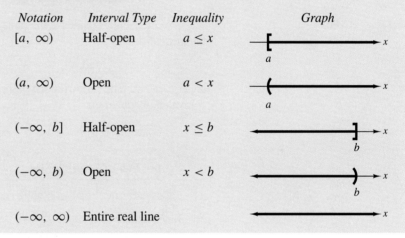

Notation	Interval Type	Inequality	Graph
$[a, \infty)$	Half-open	$a \leq x$	
(a, ∞)	Open	$a < x$	
$(-\infty, b]$	Half-open	$x \leq b$	
$(-\infty, b)$	Open	$x < b$	
$(-\infty, \infty)$	Entire real line		

The symbols ∞ (**positive infinity**) and $-\infty$ (**negative infinity**) do not represent real numbers. They are simply convenient symbols used to describe the unboundedness of an interval such as $(1, \infty)$.

EXAMPLE 1 Intervals and Inequalities

Write an inequality to represent each of the following intervals and state whether the interval is bounded or unbounded.

a. $(-3, 5]$ **b.** $(-3, \infty)$ **c.** $[0, 2]$

Solution

a. $(-3, 5]$ corresponds to $-3 < x \leq 5$.   Bounded

b. $(-3, \infty)$ corresponds to $-3 < x$.   Unbounded

c. $[0, 2]$ corresponds to $0 \leq x \leq 2$.   Bounded

Properties of Inequalities

The procedures for solving linear inequalities in one variable are much like those for solving linear equations. To isolate the variable you can make use of **properties of inequalities.** These properties are similar to the properties of equality, but there are two important exceptions. When both sides of an inequality are multiplied or divided by a negative number, the direction of the inequality symbol must be reversed. Here is an example.

$$-2 < 5 \qquad \text{Original inequality}$$
$$(-3)(-2) > (-3)(5) \qquad \text{Multiply both sides by } -3 \text{ and reverse the inequality.}$$
$$6 > -15 \qquad \text{Solution of inequality}$$

Two inequalities that have the same solution set are **equivalent.** For instance, the inequalities

$$x + 2 < 5 \qquad \text{and} \qquad x < 3$$

are equivalent. The following list describes operations that can be used to create equivalent inequalities.

Properties of Inequalities

Let a, b, c, and d be real numbers.

1. *Transitive Property*

   $$a < b \text{ and } b < c \implies a < c$$

2. *Addition of Inequalities*

   $$a < b \text{ and } c < d \implies a + c < b + d$$

3. *Addition of a Constant*

   $$a < b \implies a + c < b + c$$

4. *Multiplying by a Constant*

   $$\text{For } c > 0, a < b \implies ac < bc$$
   $$\text{For } c < 0, a < b \implies ac > bc$$

NOTE   Each of the properties above is true if the symbol $<$ is replaced by $\leq$. For instance, another form of the multiplication property would be as follows.

$$\text{For } c > 0, a \leq b \implies ac \leq bc.$$
$$\text{For } c < 0, a \leq b \implies ac \geq bc.$$

Solving a Linear Inequality

The simplest type of inequality to solve is a **linear inequality** in a single variable. For instance, $2x + 3 > 4$ is a linear inequality in x.

As you read through the following examples, pay special attention to the steps in which the inequality symbol is reversed. Remember that when you multiply or divide by a negative number, you must reverse the inequality symbol.

EXAMPLE 2   *Solving a Linear Inequality*

Solve $5x - 7 > 3x + 9$.

Solution

$5x - 7 > 3x + 9$	Original inequality
$5x > 3x + 16$	Add 7 to both sides.
$5x - 3x > 16$	Subtract $3x$ from both sides.
$2x > 16$	Combine like terms.
$x > 8$	Divide both sides by 2.

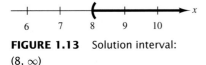

FIGURE 1.13   Solution interval: $(8, \infty)$

The solution set is all real numbers that are greater than 8, which is denoted by $(8, \infty)$. The graph is shown in Figure 1.13.

Checking the solution set of an inequality is not as simple as checking the solutions of an equation. You can, however, get an indication of the validity of a solution set by substituting a few convenient values of x to see whether the original inequality is satisfied.

EXAMPLE 3   *Solving a Linear Inequality*

Solve $1 - \dfrac{3x}{2} \geq x - 4$.

Solution

$1 - \dfrac{3x}{2} \geq x - 4$	Original inequality
$2 - 3x \geq 2x - 8$	Multiply both sides by 2.
$-3x \geq 2x - 10$	Subtract 2 from both sides.
$-5x \geq -10$	Subtract $2x$ from both sides.
$x \leq 2$	Divide both sides by -5 and reverse inequality.

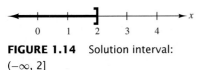

FIGURE 1.14   Solution interval: $(-\infty, 2]$

The solution set is all real numbers that are less than or equal to 2, which is denoted by $(-\infty, 2]$. The graph is shown in Figure 1.14.

Sometimes it is convenient to write two inequalities as a **double inequality.** For instance, you can write the two inequalities $-4 \leq 5x - 2$ and $5x - 2 < 7$ more simply as

$$-4 \leq 5x - 2 < 7.$$

This form allows you to solve the two given inequalities together, as demonstrated in Example 4.

EXAMPLE 4   Solving a Double Inequality

To solve the following double inequality, we isolate x as the middle term.

$-3 \leq 6x - 1 < 3$	Original inequality
$-2 \leq 6x < 4$	Add 1 to all three parts.
$-\dfrac{1}{3} \leq x < \dfrac{2}{3}$	Divide by 6 and reduce.

The solution set is all real numbers that are greater than or equal to $-\frac{1}{3}$ and less than $\frac{2}{3}$. The interval notation for this solution set is

$$\left[-\tfrac{1}{3}, \tfrac{2}{3}\right).$$ Solution set

The graph of this solution set is shown in Figure 1.15.

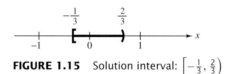

FIGURE 1.15   Solution interval: $\left[-\frac{1}{3}, \frac{2}{3}\right)$

The double inequality in Example 4 could have been solved in two parts as follows.

$-3 \leq 6x - 1$	and	$6x - 1 < 3$
$-2 \leq 6x$		$6x < 4$
$-\dfrac{1}{3} \leq x$		$x < \dfrac{2}{3}$

The solution set consists of all real numbers that satisfy *both* inequalities. In other words, the solution set is the set of all values of x for which $-\frac{1}{3} \leq x < \frac{2}{3}$.

When combining two inequalities to form a double inequality, be sure that the inequalities satisfy the Transitive Property. For instance, it is *incorrect* to combine the inequalities $3 < x$ and $x \leq -1$ as $3 < x \leq -1$. This "inequality" is obviously wrong because 3 is not less than -1.

Inequalities Involving Absolute Value

The concept of absolute value is difficult for students. Remind them that absolute value can be thought of as a distance from 0.

Solving an Absolute Value Inequality

Let x be a variable or an algebraic expression and let a be a real number such that $a \geq 0$.

1. The solutions of $|x| < a$ are all values of x that lie between $-a$ and a.

$$|x| < a \qquad \text{if and only if } -a < x < a.$$

2. The solutions of $|x| > a$ are all values of x that are less than $-a$ or greater than a.

$$|x| > a \qquad \text{if and only if } x < -a \text{ or } x > a.$$

These rules are also valid if $<$ is replaced by $\leq$ and $>$ is replaced by $\geq$.

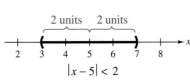

$$|x - 5| < 2$$

FIGURE 1.16

EXAMPLE 5   *Solving an Absolute Value Inequality*

Solve $|x - 5| < 2$.

Solution

$|x - 5| < 2$	Original inequality
$-2 < x - 5 < 2$	Equivalent inequalities
$-2 + 5 < x - 5 + 5 < 2 + 5$	Add 5 to all three parts.
$3 < x < 7$	Solution set

The solution set consists of all real numbers that are greater than 3 and less than 7, which is denoted by $(3, 7)$. The graph is shown in Figure 1.16.

A common error is to write $|x + 3| \geq 7$ as $-7 \geq x + 3 \geq 7$. Point out that this is incorrect because the "inequality" does not satisfy the transitive property, i.e., -7 is not greater than or equal to 7.

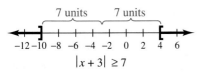

$$|x + 3| \geq 7$$

FIGURE 1.17

EXAMPLE 6   *Solving an Absolute Value Inequality*

Solve $|x + 3| \geq 7$.

Solution

$|x + 3| \geq 7$			Original inequality
$x + 3 \leq -7$	or	$x + 3 \geq 7$	Equivalent inequalities
$x + 3 - 3 \leq -7 - 3$		$x + 3 - 3 \geq 7 - 3$	Subtract 3 from both sides.
$x \leq -10$		$x \geq 4$	Solution set

The solution set is all real numbers that are less than or equal to -10 *or* greater than or equal to 4, which is denoted by $(-\infty, -10] \cup [4, \infty)$ (see Figure 1.17).

Applications

EXAMPLE 7 *Comparative Shopping*

A subcompact car can be rented from Company A for $180 per week with no extra charge for mileage. A similar car can be rented from Company B for $100 per week, plus 20 cents for each mile driven. How many miles must you drive in a week to make the rental fee for Company B more than that for Company A?

Solution

Verbal Model:

Weekly cost for Company B	$>$	Weekly cost for Company A

Labels: Miles driven in one week $= m$ (miles)
Weekly cost for Company A $= 180$ (dollars)
Weekly cost for Company B $= 100 + 0.20m$ (dollars)

Inequality: $100 + 0.2m > 180$

$$0.2m > 80$$

$$m > 400 \text{ miles}$$

If you drive more than 400 miles in a week, Company A is cheaper.

EXAMPLE 8 *Exercise Program*

A man begins an exercise and diet program that is designed to reduce his weight by at least 2 pounds per week. At the beginning of the diet the man weighs 225 pounds. Find the maximum number of weeks before the man's weight will reach (or fall below) his goal of 192 pounds.

Solution

Verbal Model:

Desired weight	$\leq$	Current weight	$-$	2 pounds per week	$\cdot$	Number of weeks

Labels: Desired weight $= 192$ (pounds)
Current weight $= 225$ (pounds)
Number of weeks $= x$ (weeks)

Inequality: $192 \leq 225 - 2x$

$$-33 \leq -2x$$

$$16.5 \geq x \text{ weeks}$$

Losing at least 2 pounds per week, it will take at most $16\frac{1}{2}$ weeks to reach his goal.

EXAMPLE 9 Accuracy of a Measurement

You go to a candy store to buy chocolates that cost $9.89 per pound. The scale that is used in the store has a state seal of approval that indicates the scale is accurate to within half an ounce. According to the scale, your purchase weighs one-half pound and costs $4.95. How much might you have been undercharged or overcharged due to an error in the scale?

Solution

To solve this problem, let x represent the *true* weight of the candy. Because the state seal indicates that the scale is accurate to within half an ounce (or $\frac{1}{32}$ of a pound), you can conclude that the difference between the exact weight (x) and the scale weight ($\frac{1}{2}$) is less than or equal to $\frac{1}{32}$ of a pound. That is,

$$\left| x - \frac{1}{2} \right| \le \frac{1}{32}.$$

You can solve this inequality as follows.

$$-\frac{1}{32} \le x - \frac{1}{2} \le \frac{1}{32}$$

$$\frac{15}{32} \le x \le \frac{17}{32}$$

$$0.46875 \le x \le 0.53125$$

In other words, your "one-half" pound of candy could have weighed as little as 0.46875 pounds (which would have cost $4.64) or as much as 0.53125 pounds (which would have cost $5.25). Thus, you could have been undercharged by as much as $0.30 or overcharged by as much as $0.31.

Group Activities Communicating Mathematically

Verbal Statements of Properties   Many people find that it is easier to memorize a verbal statement than to memorize a mathematical formula. For instance, you might remember the Pythagorean Theorem, $a^2 + b^2 = c^2$, as "the sum of the squares of the two sides is equal to the square of the hypotenuse."

Four different properties of inequalities are listed on page 138. Translate each of these mathematical statements into *verbal statements*.

$Warm\ Up$ The following warm-up exercises involve skills that were covered in earlier sections. You will use these skills in the exercise set for this section.

In Exercises 1–4, determine which of the two numbers is larger.

1. $-\frac{1}{2}, -7$

2. $-\frac{1}{3}, -\frac{1}{6}$

3. $-\pi, -3$

4. $-6, -\frac{13}{2}$

In Exercises 5–8, use inequality notation to denote the statement.

5. x is nonnegative.

6. z is strictly between -3 and 10.

7. P is no more than 2.

8. W is at least 200.

In Exercises 9 and 10, evaluate the expression for the values of x.

9. $|x - 10|, \ x = 12, \ x = 3$

10. $|2x - 3|, \ x = \frac{3}{2}, \ x = 1$

1.6 Exercises

In Exercises 1 and 2, determine whether the values of x are solutions of the inequality.

1. $5x - 12 > 0$

    (a) $x = 3$ (b) $x = -3$ (c) $x = \frac{5}{2}$ (d) $x = \frac{3}{2}$

2. $x + 1 < \dfrac{2x}{3}$

    (a) $x = 0$ (b) $x = 4$ (c) $x = -4$ (d) $x = -3$

In Exercises 3–10, match the inequality with its graph. [The graphs are labeled (a), (b), (c), (d), (e), (f), (g), and (h).]

(a)

(b)

(c)

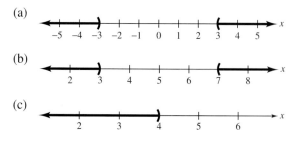

(d)

(e)

(f)

(g)

(h)

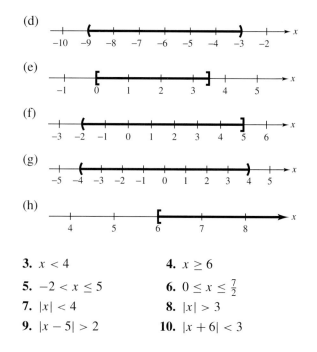

3. $x < 4$

4. $x \geq 6$

5. $-2 < x \leq 5$

6. $0 \leq x \leq \frac{7}{2}$

7. $|x| < 4$

8. $|x| > 3$

9. $|x - 5| > 2$

10. $|x + 6| < 3$

In Exercises 11–46, solve the inequality and sketch its graph.

11. $3x \geq 9$ **12.** $5x \leq 20$

13. $4x < 12$ **14.** $2x > 3$

15. $-10x < 40$ **16.** $-6x > 15$

17. $x - 5 \geq 7$ **18.** $x + 7 \leq 12$

19. $4(x + 1) < 2x + 3$ **20.** $2x + 7 < 3$

21. $2x - 1 \geq 0$ **22.** $3x + 1 \geq 2$

23. $4 - 2x < 3$ **24.** $6x - 4 \leq 2$

25. $1 < 2x + 3 < 9$

26. $-8 \leq 1 - 3(x - 2) < 13$

27. $-4 < \dfrac{2x - 3}{3} < 4$ **28.** $0 \leq \dfrac{x + 3}{2} < 5$

29. $\dfrac{3}{4} > x + 1 > \dfrac{1}{4}$ **30.** $-1 < -\dfrac{x}{3} < 1$

31. $|x| < 5$ **32.** $|2x| < 6$

33. $\left|\dfrac{x}{2}\right| > 3$ **34.** $|5x| > 10$

35. $|x - 20| \leq 4$ **36.** $|x - 7| < 6$

37. $|x - 20| \geq 4$ **38.** $|x + 14| + 3 > 17$

39. $\left|\dfrac{x - 3}{2}\right| \geq 5$ **40.** $|1 - 2x| < 5$

41. $|9 - 2x| - 2 < -1$ **42.** $\left|1 - \dfrac{2x}{3}\right| < 1$

43. $2|x + 10| \geq 9$ **44.** $3|4 - 5x| \leq 9$

45. $|x - 5| < 0$ **46.** $|x - 5| \geq 0$

In Exercises 47–52, find the domain of the expression.

47. $\sqrt{x - 5}$ **48.** $\sqrt{x - 10}$

49. $\sqrt{x + 3}$ **50.** $\sqrt[4]{6x + 15}$

51. $\sqrt[4]{7 - 2x}$ **52.** $\sqrt{3 - x}$

In Exercises 53–60, use absolute value notation to define the solution set.

53.

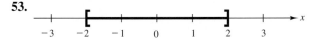

54.

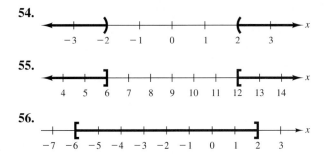

55.

56.

57. All real numbers within 10 units of 12

58. All real numbers at least 5 units from 8

59. All real numbers whose distances from -3 are more than 5

60. All real numbers whose distances from -6 are no more than 7

61. *Comparative Shopping*   You can rent a midsize car from Company A for \$250 per week with no extra charge for mileage. A similar car can be rented from Company B for \$150 per week, plus 25 cents for each mile driven. How many miles must you drive in a week to make the rental fee for Company B greater than that for Company A?

62. *Comparative Shopping*   Your department sends its copying to the photocopy center of your company. The photocopy center bills your department \$0.10 per page. You are considering buying a departmental copier for \$3000. With your own copier the cost per page would be \$0.03. The expected life of the copier is 4 years. How many copies must you make in the 4-year period to justify purchasing the copier?

63. *Interest*   For \$1000 to grow to more than \$1250 in 2 years, what must the simple interest rate be?

64. *Interest*   For \$1000 to grow to more than \$1500 in 2 years, what must the simple interest rate be?

65. *Weight Loss Program*   A person enrolls in a diet program that guarantees a loss of at least $1\frac{1}{2}$ pounds per week. The person's weight at the beginning of the program is 164 pounds. Find the maximum number of weeks that the person must be in the program before attaining a weight of 128 pounds.

66. *Salary Increase*   You accept a new job with a starting salary of $18,500. When you are employed you are told that you will receive an annual raise of at least $1250. What is the maximum number of years that you must work before your annual salary will be $23,500?

67. *Break-Even Analysis*   The revenue for selling x units of a product is $R = 115.95x$. The cost of producing x units is $C = 95x + 750$. In order to obtain a profit, the revenue must be greater than the cost.

(a) Complete the table.

x	10	20	30	40	50
R					
C					

(b) For what values of x will this product return a profit?

68. *Break-Even Analysis*   The revenue for selling x units of a product is $R = 24.55x$. The cost of producing x units is

$$C = 15.4x + 150,000.$$

In order to obtain a profit, the revenue must be greater than the cost. For what values of x will this product return a profit?

69. *Annual Operating Cost*   A utility company has a fleet of vans. The annual operating cost per van is

$$C = 0.32m + 2300$$

where m is the number of miles traveled by a van in a year. What number of miles will yield an annual operating cost that is less than $10,000?

70. *Daily Sales*   A doughnut shop at a shopping mall sells a dozen doughnuts for $2.95. Beyond the fixed costs (rent, utilities, and insurance) of $150 per day it costs $1.45 for enough materials (flour, sugar, and so on) and labor to produce a dozen doughnuts. If the daily profit varies between $50 and $200, between what levels (in dozens) do the daily sales vary?

71. *Cable Television Subscribers*   The number of cable television subscribers (in thousands) in the United States between 1980 and 1993 can be modeled by

$$\text{Subscribers} = 3184t + 16,424, \qquad 0 \le t$$

where $t = 0$ represents 1980. According to this model, when will the number of subscribers exceed 70 million (see figure)? (Source: *Television and Cable Fact Book*)

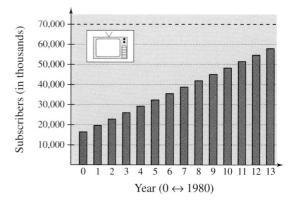

72. *Teachers' Salaries*   The average salary for elementary and secondary teachers in the United States from 1980 to 1991 can be modeled by

$$\text{Salary} = 16,020 + 1527t, \qquad 0 \le t$$

where $t = 0$ represents 1980 (see figure). According to the model, when will the average salary exceed $45,000? (Source: National Education Association)

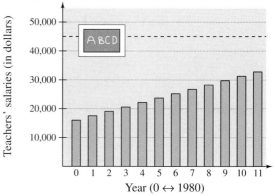

73. *Accuracy of Measurement* You buy six T-bone steaks that cost $3.98 per pound. The weight that is listed on the package is 5.72 pounds. If the scale that weighed the package is accurate to within $\frac{1}{2}$ ounce, how much money might you have been undercharged or overcharged?

74. *Accuracy of Measurement* You buy a bag of oranges that cost $0.69 per pound. The weight that is listed on the bag is 4.65 pounds. If the scale that weighed the bag is only accurate to within 1 ounce, how much money might you have been undercharged or overcharged?

75. *Human Height* The heights h of two-thirds of a population satisfy the inequality

$$|h - 68.5| \leq 2.7$$

where h is measured in inches. Determine the interval on the real number line in which these heights lie.

76. *Humidity Control* An electronic device is to be operated in a room with relative humidity h defined by

$$|h - 50| \leq 30.$$

What are the minimum and maximum relative humidities for the operation of this device?

Math Matters   Man and Mouse

A man who falls from a height of 2000 feet (without a parachute) will strike the ground with lethal force. But a mouse can fall from the same height and simply get up and walk away. Why?

The answer is that the speed at which a falling object hits the ground depends partly on the air resistance of the object, which, in turn, depends on the object's weight and surface area. If the ratio of an object's surface area to its weight is large, then its air resistance will be large. On the other hand, if the ratio of an object's surface area to its weight is small, then its air resistance will be small. This is why a parachute works—a person with a parachute has a much larger surface area (for approximately the same weight) than a person without a parachute. So how does this relate to the falling man and mouse? The falling mouse has a much greater air resistance than the falling man because the ratio of the mouse's surface area to its weight is greater than the ratio of the man's surface area to his weight. To convince yourself that the mouse's ratio is greater than the man's, try the following experiment. Find the ratio of the surface area and weight for the cubes described below.

Notice that as the cube becomes larger, the ratio of its surface area to its weight becomes smaller. (The answers are given in the back of the text.)

Length of Side	Surface Area	Volume	Density	Weight
1 ft	6 ft^2	1 ft^3	1 lb/ft^3	1 lb
2 ft	24 ft^2	8 ft^3	1 lb/ft^3	8 lb
3 ft	54 ft^2	27 ft^3	1 lb/ft^3	27 lb
4 ft	96 ft^2	64 ft^3	1 lb/ft^3	64 lb

1.7	**Other Types of Inequalities**

Polynomial Inequalities ▪ Rational Inequalities ▪ Applications

Polynomial Inequalities

To solve a polynomial inequality such as

$$x^2 - 2x - 3 < 0$$

you can use the fact that a polynomial can change signs only at its zeros (the x-values that make the polynomial equal to zero). Between two consecutive zeros a polynomial must be entirely positive or entirely negative. This means that when the real zeros of a polynomial are put in order, they divide the real number line into intervals in which the polynomial has no sign changes. These zeros are the **critical numbers** of the inequality, and the resulting intervals are the **test intervals** for the inequality. For example, the polynomial

$$x^2 - 2x - 3 = (x + 1)(x - 3)$$

has two zeros, $x = -1$ and $x = 3$, and these zeros divide the real number line into three test intervals:

$$(-\infty, -1), \quad (-1, 3), \quad \text{and} \quad (3, \infty).$$

Thus, to solve the inequality $x^2 - 2x - 3 < 0$, you need only test one value from each of these test intervals. You can use the same basic approach to determine the test intervals for any polynomial.

Finding Test Intervals for a Polynomial

To determine the intervals on which the values of a polynomial are entirely negative or entirely positive, use the following steps.

1. Find all real zeros of the polynomial, and arrange the zeros in increasing order (from smallest to largest). These zeros are the **critical numbers** of the polynomial.

2. Use the critical numbers of the polynomial to determine its **test intervals**.

3. Choose one representative x-value in each test interval and evaluate the polynomial at that value. If the value of the polynomial is negative, the polynomial will have negative values for *every* x-value in the interval. If the value of the polynomial is positive, the polynomial will have positive values for *every* x-value in the interval.

EXAMPLE 1   Finding Test Intervals for a Polynomial

Solve $x^2 - x - 6 < 0$.

Solution

By factoring the quadratic as

$$x^2 - x - 6 = (x + 2)(x - 3)$$

you can see that the critical numbers are $x = -2$ and $x = 3$. Thus, the polynomial's test intervals are

$(-\infty, -2), \quad (-2, 3), \quad$ and $\quad (3, \infty).$ Test intervals

In each test interval, choose a representative x-value and evaluate the polynomial.

Interval	x-Value	Polynomial Value	Conclusion
$(-\infty, -2)$	$x = -3$	$(-3)^2 - (-3) - 6 = 6$	Positive
$(-2, 3)$	$x = 0$	$(0)^2 - (0) - 6 = -6$	Negative
$(3, \infty)$	$x = 4$	$(4)^2 - (4) - 6 = 4$	Positive

From this, you can conclude that the polynomial is positive for all x-values in $(-\infty, -2)$ and $(3, \infty)$, and is negative for all x-values in $(-2, 3)$. This implies that the solution of the inequality $x^2 - x - 6 < 0$ is the interval $(-2, 3)$, as shown in Figure 1.18.

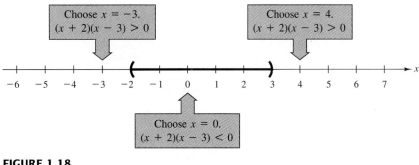

FIGURE 1.18

As with linear inequalities, you can check the reasonableness of a solution by substituting x-values into the original inequality. For instance, to check the solution found in Example 1, try substituting several x-values from the interval $(-2, 3)$ into the inequality

$$x^2 - x - 6 < 0.$$

Regardless of which x-values you choose, the inequality should be satisfied.

In Example 1, the polynomial inequality was given in standard form. Whenever this is not the case, you should begin the solution process by writing the inequality in standard form—with the polynomial on one side and zero on the other.

EXAMPLE 2   Solving a Polynomial Inequality

Solve $2x^2 + 5x > 12$.

Solution

Begin by writing the inequality in standard form.

$$2x^2 + 5x > 12 \qquad \text{Original inequality}$$

$$2x^2 + 5x - 12 > 0 \qquad \text{Write in standard form.}$$

$$(x + 4)(2x - 3) > 0 \qquad \text{Factor.}$$

Critical numbers: $x = -4, x = \frac{3}{2}$

Test intervals: $(-\infty, -4), \left(-4, \frac{3}{2}\right), \left(\frac{3}{2}, \infty\right)$

Test: Is $(x + 4)(2x - 3) > 0$?

After testing these intervals, as shown in Figure 1.19, you can see that the polynomial $2x^2 + 5x - 12$ is positive in the open intervals $(-\infty, -4)$ and $\left(\frac{3}{2}, \infty\right)$. Therefore, the solution set consists of all real numbers in the intervals $(-\infty, -4)$ and $\left(\frac{3}{2}, \infty\right)$.

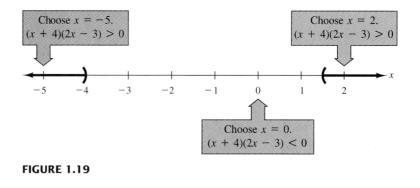

FIGURE 1.19

When solving a quadratic inequality, be sure you have accounted for the particular type of inequality symbol given in the inequality. For instance, in Example 2, note that the solution consisted of two open intervals because the original inequality contained a "less than" symbol. If the original inequality had been $2x^2 + 5x \geq 12$, the solution would have consisted of the two half-open intervals $(-\infty, -4]$ and $\left[\frac{3}{2}, \infty\right)$.

Each of the polynomial inequalities in Examples 1 and 2 has a solution set that consists of a single interval or the union of two intervals. When solving the exercises for this section, you should watch for some unusual solution sets, as illustrated in Example 3.

EXAMPLE 3   *Unusual Solution Sets*

a. The solution set of the following inequality consists of the entire set of real numbers, $(-\infty, \infty)$.

$$x^2 + 2x + 4 > 0$$

b. The solution set of the following inequality consists of the single real number $\{-1\}$.

$$x^2 + 2x + 1 \leq 0$$

c. The solution set of the following inequality is empty.

$$x^2 + 3x + 5 < 0$$

d. The solution set of the following inequality consists of all real numbers except the number 2.

$$x^2 - 4x + 4 > 0$$

Technology

Graphs of Inequalities and Graphing Utilities

Most graphing utilities can sketch the graph of an inequality. For instance, the following steps show how to sketch the graph of $x^2 - 5x < 0$ on a *TI-82*.

$\boxed{\text{Y=}}$ $Y_1 = X^2 - 5X < 0$

$\boxed{\text{GRAPH}}$

The graph produced by these steps is shown below. Notice that the graph occurs as an interval *above* the x-axis.

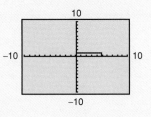

Rational Inequalities

The concepts of critical numbers and test intervals can be extended to inequalities involving rational expressions. To do this, use the fact that the value of a rational expression can change sign only at its *zeros* (the *x*-values for which its numerator is zero) and its *undefined values* (the *x*-values for which its denominator is zero). These two types of numbers make up the **critical numbers** of a rational inequality.

EXAMPLE 4   Solving a Rational Inequality

Solve $\dfrac{2x - 7}{x - 5} \leq 3$.

Solution

$$\frac{2x - 7}{x - 5} \leq 3 \qquad \text{Original inequality}$$

$$\frac{2x - 7}{x - 5} - 3 \leq 0 \qquad \text{Write in standard form.}$$

$$\frac{2x - 7 - 3x + 15}{x - 5} \leq 0 \qquad \text{Add fractions.}$$

$$\frac{-x + 8}{x - 5} \leq 0 \qquad \text{Simplify.}$$

Critical numbers: $x = 5, x = 8$

Test intervals: $(-\infty, 5), (5, 8), (8, \infty)$

Test: Is $\dfrac{-x + 8}{x - 5} \leq 0$?

After testing these intervals, as shown in Figure 1.20, you can see that the rational expression $(-x + 8)/(x - 5)$ is negative in the open intervals $(-\infty, 5)$ and $(8, \infty)$. Moreover, because $(-x + 8)(x - 5) = 0$ when $x = 8$, you can conclude that the solution set consists of all real numbers in the intervals

$$(-\infty, 5) \cup [8, \infty). \qquad \text{Solution set}$$

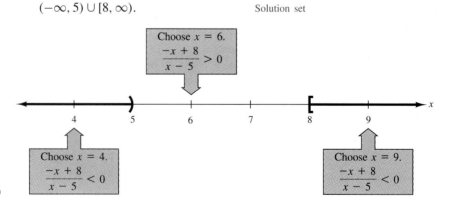

FIGURE 1.20

Applications

One common application of inequalities comes from business and involves profit, revenue, and cost. The formula that relates these three quantities is

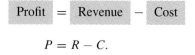

$$P = R - C.$$

EXAMPLE 5   Increasing the Profit for a Product

The marketing department of a calculator manufacturer has determined that the demand for a new model of calculator is given by

$$p = 100 - 0.00001x, \qquad 0 \le x \le 10{,}000{,}000 \qquad \text{Demand equation}$$

where the price per calculator is in dollars and x represents the number of calculators sold. (If this model is accurate, no one would be willing to pay $100 for the calculator. At the other extreme, the company couldn't *give* away more than 10 million calculators.) The revenue for selling x calculators is given by

$$R = xp = x(100 - 0.00001x) \qquad \text{Revenue equation}$$

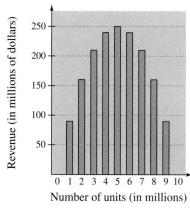

FIGURE 1.21

as shown in Figure 1.21. The total cost of producing x calculators is $10 per calculator plus a development cost of $2,500,000. Thus, the total cost is

$$C = 10x + 2{,}500{,}000. \qquad \text{Cost equation}$$

What price should the company charge per calculator to obtain a profit of at least $190,000,000?

Solution

Verbal Model: Profit   =   Revenue   −   Cost

Equation: $P = R - C$

$$P = 100x - 0.00001x^2 - (10x + 2{,}500{,}000)$$

$$P = -0.00001x^2 + 90x - 2{,}500{,}000$$

To answer the question, you must now solve the inequality

$$-0.00001x^2 + 90x - 2{,}500{,}000 \ge 190{,}000{,}000.$$

Using the techniques described in this section, you can find the solution to be $3{,}500{,}000 \le x \le 5{,}500{,}000$, as shown in Figure 1.22. The prices that correspond to these x values are given by

$$\$45.00 \le p \le \$65.00.$$

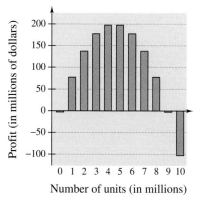

FIGURE 1.22

Another common application of inequalities is finding the domain of an expression that involves a square root, as shown in Example 6.

EXAMPLE 6   *Finding the Domain of an Expression*

Find the domain of the expression $\sqrt{64 - 4x^2}$.

Solution

Remember that the domain of an expression is the set of all x-values for which the expression is defined. Because $\sqrt{64 - 4x^2}$ is defined (has real values) only if $64 - 4x^2$ is nonnegative, the domain is given by $64 - 4x^2 \geq 0$.

$64 - 4x^2 \geq 0$	Standard form
$16 - x^2 \geq 0$	Divide both sides by 4.
$(4 - x)(4 + x) \geq 0$	Factor.

Thus, the inequality has two critical numbers: -4 and 4. You can use these two numbers to test the inequality as follows.

Critical numbers: $x = -4, x = 4$
Test intervals: $(-\infty, -4), (-4, 4), (4, \infty)$
Test: Is $(4 - x)(4 + x) \geq 0$?

A test shows that $64 - 4x^2$ is greater than or equal to 0 in the *closed interval* $[-4, 4]$. Thus, the domain of the expression $\sqrt{64 - 4x^2}$ is the interval $[-4, 4]$, as shown in Figure 1.23.

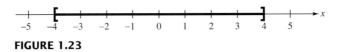

FIGURE 1.23

Group Activities   E x t e n d i n g   t h e   C o n c e p t

Profit Analysis   Consider the relationship $P = R - C$ described on page 153. Discuss why it might be beneficial to solve $P < 0$ if you owned a business. Use the situation described in Example 5 to illustrate your reasoning.

Warm Up   The following warm-up exercises involve skills that were covered in earlier sections. You will use these skills in the exercise set for this section.

In Exercises 1–10, solve the inequality.

1. $-\dfrac{y}{3} > 2$

2. $-6z < 27$

3. $-3 \leq 2x + 3 < 5$

4. $-3x + 5 \geq 20$

5. $10 > 4 - 3(x + 1)$

6. $3 < 1 + 2(x - 4) < 7$

7. $2|x| \leq 7$

8. $|x - 3| > 1$

9. $|x + 4| > 2$

10. $|2 - x| \leq 4$

1.7 Exercises

In Exercises 1–28, solve the inequality and sketch its graph.

1. $x^2 \leq 9$

2. $x^2 < 5$

3. $x^2 > 4$

4. $(x - 3)^2 \geq 1$

5. $(x + 2)^2 < 25$

6. $(x + 6)^2 \leq 8$

7. $x^2 + 4x + 4 \geq 9$

8. $x^2 - 6x + 9 < 16$

9. $x^2 + x < 6$

10. $x^2 + 2x > 3$

11. $3(x - 1)(x + 1) > 0$

12. $6(x + 2)(x - 1) < 0$

13. $x^2 + 2x - 3 < 0$

14. $x^2 - 4x - 1 > 0$

15. $4x^3 - 6x^2 < 0$

16. $4x^3 - 12x^2 > 0$

17. $x^3 - 4x \geq 0$

18. $2x^3 - x^4 \leq 0$

19. $\dfrac{1}{x} > x$

20. $\dfrac{1}{x} < 4$

21. $\dfrac{x + 6}{x + 1} < 2$

22. $\dfrac{x + 12}{x + 2} \geq 3$

23. $\dfrac{3x - 5}{x - 5} > 4$

24. $\dfrac{5 + 7x}{1 + 2x} < 4$

25. $\dfrac{4}{x + 5} > \dfrac{1}{2x + 3}$

26. $\dfrac{5}{x - 6} > \dfrac{3}{x + 2}$

27. $\dfrac{1}{x - 3} \leq \dfrac{9}{4x + 3}$

28. $\dfrac{1}{x} \geq \dfrac{1}{x + 3}$

In Exercises 29–36, find the domain of the expression.

29. $\sqrt[4]{4 - x^2}$

30. $\sqrt{x^2 - 4}$

31. $\sqrt{x^2 - 7x + 12}$

32. $\sqrt{144 - 9x^2}$

33. $\sqrt{12 - x - x^2}$

34. $\sqrt{x^2 + 4}$

35. $\sqrt{x^2 - 3x + 3}$

36. $\sqrt[4]{-x^2 + 2x - 2}$

In Exercises 37–42, use a calculator to solve the inequality. (Round each number in your answer to two decimal places.)

37. $0.4x^2 + 5.26 < 10.2$

38. $-1.3x^2 + 3.78 > 2.12$

39. $-0.5x^2 + 12.5x + 1.6 > 0$

40. $1.2x^2 + 4.8x + 3.1 < 5.3$

41. $\dfrac{1}{2.3x - 5.2} > 3.4$

42. $\dfrac{2}{3.1x - 3.7} > 5.8$

43. *Height of a Projectile*   A projectile is fired straight upward from ground level with an initial velocity of 160 feet per second. During what time period will its height exceed 384 feet?

44. *Height of a Projectile*   A projectile is fired straight upward from ground level with an initial velocity of 128 feet per second. During what time period will its height be less than 128 feet?

45. *Geometry*   A rectangular playing field (see figure) with a perimeter of 100 meters is to have an area of at least 500 square meters. Within what bounds must the length lie?

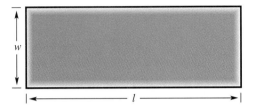

46. *Geometry*   A rectangular room with a perimeter of 50 feet is to have an area of at least 120 square feet. Within what bounds must the length lie?

47. *Company Profits*   The revenue and cost equations for a product are given by

$$R = x(50 - 0.0002x)$$
$$C = 12x + 150,000$$

where R and C are measured in dollars and x represents the number of units sold (see figure).

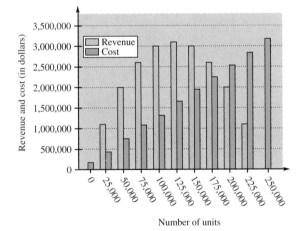

(a) How many units must be sold to obtain a profit of at least $1,650,000?

(b) The demand equation for the product is

$$p = 50 - 0.0002x$$

where p is the price per unit. What price per unit will produce a profit of at least $1,650,000?

48. *Company Profits*   The revenue and cost equations for a product are given by

$$R = x(75 - 0.0005x)$$
$$C = 30x + 250,000$$

where R and C are measured in dollars and x represents the number of units sold (see figure).

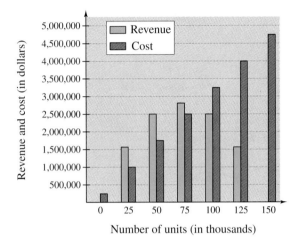

(a) How many units must be sold to obtain a profit of at least $750,000?

(b) The demand equation for the product is

$$p = 75 - 0.0005x$$

where p is the price per unit. What price per unit will produce a profit of at least $750,000?

49. *Compound Interest* P dollars, invested at interest rate r compounded annually, increases to an amount

$$A = P(1 + r)^2$$

in 2 years. If an investment of $1000 is to increase to an amount greater than $1200 in 2 years, the interest rate must be greater than what percent?

50. *Compound Interest* P dollars, invested at interest rate r compounded annually, increases to an amount

$$A = P(1 + r)^3$$

in 3 years. If an investment of $500 is to increase to an amount greater than $600 in 3 years, the interest rate must be greater than what percent?

51. *World Population*   The world population (in millions) from 1980 to 1990 can be modeled by

$$\text{Population} = 4449 + 76.7t + 0.78t^2, \quad 0 \le t$$

where $t = 0$ represents 1980 (see figure). According to this model, when will the world population exceed 6,000,000,000?   (Source: Statistical Office of United Nations)

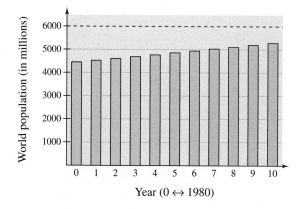

Year (0 ↔ 1980)

52. *Percent of College Graduates*   The percent of the American population that graduated from college from 1960 to 1990 can be modeled by

$$\text{Percent of graduates} = 7.34 + 0.41t + 0.002t^2, \\ 0 \le t$$

where $t = 0$ represents 1960. According to this model, when will the percent of college graduates exceed 25%?   (Source: U.S. Bureau of Census)

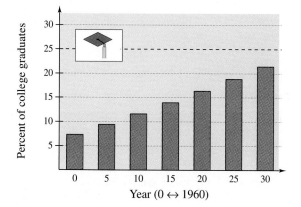

Year (0 ↔ 1960)

53. *Vehicle Registration*   The number of motor vehicle registrations (in millions) in the United States from 1960 to 1990 can be modeled by

$$\text{Registrations} = 71.78 + 4.05t + 0.002t^2, \quad 0 \le t$$

where $t = 0$ represents 1960 (see figure). According to this model, when will the number of motor vehicle registrations exceed 225,000,000?   (Source: Motor Vehicle Manufacturers Association)

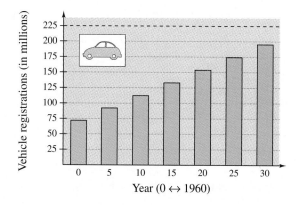

Year (0 ↔ 1960)

54. *Vehicle Registration*   The number of motor vehicle registrations (in millions) in the world from 1960 to 1990 can be modeled by

$$\text{Registrations} = 121.66 + 11.95t + 0.108t^2, \\ 0 \le t$$

where $t = 0$ represents 1960 (see figure). According to this model, when will the number of motor vehicle registrations exceed 700,000,000?   (Source: Motor Vehicle Manufacturers Association)

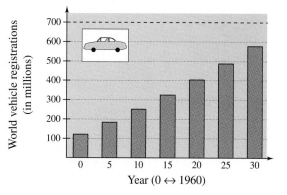

Year (0 ↔ 1960)

CHAPTER PROJECT: The Federal Deficit

Year	National Debt
1970	$370.1
1971	$397.3
1972	$426.4
1973	$457.3
1974	$474.2
1975	$533.2
1976	$620.4
1977	$698.8
1978	$771.5
1979	$826.5
1980	$907.7
1981	$997.9
1982	$1142.0
1983	$1377.2
1984	$1572.3
1985	$1823.1
1986	$2125.3
1987	$2350.3
1988	$2602.3
1989	$2857.4
1990	$3233.3
1991	$3569.3
1992	$3972.6

During each year in recent history, the federal government has had less income than expenses. The difference between expenses and income for a year is called the annual federal deficit. The national debt is the sum of the yearly deficits.

The table at the left shows the national debt (in billions) from 1970 through 1992. The national debt D (in billions) can be modeled by

$$D = 9.2t^2 - 46.7t + 480, \qquad 0 \le t$$

where $t = 0$ represents 1970. The following bar graph compares the actual debt with the debt given by the model.

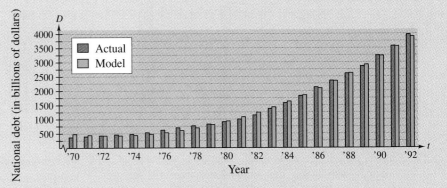

Use this information to investigate the following questions.

1. **Comparing Data**   Make a table that compares the actual debt from 1970 through 1992 with the debt given by the model. From the table, does it appear that the model could make a good predictor for the national debt?

2. **Comparing Projected Debt with Actual Debt**   Use the model to predict the national debt in 1993. Then use your school's library or some other reference source to find the actual national debt during that year. How closely did the model predict the actual debt?

3. **Projecting Future Debt**   According to this model, when will the national debt reach $7 trillion?

   a. Answer the question numerically by creating a table of values.

   b. Answer the question algebraically by solving $7000 = 9.2t^2 - 46.7t + 480$.

4. **Finding a Model**   Enter the data in the table at the left in a graphing utility and use its quadratic fit program to find a quadratic model for the data. Do you get the same model that is listed above?

5. **Research**   Use your school's library or some other reference source to find data that can be closely modeled with a quadratic model. Then use a graphing utility to find a model. Compare the model with the actual data numerically and graphically.

CHAPTER SUMMARY

After studying this chapter, you should have acquired the following skills. These skills are keyed to the Review Exercises that begin on page 160. Answers to odd-numbered Review Exercises are given in the back of the book.

- Determine whether an equation is an identity or a conditional equation. *(Section 1.1)* — **Review Exercises 1, 2**

- Determine whether a given value is a solution. *(Section 1.1)* — **Review Exercises 3, 4**

- Solve a linear equation. *(Section 1.1)* — **Review Exercises 5–10**

- Use a calculator to solve an equation. *(Section 1.1)* — **Review Exercises 11, 12**

- Solve using a linear model. *(Section 1.2)* — **Review Exercises 13–16, 19–22**

- Solve using a common formula. *(Section 1.2)* — **Review Exercises 17, 18**

- Solve a mixture problem. *(Section 1.2)* — **Review Exercises 23, 24**

- Solve a quadratic equation by factoring. *(Section 1.3)* — **Review Exercises 25–28**

- Solve an equation by extracting square roots. *(Section 1.3)* — **Review Exercises 29–32**

- Solve using a quadratic model. *(Section 1.3)* — **Review Exercises 33–36**

- Determine the number of real solutions to a quadratic equation using the discriminant. *(Section 1.4)* — **Review Exercises 37, 38**

- Solve a quadratic equation using the quadratic formula. *(Section 1.4)* — **Review Exercises 39–44, 47, 48**

- Use a calculator to solve a quadratic equation. *(Section 1.4)* — **Review Exercises 45, 46**

- Solve a polynomial equation. *(Section 1.5)* — **Review Exercises 49–52**

- Solve an equation involving radicals. *(Section 1.5)* — **Review Exercises 53–56**

- Solve an equation with absolute values. *(Section 1.5)* — **Review Exercises 57, 58**

- Solve an equation with fractions. *(Section 1.5)* — **Review Exercises 59, 60**

- Solve a real-life problem with modeling. *(Section 1.5)* — **Review Exercises 61–63**

- Solve an inequality and sketch its graph. *(Section 1.6)* — **Review Exercises 64–69**

- Find the domain of an expression. *(Section 1.6)* — **Review Exercises 70–73**

- Use an inequality to solve applications. *(Section 1.6)* — **Review Exercises 74, 75**

- Solve a polynomial inequality. *(Section 1.7)* — **Review Exercises 76–78**

- Solve a rational inequality. *(Section 1.7)* — **Review Exercises 79–81**

- Use a calculator to solve an inequality. *(Section 1.7)* — **Review Exercises 82–85**

- Model an inequality to solve a real-life problem. *(Section 1.7)* — **Review Exercises 86–91**

REVIEW EXERCISES

In Exercises 1 and 2, determine whether the equation is an identity or a conditional equation.

1. $5(x - 3) = 2x + 9$ **2.** $3(x + 2) = 3x + 6$

In Exercises 3 and 4, determine whether each given value of x is a solution of the equation.

3. $3x^2 + 7x + 5 = x^2 + 9$

   (a) $x = 0$ (b) $x = \dfrac{1}{2}$ (c) $x = -4$ (d) $x = -1$

4. $6 + \dfrac{3}{x - 4} = 5$

   (a) $x = 5$ (b) $x = 0$ (c) $x = -2$ (d) $x = 7$

In Exercises 5–10, solve the equation (if possible) and check your answer.

5. $4(x + 3) - 3 = 2(4 - 3x) - 4$

6. $\dfrac{3x - 2}{5x - 1} = \dfrac{3}{4}$

7. $(x + 3) + 2(x - 4) = 5(x + 3)$

8. $\dfrac{3}{x - 4} + \dfrac{8}{2x + 5} = \dfrac{11}{2x^2 - 3x - 20}$

9. $\dfrac{x}{x + 3} - \dfrac{4}{x + 3} + 2 = 0$

10. $7 - \dfrac{3}{x} = 8 + \dfrac{5}{x}$

In Exercises 11 and 12, use a calculator to solve the equation for x. (Round your answer to three decimal places.)

11. $0.375x - 0.75(300 - x) = 200$

12. $\dfrac{x}{0.0645} + \dfrac{x}{0.098} = 2$

13. Three consecutive even integers have a sum of 42. Find the smallest of these integers.

14. Suppose your annual salary is $24,500. You receive a 6% raise. What is your new annual salary?

15. *Monthly Profit*   The total profit for a company in June was 15% less than it was in May. The total profit for the two months was $225,392. Find the profit for each month.

16. *Study Habits*   Based on a survey of 320 math students, the most common study techniques are shown in the graph. How many students in the survey used each type of study technique? (Most students used more than one type of technique.)

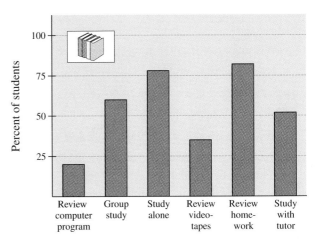

17. *Geometry*   A volleyball court is twice as long as it is wide, and its perimeter is 177 feet. Find the dimensions of the volleyball court.

18. *Geometry*   A room is 1.25 times as long as it is wide, and its perimeter is 90 feet. Find the dimensions of the room.

19. *List Price*   The price of a microwave oven has been discounted 15%. The sale price is $339.15. Find the original price of the microwave.

20. *Discount Rate*   The price of a VCR has been discounted by $85. The sale price is $257. What was the percent discount?

21. *Travel Time*   Two cars start at the same time at a given point and travel in the same direction at average speeds of 45 miles per hour and 50 miles per hour. How much time must elapse before the two cars are 10 miles apart?

22. *Projected Revenue*   From February through July, a company's revenues have totaled $435,112. If the monthly revenues continue at this rate, what will the total revenue for the year be?

23. *Mixture*   A car radiator contains 10 quarts of a 30% antifreeze solution. How many quarts will have to be replaced with pure antifreeze if the resulting 10-quart solution is to be 50% antifreeze?

24. *Mixture*   A 3-gallon acid solution contains 5% boric acid. How many gallons of 20% boric acid solution should be added to make a final solution that is 12% boric acid?

In Exercises 25–28, solve the quadratic equation by factoring. Check your solutions by substituting in the original equation.

25. $6x^2 = 5x + 4$

26. $-x^2 = 15x + 36$

27. $x^2 - 11x + 24 = 0$

28. $15 + x - 2x^2 = 0$

In Exercises 29–32, solve the equation by extracting square roots. List both the exact answer and a decimal answer that has been rounded to two decimal places.

29. $x^2 = 11$

30. $16x^2 = 25$

31. $(x + 4)^2 = 18$

32. $(x - 1)^2 = 5$

33. *Geometry*   A billboard is 12 feet longer than it is high. The billboard has 405 square feet of advertising space. What are the dimensions of the billboard? Use a diagram to help answer the question.

34. *Grand Canyon*   The Grand Canyon is 6000 feet deep at the deepest part. If a rock was dropped over the deepest part of the canyon, how long would the rock take to hit the water in the Colorado River below?

35. *Total Revenue*   The demand equation for a certain product is

$$p = 50 - 0.0001x$$

where p is the price per unit and x is the number of units sold. The total revenue for selling x units is given by

$$\text{Revenue} = xp = x(50 - 0.0001x).$$

How many units must be sold to produce a revenue of $6,000,000?

36. *Depth of an Underwater Cable*   A ship's sonar locates a cable 2000 feet from the ship (see figure). The angle between the surface of the water and a line from the ship to the cable is 45 degrees. How deep is the cable?

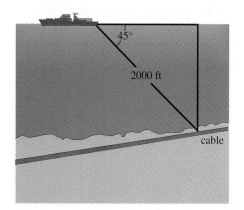

In Exercises 37 and 38, use the discriminant to determine the number of real solutions of the quadratic equation.

37. $x^2 + 11x + 24 = 0$

38. $x^2 + 5x + 12 = 0$

In Exercises 39–44, use the Quadratic Formula to solve the equation. Check your solutions by substituting in the original equation.

39. $x^2 - 12x + 30 = 0$

40. $5x^2 + 16x - 12 = 0$

41. $(y + 7)^2 = -5y$

42. $6x = 7 - 2x^2$

43. $x^2 + 6x - 3 = 0$

44. $10x^2 - 11x = 2$

In Exercises 45 and 46, use a calculator to solve the equation. (Round your answers to three decimal places.)

45. $3.6x^2 - 5.7x - 1.9 = 0$

46. $34x^2 - 296x + 47 = 0$

47. *On the Moon*   An astronaut standing on the edge of a cliff on the moon drops a rock over the cliff. If the cliff is 100 feet high, how long will it take the rock to hit the lunar surface? If the rock were dropped off a similar cliff on earth, how long would it remain in the air?

48. *Geometry*   An open box is to be made from a square piece of material by cutting 3-inch squares from the corners and turning up the sides (see figure). The volume of the finished box is to be 363 cubic inches. Find the size of the original piece of material.

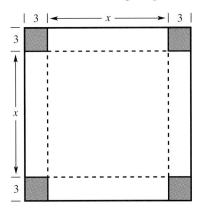

In Exercises 49–60, find all the solutions of the equation. (Check your solutions.)

49. $3x^3 - 9x^2 - 12x = 0$

50. $x^4 - 5x^2 + 4 = 0$

51. $x^4 + 3x^3 - 5x - 15 = 0$

52. $x^6 + 26x^3 - 27 = 0$

53. $2\sqrt{x} - 5 = 0$

54. $\sqrt{3x - 2} + x = 4$

55. $(x^2 - 5)^{2/3} = 9$

56. $(x^2 - 5x - 6)^{4/3} = 16$

57. $|5x + 4| = 11$

58. $|x^2 + 4x| - 2x = 8$

59. $\dfrac{5}{x + 1} + \dfrac{3}{x + 3} = 1$

60. $x + \dfrac{3}{x + 2} = 2$

61. *Sharing the Cost*   Three students are planning to share the expense of renting a condominium at a resort for 2 weeks. By adding a fourth person to the group, each person could save $50 in rental fees. How much is the rent for the 2-week period?

62. *Cash Advance*   Suppose you took out a cash advance on a credit card for $400. After 3 months, the amount you owe is $421.56. What is the annual percentage rate for this loan? (Assume that the interest is compounded monthly and that you made no payments after the first and second months.)

63. *Market Research*   The demand equation for a product is given by

$$p = 42 - \sqrt{0.001x + 2}$$

where x is the number of units demanded per day and p is the price per unit. Find the demand if the price is set at $29.95.

In Exercises 64–69, solve the inequality and sketch its graph.

64. $3(x - 1) < 2x + 8$

65. $-5 \le 2 - 4(x + 2) \le 6$

66. $-3 < \dfrac{2x + 1}{4} < 3$

67. $-1 \le -5 - 3x < 4$

68. $|x + 10| + 3 < 5$

69. $|2x - 3| - 4 > 2$

In Exercises 70–73, find the domain of the expression.

70. $\sqrt{x - 10}$

71. $\sqrt[4]{2x + 5}$

72. $\sqrt{x^2 - 15x + 54}$

73. $\sqrt{81 - 4x^2}$

74. *Break-Even Analysis*   The revenue for selling x units of a product is

$$R = 125.95x.$$

The cost of producing x units is

$$C = 92x + 1200.$$

In order to obtain a profit, the revenue must be greater than the cost. For what values of x will this product return a profit?

75. *Accuracy of Measurement*   Suppose you buy an 18-inch gold chain that costs $8.95 per inch. If the chain is measured accurately to within $\frac{1}{16}$ of an inch, how much money might you have been undercharged or overcharged?

In Exercises 76–81, solve the inequality and sketch its graph.

76. $5(x+1)(x-3) < 0$

77. $(x+4)^2 \leq 4$

78. $x^3 - 9x < 0$

79. $\dfrac{x+5}{x+8} \geq 2$

80. $\dfrac{2+3x}{4-x} < 2$

81. $\dfrac{1}{x+1} \geq \dfrac{1}{x+5}$

In Exercises 82–85, use a calculator to solve the inequality. (Round to two decimal places.)

82. $-1.2x^2 + 4.76 > 1.32$

83. $3.5x^2 + 4.9x - 6.1 < 2.4$

84. $\dfrac{1}{3.7x - 6.1} > 2.9$

85. $\dfrac{3}{5.4x - 2.7} < 8.9$

86. *Height of a Projectile*   A projectile is fired straight upward from ground level with an initial velocity of 134 feet per second. During what time period will its height exceed 276 feet?

87. *Height of a Flare*   A flare is fired straight upward from ground level with an initial velocity of 100 feet per second. During what time period will its height exceed 150 feet?

88. *Compound Interest* P dollars invested at interest rate r compounded annually increased to an amount

$$A = P(1+r)^5$$

in 5 years. If an investment of $1000 is to increase to an amount greater than $1400 in 5 years, the interest rate must be greater than what percent?

89. *Compound Interest* P dollars invested at an interest rate r compounded semiannually increased to an amount

$$A = P(1+r/2)^{2 \cdot 8}$$

in 8 years. If an investment of $2000 is to increase to an amount greater than $4200 in 8 years, the interest rate must be greater than what percent?

90. *Company Profits*   The revenue and cost equations for a product are given by

$$R = x(80 - 0.0005x)$$
$$C = 20x + 300,000$$

where R and C are measured in dollars and x represents the number of units sold. How many units must be sold to obtain a profit of at least $900,000?

91. *Price of a Product*   In Exercise 90, the revenue equation is

$$R = x(80 - 0.0005x)$$

which implies that the demand equation is

$$p = 80 - 0.0005x$$

where p is the price per unit. What price per unit should the company set to obtain a profit of at least $1,200,000?

CHAPTER TEST

Take this test as you would take a test in class. After you are done, check your work against the answers given in the back of the book.

In Exercises 1 and 2, solve the equation.

1. $3(x + 2) - 8 = 4(2 - 5x) + 7$

2. $0.875x + 0.375(300 - x) = 200$

3. In May the total profit for a company was 20% less than it was in April. The total profit for the two months was \$315,655.20. Find the profit for each month.

In Exercises 4–13, solve the equation.

4. *Factoring:* $6x^2 + 7x = 5$

5. *Factoring:* $12 + 5x - 2x^2 = 0$

6. *Extracting Roots:* $x^2 - 5 = 10$

7. *Quadratic Formula:* $(x + 5)^2 = -3x$

8. *Quadratic Formula:* $3x^2 - 11x = 2$

9. *Quadratic Formula:* $5.4x^2 - 3.2x - 2.5 = 0$

10. $|3x + 2| = 8$

11. $\sqrt{x - 3} + x = 5$

12. $x^4 - 10x^2 + 9 = 0$

13. $(x^2 - 9)^{2/3} = 9$

14. The demand equation for a product is $p = 40 - 0.0001x$, where p is the price per unit and x is the number of units sold. The total revenue for selling x units is given by $R = xp$. How many units must be sold to produce a revenue of \$2,000,000? Explain your reasoning.

In Exercises 15–18, solve the inequality and sketch its graph.

15. $-2 \le \dfrac{3x + 1}{5} \le 2$

16. $(x + 3)^2 \le 5$

17. $\dfrac{x + 3}{x + 7} > 2$

18. $3x^3 - 12x \le 0$

19. The revenue and cost equations for a product are given by

$$R = x(100 - 0.0005x) \quad \text{and} \quad C = 30x + 200,000$$

where R and C are measured in dollars and x represents the number of units sold. How many units must be sold to obtain a profit of at least \$500,000?

20. P dollars, invested at interest rate r compounded annually, increases to an amount $A = P(1 + r)^{10}$ in 10 years. If an investment of \$2000 is to increase to an amount greater than \$5000 in 10 years, the interest rate must be greater than what percent?

The Cartesian Plane and Graphs

2

- The Cartesian Plane
- Graphs of Equations
- Graphing Calculators
- Lines in the Plane
- Linear Modeling

The special icon above the graph below alerts you to the use and modeling of real data collected by the Texas Instruments Calculator-Based Laboratory (CBL) System. The chapter project provides the opportunity for further exploration of this topic and more in-depth data analysis.

One force that acts on objects falling to earth is gravity. Gravity pulls an object toward the earth. Trying to keep the object in the air is another force known as air resistance. Air resistance is the reason a leaf floats to the ground and an apple falls to the ground.

The table gives the height (in feet) of a weighted miniature parachute t seconds after its release at 4.8 feet. This data is also shown in the graph at the right.

From the graph you can see that the points appear to fall along a straight line. This line can be modeled by the equation

$$y = -2.7t + 4.96.$$

The slope of the line tells you that the miniature parachute is falling at a rate of 2.7 feet per second. The y-intercept indicates the height of the parachute when it was released.

Time (in seconds)	0	0.2	0.4	0.6	0.8	1.0	1.2
Height (in feet)	4.8	4.5	4.0	3.4	2.8	2.3	1.7

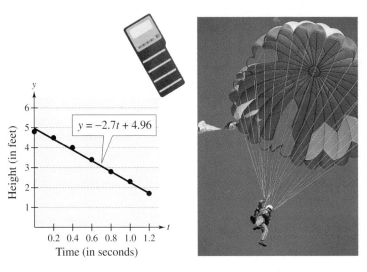

$y = -2.7t + 4.96$

The chapter project related to this information is on page 221. The project uses actual data, collected with a *Texas Instruments CBL* unit.

2.1 | The Cartesian Plane

The Cartesian Plane ▪ The Distance Between Two Points in the Plane ▪ The Midpoint Formula

The Cartesian Plane

Good graphing habits are stressed in this chapter. If graphing utilities are being used, you can more visually and effectively reinforce concepts of domain, range, symmetry, and intercepts. The basic skills learned in the chapter are transferable to graphs of both algebraic and transcendental functions later in the text.

Just as you can represent real numbers by points on the real number line, you can represent ordered pairs of real numbers by points in a **rectangular coordinate plane.** This plane is called the **Cartesian plane,** after the French mathematician René Descartes (1596–1650).

The Cartesian plane is formed by two real lines intersecting at right angles, as shown in Figure 2.1. The horizontal number line is the **x-axis** and the vertical number line is the **y-axis.** (The plural of axis is *axes.*) The point of intersection of the two axes is the **origin** and the axes separate the plane into four **quadrants.**

An overhead projector can be useful for showing how to plot points and equations: try projecting a grid onto the chalkboard, and then plot points on the chalkboard. Or try using overhead markers and graph directly on the transparency. Another option useful for graphing equations is using a viewscreen—a device that, when used with an overhead projector, will project a graphing calculator's screen image.

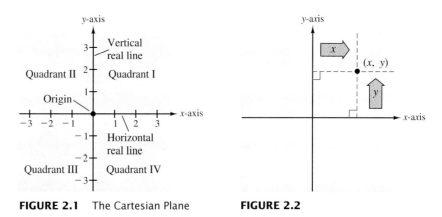

FIGURE 2.1   The Cartesian Plane   **FIGURE 2.2**

Each point in the plane corresponds to an **ordered pair** (x, y) of real numbers x and y, called the **coordinates** of the point. The first number (or **x-coordinate**) tells how far to the left or right the point is from the vertical axis, and the second number (or **y-coordinate**) tells how far up or down the point is from the horizontal axis, as shown in Figure 2.2.

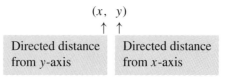

$(x, \ y)$

Directed distance from y-axis   Directed distance from x-axis

NOTE   It is customary to use the notation (x, y) to denote both a point in the plane and an open interval on the real line. The nature of a specific problem will show which of the two is being talked about.

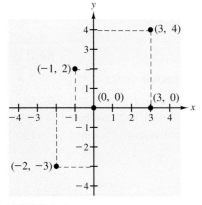

FIGURE 2.3

EXAMPLE 1   *Plotting Points in the Cartesian Plane*

Plot the points $(-1, 2)$, $(3, 4)$, $(0, 0)$, $(3, 0)$, and $(-2, -3)$ in the Cartesian plane.

Solution

To plot the point $(-1, 2)$ you can envision a vertical line through -1 on the x-axis and a horizontal line through 2 on the y-axis. The intersection of these two lines is the point $(-1, 2)$, as shown in Figure 2.3.

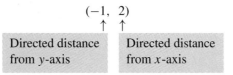

The other four points can be plotted in a similar way.

EXAMPLE 2   *Shifting Points in the Plane*

The triangle shown in Figure 2.4(a) has vertices at the points $(-1, 2)$, $(1, -4)$, and $(2, 3)$. Shift the triangle three units to the right and two units up and find the vertices of the shifted triangle.

Solution

To shift the vertices three units to the right, add 3 to each x-coordinate. To shift the vertices two units up, add 2 to each y-coordinate. See Figure 2.4(b).

Original Vertices	*Shifted Vertices*
$(-1, 2)$	$(-1 + 3, 2 + 2) = (2, 4)$
$(1, -4)$	$(1 + 3, -4 + 2) = (4, -2)$
$(2, 3)$	$(2 + 3, 3 + 2) = (5, 5)$

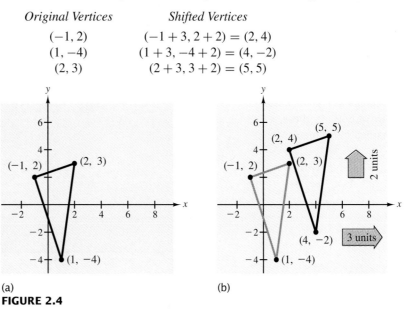

(a)

(b)

FIGURE 2.4

The value of the rectangular coordinate system is that it allows you to visualize relationships between variables x and y. Today, Descartes's ideas are in common use in virtually every scientific and business-related field.

Real Life

EXAMPLE 3   *Number of Doctorates in Mathematics*

The numbers of doctorates in mathematics granted to United States citizens by universities in the United States in the years 1978 to 1993 are given in the table. Plot these points on a rectangular coordinate system. (Source: American Mathematical Society)

Year	1978	1979	1980	1981	1982	1983	1984	1985
Degrees	634	596	578	567	519	455	433	396

Year	1986	1987	1988	1989	1990	1991	1992	1993
Degrees	386	362	363	411	401	461	430	526

Solution

The points are shown in Figure 2.5. Note that the break in the x-axis indicates that the numbers for the years prior to 1978 have been omitted.

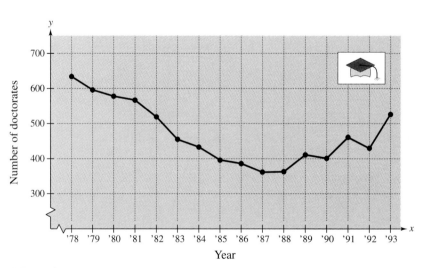

FIGURE 2.5

In 1971, nearly 25,000 bachelors degrees in mathematics were earned in the United States. Of these, 38% were earned by women. In 1991, the total number of bachelors degrees in mathematics had fallen to 14,660. Of these, 47.2% were earned by women.

The Distance Between Two Points in the Plane

You know from Section P.1 that the distance d between two points a and b on the real number line is simply

$$d = |b - a|.$$

The same "absolute value rule" is used to find the distance between two points that lie on the same *vertical* or *horizontal* line in the plane.

EXAMPLE 4   *Finding Horizontal and Vertical Distances*

a. Find the distance between the points $(1, -1)$ and $(1, 4)$.

b. Find the distance between the points $(-3, -1)$ and $(1, -1)$.

Solution

a. Because the x-coordinates are equal, you can envision a vertical line through the points $(1, -1)$ and $(1, 4)$, as shown in Figure 2.6. The distance between these two points is given by the absolute value of the difference of their y-coordinates. That is,

$$\text{Vertical distance} = |4 - (-1)| \qquad \text{Subtract } y\text{-coordinates.}$$
$$= 5. \qquad \text{Simplify.}$$

Thus, the points are five units apart.

b. Because the y-coordinates are equal, you can envision a horizontal line through the points $(-3, -1)$ and $(1, -1)$, as shown in Figure 2.6. The distance between these two points is given by the absolute value of the difference of their x-coordinates. That is,

$$\text{Horizontal distance} = |1 - (-3)| \qquad \text{Subtract } x\text{-coordinates.}$$
$$= 4. \qquad \text{Simplify.}$$

Thus, the points are four units apart.

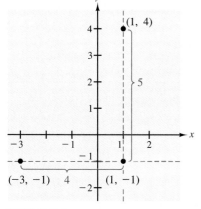

FIGURE 2.6

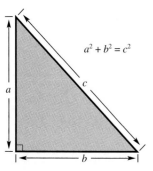

FIGURE 2.7   Pythagorean Theorem

The technique used in Example 4 can be used to develop a general formula for finding the distance between two points in the plane. This general formula will work for any two points, even if they do not lie on the same vertical or horizontal line. To develop the formula, you can use the Pythagorean Theorem, which states that for a right triangle, the hypotenuse c and sides a and b are related by the formula

$$a^2 + b^2 = c^2 \qquad \text{Pythagorean Theorem}$$

as shown in Figure 2.7. (The converse is also true. That is, if $a^2 + b^2 = c^2$, then the triangle is a right triangle.)

To develop a general formula for the distance between two points, let (x_1, y_1) and (x_2, y_2) represent two points in the plane that do not lie on the same horizontal or vertical line. With these two points, a right triangle can be formed, as shown in Figure 2.8. Note that the third vertex of the triangle is (x_1, y_2). Because (x_1, y_1) and (x_1, y_2) lie on the same vertical line, the length of the vertical side of the triangle is $|y_2 - y_1|$. Similarly, the length of the horizontal side is $|x_2 - x_1|$. Thus, by the Pythagorean Theorem, the distance between (x_1, y_1) and (x_2, y_2) is given by

$$d^2 = |x_2 - x_1|^2 + |y_2 - y_1|^2.$$

Because the distance d must be positive, choose the positive square root and write

$$d = \sqrt{|x_2 - x_1|^2 + |y_2 - y_1|^2}.$$

Finally, replacing $|x_2 - x_1|^2$ and $|y_2 - y_1|^2$ by the equivalent expressions $(x_2 - x_1)^2$ and $(y_2 - y_1)^2$ gives the following formula for the distance between two points in a rectangular coordinate plane.

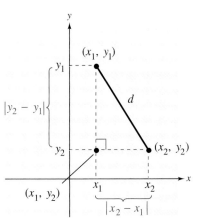

FIGURE 2.8

The Distance Formula

The distance d between two points (x_1, y_1) and (x_2, y_2) in the coordinate plane is

$$d = \sqrt{(x_2 - x_1)^2 + (y_2 - y_1)^2}.$$

EXAMPLE 5   Finding the Distance Between Two Points

Find the distance between the points $(-2, 1)$ and $(3, 4)$.

Solution

Let $(x_1, y_1) = (-2, 1)$ and $(x_2, y_2) = (3, 4)$, and apply the Distance Formula.

$$\begin{aligned}
d &= \sqrt{[3 - (-2)]^2 + (4 - 1)^2} && \text{Distance Formula} \\
&= \sqrt{5^2 + 3^2} && \text{Simplify.} \\
&= \sqrt{25 + 9} && \text{Simplify.} \\
&= \sqrt{34} && \text{Simplify.} \\
&\approx 5.83 && \text{Use a calculator.}
\end{aligned}$$

See Figure 2.9.

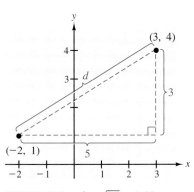

FIGURE 2.9 $d = \sqrt{34} \approx 5.83$

NOTE   In Example 5, try letting $(x_1, y_1) = (3, 4)$ and $(x_2, y_2) = (-2, 1)$ to see that the Distance Formula yields the same result.

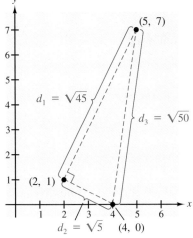

y

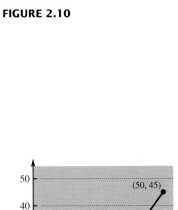

FIGURE 2.10

EXAMPLE 6   Using the Distance Formula in Geometry

Show that the points $(2, 1)$, $(4, 0)$, and $(5, 7)$ are vertices of a right triangle.

Solution

The three points are plotted in Figure 2.10. Using the Distance Formula, you can find the lengths of the three sides of the triangle.

$$d_1 = \sqrt{(5-2)^2 + (7-1)^2} = \sqrt{9+36} = \sqrt{45}$$

$$d_2 = \sqrt{(4-2)^2 + (0-1)^2} = \sqrt{4+1} = \sqrt{5}$$

$$d_3 = \sqrt{(5-4)^2 + (7-0)^2} = \sqrt{1+49} = \sqrt{50}$$

Because

$$d_1{}^2 + d_2{}^2 = 45 + 5$$
$$= 50$$
$$= d_3{}^2$$

you can conclude from the Pythagorean Theorem that the triangle is a right triangle.

The next example shows how the Distance Formula can be used to solve a real-life problem.

Real Life

EXAMPLE 7   The Length of a Football Pass

In a football game, a quarterback throws a pass from the 5-yard line, 20 yards from the sideline. The pass is caught by a wide receiver on the 45-yard line, 50 yards from the same sideline, as shown in Figure 2.11. How long was the pass?

Solution

Using the Distance Formula, you can find the distance to be

$$d = \sqrt{(50-20)^2 + (45-5)^2}$$ Distance Formula

$$= \sqrt{900 + 1600}$$ Simplify.

$$= \sqrt{2500}$$ Simplify.

$$= 50 \text{ yards}$$ Simplify.

FIGURE 2.11

NOTE   In Example 7, note that a scale was added to the goal line. When you use coordinate geometry to solve real-life problems, you are free to place the coordinate system in any way that is convenient in the problem.

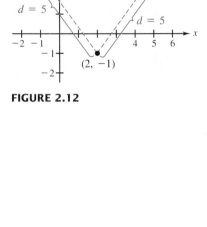

FIGURE 2.12

EXAMPLE 8   Using the Distance Formula

Find x so that the distance between $(x, 3)$ and $(2, -1)$ is 5.

Solution

Using the Distance Formula, you can write the following.

$\sqrt{(x-2)^2 + (3+1)^2} = 5$	Distance Formula
$(x^2 - 4x + 4) + 16 = 25$	Square both sides.
$x^2 - 4x - 5 = 0$	Standard form
$(x-5)(x+1) = 0$	Factor.
$x - 5 = 0 \implies x = 5$	Set 1st factor equal to 0.
$x + 1 = 0 \implies x = -1$	Set 2nd factor equal to 0.

Thus, you can conclude that each of the points $(5, 3)$ and $(-1, 3)$ lies five units from the point $(2, -1)$, as shown in Figure 2.12.

The Midpoint Formula

The following formula shows how to find the midpoint of the line segment that joins two points.

The Midpoint Formula

The **midpoint** of the line segment joining the points (x_1, y_1) and (x_2, y_2) in the coordinate plane is

$$\left(\frac{x_1 + x_2}{2}, \frac{y_1 + y_2}{2} \right).$$

EXAMPLE 9   Finding the Midpoint of a Line Segment

Find the midpoint of the line segment joining the points $(-5, -3)$ and $(9, 3)$.

Solution

Figure 2.13 shows the two given points and their midpoint. By the Midpoint Formula, you have

$$\text{Midpoint} = \left(\frac{-5+9}{2}, \frac{-3+3}{2} \right) = (2, 0).$$

FIGURE 2.13

Real Life

EXAMPLE 10   Retail Sales

A business had annual retail sales of \$240,000 in 1990 and \$312,000 in 1996. Find the sales for 1993. (Assume that the annual increase in sales followed a linear pattern. That is, assume that the points representing the sales lie on a straight line, as shown in Figure 2.14.)

Solution

To make the computations simpler, let $t = 0$ represent the year 1990 and $t = 6$ represent the year 1996. Then, if you measure the retail sales in thousands of dollars, the sales for 1990 and 1996 are represented by the points

$$(0, 240) \quad \text{and} \quad (6, 312).$$

Because 1993 is midway between 1990 and 1996 and because the growth pattern is linear, you can use the Midpoint Formula to find the 1993 sales as follows.

$$\text{Midpoint} = \left(\frac{0 + 6}{2}, \frac{240 + 312}{2}\right) = (3, 276)$$

Thus, the 1993 sales were approximately \$276,000, as indicated in Figure 2.14.

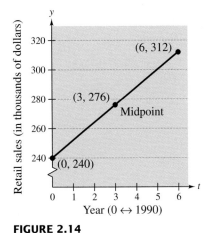

FIGURE 2.14

Group Activities   Extending the Concept

A Misleading Graph   Although graphs can help us visualize relationships between two variables, they can also be used to mislead people. The graphs shown below represent the same data points. Which of the two graphs is misleading, and why? Discuss other ways in which graphs can be misleading. Try to find another example of a misleading graph in a newspaper or magazine. Why is it misleading?

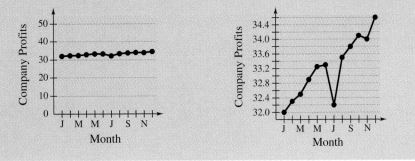

Warm Up

The following warm-up exercises involve skills that were covered in earlier sections. You will use these skills in the exercise set for this section.

In Exercises 1–6, simplify the expression.

1. $\sqrt{(2-6)^2 + [1-(-2)]^2}$

2. $\sqrt{(1-4)^2 + (-2-1)^2}$

3. $\dfrac{4 + (-2)}{2}$

4. $\dfrac{-1 + (-3)}{2}$

5. $\sqrt{18} + \sqrt{45}$

6. $\sqrt{12} + \sqrt{44}$

In Exercises 7–10, solve the equation.

7. $\sqrt{(4-x)^2 + (5-2)^2} = \sqrt{58}$

8. $\sqrt{(8-6)^2 + (y-5)^2} = 2\sqrt{5}$

9. $\dfrac{x+3}{2} = 7$

10. $\dfrac{-2+y}{2} = 1$

2.1 Exercises

In Exercises 1–4, sketch the polygon with the indicated vertices.

1. Triangle: $(-1, 1)$, $(2, -1)$, $(3, 4)$

2. Triangle: $(0, 3)$, $(-1, -2)$, $(4, 8)$

3. Square: $(2, 4)$, $(5, 1)$, $(2, -2)$, $(-1, 1)$

4. Parallelogram: $(5, 2)$, $(7, 0)$, $(1, -2)$, $(-1, 0)$

Shifting a Graph   In Exercises 5 and 6, the figure is shifted in the plane. Find the vertices of the shifted figure.

5.

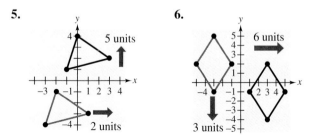

6.

In Exercises 7–10, plot the points and find the distance between them.

7. $(6, -3)$, $(6, 5)$

8. $(1, 4)$, $(8, 4)$

9. $(-3, -1)$, $(2, -1)$

10. $(-3, -4)$, $(-3, 6)$

11. Find two points in Quadrant I that are two units apart.

12. Find two points in Quadrant II that are three units apart.

In Exercises 13–16, find the length of the hypotenuse in two ways: (a) use the Pythagorean Theorem and (b) use the Distance Formula.

13.

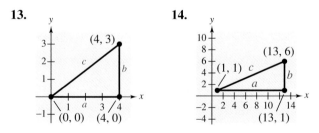

14.

15.

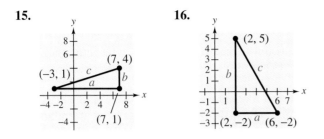

16.

In Exercises 17–28, (a) plot the points, (b) find the distance between the points, and (c) find the midpoint of the line segment joining the points.

17. $(1, 1), (9, 7)$

18. $(1, 12), (6, 0)$

19. $(-4, 10), (4, -5)$

20. $(-7, -4), (2, 8)$

21. $(-1, 2), (5, 4)$

22. $(2, 10), (10, 2)$

23. $\left(\frac{1}{2}, 1\right), \left(-\frac{5}{2}, \frac{4}{3}\right)$

24. $\left(-\frac{1}{3}, -\frac{1}{3}\right), \left(-\frac{1}{6}, -\frac{1}{2}\right)$

25. $(6.2, 5.4), (-3.7, 1.8)$

26. $(-16.8, 12.3), (5.6, 4.9)$

27. $(-36, -18), (48, -72)$

28. $(1.451, 3.051), (5.906, 11.360)$

In Exercises 29 and 30, use the Midpoint Formula to estimate the sales for 1993.

29.

Year	1991	1995
Sales	$520,000	$740,000

30.

Year	1991	1995
Sales	$4,200,000	$5,650,000

In Exercises 31–34, show that the points form the vertices of the indicated polygon.

31. Right triangle: $(4, 0), (2, 1), (-1, -5)$

32. Isosceles triangle: $(1, -3), (3, 2), (-2, 4)$

33. Rhombus: $(0, 0), (1, 2), (2, 1), (3, 3)$

34. Parallelogram: $(0, 1), (3, 7), (4, 4), (1, -2)$

In Exercises 35 and 36, find x so that the distance between the points is 13.

35. $(1, 2), (x, -10)$

36. $(-8, 0), (x, 5)$

In Exercises 37 and 38, find y so that the distance between the points is 17.

37. $(0, 0), (8, y)$

38. $(-8, 4), (7, y)$

In Exercises 39–42, state the quadrant in which (x, y) lies.

39. $x > 0$ and $y < 0$

40. $x < 0$ and $y < 0$

41. $x > 0$ and $y > 0$

42. $x < 0$ and $y > 0$

43. *Football Pass* In a football game, a quarterback throws a pass from the 15-yard line, 10 yards from the sideline. The pass is caught on the 40-yard line, 45 yards from the same sideline. How long was the pass? (Assume the 15-yard line and 40-yard line are on the same side of midfield.)

44. *Flying Distance* A plane flies in a straight line to a city that is 100 miles east and 150 miles north of the point where the plane began. How far does it fly?

Average Life Expectancy In Exercises 45 and 46, use the figure, which shows the average life expectancy for Americans from 1920 to 1995. (Source: U.S. National Center for Health Statistics)

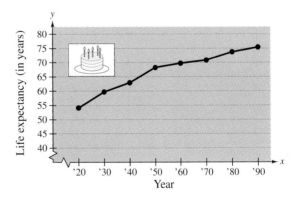

45. Estimate the average life expectancy in 1920.

46. Estimate the average life expectancy in 1990.

Gold Prices   In Exercises 47 and 48, use the figure, which shows the average price of gold from 1975 to 1992.   (Source: U.S. Bureau of Mines)

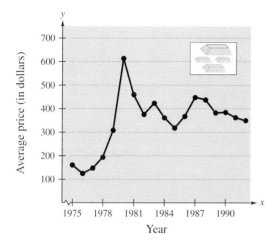

47. What is the highest price of gold shown in the graph? When did this occur?

48. What is the lowest price of gold shown in the graph? When did this occur?

Fuel Efficiency   In Exercises 49 and 50, use the figure, which shows the average fuel efficiency for automobiles in the United States from 1970 to 1991. (Source: U.S. Federal Highway Administration)

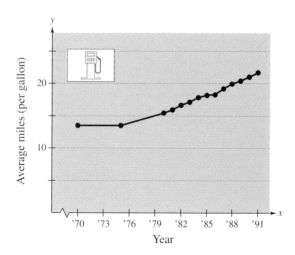

49. Estimate the percent increase in fuel efficiency from 1975 to 1988.

50. Estimate the percent increase in fuel efficiency from 1970 to 1991.

Football Attendance   In Exercises 51 and 52, use the figure, which shows the average paid attendance at NFL football games for the years 1940, 1950, 1960, 1970, 1980, and 1990.   (Source: National Football League)

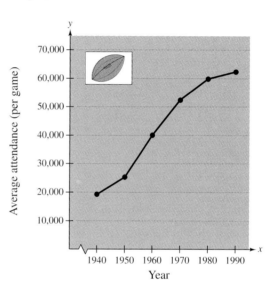

51. Estimate the increase in attendance from 1940 to 1960.

52. Estimate the increase in attendance from 1960 to 1990.

2.2 | Graphs of Equations

The Graph of an Equation ▪ Intercepts of a Graph ▪ Symmetry ▪
The Equation of a Circle

The Graph of an Equation

News magazines often show graphs comparing the rate of inflation, the federal deficit, wholesale prices, or the unemployment rate with the time of year. Industrial firms and businesses use graphs to report their monthly production and sales statistics. Such graphs provide simple geometric pictures of the ways in which certain quantities change with respect to other quantities.

Frequently, the relationship between two quantities is expressed in the form of an equation. In this section, you will study a basic procedure for sketching the graph of an equation.

For an equation in variables x and y, a point (a, b) is a **solution point** if the substitution of $x = a$ and $y = b$ satisfies the equation. Most equations have *infinitely* many solution points. For example, the equation

$$3x + y = 5$$

has solution points $(0, 5)$, $(1, 2)$, $(2, -1)$, $(3, -4)$, and so on. The set of all solution points of an equation is its **graph.**

EXAMPLE 1   *Sketching the Graph of an Equation*

Sketch the graph of $3x + y = 5$.

Solution

Begin by rewriting the equation with y isolated on the left.

$$y = 5 - 3x$$

Next, construct a table of values by choosing several values of x and calculating the corresponding values of y.

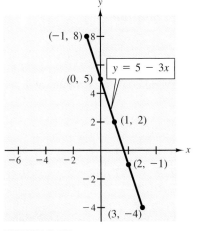

x	-1	0	1	2	3
$y = 5 - 3x$	8	5	2	-1	-4

FIGURE 2.15

Finally, plot the points given in the table and connect them as shown in Figure 2.15. It appears that the graph is a straight line. (You will study lines extensively in Section 2.4.)

The Point-Plotting Method of Graphing

1. If possible, isolate one of the variables.

2. Construct a table of several solution points.

3. Plot these points in the coordinate plane.

4. Connect the points with a smooth curve.

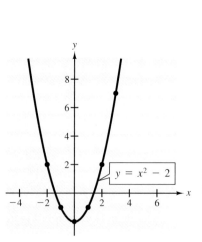

FIGURE 2.16

Step 4 of the point-plotting method can be difficult. For instance, how would you connect the four points in Figure 2.16? Without further information about the equation, any one of the three graphs in Figure 2.17 would be reasonable. These graphs show that with too few solution points, you can grossly misrepresent the graph of an equation. Throughout this course, you will study many ways to improve your graphing techniques. For now, we suggest that you plot enough points to reveal the essential behavior of the graph.

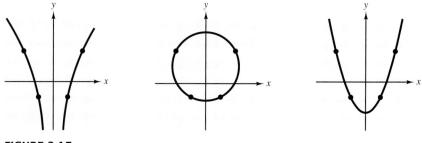

FIGURE 2.17

STUDY TIP

When constructing a table, use negative, zero, and positive values for x.

EXAMPLE 2 Sketching the Graph of an Equation

Sketch the graph of $y = x^2 - 2$.

Solution

First, construct a table of values by choosing several convenient values of x and calculating the corresponding values of y.

FIGURE 2.18

x	-2	-1	0	1	2	3
$y = x^2 - 2$	2	-1	-2	-1	2	7

Next, plot the corresponding solution points. Finally, connect the points with a smooth curve, as shown in Figure 2.18.

Intercepts of a Graph

When you are sketching a graph, two types of points that are especially useful are those for which either the y-coordinate or the x-coordinate is zero.

Definition of Intercepts

1. The **x-intercepts** of a graph are the points at which the graph intersects the x-axis. To find the x-intercepts, let y be zero and solve for x.
2. The **y-intercepts** of a graph are the points at which the graph intersects the y-axis. To find the y-intercepts, let x be zero and solve for y.

Some graphs have no intercepts and some have several. For instance, consider the three graphs in Figure 2.19.

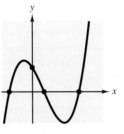

Three x-intercepts
One y-intercept
FIGURE 2.19

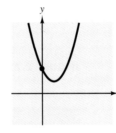

No x-intercept
One y-intercept

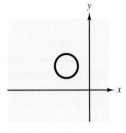

No intercepts

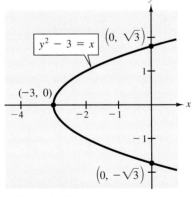

FIGURE 2.20

EXAMPLE 3   Finding x- and y-Intercepts

Find the x- and y-intercepts of the graph of $y^2 - 3 = x$.

Solution

To find the x-intercept, let $y = 0$. This produces $-3 = x$, which implies that the graph has one x-intercept, which occurs at

$$(-3, 0). \qquad\qquad x\text{-intercept}$$

To find the y-intercept, let $x = 0$. This produces $y^2 - 3 = 0$, which has two solutions: $y = \pm\sqrt{3}$. Thus, the equation has two y-intercepts, which occur at

$$\left(0, \sqrt{3}\right) \quad \text{and} \quad \left(0, -\sqrt{3}\right). \qquad y\text{-intercepts}$$

See Figure 2.20.

Symmetry

Symmetry with respect to the x-axis means that if the Cartesian plane were folded along the x-axis, the portion of the graph above the x-axis would coincide with the portion below the x-axis. Symmetry with respect to the y-axis or the origin can be described in a similar manner. Symmetry with respect to the origin means that if the Cartesian plane were rotated $180°$ about the origin, the portion of the graph to the right of the origin would coincide with the portion to the left of the origin. (See Figure 2.21.)

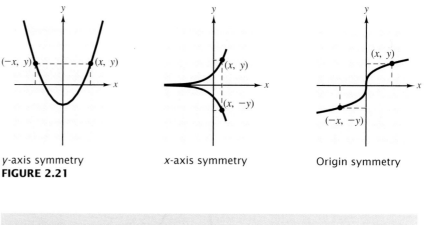

y-axis symmetry x-axis symmetry   Origin symmetry
FIGURE 2.21

Definition of Symmetry

1. A graph is **symmetric with respect to the y-axis** if, whenever (x, y) is on the graph, $(-x, y)$ is also on the graph.

2. A graph is **symmetric with respect to the x-axis** if, whenever (x, y) is on the graph, $(x, -y)$ is also on the graph.

3. A graph is **symmetric with respect to the origin** if, whenever (x, y) is on the graph, $(-x, -y)$ is also on the graph.

Suppose you apply this definition of symmetry to the graph of the equation $y = x^2 - 1$. Replacing x with $-x$ produces the following.

$y = x^2 - 1$	Given equation
$y = (-x)^2 - 1$	Replace x with $-x$.
$y = x^2 - 1$	Replacement yields equivalent equation.

Because the substitution did not change the equation, it follows that if (x, y) is a solution of the equation, then $(-x, y)$ must also be a solution. Thus, the graph of $y = x^2 - 1$ is symmetric with respect to the y-axis, as shown in Figure 2.22.

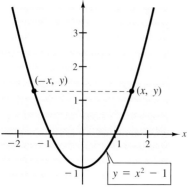

FIGURE 2.22 y-Axis Symmetry

Tests for Symmetry

1. The graph of an equation is symmetric with respect to the *y-axis* if replacing x with $-x$ yields an equivalent equation.
2. The graph of an equation is symmetric with respect to the *x-axis* if replacing y with $-y$ yields an equivalent equation.
3. The graph of an equation is symmetric with respect to the *origin* if replacing x with $-x$ *and* y with $-y$ yields an equivalent equation.

EXAMPLE 4 *Using Symmetry as a Sketching Aid*

Describe the symmetry of the graph of $x - y^2 = 1$.

Solution

Of the three tests for symmetry, the only one that is satisfied by this equation is the test for x-axis symmetry.

$x - y^2 = 1$	Original equation
$x - (-y)^2 = 1$	Replace y with $-y$.
$x - y^2 = 1$	Replacement yields equivalent equation.

Thus, the graph is symmetric with respect to the x-axis. To sketch the graph, plot the points above the x-axis and use symmetry to complete the graph, as shown in Figure 2.23.

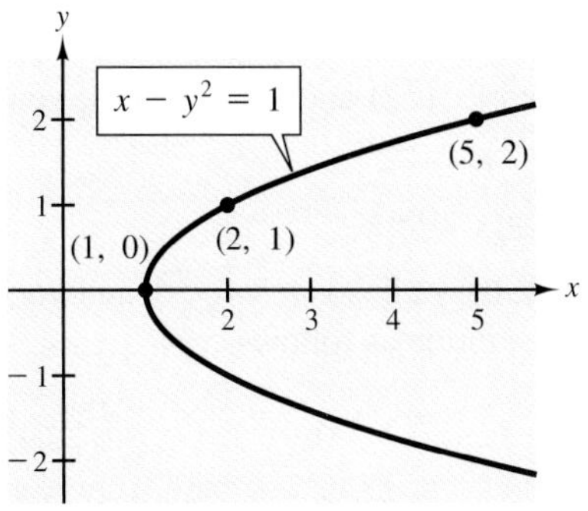

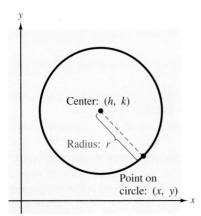

FIGURE 2.23

The Equation of a Circle

So far in this section you have studied the point-plotting method and two additional concepts (intercepts and symmetry) that can be used to streamline the graphing procedure. Another graphing aid is *equation recognition*, which is the ability to recognize the general shape of a graph simply by looking at its equation. A circle is one type of graph that is easily recognized.

Figure 2.24 shows a circle of radius r with center at the point (h, k). The point (x, y) is on this circle if and only if its distance from the center (h, k) is r. This means that a **circle** in the plane consists of all points (x, y) that are a given positive distance r from a fixed point (h, k). Using the Distance Formula, you can conclude that the point (x, y) lies on the circle if and only if

$$\sqrt{(x - h)^2 + (y - k)^2} = r.$$

By squaring both sides of this equation, you obtain the **standard form of the equation of a circle.**

FIGURE 2.24

NOTE The standard form of the equation of a circle whose center is the *origin* is simply

$$x^2 + y^2 = r^2.$$

Finding the correct h and k may be difficult for students. Rewriting the quantities in the form $(x- \ )^2$ and $(y- \ )^2$ can be helpful.

$$(x+1)^2 = [x-(-1)]^2, \quad h = -1$$
$$(y-2)^2 = [y-(2)]^2, \quad k = 2$$

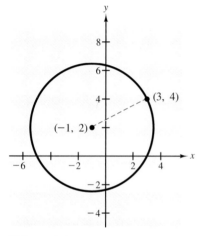

FIGURE 2.25

Standard Form of the Equation of a Circle

The **standard form of the equation of a circle** is

$$(x-h)^2 + (y-k)^2 = r^2.$$

The point (h, k) is called the **center** of the circle and the positive number r is called the **radius** of the circle.

EXAMPLE 5 *Finding an Equation of a Circle*

The point $(3, 4)$ lies on a circle whose center is at $(-1, 2)$, as shown in Figure 2.25. Find an equation for the circle.

Solution

The radius r of the circle is the distance between $(-1, 2)$ and $(3, 4)$. Thus, you have

$$r = \sqrt{[3-(-1)]^2 + (4-2)^2} = \sqrt{16+4} = \sqrt{20}.$$

Thus, the center of the circle is $(h, k) = (-1, 2)$, the radius is $r = \sqrt{20}$, and you can write the standard form of the equation of the circle as follows.

$$(x-h)^2 + (y-k)^2 = r^2 \qquad \text{Standard form}$$
$$[x-(-1)]^2 + (y-2)^2 = \left(\sqrt{20}\right)^2 \qquad \text{Let } h = -1, k = 2, \text{ and } r = \sqrt{20}.$$
$$(x+1)^2 + (y-2)^2 = 20 \qquad \text{Equation of circle}$$

If you remove the parentheses in the standard equation in Example 5, you obtain the following.

$$(x+1)^2 + (y-2)^2 = 20 \qquad \text{Standard form}$$
$$x^2 + 2x + 1 + y^2 - 4y + 4 = 20 \qquad \text{Expand terms.}$$
$$x^2 + y^2 + 2x - 4y - 15 = 0 \qquad \text{General form}$$

The last equation is in the **general form of the equation of a circle.**

$$Ax^2 + Ay^2 + Dx + Ey + F = 0, \qquad A \neq 0$$

The general form of the equation of a circle is less useful than the standard form. For instance, it is not immediately apparent from the general equation shown above that the center is $(-1, 2)$ and the radius is $\sqrt{20}$. To graph the equation of a circle, it is best to write the equation in standard form. You can do this by **completing the square,** as demonstrated in Example 6.

Activities

1. Write the equation of the circle, with center $(-1, 2)$ and radius 3, in general form.
   Answer: $x^2 + y^2 + 2x - 4y - 4 = 0$

2. Find the intercepts:
   $y = 2x^2 - 3x - 2$.
   Answer: $(0, -2), (2, 0), (-0.5, 0)$

3. Identify any symmetry:
   $y = \sqrt{4 - x^2}$.
   Answer: y-axis symmetry

EXAMPLE 6   Completing the Square to Sketch a Circle

Describe the circle given by $4x^2 + 4y^2 + 20x - 16y + 37 = 0$.

Solution

Begin by writing the given equation in standard form by completing the square for both the x-terms *and* the y-terms.

$$4x^2 + 4y^2 + 20x - 16y + 37 = 0 \qquad \text{General form}$$

$$x^2 + y^2 + 5x - 4y + \frac{37}{4} = 0 \qquad \text{Divide by 4.}$$

$$(x^2 + 5x +) + (y^2 - 4y +) = -\frac{37}{4} \qquad \text{Group terms.}$$

$$\left[x^2 + 5x + \left(\frac{5}{2}\right)^2\right] + (y^2 - 4y + 2^2) = -\frac{37}{4} + \frac{25}{4} + 4 \qquad \text{Complete square.}$$

$$\left(x + \frac{5}{2}\right)^2 + (y - 2)^2 = 1 \qquad \text{Standard form}$$

Thus, the center of the circle is $\left(-\frac{5}{2}, 2\right)$ and the radius of the circle is 1. Using this information, you can sketch the circle shown in Figure 2.26.

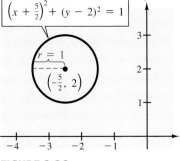

$\left(x + \frac{5}{2}\right)^2 + (y - 2)^2 = 1$

$r = 1$

$\left(-\frac{5}{2}, 2\right)$

FIGURE 2.26

Group Activities   Extending the Concept

Interpreting an Intercept   For the graph of each of the following real-life situations, find the indicated intercept and explain its practical significance.

a. *(y-intercept)* The trade-in value y of a used car x years after its purchase is $y = 9500 - 2400x$.

b. *(x-intercept)* The number of families y demanding child care services each week is related to the price x per hour of care by $y = 7850 - 20x^2$, $x \geq 0$.

Can you think of a real-life situation in which it makes sense for there to be no x-intercept?

Warm Up   The following warm-up exercises involve skills that were covered in earlier sections. You will use these skills in the exercise set for this section.

In Exercises 1 and 2, solve for y in terms of x.

1. $3x - 5y = 2$ **2.** $x^2 - 4x + 2y - 5 = 0$

In Exercises 3–6, solve the equation.

3. $x^2 - 4x + 4 = 0$ **4.** $(x - 1)(x + 5) = 0$

5. $x^3 - 9x = 0$ **6.** $x^4 - 8x^2 + 16 = 0$

In Exercises 7–10, simplify the equation.

7. $-y = (-x)^3 + 4(-x)$ **8.** $(-x)^2 + (-y)^2 = 4$

9. $y = 4(-x)^2 + 8$ **10.** $(-y)^2 = 3(-x) + 4$

2.2 Exercises

In Exercises 1–6, determine whether the points lie on the graph of the equation.

Equation		*Points*
1. $y = \sqrt{x + 4}$	(a) $(0, 2)$	(b) $(5, 3)$
2. $y = x^2 - 3x + 2$	(a) $(2, 0)$	(b) $(-2, 8)$
3. $2x - y - 3 = 0$	(a) $(1, 2)$	(b) $(1, -1)$
4. $x^2 + y^2 = 20$	(a) $(3, -2)$	(b) $(-4, 2)$
5. $x^2 y - x^2 + 4y = 0$	(a) $\left(1, \frac{1}{5}\right)$	(b) $\left(2, \frac{1}{2}\right)$
6. $y = \dfrac{1}{x^2 + 1}$	(a) $(0, 0)$	(b) $(3, 0.1)$

In Exercises 7–14, find the intercepts of the graph.

7. $y = x - 5$ **8.** $y = (x - 1)(x - 3)$

9. $y = x^2 + x - 2$ **10.** $y = 4 - x^2$

11. $y = x\sqrt{x + 2}$ **12.** $xy = 4$

13. $xy - 2y - x + 1 = 0$ **14.** $x^2 y - x^2 + 4y = 0$

In Exercises 15–22, check for symmetry.

15. $x^2 - y = 0$ **16.** $xy^2 + 10 = 0$

17. $x - y^2 = 0$ **18.** $y = \sqrt{9 - x^2}$

19. $y = x^3$ **20.** $xy = 4$

21. $y = \dfrac{x}{x^2 + 1}$ **22.** $y = x^4 - x^2 + 3$

In Exercises 23–26, use symmetry to complete the graph.

23. y-axis symmetry
$y = -x^2 + 4$

24. x-axis symmetry
$y^2 = -x + 4$

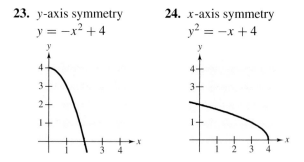

25. Origin symmetry

$y = -x^3 + x$

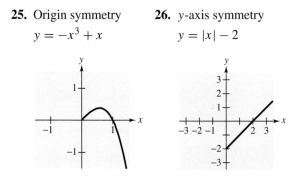

26. y-axis symmetry

$y = |x| - 2$

In Exercises 27–32, match the equation with its graph. [The graphs are labeled (a), (b), (c), (d), (e), and (f).]

(a)

(b)

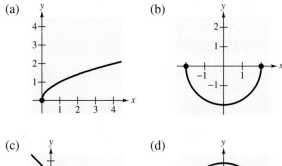

(c)

(d)

(e)

(f)

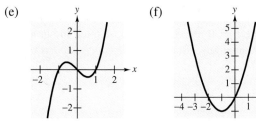

27. $y = 4 - x$

28. $y = x^2 + 2x$

29. $y = \sqrt{4 - x^2}$

30. $y = \sqrt{x}$

31. $y = x^3 - x$

32. $y = -\sqrt{4 - x^2}$

In Exercises 33–52, sketch the graph of the equation. Identify any intercepts and test for symmetry.

33. $y = -3x + 2$

34. $y = 2x - 3$

35. $y = 1 - x^2$

36. $y = x^2 - 1$

37. $y = x^2 - 4x + 3$

38. $y = -x^2 - 4x$

39. $y = x^3 + 2$

40. $y = x^3 - 1$

41. $y = x(x - 2)^2$

42. $y = \dfrac{4}{x^2 + 1}$

43. $y = \sqrt{x - 3}$

44. $y = \sqrt{1 - x}$

45. $y = \sqrt[3]{x}$

46. $y = \sqrt[3]{x + 1}$

47. $y = |x - 2|$

48. $y = 4 - |x|$

49. $x = y^2 - 1$

50. $x = y^2 - 4$

51. $x^2 + y^2 = 4$

52. $x^2 + y^2 = 16$

In Exercises 53–60, find the standard form of the equation of the specified circle.

53. Center: $(0, 0)$; radius: 3

54. Center: $(0, 0)$; radius: 5

55. Center: $(2, -1)$; radius: 4

56. Center: $\left(0, \frac{1}{3}\right)$; radius: $\frac{1}{3}$

57. Center: $(-1, 2)$; solution point: $(0, 0)$

58. Center: $(3, -2)$; solution point: $(-1, 1)$

59. Endpoints of a diameter: $(0, 0)$, $(6, 8)$

60. Endpoints of a diameter: $(-4, -1)$, $(4, 1)$

In Exercises 61–68, write the equation of the circle in standard form. Then sketch the circle.

61. $x^2 + y^2 - 2x + 6y + 6 = 0$

62. $x^2 + y^2 - 2x + 6y - 15 = 0$

63. $x^2 + y^2 - 2x + 6y + 10 = 0$

64. $3x^2 + 3y^2 - 6y - 1 = 0$

65. $2x^2 + 2y^2 - 2x - 2y - 3 = 0$

66. $4x^2 + 4y^2 - 4x + 2y - 1 = 0$

67. $16x^2 + 16y^2 + 16x + 40y - 7 = 0$

68. $x^2 + y^2 - 4x + 2y + 3 = 0$

In Exercises 69–72, find the constant C such that the ordered pair is a solution point of the equation.

69. $y = x^2 + C$ $(2, 6)$

70. $y = Cx^3$ $(-4, 8)$

71. $y = C\sqrt{x + 1}$ $(3, 8)$

72. $x + C(y + 2) = 0$ $(4, 3)$

In Exercises 73 and 74, an equation of a circle is written in standard form. Indicate the coordinates of the center of the circle and determine the radius of the circle. Rewrite the equation of the circle in general form.

73. $(x - 2)^2 + (y + 3)^2 = 16$

74. $\left(x + \frac{1}{2}\right)^2 + (y + 1)^2 = 5$

In Exercises 75 and 76, (a) sketch a graph comparing the data and the model for that data, and (b) use the model to predict y for the year 1998.

75. *Purchasing Power of the Dollar* The table gives the purchasing power of the dollar for *consumers* from 1986 to 1993. The base year for comparison is 1982, in which the purchasing power of the dollar for *producers* is $1.00. (Source: U.S. Bureau of Labor Statistics)

Year	1986	1987	1988	1989
Purchasing power	0.913	0.880	0.846	0.807

Year	1990	1991	1992	1993
Purchasing power	0.766	0.734	0.713	0.692

A mathematical model that approximates the purchasing power of the dollar is

$$y = 1.100 - 0.032t - \frac{0.016}{t}, \qquad 6 \le t \le 13$$

where y represents the purchasing power of the dollar and $t = 6$ represents 1986.

76. *Life Expectancy* The table gives the life expectancy of a child (at birth) for selected years from 1920 to 1990. (Source: Department of Health and Human Services)

Year	1920	1930	1940	1950
Life expectancy	54.1	59.7	62.9	68.2

Year	1960	1970	1980	1990
Life expectancy	69.7	70.8	73.7	75.4

A mathematical model for the life expectancy during this period is

$$y = \frac{66.715 + 1.007t}{0.996 + 0.011t}, \qquad -30 \le t \le 40$$

where y represents the life expectancy and $t = 0$ represents 1950.

77. *Earnings per Share* The earnings per share y (in dollars) for Bristol-Myers Squibb Corporation from 1984 to 1992 can be modeled by

$$y = 0.01t^2 + 0.128t + 1.03, \qquad 4 \le t \le 12$$

where $t = 0$ represents 1980. Sketch the graph of this equation. (Source: *NYSE Stock Reports*)

78. *Earnings per Share* The earnings per share y (in dollars) for Warner-Lambert Company from 1986 to 1993 can be modeled by

$$y = 0.478t - 1.195, \qquad 6 \le t \le 13$$

where $t = 0$ represents 1980. Sketch the graph of this equation. (Source: *NYSE Stock Reports*)

2.3 Graphing Calculators

Introduction ▪ Using a Graphing Calculator ▪
Using a Graphing Calculator's Special Features

Introduction

Calculators have interested mathematicians for hundreds of years. In 1642, at the age of 18, Blaise Pascal began designing a calculating machine. Over the next few years, Pascal built and sold approximately 50 machines.

In Section 2.2 you studied the point-plotting method for sketching the graph of an equation. One of the disadvantages of the point-plotting method is that to get a good idea about the shape of a graph you need to plot *many* points. By plotting only a few points, you can badly misrepresent the graph.

For instance, consider the equation $y = \frac{1}{30}x(x^4 - 10x^2 + 39)$. To graph this equation, suppose you calculated only the following five points.

x	-3	-1	0	1	3
$y = \frac{1}{30}x(x^4 - 10x^2 + 39)$	-3	-1	0	1	3

By plotting these five points, as shown in Figure 2.27(a), you might assume that the graph of the equation is a straight line. This, however, is not correct. By plotting several more points, as shown in Figure 2.27(b), you can see that the actual graph is not straight at all.

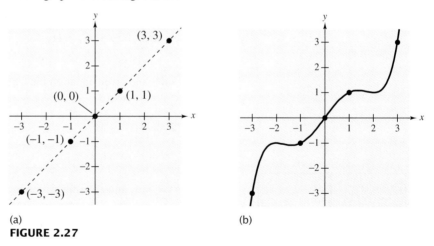

(a)   (b)

FIGURE 2.27

Thus, the point-plotting method leaves you with a dilemma. On the one hand, the method can be very inaccurate if only a few points are plotted. But, on the other hand, it is very time consuming to plot a dozen (or more) points. Technology can help you resolve this dilemma. Plotting several points (even hundreds of points) on a rectangular coordinate system is something that a computer or graphing calculator can do easily.

Using a Graphing Calculator

There are many different graphing utilities: some are graphing packages for computers and some are hand-held graphing calculators. In this section we describe only one—the *TI-82* by *Texas Instruments*. The keys for the *TI-82* are shown in Figure 2.28. The steps used to graph an equation with this calculator are summarized below. The steps used with other graphing calculators are similar.

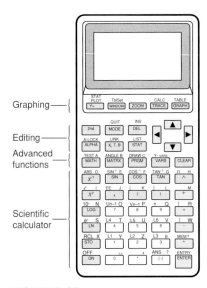

Graphing

Editing

Advanced
functions

Scientific
calculator

FIGURE 2.28

Graphing an Equation with a *TI-82* Graphing Calculator

Before performing the following steps, set your calculator so that all the standard defaults are active. For instance, all of the options at the left of the MODE screen should be highlighted.

1. Set the viewing window for the graph. (To set the standard viewing window, press ZOOM 6 .)
2. Rewrite the equation so that y is isolated on the left side of the equation.
3. Press the Y= key. Then enter the right side of the equation on the first line of the display. (The first line is labeled $Y_1 =$.)
4. Press the GRAPH key.

EXAMPLE 1   Graphing an Equation

Sketch the graph of $2y + x^3 = 4x$.

Solution

To begin, solve the given equation for y in terms of x.

$$2y + x^3 = 4x \qquad \text{Original equation}$$

$$2y = -x^3 + 4x \qquad \text{Subtract } x^3 \text{ from both sides.}$$

$$y = -\frac{1}{2}x^3 + 2x \qquad \text{Divide both sides by 2.}$$

Press the Y= key, and enter the following keystrokes.

(-) X,T,θ ∧ 3 ÷ 2 + 2 X,T,θ

The top row of the display should now be as follows.

$$Y_1 = -X\wedge 3/2 + 2X$$

Press the GRAPH key, and the screen should look like the one shown in Figure 2.29.

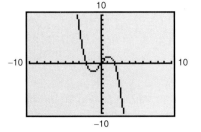

FIGURE 2.29

In Figure 2.30, notice that the calculator screen does not label the tick marks on the x-axis or the y-axis. To see what the tick marks represent, you can press WINDOW . If you set your calculator to the standard graphing defaults before working Example 1, the screen should show the following values.

Xmin=-10	The minimum x-value is -10.
Xmax=10	The maximum x-value is 10.
Xscl=1	The x-scale is one unit per tick mark.
Ymin=-10	The minimum y-value is -10.
Ymax=10	The maximum y-value is 10.
Yscl=1	The y-scale is one unit per tick mark.

These settings are summarized visually in Figure 2.30.

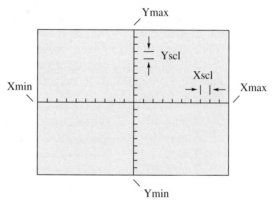

FIGURE 2.30

EXAMPLE 2 *Graphing of an Equation Involving Absolute Value*

Sketch the graph of $y = |x - 3|$.

Solution

This equation is already written with y isolated on the left side of the equation. Press the Y= key, and enter the following keystrokes.

The top row of the display should now be as follows.

$Y_1 = abs(X - 3)$

Press the GRAPH key, and the screen should look like the one shown in Figure 2.31.

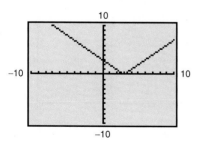

FIGURE 2.31

Using a Graphing Calculator's Special Features

To use your graphing calculator to its best advantage, you must learn how to set the viewing window, as illustrated in the next example.

EXAMPLE 3 Setting the Viewing Window

Sketch the graph of $y = x^2 + 12$.

Solution

Press Y= and enter $y = x^2 + 12$ on the first line.

X,T,θ x^2 + 12

Press the GRAPH key. If your calculator is set to the standard viewing window, nothing will appear on the screen. The reason for this is that the lowest point on the graph of $y = x^2 + 12$ occurs at the point (0, 12). Using the standard viewing window, you obtain a screen whose largest y-value is 10. In other words, none of the graph is visible on a screen whose y-values vary between -10 and 10, as shown in Figure 2.32(a). To change these settings, press WINDOW and enter the following values.

Xmin=-10	The minimum x-value is -10.
Xmax=10	The maximum x-value is 10.
Xscl=1	The x-scale is one unit per tick mark.
Ymin=-10	The minimum y-value is -10.
Ymax=30	The maximum y-value is 30.
Yscl=5	The y-scale is five units per tick mark.

Press GRAPH, and you will obtain the graph shown in Figure 2.32(b). On this graph, note that each tick mark on the y-axis represents five units because you changed the y-scale to 5. Also note that the highest point on the y-axis is now 30 because you changed the maximum value of y to 30.

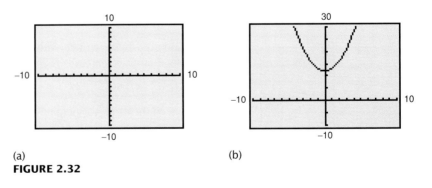

(a)   (b)

FIGURE 2.32

Point out to students how to clear a previously entered equation when they wish to enter a new equation with the $Y =$ key. On the *TI-82*, this is done by positioning the cursor on the line of the equation to be erased and pressing the CLEAR key.

You may want to point out that another use of the ZOOM key is to "zoom in," or enlarge a certain portion of the graph on the viewing window, and "zoom out," or extend the x- and y-axis ranges of the viewing window to see more of the whole graph.

NOTE If you changed the y-maximum and y-scale on your calculator as indicated in Example 3, you should return to the standard settings before working Example 4. To do this, press ZOOM 6 .

EXAMPLE 4 *Using a Square Setting*

Sketch the graph of $y = x$. The graph of this equation is a straight line that makes a 45° angle with the x-axis and y-axis. From the graph on your calculator, does the angle appear to be 45°?

Solution

Press $\boxed{Y=}$ and enter $y = x$ on the first line.

$$Y_1 = X$$

Press the $\boxed{\text{GRAPH}}$ key, and you will obtain the graph shown in Figure 2.33(a). Note that the angle the line makes with the x-axis doesn't appear to be 45°. The reason for this is that the screen is wider than it is tall. This has the effect of making the tick marks on the x-axis farther apart than the tick marks on the y-axis. To obtain the same distance between tick marks on both axes, you can change the graphing settings from "standard" to "square." To do this, press the following keys.

     Square setting

The screen should look like that shown in Figure 2.33(b). Note in this figure that the square setting has changed the viewing window so that the x-values vary between -15 and 15.

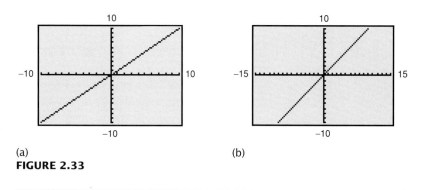

(a)   (b)

FIGURE 2.33

STUDY TIP

A square setting is useful anytime you want to sketch a graph that has a true geometric perspective. For instance, if you want the graph of $x^2 + y^2 = 25$ to look like a circle, you need to use a square setting. You also need to enter the top and bottom portions of the circle separately, as follows.

$$y = \sqrt{25 - x^2} \qquad \text{Top half}$$
$$y = -\sqrt{25 - x^2} \qquad \text{Bottom half}$$

NOTE There are many possible square settings on a graphing calculator. To create a square setting, you need the following ratio to be $\frac{2}{3}$.

$$\frac{\text{Ymax} - \text{Ymin}}{\text{Xmax} - \text{Xmin}}$$

For instance, the setting in Example 4 is square because $(\text{Ymax} - \text{Ymin}) = 20$ and $(\text{Xmax} - \text{Xmin}) = 30$.

EXAMPLE 5 *Sketching More than One Graph on the Same Screen*

Sketch the graphs of the following equations on the same screen.

$$y = -x + 5, \quad y = -x, \quad \text{and} \quad y = -x - 5$$

Solution

To begin, press $\boxed{Y=}$ and enter all three equations on the first three lines. The display should now be as follows.

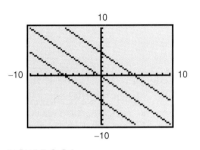

10

−10   10

−10

FIGURE 2.34

$Y_1 = \text{-X} + 5$ $\qquad$ $\boxed{(\text{-})}$ $\boxed{\text{X,T,}\theta}$ $\boxed{+}$ 5

$Y_2 = \text{-X}$ $\qquad$ $\boxed{(\text{-})}$ $\boxed{\text{X,T,}\theta}$

$Y_3 = \text{-X} - 5$ $\qquad$ $\boxed{(\text{-})}$ $\boxed{\text{X,T,}\theta}$ $\boxed{-}$ 5

Press the $\boxed{\text{GRAPH}}$ key and you will obtain the graph shown in Figure 2.34. Note that the graph of each equation is a straight line, and that the lines are parallel to each other.

Group Activities   E x p l o r i n g   w i t h   T e c h n o l o g y

Graphing For each of the following equations, describe what you must do before you can graph the equation using a graphing utility.

a. $-4y + 32 = 2x$ $\qquad$ **b.** $y = \frac{1}{3}(x^3 - 10x)$

c. $x^2 + 2y^2 = 16$ $\qquad$ **d.** $x^2 = y + 12x$

Now use a graphing utility and a standard viewing setting to graph each equation. Enlarge your view of the graph by zooming out or increasing the boundaries of your viewing window several times. For which equations are all of the x-intercepts shown in the standard viewing window?

Warm Up   The following warm-up exercises involve skills that were covered in earlier sections. You will use these skills in the exercise set for this section.

In Exercises 1–10, solve for y in terms of x.

1. $3x + y = 4$ **2.** $x - y = 0$

3. $2x + 3y = 2$ **4.** $4x - 5y = -2$

5. $3x + 4y - 5 = 0$ **6.** $-2x - 3y + 6 = 0$

7. $x^2 + y - 4 = 0$ **8.** $-2x^2 + 3y + 2 = 0$

9. $x^2 + y^2 = 4$ **10.** $x^2 - y^2 = 9$

2.3 Exercises

In Exercises 1–10, use a graphing utility to match the equation with its graph. [The graphs are labeled (a), (b), (c), (d), (e), (f), (g), (h), (i), and (j).]

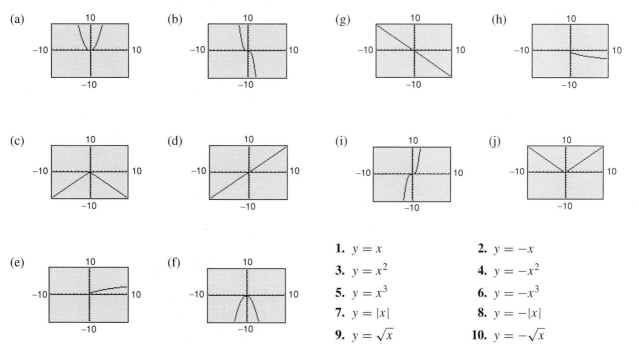

1. $y = x$ **2.** $y = -x$

3. $y = x^2$ **4.** $y = -x^2$

5. $y = x^3$ **6.** $y = -x^3$

7. $y = |x|$ **8.** $y = -|x|$

9. $y = \sqrt{x}$ **10.** $y = -\sqrt{x}$

In Exercises 11–30, use a graphing utility to graph the equation. Use a standard setting for each graph.

11. $y = x - 5$

12. $y = -x + 4$

13. $y = -\frac{1}{2}x + 3$

14. $y = \frac{2}{3}x + 1$

15. $2x - 3y = 4$

16. $x + 2y = 3$

17. $y = \frac{1}{2}x^2 - 1$

18. $y = -x^2 + 6$

19. $y = x^2 - 4x - 5$

20. $y = x^2 - 3x + 2$

21. $y = -x^2 + 2x + 1$

22. $y = -x^2 + 4x - 1$

23. $2y = x^2 + 2x - 3$

24. $3y = -x^2 - 4x + 5$

25. $y = |x + 5|$

26. $y = \frac{1}{2}|x - 6|$

27. $y = \sqrt{x^2 + 1}$

28. $y = 2\sqrt{x^2 + 2} - 4$

29. $y = \frac{1}{5}(-x^3 + 16x)$

30. $y = \frac{1}{8}(x^3 + 8x^2)$

In Exercises 31–34, use a graphing utility to graph the equation. Does the setting give a good representation of the graph? Explain.

31. $y = -2x^2 + 12x + 14$

32. $y = -x^2 + 5x + 6$

```
Xmin = -5
Xmax = 10
Xscl = 1
Ymin = -5
Ymax = 35
Yscl = 5
```

```
Xmin = -8
Xmax = 4
Xscl = 1
Ymin = -5
Ymax = 15
Yscl = 5
```

33. $y = x^3 + 6x^2$

34. $y = -x^3 + 16x$

```
Xmin = -10
Xmax = 5
Xscl = 1
Ymin = -4
Ymax = 36
Yscl = 4
```

```
Xmin = -6
Xmax = 6
Xscl = 1
Ymin = -25
Ymax = 25
Yscl = 5
```

In Exercises 35–38, find a setting on a graphing utility such that the graph of the equation agrees with the graph shown.

35. $y = -x^2 - 4x + 20$

36. $y = x^2 + 12x - 8$

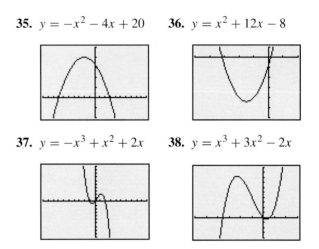

37. $y = -x^3 + x^2 + 2x$

38. $y = x^3 + 3x^2 - 2x$

In Exercises 39–42, use a graphing utility to find the number of x-intercepts of the equation.

39. $y = \frac{1}{8}(4x^2 - 32x + 65)$

40. $y = \frac{1}{4}(-4x^2 + 16x - 15)$

41. $y = 4x^3 - 20x^2 - 4x + 61$

42. $y = \frac{1}{4}(2x^3 + 6x^2 - 4x + 1)$

In Exercises 43–46, use a graphing utility to graph the equations on the same screen. Use a square setting that shows a good representation of the graphs. What geometrical shape is bounded by the graphs?

43. $y = |x| - 4$, $y = -|x| + 4$

44. $y = x + |x| - 4$, $y = x - |x| + 4$

45. $y = -\sqrt{25 - x^2}$, $y = \sqrt{25 - x^2}$

46. $y = 6$, $y = -\sqrt{3}x - 4$, $y = \sqrt{3}x - 4$

47. A mathematical model that approximates the purchasing power of the dollar is given by

$$y = 1.100 - 0.032t - \frac{0.016}{t}, \quad 5 \le t \le 13$$

where y represents the purchasing power of the dollar and $t = 5$ represents 1985.

(a) Use this model to find the purchasing power of the dollar in 1993.

(b) The purchasing power of the dollar in 1993 was 0.692. What does this tell you about the model?

48. A mathematical model for the life expectancy from 1920 to 1990 is

$$y = \frac{66.715 + 1.007t}{0.996 + 0.011t}, \quad -30 \le t \le 40$$

where y represents the life expectancy and $t = 0$ represents 1950.

(a) Using this model, predict the life expectancy in 2000 and in 2010.

(b) Are your answers reasonable? What cautions would you give when using a mathematical model for a time span that is greatly different than the period in which the data for the model was collected?

Ever Been Married?   In Exercises 49–52, use the following models, which relate ages to the percents y (in decimal form) of American males and females who have never been married.

$$y = \frac{47.07 - 0.7487x}{1 - 0.0738x + 0.00205x^2} \qquad \text{Males}$$

$$y = \frac{7.28 + 0.1864x}{1 - 0.09853x + 0.00275x^2} \qquad \text{Females}$$

In these models, x is the age of the person, and each model is valid for $20 \le x \le 50$.   (Source: U. S. Bureau of Census)

49. Use a graphing utility to graph both models on the same screen.

50. Write a paragraph describing the relationship between the two graphs that were plotted in Exercise 49.

51. Suppose that an American male is chosen at random from the population. If the person is 25 years old, what is the probability that he has never been married?

52. Suppose that an American female is chosen at random from the population. If the person is 25 years old, what is the probability that she has never been married?

Earnings and Dividends   In Exercises 53–56, use the following model, which approximates the relationship between dividends y (in dollars) per share and earnings x (in dollars) per share for the Pall Corporation between 1982 and 1989.   (Source: *Standard ASE Stock Reports*)

$$y = -0.166 + 0.502x - 0.0953x^2, \quad 0.25 \le x \le 2$$

53. Use a graphing utility to graph the model.

54. According to the model, what size dividend would the Pall Corporation pay if the earnings per share were $1.30?

55. Use the TRACE key on your graphing utility to estimate the earnings per share that would produce a dividend per share of $0.25.

(a) $1.00   (b) $1.03   (c) $1.06   (d) $1.09

56. The *payout ratio* for a stock is the ratio of the dividend per share to earnings per share. Use the model to find the payout ratio for an earnings per share of

(a) $0.75   (b) $1.00   (c) $1.25.

MID-CHAPTER QUIZ

Take this quiz as you would take a quiz in class. After you are done, check your work against the answers given in the back of the book.

In Exercises 1 and 2, (a) plot the points, (b) find the distance between the points, and (c) find the midpoint of the line segment joining the points.

1. $(-3, 2), (4, -5)$
2. $(1.3, -4.5), (-3.7, 0.7)$

3. A business had sales of \$460,000 in 1991 and \$580,000 in 1995. Estimate the sales in 1993. Explain your reasoning.

4. One plane is 200 miles due north of a transmitter and another plane is 150 miles due east of the transmitter. What is the distance between the planes?

In Exercises 5 and 6, describe the figure with the given vertices.

5. $(-1, -1), (10, 7), (2, 18)$
6. $(-3, -1), (0, 2), (4, 2), (1, -1)$

In Exercises 7–12, sketch the graph of the equation. Identify any intercepts and symmetry.

7. $y = 9 - x^2$
8. $y = x\sqrt{x + 4}$
9. $xy = 9$
10. $y = \sqrt{16 - x^2}$
11. $x^2 + y^2 = 9$
12. $y = |x - 3|$

In Exercises 13 and 14, find the standard form of the equation of the circle.

13. Center: $(2, -3)$; radius: 4
14. Center: $\left(0, -\frac{1}{2}\right)$; radius: 2

15. Write the equation $x^2 + y^2 - 2x + 4y - 4 = 0$ in standard form. Then sketch the circle.

In Exercises 16 and 17, use a graphing utility to find the number of x-intercepts of the equation.

16. $y = \frac{1}{4}(-4x^2 + 16x - 14)$
17. $y = \frac{1}{8}(4x^2 - 32x + 63)$

In Exercises 18–20, use the following model. The earnings per share y (in dollars) for Warner-Lambert Company from 1986 to 1993 can be modeled by $y = 0.478t - 1.195$, $6 \leq t \leq 13$, where $t = 0$ represents 1980. (Source: *Value Line*)

18. Determine a convenient graphing utility viewing window.

19. Use a graphing utility to graph this model.

20. Use the TRACE feature to predict the earnings per share in 1995.

2.4 Lines in the Plane

The Slope of a Line ▪ The Point-Slope Form ▪
Sketching Graphs of Lines ▪ Parallel and Perpendicular Lines

The Slope of a Line

The **slope** of a nonvertical line is a measure of the steepness of the line. Specifically, the slope represents the number of units the line rises or falls vertically for each unit of horizontal change from left to right. For instance, consider the two points (x_1, y_1) and (x_2, y_2) on the line shown in Figure 2.35. As you move from left to right along this line, a change of $y_2 - y_1$ units in the vertical direction corresponds to a change of $x_2 - x_1$ units in the horizontal direction. That is,

$$y_2 - y_1 = \text{the change in } y$$

and

$$x_2 - x_1 = \text{the change in } x.$$

The slope of the line is defined as the quotient of these two changes.

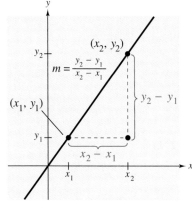

FIGURE 2.35

Definition of the Slope of a Line

The **slope** m of the nonvertical line passing through the points (x_1, y_1) and (x_2, y_2) is

$$m = \frac{y_2 - y_1}{x_2 - x_1} = \frac{\text{change in } y}{\text{change in } x}$$

where $x_1 \neq x_2$.

NOTE The change in x is sometimes called the "run" and the change in y is sometimes called the "rise."

When this formula is used, the *order of subtraction* is important. Given two points on a line, you are free to label either one of them as (x_1, y_1) and the other as (x_2, y_2). However, once this is done, you must form the numerator and denominator using the same order of subtraction.

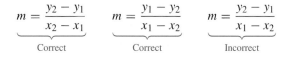

$$m = \frac{y_2 - y_1}{x_2 - x_1} \qquad m = \frac{y_1 - y_2}{x_1 - x_2} \qquad m = \frac{y_2 - y_1}{x_1 - x_2}$$

Correct   Correct   Incorrect

When working with the slope of a line, it helps to remember the following interpretations of slope.

1. A line with positive slope *rises* from left to right.

2. A line with negative slope *falls* from left to right.

3. A line with zero slope is *horizontal*.

4. A line with undefined slope is *vertical*.

EXAMPLE 1 Finding the Slope of a Line Through Two Points

Find the slopes of the lines passing through the following pairs of points.

a. $(-2, 0)$ and $(3, 1)$ **b.** $(-1, 2)$ and $(2, 2)$ **c.** $(0, 4)$ and $(1, -1)$

Solution

a. Let $(x_1, y_1) = (-2, 0)$ and $(x_2, y_2) = (3, 1)$.

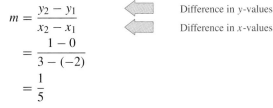

$$m = \frac{y_2 - y_1}{x_2 - x_1} \qquad \text{Difference in } y\text{-values}$$
$$\text{Difference in } x\text{-values}$$
$$= \frac{1 - 0}{3 - (-2)}$$
$$= \frac{1}{5}$$

b. The slope of the line through $(-1, 2)$ and $(2, 2)$ is

$$m = \frac{2 - 2}{2 - (-1)} = \frac{0}{3} = 0.$$

c. The slope of the line through $(0, 4)$ and $(1, -1)$ is

$$m = \frac{-1 - 4}{1 - 0} = \frac{-5}{1} = -5.$$

The graphs of the three lines are shown in Figure 2.36.

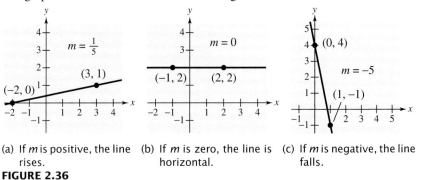

(a) If *m* is positive, the line rises.

(b) If *m* is zero, the line is horizontal.

(c) If *m* is negative, the line falls.

FIGURE 2.36

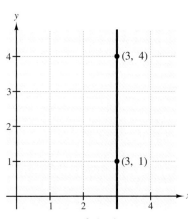

FIGURE 2.37 If the line is vertical, the slope is undefined.

The definition of slope does not apply to vertical lines. For instance, consider the points $(3, 4)$ and $(3, 1)$ on the vertical line shown in Figure 2.37. Applying the formula for slope, you have

$$\frac{1 - 4}{3 - 3} = \frac{-3}{0}. \qquad \text{Undefined division by zero}$$

Because division by zero is not defined, the slope of a vertical line is not defined.

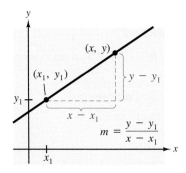

FIGURE 2.38 Any two points on a line can be used to determine the slope of the line.

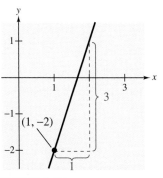

FIGURE 2.39

PROGRAMMING

In the Appendix, you will find programs for a variety of calculator models that use the two-point form to find an equation of a line. After entering the coordinates of two points, the program outputs the slope and y-intercept of the line that passes through the points.

The Point-Slope Form

If you know the slope of a line and you also know the coordinates of one point on the line, you can find an equation for the line. For instance, in Figure 2.38, let (x_1, y_1) be a given point on the line whose slope is m. If (x, y) is any *other* point on the line, it follows that

$$\frac{y - y_1}{x - x_1} = m.$$

This equation in variables x and y can be rewritten to produce the following **point-slope form** of the equation of a line.

Point-Slope Form of the Equation of a Line

The **point-slope form** of the equation of the line that passes through the point (x_1, y_1) and has a slope of m is

$$y - y_1 = m(x - x_1).$$

EXAMPLE 2   *The Point-Slope Form of the Equation of a Line*

Find an equation of the line that passes through $(1, -2)$ and has a slope of 3.

Solution

Use the point-slope form with $(x_1, y_1) = (1, -2)$ and $m = 3$.

$$y - y_1 = m(x - x_1) \qquad \text{Point-slope form}$$
$$y - (-2) = 3(x - 1) \qquad \text{Substitute } y_1 = -2, \ x_1 = 1, \text{ and } m = 3.$$
$$y + 2 = 3x - 3 \qquad \text{Simplify.}$$
$$y = 3x - 5 \qquad \text{Equation of line}$$

The graph of this line is shown in Figure 2.39.

The point-slope form can be used to find the equation of a line passing through two points (x_1, y_1) and (x_2, y_2). First, use the formula for the slope of the line passing through two points. Then, use the point-slope form to obtain the equation

$$y - y_1 = \frac{y_2 - y_1}{x_2 - x_1}(x - x_1).$$

This is sometimes called the **two-point form** of the equation of a line.

EXAMPLE 3   A Linear Model for Sales Prediction

During the first two quarters of the year, a company had sales of $3.4 million and $3.7 million, respectively.

a. Write a linear equation giving the sales y in terms of the quarter x.

b. Use the equation to predict the sales during the fourth quarter.

Solution

a. In Figure 2.40 let (1, 3.4) and (2, 3.7) be two points on the line representing the total sales. The slope of the line passing through these two points is

$$m = \frac{3.7 - 3.4}{2 - 1} = 0.3.$$

By the point-slope form, the equation of the line is as follows.

$y - y_1 = m(x - x_1)$	Point-slope form
$y - 3.4 = 0.3(x - 1)$	Substitute for y_1, m, and x_1.
$y = 0.3x - 0.3 + 3.4$	Simplify.
$y = 0.3x + 3.1$	Equation of line

b. Using the equation from part (a), the fourth quarter sales ($x = 4$) should be

$$y = 0.3(4) + 3.1 = 4.3 \text{ million dollars.}$$

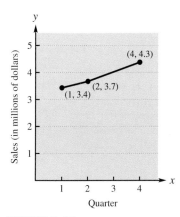

FIGURE 2.40

The estimation method illustrated in Example 3 is called **linear extrapolation.** Note in Figure 2.41(a), that for linear extrapolation, the estimated point lies to the right of the given points. When the estimated point lies *between* two given points, the procedure is called **linear interpolation,** as shown in Figure 2.41(b).

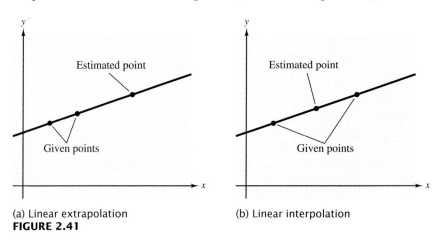

(a) Linear extrapolation

(b) Linear interpolation

FIGURE 2.41

Sketching Graphs of Lines

You have seen that to *find the equation of a line* it is convenient to use the point-slope form. This formula, however, is not particularly useful for *sketching the graph of a line*. The form that is better suited to graphing linear equations is the **slope-intercept form** of the equation of a line. To derive the slope-intercept form, write the following.

$$y - y_1 = m(x - x_1) \qquad \text{Point-slope form}$$

$$y = mx - mx_1 + y_1 \qquad \text{Solve for } y.$$

$$y = mx + b \qquad \text{Slope-intercept form}$$

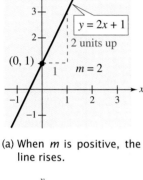

(a) When m is positive, the line rises.

Slope-Intercept Form of the Equation of a Line

The graph of the equation

$$y = mx + b$$

is a line whose slope is m and whose y-intercept is $(0, b)$.

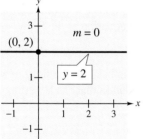

(b) When m is zero, the line is horizontal.

EXAMPLE 4 *Sketching the Graph of a Linear Equation*

Sketch the graph of each of the following linear equations.

a. $y = 2x + 1$ **b.** $y = 2$ **c.** $x + y = 2$

Solution

a. Because $b = 1$, the y-intercept is $(0, 1)$. Moreover, because the slope is $m = 2$, this line *rises* two units for each unit the line moves to the right, as shown in Figure 2.42(a).

b. By writing the equation $y = 2$ in the form

$$y = (0)x + 2$$

you can see that the y-intercept is $(0, 2)$ and the slope is zero. A zero slope implies that the line is horizontal, as shown in Figure 2.42(b).

c. By writing the equation $x + y = 2$ in slope-intercept form,

$$y = -x + 2$$

you can see that the y-intercept is $(0, 2)$. Moreover, because the slope is $m = -1$, this line *falls* one unit for each unit the line moves to the right, as shown in Figure 2.42(c).

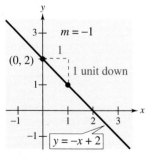

(c) When m is negative, the line falls.

FIGURE 2.42

Use a graphing utility to graph the following equations in the same window.

$y_1 = \frac{1}{4}x + 1$

$y_2 = x + 1$

$y_3 = -x + 1$

$y_4 = 3x + 1$

$y_5 = -3x + 1$

What effect does the negative x-coefficient have on the graph? What is the y-intercept of each graph?

From the slope-intercept form of the equation of a line, you can see that a horizontal line ($m = 0$) has an equation of the form

$y = (0)x + b$ or $y = b$.   Horizontal line

This is consistent with the fact that each point on a horizontal line through $(0, b)$ has a y-coordinate of b, as shown in Figure 2.43.

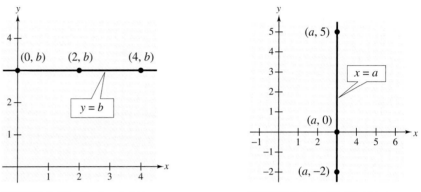

FIGURE 2.43 Horizontal Line   **FIGURE 2.44** Vertical Line

Each form of the equation of a line should be carefully explained to and learned by your students. Be sure students know and recognize the forms of vertical and horizontal lines, as that knowledge is essential when vertical and horizontal asymptotes are discussed in Section 3.7.

Discuss each form of the equation of a line and which is more efficient for different problem situations. For instance, the point-slope form of the line is more efficient for problems in which two points are given.

Similarly, each point on a vertical line through $(a, 0)$ has an x-coordinate of a, as shown in Figure 2.44. Hence, a vertical line has an equation of the form

$x = a$.   Vertical line

This equation cannot be written in the slope-intercept form because the slope of a vertical line is undefined. However, *every* line has an equation that can be written in the **general form**

$Ax + By + C = 0$ General form

where A and B are not *both* zero. If $A = 0$ (and $B \neq 0$), the equation can be reduced to the form $y = b$, which represents a horizontal line. If $B = 0$ (and $A \neq 0$), the general equation can be reduced to the form $x = a$, which represents a vertical line.

Notice that a fourth form of a line (the *two-intercept* or *intercept* form) is introduced in Exercises 49–52 of this section.

Summary of Equations of Lines

1. General form: $Ax + By + C = 0$

2. Vertical line: $x = a$

3. Horizontal line: $y = b$

4. Slope-intercept form: $y = mx + b$

5. Point-slope form: $y - y_1 = m(x - x_1)$

Parallel and Perpendicular Lines

The slope of a line is a convenient tool for determining whether two lines are parallel or perpendicular.

> ### Parallel Lines
>
> Two distinct nonvertical lines are **parallel** if and only if their slopes are equal.

EXAMPLE 5   Equations of Parallel Lines

Find an equation of the line that passes through the point $(2, -1)$ and is parallel to the line $2x - 3y = 5$, as shown in Figure 2.45.

Solution

Writing the given equation in slope-intercept form produces the following.

$$2x - 3y = 5 \qquad \text{Original equation}$$

$$3y = 2x - 5 \qquad \text{Isolate } y\text{-term.}$$

$$y = \frac{2}{3}x - \frac{5}{3} \qquad \text{Slope-intercept form}$$

Therefore, the given line has a slope of $m = \frac{2}{3}$. Because any line parallel to the given line must also have a slope of $\frac{2}{3}$, the required line through $(2, -1)$ has the following equation.

$$y - (-1) = \frac{2}{3}(x - 2) \qquad \text{Point-slope form}$$

$$y = \frac{2}{3}x - \frac{4}{3} - 1 \qquad \text{Solve for } y.$$

$$y = \frac{2}{3}x - \frac{7}{3} \qquad \text{Slope-intercept form}$$

Notice the similarity between the slope-intercept form of the original equation and the slope-intercept form of the parallel equation.

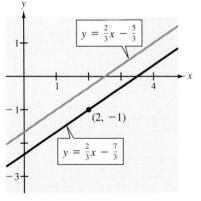

FIGURE 2.45

You have seen that two nonvertical lines are parallel if and only if they have the same slope. Two nonvertical lines are *perpendicular* if and only if their slopes are negative reciprocals of each other. For instance, the lines $y = 2x$ and $y = -\frac{1}{2}x$ are perpendicular because one has a slope of 2 and the other has a slope of $-\frac{1}{2}$.

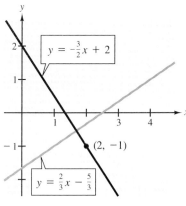

FIGURE 2.46

Perpendicular Lines

Two nonvertical lines are **perpendicular** if and only if their slopes are negative reciprocals of each other. That is,

$$m_1 = -\frac{1}{m_2}.$$

EXAMPLE 6 Equations of Perpendicular Lines

Find an equation of the line that passes through the point $(2, -1)$ and is perpendicular to the line $2x - 3y = 5$.

Solution

By writing the given line in the form $y = \frac{2}{3}x - \frac{5}{3}$, you can see that the line has a slope of $\frac{2}{3}$. Hence, any line that is perpendicular to this line must have a slope of $-\frac{3}{2}$ (because $-\frac{3}{2}$ is the negative reciprocal of $\frac{2}{3}$). Therefore, the required line through the point $(2, -1)$ has the following equation.

$$y - (-1) = -\frac{3}{2}(x - 2) \qquad \text{Point-slope form}$$

$$y = -\frac{3}{2}x + 3 - 1 \qquad \text{Solve for } y.$$

$$y = -\frac{3}{2}x + 2 \qquad \text{Slope-intercept form}$$

The graphs of both equations are shown in Figure 2.46.

Group Activities   Problem Solving

Interpreting Slope   You think the relationship between daily high temperature y (in degrees Fahrenheit) and time x (in days) is linear. You observe the high temperature on day 0 to be 60°F and the high temperature 15 days later to be 68°F.

a. Find the slope of the line passing through these two points.

b. Interpret the meaning of the slope in this situation.

c. Use linear extrapolation to predict the high temperatures on days 30, 60, 105, and 180. Are all of these predictions reasonable? Why or why not?

Warm Up   The following warm-up exercises involve skills that were covered in earlier sections. You will use these skills in the exercise set for this section.

In Exercises 1–4, simplify the expression.

1. $\dfrac{4 - (-5)}{-3 - (-1)}$

2. $\dfrac{-5 - 8}{0 - (-3)}$

3. Find $-1/m$ for $m = 4/5$.

4. Find $-1/m$ for $m = -2$.

In Exercises 5–10, solve for y in terms of x.

5. $2x - 3y = 5$

6. $4x + 2y = 0$

7. $y - (-4) = 3[x - (-1)]$

8. $y - 7 = \frac{2}{3}(x - 3)$

9. $y - (-1) = \dfrac{3 - (-1)}{2 - 4}(x - 4)$

10. $y - 5 = \dfrac{3 - 5}{0 - 2}(x - 2)$

2.4 Exercises

In Exercises 1–6, estimate the slope of the line from its graph.

1.

2.

5.

6.

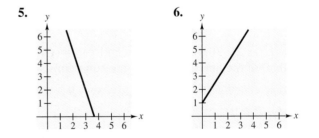

In Exercises 7 and 8, sketch the lines through the given point with the given slopes. Sketch all four lines on the same coordinate plane.

3.

4.

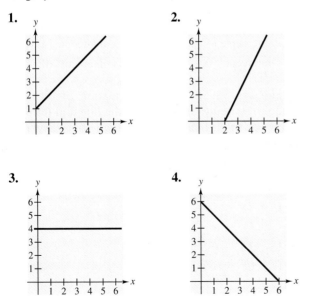

	Point		*Slopes*		
7.	$(2, 3)$	(a) 0 (b) 1	(c) 2	(d) -3
8.	$(-4, 1)$	(a) 3 (b) -3	(c) $\frac{1}{2}$	(d) undefined

In Exercises 9–14, plot the points and find the slope of the line passing through the points.

9. $(-3, -2), (1, 6)$

10. $(2, 4), (4, -4)$

11. $(-6, -1), (-6, 4)$

12. $(0, -10), (-4, 0)$

13. $(1, 2), (-2, -2)$

14. $\left(\frac{7}{8}, \frac{3}{4}\right), \left(\frac{5}{4}, -\frac{1}{4}\right)$

In Exercises 15–18, determine if the lines L_1 and L_2 passing through the pairs of points are parallel, perpendicular, or neither.

15. L_1: $(0, -1), (5, 9)$; L_2: $(0, 3), (4, 1)$

16. L_1: $(3, 6), (-6, 0)$; L_2: $(0, -1), \left(5, \frac{7}{3}\right)$

17. L_1: $(-2, -1), (1, 5)$; L_2: $(1, 3), (5, -5)$

18. L_1: $(4, 8), (-4, 2)$; L_2: $(3, -5), \left(-1, \frac{1}{3}\right)$

In Exercises 19–24, use the point on the line and the slope of the line to find three additional points that the line passes through. (Each problem has more than one correct answer.)

Point	Slope
19. $(2, 1)$	$m = 0$
20. $(-4, 1)$	m is undefined
21. $(5, -6)$	$m = 1$
22. $(10, -6)$	$m = -1$
23. $(-8, 1)$	m is undefined
24. $(-3, -1)$	$m = 0$

In Exercises 25–30, find the slope and y-intercept (if possible) of the line specified by the equation.

25. $5x - y + 3 = 0$

26. $2x + 3y - 9 = 0$

27. $5x - 2 = 0$

28. $3y + 5 = 0$

29. $7x + 6y - 30 = 0$

30. $x - y - 10 = 0$

In Exercises 31–38, find an equation of the line passing through the points and use a graphing utility to verify your answer.

31. $(5, -1), (-5, 5)$

32. $(4, 3), (-4, -4)$

33. $\left(2, \frac{1}{2}\right), \left(\frac{1}{2}, \frac{5}{4}\right)$

34. $(-1, 4), (6, 4)$

35. $(-8, 1), (-8, 7)$

36. $(1, 1), \left(6, -\frac{2}{3}\right)$

37. $(1, 0.6), (-2, -0.6)$

38. $(-8, 0.6), (2, -2.4)$

In Exercises 39–48, find an equation of the line that passes through the point and has the indicated slope. Sketch the line.

Point	Slope
39. $(0, -2)$	$m = 3$
40. $(0, 10)$	$m = -1$
41. $(-3, 6)$	$m = -2$
42. $(0, 0)$	$m = 4$
43. $(4, 0)$	$m = -\frac{1}{3}$
44. $(-2, -5)$	$m = \frac{3}{4}$
45. $(6, -1)$	m is undefined
46. $(-10, 4)$	$m = 0$
47. $\left(4, \frac{5}{2}\right)$	$m = \frac{4}{3}$
48. $\left(-\frac{1}{2}, \frac{3}{2}\right)$	$m = -3$

In Exercises 49–52, use the *intercept form* of the equation of a line

$$\frac{x}{a} + \frac{y}{b} = 1, \qquad a \neq 0, \quad b \neq 0$$

to find the equation of the line with the given intercepts $(a, 0)$ and $(0, b)$.

49. x-intercept: $(2, 0)$
  y-intercept: $(0, 3)$

50. x-intercept: $(-3, 0)$
  y-intercept: $(0, 4)$

51. x-intercept: $\left(-\frac{1}{6}, 0\right)$
  y-intercept: $\left(0, -\frac{2}{3}\right)$

52. x-intercept: $\left(\frac{2}{3}, 0\right)$
  y-intercept: $(0, -2)$

In Exercises 53–58, write an equation of the line through the point (a) parallel to the line and (b) perpendicular to the line. Use a graphing utility to verify your answer.

Point	Line
53. $(2, 1)$	$4x - 2y = 3$
54. $(-3, 2)$	$x + y = 7$
55. $(-6, 4)$	$3x + 4y = 7$
56. $\left(\frac{7}{8}, \frac{3}{4}\right)$	$5x + 3y = 0$
57. $(-1, 0)$	$y = -3$
58. $(2, 5)$	$x = 4$

59. *Temperature*   Find an equation of the line that gives the relationship between the temperature in degrees Celsius, C, and degrees Fahrenheit, F. Remember that water freezes at $0°$ Celsius ($32°$ Fahrenheit) and boils at $100°$ Celsius ($212°$ Fahrenheit).

60. *Temperature*   Use the result of Exercise 59 to complete the table. Is there a temperature for which the Fahrenheit reading is the same as the Celsius reading? If so, what is it?

C		$-10°$	$10°$			$177°$
F	$0°$			$68°$	$90°$	

61. *Simple Interest*   A person deposits P dollars in an account that pays simple interest. After 2 months the balance in the account is $759, and after 3 months the balance in the account is $763.50. Find an equation that gives the relationship between the balance A and the time t in months.

62. *Simple Interest*   Use the result of Exercise 61 to complete the table.

P			$759.00	$763.50			
t	0	1			4	5	6

63. *Fourth Quarter Sales*   During the first and second quarters of the year, a business had sales of $145,000 and $152,000. If the growth of the sales follows a linear pattern, what will the sales be during the fourth quarter?

64. *College Enrollment*   A small college had 2546 students in 1994 and 2702 students in 1996. If the enrollment follows a linear growth pattern, how many students will the college have in 2001?

65. *Annual Salary*   Suppose that your salary was $23,500 in 1992 and $26,800 in 1995. If your salary follows a linear growth pattern, what salary will you be making in 1999?

66. *Fund Raising*   In 1983, the total amount of money given by private corporate businesses to symphony orchestras was $61 million. By 1985 the total amount had increased to $79.8 million. If these contributions continued to increase in a linear pattern, how much money would have been given by private corporate businesses to symphony orchestras in 1986? (Source: Business Committee for the Arts)

67. *Turner Broadcasting Revenue*   In 1988, Turner Broadcasting System, Inc. had a total revenue of $810 million. By 1991, the total revenue had increased to $1480 million. If the total revenue continued to increase in a linear pattern, how much would the revenue have been in 1993? The actual revenue in 1993 was $1920 million. Do you think the increase in revenue was approximately linear? (Source: Turner Broadcasting System, Inc.)

68. *Outlet Malls*   In 1991, the number of outlet shopping centers in the United States was 142. By 1993, the number had increased to 249. If the number of outlet shopping centers continued to increase in a linear pattern, what would the number have been in 1994? The actual number of outlet shopping centers in 1994 was 300. Do you think the increase in the number of outlet shopping centers was approximately linear? (Source: *Value Retail News*)

Math Matters   The Pythagorean Theorem

The Pythagorean Theorem is one of the most famous theorems in mathematics. Its name comes from Pythagoras of Samos (*c.* 580–500 B.C.), a Greek mathematician who founded a school of mathematics and philosophy at Crotona in what is now southern Italy. The Pythagorean Theorem states a relationship among the three sides of a right triangle. Specifically, the theorem states that the sum of the squares of the two legs of a right triangle is equal to the square of the hypotenuse. This result is usually written as

$$a^2 + b^2 = c^2.$$ Pythagorean Theorem

One way to prove the Pythagorean Theorem is to use a geometrical argument, as indicated in the following series of figures. In the figure at the left, the areas of the three squares represent the squares of the sides of a right triangle. To prove that $a^2 + b^2 = c^2$, you need to show that the sum of the areas of the two smaller squares is equal to the area of the larger square. One way to do this is to make four copies of the given right triangle, and use the four triangles to form a square whose sides have lengths of $a + b$. By doing this in two different ways, you can conclude that $a^2 + b^2 = c^2$. Do you see why this is true?

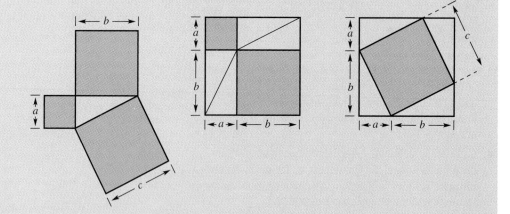

According to the *Guinness Book of World Records,* 1995, the Pythagorean Theorem is the "most proven theorem" in mathematics. This publication claims that "*The Pythagorean Proposition* published in 1940 contained 370 different proofs of Pythagoras' theorem, including one by President James Garfield."

2.5	**Linear Modeling**

Introduction ▪ Direct Variation ▪ Rates of Change ▪
Scatter Plots

Introduction

The primary objective of applied mathematics is to find equations or **mathematical models** that describe real-world phenomena. In developing a mathematical model to represent actual data, you should strive for two (often conflicting) goals—accuracy and simplicity. That is, you want the model to be simple enough to be workable, yet accurate enough to produce meaningful results.

Real Life

EXAMPLE 1   *A Mathematical Model*

The total annual amounts of advertising expenses (in billions of dollars) in the United States from 1980 to 1989 are given in the table.   (Source: McCann Erickson)

Year	1980	1981	1982	1983	1984	1985	1986	1987	1988	1989
y	$54.8	$60.4	$66.6	$75.9	$88.1	$94.8	$102.1	$109.8	$118.1	$125.6

A linear model that approximates this data is

$$y = 8.1358t + 53.009, \qquad 0 \le t \le 9$$

where y represents the advertising expenses (in billions of dollars) and t represents the year, with $t = 0$ corresponding to 1980. Plot the actual data *and* the model on the same graph. How closely does the model represent the data?

Solution

The actual data is plotted in Figure 2.47, along with the graph of the linear model. From the figure, it appears that the model is a "good fit" for the actual data. You can see how well the model fits by comparing the actual values of y with the values of y given by the model (these are labeled y^* in the table below).

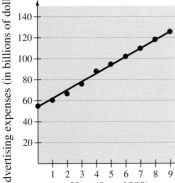

Advertising expenses (in billions of dollars)

Year (0 ↔ 1980)

FIGURE 2.47

t	0	1	2	3	4	5	6	7	8	9
y	$54.8	$60.4	$66.6	$75.9	$88.1	$94.8	$102.1	$109.8	$118.1	$125.6
y^*	$53.0	$61.1	$69.3	$77.4	$85.6	$93.7	$101.8	$110.0	$118.1	$126.2

Direct Variation

There are two basic types of linear models. The more general model has a y-intercept that is nonzero: $y = mx + b$, $b \neq 0$. The simpler one, $y = mx$, has a y-intercept that is zero. In the simpler model, y is said to **vary directly** as x, or be **proportional** to x.

Direct Variation

The following statements are equivalent.

1. y **varies directly** as x.

2. y is **directly proportional** to x.

3. $y = mx$ for some nonzero constant m.

m is the **constant of variation** or the **constant of proportionality.**

Real Life

EXAMPLE 2   State Income Tax

In Pennsylvania, the state income tax is directly proportional to *gross income*. Suppose you were working in Pennsylvania and your state income tax deduction was $31.50 for a gross monthly income of $1500.00. Find a mathematical model that gives the Pennsylvania state income tax in terms of the gross income.

Solution

Verbal Model: $\boxed{\text{State income tax}} = \boxed{m} \cdot \boxed{\text{Gross income}}$

Labels: State income tax $= y$ (dollars)
             Gross income $= x$ (dollars)
             Income tax rate $= m$ (percent in decimal form)

Equation: $y = mx$

To solve for m, substitute the given information into the equation $y = mx$, and then solve for m.

$y = mx$ Direct variation model

$31.50 = m(1500)$ Substitute $y = 31.50$ and $x = 1500$.

$0.021 = m$ Income tax rate

Thus, the equation (or model) for state income tax in Pennsylvania is $y = 0.021x$. In other words, Pennsylvania has a state income tax rate of 2.1% of the gross income. The graph of this equation is shown in Figure 2.48.

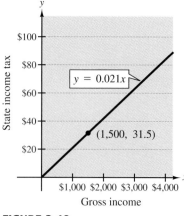

$y = 0.021x$

$(1,500, 31.5)$

FIGURE 2.48

Most measurements in the English system and metric system are directly proportional. The next example shows how to use a direct proportion to convert between miles per hour and kilometers per hour.

Real Life

EXAMPLE 3   The English and Metric Systems

You are traveling at a rate of 64 miles per hour. You switch your speedometer reading to metric units and notice that the speed is 103 kilometers per hour. Use this information to find a mathematical model that relates miles per hour to kilometers per hour.

Solution

If you let y represent the speed in miles per hour and x represent the speed in kilometers per hour, you know that y and x are related by the equation

$$y = mx.$$

You are given that $y = 64$ when $x = 103$. By substituting these values into the equation $y = mx$, you can find the value of m.

$$y = mx \qquad \text{Direct variation model}$$
$$64 = m(103) \qquad \text{Substitute } y = 64 \text{ and } x = 103.$$
$$\frac{64}{103} = m \qquad \text{Divide both sides by 103.}$$
$$0.62136 \approx m \qquad \text{Use a calculator.}$$

Thus, the conversion factor from kilometers per hour to miles per hour is approximately 0.62136, and the model is

$$y = 0.62136x.$$

The graph of this equation is shown in Figure 2.49.

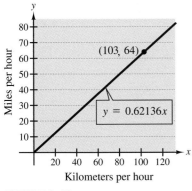

FIGURE 2.49

Once you have found a model that converts speeds from kilometers per hour to miles per hour, you can use the model to convert other speeds from the metric system to the English system, as shown in the table.

Kilometers per hour	20.0	40.0	60.0	80.0	100.0	120.0
Miles per hour	12.4	24.9	37.3	49.7	62.1	74.6

NOTE   The conversion equation $y = 0.62136x$ can be approximated by the simpler equation $y = \frac{5}{8}x$. For instance, to convert 40 kilometers per hour, divide by 8 and multiply by 5 to obtain 25 miles per hour.

Rates of Change

There are three basic types of mountain climbing: rock climbing, snow and ice climbing, and mixed climbing. Of these, rock climbing is the most popular.

A second common type of linear model is one that involves a known rate of change. In the linear equation

$$y = mx + b$$

you know that m represents the slope of the line. In real-life problems, the slope can often be interpreted as the **rate of change** of y with respect to x. Rates of change should always be listed in appropriate units of measure.

Real Life

EXAMPLE 4   *Height of a Mountain Climber*

A mountain climber is climbing up a 500-foot cliff. By 1 P.M., the mountain climber has climbed 115 feet up the cliff. By 4 P.M., the climber has reached a height of 280 feet, as shown in Figure 2.50. Find the average rate of change of the climber and use this rate of change to find the equation that relates the height of the climber to the time. Use the model to estimate the time when the climber will reach the top of the cliff.

Solution

Let y represent the height of the climber and let t represent the time. Then the two points that represent the climber's two positions are

$$(t_1, y_1) = (1, 115) \qquad \text{and} \qquad (t_2, y_2) = (4, 280).$$

Thus, the average rate of change of the climber is

$$
\begin{aligned}
\text{Average rate of change} &= \frac{y_2 - y_1}{t_2 - t_1} \\
&= \frac{280 - 115}{4 - 1} \\
&= 55 \text{ feet per hour.}
\end{aligned}
$$

Thus, an equation that relates the height of the climber to the time is

$y - y_1 = m(t - t_1)$	Point-slope form
$y - 115 = 55(t - 1)$	Substitute $y_1 = 115$, $t_1 = 1$, and $m = 55$.
$y = 55t + 60.$	Linear model

To find the time when the climber reaches the top of the cliff, let $y = 500$ and solve for t to obtain

$$t = 8.$$

Thus, continuing at the same rate, the climber will reach the top of the cliff at 8 P.M.

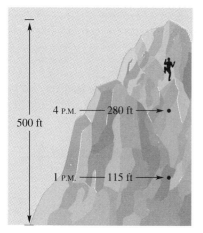

4 P.M. — 280 ft
500 ft
1 P.M. — 115 ft

FIGURE 2.50

EXAMPLE 5   *Population of Hampton, Virginia*

Between 1980 and 1990, the population of the city of Hampton, Virginia increased at an average rate of approximately 1100 people per year. In 1980, the population was 123,000. Find a mathematical model that gives the population of Hampton in terms of the year, and use the model to estimate the population in 1992.   (Source: U. S. Bureau of the Census)

Solution

Let y represent the population of Hampton, and let t represent the calendar year, with $t = 0$ corresponding to 1980. Letting $t = 0$ correspond to 1980 is convenient because you were given the population in 1980. Now, using the rate of change of 1100 people per year, you have

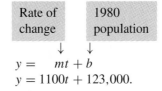

$$y = mt + b$$
$$y = 1100t + 123,000.$$

Using this model, you can predict the 1992 population to be

$$(1992\ \text{population}) = 1100(12) + 123,000 = 136,200.$$

The graph is shown in Figure 2.51. (In this particular example, the linear model is quite good—the actual population of Hampton, Virginia in 1992 was 137,000.)

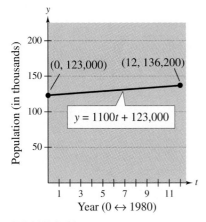

FIGURE 2.51

In Example 5, note that in the linear model the population changed by the same *amount* each year (see Figure 2.52a). If the population had changed by the same *percent* each year, the model would have been exponential, not linear (see Figure 2.52b). (You will study exponential models in Chapter 5.)

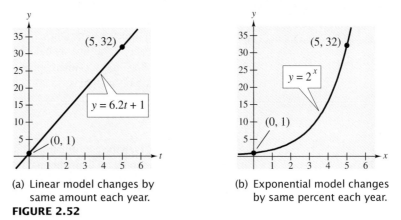

(a) Linear model changes by same amount each year.

(b) Exponential model changes by same percent each year.

FIGURE 2.52

Scatter Plots

Another type of linear modeling is a graphical approach that is commonly used in statistics. To find a mathematical model that approximates a set of actual data points, plot the points on a rectangular coordinate system. This collection of points is called a **scatter plot.** Once the points have been plotted, try to find the line that most closely represents the plotted points. (In this section, we will rely on a visual technique for fitting a line to a set of points. If you take a course in statistics, you will encounter *regression analysis* formulas that can fit a line to a set of points.)

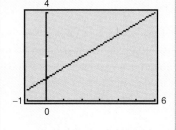

Technology

Most graphing utilities have built-in programs that will calculate the equation of the best-fitting line for a collection of points. For instance, to find the best-fitting line on a *TI-82*, enter the *x*- and *y*-values with the STAT (Edit) menu so that the *x*-values are in L_1 and the *y*-values are in L_2. Then, from the STAT (Calc) menu, choose "LinReg(ax + b)." The graph below shows the best-fitting line for the data in Example 6.

EXAMPLE 6 Fitting a Line to a Set of Points

The scatter plot in Figure 2.53(a) shows 35 different points in the plane. Find the equation of a line that approximately fits these points.

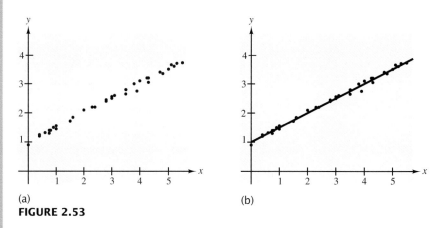

(a)   (b)

FIGURE 2.53

Solution

From Figure 2.53(a), you can see that there is no line that *exactly* fits the given points. The points, however, do appear to resemble a linear pattern. Figure 2.53(b) shows a line that appears to best describe the given points. (Notice that about as many points lie above the line as below it.) From this figure, you can see that the "best-fitting line" has a *y*-intercept at about $(0, 1)$ and has a slope of about $\frac{1}{2}$. Thus, the equation of the line is

$$y = \frac{1}{2}x + 1.$$

If you had been given the coordinates of the 35 points, you could have checked the accuracy of this model by constructing a table that compared the actual *y*-values with the *y*-values given by the model.

Alternate Group Activity

Ask each group to collect measurements of both height and forearm length (from elbow to fingertip) to the nearest $\frac{1}{4}$ inch from 10 to 20 people. Have each group plot this data (y represents height and x represents forearm length), draw the best-fitting line, and find its equation. Then ask each group to predict *your* height based on your forearm length. Supply your actual height and have each group compare its prediction to your actual height. Discuss. This activity could also be done with the entire class; if so, you may want to bring a tape measure(s) to class.

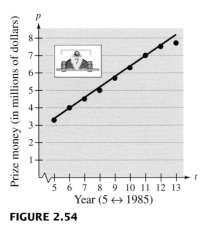

Prize money (in millions of dollars)

Year (5 ↔ 1985)

FIGURE 2.54

Real Life

EXAMPLE 7 Prize Money at the Indianapolis 500

The total prize money p (in millions of dollars) awarded at the Indianapolis 500 race from 1985 to 1993 is given in the following table. Construct a scatter plot that represents the data and find a linear model that approximates the data. (Source: Indianapolis Motor Speedway Hall of Fame)

Year	1985	1986	1987	1988	1989	1990	1991	1992	1993
p	$3.27	$4.00	$4.49	$5.03	$5.72	$6.33	$7.01	$7.53	$7.68

Solution

Let $t = 5$ represent 1985. The scatter plot for the points is shown in Figure 2.54. From the scatter plot, draw a line that approximates the data. Then, to find the equation of the line, approximate two points *on the line*: $(6, 4)$ and $(11, 7)$. The slope of this line is

$$m \approx \frac{p_2 - p_1}{t_2 - t_1} = \frac{7 - 4}{11 - 6} = 0.6.$$

Using the point-slope form, you can determine that the equation of the line is

$$p = 0.6t + 0.4.$$

To check this model, compare the actual p-values with the p-values given by the model (these are labeled p^* below).

t	5	6	7	8	9	10	11	12	13
p	$3.27	$4.00	$4.49	$5.03	$5.72	$6.33	$7.01	$7.53	$7.68
p^*	$3.4	$4.0	$4.6	$5.2	$5.8	$6.4	$7.0	$7.6	$8.2

Group Activities Extending the Concept

An Experiment Go to the library, or some other reference source, and find some data that you think describes a linear relationship. Plot the points and find the equation of a line that approximately represents the points.

Warm Up   The following warm-up exercises involve skills that were covered in earlier sections. You will use these skills in the exercise set for this section.

In Exercises 1–4, sketch the graph of the line.

1. $y = 2x$ **2.** $y = \frac{1}{2}x$

3. $y = 2x + 1$ **4.** $y = \frac{1}{2}x + 1$

In Exercises 5 and 6, find an equation of the line that has the given slope and y-intercept.

5. Slope: 1; y-intercept: $(0, 2)$ **6.** Slope: $\frac{3}{2}$; y-intercept: $(0, 3)$

In Exercises 7–10, find an equation of the line that passes through the two points.

7. $(1, 3)$ and $(6, 8)$ **8.** $(0, 4)$ and $(7, 10)$

9. $(1, 5.2)$ and $(5, 4.7)$ **10.** $(2, 6.5)$ and $(8, 3.6)$

2.5 Exercises

1. *Employment* The total numbers of employed (in thousands) in the United States from 1987 to 1993 are given by the following ordered pairs.

(1987, 121,602)
(1988, 123,378)
(1989, 125,557)
(1990, 126,424)
(1991, 126,867)
(1992, 128,548)
(1993, 129,525)

A linear model that approximates this data is

$y = 113,336.2 + 1265.0t, \quad 7 \le t \le 13$

where y represents the number of employed (in thousands) and $t = 0$ represents 1980. Plot the actual data *and* the model on the same graph. How closely does the model represent the data? (Source: U. S. Bureau of Labor Statistics)

2. *Olympic Swimming* The winning times (in minutes) in the women's 400-meter freestyle swimming event in the Olympics from 1948 to 1992 are given by the following ordered pairs.

(1948, 5.30), (1952, 5.20)
(1956, 4.91), (1960, 4.84)
(1964, 4.72), (1968, 4.53)
(1972, 4.32), (1976, 4.16)
(1980, 4.15), (1984, 4.12)
(1988, 4.06), (1992, 4.12)

A linear model that approximates this data is

$y = 5.4 - 0.03t, \quad 8 \le t \le 52$

where y represents the winning time in minutes and $t = 0$ represents 1940. Plot the actual data *and* the model on the same graph. How closely does the model represent the data? (Source: Olympic Committee)

Direct Variation In Exercises 3–8, y is proportional to x. Use the x- and y-values to find a linear model that relates y and x.

3. $x = 5$ $\qquad y = 12$

4. $x = 2$ $\qquad y = 14$

5. $x = 10$ $\qquad y = 2050$

6. $x = 6$ $\qquad y = 580$

7. $x = 12$ $\qquad y = 1.75$

8. $x = 15$ $\qquad y = 2.3$

9. *Simple Interest* The simple interest that a person receives for an investment is directly proportional to the amount of the investment. By investing $2500 in a bond issue, you obtained an interest payment of $187.50 at the end of 1 year. Find a mathematical model that gives the interest I at the end of 1 year in terms of the amount invested P.

10. *Simple Interest* The simple interest that a person receives for an investment is directly proportional to the amount of the investment. By investing $5000 in a municipal bond, you obtained interest of $337.50 at the end of 1 year. Find a mathematical model that gives the interest I for the municipal bond at the end of 1 year in terms of the amount invested P.

11. *Property Tax* Your property tax is based on the assessed value of your property. (The assessed value is often lower than the actual value of the property.) A house that has an assessed value of $50,000 has a property tax of $1840.

(a) Find a mathematical model that gives the amount of property tax y in terms of the assessed value x of the property.

(b) Use the model to find the property tax on a house that has an assessed value of $85,000.

12. *State Sales Tax* An item that sells for $145.99 has a sales tax of $10.22.

(a) Find a mathematical model that gives the amount of sales tax y in terms of the retail price x.

(b) Use the model to find the sales tax on a purchase that has a retail price of $540.50.

13. *Centimeters and Inches* On a yardstick, you notice that 13 inches is the same length as 33 centimeters.

(a) Use this information to find a mathematical model that relates centimeters to inches.

(b) Use the model to complete the table.

Inches	5	10	20	25	30
Centimeters					

14. *Liters and Gallons* You are buying gasoline and notice that 14 gallons of gasoline is the same as 53 liters.

(a) Use this information to find a linear model that relates gallons to liters.

(b) Use the model to complete the table.

Gallons	5	10	20	25	30
Liters					

In Exercises 15–20, you are given the 1994 value of a product *and* the rate at which the value is expected to change during the next 5 years. Use this information to write a linear equation that gives the dollar value of the product in terms of the year. (Let $t = 0$ represent 1994.)

	1994 Value	*Rate*
15.	$2540	$125 increase per year
16.	$156	$4.50 increase per year
17.	$20,400	$2000 decrease per year
18.	$45,000	$2800 decrease per year
19.	$154,000	$12,500 increase per year
20.	$245,000	$5600 increase per year

21. *Straight-Line Depreciation* A business purchases a piece of equipment for $875. After 5 years the equipment will have no value. Write a linear equation giving the value V of the equipment during the 5 years.

22. *Straight-Line Depreciation* A business purchases a piece of equipment for $25,000. The equipment will be replaced in 10 years, at which time its salvage value is expected to be $2000. Write a linear equation giving the value *V* of the equipment during the 10 years.

23. *Sales Price and List Price* A store is offering a 15% discount on all items. Write a linear equation giving the sale price *S* for an item with a list price *L*.

24. *Sales Price and List Price* A store is offering a 25% discount on all shirts. Write a linear equation giving the sale price *S* for a shirt with a list price *L*.

25. *Hourly Wages* A manufacturer pays its assembly line workers $11.50 per hour. In addition, workers receive a piecework rate of $0.75 per unit produced. Write a linear equation for the hourly wages *W* in terms of the number of units *x* produced per hour.

26. *Sales Commission* A salesperson receives a monthly salary of $2500 plus a commission of 7% of sales. Write a linear equation for the salesperson's monthly wage *W* in terms of the person's monthly sales *S*.

27. *Height of a Parachutist* After opening the parachute, the descent of a parachutist follows a linear model. At 2:08 P.M. the height of the parachutist is 7000 feet. At 2:10 P.M. the height is 4600 feet.

(a) Write a linear equation that gives the height of the parachutist in terms of the time *t*. (Let *t* = 0 represent 2:08 P.M. and let *t* be measured in seconds.)

(b) Use the equation to find the time when the parachutist will reach the ground.

28. *Distance Traveled by a Car* You are driving at a constant speed on a freeway. At 4:30 P.M. you drive by a sign that gives the distance to Montgomery, Alabama as 84 miles. At 4:59 P.M. you drive by another sign that says the distance to Montgomery is 56 miles.

(a) Write a linear equation that gives your distance from Montgomery in terms of *t*. (Let *t* = 0 represent 4:30 P.M. and let *t* be measured in minutes.)

(b) Use the equation to find the time when you will reach Montgomery.

In Exercises 29–32, match the description with one of the graphs. Also find the slope and describe how it is interpreted in the real-life situation. [The graphs are labeled (a), (b), (c), and (d).]

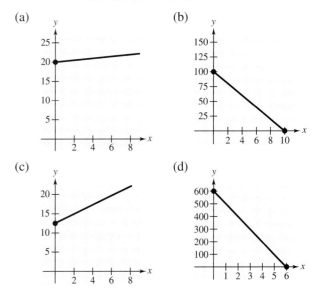

(a)   (b)

(c)   (d)

29. A person is paying $10 per week to a friend to repay a $100 loan.

30. An employee is paid $12.50 per hour plus $1.50 for each unit produced per hour.

31. A sales representative receives $20 per day for food, plus $0.25 for each mile traveled.

32. A typewriter that was purchased for $600 depreciates $100 per year.

33. *High School Enrollment* A high school had an enrollment of 1200 students in 1980. During the next 10 years the enrollment increased by approximately 50 students per year.

(a) Write a linear equation giving the enrollment *N* in terms of the year *t*. (Let *t* = 0 correspond to the year 1980.)

(b) If this constant rate of growth continues, predict the enrollment in the year 2000.

(c) Use a graphing utility to graphically confirm the result in part (b).

34. *Cost of Renting* The monthly rent in an office building is related to the office size. The rent for a 600-square-foot office is $750. The rent for a 900-square-foot office is $1150.

(a) Write a linear equation giving the monthly rent r in terms of the square footage x.

(b) Use the equation to find the monthly rent for an office that has 1300 square feet of floor space.

(c) Use a graphing utility to graphically confirm the result in part (b).

In Exercises 35–38, how well can the data be approximated by a linear model?

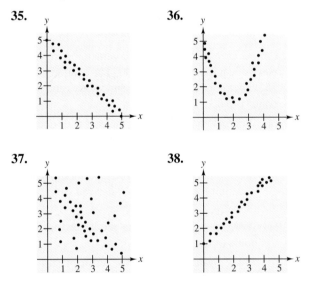

35. **36.**

37. **38.**

In Exercises 39–42, sketch the line that you think best approximates the data. Then find an equation of the line.

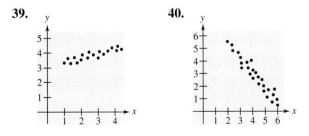

39. **40.**

41. **42.**

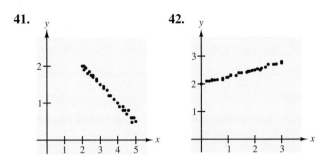

43. *Football Salaries* The average annual salary of professional football players (in thousands of dollars) from 1987 to 1992 is shown in the scatter plot. (Source: National Football League Players Association)

(a) Find the equation of the line that you think best fits this data. (Let y represent the average salary and let $t = 7$ represent 1987.)

(b) Use the equation to predict the average salaries in 1994, 1995, and 1996.

(c) Use your school's library or some other reference source to analyze the accuracy of the predictions in part (b).

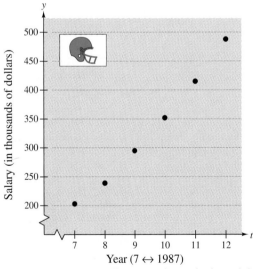

Year (7 ↔ 1987)

44. *Think About It* A linear mathematical model for predicting prize winnings in a race is based on data for only 3 years. Discuss the potential accuracy and inaccuracy of such a model.

45. *Discus Throw* The lengths of the winning men's discus throws in the Olympics from 1900 to 1992 are shown in the scatter plot. (Source: Olympic Committee)

(a) Find the equation of the line that you think best fits this data. (Let y represent the length of the winning discus throw in feet and let $t = 0$ represent 1900.)

(b) Use the equation to predict the winning men's discus throw in 1996.

(c) Use your school's library or some other reference source to analyze the accuracy of the prediction in part (b).

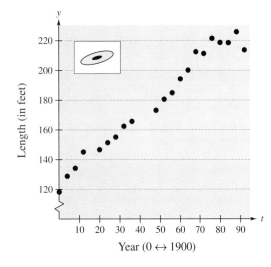

Year $(0 \leftrightarrow 1900)$

46. *Total Assets* The total assets for Boatmen's Bancshares, Inc. from 1980 to 1993 (in millions of dollars) is shown in the scatter plot. (Source: Standard OTC Stock Reports)

(a) Find the equation of the line that you think best fits this data. (Let y represent the total assets in billions of dollars and let $t = 0$ represent 1980.)

(b) Use a graphing utility to predict the assets of Boatmen's Bancshares in 1996. Compare the result with the prediction given by the equation in part (a).

(c) Use your school's library or some other reference source to analyze the accuracy of the prediction in part (b).

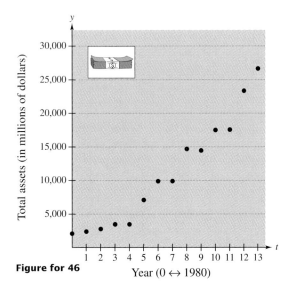

Figure for 46

Year $(0 \leftrightarrow 1980)$

47. *Contracting Purchase* A contractor purchases a piece of equipment for $36,500. The equipment costs $5.25 per hour for fuel and maintenance, and the operator is paid $11.50 per hour.

(a) Write an equation giving the cost C of operating the equipment for t hours. (Include the purchase cost.)

(b) If customers are charged $27 per hour, write an equation for the revenue R derived from t hours of use.

(c) Write an equation for the profit derived from t hours of use.

(d) *Break-Even Point* Use the result of part (c) to find the number of hours this equipment must be used to yield a profit of 0 dollars.

48. *Real Estate Purchase* A real estate office handles an apartment complex with 50 units. When the rent per unit is $380 per month, all 50 units are occupied. However, when the rent is $425 per month, the average number of occupied units drops to 47. Assume that the relationship between the monthly rent p and the demand x is linear.

(a) Write the equation of the line giving the demand x in terms of the rent p.

(b) Use this equation to predict the number of units occupied if the rent is raised to $455.

CHAPTER PROJECT: Gravity and Air Resistance

This chapter project (as well as the projects in Chapters 3, 5, and 8) utilizes data that was collected by the Texas Instruments Calculator-Based Laboratory *(CBL)* System. The *CBL* is used in conjunction with either the *TI-82* or *TI-85* calculator to collect real data in real time. This data may then be stored in the *TI-82* or *TI-85* for further analysis. It is not necessary to have access to a *CBL* to complete most of this project. However, if you do have access to a *CBL*, the last question of the project (indicated by a *CBL* icon) gives the opportunity to explore the topic further with a *CBL*. Instruction on how to use a *CBL* for your own data collection is given in the Appendix.

There are two forces that act on any falling object—gravity and air resistance. Gravity pulls an object toward the earth, and air resistance tries to keep an object from falling. However, the force of gravity is much stronger than the force of air resistance; thus, air resistance only slows the fall of an object.

In this project, you will consider the fall of a miniature parachute. The parachute is released approximately 5 feet above the ground. A motion detector attached to a *CBL* unit tracks the fall of a weight attached to the parachute.

To determine the effects of gravity and air resistance on a weighted parachute as it falls, one records the height of the falling object at timed intervals. The table shows the data collected with a *Texas Instruments CBL* unit.*

Use this information to investigate the following questions.

1. ***Predicting the Outcome***   Which of the following predictions describes how the parachute will fall immediately after its release? Give a reason for your prediction.

   (a) Air resistance will cause the parachute to fall slowly at first, then more rapidly as gravity takes over.

   (b) The parachute will fall rapidly at first, then more slowly as air resistance affects it.

   (c) Gravity and air resistance will work together to cause the parachute to fall at a constant rate.

2. ***Graphing the Data***   Input the data from the table at the left into a graphing utility, and use it to make a scatter plot of the data. How does the scatter plot compare to your prediction in Question 1?

3. ***Fitting a Model to Data***   Use the statistical features of your graphing utility to find a linear model for the data.

   (a) Write the linear model.

   (b) What is the slope of the line? Explain what the slope of the line tells you about the fall of the parachute.

   (c) What is the *y*-intercept of the line? Explain what the *y*-intercept tells you about the fall of the parachute.

4. ***Comparing Actual Data with the Model***   Enter the linear model from Question 3(a) in your graphing utility. Using the TRACE feature, find the height of the parachute at $t = 0.2$ seconds and at $t = 0.6$ seconds. How do the model values compare to the values from the table?

5. ***Extended CBL Exploration***   Use the *CBL* instructions in Appendix C to collect and analyze data about falling objects. Attach binder clips or paperweights to a parachute to vary the weight. How does the amount of weight affect a falling object? What happens when the size of the parachute decreases?

Time	Height
0	4.83262
0.10002	4.70658
0.199997	4.47971
0.300005	4.23124
0.400013	3.97196
0.499992	3.68748
0.599972	3.39579
0.699962	3.10771
0.0799942	2.81962
0.899922	2.52434
0.999902	2.25066
1.099872	1.95897
1.199872	1.67809

*Instructions for using a *CBL* to collect your own data are given in Appendix C.

CHAPTER SUMMARY

After studying this chapter, you should have acquired the following skills. These skills are keyed to the Review Exercises that begin on page 223. Answers to odd-numbered Review Exercises are given in the back of the book.

- Find the distance between two points. *(Section 2.1)* — **Review Exercises 1–4**
- Find the distance and midpoint between two points. *(Section 2.1)* — **Review Exercises 5–8**
- Use the midpoint formula to estimate sales of a company. *(Section 2.1)* — **Review Exercises 9, 10**
- Show that points on the Cartesian plane form a polygon. *(Section 2.1)* — **Review Exercises 11, 12**
- Use the distance formula to find a point. *(Section 2.1)* — **Review Exercises 13–16**
- Determine if points lie on the graph of an equation. *(Section 2.2)* — **Review Exercises 17, 18**
- Find the x- and y-intercepts of the graph of an equation. *(Section 2.2)* — **Review Exercises 19–22**
- Check for symmetry with respect to the axes and origin. *(Section 2.2)* — **Review Exercises 23–26**
- Identify intercepts and symmetry of an equation. *(Section 2.2)* — **Review Exercises 27–32**
- Find the standard form of the equation of a circle. *(Section 2.2)* — **Review Exercises 33–36**
- Write the equation of a circle in standard form. *(Section 2.2)* — **Review Exercises 37, 38**
- Use a graphing utility to match the equation with its graph. *(Section 2.3)* — **Review Exercises 39–44**
- Use a graphing utility to sketch the graph of an equation. *(Section 2.3)* — **Review Exercises 45–48**
- Write an equation so that an ordered pair is a solution. *(Section 2.2)* — **Review Exercises 49, 50**
- Use a graphing utility to find x-intercepts of an equation. *(Section 2.3)* — **Review Exercises 51, 52**
- Use a graphing utility to determine geometric shape. *(Section 2.3)* — **Review Exercises 53, 54**
- Find the slope of the line passing through two points. *(Section 2.4)* — **Review Exercises 55, 56**
- Use slope to determine if lines are parallel or perpendicular. *(Section 2.4)* — **Review Exercises 57–60**
- Find the slope and y-intercept of a line. *(Section 2.4)* — **Review Exercises 61, 62**
- Find the equation of a line passing through two points. *(Section 2.4)* — **Review Exercises 63–66**
- Find the equation of a line given a slope and point. *(Section 2.4)* — **Review Exercises 67–70**
- Write an equation of a line parallel to the given line and perpendicular to the given line. *(Section 2.4)* — **Review Exercises 71, 72**
- Use a linear model to predict sales and annual salaries. *(Section 2.5)* — **Review Exercises 73, 74**
- Use direct variation to find a model that relates y and x. *(Section 2.5)* — **Review Exercises 75–78**
- Use direct variation to relate two quantities. *(Section 2.5)* — **Review Exercises 79–82**
- Write a linear equation where the slope of the line represents a rate of change. *(Section 2.5)* — **Review Exercises 83–88**

REVIEW EXERCISES

In Exercises 1 and 2, find the distance between the points. (*Note:* In each case, the two points lie on the same horizontal or vertical line.)

1. $(5, -2), (-8, -2)$ **2.** $(3, 7), (3, 11)$

In Exercises 3 and 4, (a) find the lengths of the two sides of the right triangle and use the Pythagorean Theorem to find the length of the hypotenuse, and (b) use the Distance Formula to find the length of the hypotenuse of the triangle.

3.   **4.**

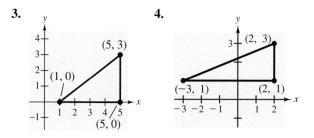

In Exercises 5–8, (a) plot the points, (b) find the distance between the points, and (c) find the midpoint of the line segment joining the points.

5. $(-9, 3), (5, 7)$ **6.** $(3, 2), (-3, -5)$

7. $(-6.7, -3.9), (5.1, 8.2)$

8. $(3.45, 6.55), (-1.06, -3.87)$

In Exercises 9 and 10, use the Midpoint Formula to estimate the sales of a company for 1995 given the sales in 1993 and 1997.

9.

Year	1993	1997
Sales	$640,000	$810,000

10.

Year	1993	1997
Sales	$3,250,000	$5,690,000

In Exercises 11 and 12, show that the points form the vertices of the indicated polygon.

11. Parallelogram: $(1, 1), (8, 2), (9, 5), (2, 4)$

12. Right Triangle: $(-1, -1), (10, 7), (2, 18)$

In Exercises 13 and 14, find x so that the distance between the points is 25.

13. $(10, 10), (x, -5)$ **14.** $(x, -5), (-15, 10)$

In Exercises 15 and 16, find y so that the distance between the points is 13.

15. $(6, y), (-6, -2)$ **16.** $(5, 1), (17, y)$

In Exercises 17 and 18, determine whether the points lie on the graph of the equation.

17. $y = x^2 - 7x + 10$ (a) $(1, 4)$ (b) $(2, 0)$

18. $y = \sqrt{25 - x^2}$ (a) $(3, -2)$ (b) $(0, 5)$

In Exercises 19–22, find the x- and y-intercepts of the graph of the equation.

19. $y = (x + 3)(x - 2)$ **20.** $y = x^2 - 9$

21. $y = x\sqrt{x + 3}$ **22.** $xy + x - 2y + 4 = 0$

In Exercises 23–26, check for symmetry with respect to both axes and the origin.

23. $y = \dfrac{x}{x^2 - 2}$ **24.** $x^2 + y^2 = 9$

25. $y = x^3 + 1$ **26.** $y^2 = x$

In Exercises 27–32, sketch the graph of the equation. Identify any intercepts and test for symmetry.

27. $y = x^2 + 3$ **28.** $x^2 + y^2 = 25$

29. $y = 3x - 4$ **30.** $y = \sqrt{9 - x}$

31. $y = x^3 + 1$ **32.** $y = |x - 3|$

In Exercises 33–36, find the standard form of the equation of the specified circle.

33. Center: $(1, 3)$; radius: 5

34. Center: $(-1, 0)$; solution point: $(2, 4)$

35. Endpoints of the diameter: $(-1, -2)$, $(3, 1)$

36. Center: $\left(\frac{1}{2}, -\frac{1}{3}\right)$; radius: $\frac{3}{2}$

In Exercises 37 and 38, write the equation of the circle in standard form and sketch its graph.

37. $x^2 + y^2 - 6x + 4y - 3 = 0$

38. $4x^2 + 4y^2 + 8x - 16y - 44 = 0$

In Exercises 39–44, use a graphing utility to match the equation with its graph. [The graphs are labeled (a), (b), (c), (d), (e), and (f).]

(a)

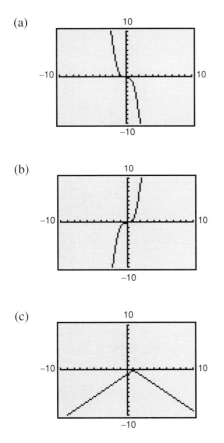

(b)

(c)

(d)

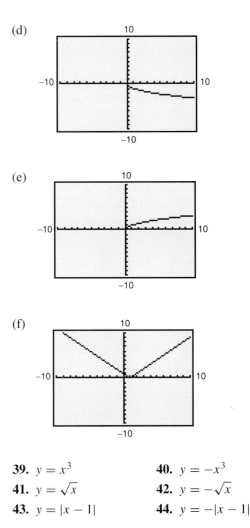

(e)

(f)

39. $y = x^3$ **40.** $y = -x^3$

41. $y = \sqrt{x}$ **42.** $y = -\sqrt{x}$

43. $y = |x - 1|$ **44.** $y = -|x - 1|$

In Exercises 45–48, use a graphing utility to sketch the graph of the equation. Use a standard setting for each graph.

45. $y = -2x + 3$ **46.** $4y = x^2 + 5x - 2$

47. $y = |x + 4|$ **48.** $y = \frac{1}{3}(x^3 - 3x)$

In Exercises 49 and 50, find the constant C such that the ordered pair is a solution point of the equation.

49. $y = Cx^2$ $(2, 12)$

50. $x + C(y + 3) = 0$ $(-3, 4)$

In Exercises 51 and 52, use a graphing utility to find the number of x-intercepts of the equation.

51. $y = \frac{1}{3}(3x^2 - 27x + 55)$

52. $y = \frac{1}{4}(-8x^2 + 32x - 9)$

In Exercises 53 and 54, use a graphing utility to sketch the graphs on the same screen. Using a square setting, what geometrical shape is bounded by the graphs?

53. $y = -\sqrt{16 - x^2}$, $y = \sqrt{16 - x^2}$

54. $y = |x| - 6$, $y = -|x| + 6$

In Exercises 55 and 56, plot the points and find the slope of the line passing through the two points.

55. $(1, 5), (-3, -6)$ **56.** $(0, -5), (3, 2)$

In Exercises 57–60, determine whether the lines L_1 and L_2 passing through the pairs of points are parallel, perpendicular, or neither.

57. L_1: $(0, 3), (-2, 1)$; L_2: $(-8, -3), (4, 9)$

58. L_1: $(-3, -1), (2, 5)$; L_2: $(2, 1), (8, 6)$

59. L_1: $(3, 6), (-1, -5)$; L_2: $(-2, 3), (4, 7)$

60. L_1: $(-1, 2), (-1, 4)$; L_2: $(7, 3), (4, 7)$

In Exercises 61 and 62, find the slope and y-intercept (if possible) of the line specified by the equation.

61. $5x - 4y + 11 = 0$ **62.** $3y - 2 = 0$

In Exercises 63–66, find an equation for the line passing through the points and sketch the line.

63. $(3, 7), (2, -1)$ **64.** $(3, -2), (-1, -2)$

65. $(3, 4), (3, -2)$ **66.** $(-1, 5), (2, -3)$

In Exercises 67–70, find an equation of the line that passes through the point and has the indicated slope. Sketch the graph of the line.

67. $(0, -5)$ $m = \frac{3}{2}$

68. $(-2, 6)$ $m = 0$

69. $(3, 0)$ $m = -\frac{2}{3}$

70. $(5, 4)$ m is undefined

In Exercises 71 and 72, write an equation of the line through the point (a) parallel to the given line and (b) perpendicular to the line.

71. $(3, -2)$ $5x - 4y = 8$

72. $(-8, 3)$ $2x + 3y = 5$

73. *Fourth Quarter Sales*   During the second and third quarters of the year, a business had sales of $160,000 and $185,000. If the growth of the sales follows a linear pattern, what will sales be during the fourth quarter?

74. *Annual Salary*   Suppose that your annual salary was $22,000 in 1993 and $24,500 in 1994. If your salary follows a linear growth pattern, what salary will you be making in 1998?

Direct Variation   In Exercises 75–78, assume that y is proportional to x. Use the x-value and y-value to find a linear model that relates y and x.

75. $x = 3$ $y = 7$

76. $x = 5$ $y = 3.8$

77. $x = 10$ $y = 3480$

78. $x = 14$ $y = 1.95$

79. *Simple Interest*   The simple interest that a person receives for an investment is directly proportional to the amount of the investment. Suppose that by investing $7000 in a municipal bond, you obtained an interest of $438.50 at the end of 1 year. Find a mathematical model that gives the amount of interest I for this municipal bond at the end of 1 year in terms of the amount invested.

80. *Property Tax*   The property tax in a certain city is based on the assessed value of the property. A house that has an assessed value of $40,000 has a property tax of $1620. Find a mathematical model that gives the amount of property tax y in terms of the assessed value of the property x. Use the model to find the property tax on a house that has an assessed value of $65,000.

81. *Grams and Ounces*   Suppose that you are buying cherry pie filling and notice that 21 ounces of filling is the same as 595 grams. Use this information to find a linear model that relates ounces to grams. Use the model to complete the table.

Ounces	6	8	12	16	24
Grams					

82. *Kilometers and Miles*   Suppose you are driving on the highway and you notice that a billboard indicates that it is 2.5 miles or 4 kilometers to the next restaurant of a national fast food chain. Use this information to find a linear model that relates miles to kilometers. Use this model to complete the table.

Miles	2	5	10	12
Kilometers				

83. *Dollar Value*   The dollar value of a product in 1994 is $85 and the item is expected to increase in value at a rate of $3.75 per year. Write a linear equation that gives the dollar value of the product in terms of the year. Use this model to estimate the dollar value of the product in 1999. (Let $t = 4$ represent 1994.)

84. *Straight-Line Depreciation*   A small business purchases a piece of equipment for $30,000. After 10 years, the equipment will have to be replaced. Its salvage value at that time is expected to be $2500. Write a linear equation giving the value V of the equipment during the 10 years it will be used.

85. *Cost of Renting*   The monthly rent in an office building is related to the size of the office that is rented. The monthly rent for an office that has 800 square feet of floor space is $975. The monthly rent for an office that has 1100 square feet of floor space is $1350. Write a linear equation giving the monthly rent r in terms of the number of square feet of floor space x. Use the result to find the monthly rent for an office that has 1400 square feet of floor space.

86. *Hourly Wages*   A manufacturer pays its assembly line workers $9.50 per hour. In addition, workers receive a piecework rate of $0.60 per unit produced. Write a linear equation for the hourly wages W in terms of the number of units x produced per hour.

87. *Vehicle Sales*   During November 1–10, 1989, the ten major United States automakers sold cars and trucks at an average daily rate of 26,194. During the same period in 1990, they sold cars and trucks at an average daily rate of 26,817. Write a linear equation for the average daily sales rate of vehicles during November 1–10. (Let $t = 0$ represent 1989.) Use the result to estimate the average daily sales rate for November 1–10, 1991. (Source: *Dallas Times Herald,* November 15, 1990)

88. *Contracting Purchase*   A contractor purchases a piece of equipment for $24,500. The equipment requires an average expenditure of $5.75 per hour for fuel and maintenance, and the operator is paid $9.50 per hour.

(a) Write a linear equation giving the total cost C of operating this equipment for t hours. (Include the purchase price of the equipment.)

(b) If customers are charged $35 per hour of machine use, write an equation for the revenue R derived from t hours of use.

(c) Use the formula for profit ($P = R - C$) to write an equation for the profit derived from t hours of use.

(d) *Break-Even Point*   Use the result of part (c) to find the number of hours this equipment must be used to yield a profit of 0 dollars.

CHAPTER TEST

Take this test as you would take a test in class. After you are done, check your work against the answers given in the back of the book.

In Exercises 1 and 2, find the distance between the points and the midpoint of the line segment connecting the points.

1. $(-3, 2), (5, -2)$ **2.** $(3.25, 7.05), (-2.37, 1.62)$

3. Find the intercepts of the graph of $y = (x + 5)(x - 3)$.

4. Find the intercepts of the graph of $y = x\sqrt{x - 2}$.

5. Describe the symmetry of the graph of $x^2 + y^2 = 16$.

6. Describe the symmetry of the graph of $y = \dfrac{x}{x^2 - 4}$.

7. Find an equation of the line through $(3, 4)$ and $(-2, 4)$.

8. Find an equation of the line through $(4, -3)$ with a slope of $\frac{3}{4}$.

9. Find an equation of the line through $(-1, 5)$ and $(2, 1)$.

10. Find an equation of the vertical line through $(2, -7)$.

11. Find an equation of the line with intercepts $(3, 0)$ and $(0, -4)$.

In Exercises 12–15, match the equation with its graph. [The graphs are labeled (a), (b), (c), and (d).]

12. $y = \sqrt{x}$ **13.** $y = -\sqrt{x - 2}$

14. $y = |x - 2|$ **15.** $y = -x^3$

16. Write the equation of the circle in standard form and sketch its graph.

$$x^2 + y^2 - 4x - 2y - 4 = 0$$

17. During the first and second quarters of the year, a business had sales of $150,000 and $185,000, respectively. If the sales follow a linear pattern, what will sales be in the third quarter?

18. An assembly-line worker is paid $8.75 per hour, plus a piecework rate of $0.55 per unit. Write a linear equation for the hourly wages W in terms of the number of units x produced per hour.

19. A business purchases a piece of equipment for $25,000. After 5 years the equipment will be worth only $5000. Write a linear equation that gives the value V of the equipment during the 5 years.

20. A store is offering a 30% discount on all summer apparel. Write a linear equation that gives the sale price S of an item in terms of its list price L.

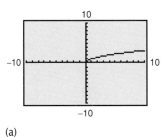

(a)

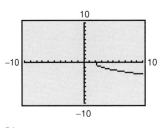

(b)

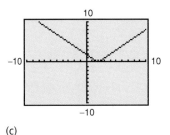

(c)

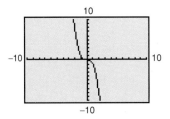

(d)
Figures for 12–15

CUMULATIVE TEST: CHAPTERS P–2

Take this test as you would take a test in class. After you are done, check your work against the answers given in the back of the book.

In Exercises 1–4, simplify the expression.

1. $5\sqrt{x} - 11\sqrt{x}$

2. $7^{1/3} \cdot 7^{5/3}$

3. $\sqrt{24x^3}$

4. $\dfrac{1}{3 - \sqrt{5}}$

In Exercises 5–10, solve the equation.

5. $4y^2 - 64 = 0$

6. $2(x + 4) - 5 = 11 + 5(2 - x)$

7. $x^4 - 17x^2 + 16 = 0$

8. $|2x - 3| = 5$

9. $\sqrt{y + 3} + y = 3$

10. $2x^2 + x = 5$

11. You deposit $5000 in an account that pays 8.5%, compounded monthly. Find the balance after 10 years.

12. You just bought a new camcorder. The manufacturer made 300,000 camcorders of your model and, of these, 20,000 have defective zoom lenses. What is the probability that your new camcorder has a defective zoom lens? Explain what this tells you about the probability that your new camcorder has a properly working zoom lens.

In Exercises 13 and 14, solve the inequality.

13. $-3 \leq \dfrac{2x - 5}{4} \leq 3$

14. $(x + 4)^2 \leq 4$

15. Find an equation of the line passing through $(3, -2)$ and $(-1, 5)$.

16. Find an equation of the line with slope $\frac{2}{3}$ passing through $(2, -1)$.

17. Find an equation of the line with zero slope passing through $(-1, -3)$.

18. Write the equation of the circle in standard form and sketch its graph.

$$x^2 + y^2 - 6x + 4y - 3 = 0$$

19. During the second and third quarters of the year, a business had sales of $210,000 and $230,000, respectively. If the sales growth follows a linear pattern, what will the sales be during the fourth quarter?

20. The revenue and cost equations for a product are given by

$$R = x(100 - 0.001x) \quad \text{and} \quad C = 20x + 30,000$$

where R and C are measured in dollars and x represents the number of units sold. How many units must be sold for the revenue to equal the cost?

Functions and Graphs

3

A function that gives the height s of a falling object in terms of the time t is called a position function. If air resistance is not considered, the position of a falling object can be modeled by

$$s(t) = \tfrac{1}{2}gt^2 + v_0 t + s_0$$

where g is the acceleration due to gravity, v_0 is the initial velocity, and s_0 is the initial height.

The value of g depends on where the object is dropped. On Earth's surface, g is approximately -32 feet per second per second.

To discover experimentally the value of g, you can record the heights of a falling object for several times, as indicated in the partial table. A set of 23 points is shown in the scatter plot. The model for this data is $s(t) = -15.45t^2 - 1.301t + 5.234$.

Time (seconds)	0.0	0.1	0.2	0.3	0.4
Height (feet)	5.24	4.95	4.36	3.45	2.26

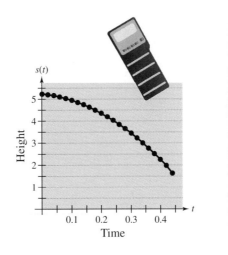

The chapter project related to this information is on page 312. The project uses actual data, collected with a *Texas Instruments CBL System.*

229

3.1	**Functions**
	Introduction to Functions ▪ Function Notation ▪ Finding the Domain of a Function ▪ Applications

Introduction to Functions

Many everyday phenomena involve two quantities that are related to each other by some rule of correspondence. Here are some examples.

1. The simple interest I earned on $1000 for 1 year is related to the annual percentage rate r by the formula $I = 1000r$.

2. The distance d traveled on a bicycle in 2 hours is related to the speed s of the bicycle by the formula $d = 2s$.

3. The area A of a circle is related to its radius r by the formula $A = \pi r^2$.

Not all correspondences between two quantities have simple mathematical formulas. For instance, people commonly match up NFL starting quarterbacks with touchdown passes and hours of the day with temperature. In each of these cases, however, there is some rule of correspondence that matches each item from one set with exactly one item from a different set. Such a rule of correspondence is called a **function.**

Definition of a Function

A **function** f from a set A to a set B is a rule of correspondence that assigns to each element x in the set A exactly one element y in the set B. The set A is the **domain** (or set of inputs) of the function f, and the set B contains the **range** (or set of outputs).

To get a better idea of this definition, look at the function illustrated in Figure 3.1. This function can be represented by the following set of ordered pairs.

$$\{(1, 9°), (2, 13°), (3, 15°), (4, 15°), (5, 12°), (6, 4°)\}$$

In each ordered pair, the first coordinate is the input and the second coordinate is the output. In this example, note the following characteristics of a function.

1. Each element in A must be matched with an element of B.

2. Some elements in B may not be matched with any element in A.

3. Two or more elements of A may be matched with the same element of B.

The converse of the third statement is not true. That is, an element of A (the domain) cannot be matched with two different elements of B.

Hours of the day | Celsius temperature

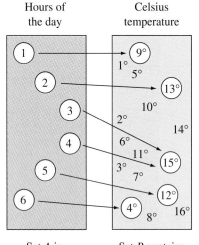

Set A is the domain. Input 1, 2, 3, 4, 5, 6

Set B contains the range. Output: 4°, 9°, 12°, 13°, 15°

FIGURE 3.1 Function from Set A to Set B

In the following two examples, you are asked to decide whether different correspondences are functions. To do this, you must decide whether each element in the domain A is matched with exactly one element in the range B. If any element in A is matched with two or more elements in B, the correspondence is not a function.

EXAMPLE 1   Testing for Functions

Let $A = \{a, b, c\}$ and $B = \{1, 2, 3, 4, 5\}$. Which of the following sets of ordered pairs or figures represent functions from set A to set B?

a. $\{(a, 2), (b, 3), (c, 4)\}$ **b.** $\{(a, 4), (b, 5)\}$

c.   **d.**

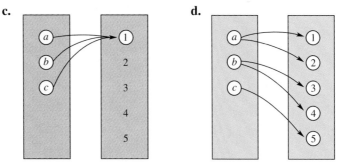

Solution

a.   This collection of ordered pairs *does* represent a function from A to B. Each element of A is matched with exactly one element of B.

b.   This collection of ordered pairs *does not* represent a function from A to B. Not every element of A is matched with an element of B.

c.   This figure *does* represent a function from A to B. It does not matter that each element of A is matched with the same element of B.

d.   This figure *does not* represent a function from A to B. The element a in A is matched with *two* elements, 1 and 2, of B. This is also true of the element b.

Representing functions by sets of ordered pairs is a common practice in *discrete mathematics*. In algebra, however, it is more common to represent functions by equations or formulas involving two variables. For instance, the equation

$$y = x^2 \qquad \text{\small y is a function of x.}$$

represents the variable y as a function of the variable x. In this equation, x is the **independent variable** and y is the **dependent variable.** The domain of the function is the set of all values taken on by the independent variable x, and the range of the function is the set of all values taken on by the dependent variable y.

EXAMPLE 2   *Testing for Functions Represented by Equations*

Which of the following equations represent y as a function of x?

a. $x^2 + y = 1$ **b.** $-x + y^2 = 1$

Solution

To determine whether y is a function of x, try to solve for y in terms of x.

a. Solving for y yields the following.

$$x^2 + y = 1 \qquad\qquad \text{Original equation}$$
$$y = 1 - x^2 \qquad\qquad \text{Solve for } y.$$

To each value of x there corresponds exactly one value of y. Thus, y *is* a function of x.

b. Solving for y yields the following.

$$-x + y^2 = 1 \qquad\qquad \text{Original equation}$$
$$y^2 = 1 + x \qquad\qquad \text{Add } x \text{ to both sides.}$$
$$y = \pm\sqrt{1 + x} \qquad\qquad \text{Solve for } y.$$

In Example 2(b), you may also want to consider relationships that cannot readily be solved for y, such as Exercise 38 in Section 3.1.

The $\pm$ indicates that to a given value of x there correspond two values of y. Thus, y *is not* a function of x.

Function Notation

When an equation is used to represent a function, it is convenient to name the function so that it can be referenced easily. For example, you know that the equation $y = 1 - x^2$ describes y as a function of x. Suppose you give this function the name "f." Then you can use the following **function notation.**

Input	*Output*	*Equation*
x	$f(x)$	$f(x) = 1 - x^2$

Understanding the concept of functions is essential. Be sure students understand functional notation. Frequently $f(x)$ is misinterpreted as "f times x" rather than "f of x".

The symbol $f(x)$ is read as the **value of f at x** or simply **f of x.** The symbol $f(x)$ corresponds to the y-value for a given x. Thus, you can write $y = f(x)$. Keep in mind that f is the *name* of the function, whereas $f(x)$ is the *value* of the function at x. For instance, the function given by

$$f(x) = 3 - 2x$$

has *function values* denoted by $f(-1)$, $f(0)$, $f(2)$, and so on. To find these values, substitute the specified input values into the given equation.

For $x = -1$, $f(-1) = 3 - 2(-1) = 3 + 2 = 5.$
For $x = 0$, $f(0) = 3 - 2(0) = 3 - 0 = 3.$
For $x = 2$, $f(2) = 3 - 2(2) = 3 - 4 = -1.$

Although f is often used as a convenient function name and x is often used as the independent variable, you can use other letters. For instance,

$$f(x) = x^2 - 4x + 7, \quad f(t) = t^2 - 4t + 7, \quad \text{and} \quad g(s) = s^2 - 4s + 7$$

all define the same function. In fact, the role of the independent variable in a function is simply that of a "placeholder." Consequently, the above function could be described by the form

$$f() = ()^2 - 4() + 7.$$

EXAMPLE 3 Evaluating a Function

Let $g(x) = -x^2 + 4x + 1$ and find the following.

a. $g(2)$ **b.** $g(t)$ **c.** $g(x+2)$

Solution

a. Replacing x with 2 in $g(x) = -x^2 + 4x + 1$ yields the following.

$$g(2) = -(2)^2 + 4(2) + 1 = -4 + 8 + 1 = 5$$

b. Replacing x with t yields the following.

$$g(t) = -(t)^2 + 4(t) + 1 = -t^2 + 4t + 1$$

> **NOTE** In Example 3, note that $g(x+2)$ is not equal to $g(x) + g(2)$. In general, $g(u+v) \neq g(u) + g(v)$.

c. Replacing x with $x + 2$ yields the following.

$$g(x+2) = -(x+2)^2 + 4(x+2) + 1$$
$$= -(x^2 + 4x + 4) + 4x + 8 + 1$$
$$= -x^2 - 4x - 4 + 4x + 8 + 1$$
$$= -x^2 + 5$$

EXAMPLE 4 A Function Defined by Two Equations

Evaluate the following function when $x = -1, 0,$ and 1.

$$f(x) = \begin{cases} x^2 + 1, & x < 0 \\ x - 1, & x \geq 0 \end{cases}$$

Solution

Because $x = -1$ is less than 0, use $f(x) = x^2 + 1$ to obtain

$$f(-1) = (-1)^2 + 1 = 2.$$

> **NOTE** A function defined by two or more equations over a specified domain is called a *piecewise-defined* function.

For $x = 0$, use $f(x) = x - 1$ to obtain

$$f(0) = (0) - 1 = -1.$$

For $x = 1$, use $f(x) = x - 1$ to obtain $f(1) = (1) - 1 = 0.$

DISCOVERY

Use a graphing utility to graph $y = \sqrt{4 - x^2}$. What is the domain of this function? Then graph $y = \sqrt{x^2 - 4}$. What is the domain of this function? Do the domains of these two functions overlap? If so, for what values?

Finding the Domain of a Function

The domain of a function can be described explicitly or it can be *implied* by the expression used to define the function. The **implied domain** is the set of all real numbers for which the expression is defined. For instance, the function given by

$$f(x) = \frac{1}{x^2 - 4}$$

has an implied domain that consists of all real x other than $x = \pm 2$. These two values are excluded from the domain because division by zero is undefined. Another common type of implied domain is that used to avoid even roots of negative numbers. For example, the function given by

$$f(x) = \sqrt{x}$$

is defined only for $x \geq 0$. Hence, its implied domain is the interval $[0, \infty)$. In general, the domain of a function *excludes* values that would cause division by zero *or* result in the even root of a negative number.

EXAMPLE 5 *Finding the Domain of a Function*

Find the domain of each of the following functions.

a. $f: \{(-3, 0), (-1, 4), (0, 2), (2, 2), (4, -1)\}$

b. $g(x) = \dfrac{1}{x + 5}$

c. Volume of a sphere: $V = \frac{4}{3}\pi r^3$

d. $h(x) = \sqrt{4 - x^2}$

Solution

a. The domain of f consists of all first coordinates in the set of ordered pairs.

Domain $= \{-3, -1, 0, 2, 4\}$.

b. Excluding x-values that yield zero in the denominator, the domain of g is the set of all real numbers $x \neq -5$.

c. Because this function represents the volume of a sphere, the values of the radius r must be positive. Thus, the domain is the set of all real numbers r such that $r \geq 0$.

d. This function is defined only for x-values for which $4 - x^2 \geq 0$. Using the methods described in Section 1.7, you can conclude that $-2 \leq x \leq 2$. Thus, the domain is the interval $[-2, 2]$.

NOTE In Example 5(c), note that the domain of a function may be implied by the physical context. For instance, from the equation $V = \frac{4}{3}\pi r^3$, you would have no reason to restrict r to nonnegative values, but the physical context implies that a sphere cannot have a negative radius.

Applications

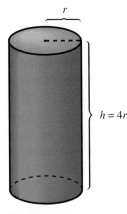

r

$h = 4r$

FIGURE 3.2

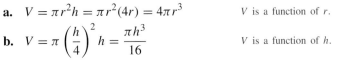

EXAMPLE 6   *The Dimensions of a Container*

You work in the marketing department of a soft-drink company and are experimenting with a new soft-drink can that is slightly narrower and taller than a standard can. For your experimental can, the ratio of the height to the radius is 4, as shown in Figure 3.2.

a.   Express the volume of the can as a function of the radius r.

b.   Express the volume of the can as a function of the height h.

Solution

a. $V = \pi r^2 h = \pi r^2 (4r) = 4\pi r^3$ V is a function of r.

b. $V = \pi \left(\dfrac{h}{4}\right)^2 h = \dfrac{\pi h^3}{16}$ V is a function of h.

EXAMPLE 7   *The Path of a Baseball*

A baseball is hit 3 feet above ground at a velocity of 100 feet per second and at an angle of $45°$. The path of the baseball is given by the function

$$y = -0.0032x^2 + x + 3$$

where y and x are measured in feet, as shown in Figure 3.3. Will the baseball clear a 10-foot fence located 300 feet from home plate?

Solution

When $x = 300$, the height of the baseball is given by

$$y = -0.0032(300^2) + 300 + 3 = 15 \text{ feet.}$$

Thus, the ball will clear the fence.

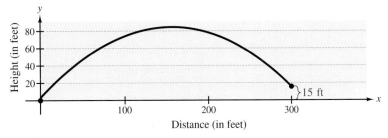

15 ft

FIGURE 3.3

In 1988, Americans made about 40.7 million trips to foreign countries. By 1993, this number had increased to about 45.5 million. The most visited country was Mexico.

EXAMPLE 8   U.S. Travelers Abroad

The amount of money spent by U.S. travelers in other countries increased in a linear pattern from 1980 to 1983, as shown in Figure 3.4. Then, in 1984, the amount of money spent took a sharp jump and until 1988 increased in a *different* linear pattern. These two patterns can be approximated by the function

$$C = \begin{cases} 10{,}479 + 917.1t, & 0 \le t \le 3 \\ 12{,}808 + 2350.4t, & 4 \le t \le 8 \end{cases}$$

where C represents the amount of money spent (in millions of dollars) and t represents the calendar year, with $t = 0$ corresponding to 1980. Use this function to approximate the total amount of money spent by U.S. travelers abroad between 1980 and 1988.   (Source: U.S. Bureau of Economic Analysis)

Solution

From 1980 to 1983, use the formula $C = 10{,}479 + 917.1t$ to approximate the money spent.

Year	1980	1981	1982	1983
t	0	1	2	3
Money Spent	$10,479	$11,396	$12,313	$13,230

From 1984 to 1988, use the formula $C = 12{,}808 + 2350.4t$ to approximate the money spent.

Year	1984	1985	1986	1987	1988
t	4	5	6	7	8
Money Spent	$22,210	$24,560	$26,910	$29,261	$31,611

To find the total amount spent, add the amounts for each of the years, as shown below.

$$10{,}479 + 11{,}396 + 12{,}313 + 13{,}230 + 22{,}210 +$$
$$24{,}560 + 26{,}910 + 29{,}261 + 31{,}611 = 181{,}970$$

Because the money spent is measured in millions of dollars, you can conclude that the total amount spent by U.S. travelers abroad from 1980 through 1988 was approximately

$$\text{Total spent} \approx \$181{,}970{,}000{,}000.$$

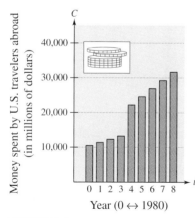

FIGURE 3.4

Summary of Function Terminology

Function: A **function** is a relationship between two variables such that to each value of the independent variable there corresponds exactly one value of the dependent variable.

Function Notation: $y = f(x)$

   f is the **name** of the function.

   y is the **dependent variable.**

   x is the **independent variable.**

   $f(x)$ is the **value of the function at x.**

Domain: The **domain** of a function is the set of all values (inputs) of the independent variable for which the function is defined. If x is in the domain of f, we say that f is **defined** at x. If x is not in the domain of f, we say that f is **undefined** at x.

Range: The **range** of a function is the set of all values (outputs) assumed by the dependent variable (that is, the set of all function values).

Implied Domain: If f is defined by an algebraic expression and the domain is not specified, the **implied domain** consists of all real numbers for which the expression is defined.

Group Activities   Extending the Concept

Determining Relationships That Are Functions   Write two statements describing relationships in everyday life that *are* functions and two that *are not* functions. Here are a couple of examples.

a. The statement, "Your happiness is a function of the grade you receive in this course" *is not* a correct mathematical use of the word "function." The word "happiness" is ambiguous.

b. The statement, "Your federal income tax is a function of your adjusted gross income" *is* a correct mathematical use of the word "function." Once you have determined your adjusted gross income, your income tax can be determined.

Warm Up The following warm-up exercises involve skills that were covered in earlier sections. You will use these skills in the exercise set for this section.

In Exercises 1–4, simplify the expression.

1. $2(-3)^3 + 4(-3) - 7$

2. $4(-1)^2 - 5(-1) + 4$

3. $(x+1)^2 + 3(x+1) - 4 - (x^2 + 3x - 4)$

4. $(x-2)^2 - 4(x-2) - (x^2 - 4)$

In Exercises 5 and 6, solve for y in terms of x.

5. $2x + 5y - 7 = 0$

6. $y^2 = x^2$

In Exercises 7–10, solve the inequality.

7. $x^2 - 4 \geq 0$

8. $9 - x^2 \geq 0$

9. $x^2 + 2x + 1 \geq 0$

10. $x^2 - 3x + 2 \geq 0$

3.1 Exercises

In Exercises 1–4, evaluate the function.

1. $f(x) = 6 - 4x$

   (a) $f(3) = 6 - 4(\quad)$

   (b) $f(-7) = 6 - 4(\quad)$

   (c) $f(t) = 6 - 4(\quad)$

   (d) $f(c+1) = 6 - 4(\quad)$

2. $f(s) = \dfrac{1}{s+1}$

   (a) $f(4) = \dfrac{1}{(\quad) + 1}$

   (b) $f(0) = \dfrac{1}{(\quad) + 1}$

   (c) $f(4x) = \dfrac{1}{(\quad) + 1}$

   (d) $f(x+1) = \dfrac{1}{(\quad) + 1}$

3. $g(x) = \dfrac{1}{x^2 - 2x}$

   (a) $g(1) = \dfrac{1}{(\quad)^2 - 2(\quad)}$

   (b) $g(-3) = \dfrac{1}{(\quad)^2 - 2(\quad)}$

   (c) $g(t) = \dfrac{1}{(\quad)^2 - 2(\quad)}$

   (d) $g(t+1) = \dfrac{1}{(\quad)^2 - 2(\quad)}$

4. $f(t) = \sqrt{25 - t^2}$

   (a) $f(3) = \sqrt{25 - (\quad)^2}$

   (b) $f(5) = \sqrt{25 - (\quad)^2}$

   (c) $f(x+5) = \sqrt{25 - (\quad)^2}$

   (d) $f(2x) = \sqrt{25 - (\quad)^2}$

In Exercises 5–16, evaluate the function and simplify the results.

5. $f(x) = 2x - 3$

  (a) $f(1)$ (b) $f(-3)$

  (c) $f(x - 1)$ (d) $f(\frac{1}{4})$

6. $g(y) = 7 - 3y$

  (a) $g(0)$ (b) $g(\frac{7}{3})$

  (c) $g(s)$ (d) $g(s + 2)$

7. $h(t) = t^2 - 2t$

  (a) $h(2)$ (b) $h(-1)$

  (c) $h(x + 2)$ (d) $h(1.5)$

8. $V(r) = \frac{4}{3}\pi r^3$

  (a) $V(3)$ (b) $V(0)$

  (c) $V(\frac{3}{2})$ (d) $V(2r)$

9. $f(y) = 3 - \sqrt{y}$

  (a) $f(4)$ (b) $f(100)$

  (c) $f(4x^2)$ (d) $f(0.25)$

10. $f(x) = \sqrt{x + 8} + 2$

  (a) $f(-8)$ (b) $f(1)$

  (c) $f(x - 8)$ (d) $f(x + 8)$

11. $q(x) = \dfrac{1}{x^2 - 9}$

  (a) $q(4)$ (b) $q(0)$

  (c) $q(3)$ (d) $q(y + 3)$

12. $q(t) = \dfrac{2t^2 + 3}{t^2}$

  (a) $q(2)$ (b) $q(0)$

  (c) $q(x)$ (d) $q(-x)$

13. $f(x) = \dfrac{|x|}{x}$

  (a) $f(2)$ (b) $f(-2)$

  (c) $f(x^2)$ (d) $f(x - 1)$

14. $f(x) = |x| + 4$

  (a) $f(2)$ (b) $f(-2)$

  (c) $f(x^2)$ (d) $f(x + 2)$

15. $f(x) = \begin{cases} 2x + 1, & x < 0 \\ 2x + 2, & x \geq 0 \end{cases}$

  (a) $f(-1)$ (b) $f(0)$

  (c) $f(1)$ (d) $f(2)$

16. $f(x) = \begin{cases} x^2 + 2, & x \leq 1 \\ 2x^2 + 2, & x > 1 \end{cases}$

  (a) $f(-2)$ (b) $f(0)$

  (c) $f(1)$ (d) $f(2)$

In Exercises 17–22, find all real values of x such that $f(x) = 0$.

17. $f(x) = 15 - 3x$ **18.** $f(x) = \dfrac{3x - 4}{5}$

19. $f(x) = x^2 - 9$ **20.** $f(x) = x^3 - x$

21. $f(x) = \dfrac{3}{x - 1} + \dfrac{4}{x - 2}$

22. $f(x) = 2 + \dfrac{3}{x}$

In Exercises 23–32, find the domain of the function.

23. $g(x) = 1 - 2x^2$ **24.** $f(x) = 5x^2 + 2x - 1$

25. $h(t) = \dfrac{4}{t}$ **26.** $s(y) = \dfrac{3y}{y + 5}$

27. $g(y) = \sqrt{y - 10}$ **28.** $f(t) = \sqrt[3]{t + 4}$

29. $f(x) = \sqrt[4]{1 - x^2}$ **30.** $h(x) = \dfrac{10}{x^2 - 2x}$

31. $g(x) = \dfrac{1}{x} - \dfrac{3}{x + 2}$ **32.** $f(s) = \dfrac{\sqrt{s - 1}}{s - 4}$

In Exercises 33–42, decide whether the equation represents y as a function of x.

33. $x^2 + y^2 = 4$ **34.** $x = y^2$

35. $x^2 + y = 4$ **36.** $x + y^2 = 4$

37. $2x + 3y = 4$

38. $x^2 + y^2 - 2x - 4y + 1 = 0$

39. $y^2 = x^2 - 1$ **40.** $y = \sqrt{x + 5}$

41. $x^2y - x^2 + 4y = 0$ **42.** $xy - y - x - 2 = 0$

In Exercises 43–46, decide whether the set of ordered pairs represents a function from A to B.

$A = \{0, 1, 2, 3\}$ and $B = \{-2, -1, 0, 1, 2\}$

Give reasons for your answers.

43. $\{(0, 1), (1, -2), (2, 0), (3, 2)\}$

44. $\{(0, -1), (2, 2), (1, -2), (3, 0), (1, 1)\}$

45. $\{(0, 0), (1, 0), (2, 0), (3, 0)\}$

46. $\{(0, 2), (3, 0), (1, 1)\}$

In Exercises 47–50, decide whether the set of ordered pairs represents a function from A to B.

$A = \{a, b, c\}$ and $B = \{0, 1, 2, 3\}$

Give reasons for your answers.

47. $\{(a, 1), (c, 2), (c, 3), (b, 3)\}$

48. $\{(a, 1), (b, 2), (c, 3)\}$

49. $\{(1, a), (0, a), (2, c), (3, b)\}$

50. $\{(c, 0), (b, 0), (a, 3)\}$

In Exercises 51–54, decide whether the set of ordered pairs represents a function from A to B.

$A = \{a, b, c\}$ and $B = \{1, 2, 3, 4\}$

Give reasons for your answers.

51. **52.**

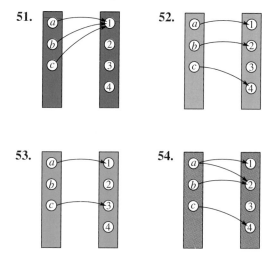

53. **54.**

In Exercises 55–58, the domain of f is the set

$A = \{-2, -1, 0, 1, 2\}$.

Write the function as a set of ordered pairs.

55. $f(x) = x^2$

56. $f(x) = \dfrac{2x}{x^2 + 1}$

57. $f(x) = \sqrt{x + 2}$

58. $f(x) = |x + 1|$

59. *Volume of a Box* An open box is to be made from a square piece of material 12 inches on a side by cutting equal squares from the corners and turning up the sides (see figure).

(a) Write the volume V as a function of x.

(b) What is the domain of this function?

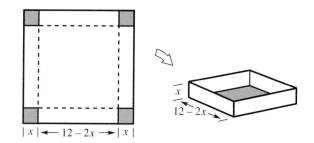

60. *Volume of a Package* A rectangular package to be sent by a postal service can have a maximum combined length and girth (perimeter of a cross section) of 108 inches (see figure).

(a) Write the volume as a function of x.

(b) What is the domain of this function?

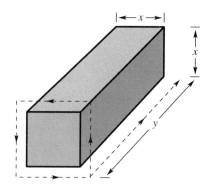

61. *Height of a Balloon*   A balloon carrying a transmitter ascends vertically from a point 2000 feet from the receiving station (see figure). Let d be the distance between the balloon and the receiving station. Express the height of the balloon as a function of d. What is the domain of this function?

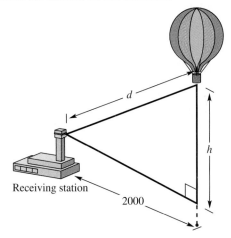

Receiving station

2000

62. *Price of Mobile Homes*   The average price of a new mobile home in the United States from 1985 to 1993 can be approximated by the model

$$p = \begin{cases} 20{,}568 + 9t^3, & 5 \le t \le 9 \\ 90{,}530 - 11{,}770t + 550t^2, & 10 \le t \le 13 \end{cases}$$

where p is the price in dollars and t is the year, with $t = 5$ corresponding to 1985 (see figure). Find the average prices of mobile homes in 1986 and in 1991. (Source: U.S. Bureau of Census)

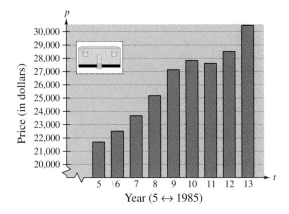

Year (5 ↔ 1985)

63. *Balance in an Account*   A person deposits $2000 in an account that pays 6% interest compounded quarterly.

(a) Write the balance A of the account in terms of the number of years t that the principal is left in the account.

(b) What is the domain of this function?

64. *Unemployment Rates*   The average annual unemployment rate in the United States from 1986 to 1993 is given by the function

$$r = \begin{cases} 18.5 - 2.83t + 0.15t^2, & 6 \le t \le 9 \\ -57.14 + 10.81t - 0.45t^2, & 10 \le t \le 13 \end{cases}$$

where r is the unemployment rate (as a percent) and $t = 6$ represents 1986. Use this function to find the average annual unemployment rates in 1989 and in 1992.   (Source: U.S. Bureau of Labor Statistics)

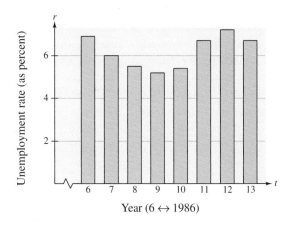

Year (6 ↔ 1986)

65. *Revenue, Cost, and Profit*   A company produces a product for which the variable cost is $12.30 per unit and the fixed costs are $98,000. The product sells for $17.98. Let x be the number of units produced.

(a) Write the total cost C as a function of the number of units produced.

(b) Write the revenue R as a function of the number of units produced.

(c) Write the profit P as a function of the number of units produced.

66. *Revenue, Cost, and Profit*   A company produces a product for which the variable cost is $15.95 per unit and the fixed costs are $110,000. The product sells for $23.79. Let x be the number of units produced.

(a) Write the total cost C as a function of the number of units produced.

(b) Write the revenue R as a function of the number of units produced.

(c) Write the profit P as a function of the number of units produced.

67. *Average Cost*   The inventor of a new game believes that the variable cost for producing the game is $0.95 per unit and the fixed costs are $6000. The inventor sells each game for $1.69. Let x be the number of games sold.

(a) Write the total cost C as a function of the number of games sold.

(b) Write the average cost per unit $\overline{C} = C/x$ as a function of x.

68. *Average Cost*   A manufacturer determines that the variable cost for a new product is $1.15 per unit and the fixed costs are $35,000. The product is to be sold for $2.19. Let x be the number of units sold.

(a) Write the total cost C as a function of the number of units sold.

(b) Write the average cost per unit $\overline{C} = C/x$ as a function of x.

69. *Charter Bus Fares*   For groups of 80 or more people, a charter bus company determines the rate per person as Rate $= 8 - 0.05(n - 80), n \geq 80$, where the rate is given in dollars and n is the number of people.

(a) Express the total revenue R for the bus company as a function of n.

(b) Complete the table.

n	90	100	110	120	130	140	150
R							

(c) Write a paragraph analyzing the data in the table.

70. *Apartment Rentals*   An apartment complex finds that apartment occupancy depends on the monthly rent, according to the model Rent $= 1700 - 10x, 0 \leq x \leq 150$, where rent is given in dollars and x is the number of rented apartments.

(a) Express the total revenue R for the apartment complex as a function of x.

(b) Complete the table.

x	25	50	75	100	125
R					

(c) Write a paragraph analyzing the data in the table.

CAREER INTERVIEW

Stephen T. King

Project Manager

Congress Construction Company, Inc.

Danvers, MA 01923

Congress Construction Company works in construction of buildings such as nursing homes, office buildings, schools, and manufacturing plants. As project manager, I oversee all aspects of construction in the office and at job sites including the interviewing and hiring of subcontractors (about 25 per site) and the quantifying and ordering of all materials needed for the job.

To order construction materials, I estimate quantities from the building's blueprints. For example, I count the number of doors and windows as well as estimate required amounts of items such as concrete, bricks, lumber, concrete blocks, and shingles. Because the price of the required concrete depends on its volume, I use dimensions from the blueprints to estimate the volume of each concrete component of the building. The number of bricks needed to build a wall is a function of the wall's area; for example, if modular bricks are being used, I estimate 7.5 bricks per square foot of wall. As you can see, estimating material quantities requires a variety of skills from both geometry and algebra.

<table>
<tr><td>**3.2**</td><td>## Graphs of Functions</td></tr>
</table>

The Graph of a Function ▪ Increasing and Decreasing Functions ▪
Step Functions ▪ Even and Odd Functions ▪ Common Graphs

The Graph of a Function

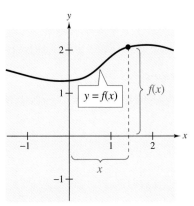

In Section 3.1 you studied functions from an algebraic point of view. In this section, you will study functions from a geometric perspective. The **graph of a function** f is the collection of ordered pairs $(x, f(x))$ such that x is in the domain of f. As you study this section, remember that

$$x = \text{ the directed distance from the } y\text{-axis}$$
$$f(x) = \text{ the directed distance from the } x\text{-axis}$$

as shown in Figure 3.5. If the graph of a function has an x-intercept at $(a, 0)$, then a is a **zero** of the function. In other words, the zeros of a function are the values of x for which $f(x) = 0$. For instance, the function given by $f(x) = x^2 - 4$ has two zeros: -2 and 2.

The **range** of a function (the set of values assumed by the dependent variable) is often more easily determined graphically than algebraically. This technique is illustrated in Example 1.

FIGURE 3.5

EXAMPLE 1   *Finding the Domain and Range of a Function*

Use the graph of the function f, shown in Figure 3.6, to find the following.

a. The domain of f.

b. The function values $f(-1)$ and $f(2)$.

c. The range of f.

Solution

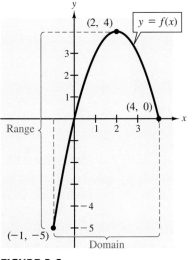

a. Because the graph does not extend beyond $x = -1$ (on the left) and $x = 4$ (on the right), the domain of f is all x in the interval $[-1, \ 4]$.

b. Because $(-1, -5)$ is a point on the graph of f, it follows that

$$f(-1) = -5.$$

Similarly, because $(2, 4)$ is a point on the graph of f, it follows that

$$f(2) = 4.$$

c. Because the graph does not extend below $f(-1) = -5$ nor above $f(2) = 4$, the range of f is the interval $[-5, 4]$.

FIGURE 3.6

By the definition of a function, at most one y-value corresponds to a given x-value. It follows, then, that a vertical line can intersect the graph of a function at most once. This observation provides a convenient visual test for functions.

Vertical Line Test for Functions

A set of points in a coordinate plane is the graph of y as a function of x if and only if no vertical line intersects the graph at more than one point.

EXAMPLE 2   *Vertical Line Test for Functions*

Use the Vertical Line Test to decide whether the graphs in Figure 3.7 represent y as a function of x.

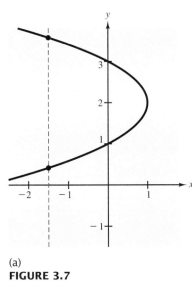 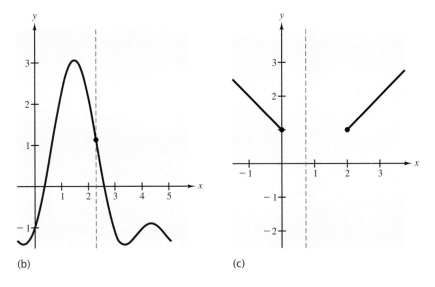

(a)   (b)   (c)

FIGURE 3.7

Solution

a. This *is not* a graph of y as a function of x because we can find a vertical line that intersects the graph twice.

b. This *is* a graph of y as a function of x because every vertical line intersects the graph at most once.

c. This *is* a graph of y as a function of x. (Note that if a vertical line does not intersect the graph, it simply means that the function is undefined for that particular value of x.)

Increasing and Decreasing Functions

The more you know about the graph of a function, the more you know about the function itself. Consider the graph shown in Figure 3.8. As you move from *left to right*, this graph decreases, then is constant, and then increases.

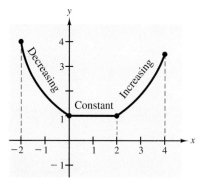

FIGURE 3.8

Increasing, Decreasing, and Constant Functions

A function f is **increasing** on an interval if, for any x_1 and x_2 in the interval, $x_1 < x_2$ implies $f(x_1) < f(x_2)$.

A function f is **decreasing** on an interval if, for any x_1 and x_2 in the interval, $x_1 < x_2$ implies $f(x_1) > f(x_2)$.

A function f is **constant** on an interval if, for any x_1 and x_2 in the interval, $f(x_1) = f(x_2)$.

EXAMPLE 3   *Increasing and Decreasing Functions*

In Figure 3.9, describe the increasing or decreasing behavior of the function.

Solution

a. This function is increasing over the entire real line.

b. This function is increasing on the interval $(-\infty, -1)$, decreasing on the interval $(-1, 1)$, and increasing on the interval $(1, \infty)$.

c. This function is increasing on the interval $(-\infty, 0)$, constant on the interval $(0, 2)$, and decreasing on the interval $(2, \infty)$.

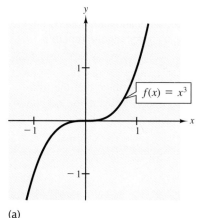

(a)
FIGURE 3.9

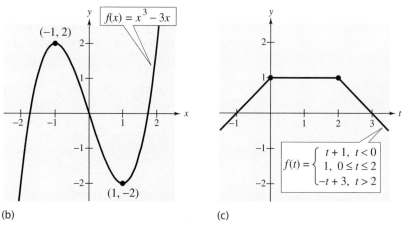

(b)

(c)

To be able to sketch an accurate graph of a function, it is important that you know the points at which the function changes its increasing, decreasing, or constant behavior. These points often identify *maximum* or *minimum* values of the function.

Real Life

EXAMPLE 4 *The Price of Diamonds*

During the 1980s, the average price of a 1-carat polished diamond decreased and then increased according to the model

$$C = -0.7t^3 + 16.25t^2 - 106t + 388, \qquad 2 \le t \le 10$$

where C is the average price in dollars (on the Antwerp Index) and t represents the calendar year, with $t = 2$ corresponding to 1982. According to this model, during which years was the price of diamonds decreasing? During which years was the price of diamonds increasing? Approximate the minimum price of a 1-carat diamond between 1982 and 1990. (Source: Diamond High Council)

Solution

To solve this problem, sketch an accurate graph of the function, as shown in Figure 3.10. From the graph, you can see that the price of diamonds decreased from 1982 until late 1984. Then, from late 1984 to 1990, the price increased. The minimum price during the 8-year period was approximately $175.

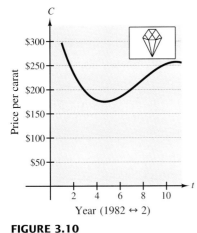

FIGURE 3.10

Technology

A graphing utility is useful for determining the minimum and maximum values of a function over a closed interval. For instance, try graphing the function

$$C = -0.7t^3 + 16.25t^2 - 106t + 388, \qquad 2 \le t \le 10$$

as shown below. By using the TRACE key, you can determine that the minimum value occurs when $x \approx 4.6$.

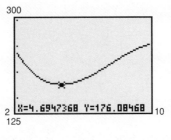

Step Functions

EXAMPLE 5 The Greatest Integer Function

The **greatest integer function** is denoted by $[[x]]$ and is defined by

$$[[x]] = \text{the greatest integer less than or equal to } x.$$

The graph of this function is shown in Figure 3.11. Note that the graph of the greatest integer function jumps vertically one unit at each integer and is constant (a horizontal line segment) between each pair of consecutive integers. Because of the jumps in its graph, the greatest integer function is an example of a category of functions called **step functions.** Some values of the greatest integer function are as follows.

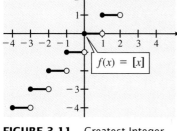

$$[[-1]] = -1 \qquad [[-0.5]] = -1$$
$$[[0]] = 0 \qquad [[0.5]] = 0$$
$$[[1]] = 1 \qquad [[1.5]] = 1$$

FIGURE 3.11 Greatest Integer Function

The range of the greatest integer function is the set of all integers.

Demonstrate the real-life nature of step functions by discussing Exercises 64–67 in Section 3.2. If writing is a part of your course, this section provides a good opportunity for students to find other examples of step functions and write a brief essay on their application.

NOTE If you use a graphing utility to sketch a step function, you should set the utility to *Dot* mode rather than *Connected* mode.

Real Life

EXAMPLE 6 The Price of a Telephone Call

The cost of a telephone call between Los Angeles and San Francisco is $0.50 for the first minute and $0.36 for each additional minute (or portion of a minute). The greatest integer function can be used to create a model for the cost of this call, as follows.

$$C = 0.50 + 0.36[[t]], \qquad t > 0$$

where C is the total cost of the call in dollars and t is the length of the call in minutes. Sketch the graph of this function.

Solution

For calls up to 1 minute, the cost is $0.50. For calls between 1 and 2 minutes, the cost is $0.86, and so on.

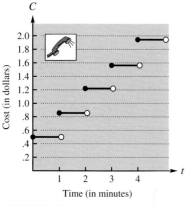

FIGURE 3.12

Length of Call	$0 \leq t < 1$	$1 \leq t < 2$	$2 \leq t < 3$	$3 \leq t < 4$	$4 \leq t < 5$
Cost of Call	$0.50	$0.86	$1.22	$1.58	$1.94

Using these values, you can sketch the graph shown in Figure 3.12.

Even and Odd Functions

In Section 2.2, you studied different types of symmetry of a graph. In the terminology of functions, a function is said to be **even** if its graph is symmetric with respect to the y-axis and to be **odd** if its graph is symmetric with respect to the origin. The symmetry tests in Section 2.2 yield the following tests for even and odd functions.

Test for Even and Odd Functions

A function given by $y = f(x)$ is **even** if, for each x in the domain of f,

$$f(-x) = f(x).$$

A function given by $y = f(x)$ is **odd** if, for each x in the domain of f,

$$f(-x) = -f(x).$$

The Cartesian coordinate plane named after René Descartes was developed independently by another French mathematician, Pierre Fermat. Fermat's *Introduction to Loci*, written about 1629, was clearer and more systematic than Descartes's *La géométrie*. However, Fermat's work was not published during his lifetime. Consequently, Descartes received the credit for the development of the coordinate plane with the now familiar x and y axes.

EXAMPLE 7   Even and Odd Functions

a.   The function $g(x) = x^3 - x$ is odd because

$$g(-x) = (-x)^3 - (-x) = -x^3 + x = -(x^3 - x) = -g(x).$$

b.   The function $h(x) = (x)^2 + 1$ is even because

$$h(-x) = (-x)^2 + 1 = x^2 + 1 = h(x).$$

The graphs of the two functions are shown in Figure 3.13.

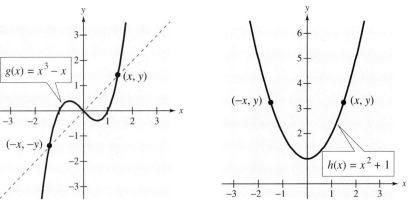

(a) Odd function (symmetric to origin)   (b) Even function (symmetric to y-axis)

FIGURE 3.13

Common Graphs

When discussing this summary of graphs of common functions, emphasize that the graph of a function is related to a "family" of graphs. If students learn these "families," graphing will be much easier. You can reinforce this concept by discovery methods such as graphing $f(x) = x^2$, $f(x) = x^2 + 2$, $f(x) = (x - 1)^2$, and $f(x) = (x - 1)^2 + 2$ and noting similarities and differences.

Figure 3.14 shows the graphs of six common functions. You need to be familiar with these graphs.

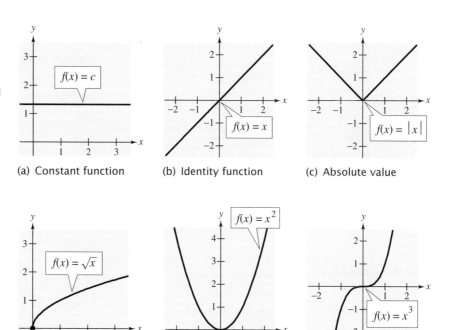

(a) Constant function

(b) Identity function

(c) Absolute value

(d) Square root

(e) Squaring function

(f) Cubing function

FIGURE 3.14

Activities

1. Determine whether the function is even, odd, or neither: $f(x) = x^3 - x + 1$.

   Answer: Neither

2. Does the graph of a circle pass the vertical line test for functions?

   Answer: No

3. Use the graph of $f(x) = x^3 - 3x$ to determine the open interval(s) where the graph is decreasing.

   Answer: $(-1, 1)$

Group Activities   Extending the Concept

Increasing and Decreasing Functions   Use your school's library or some other reference source to find examples of three different functions that represent quantities between 1983 and 1993. Find one that decreased during the decade, one that increased, and one that was constant. For instance, the value of the dollar decreased, the population of the United States increased, and the land size of the United States remained constant. Can you find three other examples? If the examples you find appear to represent *linear growth or decline*, use the methods described in Section 2.5 to find a linear function $f(x) = a + bx$ that approximates the data.

Warm Up

The following warm-up exercises involve skills that were covered in earlier sections. You will use these skills in the exercise set for this section.

1. Find $f(2)$ for $f(x) = -x^3 + 5x$.

2. Find $f(6)$ for $f(x) = x^2 - 6x$.

3. Find $f(-x)$ for $f(x) = 3/x$.

4. Find $f(-x)$ for $f(x) = x^2 + 3$.

In Exercises 5 and 6, solve the equation.

5. $x^3 - 16x = 0$

6. $2x^2 - 3x + 1 = 0$

In Exercises 7–10, find the domain of the function.

7. $g(x) = 4(x - 4)^{-1}$

8. $f(x) = 2x/(x^2 - 9x + 20)$

9. $h(t) = \sqrt[4]{5 - 3t}$

10. $f(t) = t^3 + 3t - 5$

3.2 **Exercises**

In Exercises 1–8, find the range of the function.

1. $f(x) = \sqrt{x - 1}$

2. $f(x) = 4 - x^2$

5. $f(x) = \sqrt{25 - x^2}$

6. $f(x) = \dfrac{|x|}{x}$

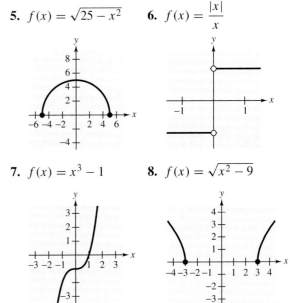

3. $f(x) = \sqrt{x^2 - 4}$

4. $f(x) = |x - 2|$

7. $f(x) = x^3 - 1$

8. $f(x) = \sqrt{x^2 - 9}$

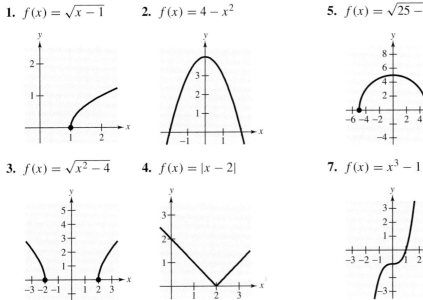

In Exercises 9–14, use the Vertical Line Test to decide whether y is a function of x.

9. $y = x^2$

10. $y = x^3 - 1$

11. $x - y^2 = 0$

12. $x^2 + y^2 = 9$

13. $x^2 = xy - 1$

14. $x = |y|$

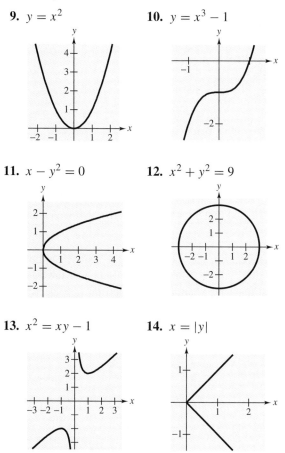

In Exercises 15–22, describe the increasing and decreasing behavior of the function.

15. $f(x) = 2x$

16. $f(x) = x^2 - 2x$

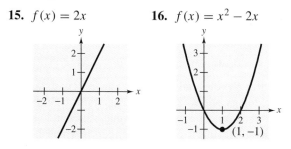

17. $f(x) = x^3 - 3x^2$

18. $f(x) = \sqrt{x^2 - 4}$

19. $f(x) = 3x^4 - 6x^2$

20. $f(x) = x^{2/3}$

21. $y = x\sqrt{x + 3}$

22. $y = |x + 1| + |x - 1|$

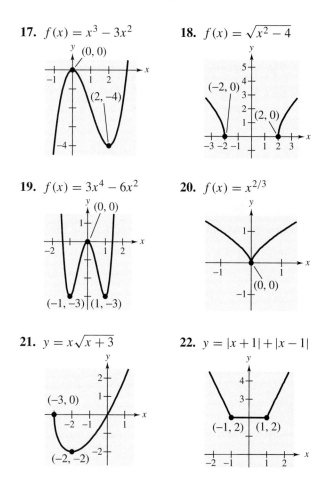

In Exercises 23–28, decide whether the function is even, odd, or neither.

23. $f(x) = x^6 - 2x^2 + 3$

24. $h(x) = x^3 - 5$

25. $g(x) = x^3 - 5x$

26. $f(x) = x\sqrt{1 - x^2}$

27. $f(t) = t^2 + 2t - 3$

28. $g(s) = 4s^{2/3}$

In Exercises 29–40, sketch the graph of the function and determine whether the function is even, odd, or neither.

29. $f(x) = 3$

30. $g(x) = x$

31. $f(x) = 5 - 3x$

32. $h(x) = x^2 - 4$

33. $g(s) = \dfrac{s^3}{4}$

34. $f(t) = -t^4$

35. $f(x) = \sqrt{1-x}$

36. $f(x) = x^{3/2}$

37. $g(t) = \sqrt[3]{t-1}$

38. $f(x) = |x+2|$

39. $f(x) = \begin{cases} x+3, & x \le 0 \\ 3, & 0 < x \le 2 \\ 2x-1, & x > 2 \end{cases}$

40. $f(x) = \begin{cases} 2x+1, & x \le -1 \\ x^2 - 2, & x > -1 \end{cases}$

In Exercises 41–54, sketch the graph of the function.

41. $f(x) = 4 - x$

42. $f(x) = 4x + 2$

43. $f(x) = x^2 - 9$

44. $f(x) = x^2 - 4x$

45. $f(x) = 1 - x^4$

46. $f(x) = \sqrt{x+2}$

47. $f(x) = x^2 + 1$

48. $f(x) = -1(1 + |x|)$

49. $f(x) = -5$

50. $f(x) = \frac{1}{2}(2 + |x|)$

51. $f(x) = -[[x]]$

52. $f(x) = 2[[x]]$

53. $f(x) = [[x-1]]$

54. $f(x) = [[x+1]]$

In Exercises 55–58, use a graphing utility to graph the function and then estimate the open intervals on which the function is increasing or decreasing.

55. $f(x) = x^2 - 4x + 1$

56. $f(x) = -x^2 + 6x + 3$

57. $f(x) = x^3 - 3x^2$

58. $f(x) = -x^3 + 3x + 1$

59. *Population of Oklahoma* The population of the state of Oklahoma fell during the latter part of the 1980s and then began to increase. A model that approximates the population of Oklahoma is

$$P = 7.8t^2 - 151t + 3878, \qquad 6 \le t \le 13$$

where P represents the population in thousands and $t = 6$ represents 1986. Use the figure to estimate the years that the population was decreasing and the years that the population was increasing. (Source: U.S. Census Bureau)

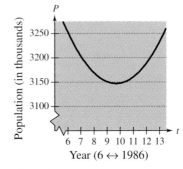

Figure for 59

60. *Bank Deposits* From 1980 to 1993, the total deposits for a bank can be approximated by the function

$$D = -0.0027t^3 + 0.058t^2 - 0.32t + 1.45,$$
$$0 \le t \le 13$$

where D is the total annual deposits in billions of dollars and $t = 0$ represents 1980. Use the figure to estimate the years that the deposits were decreasing and the years that the deposits were increasing.

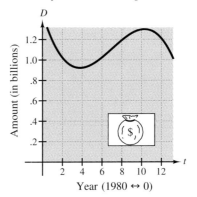

61. *Company Revenue* A company determines that the total revenue for the years from 1985 to 1995 can be approximated by the function

$$R = -0.045t^3 + 0.75t^2 - 0.7t + 2.54, \quad 5 \le t \le 15$$

where R is the total revenue in hundreds of thousands of dollars and $t = 5$ represents 1985. Graph the revenue function with a graphing utility and use the TRACE feature to estimate the years in which the revenue was increasing and the years in which the revenue was decreasing.

62. *Maximum Profit* The marketing department of a company estimates that the demand for a product is given by $p = 100 - 0.0001x$, where p is the price per unit and x is the number of units. The cost of producing x units is given by $C = 350,000 + 30x$, and the profit for producing x units is given by

$$P = R - C = xp - C.$$

Sketch the graph of the profit function and estimate the number of units that would produce a maximum profit.

63. *Maximum Profit* The marketing department of a company estimates that the demand for a product is given by $p = 120 - 0.0001x$, where p is the price per unit and x is the number of units. The cost of producing x units is given by $C = 450,000 + 50x$, and the profit for producing x units is given by

$$P = R - C = xp - C.$$

Sketch the graph of the profit function and estimate the number of units that would produce a maximum profit.

64. *Cost of a Phone Call* Suppose that the cost of a telephone call between Denver and Boise is $0.60 for the first minute and $0.42 for each additional minute (or portion of a minute). A model for the total cost of the phone call is

$$C = 0.60 + 0.42[\![x]\!], \qquad 0 < x$$

where C is the total cost of the call in dollars and x is the length of the call in minutes. Sketch the graph of this function.

65. *Cost of a Phone Call* Suppose the cost of a telephone call between Dallas and New York City is $0.75 for the first minute and $0.46 for each additional minute (or portion of a minute). A model for the total cost of the phone call is

$$C = 0.75 + 0.46[\![x]\!], \qquad 0 < x$$

where C is the total cost of the call in dollars and x is the length of the call in minutes. Sketch the graph of this function.

66. *Cost of Overnight Delivery* Suppose that the cost of sending an overnight package from New York to Atlanta is $9.80 for the first pound and $2.50 for each additional pound (or portion of a pound). A model for the total cost of sending the package is

$$C = 9.8 + 2.5[\![x]\!], \qquad 0 < x$$

where C is the total cost in dollars and x is the weight of the package in pounds. Sketch the graph of this function.

67. *Cost of Overnight Delivery* Suppose that the cost of sending an overnight package from Los Angeles to Miami is $10.75 for the first pound and $3.95 for each additional pound (or portion of a pound). A model for the total cost of sending the package is

$$C = 10.75 + 3.95[\![x]\!], \qquad 0 < x$$

where C is the total cost in dollars and x is the weight in pounds. Sketch the graph of this function.

68. *Research and Development* The total amount spent on basic research and development in the United States from 1960 to 1988 can be approximated by the model

$$C = \begin{cases} 13.24 + 1.35t, & 0 \le t \le 15.75 \\ -83.15 + 7.47t, & 15.75 \le t \le 28 \end{cases}$$

where C is the amount in billions of dollars and t is the year, with $t = 0$ corresponding to 1960. Sketch the graph of this function. (Source: National Science Foundation)

69. *Grade Level Salaries* The 1988 average salary for federal employees can be approximated by the model

$$S = \begin{cases} 6874.7 + 2162.1x, & x = 1, 2, \ldots, 12 \\ -66,957 + 8706.9x, & x = 13, 14, 15, 16 \end{cases}$$

where S is the salary in dollars and x represents the "GS" level. Sketch a *bar graph* that represents this function. (Source: U.S. Office of Personnel Management)

3.3 | Translations and Combinations of Functions

Shifting, Reflecting, and Stretching Graphs ▪ Arithmetic Combinations of Functions ▪ Composition of Functions ▪ Applications

Shifting, Reflecting, and Stretching Graphs

Many functions have graphs that are simple transformations of the common graphs that are summarized on page 249. For example, you can obtain the graph of $h(x) = x^2 + 2$ by shifting the graph of $f(x) = x^2$ *up* two units, as shown in Figure 3.15. In function notation, h and f are related as follows.

$$h(x) = x^2 + 2 = f(x) + 2 \qquad \text{Upward shift of 2}$$

Similarly, you can obtain the graph of $g(x) = (x - 2)^2$ by shifting the graph of $f(x) = x^2$ to the *right* two units, as shown in Figure 3.16. In this case, the functions g and f have the following relationship.

$$g(x) = (x - 2)^2 = f(x - 2) \qquad \text{Right shift of 2}$$

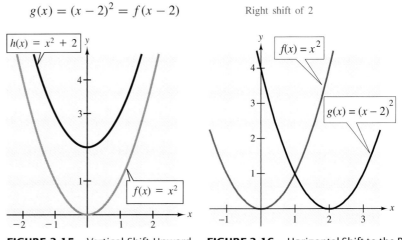

FIGURE 3.15 Vertical Shift Upward   **FIGURE 3.16** Horizontal Shift to the Right

NOTE In items 3 and 4, be sure you see that $h(x) = f(x - c)$ corresponds to a *right* shift and $h(x) = f(x + c)$ corresponds to a *left* shift.

Vertical and Horizontal Shifts

Let c be a positive real number. **Vertical** and **horizontal shifts** in the graph of $y = f(x)$ are represented as follows.

1. Vertical shift c units **upward:** $\qquad h(x) = f(x) + c$

2. Vertical shift c units **downward:** $\qquad h(x) = f(x) - c$

3. Horizontal shift c units to the **right:** $\qquad h(x) = f(x - c)$

4. Horizontal shift c units to the **left:** $\qquad h(x) = f(x + c)$

Some graphs can be obtained from a combination of vertical and horizontal shifts. This is demonstrated in Example 1(b).

NOTE   Vertical and horizontal shifts such as those shown in Example 1 generate a *family of functions*, each with the same shape but at different locations in the plane.

EXAMPLE 1   Shifts in the Graph of a Function

Use the graph of $f(x) = x^3$ to sketch the graphs of the following functions.

a. $g(x) = x^3 + 1$ **b.** $h(x) = (x + 2)^3 + 1$

Solution

Relative to the graph of $f(x) = x^3$, the graph of $g(x) = x^3 + 1$ is an upward shift of one unit, and the graph of $h(x) = (x + 2)^3 + 1$ involves a left shift of two units *and* an upward shift of one unit. The graphs of both functions are compared with the graph of $f(x) = x^3$ in Figure 3.17.

(a) Vertical shift: 1 up   (b) Horizontal shift: 2 left; vertical shift: 1 up
FIGURE 3.17

Technology

Graphing utilities are ideal tools for exploring translations of functions. Graph f, g, and h on the same screen. Before looking at the graphs, try to predict how the graphs of g and h relate to the graph of f.

a. $f(x) = x^2$, $g(x) = (x - 4)^2$, $h(x) = (x - 4)^2 + 3$

b. $f(x) = x^2$, $g(x) = (x + 1)^2$, $h(x) = (x + 1)^2 - 2$

c. $f(x) = x^2$, $g(x) = (x + 4)^2$, $h(x) = (x + 4)^2 + 2$

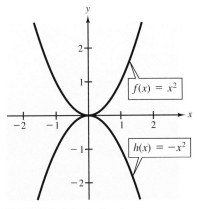

FIGURE 3.18 Reflection

The second common type of transformation is a **reflection.** For instance, if you consider the x-axis to be a mirror, the graph of $h(x) = -x^2$ is the mirror image (or reflection) of the graph of $f(x) = x^2$, as shown in Figure 3.18.

Reflections in the Coordinate Axes

Reflections in the coordinate axes of the graph of $y = f(x)$ are represented as follows.

1. **Reflection in the x-axis:** $\quad h(x) = -f(x)$

2. **Reflection in the y-axis:** $\quad h(x) = f(-x)$

EXAMPLE 2 *Reflections and Shifts*

Compare the graphs of the following with the graph of $f(x) = \sqrt{x}$.

a. $g(x) = -\sqrt{x}$ **b.** $h(x) = \sqrt{-x}$

Solution

a. The graph of g is a reflection of the graph of f in the x-axis because

$$g(x) = -\sqrt{x} = -f(x).$$

b. The graph of h is a reflection of the graph of f in the y-axis because

$$h(x) = \sqrt{-x} = f(-x).$$

The graphs of both functions are shown in Figure 3.19.

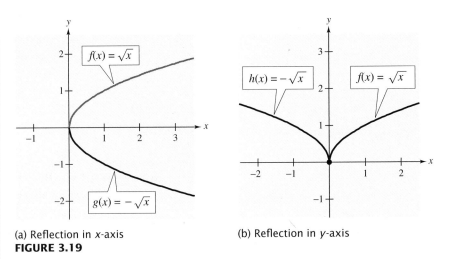

(a) Reflection in x-axis   (b) Reflection in y-axis

FIGURE 3.19

PROGRAMMING

In the Appendix, you will find programs for a variety of calculator models that will give you practice working with reflections, horizontal shifts, and vertical shifts. These programs will sketch a graph of the function

$$y = R(x + H)^2 + V$$

where $R = \pm 1$, H is an integer between -6 and 6, and V is an integer between -3 and 3. Each time you run the program, different values of R, H, and V are possible. From the graph, you should be able to determine the values of R, H and V. After you have determined the values, press ENTER to see the answer. (To look at the graph again, press GRAPH.)

DISCOVERY

Use a graphing utility to graph $f(x) = 2x^2$. Compare this graph with the graph of $h(x) = x^2$. Describe the effect of multiplying x^2 by a number greater than 1. Then graph $g(x) = \frac{1}{2}x^2$. Compare this with the graph of $h(x) = x^2$. Describe the effect of multiplying x^2 by a number less than 1. Can you think of an easy way to remember this generalization?

Horizontal shifts, vertical shifts, and reflections are **rigid** transformations because the basic shape of the graph is unchanged. These transformations change only the *position* of the graph in the xy-plane. **Nonrigid** transformations are those that cause a *distortion*—a change in the shape of the original graph. For instance, a nonrigid transformation of the graph of $y = f(x)$ is represented by $g(x) = cf(x)$, where the transformation is a **vertical stretch** if $c > 1$ and a **vertical shrink** if $0 < c < 1$.

EXAMPLE 3   Nonrigid Transformations

Compare the graphs of the following with the graph of $f(x) = |x|$.

a. $h(x) = 3|x|$ **b.** $g(x) = \frac{1}{3}|x|$

Solution

a. Relative to the graph of $f(x) = |x|$, the graph of

$$h(x) = 3|x| = 3f(x)$$

is a vertical stretch (multiply each y-value by 3) of the graph of f.

b. Similarly, the equation

$$g(x) = \frac{1}{3}|x| = \frac{1}{3}f(x)$$

indicates that the graph of g is a vertical shrink of the graph of f.
The graphs of both functions are shown in Figure 3.20.

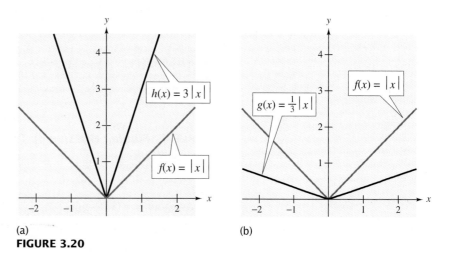

(a)   (b)

FIGURE 3.20

Arithmetic Combinations of Functions

Just as two real numbers can be combined by the operations of addition, subtraction, multiplication, and division to form other real numbers, two *functions* can be combined to create new functions. For example, the functions

$$f(x) = 2x - 3 \quad \text{and} \quad g(x) = x^2 - 1$$

can be combined to form the sum, difference, product, and quotient of f and g as follows.

$$f(x) + g(x) = (2x - 3) + (x^2 - 1) = x^2 + 2x - 4 \qquad \text{Sum}$$

$$f(x) - g(x) = (2x - 3) - (x^2 - 1) = -x^2 + 2x - 2 \qquad \text{Difference}$$

$$f(x)g(x) = (2x - 3)(x^2 - 1) = 2x^3 - 3x^2 - 2x + 3 \qquad \text{Product}$$

$$\frac{f(x)}{g(x)} = \frac{2x - 3}{x^2 - 1}, \quad x \neq \pm 1 \qquad \text{Quotient}$$

The domain of an arithmetic combination of functions f and g consists of all real numbers that are common to the domains of f and g. In the case of the quotient $f(x)/g(x)$, there is the further restriction that $g(x) \neq 0$.

Sum, Difference, Product, and Quotient of Functions

Let f and g be two functions with overlapping domains. Then, for all x common to both domains, the **sum, difference, product,** and **quotient** of f and g are defined as follows.

1. *Sum:* $\qquad (f + g)(x) = f(x) + g(x)$

2. *Difference:* $\qquad (f - g)(x) = f(x) - g(x)$

3. *Product:* $\qquad (fg)(x) = f(x) \cdot g(x)$

4. *Quotient:* $\qquad \left(\dfrac{f}{g}\right)(x) = \dfrac{f(x)}{g(x)}, \quad g(x) \neq 0$

EXAMPLE 4   *Finding the Sum of Two Functions*

Given $f(x) = 2x + 1$ and $g(x) = x^2 + 2x - 1$, find $(f + g)(x)$.

Solution

$$(f + g)(x) = f(x) + g(x)$$
$$= (2x + 1) + (x^2 + 2x - 1)$$
$$= x^2 + 4x$$

EXAMPLE 5   Finding the Difference of Two Functions

Given $f(x) = 2x + 1$ and $g(x) = x^2 + 2x - 1$, find $(f - g)(x)$. Then evaluate the difference when $x = 2$.

Solution

The difference of the functions f and g is given by

$$(f - g)(x) = f(x) - g(x)$$
$$= (2x + 1) - (x^2 + 2x - 1)$$
$$= -x^2 + 2.$$

When $x = 2$, the value of this difference is

$$(f - g)(2) = -(2)^2 + 2 = -2.$$

In Examples 4 and 5, both f and g have domains that consist of all real numbers. Thus, the domains of $(f + g)$ and $(f - g)$ are also the set of all real numbers. Remember that any restrictions on the domains of f and g must be considered when forming the sum, difference, product, or quotient of f and g.

EXAMPLE 6   The Quotient of Two Functions

Find the domains of $(f/g)(x)$ and $(g/f)(x)$ for the functions

$$f(x) = \sqrt{x} \qquad \text{and} \qquad g(x) = \sqrt{4 - x^2}.$$

Solution

The quotient of f and g is given by

$$\left(\frac{f}{g}\right)(x) = \frac{f(x)}{g(x)} = \frac{\sqrt{x}}{\sqrt{4 - x^2}}$$

and the quotient of g and f is given by

$$\left(\frac{g}{f}\right)(x) = \frac{g(x)}{f(x)} = \frac{\sqrt{4 - x^2}}{\sqrt{x}}.$$

The domain of f is $[0, \infty)$ and the domain of g is $[-2, 2]$. The intersection of these two domains is $[0, 2]$, which implies that the domains of f/g and g/f are as follows.

$$\text{Domain of } \frac{f}{g} : \ [0, 2) \qquad \text{Domain of } \frac{g}{f} : \ (0, 2]$$

Can you see why these two domains differ slightly?

Composition of Functions

Another way of combining two functions is to form the **composition** of one with the other. For instance, if $f(x) = x^2$ and $g(x) = x + 1$, the composition of f with g is given by

$$f(g(x)) = f(x + 1) = (x + 1)^2.$$

This composition is denoted as $f \circ g$.

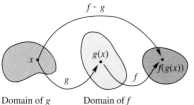

Domain of g Domain of f

FIGURE 3.21

Definition of Composition of Two Functions

The **composition** of the functions f and g is given by

$$(f \circ g)(x) = f(g(x)).$$

The domain of $(f \circ g)$ is the set of all x in the domain of g such that $g(x)$ is in the domain of f. (See Figure 3.21.)

EXAMPLE 7   Composition of Functions

Given $f(x) = x + 2$ and $g(x) = 4 - x^2$, find the following.

a. $(f \circ g)(x)$ **b.** $(g \circ f)(x)$

Solution

a. The composition of f with g is as follows.

$$
\begin{aligned}
(f \circ g)(x) &= f(g(x)) && \text{Definition of } f \circ g \\
&= f(4 - x^2) && \text{Definition of } g(x) \\
&= (4 - x^2) + 2 && \text{Definition of } f(x) \\
&= -x^2 + 6 && \text{Simplify.}
\end{aligned}
$$

b. The composition of g with f is as follows.

$$
\begin{aligned}
(g \circ f)(x) &= g(f(x)) && \text{Definition of } g \circ f \\
&= g(x + 2) && \text{Definition of } f(x) \\
&= 4 - (x + 2)^2 && \text{Definition of } g(x) \\
&= 4 - (x^2 + 4x + 4) && \text{Expand.} \\
&= -x^2 - 4x && \text{Simplify.}
\end{aligned}
$$

Note that, in this case, $(f \circ g)(x) \neq (g \circ f)(x)$.

Applications

EXAMPLE 8   Political Makeup of the U.S. Senate

Consider three functions R, D, and I that represent the numbers of Republicans, Democrats, and Independents in the U.S. Senate from 1965 to 1995. Sketch the graphs of R, D, and I, and the sum of R, D, and I, in the same coordinate plane. The numbers of Republicans and Democrats in the Senate are shown below.

Year	Republicans	Democrats	Year	Republicans	Democrats
1965	32	68	1981	53	46
1967	36	64	1983	54	46
1969	43	57	1985	53	47
1971	44	54	1987	45	55
1973	42	56	1989	45	55
1975	37	60	1991	44	56
1977	38	61	1993	43	57
1979	41	58	1995	53	47

Solution

The graphs of R, D, and I are shown in Figure 3.22. Note that the sum of R, D, and I is the constant function $R + D + I = 100$. This follows from the fact that the number of senators in the United States is 100 (two from each state).

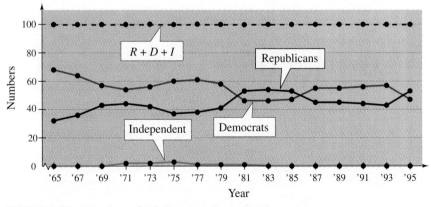

FIGURE 3.22   Number of U.S. Senators (by political party)

Activities

1. Find $(f + g)(-1)$ for $f(x) = 3x^2 + 2$, $g(x) = 2x$.

   Answer: 3

2. Given $f(x) = 3x^2 + 2$ and $g(x) = 2x$, find $f \circ g$.

   Answer: $(f \circ g)(x) = 12x^2 + 2$

3. Find two functions f and g such that $(f \circ g)(x) = h(x)$.

   $h(x) = \dfrac{1}{\sqrt{3x + 1}}$.

   Answers are not unique:

   $f(x) = \dfrac{1}{\sqrt{x}}$ and $g(x) = 3x + 1$

Group Activity Suggestion

Depending on the availability of graphing utilities in your class, you may want to consider the following suggestions.

1. If everyone in your class uses the same model of graphing utility, divide the class into groups of four so that each person in the group is responsible for one translation.

2. If you require graphing utilities but not everyone has the same model, divide the class based on graphing utilities so each group has its own built-in support system.

Real Life

EXAMPLE 9   *Bacteria Count*

The number of bacteria in a refrigerated food is given by

$$N(T) = 20T^2 - 80T + 500, \qquad 2 \le T \le 14$$

where T is the temperature of the food. When the food is removed from refrigeration, the temperature is given by

$$T(t) = 4t + 2, \qquad 0 \le t \le 3$$

where t is the time in hours. Find (a) the composite $N(T(t))$, (b) the number of bacteria in the food when $t = 2$ hours, and (c) the time when the bacteria count reaches 2000.

Solution

a. $N(T(t)) = 20(4t + 2)^2 - 80(4t + 2) + 500$

   $\qquad = 20(16t^2 + 16t + 4) - 320t - 160 + 500$

   $\qquad = 320t^2 + 320t + 80 - 320t - 160 + 500$

   $\qquad = 320t^2 + 420$

b. When $t = 2$, the number of bacteria is

   $N = 320(2)^2 + 420 = 1280 + 420 = 1700.$

c. The bacteria count will reach $N = 2000$ when $320t^2 + 420 = 2000$. By solving this equation, you can determine that the bacteria count will reach 2000 when

   $t \approx 2.2$ hours.

Group Activities   Exploring with Technology

3. If you do not require graphing utilities in your class but some students use them, divide the class so that at least one student owning a graphing utilitiy is in the group. Consider asking the graphing utility owner to conduct a demonstration for the rest of the group and then work cooperatively.

Constructing Translations   Use a graphing utility to graph $f(x) = 3 - x^3$. Decide how to alter this function to produce each of the following translation descriptions. Graph each translation on the same screen with f; confirm that the translation moves f as described.

a. The graph of f shifted upward four units

b. The graph of f shifted to the left two units

c. The graph of f shifted downward two units and to the right one unit

d. The graph of f reflected in the x-axis

Warm Up   The following warm-up exercises involve skills that were covered in earlier sections. You will use these skills in the exercise set for this section.

In Exercises 1–10, perform the indicated operations and simplify the result.

1. $\dfrac{1}{x} + \dfrac{1}{1-x}$

2. $\dfrac{2}{x+3} - \dfrac{2}{x-3}$

3. $\dfrac{3}{x-2} - \dfrac{2}{x(x-2)}$

4. $\dfrac{x}{x-5} + \dfrac{1}{3}$

5. $(x-1)\left(\dfrac{1}{\sqrt{x^2-1}}\right)$

6. $\left(\dfrac{x}{x^2-4}\right)\left(\dfrac{x^2-x-2}{x^2}\right)$

7. $(x^2-4) \div \left(\dfrac{x+2}{5}\right)$

8. $\left(\dfrac{x}{x^2+3x-10}\right) \div \left(\dfrac{x^2+3x}{x^2+6x+5}\right)$

9. $\dfrac{(1/x)+5}{3-(1/x)}$

10. $\dfrac{(x/4)-(4/x)}{x-4}$

3.3 Exercises

In Exercises 1–8, use the graph of $f(x) = x^2$ to sketch the graph of the function.

1. $g(x) = x^2 + 3$

2. $g(x) = x^2 - 2$

3. $g(x) = (x+3)^2$

4. $g(x) = (x-4)^2$

5. $g(x) = (x-2)^2 + 2$

6. $g(x) = (x+1)^2 - 3$

7. $g(x) = -x^2 + 1$

8. $g(x) = -(x-2)^2$

In Exercises 9–14, use the graph of $f(x) = \sqrt{x}$ to sketch the graph of the function.

9. $y = \sqrt{x} + 2$

10. $y = -\sqrt{x}$

11. $y = \sqrt{x-2}$

12. $y = \sqrt{x+3}$

13. $y = 2 - \sqrt{x-4}$

14. $y = \sqrt{2x}$

In Exercises 15–20, use the graph of $f(x) = \sqrt[3]{x}$ to sketch the graph of the function.

15. $y = \sqrt[3]{x} - 1$

16. $y = \sqrt[3]{x+1}$

17. $y = \sqrt[3]{x-1}$

18. $y = -\sqrt[3]{x-2}$

19. $y = \sqrt[3]{x+1} - 1$

20. $y = \frac{1}{2}\sqrt[3]{x}$

21. Use a graphing utility to graph f for $c = -2, 0$, and 2 on the same screen.

(a) $f(x) = \frac{1}{2}x + c$

(b) $f(x) = \frac{1}{2}(x-c)$

(c) $f(x) = \frac{1}{2}(cx)$

In each case, compare the three graphs with the graph of $y = \frac{1}{2}x$.

22. Use a graphing utility to graph f for $c = -2, 0$, and 2 on the same screen.

(a) $f(x) = x^3 + c$

(b) $f(x) = (x-c)^3$

(c) $f(x) = (x-2)^3 + c$

In each case, compare the three graphs with the graph of $y = x^3$.

23. Use the graph of $f(x) = x^2$ to write formulas for the functions whose graphs are shown.

(a)   (b)

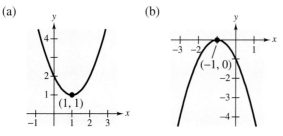

24. Use the graph of $f(x) = x^3$ to write formulas for the functions whose graphs are shown.

(a)   (b)

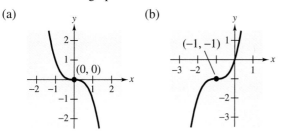

In Exercises 25–32, find the sum, difference, product, and quotient of f and g. What is the domain of f/g?

25. $f(x) = x + 1$, $g(x) = x - 1$

26. $f(x) = 2x - 5$, $g(x) = 1 - x$

27. $f(x) = x^2$, $g(x) = 1 - x$

28. $f(x) = 2x - 5$, $g(x) = 5$

29. $f(x) = x^2 + 5$, $g(x) = \sqrt{1 - x}$

30. $f(x) = \sqrt{x^2 - 4}$, $g(x) = \dfrac{x^2}{x^2 + 1}$

31. $f(x) = \dfrac{1}{x}$, $g(x) = \dfrac{1}{x^2}$

32. $f(x) = \dfrac{x}{x + 1}$, $g(x) = x^3$

In Exercises 33–44, evaluate the function for $f(x) = x^2 + 1$ and $g(x) = x - 4$.

33. $(f + g)(3)$ **34.** $(f - g)(-2)$

35. $(f - g)(2t)$ **36.** $(f + g)(t - 1)$

37. $(fg)(4)$ **38.** $(fg)(-6)$

39. $\left(\dfrac{f}{g}\right)(5)$ **40.** $\left(\dfrac{f}{g}\right)(0)$

41. $(f - g)(0)$ **42.** $(f + g)(1)$

43. $\left(\dfrac{f}{g}\right)(-1) - g(3)$ **44.** $(2f)(5)$

In Exercises 45–48, find (a) $f \circ g$ and (b) $f \circ f$.

45. $f(x) = x^2$, $g(x) = x - 1$

46. $f(x) = 3x$, $g(x) = 2x + 1$

47. $f(x) = 3x + 5$, $g(x) = 5 - x$

48. $f(x) = x^3$, $g(x) = \dfrac{1}{x}$

In Exercises 49–56, find (a) $f \circ g$ and (b) $g \circ f$.

49. $f(x) = \sqrt{x + 4}$, $g(x) = x^2$

50. $f(x) = \sqrt[3]{x - 1}$, $g(x) = x^3 + 1$

51. $f(x) = \frac{1}{3}x - 3$, $g(x) = 3x + 1$

52. $f(x) = x^4$, $g(x) = x^4$

53. $f(x) = \sqrt{x}$, $g(x) = \sqrt{x}$

54. $f(x) = 2x - 3$, $g(x) = 2x - 3$

55. $f(x) = |x|$, $g(x) = x + 6$

56. $f(x) = x^{2/3}$, $g(x) = x^6$

In Exercises 57–60, use the graphs of f and g to evaluate the functions.

(a)   (b)

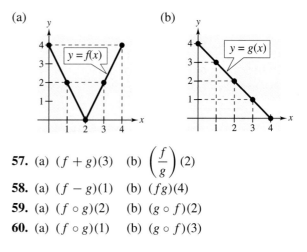

57. (a) $(f + g)(3)$ (b) $\left(\dfrac{f}{g}\right)(2)$

58. (a) $(f - g)(1)$ (b) $(fg)(4)$

59. (a) $(f \circ g)(2)$ (b) $(g \circ f)(2)$

60. (a) $(f \circ g)(1)$ (b) $(g \circ f)(3)$

In Exercises 61–64, determine the domain of (a) f, (b) g, and (c) $f \circ g$.

61. $f(x) = \sqrt{x}, \quad g(x) = x^2 + 1$

62. $f(x) = \dfrac{1}{x}, \quad g(x) = x + 3$

63. $f(x) = \dfrac{3}{x^2 - 1}, \quad g(x) = x + 1$

64. $f(x) = 2x + 3, \quad g(x) = \dfrac{x}{2}$

65. *Comparing Profits*   A company has two manufacturing plants, one in New Jersey and the other in California. From 1990 to 1995, the profits for the manufacturing plant in New Jersey have been decreasing according to the function

$$P_1 = 14.82 - 0.43t, \qquad t = 0, 1, 2, 3, 4, 5$$

where P_1 represents the profit in millions of dollars and $t = 0$ represents 1990. On the other hand, the profits for the manufacturing plant in California have been increasing according to the function

$$P_2 = 16.16 + 0.56t, \qquad t = 0, 1, 2, 3, 4, 5.$$

Write a function that represents the overall company profits during the 6-year period. Use the *stacked bar graph* in the figure, which represents the total profit for the company during this 6-year period, to determine whether the overall company profits have been increasing or decreasing.

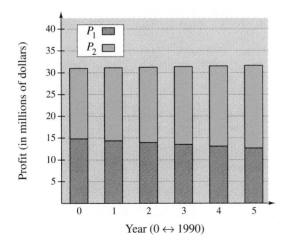

Year (0 ↔ 1990)

66. *Comparing Sales*   You own two fast-food restaurants. During the years 1990 to 1995, the sales for the first restaurant have been decreasing according to the function

$$R_1 = 480 - 8t - 0.8t^2, \qquad t = 0, 1, 2, 3, 4, 5$$

where R_1 represents the sales for the first restaurant (in thousands of dollars) and $t = 0$ represents 1990. During the same 6-year period, the sales for the second restaurant have been increasing according to the function

$$R_2 = 253.9 + 0.78t, \qquad t = 0, 1, 2, 3, 4, 5.$$

Write a function that represents the total sales for the two restaurants. Use the *stacked bar graph* in the figure, which represents the total sales during this 6-year period, to determine whether the total sales have been increasing or decreasing.

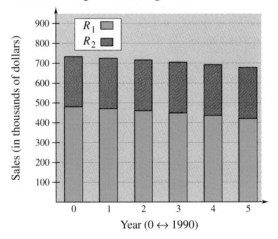

Year (0 ↔ 1990)

67. *Bacteria Count*   The number of bacteria in a refrigerated food product is given by

$$N(T) = 10T^2 - 20T + 600, \qquad 1 \le T \le 20$$

where T is the temperature of the food. When the food is removed from the refrigerator, the temperature is given by

$$T(t) = 3t + 1$$

where t is the time in hours. Find (a) the composite $N(T(t))$ and (b) the time when the bacteria count reaches 1500.

68. *Bacteria Count* The number of bacteria in a refrigerated food product is given by

$$N(T) = 25T^2 - 50T + 300, \qquad 2 \leq T \leq 20$$

where T is the temperature of the food. When the food is removed from the refrigerator, the temperature is given by

$$T(t) = 2t + 1$$

where t is the time in hours. Find (a) the composite $N(T(t))$ and (b) the time when the bacteria count reaches 750.

Price-Earnings Ratio In Exercises 69 and 70, the average annual price-earnings ratio for a corporation's stock is defined as the average price of the stock divided by the earnings per share. The average price of a corporation's stock is given as the function P and the earnings per share is given as the function E. Find the price-earnings ratio, P/E, for the years from 1985 to 1993.

69. *McDonald's Corporation*

Year	P	E
1985	$ 7.21	$0.55
1986	$10.42	$0.62
1987	$12.74	$0.72
1988	$11.35	$0.86
1989	$14.41	$0.98
1990	$15.51	$1.10
1991	$16.76	$1.18
1992	$22.10	$1.30
1993	$26.13	$1.46

70. *Maytag Corporation*

Year	P	E
1985	$13.73	$1.32
1986	$22.20	$1.51
1987	$25.79	$1.91
1988	$23.01	$1.77
1989	$21.34	$1.27
1990	$15.51	$0.94
1991	$14.01	$0.87
1992	$16.31	$0.62
1993	$15.41	$0.48

71. *Troubled Waters* A pebble is dropped into a calm pond, causing ripples in the form of concentric circles (see figure). The radius (in feet) of the outer ripple is given by $r(t) = 0.6t$, where t is time in seconds after the pebble strikes the water. The area of the circle is given by the function $A(r) = \pi r^2$. Find and interpret $(A \circ r)(t)$.

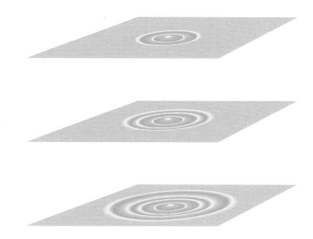

72. *Cost* The weekly cost of producing x units in a manufacturing process is given by the function

$$C(x) = 60x + 750.$$

The number of units produced in t hours is given by $x(t) = 50t$. Find and interpret $(C \circ x)(t)$.

73. *Cost* The weekly cost of producing x units in a manufacturing process is given by the function

$$C(x) = 70x + 375.$$

The number of units produced in t hours is given by $x(t) = 40t$. Find and interpret $(C \circ x)(t)$.

74. Find the domain of $(f/g)(x)$ and $(g/f)(x)$ for the functions

$$f(x) = \sqrt{x} \text{ and } g(x) = \sqrt{9 - x^2}.$$

Why do the two domains differ?

<table>
<tr><td>**3.4**</td><td>## Inverse Functions</td></tr>
</table>

The Inverse of a Function ▪ Finding the Inverse of a Function ▪ The Graph of the Inverse of a Function

The Inverse of a Function

Recall from Section 3.1 that a function can be represented by a set of ordered pairs. For instance, the function $f(x) = x + 4$ from the set $A = \{1, 2, 3, 4\}$ to the set $B = \{5, 6, 7, 8\}$ can be written as follows.

$$f(x) = x + 4 : \quad \{(1, 5), (2, 6), (3, 7), (4, 8)\}$$

By interchanging the first and second coordinates of each of these ordered pairs, you can form the **inverse function** of f, which is denoted by f^{-1}. It is a function from the set B to the set A, and can be written as follows.

$$f^{-1}(x) = x - 4 : \quad \{(5, 1), (6, 2), (7, 3), (8, 4)\}$$

Note that the domain of f is equal to the range of f^{-1}, and vice versa, as shown in Figure 3.23. Also note that the functions f and f^{-1} have the effect of "undoing" each other. In other words, when you form the composition of f with f^{-1} or the composition of f^{-1} with f, you obtain identity functions, as follows.

$$f(f^{-1}(x)) = f(x - 4) = (x - 4) + 4 = x$$
$$f^{-1}(f(x)) = f^{-1}(x + 4) = (x + 4) - 4 = x$$

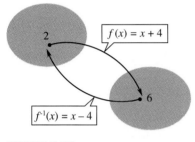

FIGURE 3.23

EXAMPLE 1   *Finding Inverse Functions Informally*

Find the inverse of $f(x) = 4x$. Then verify that both $f(f^{-1}(x))$ and $f^{-1}(f(x))$ are equal to the identity function.

Solution

The given function *multiplies* each input by 4. To "undo" this function, you need to *divide* each input by 4. Thus, the inverse function of $f(x) = 4x$ is

$$f^{-1}(x) = \frac{x}{4}.$$

You can verify that both $f(f^{-1}(x))$ and $f^{-1}(f(x))$ are equal to the identity function as follows.

$$f(f^{-1}(x)) = f\left(\frac{x}{4}\right) = 4\left(\frac{x}{4}\right) = x$$

$$f^{-1}(f(x)) = f^{-1}(4x) = \frac{4x}{4} = x$$

EXAMPLE 2 *Finding Inverse Functions Informally*

Find the inverse of $f(x) = x - 6$. Then verify that both $f(f^{-1}(x))$ and $f^{-1}(f(x))$ are equal to the identity function.

Solution

The given function *subtracts* 6 from each input. To "undo" this function, you need to *add* 6 to each input. Thus, the inverse function of $f(x) = x - 6$ is

$$f^{-1}(x) = x + 6.$$

You can verify that both $f(f^{-1}(x))$ and $f^{-1}(f(x))$ are equal to the identity function as follows.

$$f(f^{-1}(x)) = f(x + 6) = (x + 6) - 6 = x$$
$$f^{-1}(f(x)) = f^{-1}(x - 6) = (x - 6) + 6 = x$$

The formal definition of the inverse of a function is as follows.

Definition of the Inverse of a Function

Let f and g be two functions such that

$$f(g(x)) = x \qquad \text{for every } x \text{ in the domain of } g$$

and

$$g(f(x)) = x \qquad \text{for every } x \text{ in the domain of } f.$$

Under these conditions, the function g is the **inverse** of the function f. The function g is denoted by f^{-1} (read "f-inverse"). Thus,

$$f(f^{-1}(x)) = x \qquad \text{and} \qquad f^{-1}(f(x)) = x.$$

The domain of f must be equal to the range of f^{-1}, and the range of f must be equal to the domain of f^{-1}.

NOTE Don't be confused by the use of -1 to denote the inverse function f^{-1}. In this text, whenever we write f^{-1}, we will *always* be referring to the inverse of the function f and *not* to the reciprocal of $f(x)$.

If the function g is the inverse of the function f, it must also be true that the function f is the inverse of the function g. For this reason, you can say that the functions f and g are *inverses of each other*.

EXAMPLE 3   *Verifying Inverse Functions*

Show that the following functions are inverses of each other.

$$f(x) = 2x^3 - 1 \qquad \text{and} \qquad g(x) = \sqrt[3]{\frac{x+1}{2}}$$

Solution

$$\begin{aligned}
f(g(x)) = f\left(\sqrt[3]{\frac{x+1}{2}}\right) &= 2\left(\sqrt[3]{\frac{x+1}{2}}\right)^3 - 1 \\
&= 2\left(\frac{x+1}{2}\right) - 1 \\
&= x + 1 - 1 \\
&= x
\end{aligned}$$

$$\begin{aligned}
g(f(x)) = g(2x^3 - 1) &= \sqrt[3]{\frac{(2x^3 - 1) + 1}{2}} \\
&= \sqrt[3]{\frac{2x^3}{2}} \\
&= \sqrt[3]{x^3} \\
&= x
\end{aligned}$$

EXAMPLE 4   *Verifying Inverse Functions*

Which of the functions

$$g(x) = \frac{x-2}{5} \qquad \text{and} \qquad h(x) = \frac{5}{x} + 2$$

is the inverse of the function $f(x) = \dfrac{5}{x-2}$?

Solution

By forming the composition of f with g, you can see that

$$f(g(x)) = f\left(\frac{x-2}{5}\right) = \frac{5}{[(x-2)/5] - 2} = \frac{25}{x-12} \neq x.$$

Because this composition is not equal to the identity function x, it follows that g *is not* the inverse of f. By forming the composition of f with h, you have

$$f(h(x)) = f\left(\frac{5}{x} + 2\right) = \frac{5}{(5/x) + 2 - 2} = \frac{5}{5/x} = x.$$

Thus, it appears that h *is* the inverse of f. You can confirm this by showing that the composition of h with f is also equal to the identity function. (Try doing this.)

Finding the Inverse of a Function

For simple functions (such as the ones in Examples 1 and 2) you can find inverse functions by inspection. For more complicated functions, however, it is best to use the following guidelines. The key step in these guidelines is switching the roles of x and y. This step corresponds to the fact that inverse functions have ordered pairs with the coordinates reversed.

STUDY TIP

Note in Step 3 of the guidelines for finding the inverse of a function that it is possible that a function has no inverse. For instance, the function $f(x) = x^2$ has no inverse function.

Finding the Inverse of a Function

1. In the equation for $f(x)$, replace $f(x)$ by y.

2. Interchange the roles of x and y.

3. If the new equation does not represent y as a function of x, the function f does not have an inverse function. If the new equation does represent y as a function of x, solve the new equation for y.

4. Replace y by $f^{-1}(x)$.

5. Verify that f and f^{-1} are inverses of each other by showing that the domain of f is equal to the range of f^{-1}, the range of f is equal to the domain of f^{-1}, and $f(f^{-1}(x)) = x = f^{-1}(f(x))$.

EXAMPLE 5   Finding the Inverse of a Function

Find the inverse of $f(x) = \dfrac{5 - 3x}{2}$.

Solution

$$f(x) = \frac{5 - 3x}{2} \qquad \text{Given function}$$

$$y = \frac{5 - 3x}{2} \qquad \text{Replace } f(x) \text{ by } y.$$

$$x = \frac{5 - 3y}{2} \qquad \text{Interchange } x \text{ and } y.$$

$$2x = 5 - 3y \qquad \text{Multiply both sides by 2.}$$

$$3y = 5 - 2x \qquad \text{Isolate the } y\text{-term.}$$

$$y = \frac{5 - 2x}{3} \qquad \text{Solve for } y.$$

$$f^{-1}(x) = \frac{5 - 2x}{3} \qquad \text{Replace } y \text{ by } f^{-1}(x).$$

Note that both f and f^{-1} have domains and ranges that consist of the entire set of real numbers. Check that $f(f^{-1}(x)) = x$ and $f^{-1}(f(x)) = x$.

The Graph of the Inverse of a Function

The graphs of a function f and its inverse f^{-1} are related to each other in the following way. If the point (a, b) lies on the graph of f, then the point (b, a) must lie on the graph of f^{-1}, and vice versa. This means that the graph of f^{-1} is a *reflection* of the graph of f in the line $y = x$, as shown in Figure 3.24.

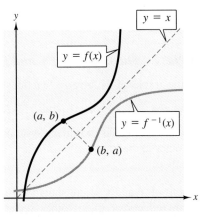

FIGURE 3.24   The graph of f^{-1} is a reflection of the graph of f in the line $y = x$.

EXAMPLE 6   The Graphs of f and f⁻¹

Sketch the graphs of the inverse functions

$$f(x) = 2x - 3 \qquad \text{and} \qquad f^{-1}(x) = \tfrac{1}{2}(x + 3)$$

on the same rectangular coordinate system and show that the graphs are reflections of each other in the line $y = x$.

Solution

The graphs of f and f^{-1} are shown in Figure 3.25. Visually, it appears that the graphs are reflections of each other in the line $y = x$. You can further verify this reflective property by testing a few points on each graph. Note in the following list that if the point (a, b) is on the graph of f, then the point (b, a) is on the graph of f^{-1}.

$f(x) = 2x - 3$	$f^{-1}(x) = \tfrac{1}{2}(x + 3)$
$(-1, -5)$	$(-5, -1)$
$(0, -3)$	$(-3, 0)$
$(1, -1)$	$(-1, 1)$
$(2, 1)$	$(1, 2)$
$(3, 3)$	$(3, 3)$

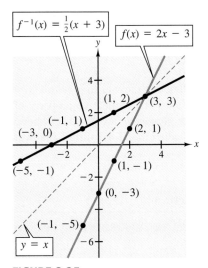

FIGURE 3.25

In the study tip on page 270, we mentioned that the function

$$f(x) = x^2$$

It is easy to show graphically the inverse of a function by using a graphing utility to plot the function, the line $y = x$, and the inverse of the function. Also consider using the graphing utility program given on page 271 to illustrate.

has no inverse. What we really meant is that *assuming the domain of f is the entire real line*, the function $f(x) = x^2$ has no inverse. If, however, we restrict the domain of f to the nonnegative real numbers, then f does have an inverse, as demonstrated in Example 7.

EXAMPLE 7   The Graphs of f and f⁻¹

Sketch the graphs of the inverse functions

$$f(x) = x^2, \quad x \geq 0, \quad \text{and} \quad f^{-1}(x) = \sqrt{x}$$

on the same rectangular coordinate system and show that the graphs are reflections of each other in the line $y = x$.

Solution

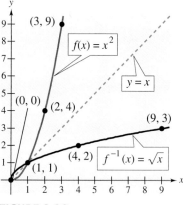

FIGURE 3.26

The graphs of f and f^{-1} are shown in Figure 3.26. Visually, it appears that the graphs are reflections of each other in the line $y = x$. You can further verify this reflective property by testing a few points on each graph. Note in the following list that if the point (a, b) is on the graph of f, then the point (b, a) is on the graph of f^{-1}.

$$
\begin{array}{cc}
f(x) = x^2, \quad x \geq 0 & f^{-1}(x) = \sqrt{x} \\
(0, 0) & (0, 0) \\
(1, 1) & (1, 1) \\
(2, 4) & (4, 2) \\
(3, 9) & (9, 3)
\end{array}
$$

Try showing that $f(f^{-1}(x)) = x$ and $f^{-1}(f(x)) = x$.

Activities

1. Given $f(x) = 5x - 7$, find $f^{-1}(x)$.

   Answer: $f^{-1}(x) = \dfrac{x + 7}{5}$

2. Show that f and g are inverse functions by showing $f(g(x)) = x$ and $g(f(x)) = x$.

   $f(x) = 3x^3 + 1$

   $g(x) = \sqrt[3]{\dfrac{x - 1}{3}}$

3. Describe the graphs of functions that have inverses, and show how the graph of a function and its inverse are related.

The guidelines for finding the inverse of a function include an *algebraic* test for determining whether a function has an inverse. The reflective property of the graphs of inverse functions gives you a nice *geometric* test for determining whether a function has an inverse. This test is called the **Horizontal Line Test** for inverse functions.

Horizontal Line Test for Inverse Functions

A function f has an inverse function if and only if no *horizontal* line intersects the graph of f at more than one point.

EXAMPLE 8   *Applying the Horizontal Line Test*

a. The graph of the function $f(x) = x^3 - 1$ is shown in Figure 3.27(a). Because no horizontal line intersects the graph of f at more than one point, you can conclude that f *does* possess an inverse function.

b. The graph of the function $f(x) = x^2 - 1$ is shown in Figure 3.27(b). Because it is possible to find a horizontal line that intersects the graph of f at more than one point, you can conclude that f *does not* possess an inverse function.

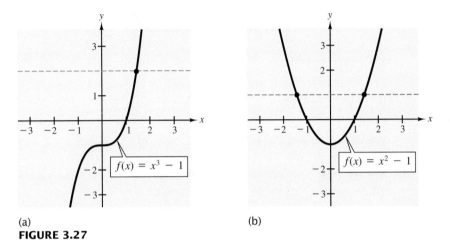

(a)   (b)

FIGURE 3.27

Group Activities   You Be the Instructor

Error Analysis   One of your students has handed in the following quiz. Find the error(s) in each solution and discuss how to explain each error to your student.

1. Find the inverse f^{-1} of $y = \sqrt{2x - 5}$.

$f(x) = \sqrt{2x - 5}$, so

$f^{-1}(x) = \dfrac{1}{\sqrt{2x - 5}}$

2. Find the inverse f^{-1} of $y = \frac{3}{5}x + \frac{1}{3}$.

$f(x) = \frac{3}{5}x + \frac{1}{3}$, so

$f^{-1}(x) = \frac{5}{3}x - 3$

Warm Up

The following warm-up exercises involve skills that were covered in earlier sections. You will use these skills in the exercise set for this section.

In Exercises 1–4, find the domain of the function.

1. $f(x) = \sqrt[3]{x+1}$

2. $f(x) = \sqrt{x+1}$

3. $g(x) = \dfrac{2}{x^2 - 2x}$

4. $h(x) = \dfrac{x}{3x+5}$

In Exercises 5–8, simplify the expression.

5. $2\left(\dfrac{x+5}{2}\right) - 5$

6. $7 - 10\left(\dfrac{7-x}{10}\right)$

7. $\sqrt[3]{2\left(\dfrac{x^3}{2} - 2\right) + 4}$

8. $\sqrt[5]{(x+2)^5} - 2$

In Exercises 9 and 10, solve for x in terms of y.

9. $y = \dfrac{2x-6}{3}$

10. $y = \sqrt[3]{2x-4}$

3.4 Exercises

In Exercises 1–10, show that f and g are inverses algebraically *and* using a graphing utility.

1. $f(x) = 2x$, $\qquad g(x) = \dfrac{x}{2}$

2. $f(x) = x - 5$, $\qquad g(x) = x + 5$

3. $f(x) = 5x + 1$, $\qquad g(x) = \dfrac{x-1}{5}$

4. $f(x) = 3 - 4x$, $\qquad g(x) = \dfrac{3-x}{4}$

5. $f(x) = x^3$, $\qquad g(x) = \sqrt[3]{x}$

6. $f(x) = \dfrac{1}{x}$, $\qquad g(x) = \dfrac{1}{x}$

7. $f(x) = \sqrt{x-4}$, $\qquad g(x) = x^2 + 4, \ x \geq 0$

8. $f(x) = 9 - x^2, \ x \geq 0$, $\qquad g(x) = \sqrt{9-x}, x \leq 9$

9. $f(x) = 1 - x^3$, $\qquad g(x) = \sqrt[3]{1-x}$

10. $f(x) = \dfrac{1}{1+x}, \ x \geq 0$

$\qquad g(x) = \dfrac{1-x}{x}, \ 0 < x \leq 1$

In Exercises 11–20, use the graph of $y = f(x)$ to determine whether the function has an inverse.

11.

12.

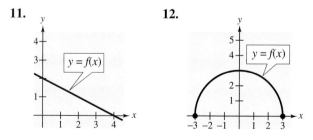

13.

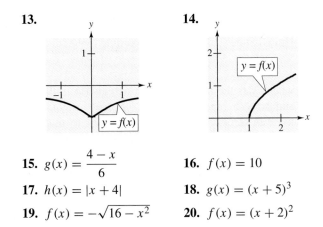

14.

15. $g(x) = \dfrac{4 - x}{6}$

16. $f(x) = 10$

17. $h(x) = |x + 4|$

18. $g(x) = (x + 5)^3$

19. $f(x) = -\sqrt{16 - x^2}$

20. $f(x) = (x + 2)^2$

In Exercises 21–30, find the inverse of the function f. Then, using a graphing utility, graph both f and f^{-1} on the same screen.

21. $f(x) = 2x - 3$

22. $f(x) = 3x$

23. $f(x) = x^5$

24. $f(x) = x^3 + 1$

25. $f(x) = \sqrt{x}$

26. $f(x) = x^2,\ x \geq 0$

27. $f(x) = \sqrt{4 - x^2},\ 0 \leq x \leq 2$

28. $f(x) = \dfrac{4}{x}$

29. $f(x) = \sqrt[3]{x - 1}$

30. $f(x) = x^{3/5}$

In Exercises 31–46, determine whether the function has an inverse. If it does, find its inverse.

31. $f(x) = x^4$

32. $f(x) = \dfrac{1}{x^2}$

33. $g(x) = \dfrac{x}{8}$

34. $f(x) = 3x + 5$

35. $p(x) = -4$

36. $f(x) = \dfrac{3x + 4}{5}$

37. $f(x) = (x + 3)^2,\ x \geq -3$

38. $q(x) = (x - 5)^2$

39. $h(x) = \dfrac{1}{x}$

40. $f(x) = |x - 2|,\ x \leq 2$

41. $f(x) = \sqrt{2x + 3}$

42. $f(x) = \sqrt{x - 2}$

43. $g(x) = x^2 - x^4$

44. $f(x) = \dfrac{x^2}{x^2 + 1}$

45. $f(x) = 25 - x^2,\ x \leq 0$

46. $f(x) = 36 + x^2,\ x \leq 0$

In Exercises 47 and 48, use the graph of f to complete the table and to sketch the graph of f^{-1}.

47.

x	0	1	2	3	4
$f^{-1}(x)$					

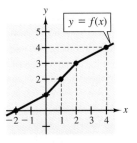

48.

x	0	2	4	6
$f^{-1}(x)$				

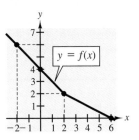

In Exercises 49–52, use the functions

$$f(x) = \tfrac{1}{8}x - 3 \text{ and } g(x) = x^3$$

to find the value.

49. $(f^{-1} \circ g^{-1})(1)$

50. $(g^{-1} \circ f^{-1})(-3)$

51. $(f^{-1} \circ f^{-1})(6)$

52. $(g^{-1} \circ g^{-1})(-4)$

In Exercises 53–56, use the functions

$$f(x) = x + 4 \text{ and } g(x) = 2x - 5$$

to find the composition of functions.

53. $g^{-1} \circ f^{-1}$ **54.** $f^{-1} \circ g^{-1}$

55. $(f \circ g)^{-1}$ **56.** $(g \circ f)^{-1}$

57. *Soccer Goals*   From 1984 to 1990, the average number of soccer goals per game in the Major Soccer League decreased. A function that approximates the average number of goals per game is

$$y = 10.88 - 0.3t, \qquad t = 4, 5, 6, 7, 8, 9$$

where y represents the average number of goals per game and $t = 4$ represents the 1984–85 season (see figure). Find the inverse of this function, and use f^{-1} to find the playing season that had an average of nine goals per game.   (Source: Major Soccer League)

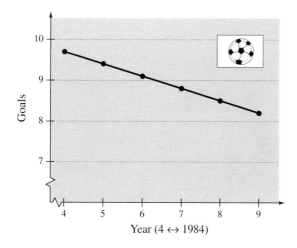

Goals

Year (4 ↔ 1984)

58. *Reasoning*   You are teaching another student how to find the inverse function of a one-to-one function. The student you are helping states that interchanging the roles of x and y is "cheating." Discuss how you would use the graph of $f(x) = x^2$, $x \geq 0$ and $f^{-1}(x) = \sqrt{x}$ to justify that particular step in the process of finding an inverse function.

59. *Automated Teller Machines (ATMs)*   From 1985 to 1993, the annual number of ATM transactions in the United States has increased according to the function

$$y = 2615 + 30.895t^2, \qquad 5 \leq t \leq 13$$

where y is the number of transactions (in millions) and $t = 5$ represents 1985. Solve the equation for t in terms of y and use the result to find the year in which there were 5,117,000,000 transactions. (Source: *Bank Network News*)

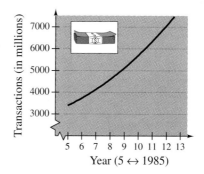

Transactions (in millions)

Year (5 ↔ 1985)

60. *Median Family Income*   The median family income in the United States increased from 1970 to 1992 according to the function

$$y = (92.311 + 3.98t)^2, \qquad 0 \leq t \leq 22$$

where y is the median income in dollars and $t = 0$ represents 1970. Solve the equation for t in terms of y and use the result to find the year in which the median family income was $29,553.   (Source: U.S. Census Bureau)

61. *Earnings-Dividend Ratio*   From 1983 to 1993, the earnings per share for the Coca-Cola Corporation were approximately related to the dividends per share by the function

$$f(x) = \sqrt{0.214 + 0.169x}, \qquad 0.22 \leq x \leq 1.75$$

where $f(x)$ represents the dividends per share (in dollars) and x represents the earnings per share (in dollars). In 1992, Coca-Cola paid dividends of $0.56 per share. What were the earnings per share in 1992? (Source: Coca-Cola)

MID-CHAPTER QUIZ

Take this quiz as you would take a quiz in class. After you are done, check your work against the answers given in the back of the book.

In Exercises 1–4, find the value of $f(x) = 2x^2 - x + 1$.

1. $f(3)$ **2.** $f(-1)$ **3.** $f(0)$ **4.** $f(a)$

In Exercises 5 and 6, find all real values of x such that $f(x) = 0$.

5. $f(x) = x^3 - 4x$ **6.** $f(x) = 3 - \dfrac{4}{x}$

7. A company produces a product for which the variable cost is $11.40 per unit and the fixed costs are $85,000. The product sells for $16.89. Let x be the number of units produced. Write the total profit P as a function of x.

In Exercises 8 and 9, find the domain of the function.

8. $f(x) = \dfrac{3}{x^2 - x}$ **9.** $h(x) = \sqrt{5 - x}$

In Exercises 10–13, use a graphing utility to graph the function. Then estimate the open intervals on which the function is increasing or decreasing.

10. $f(x) = -x^2 + 5x + 3$ **11.** $f(x) = 2x^3 - 5x^2$

12. $f(x) = x^2 + 2$ **13.** $f(x) = \sqrt{x + 1}$

In Exercises 14 and 15, compare the graphs of $f(x) = \sqrt[3]{x}$ and g.

14. $g(x) = \sqrt[3]{x - 2}$ **15.** $g(x) = -\sqrt[3]{x + 1}$

In Exercises 16 and 17, find $(f \circ g)(x)$.

16. $f(x) = x^2,\ g(x) = x - 1$ **17.** $f(x) = \dfrac{1}{x},\ g(x) = x^3$

In Exercises 18 and 19, compare the graphs of f and f^{-1}.

18. $f(x) = x^2,\quad x \geq 0$ **19.** $f(x) = \sqrt[3]{x - 1}$

20. *Company Income*   The income y for a company can be modeled by $y = (96.5 + 3.75t)^2$, $4 \leq t \leq 14$, where $t = 4$ represents 1984. Solve the model for t and use the result to find the year in which the income was $16,200.

3.5 Quadratic Functions

The Graph of a Quadratic Function ▪
The Standard Form of a Quadratic Function ▪ Applications

The Graph of a Quadratic Function

In this and the next two sections, you will study the graphs of polynomial functions and rational functions.

Definition of Polynomial Function

Let n be a nonnegative integer and let $a_n, a_{n-1}, \ldots, a_2, a_1, a_0$ be real numbers with $a_n \neq 0$. The function given by

$$f(x) = a_n x^n + a_{n-1} x^{n-1} + \cdots + a_2 x^2 + a_1 x + a_0$$

is called a **polynomial function of x with degree n.**

Polynomial functions are classified by degree. For instance, the polynomial function

$$f(x) = a, \quad a \neq 0 \qquad\qquad \text{Constant function}$$

has degree 0 and is called a **constant function.** In Chapter 2, you learned that the graph of this type of function is a horizontal line. The polynomial function

$$f(x) = ax + b, \quad a \neq 0 \qquad\qquad \text{Linear function}$$

has degree 1 and is called a **linear function.** In Chapter 2, you learned that the graph of the linear function $f(x) = ax + b$ is a line whose slope is a and whose y-intercept is $(0, b)$. In this section you will study second-degree polynomial functions, which are called **quadratic functions.**

Definition of Quadratic Function

Let a, b, and c be real numbers with $a \neq 0$. The function of x given by

$$f(x) = ax^2 + bx + c \qquad\qquad \text{Quadratic function}$$

is called a **quadratic function.**

NOTE The graph of a quadratic function is a "∪"-shaped curve that is called a **parabola.**

All parabolas are symmetric with respect to a line called the **axis of symmetry,** or simply the **axis** of the parabola. The point where the axis intersects the parabola is the **vertex** of the parabola, as shown in Figure 3.28. If the leading coefficient is positive, the graph of $f(x) = ax^2 + bx + c$ is a parabola that opens upward, and if the leading coefficient is negative, the graph of $f(x) = ax^2 + bx + c$ is a parabola that opens downward.

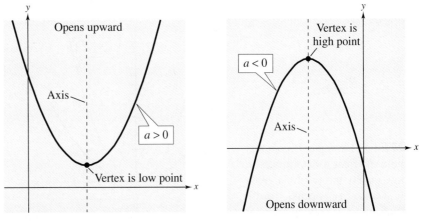

FIGURE 3.28

The simplest type of quadratic function is $f(x) = ax^2$. Its graph is a parabola whose vertex is $(0, 0)$. If $a > 0$, the vertex is the *minimum* point on the graph, and if $a < 0$, the vertex is the *maximum* point on the graph, as shown in Figure 3.29.

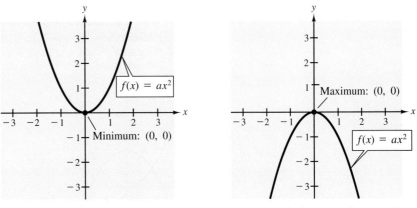

$a > 0$: Parabola opens upward $a < 0$: Parabola opens downward
FIGURE 3.29

When sketching the graph of $f(x) = ax^2$, it is helpful to use the graph of $y = x^2$ as a reference, as discussed in Section 3.3.

EXAMPLE 1   Sketching the Graph of a Quadratic Function

a. Compared with $y = x^2$, each output of $f(x) = \frac{1}{3}x^2$ "shrinks" by a factor of $\frac{1}{3}$, creating the broader parabola shown in Figure 3.30(a).

b. Compared with $y = x^2$, each output of $g(x) = 2x^2$ "stretches" by a factor of 2, creating the narrower parabola shown in Figure 3.30(b).

NOTE In Example 1, note that the coefficient a determines how widely the parabola given by $f(x) = ax^2$ opens. If $|a|$ is small, the parabola opens more widely than if $|a|$ is large.

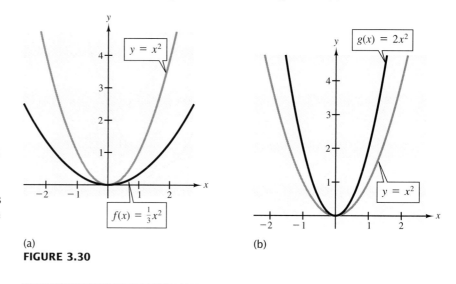

(a)   (b)

FIGURE 3.30

Recall from Section 3.3 that the graphs of $y = f(x \pm c)$, $y = f(x) \pm c$, and $y = -f(x)$ are rigid transformations of the graph of $y = f(x)$. For instance, in Figure 3.31, notice how the graph of $y = x^2$ can be transformed to produce the graphs of $f(x) = -x^2 + 1$ and $g(x) = (x + 2)^2 - 3$.

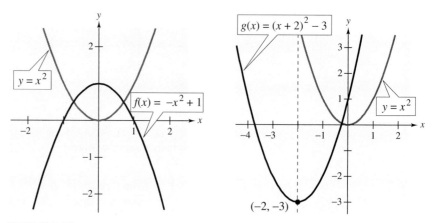

FIGURE 3.31

The Standard Form of a Quadratic Function

The **standard form** of a quadratic function is

$$f(x) = a(x - h)^2 + k.$$

This form is especially convenient for sketching a parabola because it identifies the vertex of the parabola.

Standard Form of a Quadratic Function

The quadratic function

$$f(x) = a(x - h)^2 + k, \qquad a \neq 0$$

is said to be in **standard form.** The graph of f is a parabola whose axis is the vertical line $x = h$ and whose vertex is the point (h, k). If $a > 0$, the parabola opens upward, and if $a < 0$, the parabola opens downward.

To write a quadratic function in standard form, you can use the process of *completing the square*, as illustrated in Example 2.

EXAMPLE 2   *Writing a Quadratic Function in Standard Form*

Sketch the graph of $f(x) = 2x^2 + 8x + 7$ and identify the vertex.

Solution

Begin by writing the quadratic function in standard form. Notice that the first step in completing the square is to factor out any coefficient of x^2 that is not 1.

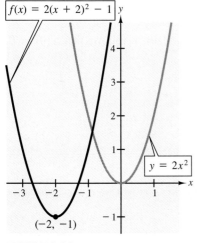

$$f(x) = 2x^2 + 8x + 7 \qquad \text{Given form}$$

$$= 2(x^2 + 4x) + 7 \qquad \text{Factor 2 out of } x \text{ terms.}$$

$$= 2(x^2 + 4x + 4 - 4) + 7 \qquad \text{Add and subtract 4 within parentheses.}$$

$$\underset{2^2}{\underbrace{\qquad}}$$

$$= 2(x^2 + 4x + 4) - 2(4) + 7 \qquad \text{Regroup terms.}$$

$$= 2(x^2 + 4x + 4) - 8 + 7 \qquad \text{Simplify.}$$

$$= 2(x + 2)^2 - 1 \qquad \text{Standard form}$$

From this form, you can see that the graph of f is a parabola that opens upward with vertex $(-2, -1)$. This corresponds to a left shift of two units and a downward shift of one unit relative to the graph of $y = 2x^2$, as shown in Figure 3.32.

$f(x) = 2(x + 2)^2 - 1$

$y = 2x^2$

$(-2, -1)$

FIGURE 3.32

EXAMPLE 3 Writing a Quadratic Function in Standard Form

Sketch the graph of $f(x) = -x^2 + 6x - 8$ and identify the vertex.

Solution

As in Example 2, we begin by writing the quadratic function in standard form.

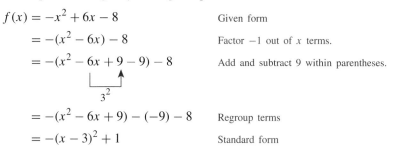

$$f(x) = -x^2 + 6x - 8 \qquad \text{Given form}$$
$$= -(x^2 - 6x) - 8 \qquad \text{Factor } -1 \text{ out of } x \text{ terms.}$$
$$= -(x^2 - 6x + 9 - 9) - 8 \qquad \text{Add and subtract 9 within parentheses.}$$

$$3^2$$

$$= -(x^2 - 6x + 9) - (-9) - 8 \qquad \text{Regroup terms}$$
$$= -(x - 3)^2 + 1 \qquad \text{Standard form}$$

Thus, the graph of f is a parabola that opens downward with vertex at $(3, 1)$, as shown in Figure 3.33.

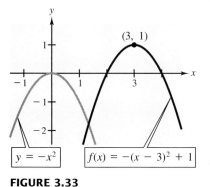

$y = -x^2$ $\quad$ $f(x) = -(x - 3)^2 + 1$

FIGURE 3.33

EXAMPLE 4 Finding the Equation of a Parabola

Find an equation for the parabola whose vertex is $(1, 2)$ and passes through the point $(0, 0)$, as shown in Figure 3.34.

Solution

Because the parabola has a vertex at $(h, k) = (1, 2)$, the equation must have the form

$$f(x) = a(x - 1)^2 + 2. \qquad \text{Standard form}$$

Because the parabola passes through the point $(0, 0)$, it follows that $f(0) = 0$. Thus, you obtain

$$0 = a(0 - 1)^2 + 2 \qquad \Longrightarrow \qquad a = -2$$

which implies that the equation is

$$f(x) = -2(x - 1)^2 + 2 = -2x^2 + 4x.$$

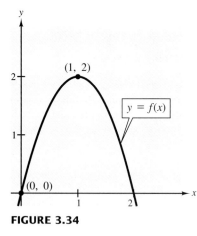

$y = f(x)$

FIGURE 3.34

To find the x-intercepts of the graph of $f(x) = ax^2 + bx + c$, you must solve the equation

$$ax^2 + bx + c = 0.$$

If $ax^2 + bx + c$ does not factor, you can use the Quadratic Formula to find the x-intercepts. Remember, however, that a parabola may have no x-intercepts.

Applications

Many applications involve finding the maximum or minimum value of a quadratic function. By writing the quadratic function $f(x) = ax^2 + bx + c$ in standard form, you can determine that the vertex occurs when $x = -b/2a$.

Real Life

EXAMPLE 5 *The Maximum Height of a Baseball*

Baseball is often called America's *national pastime.* It was invented in the United States in the mid-1800's. The National League was founded in 1876, and the American League was founded in 1900. Each year, the winners of these two leagues play in the World Series.

A baseball is hit 3 feet above ground at a velocity of 100 feet per second and at an angle of 45° with respect to the ground. The path of the baseball is given by the function

$$f(x) = -0.0032x^2 + x + 3$$

where $f(x)$ is the height of the baseball (in feet) and x is the distance from home plate (in feet). What is the maximum height reached by the baseball?

Solution

For this quadratic function, you have

$$f(x) = ax^2 + bx + c = -0.0032x^2 + x + 3.$$

Thus, $a = -0.0032$ and $b = 1$. Because the function has a maximum when $x = -b/2a$, you can conclude that the baseball reaches its maximum height when it is

$$x = -\frac{b}{2a} = -\frac{1}{2(-0.0032)} = 156.25 \text{ feet}$$

from home plate. At this distance, the maximum height is

$$f(156.25) = -0.0032(156.25)^2 + 156.25 + 3 = 81.125 \text{ feet}.$$

The path of the baseball is shown in Figure 3.35.

Activities

1. Describe the effect a has on the graph of $f(x) = ax^2 + bx + c$.

2. Rewrite the quadratic function $f(x) = -x^2 + 4x - 1$ in standard form.

   Answer: $f(x) = -(x - 2)^2 + 3$

3. Find the quadratic function whose graph has vertex $(3, -1)$ and passes through the point $(0, 2)$.

   Answer: $f(x) = \frac{1}{3}x^2 - 2x + 2$

4. Fit a quadratic model to the points $(0, -2)$, $(2, 0)$, and $(-0.5, 0)$.

   Answer: $y = 2x^2 - 3x - 2$

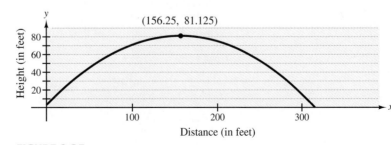

FIGURE 3.35

EXAMPLE 6 Charitable Contributions

According to a survey conducted by *Independent Sector*, the percent of income that Americans give to charities is related to the amount of income. For families with annual incomes of $100,000 or less, the percent can be modeled by

$$P = 0.0014x^2 - 0.1529x + 5.855, \qquad 5 \le x \le 100$$

where x is the annual income in thousands of dollars. According to this model, what income level corresponds to the minimum percent of charitable contributions?

Solution

There are two ways to answer this question. One is to sketch the graph of the quadratic function, as shown in Figure 3.36. From this graph, it appears that the minimum percent corresponds to an income level of about $55,000. The other way to answer the question is to use the fact that the minimum point of the parabola occurs when $x = -b/2a$.

$$x = -\frac{b}{2a} = -\frac{-0.1529}{2(0.0014)} \approx 54.6.$$

From this x-value, you can conclude that the minimum percent corresponds to an income level of about $54,600.

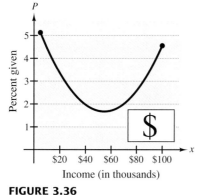

Income (in thousands)

FIGURE 3.36

Group Activities   Exploring with Technology

Approximating Intercepts and Extreme Points   A graphing utility with ZOOM and TRACE features can be a useful tool in approximating the intercepts and vertex coordinates of quadratic functions to acceptable degrees of accuracy. Use a graphing utility to locate and approximate the intercepts and vertices of the following quadratic functions to two decimal places. Confirm the coordinates of the vertices algebraically.

a. $y = x^2 - 3x + 4$

b. $y = x^2 - 17x + 64$

c. $y = 7.52x^2 + 4.16x$

d. $y = -\frac{1}{2}x^2 + 5x - \frac{13}{2}$

e. $y = \frac{x^2}{19} + \frac{13x}{3} + \frac{4}{21}$

f. $y = 14x^2 - 5x + 2$

Warm Up

The following warm-up exercises involve skills that were covered in earlier sections. You will use these skills in the exercise set for this section.

In Exercises 1–4, solve the equation by factoring.

1. $2x^2 + 11x - 6 = 0$ **2.** $5x^2 - 12x - 9 = 0$

3. $3 + x - 2x^2 = 0$ **4.** $x^2 + 20x + 100 = 0$

In Exercises 5–10, use the Quadratic Formula to solve the equation.

5. $x^2 - 6x + 4 = 0$ **6.** $x^2 + 4x + 1 = 0$

7. $2x^2 - 16x + 25 = 0$ **8.** $3x^2 + 30x + 74 = 0$

9. $x^2 + 3x + 1 = 0$ **10.** $x^2 + 3x - 3 = 0$

3.5 Exercises

In Exercises 1–8, match the quadratic function with its graph. [The graphs are labeled (a), (b), (c), (d), (e), (f), (g), and (h).]

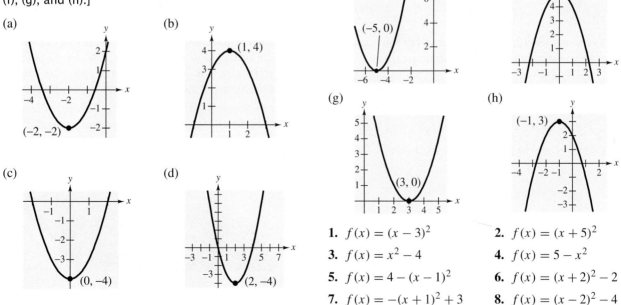

1. $f(x) = (x - 3)^2$ **2.** $f(x) = (x + 5)^2$

3. $f(x) = x^2 - 4$ **4.** $f(x) = 5 - x^2$

5. $f(x) = 4 - (x - 1)^2$ **6.** $f(x) = (x + 2)^2 - 2$

7. $f(x) = -(x + 1)^2 + 3$ **8.** $f(x) = (x - 2)^2 - 4$

In Exercises 9–14, find an equation of the parabola.

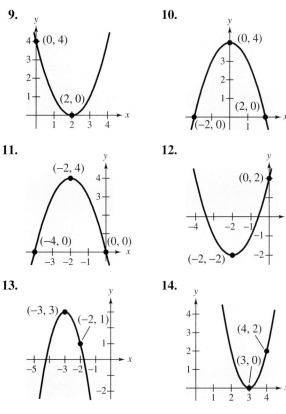

9.

10.

11.

12.

13.

14.

In Exercises 15–32, sketch the graph of the quadratic function. Identify the vertex and intercepts.

15. $f(x) = x^2 - 5$

16. $f(x) = \frac{1}{2}x^2 - 4$

17. $f(x) = 16 - x^2$

18. $h(x) = 25 - x^2$

19. $f(x) = (x + 5)^2 - 6$

20. $f(x) = (x - 6)^2 + 3$

21. $h(x) = x^2 - 8x + 16$

22. $g(x) = x^2 + 2x + 1$

23. $f(x) = -(x^2 + 2x - 3)$

24. $g(x) = x^2 + 8x + 11$

25. $f(x) = x^2 - x + \frac{5}{4}$

26. $f(x) = x^2 + 3x + \frac{1}{4}$

27. $f(x) = -x^2 + 2x + 5$

28. $f(x) = -x^2 - 4x + 1$

29. $h(x) = 4x^2 - 4x + 21$

30. $f(x) = 2x^2 - x + 1$

31. $f(x) = 2x^2 - 16x + 31$

32. $g(x) = \frac{1}{2}(x^2 + 4x - 2)$

In Exercises 33–36, find the quadratic function that has the indicated vertex and whose graph passes through the point.

33. Vertex: $(3, 4)$; point: $(1, 2)$

34. Vertex: $(2, 3)$; point: $(0, 2)$

35. Vertex: $(5, 12)$; point: $(7, 15)$

36. Vertex: $(-2, -2)$; point: $(-1, 0)$

In Exercises 37–42, find two quadratic functions whose graphs have the given x-intercepts. Find one function that has a graph that opens upward and another that has a graph that opens downward. (Each exercise has many correct answers.)

37. $(-1, 0)$, $(3, 0)$

38. $\left(-\frac{5}{2}, 0\right)$, $(2, 0)$

39. $(0, 0)$, $(10, 0)$

40. $(4, 0)$, $(8, 0)$

41. $(-3, 0)$, $\left(-\frac{1}{2}, 0\right)$

42. $(-5, 0)$, $(5, 0)$

43. *Maximum Area*   The perimeter of a rectangle is 100 feet. Let x represent the width of the rectangle and write a quadratic function that expresses the area of the rectangle in terms of its width. Of all possible rectangles with perimeters of 100 feet, what are the dimensions of the one that has the greatest area?

44. *Maximum Area*   The perimeter of a rectangle is 400 feet. Let x represent the width of the rectangle and write a quadratic function that expresses the area of the rectangle in terms of its width. Of all possible rectangles with perimeters of 400 feet, what are the dimensions of the one that has the greatest area?

45. *Maximum Area*   A rancher has 200 feet of fencing with which to enclose two adjacent rectangular corrals (see figure). What dimensions will produce a maximum enclosed area?

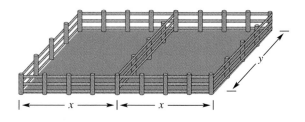

46. *Maximum Area*   An indoor physical-fitness room consists of a rectangular region with a semicircle on each end (see figure). The perimeter of the room is to be a 200-meter running track. What dimensions will produce a maximum area of the rectangle?

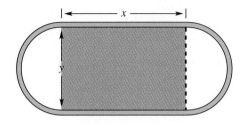

Maximum Revenue   In Exercises 47 and 48, find the number of units that produces a maximum revenue. The revenue R is measured in dollars and x is the number of units produced.

47. $R = 900x - 0.01x^2$

48. $R = 100x - 0.0002x^2$

49. *Minimum Cost*   A manufacturer of lighting fixtures has daily production costs of

$$C = 800 - 10x + 0.25x^2$$

where C is the total cost in dollars and x is the number of units produced. How many fixtures should be produced each day to yield a minimum cost?

50. *Minimum Cost*   A textile manufacturer has daily production costs of

$$C = 10,000 - 10x + 0.045x^2$$

where C is the total cost in dollars and x is the number of units produced. How many units should be produced each day to yield a minimum cost?

51. *Maximum Profit*   Let x be the amount (in hundreds of dollars) a company spends on advertising, and let P be the profit, where

$$P = 230 + 20x - 0.5x^2.$$

What expenditure for advertising results in the maximum profit?

52. *Maximum Profit*   The profit for a company is given by

$$P = -0.0002x^2 + 140x - 250,000$$

where x is the number of units produced. What production level will yield a maximum profit?

53. *Maximum Height of a Diver*   The path of a diver is given by

$$y = -\frac{4}{9}x^2 + \frac{24}{9}x + 10$$

where y is the height in feet and x is the horizontal distance from the end of the diving board in feet (see figure). Use a graphing utility and the TRACE feature to find the maximum height of the diver.

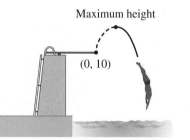

Maximum height

(0, 10)

54. *Maximum Height*   The winning women's shot put in the 1988 Summer Olympics was thrown by Natalya Lisovskaya of the Soviet Union. The path of her winning toss was approximately given by

$$y = -0.01464x^2 + x + 5$$

where y is the height of the shot put in feet and x is the horizontal distance in feet. Use a graphing utility and the TRACE feature to find how long the winning toss was and the maximum height of the shot put.

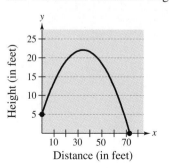

Distance (in feet)

55. *Nobel Prize* Nobel Prizes are awarded in Swedish currency (krona). During the 1980s, the dollar value of a Nobel Prize fell slightly, then rose according to the model

$$N = 11.577t^2 - 60.03t + 230, \qquad 0 \le t \le 12$$

where N is the approximate dollar value (in thousands of dollars) of a Nobel Prize and t represents the calendar year, with $t = 0$ corresponding to 1980. Use a graphing utility to determine the year, from 1980 to 1992, when the dollar value of the Nobel Prize was at its lowest. (Source: Swedish Embassy)

56. *Unemployment Rate* The annual average unemployment rate for married men from 1990 to 1993 can be approximated by the model

$$P = -0.4t^2 + 1.56t + 3.36, \qquad 0 \le t \le 3$$

where P is the percent of married men who were unemployed and t represents the calendar year, with $t = 0$ corresponding to 1990. Use a graphing utility to determine the year in which the unemployment rate was at its highest. (Source: U.S. Department of Labor)

Math Matters   Circles and Pi

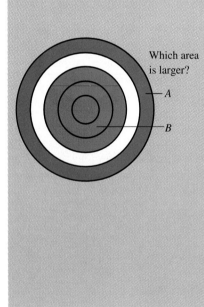

Which area is larger?

— A

— B

Pi is a special number that represents a relationship that is found in *all circles*. The relationship is this: The ratio of the circumference of *any* circle to the diameter of the circle always produces the same number—the number denoted by the Greek symbol π.

Calculation of the decimal representation of the number π has consumed the time of many mathematicians (and computers). Because the number π is irrational, its decimal representation does not repeat or terminate. Here is a decimal representation of π that is accurate to 50 decimal places.

$$\pi = 3.14159265358979323846264338327950288419716939937511\ldots$$

The greatest number of decimal places (as of 1994) to which π has been calculated is 2,260,321,336. Two brothers, Gregory Volfovich and David Volfovich Chudnovsky, arrived at this number on their home-made supercomputer, *m zero*, in New York City in the summer of 1991.

The number π is used in most formulas that deal with circles. For instance, the area of a circle is given by

$$A = \pi r^2$$

where r is the radius of the circle. Use this formula to determine which of the colored regions inside the circle at the left has the greater area. Is the area of the red circle larger than that of the blue ring, or do both regions have the same area? (The answer is given in the back of the book.)

<table>
<tr><td>**3.6**</td><td>**Polynomial Functions of Higher Degree**</td></tr>
</table>

Graphs of Polynomial Functions ▪ The Leading Coefficient Test ▪
Zeros of Polynomial Functions ▪ The Intermediate Value Theorem

Graphs of Polynomial Functions

In this section, you will study basic features of the graphs of polynomial functions. The first feature is that the graph of a polynomial function is **continuous.**
Essentially, this means that the graph of a polynomial function has no breaks, as
shown in Figure 3.37(a).

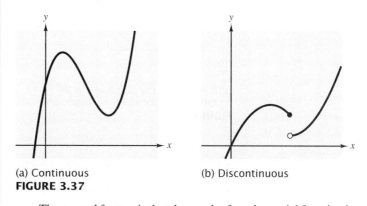

(a) Continuous   (b) Discontinuous
FIGURE 3.37

The second feature is that the graph of a polynomial function has only smooth,
rounded turns, as shown in Figure 3.38(a). A polynomial function cannot have a
sharp turn. For instance, the function $f(x) = |x|$, which has a sharp turn at the
point $(0, 0)$, as shown in Figure 3.38(b), is not a polynomial function.

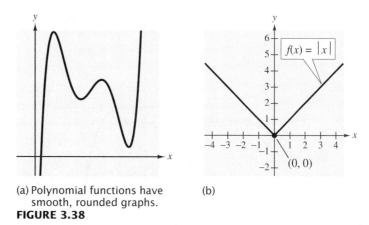

(a) Polynomial functions have   (b)
    smooth, rounded graphs.
FIGURE 3.38

The polynomial functions that have the simplest graphs are monomials of the form $f(x) = x^n$, where n is an integer greater than zero. From Figure 3.39, you can see that when n is *even* the graph is similar to the graph of $f(x) = x^2$ and when n is *odd* the graph is similar to the graph of $f(x) = x^3$. Moreover, the greater the value of n, the flatter the graph near the origin.

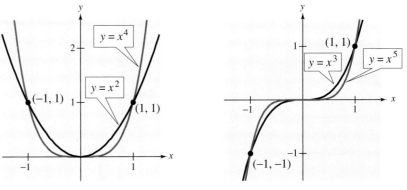

(a) If n is even, the graph of $y = x^n$ touches the axis at the x-intercept.

(b) If n is odd, the graph of $y = x^n$ *crosses* the axis at the x-intercept.

FIGURE 3.39

EXAMPLE 1   *Sketching Transformations of Monomial Functions*

a. Because the degree of $f(x) = -x^5$ is odd, its graph is similar to the graph of $y = x^3$. In Figure 3.40(a), note that the negative coefficient has the effect of reflecting the graph about the x-axis.

b. The graph of $h(x) = (x + 1)^4$ is a left shift, by one unit, of the graph of $y = x^4$, as shown in Figure 3.40(b).

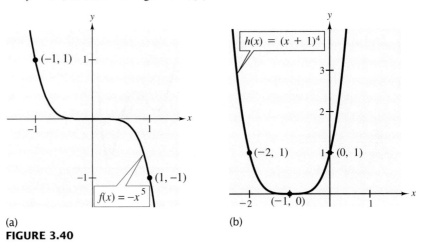

(a)

(b)

FIGURE 3.40

The Leading Coefficient Test

In Example 1, note that both graphs eventually rise or fall without bound as x moves to the right. Whether the graph of a polynomial eventually rises or falls can be determined by the function's degree (even or odd) and by its leading coefficient, as indicated in the **Leading Coefficient Test.**

Leading Coefficient Test

As x moves without bound to the left or to the right, the graph of the polynomial function $f(x) = a_n x^n + \cdots + a_1 x + a_0$ eventually rises or falls in the following manner.

1. When n is *odd*:

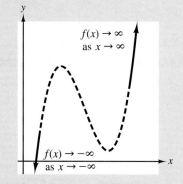

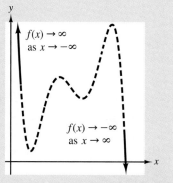

If the leading coefficient is positive $(a_n > 0)$, the graph falls to the left and rises to the right.

If the leading coefficient is negative $(a_n < 0)$, the graph rises to the left and falls to the right.

2. When n is *even*:

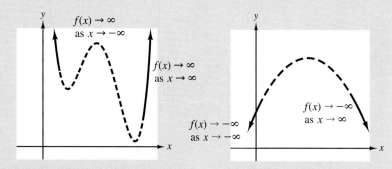

If the leading coefficient is positive $(a_n > 0)$, the graph rises to the left and right.

If the leading coefficient is negative $(a_n < 0)$, the graph falls to the left and right.

NOTE   The dashed portions of the graphs indicate that the test determines *only* the right and left behavior of the graph.

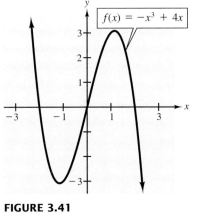

FIGURE 3.41

EXAMPLE 2 Applying the Leading Coefficient Test

Describe the right and left behavior of the graph of

$$f(x) = -x^3 + 4x.$$

Solution

Because the degree is odd and the leading coefficient is negative, the graph rises to the left and falls to the right, as shown in Figure 3.41.

In Example 2, note that the Leading Coefficient Test only tells you whether the graph *eventually* rises or falls to the right or left. Other characteristics of the graph, such as intercepts and minimum and maximum points, must be determined by means of other tests.

EXAMPLE 3 Applying the Leading Coefficient Test

Describe the right and left behavior of the graphs of the following functions.

a. $f(x) = x^4 - 5x^2 + 4$ **b.** $f(x) = x^5 - x$

Solution

a. Because the degree is even and the leading coefficient is positive, the graph rises to the left and right, as shown in Figure 3.42(a).

b. Because the degree is odd and the leading coefficient is positive, the graph falls to the left and rises to the right, as shown in Figure 3.42(b).

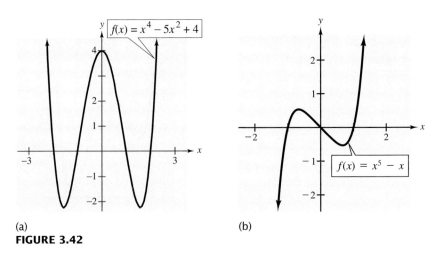

(a)   (b)

FIGURE 3.42

Zeros of Polynomial Functions

It can be shown that for a polynomial function f of degree n, the following statements are true. (Remember that the **zeros** of a function are the x-values for which the function is zero.)

1. The graph of f has, at most, $n - 1$ turning points. (Turning points are points at which the graph changes from increasing to decreasing or vice versa.)

2. The function f has, at most, n real zeros. (You will study this result in detail in Section 4.5 on the Fundamental Theorem of Algebra.)

Finding the zeros of polynomial functions is one of the most important problems in algebra. There is a strong interplay between graphical and algebraic approaches to this problem. Sometimes you can use information about the graph of a function to help find its zeros, and in other cases you can use information about the zeros of a function to help sketch its graph.

> **Real Zeros of Polynomial Functions**
>
> If f is a polynomial function and a is a real number, the following statements are equivalent.
>
> 1. $x = a$ is a *zero* of the function f.
>
> 2. $x = a$ is a *solution* of the polynomial equation $f(x) = 0$.
>
> 3. $(x - a)$ is a *factor* of the polynomial $f(x)$.
>
> 4. $(a, 0)$ is an *x-intercept* of the graph of f.

NOTE   In the equivalent statements at the right, notice that finding zeros of polynomial functions is closely related to factoring and finding x-intercepts.

EXAMPLE 4   *Finding Zeros of a Polynomial Function*

Find all real zeros of $f(x) = x^3 - x^2 - 2x$.

Solution

By factoring, you obtain the following.

$$f(x) = x^3 - x^2 - 2x \qquad \text{Original function}$$

$$= x(x^2 - x - 2) \qquad \text{Remove common monomial factor.}$$

$$= x(x - 2)(x + 1) \qquad \text{Factor completely.}$$

Thus, the real zeros are $x = 0$, $x = 2$, and $x = -1$, and the corresponding x-intercepts are $(0, 0)$, $(2, 0)$, and $(-1, 0)$, as shown in Figure 3.43. In the figure, note that the graph has two turning points. This is consistent with the fact that a third-degree polynomial can have *at most* two turning points.

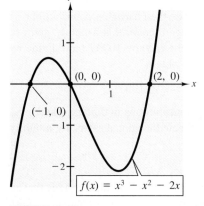

FIGURE 3.43

EXAMPLE 5 Finding Zeros of a Polynomial Function

Find all real zeros of $f(x) = -2x^4 + 2x^2$.

Solution

In this case, the polynomial factors as follows.

$$f(x) = -2x^4 + 2x^2 \qquad \text{Original function}$$
$$= -2x^2(x^2 - 1) \qquad \text{Remove common monomial factor.}$$
$$= -2x^2(x - 1)(x + 1) \qquad \text{Factor completely.}$$

Thus, the real zeros are $x = 0$, $x = 1$, and $x = -1$, and the corresponding x-intercepts are $(0, 0)$, $(1, 0)$, and $(-1, 0)$, as shown in Figure 3.44. Note in the figure that the graph has three turning points, which is consistent with the fact that a fourth-degree polynomial can have at most three turning points.

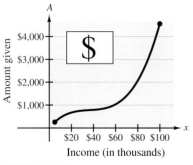

$f(x) = -2x^4 + 2x^2$

$(-1, 0)$

$(1, 0)$

$(0, 0)$

FIGURE 3.44

In Example 5, the real zero arising from $-2x^2 = 0$ is called a **repeated zero.** In general, a factor $(x - a)^k$ yields a repeated zero $x = a$ of **multiplicity** k. If k is odd, the graph *crosses* the x-axis at $x = a$. If k is even, the graph *touches* the x-axis (but does not cross the x-axis) at $x = a$. Note how this occurs in Figure 3.44.

Real Life

EXAMPLE 6 Charitable Contributions Revisited

Example 6 in Section 3.5 discussed the model

$$P = 0.0014x^2 - 0.1529x + 5.855, \qquad 5 \le x \le 100$$

where P is the percent of annual income given and x is the annual income in thousands of dollars. Note that this model gives the charitable contributions as a *percent* of annual income. To find the average *amount* that a family gives to charity, you can multiply the given model by the income $1000x$ (and divide by 100 to change from percent to decimal form) to obtain

$$A = 0.014x^3 - 1.529x^2 + 58.55x, \qquad 5 \le x \le 100$$

where A represents the amount of charitable contributions in dollars. Sketch the graph of this function and use the graph to estimate the annual salary for a family that gives $1000 a year to charities.

Solution

The graph of this function is shown in Figure 3.45. From the graph you see that an average contribution of $1000 corresponds to an annual income of about $59,000.

Income (in thousands)

FIGURE 3.45

Example 7 shows how the Leading Coefficient Test and zeros of polynomial functions can be used as sketching aids.

Technology

Example 7 uses an "algebraic approach" to describe the graph of the function. A graphing utility is a valuable comple-ment to this approach. Re-member that the most impor-tant part of using a graphing utility is to find a viewing win-dow that shows all important parts of the graph. For in-stance, the graph below shows the important parts of the graph of the function in Example 7.

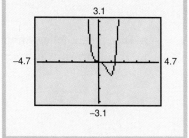

EXAMPLE 7   *Sketching the Graph of a Polynomial Function*

Sketch the graph of $f(x) = 3x^4 - 4x^3$.

Solution

Because the leading coefficient is positive and the degree is even, you know that the graph eventually rises to the left and to the right, as shown in Figure 3.46(a). By factoring

$$f(x) = 3x^4 - 4x^3 = x^3(3x - 4)$$

you can see that the zeros of f are $x = 0$ and $x = \frac{4}{3}$ (both of odd multiplicity). Thus, the x-intercepts occur at $(0, 0)$ and $\left(\frac{4}{3}, 0\right)$. To sketch the graph by hand, find a few additional points, as shown in the table. Then plot the points and complete the graph, as shown in Figure 3.46(b).

x	-1	0.5	1	1.5
$f(x)$	7	-0.3125	-1	1.6875

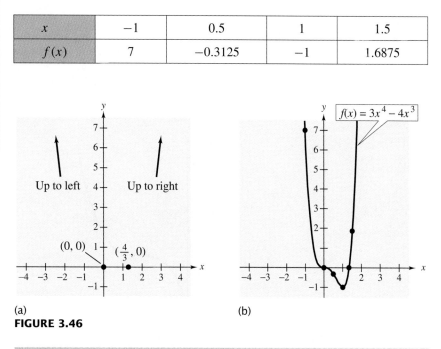

(a)

(b)

FIGURE 3.46

The Intermediate Value Theorem

The next theorem, called the **Intermediate Value Theorem,** tells you of the existence of real zeros of polynomial functions. The theorem implies that if $(a, f(a))$ and $(b, f(b))$ are two points on the graph of a polynomial such that $f(a) \neq f(b)$, then for any number d between $f(a)$ and $f(b)$ there must be a number c between a and b such that $f(c) = d$. (See Figure 3.47.)

Intermediate Value Theorem

Let a and b be real numbers such that $a < b$. If f is a polynomial function such that $f(a) \neq f(b)$, then, in the interval $[a, b]$, f takes on every value between $f(a)$ and $f(b)$.

The Intermediate Value Theorem helps you locate the real zeros of a polynomial function in the following way. If you can find a value $x = a$ where a polynomial function is positive, and another value $x = b$ where it is negative, you can conclude that the function has at least one real zero between these two values. For example, the function

$$f(x) = x^3 + x^2 + 1$$

is negative when $x = -2$ and positive when $x = -1$. Therefore, it follows from the Intermediate Value Theorem that f must have a real zero somewhere between -2 and -1, as shown in Figure 3.48.

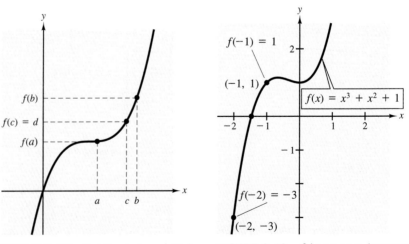

FIGURE 3.47 If d lies between $f(a)$ and $f(b)$, there exists c between a and b such that $f(c) = d$.

FIGURE 3.48 f has a zero between -2 and -1.

EXAMPLE 8   *Approximating a Zero of a Polynomial Function*

Use the Intermediate Value Theorem to approximate the real zero of

$$f(x) = x^3 - x^2 + 1.$$

Solution

Begin by computing a few function values, as follows.

x	-2	-1	0	1
$f(x)$	-11	-1	1	1

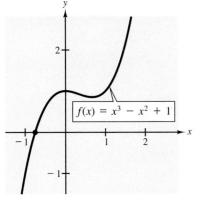

Because $f(-1)$ is negative and $f(0)$ is positive, you can apply the Intermediate Value Theorem to conclude that the function has a zero between -1 and 0. To pinpoint this zero more closely, divide the interval $[-1, 0]$ into tenths and evaluate the function at each point. When you do this, you will find that

$$f(-0.8) = -0.152$$

and

$$f(-0.7) = 0.167.$$

FIGURE 3.49   *f* has a zero between -0.8 and -0.7.

Thus, f must have a zero between -0.8 and -0.7, as shown in Figure 3.49. By continuing this process you can approximate this zero to any desired accuracy.

NOTE   The approximation process adapts very well to a graphing utility. By repeatedly using the ZOOM and TRACE features, you can find that the real zero of $f(x) = x^3 - x^2 + 1$ occurs between -0.755 and -0.754.

Group Activities   You Be the Instructor

Creating Polynomial Functions   Suppose you are writing a quiz for your algebra class and want to make up several polynomial functions for your students to investigate. Discuss how you could find polynomial functions that have reasonably simple zeros. Then use the methods you have discussed to find polynomial functions that have the following zeros.

a. $(2, 4, -3)$ **b.** $\left(3, -\frac{2}{3}, \frac{3}{4}\right)$ **c.** $(-3, -3, -3, -3)$ **d.** $(1, -2, 3, -4)$

Warm Up   The following warm-up exercises involve skills that were covered in earlier sections. You will use these skills in the exercise set for this section.

In Exercises 1–6, factor the expression completely.

1. $12x^2 + 7x - 10$

2. $25x^3 - 60x^2 + 36x$

3. $12z^4 + 17z^3 + 5z^2$

4. $y^3 + 125$

5. $x^3 + 3x^2 - 4x - 12$

6. $x^3 + 2x^2 + 3x + 6$

In Exercises 7–10, find all real solutions of the equation.

7. $5x^2 + 8 = 0$

8. $x^2 - 6x + 4 = 0$

9. $4x^2 + 4x - 11 = 0$

10. $x^4 - 18x^2 + 81 = 0$

3.6 Exercises

In Exercises 1–8, match the polynomial function with its graph. [The graphs are labeled (a), (b), (c), (d), (e), (f), (g), and (h).]

(a)
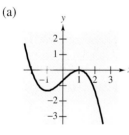

(b)
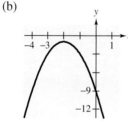

(e)
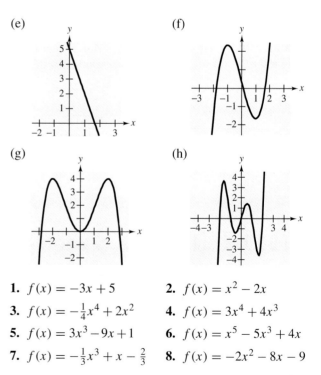

(f)

(c)
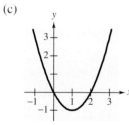

(d)
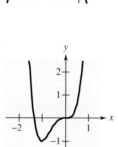

(g)

(h)

1. $f(x) = -3x + 5$

2. $f(x) = x^2 - 2x$

3. $f(x) = -\frac{1}{4}x^4 + 2x^2$

4. $f(x) = 3x^4 + 4x^3$

5. $f(x) = 3x^3 - 9x + 1$

6. $f(x) = x^5 - 5x^3 + 4x$

7. $f(x) = -\frac{1}{3}x^3 + x - \frac{2}{3}$

8. $f(x) = -2x^2 - 8x - 9$

In Exercises 9–18, determine the right-hand and left-hand behavior of the graph of the function.

9. $f(x) = 2x^2 - 3x + 1$

10. $f(x) = \frac{1}{3}x^3 + 5x$

11. $g(x) = 5 - \frac{7}{2}x - 3x^2$

12. $f(x) = -2.1x^5 + 4x^3 - 2$

13. $f(x) = 2x^5 - 5x + 7.5$

14. $h(x) = 1 - x^6$

15. $f(x) = 6 - 2x + 4x^2 - 5x^3$

16. $f(x) = \dfrac{3x^4 - 2x + 5}{4}$

17. $h(t) = -\frac{2}{3}(t^2 - 5t + 3)$

18. $f(s) = -\frac{7}{8}(s^3 + 5s^2 - 7s + 1)$

▦ *Algebraic and Graphical Approaches* In Exercises 19–34, find all real zeros of the function algebraically. Then use a graphing utility to confirm your results.

19. $f(x) = x^2 - 25$

20. $f(x) = 49 - x^2$

21. $h(t) = t^2 - 6t + 9$

22. $f(x) = x^2 + 10x + 25$

23. $f(x) = x^2 + x - 2$

24. $f(x) = \frac{1}{2}x^2 + \frac{5}{2}x - \frac{3}{2}$

25. $f(x) = 3x^2 - 12x + 3$

26. $g(x) = 5(x^2 - 2x - 1)$

27. $f(t) = t^3 - 4t^2 + 4t$

28. $f(x) = x^4 - x^3 - 20x^2$

29. $g(t) = \frac{1}{2}t^4 - \frac{1}{2}$

30. $f(x) = x^5 + x^3 - 6x$

31. $f(x) = 2x^4 - 2x^2 - 40$

32. $g(t) = t^5 - 6t^3 + 9t$

33. $f(x) = 5x^4 + 15x^2 + 10$

34. $f(x) = x^3 - 4x^2 - 25x + 100$

In Exercises 35–38, use the graph of $y = x^3$ to sketch the graph of the function.

35. $f(x) = (x - 2)^3$

36. $f(x) = x^3 - 2$

37. $f(x) = (x - 2)^3 - 2$

38. $f(x) = -\frac{1}{2}x^3$

In Exercises 39–42, use the graph of $y = x^4$ to sketch the graph of the function.

39. $f(x) = (x + 3)^4$

40. $f(x) = x^4 - 3$

41. $f(x) = 4 - x^4$

42. $f(x) = \frac{1}{2}(x - 1)^4$

▦ *Analyzing a Graph* In Exercises 43–54, analyze the graph of the function algebraically and use the results to sketch the graph *by hand*. Then use a graphing utility to confirm your sketch.

43. $f(x) = -\frac{3}{2}$

44. $h(x) = \frac{1}{3}x - 3$

45. $f(t) = \frac{1}{4}(t^2 - 2t + 15)$

46. $g(x) = -x^2 + 10x - 16$

47. $f(x) = x^3 - 3x^2$

48. $f(x) = 1 - x^3$

49. $f(x) = x^3 - 4x$

50. $f(x) = \frac{1}{4}x^4 - 2x^2$

51. $g(t) = -\frac{1}{4}(t - 2)^2(t + 2)^2$

52. $f(x) = x^2(x - 4)$

53. $f(x) = 1 - x^6$

54. $g(x) = 1 - (x + 1)^6$

▦ In Exercises 55–58, follow the procedure given in Example 8 to approximate the zero of $f(x)$ in the interval $[a, b]$. Give your approximation to the nearest tenth. (If you have a graphing utility, use it to help approximate the zero.)

55. $f(x) = x^3 + x - 1$, $\quad [0, 1]$

56. $f(x) = x^5 + x + 1$, $\quad [-1, 0]$

57. $f(x) = x^4 - 10x^2 - 11$, $\quad [3, 4]$

58. $f(x) = -x^3 + 3x^2 + 9x - 2$, $\quad [4, 5]$

59. *Hockey Players' Weights*   From 1971 to 1990, the average weight of a hockey player in the National Hockey League increased by about 11 pounds. A model that approximates the average weight of a hockey player is

$$W = 188.36 + 0.202t + 0.0042t^3, \quad -9 \le t \le 10$$

where W is the weight in pounds and t represents the playing season, with $t = 0$ corresponding to the 1980–81 playing season.   (Source: National Hockey League)

(a) Use a graphing utility to graph the model.

(b) Estimate the playing season in which the average weight was 190 pounds.

60. *Total Investment Capital*   From 1982 to 1989, the total investment capital for Northwest Natural Gas Company increased by about 200 million dollars. A model that approximates the total investment capital for this corporation is

$$C = 348.4 - 18.46t + 2.55t^2 + 0.177t^3, \quad 2 \le t \le 9$$

where C is the total investment capital (in millions of dollars) and $t = 2$ represents 1982.   (Source: Northwest Natural Gas Company)

(a) Use a graphing utility to graph the model.

(b) Estimate the year (between 1982 and 1989) in which the total investment capital was the least.

61. *Advertising Expenses*   The total revenue for a soft-drink company is related to its advertising expense by the function

$$R = \frac{1}{50,000}(-x^3 + 600x^2), \quad 0 \le x \le 400$$

where R is the total revenue in millions of dollars and x is the amount spent on advertising (in tens of thousands of dollars). Use the graph of this function to estimate the point on the graph at which the function is increasing most rapidly. This point is called the *point of diminishing returns* because any expenses above this amount will yield less return per dollar invested in advertising.

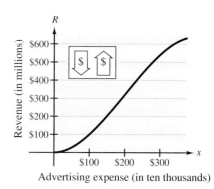

Figure for 61

62. *Advertising Expenses*   The total revenue for a hotel corporation is related to its advertising expense by the function

$$R = -0.148x^3 + 4.889x^2 - 17.778x + 125.185, \quad 0 \le x \le 20$$

where R is the total revenue in millions of dollars and x is the amount spent on advertising (in millions of dollars). Use the accompanying graph of this function to estimate the point on the graph at which the function is increasing most rapidly. This point is called the *point of diminishing returns* because any expenses above this amount will yield less return per dollar invested in advertising.

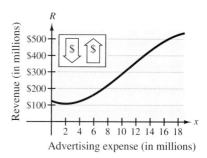

63. *Think About It*   Use a graphing utility to graph the following functions: $f(x) = x^2$; $g(x) = x^4$, $h(x) = x^6$. Do the three functions have a common shape? Are their graphs identical? Why or why not?

3.7 **Rational Functions**

Introduction ▪ Horizontal and Vertical Asymptotes ▪
Sketching the Graph of a Rational Function ▪ Applications

Introduction

A **rational function** is one that can be written in the form

$$f(x) = \frac{p(x)}{q(x)}$$

where $p(x)$ and $q(x)$ are polynomials and $q(x)$ is not the zero polynomial. In this section we assume $p(x)$ and $q(x)$ have no common factors. Unlike polynomial functions, whose domains consist of all real numbers, rational functions often have restricted domains. In general, the *domain* of a rational function of x includes all real numbers except x-values that make the denominator zero.

EXAMPLE 1   Finding the Domain of a Rational Function

Find the domain of f and discuss its behavior near any excluded x-values.

$$f(x) = \frac{1}{x}$$

Solution

The domain of f is all real numbers except $x = 0$. To determine the behavior of f near this x-value, evaluate $f(x)$ to the left and right of $x = 0$.

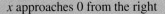

x	-1	-0.5	-0.1	-0.01	-0.001	$\longrightarrow 0$
$f(x)$	-1	-2	-10	-100	-1000	$\longrightarrow -\infty$

x approaches 0 from the right

x	$0 \longleftarrow$	0.001	0.01	0.1	0.5	1
$f(x)$	$\infty \longleftarrow$	1000	100	10	2	1

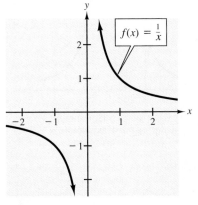

FIGURE 3.50

Note that as x approaches 0 *from the left*, $f(x)$ decreases without bound, whereas as x approaches 0 *from the right*, $f(x)$ increases without bound. The graph of f is shown in Figure 3.50.

Horizontal and Vertical Asymptotes

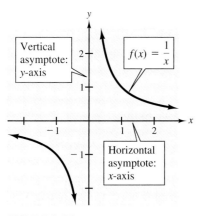

FIGURE 3.51

In Example 1, the behavior of $f(x) = 1/x$ near $x = 0$ is described by saying that the line $x = 0$ is a **vertical asymptote** of the graph of f, as shown in Figure 3.51. In this figure, note that the graph of f also has a **horizontal asymptote**—the line $y = 0$. This means that the values of $f(x) = 1/x$ approach zero as x increases or decreases without bound.

Definition of Vertical and Horizontal Asymptotes

1. The line $x = a$ is a **vertical asymptote** of the graph of f if

$$f(x) \longrightarrow \infty \quad \text{or} \quad f(x) \longrightarrow -\infty$$

as $x \longrightarrow a$, either from the right or from the left.

2. The line $y = b$ is a **horizontal asymptote** of the graph of f if

$$f(x) \longrightarrow b$$

as $x \longrightarrow \infty$ or $x \longrightarrow -\infty$.

The graph of a rational function can never intersect its vertical asymptote. It may or may not intersect its horizontal asymptote. In either case, the distance between the horizontal asymptote and the points on the graph must approach zero (as $x \longrightarrow \infty$ or $x \longrightarrow -\infty$). Figure 3.52 shows the horizontal and vertical asymptotes of the graphs of three rational functions.

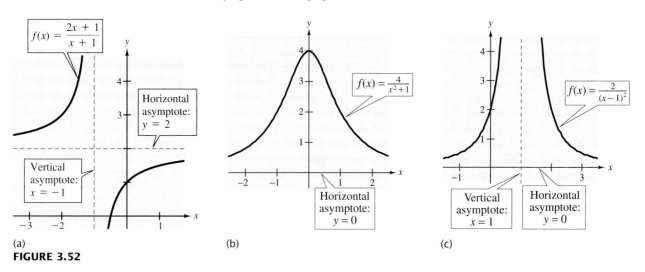

(a)

(b)

(c)

FIGURE 3.52

NOTE The graphs of $f(x) = 1/x$ (Figure 3.51) and $f(x) = (2x + 1)/(x + 1)$ (Figure 3.52a) are called **hyperbolas.**

You may want to remind students that they are writing equations of lines when they are finding equations of asymptotes. The vertical asymptote $x = 2$, for example, is sometimes written *incorrectly* as $VA = 2$ or even as the point $(2, 0)$.

Activities

1. Which of the following functions has $x = 0$ as a vertical asymptote? Discuss your answers.

   a. $f(x) = \dfrac{3 + x}{x}$

   b. $f(x) = \dfrac{x}{4 - x^2}$

   c. $f(x) = \dfrac{3x^2 - x}{x^3 - x}$

   d. $f(x) = \dfrac{x^3 - x^2 + 1}{x}$

   Answer: a and d. In a and d, $x = 0$ is the zero of the denominator. In c, the function is not reduced. The numerator and denominator have the common factor of x. Hence, $x = 0$ is not a vertical asymptote.

2. Find a function that has $x = 3$ as a vertical asymptote.

   Answers are not unique:
   $f(x) = \dfrac{5x}{x - 3}$

Horizontal Asymptote of a Rational Function

Let f be the rational function given by

$$f(x) = \frac{a_n x^n + a_{n-1} x^{n-1} + \cdots + a_1 x + a_0}{b_m x^m + b_{m-1} x^{m-1} + \cdots + b_1 x + b_0}, \quad a_n \neq 0, b_n \neq 0.$$

1. If $n < m$, the x-axis is a horizontal asymptote.

2. If $n = m$, the line $y = a_n / b_m$ is a horizontal asymptote.

3. If $n > m$, there is no horizontal asymptote.

EXAMPLE 2   Horizontal Asymptotes of Rational Functions

a. $f(x) = \dfrac{2x}{3x^2 + 1}$ See Figure 3.53(a).

Because the degree of the numerator is *less than* the degree of the denominator, the graph of f has the x-axis as a horizontal asymptote.

b. $g(x) = \dfrac{2x^2}{3x^2 + 1}$ See Figure 3.53(b).

Because the degree of the numerator is *equal to* the degree of the denominator, the graph of g has the line $y = \frac{2}{3}$ as a horizontal asymptote.

c. $h(x) = \dfrac{2x^3}{3x^2 + 1}$ See Figure 3.53(c).

Because the degree of the numerator is *greater than* the degree of the denominator, the graph of h has no horizontal asymptote.

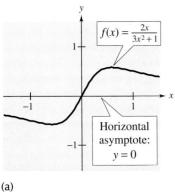

(a)

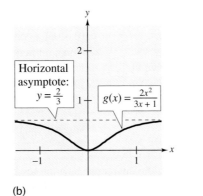

(b)

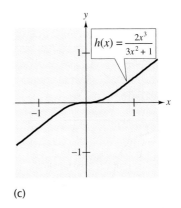

(c)

FIGURE 3.53

Sketching the Graph of a Rational Function

Guidelines for Graphing Rational Functions

Let $f(x) = p(x)/q(x)$, where $p(x)$ and $q(x)$ have no common factors.

1. Find and plot the y-intercept (if any) by evaluating $f(0)$.

2. Find the zeros of the numerator (if any) by solving the equation $p(x) = 0$. Then plot the corresponding x-intercepts.

3. Find the zeros of the denominator (if any) by solving the equation $q(x) = 0$. Then sketch the corresponding vertical asymptotes.

4. Find and sketch the horizontal asymptote (if any) by using the rule for finding the horizontal asymptote of a rational function.

5. Plot at least one point both *between and beyond* each x-intercept and vertical asymptote.

6. Use smooth curves to complete the graph between and beyond the vertical asymptotes.

Most graphing utilities do not produce good graphs of rational functions (the presence of vertical asymptotes is a problem). To obtain a reasonable graph, we suggest setting the calculator in its DOT mode.

NOTE Testing for symmetry can be useful, especially for simple rational functions. For example, the graph of $f(x) = 1/x$ is symmetrical with respect to the origin, and the graph of $g(x) = 1/x^2$ is symmetrical with respect to the y-axis.

EXAMPLE 3   Sketching the Graph of a Rational Function

Sketch the graph of $g(x) = \dfrac{3}{x-2}$.

Solution

Begin by noting that the numerator and denominator have no common factors.

y-intercept: $\left(0, -\frac{3}{2}\right)$, because $g(0) = -\frac{3}{2}$
x-intercept: None, numerator has no zeros.
Vertical asymptote: $x = 2$, zero of denominator
Horizontal asymptote: $y = 0$, degree of $p(x) <$ degree of $q(x)$
Additional points:

x	-4	1	3	5
$g(x)$	-0.5	-3	3	1

By plotting the intercepts, asymptotes, and a few additional points, you can obtain the graph shown in Figure 3.54. In the figure, note that the graph of g is a vertical stretch and a right shift of the graph of $y = 1/x$.

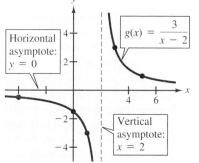

Horizontal asymptote: $y = 0$

$g(x) = \dfrac{3}{x-2}$

Vertical asymptote: $x = 2$

FIGURE 3.54

EXAMPLE 4   *Sketching the Graph of a Rational Function*

Sketch the graph of

$$f(x) = \frac{2x - 1}{x}.$$

Solution

Begin by noting that the numerator and denominator have no common factors.

y-intercept:	None, because $f(0)$ is not defined
x-intercept:	$\left(\frac{1}{2}, 0\right)$, from $2x - 1 = 0$
Vertical asymptote:	$x = 0$, zero of denominator
Horizontal asymptote:	$y = 2$, degree of $p(x) =$ degree of $q(x)$
Additional points:	

x	-4	-1	$\frac{1}{4}$	4
$f(x)$	2.25	3	-2	1.75

By plotting the intercepts, asymptotes, and a few additional points, you can obtain the graph shown in Figure 3.55. Confirm the graph with your graphing utility.

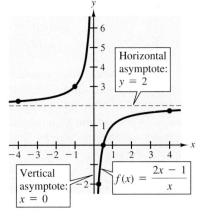

FIGURE 3.55

EXAMPLE 5   *Sketching the Graph of a Rational Function*

Sketch the graph of $f(x) = \dfrac{x}{x^2 - x - 2}$.

Solution

By factoring the denominator, you have

$$f(x) = \frac{x}{x^2 - x - 2} = \frac{x}{(x + 1)(x - 2)}.$$

Thus, the numerator and denominator have no common factors.

y-intercept:	$(0, 0)$
x-intercept:	$(0, 0)$
Vertical asymptote:	$x = -1, x = 2$, zeros of denominator
Horizontal asymptote:	$y = 0$, degree of $p(x) <$ degree of $q(x)$
Additional points:	

x	-3	-0.5	1	3
$f(x)$	-0.3	0.4	-0.5	0.75

By plotting the intercepts, asymptotes, and a few additional points, you can obtain the graph shown in Figure 3.56. Confirm the graph with your graphing utility.

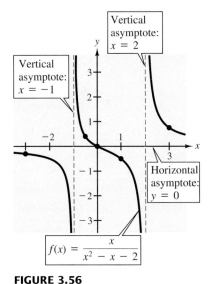

FIGURE 3.56

Applications

There are many examples of asymptotic behavior in business and biology. For instance, the following two examples describe the asymptotic behavior related to the cost of removing smokestack emissions *and* the average cost of producing a product.

EXAMPLE 6 *Cost-Benefit Model*

A utility company burns coal to generate electricity. The cost of removing a certain *percent* of the pollutants from the stack emission is typically not a linear function. That is, if it costs C dollars to remove 25% of the pollutants, it would cost more than $2C$ dollars to remove 50% of the pollutants. As the percent of removed pollutants approaches 100%, the cost tends to become prohibitive. Suppose that the cost C of removing p percent of the smokestack pollutants is given by

$$C = \frac{80,000p}{100 - p}.$$

Suppose that you are a member of a state legislature that is considering a law that will require utility companies to remove 90% of the pollutants from their smokestack emissions. If the current law requires 85% removal, how much additional expense is the new law asking the utility company to incur?

Solution

The graph of this function is shown in Figure 3.57. Note that the graph has a vertical asymptote at $p = 100$. Because the current law requires 85% removal, the current cost to the utility company is

$$C = \frac{80,000(85)}{100 - 85} \qquad \text{Substitute 85 for } p.$$
$$\approx \$453,333. \qquad \text{Use a calculator.}$$

If the new law increases the percent removal to 90%, the cost to the utility company will be

$$C = \frac{80,000(90)}{100 - 90} \qquad \text{Substitute 90 for } p.$$
$$\approx \$720,000. \qquad \text{Use a calculator.}$$

Therefore, the new law would require the utility company to spend an additional

$$\$720,000 - \$453,333 = \$266,667.$$

How much would it cost to remove another 2% of the pollutants?

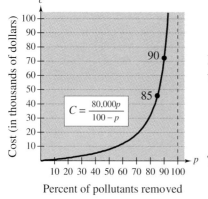

FIGURE 3.57

Cost (in thousands of dollars)

Percent of pollutants removed

$$C = \frac{80,000p}{100 - p}$$

EXAMPLE 7   *Average Cost of Producing a Product*

A business has a cost of $C = 0.5x + 5000$, where C is measured in dollars and x is the number of units produced. The *average cost per unit* is given by

$$\overline{C} = \frac{C}{x} = \frac{0.5x + 5000}{x}.$$

Find the average cost per unit when $x = 1000$, $10{,}000$, and $100{,}000$. What is the horizontal asymptote for this function, and what does it represent?

Solution

When $x = 1000$, the average cost per unit is

$$\overline{C} = \frac{0.5(1000) + 5000}{1000} = \$5.50.$$

When $x = 10{,}000$, the average cost per unit is

$$\overline{C} = \frac{0.5(10{,}000) + 5000}{10{,}000} = \$1.00.$$

When $x = 100{,}000$, the average cost per unit is

$$\overline{C} = \frac{0.5(100{,}000) + 5000}{100{,}000} = \$0.55.$$

As shown in Figure 3.58, the horizontal asymptote is $C = 0.50$. This line represents the least possible unit cost for the product. This example points out one of the major problems of a small business—it is difficult to have competitively low prices with low levels of production.

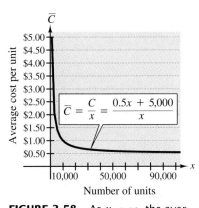

FIGURE 3.58   As $x \to \infty$, the average cost per unit approaches $0.50.

Group Activities

Extending the Concept

Constructing Rational Functions   A model for the population y (in millions) of the United States from 1960 to 1990 is $y = 2.314x + 179.759$, $0 \le x \le 30$, where x represents the calendar year, with $x = 0$ corresponding to 1960   (Source: U.S Bureau of Census). A model for the land area A (in millions of acres) of the United States from 1960 to 1990 is $A = 2316.013$. From these functions, construct a rational function for per capita land area. Sketch a graph of the rational function. If this model continues to be accurate for the years beyond 1990, when will per capita land area drop below 8 acres? Find and interpret any asymptotes of the function.

Warm Up   The following warm-up exercises involve skills that were covered in earlier sections. You will use these skills in the exercise set for this section.

In Exercises 1–6, factor the polynomial.

1. $x^2 - 4x$

2. $2x^3 - 6x$

3. $x^2 - 3x - 10$

4. $x^2 - 7x + 10$

5. $x^3 + 4x^2 + 3x$

6. $x^3 - 4x^2 - 2x + 8$

In Exercises 7–10, sketch the graph of the equation.

7. $y = 2$

8. $x = -1$

9. $y = x - 2$

10. $y = -x + 1$

3.7 Exercises

In Exercises 1–8, find the domain of the function.

1. $f(x) = \dfrac{1}{x^2}$

2. $f(x) = \dfrac{4}{(x-2)^3}$

3. $f(x) = \dfrac{2+x}{2-x}$

4. $f(x) = \dfrac{1-5x}{1+2x}$

5. $f(x) = \dfrac{3x^2+1}{x^2+9}$

6. $f(x) = \dfrac{3x^2+x-5}{x^2+1}$

7. $f(x) = \dfrac{5x^4}{x^2+1}$

8. $f(x) = \dfrac{2x}{x+1}$

In Exercises 9–14, match the function with its graph. [The graphs are labeled (a), (b), (c), (d), (e), and (f).]

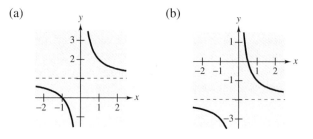

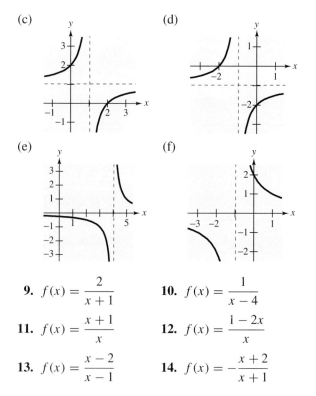

9. $f(x) = \dfrac{2}{x+1}$

10. $f(x) = \dfrac{1}{x-4}$

11. $f(x) = \dfrac{x+1}{x}$

12. $f(x) = \dfrac{1-2x}{x}$

13. $f(x) = \dfrac{x-2}{x-1}$

14. $f(x) = -\dfrac{x+2}{x+1}$

In Exercises 15–18, compare the graph of $f(x) = 1/x$ with the graph of g.

15. $g(x) = f(x) + 1 = \dfrac{1}{x} + 1$

16. $g(x) = f(x - 1) = \dfrac{1}{x - 1}$

17. $g(x) = -f(x) = -\dfrac{1}{x}$

18. $g(x) = f(x + 2) = \dfrac{1}{x + 2}$

In Exercises 19–22, compare the graph of $f(x) = 4/x^2$ with the graph of g.

19. $g(x) = f(x) - 2 = \dfrac{4}{x^2} - 2$

20. $g(x) = f(x - 2) = \dfrac{4}{(x - 2)^2}$

21. $g(x) = -f(x) = -\dfrac{4}{x^2}$

22. $g(x) = \dfrac{1}{4}f(x) = \dfrac{1}{x^2}$

In Exercises 23–26, compare the graph of $f(x) = 8/x^3$ with the graph of g.

23. $g(x) = f(x) + 1 = \dfrac{8}{x^3} + 1$

24. $g(x) = f(x + 2) = \dfrac{8}{(x + 2)^3}$

25. $g(x) = -f(x) = -\dfrac{8}{x^3}$

26. $g(x) = \dfrac{1}{8}f(x) = \dfrac{1}{x^3}$

In Exercises 27–50, sketch the graph of the rational function. As sketching aids, check for intercepts, vertical asymptotes, and horizontal asymptotes.

27. $f(x) = \dfrac{1}{x + 2}$

28. $f(x) = \dfrac{1}{x - 3}$

29. $h(x) = \dfrac{-1}{x + 2}$

30. $g(x) = \dfrac{1}{3 - x}$

31. $f(x) = \dfrac{x + 1}{x + 2}$

32. $f(x) = \dfrac{x - 2}{x - 3}$

33. $f(x) = \dfrac{2 + x}{1 - x}$

34. $f(x) = \dfrac{3 - x}{2 - x}$

35. $f(t) = \dfrac{3t + 1}{t}$

36. $f(t) = \dfrac{1 - 2t}{t}$

37. $g(x) = \dfrac{1}{x + 2} + 2$

38. $h(x) = \dfrac{1}{x - 3} + 1$

39. $C(x) = \dfrac{5 + 2x}{1 + x}$

40. $P(x) = \dfrac{1 - 3x}{1 - x}$

41. $f(x) = \dfrac{x^2}{x^2 + 9}$

42. $f(x) = 2 - \dfrac{3}{x^2}$

43. $h(x) = \dfrac{x^2}{x^2 - 9}$

44. $g(x) = \dfrac{x}{x^2 - 9}$

45. $g(x) = \dfrac{s}{s^2 + 1}$

46. $h(t) = \dfrac{4}{t^2 + 1}$

47. $f(x) = -\dfrac{1}{(x - 2)^2}$

48. $g(x) = -\dfrac{x}{(x - 2)^2}$

49. $f(x) = \dfrac{3x}{x^2 - x - 2}$

50. $f(x) = \dfrac{2x}{x^2 + x - 2}$

51. *Seizure of Illegal Drugs* The cost in millions of dollars for the federal government to seize p percent of a certain illegal drug as it enters the country is

$$C = \dfrac{528p}{100 - p}, \qquad 0 \le p < 100.$$

(a) Find the cost of seizing 25%.

(b) Find the cost of seizing 50%.

(c) Find the cost of seizing 75%.

(d) According to this model, would it be possible to seize 100% of the drug? Explain.

52. *Smokestack Emission* The cost in dollars of removing p percent of the air pollutants in the stack emission of a utility company that burns coal to generate electricity is

$$C = \dfrac{80,000p}{100 - p}, \qquad 0 \le p < 100.$$

(a) Find the cost of removing 15%.

(b) Find the cost of removing 50%.

(c) Find the cost of removing 90%.

(d) According to this model, would it be possible to remove 100% of the pollutants? Explain.

53. *Population of Deer*   The game commission introduces 50 deer into newly acquired state game lands. The population of the herd is given by

$$N = \frac{10(5 + 3t)}{1 + 0.04t}, \qquad 0 \le t$$

where t is time in years.

(a) Find the population when t is 5, 10, and 25.

(b) What is the limiting size of the herd as time progresses?

54. *Population of Elk*   The game commission introduces 40 elk into newly acquired state game lands. The population of the herd is given by

$$N = \frac{10(4 + 2t)}{1 + 0.03t}, \qquad 0 \le t$$

where t is time in years.

(a) Find the population when t is 5, 10, and 25.

(b) What is the limiting size of the herd as time progresses?

55. *Average Cost*   The cost of producing x units is

$$C = 150,000 + 0.25x$$

and therefore the average cost per unit is

$$\overline{C} = \frac{C}{x} = \frac{150,000 + 0.25x}{x}, \qquad 0 < x.$$

(a) Sketch the graph of this average cost function.

(b) Find the average cost of producing $x = 1000$, $x = 10,000$, and $x = 100,000$ units. What can you conclude?

56. *Average Cost*   The cost of producing x units is

$$C = 250,000 + 3x$$

and therefore the average cost per unit is

$$\overline{C} = \frac{C}{x} = \frac{250,000 + 3x}{x}, \qquad 0 < x.$$

(a) Sketch the graph of this average cost function.

(b) Find the average cost of producing $x = 1000$, $x = 10,000$, and $x = 100,000$ units. What can you conclude?

57. *Human Memory Model*   Psychologists have developed mathematical models to predict memory performance as a function of the number of trials n of a certain task. Consider the learning curve

$$P = \frac{0.5 + 0.9(n - 1)}{1 + 0.9(n - 1)}, \qquad 0 < n$$

where P is the percent of correct responses after n trials.

(a) Complete the following table.

n	1	2	3	4	5	6	7	8	9	10
P										

(b) According to this model, what is the limiting percent of correct responses as n increases?

58. *Human Memory Model*   Consider the learning curve

$$P = \frac{0.6 + 0.85(n - 1)}{1 + 0.85(n - 1)}, \qquad 0 < n$$

where P is the percent of correct responses after n trials.

(a) Complete the following table.

n	1	2	3	4	5	6	7	8	9	10
P										

(b) According to this model, what is the limiting percent of correct responses as n increases?

59. *Average Recycling Cost*   The cost of recycling a waste product is

$$C = 350,000 + 5x, \qquad 0 < x$$

where x is the number of pounds of waste. The average recyclying cost is

$$\overline{C} = \frac{C}{x} = \frac{350,000 + 5x}{x}, \qquad 0 < x.$$

(a) Use a graphing utility to sketch the graph of $\overline{C}$.

(b) Find the average cost of recycling for $x = 1000$, $x = 10,000$, and $x = 100,000$ pounds. What can you conclude?

60. *Average Recyclying Cost*   The cost of recycling a waste product is

$$C = 600,000 + 10x, \quad 0 < x$$

where x is the number of pounds of waste. The average recycling cost is

$$\overline{C} = \frac{C}{x} = \frac{600,000 + 10x}{x}, \quad 0 < x.$$

(a) Use a graphing utility to sketch the graph of $\overline{C}$.

(b) Find the average cost of recycling for $x = 1000$, $x = 10,000$, and $x = 100,000$ pounds. What can you conclude?

61. *Sale of Long-Playing Albums*   The sales of long-playing albums in the United States from 1975 to 1991 can be approximated by the model

$$L = \frac{178.77 - 8.49t}{1 - 0.117t + 0.005t^2}, \quad 5 \le t \le 21$$

where L is the number of albums sold (in millions) and $t = 5$ represents 1975. From the figure, notice the drop in sales of albums after about 1980. How would you account for this drop in sales?   (Source: Recording Industry Association of America)

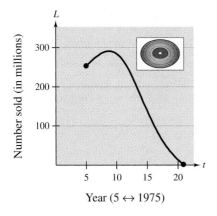

Year (5 ↔ 1975)

62. *1500-Meter Run*   The winning times for the 1500-meter run at the Olympics from 1900 to 1992 are shown in the figure. A *quadratic* model that approximates these times is

$$y = 4.13 - 0.0109t + 0.00006t^2$$

where y is the winning time and $t = 0$ represents 1900. How well did this model predict the winning time for 1996? Does this model have a horizontal asymptote? From the winning times, do you think that a good model should have a horizontal asymptote?

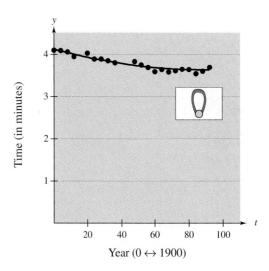

Year (0 ↔ 1900)

63. The concentration of a medication in the bloodstream t hours after injection into muscle tissue is given by

$$C(t) = \frac{2t + 1}{t^2 + 4}, \quad 0 < t.$$

(a) Use a graphing utility to graph the function. Estimate when the concentration is greatest.

(b) Does this function have a horizontal asymptote? If so, discuss the meaning of the asymptote in terms of the concentration of the medication.

64. The concentration of a medication in the bloodstream t minutes after sublingual (under the tongue) application is given by

$$C(t) = \frac{3t - 1}{2t^2 + 5}, \quad 0 < t.$$

(a) Use a graphing utility to graph the function. Estimate when the concentration is greatest.

(b) Does this function have a horizontal asymptote? If so, discuss the meaning of the asymptote in terms of the concentration of the medication.

CHAPTER PROJECT: Gravitational Force

Time	Height
0.0	5.23594
0.02	5.20353
0.04	5.16031
0.06	5.0991
0.08	5.02707
0.099996	4.95146
0.119996	4.85062
0.139992	4.74979
0.159988	4.63096
0.179988	4.50132
0.199984	4.35728
0.219984	4.19523
0.23998	4.02958
0.25998	3.84593
0.27998	3.65507
0.299976	3.44981
0.319972	3.23375
0.339961	3.01048
0.359961	2.76921
0.379951	2.52074
0.399941	2.25786
0.419941	1.98058
0.439941	1.63488

A function that gives the height s of a falling object in terms of the time t is called a *position function*. If air resistance is not considered, the position of a falling object can be modeled by

$$s(t) = \tfrac{1}{2}gt^2 + v_0 t + s_0$$

where g is the acceleration due to gravity, v_0 is the initial velocity, and s_0 is the initial height. The value of g depends on where the object is dropped. On Earth's surface, g is approximately -32 feet per second per second.

To discover the value of g experimentally, you can record the heights of a falling object for several times. The table at the left shows the data collected with a *Texas Instruments CBL* unit.*

Use this information to investigate the following questions.

1. **Fitting a Model to Data**   Enter the data given in the table in a graphing utility, and use its quadratic fit feature to find a quadratic model for the data. Does the model you obtain agree with that given on page 229?

2. **Comparing Actual Data with a Model Numerically**   Copy the table shown at the left. Then add another column to the table listing the heights predicted by the model. How well does the model fit the actual data?

3. **Comparing Actual Data with a Model Graphically**   Use a graphing utility to make a scatter plot of the data shown in the table. Then graph the model on the same screen. How well does the model fit the actual data?

4. **Experimental Value of Gravity**   Use the model found in Question 1 to answer the following.

   (a) What is the experimental value of g? What are the units for g?

   (b) What is the initial velocity of the falling object? What are the units for this velocity? Why is the velocity negative?

   (c) What is the initial height of the falling object? What are the units for this height?

   (d) Use the units found in parts (a), (b), and (c) to analyze the units of the position function $s(t) = \tfrac{1}{2}gt^2 + v_0 t + s_0$. What are the units for s, g, t, v_0, and s_0?

5. **Extended CBL Exploration**   The experimental value of g found in Question 4 is not equal to the theoretical value of g on Earth's surface. Use the *CBL* instructions in Appendix C to collect and analyze data for different types of falling objects. For instance, you might use a baseball, a basketball, and a ping-pong ball. Do the experimental values of g depend on the type of ball you use? If so, why—does air resistance come into play? Which of the objects used is least affected by air resistance?

*Instructions for using a *CBL* to collect your own data are given in Appendix C.

CHAPTER SUMMARY

After studying this chapter, you should have acquired the following skills. These skills are keyed to the Review Exercises that begin on page 314. Answers to odd-numbered Review Exercises are given in the back of the book.

- Evaluate a function. *(Section 3.1)* — **Review Exercises 1–4**

- Find a value for which a function is zero. *(Section 3.1)* — **Review Exercises 5, 6**

- Find the domain of a function. *(Section 3.1)* — **Review Exercises 7–12, 17, 18**

- Determine if an equation or a set of ordered pairs represents a function. *(Section 3.1)* — **Review Exercises 13–16**

- Determine the range and domain of a function, the intervals over which it is increasing, decreasing, or constant, and whether it is even, odd, or neither. *(Section 3.2)* — **Review Exercises 19–22**

- Sketch the graph of a function. *(Section 3.2)* — **Review Exercises 23–33**

- Use translations to sketch the graph of a function. *(Section 3.3)* — **Review Exercises 34–39**

- Determine the sum, difference, product, or quotient of two functions. *(Section 3.3)* — **Review Exercises 40–47, 52**

- Evaluate the composition of two functions. *(Section 3.3)* — **Review Exercises 48–51, 53**

- Show that two functions are inverses of each other. *(Section 3.4)* — **Review Exercises 54, 55**

- Determine if a function has an inverse and graph both in the same coordinate plane. *(Section 3.4)* — **Review Exercises 56–61**

- Sketch the graph of a quadratic function and identify its vertex and intercepts. *(Section 3.5)* — **Review Exercises 62–65**

- Find a quadratic function given its vertex and a point. *(Section 3.5)* — **Review Exercises 66, 67**

- Use the standard form of a quadratic function to determine minimum or maximum values. *(Section 3.5)* — **Review Exercises 68–71**

- Determine the right- and left-hand behavior of a graph of a function. *(Section 3.6)* — **Review Exercises 72–75**

- Find all real zeros of a function. *(Section 3.6)* — **Review Exercises 76–79**

- Use the Intermediate Value Theorem to approximate the zero of a function. *(Section 3.6)* — **Review Exercises 80, 81**

- Find the domain of a rational function and identify any asymptotes. *(Section 3.7)* — **Review Exercises 82–85**

- Sketch a graph of a rational function using intercepts, symmetry, and asymptotes. *(Section 3.7)* — **Review Exercises 86–89**

- Use rational functions to model applications. *(Section 3.7)* — **Review Exercises 90–94**

REVIEW EXERCISES

In Exercises 1–4, evaluate the function and simplify the results.

1. $f(x) = 3x - 5$

    (a) $f(1)$ (b) $f(-2)$ (c) $f(m)$ (d) $f(x+1)$

2. $f(x) = \sqrt{x+9} - 3$

    (a) $f(7)$ (b) $f(0)$ (c) $f(-5)$ (d) $f(x+2)$

3. $f(x) = |x| + 5$

    (a) $f(0)$ (b) $f(-3)$ (c) $f\left(\dfrac{1}{2}\right)$ (d) $f(-x^2)$

4. $f(x) = \begin{cases} x^2 + 1, & x \le 2 \\ x + 3, & x > 2 \end{cases}$

    (a) $f(0)$ (b) $f(2)$ (c) $f(3)$ (d) $f(-5)$

In Exercises 5 and 6, find all real values of x such that $f(x) = 0$.

5. $f(x) = \dfrac{2x+7}{3}$

6. $f(x) = x^3 - 4x$

In Exercises 7–12, find the domain of the function.

7. $f(x) = 3x^2 + 8x + 4$

8. $g(t) = \dfrac{5}{t^2 - 9}$

9. $h(x) = \sqrt{x+9}$

10. $f(t) = \sqrt[3]{t-5}$

11. $g(t) = \dfrac{\sqrt{t-3}}{t-5}$

12. $h(x) = \sqrt[4]{9-x^2}$

In Exercises 13 and 14, decide whether the equation represents y as a function of x.

13. $3x - 4y = 12$

14. $y^2 = x^2 - 9$

In Exercises 15 and 16, decide whether the set represents a function from A to B.

$A = \{1, 2, 3\}$ $B = \{-3, -4, -7\}$

Give reasons for your answer.

15. $\{(1, -3), (2, -7), (3, -3)\}$

16. $\{(1, -4), (2, -3), (3, -9)\}$

17. *Volume of a Box* An open box is to be made from a square piece of material 20 inches on a side by cutting equal squares from the corners and turning up the sides (see figure).

  (a) Write the volume of the box as a function of x.

  (b) What is the domain of this function?

  (c) Sketch the graph of this function.

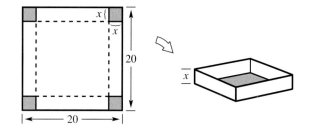

18. *Balance in an Account* A person deposits $5000 in an account that pays 6.25% interest compounded quarterly.

  (a) Write the balance of the account in terms of the time t that the principal is left in the account.

  (b) What is the domain of this function?

In Exercises 19–22, (a) determine the range and domain of the function, (b) determine the intervals over which the function is increasing, decreasing, or constant, and (c) determine if the function is even, odd, or neither.

19. $f(x) = x^2 + 1$

20. $f(x) = \sqrt{x^2 - 9}$

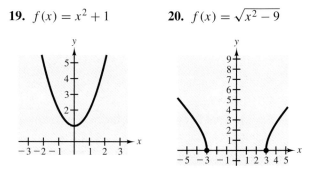

21. $f(x) = x^3 - 4x^2$

22. $f(x) = |x - 2|$

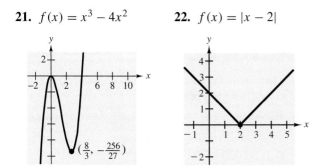

$\left(\frac{8}{3}, -\frac{256}{27}\right)$

In Exercises 23–32, sketch the graph of the function.

23. $f(x) = 3 - 2x$

24. $f(x) = |x + 3|$

25. $g(x) = \sqrt{x^2 - 16}$

26. $h(x) = [\![x]\!]$

27. $h(x) = 2[\![x]\!] + 1$

28. $f(x) = 3$

29. $g(x) = \begin{cases} x + 2, & x < 0 \\ 2, & x = 0 \\ x^2 + 2, & x > 0 \end{cases}$

30. $g(x) = \begin{cases} 3x + 1, & x < -1 \\ x^2 - 3, & x \geq -1 \end{cases}$

31. $h(x) = x^2 - 3x$

32. $f(x) = \sqrt{9 - x^2}$

33. *Cost of a Phone Call* The cost of a phone call between Dallas and Philadelphia is \$0.70 for the first minute and \$0.38 for each additional minute (or portion of a minute). A model for the total cost of the phone call is

$$C = 0.70 + 0.38[\![x]\!], \qquad 0 < x$$

where C is the total cost of the call in dollars and x is the length of the call in minutes. Sketch the graph of this function.

In Exercises 34 and 35, use the graph of $f(x) = x^2$ to sketch the graph of the function.

34. $g(x) = -(x - 1)^2 - 2$

35. $g(x) = -x^2 + 3$

In Exercises 36 and 37, use the graph of $f(x) = \sqrt{x}$ to sketch the graph of the function.

36. $h(x) = -\sqrt{x - 2}$

37. $h(x) = \sqrt{x} + 2$

In Exercises 38 and 39, use the graph of $f(x) = \sqrt[3]{x}$ to sketch the graph of the function.

38. $y = \sqrt[3]{x + 2}$

39. $y = 2\sqrt[3]{x}$

In Exercises 40–43, find $(f + g)(x)$, $(f - g)(x)$, $(fg)(x)$, and $(f/g)(x)$. What is the domain of f/g?

40. $f(x) = x + 2$, $g(x) = x - 3$

41. $f(x) = x^2 + 1$, $g(x) = 8 - x$

42. $f(x) = \sqrt{x^2 - 4}$, $g(x) = 3$

43. $f(x) = 3x + 2$, $g(x) = x^2 - 4x$

In Exercises 44–47, evaluate the function for $f(x) = x^2 - 1$ and $g(x) = x + 2$.

44. $(f + g)(2)$

45. $(f - g)(-1)$

46. $(fg)(3)$

47. $\left(\dfrac{f}{g}\right)(0)$

In Exercises 48–51, find (a) $f \circ g$ and (b) $g \circ f$.

48. $f(x) = x^2$, $g(x) = x + 3$

49. $f(x) = 2x - 5$, $g(x) = x^2 + 2$

50. $f(x) = |x| + 2$, $g(x) = x - 5$

51. $f(x) = \dfrac{1}{x}$, $g(x) = 3x + x^2$

52. *Company Sales* You own two dry cleaning establishments. From 1990 to 1994, the sales for one of the establishments have been decreasing according to the function

$$R_1 = 389.2 - 3.6t - 0.3t^2, \qquad t = 0, 1, 2, 3, 4$$

where R_1 represents the sales (in thousands of dollars) and $t = 0$ represents 1990. During the same 5-year period, the sales for the second establishment have been increasing according to the function

$$R_2 = 204.92 + 0.82t, \qquad t = 0, 1, 2, 3, 4.$$

Write a function that represents the total sales for the two establishments. Make a stacked bar graph to represent the total sales during this 5-year period. Are total sales increasing or decreasing?

53. *Cost* The daily cost of producing x units in a manufacturing process is given by the function $C(x) = 30x + 800$. The number of units produced in t hours is given by $x(t) = 150t$. Find and interpret $(C \circ x)(t)$.

In Exercises 54 and 55, show that f and g are inverses of each other.

54. $f(x) = 3x + 5$, $\quad g(x) = \dfrac{x - 5}{3}$

55. $f(x) = \sqrt[3]{x - 3}$, $\quad g(x) = x^3 + 3$

In Exercises 56–61, determine whether the function has an inverse. If it does, find the inverse and graph f and f^{-1} in the same coordinate plane.

56. $f(x) = 3x^2$

57. $f(x) = \sqrt[3]{x + 1}$

58. $f(x) = \dfrac{1}{x}$

59. $f(x) = x^4$

60. $f(x) = \dfrac{2x - 5}{3}$

61. $f(x) = \dfrac{x^2}{x^2 - 9}$

In Exercises 62–65, sketch the graph of the quadratic function. Identify the vertex and intercepts.

62. $f(x) = (x - 2)^2 + 5$

63. $g(x) = -x^2 + 6x + 8$

64. $h(x) = 3x^2 - 12x + 11$

65. $f(x) = \dfrac{1}{3}(x^2 + 5x - 4)$

In Exercises 66 and 67, find the quadratic function that has the indicated vertex and whose graph passes through the point.

66. Vertex: $(1, -4)$; point: $(2, -3)$

67. Vertex: $(2, 3)$; point: $(-1, 6)$

68. *Maximum Area* The perimeter of a rectangle is 200 feet. Let x represent the width of the rectangle and write a quadratic function that expresses the area of the rectangle in terms of its width. Of all possible rectangles with perimeters of 200 feet, what are the dimensions of the one that has the greatest area?

69. *Maximum Revenue* Find the number of units that produce a maximum revenue R for

$$R = 800x - 0.01x^2$$

where R is the total revenue in dollars and x is the number of units produced.

70. *Minimum Cost* A manufacturer has daily production costs of

$$C = 20{,}000 - 40x + 0.055x^2$$

where C is the total cost in dollars and x is the number of units produced.

(a) Use a graphing utility to graph the cost function.

(b) Graphically estimate the number of units that should be produced to yield a minimum cost.

(c) Explain how to confirm the result of part (b) algebraically.

71. *Maximum Profit* The profit for a company is given by

$$P = -0.0001x^2 + 130x - 300{,}000$$

where x is the number of units produced.

(a) Use a graphing utility to graph the profit function.

(b) Graphically estimate the number of units that should be produced to yield a maximum profit.

(c) Explain how to confirm the result of part (b) algebraically.

In Exercises 72–75, determine the right- and left-hand behavior of the graph of the function.

72. $f(x) = \dfrac{1}{2}x^3 + 2x$

73. $f(x) = -x^2 + 6x + 9$

74. $f(x) = -x^5 + 3x^2 + 11$

75. $f(x) = \dfrac{3}{4}(x^4 + 3x^2 + 2)$

In Exercises 76–79, find all real zeros of the function.

76. $f(x) = 25 - x^2$

77. $f(x) = x^4 - 6x^2 + 8$

78. $f(x) = x^3 - 7x^2 + 10x$

79. $f(x) = x^3 - 6x^2 - 3x + 18$

In Exercises 80 and 81, use the Intermediate Value Theorem to approximate the zero of $f(x)$ in the interval $[a, b]$. Give your approximation to the nearest tenth.

80. $f(x) = x^3 - 4x + 3$ $[-3, -2]$

81. $f(x) = x^5 + 5x^2 + x - 1$ $[0, 1]$

In Exercises 82–85, find the domain of the function and identify any horizontal or vertical asymptotes.

82. $f(x) = \dfrac{4}{x + 3}$

83. $f(x) = \dfrac{2x^2 + 5x - 3}{x^2 + 2}$

84. $f(x) = \dfrac{x^2}{x^2 - 4}$

85. $f(x) = \dfrac{1}{(x - 3)^2}$

In Exercises 86–89, sketch the graph of the rational function. As sketching aids, check for intercepts, symmetry, vertical asymptotes, and horizontal asymptotes.

86. $P(x) = \dfrac{3 - x}{x + 2}$

87. $f(x) = \dfrac{4}{(x - 1)^2}$

88. $g(x) = \dfrac{1}{x + 3} + 2$

89. $h(x) = -\dfrac{5}{x^2}$

90. *Average Cost* The cost of producing x units is $C = 100{,}000 + 0.9x$, and therefore the average cost per unit is

$$\overline{C} = \frac{C}{x} = \frac{100{,}000 + 0.9x}{x}, \qquad 0 < x.$$

Sketch the graph of this average cost function and find the average cost of producing $x = 1000$, $x = 10{,}000$, and $x = 100{,}000$ units.

91. *Population of Fish* The Parks and Wildlife Commission introduces 50,000 game fish into a large lake. The population of the fish (in thousands) is

$$N = \frac{20(4 + 3t)}{1 + 0.05t}, \qquad 0 \le t$$

where t is time in years.

(a) Find the population when $t = 5$, 10, and 25.

(b) What is the limiting number of this population of fish in the lake as time progresses?

92. *Human Memory Model* Consider the learning curve

$$P = \frac{0.5 + 0.9(n - 1)}{1 + 0.9(n - 1)}, \qquad 0 < n$$

where P is the percent of correct responses after n trials. Complete the table for this model.

n	1	2	3	4	5	6	7	8	9	10
P										

93. *Smokestack Emission* The cost in dollars of removing p percent of the air pollutants in the stack emission of a utility company that burns coal to generate electricity is

$$C = \frac{95{,}000p}{100 - p}, \qquad 0 \le p \le 100.$$

(a) Find the cost of removing 25%.

(b) Find the cost of removing 60%.

(c) Find the cost of removing 99%.

(d) According to this model, would it be possible to remove 100% of the pollutants? Explain.

94. *Average Recycling Cost* The cost of recycling a waste product is

$$C = 450{,}000 + 6x, \qquad 0 < x$$

where x is the number of pounds of waste. The average recycling cost is

$$\overline{C} = \frac{C}{x} = \frac{450{,}000 + 6x}{x}, \qquad 0 < x.$$

(a) Sketch the graph of $\overline{C}$.

(b) Find the average cost of recycling for $x = 1000$ pounds, $x = 10{,}000$ pounds, and $x = 100{,}000$ pounds. What can you conclude?

CHAPTER TEST

Take this test as you would take a test in class. After you are done, check your work against the answers given in the back of the book.

In Exercises 1–4, find the domain of the function.

1. $f(x) = 2x^2 - 3x + 8$ **2.** $g(x) = \sqrt{x - 7}$

3. $g(t) = \dfrac{\sqrt{t - 2}}{t - 7}$ **4.** $h(x) = \dfrac{3}{x^2 - 4}$

In Exercises 5–7, decide whether the statement is true or false. Explain.

5. The equation $2x - 3y = 5$ identifies y as a function of x.

6. The equation $y = \pm\sqrt{x^2 - 16}$ identifies y as a function of x.

7. If $A = \{3, 4, 5\}$ and $B = \{-1, -2, -3\}$, the set $\{(3, -9), (4, -2), (5, -3)\}$ represents a function from A to B.

In Exercises 8 and 9, (a) find the range and domain of the function, (b) determine the intervals over which the function is increasing, decreasing, or constant, and (c) determine whether the function is even or odd.

8. $f(x) = x^2 + 2$ (See figure.)   **9.** $g(x) = \sqrt{x^2 - 4}$ (See figure.)

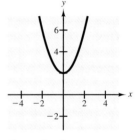

Figure for 8

In Exercises 10–12, sketch the graph of the function.

10. $f(x) = \dfrac{3}{x + 2}$ **11.** $g(x) = \begin{cases} x + 1, & x < 0 \\ 1, & x = 0 \\ x^2 + 1, & x > 0 \end{cases}$ **12.** $h(x) = (x - 3)^2 + 4$

13. Find $f^{-1}(x)$ for the function $f(x) = x^3 - 5$.

In Exercises 14–19, use $f(x) = x^2 + 2$ and $g(x) = 2x - 1$ to find the function.

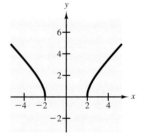

Figure for 9

14. $(f + g)(x)$ **15.** $(f - g)(x)$ **16.** $(fg)(x)$

17. $\left(\dfrac{f}{g}\right)(x)$ **18.** $(f \circ g)(x)$ **19.** $(g \circ g)(x)$

20. The profit for a company is

$$P = 300 + 10x - 0.002x^2$$

where x is the number of units produced. What production level will yield a maximum profit?

Zeros of Polynomial Functions

- Polynomial Division and Synthetic Division
- Real Zeros of Polynomial Functions
- Approximating Zeros of Polynomial Functions
- Complex Numbers
- The Fundamental Theorem of Algebra

Diabetes is one of the leading causes of death in the United States. There are two basic types of diabetes. The first is caused by a lack of insulin and represents about 20% of all diabetes cases. The second type is not related to insulin production and is more common among people over the age of 40.

Diabetes affects a person's ability to use glucose (or sugar)—hence, sugar tends to build up in the blood. A test that is used to diagnose diabetes is the Glucose Tolerance Test. With this test, a person is given a dose of sugar water and the person's blood sugar level is monitored over several hours. The partial results for a person who tested positive for diabetes is shown in the table at the right. This data can be modeled with a seventh-degree polynomial, as shown in the graph at the right.

Time	0	0.5	1	2	3	4	5
Sugar	95	175	160	120	92	80	90

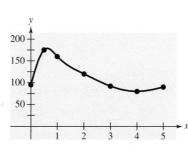

The chapter project related to this information is on page 366.

319

4.1 Polynomial Division and Synthetic Division

Long Division of Polynomials ▪ Synthetic Division ▪
Remainder and Factor Theorems ▪ Application

Long Division of Polynomials

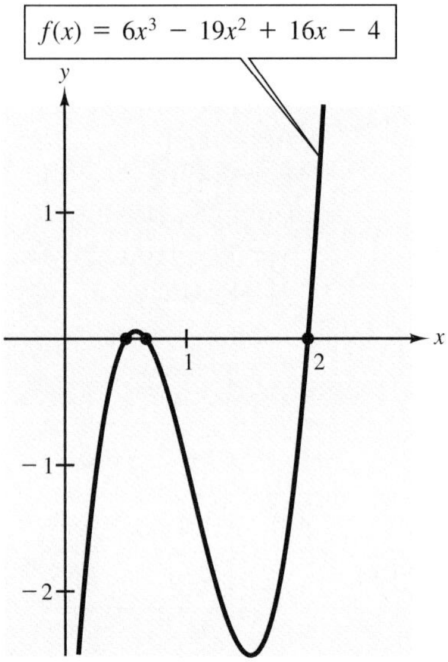

In this section, you will study two procedures for *dividing* polynomials. These procedures are especially valuable in factoring and finding the zeros of polynomial functions. To begin, suppose you are given the graph of

$$f(x) = 6x^3 - 19x^2 + 16x - 4.$$

$$\boxed{f(x) = 6x^3 - 19x^2 + 16x - 4}$$

Notice that a zero of f occurs at $x = 2$, as shown in Figure 4.1. Because $x = 2$ is a zero of the polynomial function f, you know that $(x - 2)$ is a factor of $f(x)$. This means that there exists a second-degree polynomial $q(x)$ such that

$$f(x) = (x - 2) \cdot q(x).$$

To find $q(x)$, you can use **long division,** as illustrated in Example 1.

FIGURE 4.1

EXAMPLE 1 Long Division of Polynomials

Divide the polynomial $6x^3 - 19x^2 + 16x - 4$ by $x - 2$, and use the result to factor the polynomial completely.

Solution

Partial quotients

$$
\begin{array}{r}
6x^2 - 7x + 2 \\
x - 2 \overline{)\, 6x^3 - 19x^2 + 16x - 4} \\
\underline{6x^3 - 12x^2} \\
-7x^2 + 16x \\
\underline{-7x^2 + 14x} \\
2x - 4 \\
\underline{2x - 4} \\
0
\end{array}
$$

Multiply: $6x^2(x - 2)$.
Subtract.
Multiply: $-7x(x - 2)$.
Subtract.
Multiply: $2(x - 2)$.
Subtract.

NOTE Note that the factorization shown in Example 1 agrees with the graph shown in Figure 4.1 in that the three x-intercepts occur at $x = 2$, $x = \frac{1}{2}$, and $x = \frac{2}{3}$.

From this division, you can conclude that

$$6x^3 - 19x^2 + 16x - 4 = (x - 2)(6x^2 - 7x + 2)$$

and by factoring the quadratic $6x^2 - 7x + 2$, you have

$$6x^3 - 19x^2 + 16x - 4 = (x - 2)(2x - 1)(3x - 2).$$

In Example 1, $x - 2$ is a factor of the polynomial $6x^3 - 19x^2 + 16x - 4$, and the long division process produces a remainder of zero. Often, long division will produce a nonzero remainder. For instance, if you divide $x^2 + 3x + 5$ by $x + 1$, you obtain the following.

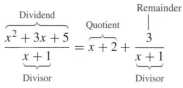

In fractional form, you can write this result as follows.

$$\underbrace{\frac{x^2 + 3x + 5}{x + 1}}_{} = \overbrace{x + 2}^{} + \frac{3}{\underbrace{x + 1}_{}}$$

Dividend / Quotient / Remainder / Divisor

This example illustrates the following well-known theorem called the **Division Algorithm.**

The Division Algorithm

If $f(x)$ and $d(x)$ are polynomials such that $d(x) \neq 0$, and the degree of $d(x)$ is less than or equal to the degree of $f(x)$, there exist unique polynomials $q(x)$ and $r(x)$ such that

$$f(x) = d(x)q(x) + r(x)$$

Dividend   Quotient
    Divisor   Remainder

where $r(x) = 0$ *or* the degree of $r(x)$ is less than the degree of $d(x)$. If the remainder $r(x)$ is zero, $d(x)$ **divides evenly** into $f(x)$.

The Division Algorithm can also be written as

$$\frac{f(x)}{d(x)} = q(x) + \frac{r(x)}{d(x)}.$$

In the Division Algorithm, the rational expression $f(x)/d(x)$ is **improper** because the degree of $f(x)$ is greater than or equal to the degree of $d(x)$. On the other hand, the rational expression $r(x)/d(x)$ is **proper** because the degree of $r(x)$ is less than the degree of $d(x)$.

EXAMPLE 2 Long Division of Polynomials

Divide $x^3 - 1$ by $x - 1$.

Solution

Because there is no x^2-term or x-term in the dividend, you need to line up the subtraction by using zero coefficients (or leaving spaces) for the missing terms.

$$
\begin{array}{r}
x^2 +\ x + 1 \\
x - 1\ \overline{)\ x^3 + 0x^2 + 0x - 1} \\
\underline{x^3 -\ x^2} \\
x^2 \\
\underline{x^2 - x} \\
x - 1 \\
\underline{x - 1} \\
0
\end{array}
$$

Thus, $x - 1$ divides evenly into $x^3 - 1$ and you can write

$$
\frac{x^3 - 1}{x - 1} = x^2 + x + 1.
$$

NOTE You can check the result of a division problem by multiplying. For instance, in Example 2, try checking that $(x - 1)(x^2 + x + 1) = x^3 - 1$.

EXAMPLE 3 Long Division of Polynomials

Divide $2x^4 + 4x^3 - 5x^2 + 3x - 2$ by $x^2 + 2x - 3$.

Solution

$$
\begin{array}{r}
2x^2 \qquad\ + 1 \\
x^2 + 2x - 3\ \overline{)\ 2x^4 + 4x^3 - 5x^2 + 3x - 2} \\
\underline{2x^4 + 4x^3 - 6x^2} \\
x^2 + 3x - 2 \\
\underline{x^2 + 2x - 3} \\
x + 1
\end{array}
$$

Note that the first subtraction eliminated two terms from the dividend. When this happens, the quotient skips a term. Thus, you can write

$$
\frac{2x^4 + 4x^3 - 5x^2 + 3x - 2}{x^2 + 2x - 3} = 2x^2 + 1 + \frac{x + 1}{x^2 + 2x - 3}.
$$

Synthetic Division

There is a nice shortcut for long division by polynomials of the form $x - k$. The shortcut is called **synthetic division.** We summarize the pattern for synthetic division of a cubic polynomial as follows. (The pattern for higher-degree polynomials is similar.)

Synthetic Division (for a Cubic Polynomial)

To divide $ax^3 + bx^2 + cx + d$ by $x - k$, use the following pattern.

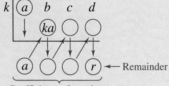

Vertical pattern: Add terms.
Diagonal pattern: Multiply by k.

NOTE   Synthetic division works *only* for divisors of the form $x - k$. You cannot use synthetic division to divide a polynomial by a quadratic such as $x^2 - 3$.

EXAMPLE 4   *Using Synthetic Division*

Use synthetic division to divide $x^4 - 10x^2 - 2x + 4$ by $x + 3$.

Solution

You should set up the array as follows. Note that a zero is included for each missing term in the dividend.

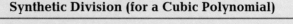

Thus, you have

$$\frac{x^4 - 10x^2 - 2x + 4}{x + 3} = x^3 - 3x^2 - x + 1 + \frac{1}{x + 3}.$$

Remainder and Factor Theorems

The remainder obtained in the synthetic division process has an important interpretation, as in the Remainder Theorem.

The Remainder Theorem

If a polynomial $f(x)$ is divided by $x - k$, the remainder is

$$r = f(k).$$

The Remainder Theorem tells you that synthetic division can be used to evaluate a polynomial function. That is, to evaluate a polynomial function $f(x)$ when $x = k$, divide $f(x)$ by $x - k$. The remainder will be $f(k)$, as illustrated in Example 5.

Technology

With the *TI-82* you can evaluate a function by *storing* the x-value. Try the following keystrokes.

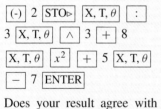

Does your result agree with the solution in Example 5?

EXAMPLE 5   Using the Remainder Theorem

Use the Remainder Theorem to evaluate the following function at $x = -2$.

$$f(x) = 3x^3 + 8x^2 + 5x - 7$$

Solution

Using synthetic division, you obtain the following.

$$
\begin{array}{r|rrrr}
-2 & 3 & 8 & 5 & -7 \\
   &   & -6 & -4 & -2 \\
\hline
   & 3 & 2 & 1 & -9
\end{array}
$$

Because the remainder is $r = -9$, you can conclude that

$$f(-2) = -9.$$

This means that $(-2, -9)$ is a point on the graph of f. Try checking this by substituting $x = -2$ in the original function.

Another important theorem is the Factor Theorem, which is stated below.

Factor Theorem

A polynomial $f(x)$ has a factor $(x - k)$ if and only if $f(k) = 0$.

EXAMPLE 6   *Factoring a Polynomial*

Show that $(x - 2)$ and $(x + 3)$ are factors of the polynomial

$$f(x) = 2x^4 + 7x^3 - 4x^2 - 27x - 18.$$

Then find the remaining factors of $f(x)$.

Solution

Using synthetic division with 2 and -3 *successively*, you obtain the following.

$$
\begin{array}{r|rrrrr}
2 & 2 & 7 & -4 & -27 & -18 \\
  &   & 4 & 22 & 36 & 18 \\
\hline
  & 2 & 11 & 18 & 9 & 0
\end{array}
$$

0 remainder
$(x - 2)$ is a factor.

$$
\begin{array}{r|rrrr}
-3 & 2 & 11 & 18 & 9 \\
   &   & -6 & -15 & -9 \\
\hline
   & 2 & 5 & 3 & 0
\end{array}
$$

0 remainder
$(x + 3)$ is a factor.

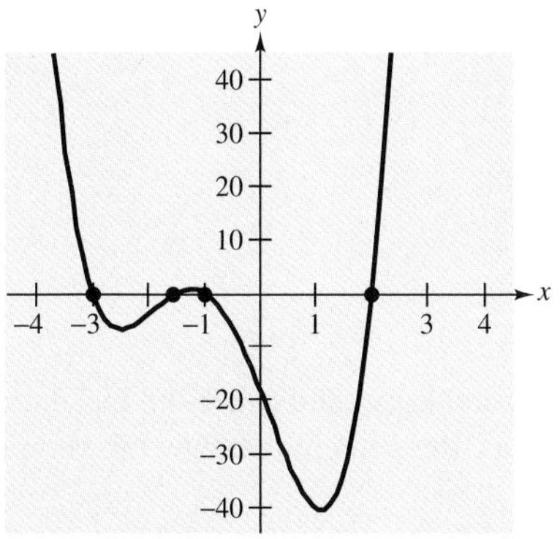

Because the resulting quadratic expression factors as

$$2x^2 + 5x + 3 = (2x + 3)(x + 1)$$

the complete factorization of $f(x)$ is

$$f(x) = (x - 2)(x + 3)(2x + 3)(x + 1).$$

Note that this factorization implies that f has four real zeros:

$$2, -3, -\tfrac{3}{2}, \text{ and } -1.$$

This is confirmed by the graph of f, which is shown in Figure 4.2.

FIGURE 4.2

In summary, the remainder r, obtained in the synthetic division of $f(x)$ by $x - k$, provides the following information.

1. The remainder r gives the value of f at $x = k$. That is, $r = f(k)$.
2. If $r = 0$, $(x - k)$ is a factor of $f(x)$.
3. If $r = 0$, $(k, 0)$ is an x-intercept of the graph of f.

NOTE   Throughout this text, we have emphasized the importance of developing several problem-solving strategies. In the exercises for this section, try using more than one strategy to solve several of the exercises. For instance, if you find that $x - k$ divides evenly into $f(x)$, try sketching the graph of f. You should find that $(k, 0)$ is an x-intercept of the graph.

Application

Real Life

EXAMPLE 7 Take-Home Pay

The 1990 monthly take-home pay for an employee who was married and claimed four deductions is given by the function

$$y = -0.00002469x^2 + 0.85534x + 32.98, \qquad 500 \leq x \leq 5000$$

where y represents the take-home pay and x represents the monthly salary. Find a function that gives the take-home pay as a *percent* of the monthly salary. (Source: Hooper International Accounting Software, based on a state and local income tax rate of 3.1%)

Solution

Because the monthly salary is given by x and the take-home pay is given by y, the percent of monthly salary that the person takes home is

$$
\begin{aligned}
P &= \frac{y}{x} \\
&= \frac{-0.00002469x^2 + 0.85534x + 32.98}{x} \\
&= -0.00002469x + 0.85534 + \frac{32.98}{x}.
\end{aligned}
$$

The graphs of these functions are shown in Figure 4.3(a) and (b). Note in Figure 4.3(b) that as a person's monthly salary increases, the percent that they take home decreases.

The difference between gross pay and take-home pay results primarily from federal deductions for income tax, Social Security tax, and Medicare tax. Because the federal income tax is graduated, the percent that is deducted depends on a person's income. People who earn more do not simply pay more tax—they pay *much* more tax because the tax rate increases.

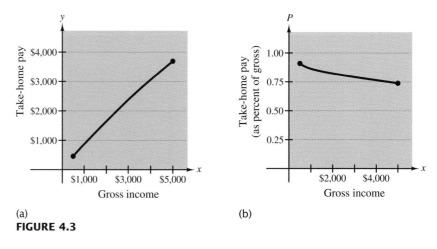

(a)   (b)

FIGURE 4.3

Group Activities

Extending the Concept

Analyzing a Slant Asymptote By use of long division, the rational function given by

$$f(x) = \frac{x^2 + 2x + 3}{x - 1}$$

can be written as

$$f(x) = \frac{x^2 + 2x + 3}{x - 1} = x + 3 + \frac{6}{x - 1}.$$

As x approaches $-\infty$ or ∞, the value of the fraction $6/(x-1)$ is insignificant and the graph of f is close to the graph of $y = x + 3$ (see figure). The line

$$y = x + 3 \qquad\qquad \text{Slant asymptote}$$

is called a **slant asymptote** of the graph of f. Complete the table to show that the values of $f(x)$ and y are nearly the same for large values of x.

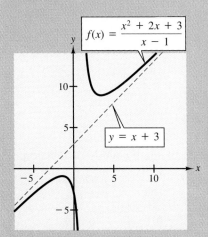

$$f(x) = \frac{x^2 + 2x + 3}{x - 1}$$

$$y = x + 3$$

x	$-10{,}000$	-1000	-100	100	1000	$10{,}000$
$f(x)$						
y						

Use long division to show that

$$g(x) = \frac{2x^3 - 5}{x^2}$$

also has a slant asymptote. What is the equation of this slant asymptote? Graph g and complete a table for g similar to the one above for f. What can you conclude?

Warm Up   The following warm-up exercises involve skills that were covered in earlier sections. You will use these skills in the exercise set for this section.

In Exercises 1–4, write the expression in standard polynomial form.

1. $(x - 1)(x^2 + 2) + 5$ **2.** $(x^2 - 3)(2x + 4) + 8$

3. $(x^2 + 1)(x^2 - 2x + 3) - 10$ **4.** $(x + 6)(2x^3 - 3x) - 5$

In Exercises 5 and 6, factor the polynomial.

5. $x^2 - 4x + 3$ **6.** $4x^3 - 10x^2 + 6x$

In Exercises 7–10, find a polynomial function that has the given zeros.

7. $0, 3, 4$ **8.** $-6, 1$

9. $-3, 1 + \sqrt{2}, 1 - \sqrt{2}$ **10.** $1, -2, 2 + \sqrt{3}, 2 - \sqrt{3}$

4.1 Exercises

In Exercises 1–14, divide by long division.

Dividend	Divisor
1. $2x^2 + 10x + 12$	$x + 3$
2. $5x^2 - 17x - 12$	$x - 4$
3. $4x^3 - 7x^2 - 11x + 5$	$4x + 5$
4. $6x^3 - 16x^2 + 17x - 6$	$3x - 2$
5. $x^4 + 5x^3 + 6x^2 - x - 2$	$x + 2$
6. $x^3 + 4x^2 - 3x - 12$	$x^2 - 3$
7. $7x + 3$	$x + 2$
8. $8x - 5$	$2x + 1$
9. $6x^3 + 10x^2 + x + 8$	$2x^2 + 1$
10. $x^3 - 9$	$x^2 + 1$
11. $x^3 - 4x^2 + 5x - 2$	$x - 1$
12. $x^3 - x^2 + 2x - 8$	$x - 2$
13. $x^4 + 3x^2 + 1$	$x^2 - 2x + 3$
14. $x^5 + 7$	$x^3 - 1$

In Exercises 15–32, divide by synthetic division.

Dividend	Divisor
15. $3x^3 - 17x^2 + 15x - 25$	$x - 5$
16. $5x^3 + 18x^2 + 7x - 6$	$x + 3$
17. $4x^3 - 9x + 8x^2 - 18$	$x + 2$
18. $9x^3 - 16x - 18x^2 + 32$	$x - 2$
19. $-x^3 + 75x - 250$	$x + 10$
20. $3x^3 - 16x^2 - 72$	$x - 6$
21. $5x^3 - 6x^2 + 8$	$x - 4$
22. $5x^3 + 6x + 8$	$x + 2$
23. $10x^4 - 50x^3 - 800$	$x - 6$
24. $x^5 - 13x^4 - 120x + 80$	$x + 3$
25. $x^3 + 512$	$x + 8$
26. $5x^3$	$x + 3$
27. $-3x^4$	$x - 2$
28. $-3x^4$	$x + 2$

Dividend	*Divisor*
29. $5 - 3x + 2x^2 - x^3$	$x + 1$
30. $180x - x^4$	$x - 6$
31. $4x^3 + 16x^2 - 23x - 15$	$x + \frac{1}{2}$
32. $3x^3 - 4x^2 + 5$	$x - \frac{3}{2}$

In Exercises 33–38, use synthetic division to show that x is a solution of the third-degree polynomial equation, and use the result to factor the polynomial completely.

33. $x^3 - 7x + 6 = 0$, $x = 2$

34. $x^3 - 28x - 48 = 0$, $x = -4$

35. $2x^3 - 15x^2 + 27x - 10 = 0$, $x = \frac{1}{2}$

36. $48x^3 - 80x^2 + 41x - 6 = 0$, $x = \frac{2}{3}$

37. $x^3 + 2x^2 - 3x - 6 = 0$, $x = \sqrt{3}$

38. $x^3 + 2x^2 - 2x - 4$, $x = \sqrt{2}$

In Exercises 39–42, express the function in the form

$$f(x) = (x - k)q(x) + r$$

for the given value of k, and demonstrate that $f(k) = r$.

39. $f(x) = x^3 - x^2 - 14x + 11$, $k = 4$

40. $f(x) = \frac{1}{3}(15x^4 + 10x^3 - 6x^2 + 17x + 14)$,

  $k = -\frac{2}{3}$

41. $f(x) = x^3 + 3x^2 - 2x - 14$, $k = \sqrt{2}$

42. $f(x) = 4x^3 - 6x^2 - 12x - 4$, $k = 1 - \sqrt{3}$

In Exercises 43–48, use synthetic division to find the required function values.

43. $f(x) = 4x^3 - 13x + 10$

  (a) $f(1)$ (b) $f(-2)$

  (c) $f(\frac{1}{2})$ (d) $f(8)$

44. $g(x) = x^6 - 4x^4 + 3x^2 + 2$

  (a) $g(2)$ (b) $g(-4)$

  (c) $g(3)$ (d) $g(-1)$

45. $h(x) = 3x^3 + 5x^2 - 10x + 1$

  (a) $h(3)$ (b) $h(\frac{1}{3})$

  (c) $h(-2)$ (d) $h(-5)$

46. $f(x) = 0.4x^4 - 1.6x^3 + 0.7x^2 - 2$

  (a) $f(1)$ (b) $f(-2)$

  (c) $f(5)$ (d) $f(-10)$

47. $f(x) = x^3 - 2x^2 - 11x + 52$

  (a) $f(5)$ (b) $f(-4)$

  (c) $f(1.2)$ (d) $f(2)$

48. $g(x) = x^3 - x^2 + 25x - 25$

  (a) $g(5)$ (b) $g(\frac{1}{5})$

  (c) $g(-1.5)$ (d) $g(-1)$

In Exercises 49–52, match the function with its graph and use the result to find all real solutions of $f(x) = 0$. [The graphs are labeled (a), (b), (c), and (d).]

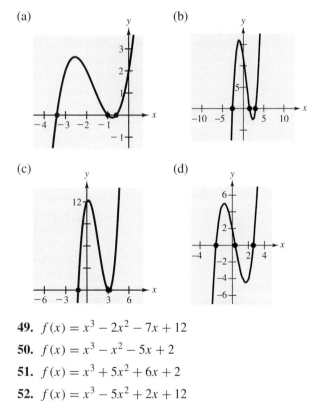

(a)   (b)

(c)   (d)

49. $f(x) = x^3 - 2x^2 - 7x + 12$

50. $f(x) = x^3 - x^2 - 5x + 2$

51. $f(x) = x^3 + 5x^2 + 6x + 2$

52. $f(x) = x^3 - 5x^2 + 2x + 12$

In Exercises 53–58, simplify the rational function.

53. $\dfrac{4x^3 - 8x^2 + x + 3}{2x - 3}$

54. $\dfrac{x^3 + x^2 - 64x - 64}{x + 8}$

55. $\dfrac{x^3 + 3x^2 - x - 3}{x + 1}$

56. $\dfrac{2x^3 + 3x^2 - 3x - 2}{x - 1}$

57. $\dfrac{x^4 + 6x^3 + 11x^2 + 6x}{x^2 + 3x + 2}$

58. $\dfrac{x^4 + 9x^3 - 5x^2 - 36x + 4}{x^2 - 4}$

59. *Dimensions of a Room* A rectangular room has a volume of

$$x^3 + 11x^2 + 34x + 24$$

cubic feet. The height of the room is $x + 1$ (see figure). Find the number of square feet of floor space in the room.

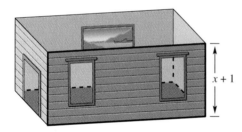

60. *Dimensions of a House* A rectangular house has a volume of

$$x^3 + 55x^2 + 650x + 2000$$

cubic feet (the space in the attic is not counted). The height of the house is $x + 5$ (see figure). Find the number of square feet of floor space *on the first floor* of the house.

61. *Profit* A company making compact discs estimates that the profit from selling a particular disc is

$$P = -153.6x^3 + 5760x^2 - 100{,}000,$$
$$0 \le x \le 35$$

where P is the profit in dollars and x is the advertising expense (in tens of thousands of dollars). For this disc, the advertising expense was \$300,000 ($x = 30$) and the profit was \$936,800.

(a) From the graph shown in the figure, it appears that the company could have obtained the same profit by spending less on advertising. Use the graph to estimate another amount that the company could have spent that would have produced the same profit.

(b) Use synthetic division with $x = 30$ to algebraically confirm the result of part (a).

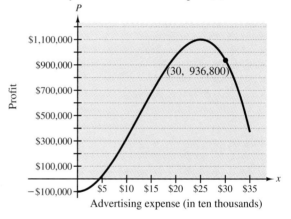

62. *Profit* A company that produces calculators estimates that the profit from selling a particular model of calculator is

$$P = -125x^3 + 6450x^2 - 150{,}000, \qquad 0 \le x \le 50$$

where P is the profit in dollars and x is the advertising expense (in tens of thousands of dollars). For this model of calculator, the advertising expense was \$400,000 ($x = 40$) and the profit was \$2,170,000.

(a) Use a graphing utility to graph the profit function.

(b) Could the company have obtained the same profit by spending less on advertising? Explain your reasoning.

4.2 | Real Zeros of Polynomial Functions

Descartes's Rule of Signs ▪ The Rational Zero Test ▪
Bounds for Real Zeros of Polynomial Functions

Descartes's Rule of Signs

In Section 3.6, you learned that an nth-degree polynomial function can have *at most n real zeros*. Of course, many nth-degree polynomials do not have that many zeros. For instance, $f(x) = x^2 + 1$ has no real zeros, and $f(x) = x^3 + 1$ has only one real zero. The following theorem, called **Descartes's Rule of Signs,** sheds more light on the number of real zeros of a polynomial.

NOTE When there is only one variation in sign, Descartes's Rule of Signs guarantees the existence of exactly one positive (or negative) real zero.

Descartes's Rule of Signs

Let $f(x) = a_n x^n + a_{n-1} x^{n-1} + \cdots + a_2 x^2 + a_1 x + a_0$ be a polynomial with real coefficients and $a_0 \neq 0$.

1. The number of *positive real zeros* of f is either equal to the number of variations in sign of $f(x)$ or less than that number by an even integer.

2. The number of *negative real zeros* of f is either equal to the number of variations in sign of $f(-x)$ or less than that number by an even integer.

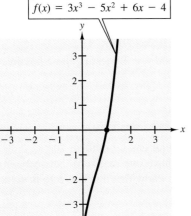

$f(x) = 3x^3 - 5x^2 + 6x - 4$

FIGURE 4.4

A **variation in sign** means that two consecutive coefficients have opposite signs. For example, the polynomial

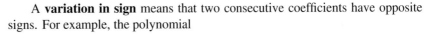

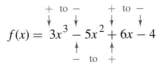

$$f(x) = 3x^3 - 5x^2 + 6x - 4$$

has *three* variations in sign, whereas

$$f(-x) = 3(-x)^3 - 5(-x)^2 + 6(-x) - 4$$
$$= -3x^3 - 5x^2 - 6x - 4$$

has no variations in sign. Thus, from Descartes's Rule of Signs, the polynomial $f(x) = 3x^3 - 5x^2 + 6x - 4$ has either three positive real zeros or one positive real zero, and has no negative real zeros. From the graph in Figure 4.4, you can see that the function has only one real zero (it is a positive number, near 1).

The Rational Zero Test

The **Rational Zero Test** relates the possible rational zeros of a polynomial (having integer coefficients) to the leading coefficient and to the constant term of the polynomial.

The Rational Zero Test

If the polynomial $f(x) = a_n x^n + a_{n-1} x^{n-1} + \cdots + a_2 x^2 + a_1 x + a_0$ has *integer* coefficients, then every rational zero of f has the form

$$\text{Rational zero} = \frac{p}{q}$$

where p and q have no common factors other than 1, and

p = a factor of the constant term a_0

q = a factor of the leading coefficient a_n.

NOTE When the leading coefficient is 1, the possible rational zeros are simply the factors of the constant term.

To use the Rational Zero Test, you should first list all rational numbers whose numerators are factors of the constant term and whose denominators are factors of the leading coefficient.

$$\text{Possible rational zeros} = \frac{\text{factors of constant term}}{\text{factors of leading coefficient}}$$

Having formed this list of *possible rational zeros*, use a trial-and-error method to determine which, if any, are actual zeros of the polynomial.

EXAMPLE 1 Rational Zero Test with Leading Coefficient of 1

Find the rational zeros of $f(x) = x^3 + x + 1$.

Solution

The only factors of the leading coefficient are 1 and -1. By testing these possible zeros, you can see that neither works.

$$f(1) = (1)^3 + 1 + 1 = 3$$
$$f(-1) = (-1)^3 + (-1) + 1 = -1$$

Thus, you can conclude that the given polynomial has *no* rational zeros. Note from the graph of f in Figure 4.5 that f does have one real zero (between -1 and 0). However, by the Rational Zero Test, you know that this real zero is *not* a rational number.

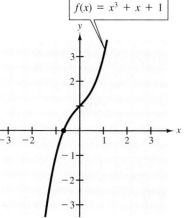

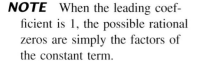
$f(x) = x^3 + x + 1$

FIGURE 4.5

EXAMPLE 2   *Rational Zero Test with Leading Coefficient of 1*

Find the rational zeros of

$$f(x) = x^4 - x^3 + x^2 - 3x - 6.$$

Solution

Because the leading coefficient is 1, the possible rational zeros are the factors of the constant term.

Possible rational zeros: $\pm 1, \ \pm 2, \ \pm 3, \ \pm 6$

A test of these possible zeros shows that $x = -1$ and $x = 2$ are the only two that work. Check the others to be sure.

If the leading coefficient of a polynomial is not 1, the list of possible rational zeros can increase dramatically. In such cases the search can be shortened in several ways: (1) a programmable calculator can be used to speed up the calculations; (2) a graph, either by hand or with a graphing utility, can give a good estimate of the locations of the zeros; (3) synthetic division can be used to test the possible rational zeros.

To see how to use synthetic division to test the possible rational zeros, let's take another look at the function

$$f(x) = x^4 - x^3 + x^2 - 3x - 6$$

given in Example 2. To test that $x = -1$ and $x = 2$ are zeros of f, you can apply synthetic division, as follows.

$$
\begin{array}{r|rrrrr}
-1 & 1 & -1 & 1 & -3 & -6 \\
   &   & -1 & 2 & -3 & 6 \\
\hline
   & 1 & -2 & 3 & -6 & 0
\end{array}
$$

$$
\begin{array}{r|rrrr}
2 & 1 & -2 & 3 & -6 \\
  &   & 2 & 0 & 6 \\
\hline
  & 1 & 0 & 3 & 0
\end{array}
$$

Thus, you have

$$f(x) = (x + 1)(x - 2)(x^2 + 3).$$

Because the factor $(x^2 + 3)$ produces no real zeros, you can conclude that $x = -1$ and $x = 2$ are the only *real* zeros of f.

Finding the first zero is often the hardest part. After that, the search is simplified by using the lower-degree polynomial obtained in synthetic division.

EXAMPLE 3   *Using the Rational Zero Test*

Find the rational zeros of $f(x) = 2x^3 + 3x^2 - 8x + 3$.

Solution

The leading coefficient is 2 and the constant term is 3.

$$\textit{Possible rational zeros: } \frac{\text{Factors of 3}}{\text{Factors of 2}} = \frac{\pm 1, \ \pm 3}{\pm 1, \pm 2} = \pm 1, \ \pm 3, \ \pm\frac{1}{2}, \ \pm\frac{3}{2}$$

By synthetic division, you can determine that $x = 1$ is a zero.

$$
\begin{array}{r|rrrr}
1 & 2 & 3 & -8 & 3 \\
  &   & 2 & 5 & -3 \\
\hline
  & 2 & 5 & -3 & 0
\end{array}
$$

Thus, $f(x)$ factors as

$$f(x) = (x - 1)(2x^2 + 5x - 3) = (x - 1)(2x - 1)(x + 3)$$

and you can conclude that the zeros of f are $x = 1$, $x = \frac{1}{2}$, and $x = -3$.

EXAMPLE 4   *Using the Rational Zero Test*

Find all the real zeros of $f(x) = 10x^3 - 15x^2 - 16x + 12$.

Solution

The leading coefficient is 10 and the constant term is 12.

$$\textit{Possible rational zeros: } \frac{\text{Factors of 12}}{\text{Factors of 10}} = \frac{\pm 1, \ \pm 2, \ \pm 3, \ \pm 4, \ \pm 6, \ \pm 12}{\pm 1, \ \pm 2, \ \pm 5, \ \pm 10}$$

With so many possibilities (32, in fact), it is worth your time to stop and sketch a graph. From Figure 4.6, it looks like three reasonable choices would be $x = -\frac{6}{5}$, $x = \frac{1}{2}$, and $x = 2$. Testing these by synthetic division shows that only $x = 2$ works. Thus, you have

$$f(x) = (x - 2)(10x^2 + 5x - 6).$$

Using the Quadratic Formula, you find that the two additional zeros are irrational numbers.

$$x = \frac{-5 + \sqrt{265}}{20} \approx 0.5639 \qquad \text{and} \qquad x = \frac{-5 - \sqrt{265}}{20} \approx -1.0639$$

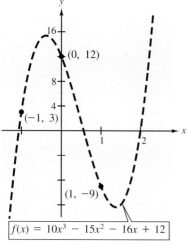

$$f(x) = 10x^3 - 15x^2 - 16x + 12$$

FIGURE 4.6

Bounds for Real Zeros of Polynomial Functions

The third test for zeros of a polynomial function is related to the sign pattern in the last row of the synthetic division tableau. This test can give you an upper or a lower bound of the real zeros of f.

Upper and Lower Bound Rule

Let $f(x)$ be a polynomial with real coefficients and a positive leading coefficient. Suppose $f(x)$ is divided by $x - c$, using synthetic division.

1. If $c > 0$ and each number in the last row is either positive or zero, then c is an *upper bound* for the real zeros of f.

2. If $c < 0$ and the numbers in the last row are alternately positive and negative (zero entries count as positive or negative), then c is a *lower bound* for the real zeros of f.

NOTE   A real number b is an **upper bound** for the real zeros of f if no zeros are greater than b. Similarly, b is a **lower bound** if no real zeros of f are less than b.

DISCOVERY

Graph $f(x) = 6x^3 - 4x^2 + 3x - 2$. Notice that the graph intersects the x-axis at the point $\left(\frac{2}{3}, 0\right)$. How does this relate to the real zero found in Example 5?

Graph $g(x) = x^4 - 5x^3 + 3x^2 + x$. How many times does the graph intersect the x-axis? How many real zeros does g have?

EXAMPLE 5   *Finding the Zeros of a Polynomial Function*

Find the real zeros of $f(x) = 6x^3 - 4x^2 + 3x - 2$.

Solution

The possible real zeros are as follows.

$$\frac{\text{Factors of 2}}{\text{Factors of 6}} = \frac{\pm 1, \ \pm 2}{\pm 1, \ \pm 2, \ \pm 3, \ \pm 6} = \pm 1, \ \pm\frac{1}{2}, \ \pm\frac{1}{3}, \ \pm\frac{1}{6}, \ \pm\frac{2}{3}, \ \pm 2$$

Because $f(x)$ has three variations in sign and $f(-x)$ has none, you can apply Descartes's Rule of Signs to conclude that there are three positive real zeros or one positive real zero, and no negative zeros. Trying $x = 1$ produces the following.

$$
\begin{array}{r|rrrr}
1 & 6 & -4 & 3 & -2 \\
  &   & 6 & 2 & 5 \\
\hline
  & 6 & 2 & 5 & 3
\end{array}
$$

Thus, $x = 1$ is not a zero, but because the last row has all positive entries, you know that $x = 1$ is an upper bound for the real zeros. Thus, you can restrict the search to zeros between 0 and 1. By trial and error, you can determine that $x = \frac{2}{3}$ is a zero. Thus,

$$f(x) = \left(x - \tfrac{2}{3}\right)\left(6x^2 + 3\right).$$

Because $6x^2 + 3$ has no real zeros, it follows that $x = \frac{2}{3}$ is the only real zero.

Before concluding this section, we list two additional hints that can help you find the real zeros of a polynomial.

1. If the terms of $f(x)$ have a common monomial factor, it should be factored out before applying the tests in this section. For instance, by writing

$$f(x) = x^4 - 5x^3 + 3x^2 + x$$
$$= x(x^3 - 5x^2 + 3x + 1)$$

you can see that $x = 0$ is a zero of f and that the remaining zeros can be obtained by analyzing the cubic factor.

2. If you are able to find all but two zeros of $f(x)$, you are home free because you can always use the Quadratic Formula on the remaining quadratic factor. For instance, if you succeeded in writing

$$f(x) = x^4 - 5x^3 + 3x^2 + x$$
$$= x(x - 1)(x^2 - 4x - 1)$$

you can apply the Quadratic Formula to $x^2 - 4x - 1$ to conclude that the two remaining zeros are

$$x = 2 + \sqrt{5} \quad \text{and} \quad x = 2 - \sqrt{5}.$$

Activities

1. Use Descartes's Rule of Signs to analyze the polynomial $f(x) = 3x^4 - 4x^3 + x - 2$ for the number of positive and negative real zeros of f.

   Answer: either three or one positive real zero(s) and exactly one negative real zero

2. List the possible rational zeros of the function $f(x) = 3x^4 - 4x^3 + x - 2$.

   Answer: $\pm\frac{2}{3}, \pm\frac{1}{3}, \pm 2, \pm 1$

3. Use synthetic division to determine if the x-value is an upper bound, a lower bound, or neither.
   $f(x) = 3x^4 - 5x^3 + x^2 - 5x - 2$, $x = -1$

   Answer: $x = -1$ is a lower bound.

Group Activities

Extending the Concept

Group Activity Suggestion

Consider dividing the class into four groups, assigning one of the four questions to each group for discussion. Then have a spokesperson from each group share the group's findings with the entire class.

Comparing Real Zeros and Rational Zeros   Discuss the meanings of *real zeros* and *rational zeros* of a polynomial function and compare these two types of zeros. Then answer the following questions.

a. Is it possible for a polynomial function to have no rational zeros but to have real zeros? If so, give an example.

b. If a polynomial function has three real zeros, and only one of them is a rational number, must the other two zeros be irrational numbers?

c. Consider a cubic polynomial function

$$f(x) = ax^3 + bx^2 + cx + d$$

where $a \neq 0$. Is it possible that f has no real zeros? Is it possible that f has no rational zeros?

d. Is it possible that a second-degree polynomial function with integer coefficients has one rational zero and one irrational zero? If so, given an example.

Warm Up

In Exercises 1 and 2, find a polynomial function with integer coefficients having the given zeros.

1. $-1, \frac{2}{3}, 3$ **2.** $-2, 0, \frac{3}{4}, 2$

In Exercises 3 and 4, divide by synthetic division.

3. $\dfrac{x^5 - 9x^3 + 5x + 18}{x + 3}$ **4.** $\dfrac{3x^4 + 17x^3 + 10x^2 - 9x - 8}{x + \frac{2}{3}}$

In Exercises 5–8, use the given zero to find all the real zeros of f.

5. $f(x) = 2x^3 + 11x^2 + 2x - 4, \; x = \frac{1}{2}$

6. $f(x) = 6x^3 - 47x^2 - 124x - 60, \; x = 10$

7. $f(x) = 4x^3 - 13x^2 - 4x + 6, \; x = -\frac{3}{4}$

8. $f(x) = 10x^3 + 51x^2 + 48x - 28, \; x = \frac{2}{5}$

In Exercises 9 and 10, find all real solutions.

9. $x^4 - 3x^2 + 2 = 0$ **10.** $x^4 - 7x^2 + 12 = 0$

4.2 Exercises

In Exercises 1–10, use Descartes's Rule of Signs to determine the possible number of positive and negative zeros of the function.

1. $f(x) = x^3 + 3$ **2.** $g(x) = x^3 + 3x^2$

3. $h(x) = 3x^4 + 2x^2 + 1$ **4.** $h(x) = 2x^4 - 3x + 2$

5. $g(x) = 2x^3 - 3x^2 - 3$

6. $f(x) = 4x^3 - 3x^2 + 2x - 1$

7. $f(x) = -5x^3 + x^2 - x + 5$

8. $f(x) = 3x^3 + 2x^2 + x + 3$

9. $g(x) = 5x^5 + 10x$ **10.** $h(x) = 4x^2 - 8x + 3$

In Exercises 11–16, use the Rational Zero Test to list all possible rational zeros of f. Then use a graphing utility to graph the function. Use the graph to help determine which of the possible rational zeros are actual zeros of the function.

11. $f(x) = x^3 + x^2 - 4x - 4$

12. $f(x) = -3x^3 + 20x^2 - 36x + 16$

13. $f(x) = -4x^3 + 15x^2 - 8x - 3$

14. $f(x) = 4x^3 - 12x^2 - x + 15$

15. $f(x) = 4x^4 - 17x^2 + 4$

16. $f(x) = -2x^4 + 13x^3 - 21x^2 + 2x + 8$

In Exercises 17–20, use synthetic division to determine whether the given x-value is an upper bound of the zeros of f, a lower bound of the zeros of f, or neither.

17. $f(x) = x^4 - 4x^3 + 15$

    (a) $x = 4$ (b) $x = -1$ (c) $x = 3$

18. $f(x) = 2x^3 - 3x^2 - 12x + 8$

    (a) $x = 2$ (b) $x = 4$ (c) $x = -1$

19. $f(x) = x^4 - 4x^3 + 16x - 16$

    (a) $x = -1$ (b) $x = -3$ (c) $x = 5$

20. $f(x) = 2x^4 - 8x + 3$

    (a) $x = 1$ (b) $x = 3$ (c) $x = -4$

In Exercises 21–36, find the real zeros of the function.

21. $f(x) = x^3 - 6x^2 + 11x - 6$

22. $f(x) = x^3 - 7x - 6$

23. $g(x) = x^3 - 4x^2 - x + 4$

24. $h(x) = x^3 - 9x^2 + 20x - 12$

25. $h(t) = t^3 + 12t^2 + 21t + 10$

26. $f(x) = x^3 + 6x^2 + 12x + 8$

27. $f(x) = x^3 - 4x^2 + 5x - 2$

28. $p(x) = x^3 - 9x^2 + 27x - 27$

29. $C(x) = 2x^3 + 3x^2 - 1$

30. $f(x) = 3x^3 - 19x^2 + 33x - 9$

31. $f(x) = 4x^3 - 3x - 1$

32. $f(z) = 12z^3 - 4z^2 - 27z + 9$

33. $f(y) = 4y^3 + 3y^2 + 8y + 6$

34. $g(x) = 3x^3 - 2x^2 + 15x - 10$

35. $f(x) = x^4 - 3x^2 + 2$

36. $P(t) = t^4 - 7t^2 + 12$

In Exercises 37–44, find all real solutions of the polynomial equation.

37. $z^4 - z^3 - 2z - 4 = 0$

38. $x^4 - x^3 - 29x^2 - x - 30 = 0$

39. $x^4 - 13x^2 - 12x = 0$

40. $2y^4 + 7y^3 - 26y^2 + 23y - 6 = 0$

41. $2x^4 - 11x^3 - 6x^2 + 64x + 32 = 0$

42. $x^5 - x^4 - 3x^3 + 5x^2 - 2x = 0$

43. $x^5 - 7x^4 + 10x^3 + 14x^2 - 24x = 0$

44. $6x^4 - 11x^3 - 51x^2 + 99x - 27 = 0$

In Exercises 45–48, (a) list the possible rational zeros of f, (b) sketch the graph of f so that some of the possible zeros in part (a) can be discarded, and then (c) determine all real zeros of f.

45. $f(x) = 32x^3 - 52x^2 + 17x + 3$

46. $f(x) = 6x^3 - x^2 - 13x + 8$

47. $f(x) = 4x^3 + 7x^2 - 11x - 18$

48. $f(x) = 2x^3 + 5x^2 - 21x - 10$

In Exercises 49–52, find the rational zeros of the polynomial function.

49. $P(x) = x^4 - \frac{25}{4}x^2 + 9$

50. $f(x) = x^3 - \frac{3}{2}x^2 - \frac{23}{2}x + 6$

51. $f(x) = x^3 - \frac{1}{4}x^2 - x + \frac{1}{4}$

52. $f(z) = z^3 + \frac{11}{6}z^2 - \frac{1}{2}z - \frac{1}{3}$

In Exercises 53–56, match the given cubic equation with the number of rational and irrational zeros given in (a), (b), (c), and (d) below.

(a) Rational zeros: 0   (b) Rational zeros: 3
    Irrational zeros: 1   Irrational zeros: 0
(c) Rational zeros: 1   (d) Rational zeros: 1
    Irrational zeros: 2   Irrational zeros: 0

53. $f(x) = x^3 - 1$ **54.** $f(x) = x^3 - 2$

55. $f(x) = x^3 - x$ **56.** $f(x) = x^3 - 2x$

57. *Dimensions of a Box* An open box is to be made from a rectangular piece of material, 12 inches by 10 inches, by cutting equal squares from the corners and turning up the sides. Find the dimensions of the box, given that the volume is to be 96 cubic inches.

58. *Dimensions of a Box* An open box is to be made from a rectangular piece of material, 9 inches by 5 inches, by cutting equal squares from the corners and turning up the sides (see figure). Find the dimensions of the box, given that the volume is to be 18 cubic inches.

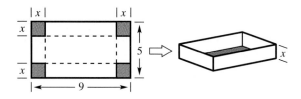

59. *Dimensions of a Package* A rectangular package to be sent by a postal service can have a maximum combined length and girth (perimeter of a cross section) of 108 inches (see figure). Find the dimensions of the package, given that the volume is 11,664 cubic inches.

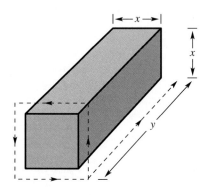

60. *Dimensions of a Package* A rectangular package to be sent by a postal service can have a maximum combined length and girth (perimeter of a cross section) of 120 inches. Find the dimensions of the package, given that the volume is 16,000 cubic inches.

61. *Dimensions of a Box* An open box is to be made from a rectangular piece of material, 10 inches by 8 inches, by cutting equal squares from the corners and turning up the sides. Find the dimensions of the box, if the volume is to be 52.5 cubic inches.

62. *Demand Function* A company that produces cellular pagers estimates that the demand for a particular model of pager is

$$D = -x^3 + 54x^2 - 140x - 3000, \quad 10 \leq x \leq 50$$

where D is the demand for the pager and x is the price of the pager.

(a) Use a graphing utility to graph D. Use the TRACE option to determine for what values of x the demand will be 14,400 pagers.

(b) You may also determine the price x of the pagers that will yield a demand of 14,400 by setting D equal to 14,400 and solving for x with a graphing utility. Discuss this alternative method of solution. How many real zeros does Descartes's Rule of Signs indicate there may be? Of the actual solutions, what price would you recommend the company charge for the pagers?

4.3 Approximating Zeros of Polynomial Functions

Approximating Zeros of Polynomial Functions ▪ Application

Approximating Zeros of Polynomial Functions

There are several different techniques for approximating the zeros of a polynomial function. All such techniques are better suited to computers or graphing utilities than they are to "hand calculations." In this section, you will study two techniques that can be used with a graphing utility. The first is called the **zoom-and-trace** technique.

Zoom-and-Trace Technique

To approximate a real zero of a function with a graphing utility, use the following steps.

1. Graph the function so that the real zero you want to approximate appears as an x-intercept on the screen.

2. Move the cursor near the x-intercept and use the ZOOM feature to zoom in to get a better look at the intercept.

3. Use the TRACE feature to find the x-values that occur just before and just after the x-intercept. If the difference in these values is sufficiently small, use their average as the approximation. If not, continue zooming in until the approximation reaches the desired accuracy.

To help you visually determine when you have zoomed in enough times to reach the desired accuracy, you can set the X-scale of the viewing window to the accuracy you need and zoom in repeatedly.

- To approximate the zero to the nearest hundredth, set the X-scale to 0.01.
- To approximate the zero to the nearest thousandth, set the X-scale to 0.001.
- To approximate the zero to the nearest ten-thousandth, set the X-scale to 0.0001.

NOTE The amount that a graphing utility zooms in is determined by the *zoom factor*. The zoom factor is an integer that gives the ratio of the larger screen to the smaller screen. For instance, if you zoom in with a zoom factor of 2, you will obtain a screen in which the x- and y-values are half their original values. In this text, we use a zoom factor of 4.

EXAMPLE 1 *Approximating the Zeros of a Polynomial Function*

Approximate the real zeros of the polynomial function

$$f(x) = x^3 + 4x + 2.$$

Use an accuracy of 0.001.

Solution

To begin, use a graphing utility to graph the function, as shown in Figure 4.7. Set the X-scale to 0.001 and zoom in several times until the tick marks on the x-axis become visible. The final screen should be similar to the one shown in Figure 4.8. At this point, you can use the TRACE feature to determine that the x-values just to the left and right of the x-intercept are

$$x \approx -0.4735 \quad \text{and} \quad x \approx -0.4733.$$

Thus, to the nearest thousandth, you can approximate the zero of the function to be

$$x \approx -0.473.$$

To check this, try substituting -0.473 into the function. You should obtain a result that is approximately zero.

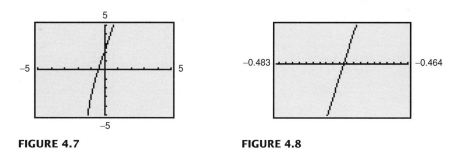

FIGURE 4.7   FIGURE 4.8

In Example 1, the cubic polynomial function has only one real zero. Remember that functions can have two or more real zeros. In such cases, you can use the zoom-and-trace technique for each zero separately. For instance, the function

$$f(x) = x^3 - 4x^2 + x + 2$$

has three real zeros, as shown in Figure 4.9. Using a zoom-and-trace approach for each real zero, you can approximate the real zeros to be

$$-0.562, \quad 1.000, \quad \text{and} \quad 3.562.$$

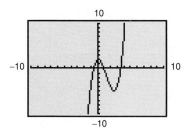

FIGURE 4.9

The second technique that can be used with some graphing utilities is to use the graphing utility's **solve key.** The name of this key or feature differs with different calculators. On a *TI-82*, it is called the root feature.

EXAMPLE 2   *Approximating the Zeros of a Polynomial Function*

Approximate the real zeros of $f(x) = x^3 - 2x^2 - x + 1$.

Solution

To begin, use a graphing utility to graph the function, as shown in Figure 4.10. Notice that the graph has three x-intercepts. To approximate the leftmost intercept with a *TI-82*, find an appropriate viewing window and use the root feature. After this has been done, the calculator should display an approximation of

$$x \approx -0.8019377,$$

which is accurate to seven decimal places. By repeating this process, you can determine that the other two zeros are $x \approx 0.556$ and $x \approx 2.247$.

Technology

On a *TI-82*, you can approximate the x-intercepts of a graph with the root feature. To do this, use the following keystrokes.

[CALC] 2

Cursor left of x. [ENTER]

Cursor right of x. [ENTER]

Cursor near x. [ENTER]

At this point, the calculator will display an approximation of the x-intercept.

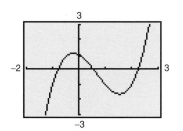

Find an appropriate viewing window, then use the *root* feature.

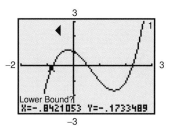

Cursor to the left of the intercept and press "Enter."

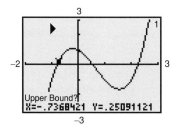

Cursor to the right of the intercept and press "Enter."
FIGURE 4.10

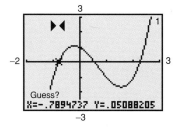

Cursor near the intercept and press "Enter."

You may be wondering why we spend so much time in algebra trying to find the *zeros* of a function. The reason is that if you have a technique that will allow you to solve the equation $f(x) = 0$, you can use the same technique to solve the more general equation

$$f(x) = c$$

where c is any real number. This procedure is demonstrated in Example 3.

EXAMPLE 3   Solving the Equation f(x) = c

Find a value of x such that $f(x) = 30$ for the function

$$f(x) = x^3 - 4x + 4.$$

Solution

The graph of $f(x) = x^3 - 4x + 4$ is shown in Figure 4.11. Note from the graph that $f(x) = 30$ when x is about 3.5. To use the zoom-and-trace technique to approximate this x-value more closely, consider the equation

$$x^3 - 4x + 4 = 30$$
$$x^3 - 4x - 26 = 0.$$

Thus, the *solutions* of the equation $f(x) = 30$ are precisely the same x-values as the *zeros* of $g(x) = x^3 - 4x - 26$. Using the graph of g, as shown in Figure 4.12, you can approximate the zero of g to be

$$x \approx 3.41.$$

You can check this value by substituting $x = 3.41$ into the original function.

$$f(3.41) = (3.41)^3 - 4(3.41) + 4$$
$$\approx 30.01 \ \checkmark$$

Remember that with decimal approximations, a check will usually not produce an exact value.

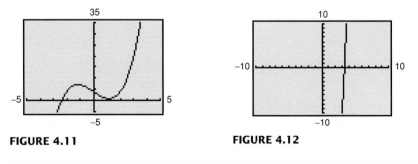

FIGURE 4.11   **FIGURE 4.12**

Application

EXAMPLE 4 *Profit and Advertising Expenses*

A company that produces sports clothes estimates that the profit from selling a particular line of sportswear is given by

$$P = -140x^3 + 7520x^2 - 400,000, \qquad 0 \le x \le 50$$

where P is the profit (in dollars) and x is the advertising expense (in tens of thousands of dollars). According to this model, how much money should the company spend to obtain a profit of $2,750,000?

Solution

From Figure 4.13, it appears that there are two different values of x between 0 and 50 that will produce a profit of $2,750,000. However, because of the context of the problem, it is clear that the better answer is the smaller of the two numbers. Thus, to solve the equation

$$-140x^3 + 7520x^2 - 400,000 = 2,750,000$$
$$-140x^3 + 7520x^2 - 3,150,000 = 0$$

you can find the zeros of the function $g(x) = -140x^3 + 7520x^2 - 3,150,000$. Using the zoom-and-trace technique, you can find that the leftmost zero is

$$x \approx 32.8.$$

You can check this solution by substituting $x = 32.8$ into the original function, as follows.

$$P = -140(32.8)^3 + 7520(32.8)^2 - 400,000 \approx \$2,750,059$$

Thus, the company should plan to spend about $328,000 in advertising for the line of sportswear.

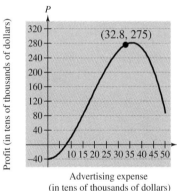

Profit (in tens of thousands of dollars)

(32.8, 275)

Advertising expense
(in tens of thousands of dollars)

FIGURE 4.13

Group Activities   Exploring with Technology

Approximating Zeros   Find a fourth-degree polynomial function that has four irrational zeros. Then use a graphing utility to approximate each of the zeros.

Warm Up   The following warm-up exercises involve skills that were covered in earlier sections. You will use these skills in the exercise set for this section.

In Exercises 1–6, use a graphing utility to graph the function.

1. $f(x) = x^3 - 2x^2 + 1$

2. $f(x) = -x^3 + 3x + 2$

3. $f(x) = -2x^3 - 3x^2 + x + 4$

4. $f(x) = 3x^3 - 4x + 10$

5. $f(x) = x^4 - 5x^2 + x - 1$

6. $f(x) = -x^4 + 2x + 3$

In Exercises 7–10, evaluate the function at the indicated x-value.

7. $f(x) = x^3 + 2x^2 + 3, \quad x = 1.5$

8. $f(x) = 2x^3 - 6x + 10, \quad x = -2.6$

9. $f(x) = -x^3 + 7x^2 - 3x - 12, \quad x = 3.4$

10. $f(x) = -2x^3 - 3x^2 + x - 2, \quad x = -3.1$

4.3 Exercises

In Exercises 1–6, match the function with its graph. Then approximate the real zeros of the function to three decimal places. [The graphs are labeled (a), (b), (c), (d), (e), and (f).]

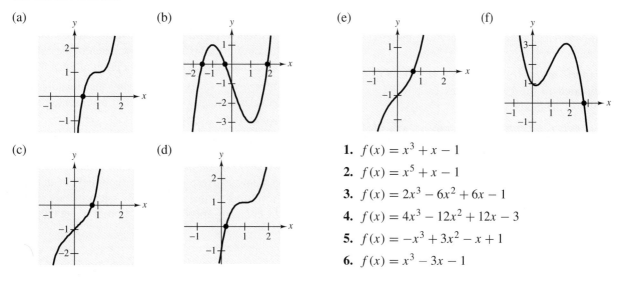

(a)

(b)

(c)

(d)

(e)

(f)

1. $f(x) = x^3 + x - 1$

2. $f(x) = x^5 + x - 1$

3. $f(x) = 2x^3 - 6x^2 + 6x - 1$

4. $f(x) = 4x^3 - 12x^2 + 12x - 3$

5. $f(x) = -x^3 + 3x^2 - x + 1$

6. $f(x) = x^3 - 3x - 1$

In Exercises 7–20, use a graphing utility to approximate the real zeros of the function. Use an accuracy of 0.001.

7. $f(x) = x^4 - x - 3$ **8.** $f(x) = x^3 + 2x + 1$

9. $f(x) = x^3 - 27x - 27$

10. $f(x) = x^3 - 3.9x^2 + 4.79x - 1.881$

11. $f(x) = 4x^3 + 14x - 8$

12. $f(x) = 4x^3 - 14x^2 - 2$

13. $f(x) = -x^3 + 2x^2 + 4x + 5$

14. $f(x) = x^3 - 4x^2 - 3x - 9$

15. $f(x) = x^4 + x - 3$ **16.** $f(x) = x^3 - 3x^2 - 7$

17. $f(x) = -x^3 - 4x + 1$

18. $f(x) = -x^4 + 2x^3 + 4$

19. $f(x) = 7x^4 - 42x^3 + 43x^2 + 216x - 324$

20. $f(x) = 3x^4 - 12x^3 + 27x^2 + 4x - 4$

21. *Cost of Dental Care* From 1970 to 1993, the cost of dental care in the United States rose according to the function

$$C = \sqrt{1.9t^3 + 13.4t^2 + 128.6t + 1536.8},$$
$$0 \le t \le 23$$

where C is the dental care index and $t = 0$ represents 1970. In 1981, the dental care index was about 84. According to this model, when did the cost of dental care reach a level that was *twice* the 1981 cost? (Source: U.S. Bureau of Labor Statistics)

22. *Imports to the United States* The value of goods imported to the United States from 1984 to 1993 follows the model

$$I = 0.232x^4 - 7.734x^3 + 91.70x^2 - 428.12x +$$
$$1008.8, \quad 4 \le t \le 13$$

where I is the annual value of goods imported (in billions of dollars) and $t = 4$ represents 1984. According to this model, when did the annual value of imports reach \$700 billion? (Source: U.S. General Imports and Imports for Consumption)

23. *Medicine* The concentration of a certain chemical in the bloodstream t hours after injection into muscle tissue is given by

$$C = \frac{3t^2 + t}{t^3 + 50}, \quad 0 \le t.$$

The concentration is greatest when

$$3t^4 + 2t^3 - 300t - 50 = 0.$$

Approximate this time to the nearest tenth of an hour.

24. *Transportation Cost* The transportation cost C of the components used in manufacturing a certain product is given by

$$C = 100 \left(\frac{200}{x^2} + \frac{x}{x + 30} \right), \quad 1 \le x$$

where C is measured in thousands of dollars and x is the order size in hundreds. The cost is a minimum when

$$3x^3 - 40x^2 - 2400x - 36{,}000 = 0.$$

Approximate the optimal order size to the nearest hundred units.

25. *Advertising Costs* A company that produces portable cassette players estimates that the profit from selling a particular model is given by

$$P = -76x^3 + 4830x^2 - 320{,}000, \quad 0 \le x \le 60$$

where P is the profit (in dollars) and x is the advertising expense (in tens of thousands of dollars). According to this model, how much money should the company spend to obtain a profit of \$2,500,000?

26. *Advertising Costs* A company that manufactures bicycles estimates that the profit from selling a particular model is given by

$$P = -45x^3 + 2500x^2 - 275{,}000, \quad 0 \le x \le 50$$

where P is the profit (in dollars) and x is the advertising expense (in tens of thousands of dollars). According to this model, how much money should the company spend to obtain a profit of \$800,000?

MID-CHAPTER QUIZ

Take this quiz as you would take a quiz in class. After you are done, check your work against the answers given in the back of the book.

In Exercises 1 and 2, completely factor the polynomial. (One factor is given.)

1. $2x^3 - x^2 - 13x - 6$, $(x + 2)$ **2.** $3x^3 - 5x^2 - 58x + 40$, $\left(x - \frac{2}{3}\right)$

In Exercise 3 and 4, use synthetic division to evaluate $f(2)$.

3. $f(x) = 3x^3 - 5x^2 + 9$ **4.** $f(x) = 0.3x^4 - 1.8x^3 + 0.7x^2 - 3$

5. Simplify: $\dfrac{2x^4 + 9x^3 - 32x^2 - 99x + 180}{x^2 + 2x - 15}$

6. The profit for a company is $P = -95x^3 + 5650x^2 - 250,000$, $0 \le x \le 55$, where x is the advertising expense (in tens of thousands of dollars). What is the profit for an advertising expense of $450,000? Use a graphing utility to approximate another advertising expense that would yield the same profit.

In Exercises 7–14, find all real zeros of the function.

7. $f(x) = -2x^3 - 7x^2 + 10x + 35$ **8.** $f(x) = 4x^4 - 37x^2 + 9$

9. $f(x) = 3x^4 + 4x^3 - 3x - 4$ **10.** $f(x) = 2x^3 - 3x^2 + 2x - 3$

11. $f(x) = x^4 - 5x^2 + 4$ **12.** $f(x) = 3x^3 - 4x^2 + 1$

13. $f(x) = x^4 + x^3 + x^2 + 3x - 6$ **14.** $f(x) = 5x^3 - 7x^2 + 2$

In Exercises 15 and 16, use Descartes's Rule of Signs to determine the possible number of positive and negative zeros. Then use a graphing utility to determine the actual number of positive and negative zeros.

15. $f(x) = -2x^3 - x^2 + 11x - 5$ **16.** $f(x) = x^4 - 19x^2 + 48$

In Exercises 17–20, an open box is to be made from a 12-inch-by-14-inch piece of material by cutting equal squares from the corners and turning up the sides. The volume of the box is to be 160 cubic inches.

17. Write an equation that will allow you to determine the dimensions of the box.

18. Find the dimensions of the box algebraically.

19. Use a graphing utility to find the dimensions of the box.

20. When solving this problem algebraically or graphically, not all possible zeros are used. Explain.

4.4 | Complex Numbers

The Imaginary Unit *i* ▪ Operations with Complex Numbers ▪
Complex Conjugates ▪ Complex Solutions ▪ Applications

The Imaginary Unit *i*

In Section 1.4, you learned that some quadratic equations have no real solutions.
For instance, the quadratic equation

$$x^2 + 1 = 0 \qquad \text{Equation with no real solution}$$

has no real solution because there is no real number x that can be squared to
produce -1. To overcome this deficiency, mathematicians created an expanded
system of numbers using the **imaginary unit *i*,** defined as

$$i = \sqrt{-1} \qquad \text{Imaginary unit}$$

where $i^2 = -1$. By adding real numbers to real multiples of this imaginary unit,
we obtain the set of **complex numbers.** Each complex number can be written in
the **standard form,** $a + bi$.

Definition of a Complex Number

For real numbers a and b, the number

$$a + bi$$

is a **complex number.** If $a = 0$ and $b \neq 0$, the complex number bi is an
imaginary number.

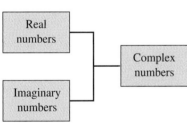

FIGURE 4.14

The set of real numbers is a subset of the set of complex numbers, as shown in
Figure 4.14. This is true because every real number a can be written as a complex
number using $b = 0$. That is, for every real number a, we can write $a = a + 0i$.

Equality of Complex Numbers

Two complex numbers $a + bi$ and $c + di$, written in standard form, are
equal to each other

$$a + bi = c + di \qquad \text{Equality of two complex numbers}$$

if and only if $a = c$ and $b = d$.

Operations with Complex Numbers

To add (or subtract) two complex numbers, you add (or subtract) the real and imaginary parts of the numbers separately.

Addition and Subtraction of Complex Numbers

If $a + bi$ and $c + di$ are two complex numbers written in standard form, their sum and difference are defined as follows.

$$\text{Sum:} \quad (a + bi) + (c + di) = (a + c) + (b + d)i$$
$$\text{Difference:} \quad (a + bi) - (c + di) = (a - c) + (b - d)i$$

The **additive identity** in the complex number system is zero (the same as in the real number system). Furthermore, the **additive inverse** of the complex number $a + bi$ is

$$-(a + bi) = -a - bi. \qquad \text{Additive inverse}$$

Thus, you have

$$(a + bi) + (-a - bi) = 0 + 0i = 0.$$

EXAMPLE 1   *Adding and Subtracting Complex Numbers*

a.
$$
\begin{aligned}
(3 - i) + (2 + 3i) &= 3 - i + 2 + 3i && \text{Remove parentheses.} \\
&= 3 + 2 - i + 3i && \text{Group real and imaginary terms.} \\
&= (3 + 2) + (-1 + 3)i \\
&= 5 + 2i && \text{Standard form}
\end{aligned}
$$

b.
$$
\begin{aligned}
2i + (-4 - 2i) &= 2i - 4 - 2i && \text{Remove parentheses.} \\
&= -4 + 2i - 2i && \text{Group real and imaginary terms.} \\
&= -4 && \text{Standard form}
\end{aligned}
$$

c.
$$
\begin{aligned}
3 - (-2 + 3i) + (-5 + i) &= 3 + 2 - 3i - 5 + i \\
&= 3 + 2 - 5 - 3i + i \\
&= 0 - 2i \\
&= -2i
\end{aligned}
$$

NOTE   Note in Example 1(b) that the sum of two complex numbers can be a real number.

DISCOVERY

Complete the table:

$i^1 = i$ $i^5 =$ $i^9 =$

$i^2 = -1$ $i^6 =$ $i^{10} =$

$i^3 = -i$ $i^7 =$ $i^{11} =$

$i^4 = 1$ $i^8 =$ $i^{12} =$

What pattern do you see? Write a brief description of how you would find i raised to a positive integer power.

Many of the properties of real numbers are valid for complex numbers as well. Here are some examples.

Associative Property of Addition and Multiplication

Commutative Property of Addition and Multiplication

Distributive Property of Multiplication Over Addition

Notice how these properties are used when two complex numbers are multiplied.

$$(a + bi)(c + di) = a(c + di) + bi(c + di) \qquad \text{Distributive}$$
$$= ac + (ad)i + (bc)i + (bd)i^2 \qquad \text{Distributive}$$
$$= ac + (ad)i + (bc)i + (bd)(-1) \qquad \text{Definition of } i$$
$$= ac - bd + (ad)i + (bc)i \qquad \text{Commutative}$$
$$= (ac - bd) + (ad + bc)i \qquad \text{Associative}$$

Rather than trying to memorize this multiplication rule, we suggest that you simply remember how the distributive property is used to multiply two complex numbers. The procedure is similar to multiplying two polynomials and combining like terms (as in the FOIL Method).

EXAMPLE 2 Multiplying Complex Numbers

a. $(i)(-3i) = -3i^2$ Multiply.

$\qquad\qquad\quad = -3(-1)$ $i^2 = -1$

$\qquad\qquad\quad = 3$ Simplify.

b. $(2 - i)(4 + 3i) = 8 + 6i - 4i - 3i^2$ Product of binomials

$\qquad\qquad\qquad = 8 + 6i - 4i - 3(-1)$ $i^2 = -1$

$\qquad\qquad\qquad = 8 + 3 + 6i - 4i$ Collect terms.

$\qquad\qquad\qquad = 11 + 2i$ Standard form

c. $(3 + 2i)(3 - 2i) = 9 - 6i + 6i - 4i^2$ Product of binomials

$\qquad\qquad\qquad = 9 - 4(-1)$ $i^2 = -1$

$\qquad\qquad\qquad = 9 + 4$ Simplify.

$\qquad\qquad\qquad = 13$ Simplify.

d. $(3 + 2i)^2 = 9 + 6i + 6i - 4i^2$ Product of binomials

$\qquad\qquad\quad = 9 - 4(-1) + 12i$ $i^2 = -1$

$\qquad\qquad\quad = 9 + 4 + 12i$ Simplify.

$\qquad\qquad\quad = 13 + 12i$ Simplify.

Complex Conjugates

Notice in Example 2(c) that the product of two complex numbers can be a real number. This occurs with pairs of complex numbers of the form $a + bi$ and $a - bi$, called **complex conjugates.** In general, the product of two complex conjugates can be written as follows.

$$(a + bi)(a - bi) = a^2 - abi + abi - b^2i^2$$
$$= a^2 - b^2(-1)$$
$$= a^2 + b^2$$

Complex conjugates can be used to divide one complex number by another. That is, to find the quotient of $a + bi$ and $c + di$, multiply the numerator and denominator by the conjugate of the denominator to obtain

$$\frac{a + bi}{c + di} = \frac{a + bi}{c + di}\left(\frac{c - di}{c - di}\right) = \frac{(ac + bd) + (bc - ad)i}{c^2 + d^2}.$$

EXAMPLE 3   *Dividing Complex Numbers*

$$\frac{1}{1 + i} = \frac{1}{1 + i}\left(\frac{1 - i}{1 - i}\right) \qquad \text{Multiply by conjugate.}$$

$$= \frac{1 - i}{1^2 - i^2} \qquad \text{Product of conjugates}$$

$$= \frac{1 - i}{1 - (-1)} \qquad i^2 = -1$$

$$= \frac{1 - i}{2} \qquad \text{Simplify.}$$

$$= \frac{1}{2} - \frac{1}{2}i \qquad \text{Standard form}$$

EXAMPLE 4   *Dividing Complex Numbers*

$$\frac{2 + 3i}{4 - 2i} = \frac{2 + 3i}{4 - 2i}\left(\frac{4 + 2i}{4 + 2i}\right) \qquad \text{Multiply by conjugate.}$$

$$= \frac{8 + 4i + 12i + 6i^2}{16 - 4i^2} \qquad \text{Expand.}$$

$$= \frac{8 - 6 + 16i}{16 + 4} \qquad i^2 = -1$$

$$= \frac{1}{20}(2 + 16i) \qquad \text{Simplify.}$$

$$= \frac{1}{10} + \frac{4}{5}i \qquad \text{Standard form}$$

Complex Solutions

Remind students that complex zeros of polynomial functions with real coefficients *always* occur in conjugate pairs.

When using the Quadratic Formula to solve a quadratic equation, you often obtain a result such as $\sqrt{-3}$, which you know is not a real number. By factoring out $i = \sqrt{-1}$, you can write this number in standard form.

$$\sqrt{-3} = \sqrt{3(-1)} = \sqrt{3}\sqrt{-1} = \sqrt{3}\,i$$

The number $\sqrt{3}\,i$ is called the principal square root of -3.

Principal Square Root of a Negative Number

If a is a positive number, the **principal square root** of the negative number $-a$ is defined as

$$\sqrt{-a} = \sqrt{a}\,i.$$

STUDY TIP

The definition of principal square root uses the rule

$$\sqrt{ab} = \sqrt{a}\sqrt{b}$$

for $a > 0$ and $b < 0$. This rule is not valid if *both a* and *b* are negative. For example,

$$\sqrt{-5}\sqrt{-5} = 5i^2 = -5$$

whereas

$$\sqrt{(-5)(-5)} = \sqrt{25} = 5.$$

To avoid problems with multiplying square roots of negative numbers, be sure to convert to standard form *before* multiplying.

EXAMPLE 5   *Writing Complex Numbers in Standard Form*

a. $\sqrt{-3}\sqrt{-12} = \sqrt{3}\,i\sqrt{12}\,i = \sqrt{36}\,i^2 = 6(-1) = -6$

b. $\sqrt{-48} - \sqrt{-27} = \sqrt{48}\,i - \sqrt{27}\,i = 4\sqrt{3}\,i - 3\sqrt{3}\,i = \sqrt{3}\,i$

c. $\left(-1 + \sqrt{-3}\right)^2 = \left(-1 + \sqrt{3}\,i\right)^2$

$$= (-1)^2 - 2\sqrt{3}\,i + \left(\sqrt{3}\right)^2 (i^2)$$

$$= 1 - 2\sqrt{3}\,i + 3(-1)$$

$$= -2 - 2\sqrt{3}\,i$$

EXAMPLE 6   *Complex Solutions of a Quadratic Equation*

Solve the equation $3x^2 - 2x + 5 = 0$.

Solution

$$x = \frac{-(-2) \pm \sqrt{(-2)^2 - 4(3)(5)}}{2(3)} \qquad \text{Quadratic Formula}$$

$$= \frac{2 \pm \sqrt{-56}}{6} \qquad \text{Simplify.}$$

$$= \frac{2 \pm 2\sqrt{14}\,i}{6} \qquad \text{Write in } i\text{-form.}$$

$$= \frac{1}{3} \pm \frac{\sqrt{14}}{3}\,i \qquad \text{Standard form}$$

A common error in working with complex numbers occurs when a negative sign precedes the imaginary number. For instance, $-\sqrt{-9}$ is often incorrectly simplified as 3.

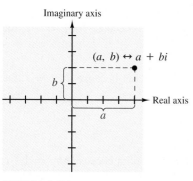

FIGURE 4.15

Applications

Most applications involving complex numbers are either theoretical (see the next section) or very technical, and are therefore not appropriate for inclusion in this text. However, to give you some idea of how complex numbers can be used in applications, we give a general description of their use in **fractal geometry.**

To begin, consider a coordinate system called the **complex plane.** Just as every real number corresponds to a point on the real line, every complex number corresponds to a point in the complex plane, as shown in Figure 4.15. In this figure, note that the vertical axis is the **imaginary axis** and the horizontal axis is the **real axis.** The point that corresponds to the complex number $a + bi$ is (a, b).

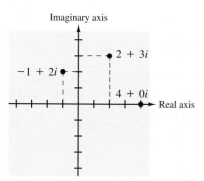

FIGURE 4.16

EXAMPLE 7 *Plotting Complex Numbers in the Complex Plane*

Plot the following complex numbers in the complex plane.

a. $2 + 3i$ **b.** $-1 + 2i$ **c.** 4

Solution

a. To plot the complex number $2 + 3i$, move (from the origin) two units to the right on the real axis and then three units up, as shown in Figure 4.16. In other words, plotting the complex number $2 + 3i$ in the complex plane is comparable to plotting the point $(2, 3)$ in the Cartesian plane.

b. The complex number $-1 + 2i$ corresponds to the point $(-1, 2)$, as shown in Figure 4.16.

c. The complex number 4 corresponds to the point $(4, 0)$, as shown in Figure 4.16.

In the hands of a person who understands "fractal geometry," the complex plane can become an easel on which stunning pictures, called **fractals,** can be drawn. The most famous such picture is called the **Mandelbrot Set,** named after the Polish-born mathematician Benoit Mandelbrot. To draw the Mandelbrot Set, consider the following sequence of numbers.

$$c, \qquad c^2 + c, \qquad (c^2 + c)^2 + c, \qquad ((c^2 + c)^2 + c)^2 + c, \qquad \dots$$

The behavior of this sequence depends on the value of the complex number c. For some values of c this sequence is **bounded,** and for other values it is **unbounded.** If the sequence is bounded, the complex number c is in the Mandelbrot Set, and if the sequence is unbounded, the complex number c is not in the Mandelbrot Set.

EXAMPLE 8 *Members of the Mandelbrot Set*

a. The complex number -2 is in the Mandelbrot Set because for $c = -2$, the corresponding Mandelbrot sequence is $-2, 2, 2, 2, 2, 2, \ldots$, which is bounded.

b. The complex number i is also in the Mandelbrot Set because for $c = i$, the corresponding Mandelbrot sequence is

$$i, \quad -1+i, \quad -i, \quad -1+i, \quad -i, \quad -1+i, \quad \ldots$$

which is bounded.

c. The complex number $1 + i$ is *not* in the Mandelbrot Set because for $c = 1 + i$, the corresponding Mandelbrot sequence is

$$1 + i, \ 1 + 3i, \ -7 + 7i, \ 1 - 97i, \ -9407 - 193i,$$
$$88454401 + 3631103i, \ \ldots$$

which is unbounded.

With this definition, a picture of the Mandelbrot Set would have only two colors. One color for points that are in the set (the sequence is bounded) and one color for points that are outside the set (the sequence is unbounded). Figure 4.17 shows a black and yellow picture of the Mandelbrot Set. The points that are black are in the Mandelbrot Set and the points that are yellow are not.

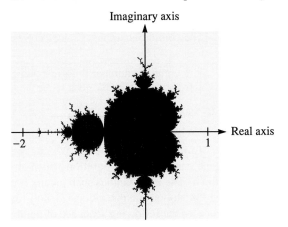

FIGURE 4.17 Mandelbrot Set

To add more interest to the picture, computer scientists discovered that the points that are not in the Mandelbrot Set can be assigned a variety of colors, depending on "how quickly" their sequences diverge. Figure 4.18 shows three different appendages of the Mandelbrot Set using a spectrum of colors. (The colored portions of the picture represent points that are *not* in the Mandelbrot Set.)

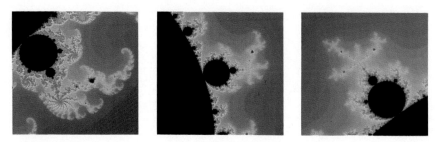

FIGURE 4.18

Figures 4.19, 4.20, and 4.21 show other types of fractal sets. From these pictures, you can see why fractals have fascinated people since their discovery (around 1980).

FIGURE 4.19

FIGURE 4.20

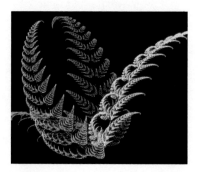

FIGURE 4.21

Group Activities You Be the Instructor

Error Analysis One of your students has handed in the following quiz. Find the error(s) in each solution and discuss how to explain each error to your student.

1. Write $\dfrac{5}{3-2i}$ in standard form.

$$\frac{5}{3-2i}\cdot\frac{3+2i}{3+2i} \quad \frac{15+10i}{9-4}=3+2i$$

2. Multiply $\left(\sqrt{-4}+3\right)\left(i-\sqrt{-3}\right)$.

$$\left(\sqrt{-4}+3\right)\left(i-\sqrt{-3}\right)=i\sqrt{-4}-\sqrt{-4}\sqrt{-3}+3i-3\sqrt{-3}$$
$$=-2i-\sqrt{12}+3i-3i\sqrt{3}$$
$$=\left(1-3\sqrt{3}\right)i-2\sqrt{3}$$

Warm Up The following warm-up exercises involve skills that were covered in earlier sections. You will use these skills in the exercise set for this section.

In Exercises 1–8, simplify the expression.

1. $\sqrt{12}$

2. $\sqrt{500}$

3. $\sqrt{20} - \sqrt{5}$

4. $\sqrt{27} - \sqrt{243}$

5. $\sqrt{24}\sqrt{6}$

6. $2\sqrt{18}\sqrt{32}$

7. $\dfrac{1}{\sqrt{3}}$

8. $\dfrac{2}{\sqrt{2}}$

In Exercises 9 and 10, solve the quadratic equation.

9. $x^2 + x - 1 = 0$

10. $x^2 + 2x - 1 = 0$

4.4 Exercises

1. Write out the first 16 positive integer powers of i (i, i^2, i^3, . . . , i^{16}), and express each as i, $-i$, 1, or -1.

2. Express each of the powers of i as i, $-i$, 1, or -1.

(a) i^{40} (b) i^{25} (c) i^{50} (d) i^{67}

In Exercises 3–6, find the real numbers a and b so that the equation is true.

3. $a + bi = -10 + 6i$

4. $a + bi = 13 + 4i$

5. $(a - 1) + (b + 3)i = 5 + 8i$

6. $(a + 6) + 2bi = 6 - 5i$

In Exercises 7–18, write the complex number in standard form and find its complex conjugate.

7. $4 + \sqrt{-9}$

8. $3 + \sqrt{-16}$

9. $2 - \sqrt{-27}$

10. $1 + \sqrt{-8}$

11. $\sqrt{-75}$

12. 45

13. $-6i + i^2$

14. $4i^2 - 2i^3$

15. $-5i^5$

16. $(-i)^3$

17. 8

18. $\left(\sqrt{-4}\right)^2 - 5$

In Exercises 19–56, perform the indicated operation and write the result in standard form.

19. $(5 + i) + (6 - 2i)$

20. $(13 - 2i) + (-5 + 6i)$

21. $(8 - i) - (4 - i)$

22. $(3 + 2i) - (6 + 13i)$

23. $\left(-2 + \sqrt{-8}\right) + \left(5 - \sqrt{-50}\right)$

24. $\left(8 + \sqrt{-18}\right) - \left(4 + 3\sqrt{2}\,i\right)$

25. $-\left(\frac{3}{2} + \frac{5}{2}i\right) + \left(\frac{5}{3} + \frac{11}{3}i\right)$

26. $(1.6 + 3.2i) + (-5.8 + 4.3i)$

27. $\sqrt{-6} \cdot \sqrt{-2}$

28. $\sqrt{-5} \cdot \sqrt{-10}$

29. $\left(\sqrt{-10}\right)^2$

30. $\left(\sqrt{-75}\right)^3$

31. $(1 + i)(3 - 2i)$

32. $(6 - 2i)(2 - 3i)$

33. $(4 + 5i)(4 - 5i)$

34. $(6 + 7i)(6 - 7i)$

35. $6i(5 - 2i)$

36. $-8i(9 + 4i)$

37. $-8(5 - 2i)$

38. $\left(\sqrt{5} - \sqrt{3}\,i\right)\left(\sqrt{5} + \sqrt{3}\,i\right)$

39. $\left(\sqrt{14} + \sqrt{10}\,i\right)\left(\sqrt{14} - \sqrt{10}\,i\right)$

40. $\left(3 + \sqrt{-5}\right)\left(7 - \sqrt{-10}\right)$

41. $(4 + 5i)^2$

42. $(2 - 3i)^3$

43. $\dfrac{4}{4 - 5i}$

44. $\dfrac{3}{1 - i}$

45. $\dfrac{2 + i}{2 - i}$

46. $\dfrac{8 - 7i}{1 - 2i}$

47. $\dfrac{6 - 7i}{i}$

48. $\dfrac{8 + 20i}{2i}$

49. $\dfrac{1}{(2i)^3}$

50. $\dfrac{1}{(4 - 5i)^2}$

51. $\dfrac{5}{(1 + i)^3}$

52. $\dfrac{(2 - 3i)(5i)}{2 + 3i}$

53. $\dfrac{(21 - 7i)(4 + 3i)}{2 - 5i}$

54. $\dfrac{1}{i^3}$

55. $(2 + 3i)^2 + (2 - 3i)^2$

56. $(1 - 2i)^2 - (1 + 2i)^2$

In Exercises 57–64, solve the equation.

57. $x^2 - 2x + 2 = 0$

58. $x^2 + 6x + 10 = 0$

59. $4x^2 + 16x + 17 = 0$

60. $9x^2 - 6x + 37 = 0$

61. $4x^2 + 16x + 15 = 0$

62. $9x^2 - 6x + 35 = 0$

63. $16t^2 - 4t + 3 = 0$

64. $5s^2 + 6s + 3 = 0$

In Exercises 65–70, plot the complex number.

65. $-2 + i$

66. i

67. 3

68. $-2 - 3i$

69. $1 - 2i$

70. $-2i$

In Exercises 71–76, decide whether the number is in the Mandelbrot Set. Explain your reasoning.

71. $c = 0$

72. $c = 2$

73. $c = \frac{1}{2}i$

74. $c = -i$

75. $c = 1$

76. $c = -1$

Math Matters

Acceptance of Imaginary Numbers

Imaginary numbers were given the name "imaginary" because many mathematicians had a difficult time accepting such numbers. The following excerpt, written by the famous mathematician Carl Gauss (1777–1855), shows that, even in the early 1800s, some people were hesitant to accept the use of imaginary numbers.

"Our general arithmetic, which far surpasses the extent of the geometry of the ancients, is entirely the creation of modern times. Starting with the notion of whole numbers, it has gradually enlarged its domain. To whole numbers have been added fractions; to rational numbers have been added irrational; to positive numbers have been added negative; and to real numbers have been added imaginary numbers.

Carl Friedrich Gauss

This advance, however, has always been made with timid steps. The early algebraists called negative solutions of equations false solutions (and this is indeed the case when the problem to which they relate has been stated in such a way that negative solutions have no meaning). The reality of negative numbers is sufficiently justified since there are many cases where they have meaningful interpretations. This has long been admitted, but the imaginary numbers (formerly and occasionally now called impossible numbers) are still rather tolerated than fully accepted."

| 4.5 | **The Fundamental Theorem of Algebra** |

The Fundamental Theorem of Algebra ■ Conjugate Pairs ■
Factoring a Polynomial

The Fundamental Theorem of Algebra

You have been using the fact that an nth-degree polynomial can have at most n real zeros. In the complex number system, this statement can be improved. That is, in the complex number system, every nth-degree polynomial function has *precisely* n zeros. This important result is derived from the **Fundamental Theorem of Algebra,** first proved by the famous German mathematician Carl Friedrich Gauss (1777–1855).

The Fundamental Theorem of Algebra

If $f(x)$ is a polynomial of degree n, where $n > 0$, then f has at least one zero in the complex number system.

Jean Le Rond d'Alembert (1717–1783) worked independently of Carl Gauss trying to prove the Fundamental Theorem of Algebra. His efforts were such that in France, the Fundamental Theorem of Algebra is frequently known as the theorem of d'Alembert.

Using the Fundamental Theorem of Algebra and the equivalence of zeros and factors, you obtain the following theorem.

Linear Factorization Theorem

If $f(x)$ is a polynomial of degree n

$$f(x) = a_n x^n + a_{n-1} x^{x-1} + \cdots + a_1 x + a_0$$

where $n > 0$, then f has precisely n linear factors

$$f(x) = a_n (x - c_1)(x - c_2) \cdots (x - c_n)$$

where $c_1, c_2, \ldots, c_n$ are complex numbers and a_n is the leading coefficient of $f(x)$.

Note that neither the Fundamental Theorem of Algebra nor the Linear Factorization Theorem tells you *how* to find the zeros or factors of a polynomial. Such theorems are called **existence theorems.** To find the zeros of a polynomial function, you still rely on the techniques developed in the earlier parts of the text.

Remember that the n zeros of a polynomial function can be real or complex, and they may be repeated. Example 1 illustrates several cases.

EXAMPLE 1   Zeros of Polynomial Functions

a. The first-degree polynomial $f(x) = x - 2$ has exactly *one* zero: $x = 2$.

b. Counting multiplicity, the second-degree polynomial function

$$f(x) = x^2 - 6x + 9 = (x - 3)(x - 3)$$

has exactly *two* zeros: $x = 3$ and $x = 3$.

c. The third-degree polynomial function

$$f(x) = x^3 + 4x = x(x - 2i)(x + 2i)$$

has exactly *three* zeros: $x = 0$, $x = 2i$, and $x = -2i$.

d. The fourth-degree polynomial function

$$f(x) = x^4 - 1 = (x - 1)(x + 1)(x - i)(x + i)$$

has exactly *four* zeros: $x = 1$, $x = -1$, $x = i$, and $x = -i$.

Technology

Remember that when you use a graphing utility to locate the zeros of a function, the only zeros that appear as x-intercepts are the *real zeros*. Compare the graphs below with the four polynomial functions in Example 1. Which zeros appear on the graphs?

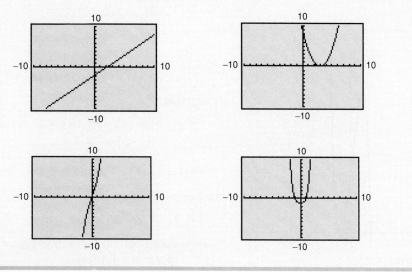

Example 2 shows how you can use the methods described in Sections 4.1 and 4.2 (Descartes's Rule of Signs, Rational Zero Test, synthetic division, and factoring) to find all the zeros of a polynomial function, including complex zeros.

EXAMPLE 2 *Finding the Zeros of a Polynomial Function*

Write the polynomial function $f(x) = x^5 + x^3 + 2x^2 - 12x + 8$ as the product of linear factors and list all of its zeros.

Solution

Descartes's Rule of Signs indicates two or no positive real zeros and one negative real zero. Moreover, the possible rational zeros are ± 1, ± 2, ± 4, and ± 8. Synthetic division produces the following.

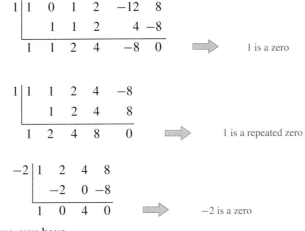

$$1 \begin{array}{|rrrrrr} 1 & 0 & 1 & 2 & -12 & 8 \\ & 1 & 1 & 2 & 4 & -8 \\ \hline 1 & 1 & 2 & 4 & -8 & 0 \end{array} \implies \text{1 is a zero}$$

$$1 \begin{array}{|rrrrr} 1 & 1 & 2 & 4 & -8 \\ & 1 & 2 & 4 & 8 \\ \hline 1 & 2 & 4 & 8 & 0 \end{array} \implies \text{1 is a repeated zero}$$

$$-2 \begin{array}{|rrrr} 1 & 2 & 4 & 8 \\ & -2 & 0 & -8 \\ \hline 1 & 0 & 4 & 0 \end{array} \implies \text{-2 is a zero}$$

Thus, you have

$$f(x) = x^5 + x^3 + 2x^2 - 12x + 8$$
$$= (x - 1)(x - 1)(x + 2)(x^2 + 4).$$

By factoring $x^2 + 4$ as

$$x^2 - (-4) = \left(x - \sqrt{-4}\right)\left(x + \sqrt{-4}\right) = (x - 2i)(x + 2i)$$

you obtain

$$f(x) = (x - 1)(x - 1)(x + 2)(x - 2i)(x + 2i)$$

which gives the following five zeros of f.

$$1, \quad 1, \quad -2, \quad 2i, \quad \text{and} \quad -2i$$

Note from the graph of f shown in Figure 4.22 that the *real* zeros are the only ones that appear as x-intercepts.

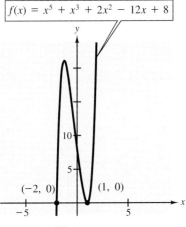

$f(x) = x^5 + x^3 + 2x^2 - 12x + 8$

FIGURE 4.22

Conjugate Pairs

In Example 2, note that the two complex zeros are **conjugates.** That is, they are of the form $a + bi$ and $a - bi$.

Complex Zeros Occur in Conjugate Pairs

Let $f(x)$ be a polynomial function that has real coefficients. If $a + bi$, where $b \neq 0$, is a zero of the function, then the conjugate $a - bi$ is also a zero of the function.

NOTE   Be sure you see that this result is true only if the polynomial function has *real coefficients*. For instance, the result applies to the function $f(x) = x^2 + 1$, but not to the function $g(x) = x - i$.

EXAMPLE 3   *Finding a Polynomial with Given Zeros*

Find a *fourth-degree* polynomial function, with real coefficients, that has $-1, -1$, and $3i$ as zeros.

Solution

Because $3i$ is a zero *and* the polynomial is stated to have real coefficients, you know that the conjugate $-3i$ must also be a zero. Thus, from the Linear Factorization Theorem, $f(x)$ can be written as

$$f(x) = a(x + 1)(x + 1)(x - 3i)(x + 3i).$$

For simplicity, let $a = 1$, and obtain

$$f(x) = (x^2 + 2x + 1)(x^2 + 9)$$
$$= x^4 + 2x^3 + 10x^2 + 18x + 9.$$

Factoring a Polynomial

The Linear Factorization Theorem shows that you can write any nth-degree polynomial as the product of n linear factors.

$$f(x) = a(x - c_1)(x - c_2)(x - c_3) \cdots (x - c_n)$$

However, this result includes the possibility that some of the values of c_i are complex. The following result implies that even if you do not want to get involved with "complex factors," you can still write $f(x)$ as the product of linear and/or quadratic factors.

Factors of a Polynomial

Every polynomial of degree $n > 0$ with real coefficients can be written as the product of linear and quadratic factors with real coefficients, where the quadratic factors have no real zeros.

A quadratic factor with no real zeros is said to be **irreducible over the reals.** Be sure you see that this is not the same as being *irreducible over the rationals.* For example, the quadratic

$$x^2 + 1 = (x - i)(x + i)$$

is irreducible over the reals (and therefore over the rationals). On the other hand, the quadratic

$$x^2 - 2 = \left(x - \sqrt{2}\right)\left(x + \sqrt{2}\right)$$

is irreducible over the rationals, but it is *reducible* over the reals.

EXAMPLE 4   *Factoring a Polynomial*

Write the polynomial $f(x) = x^4 - x^2 - 20$

a. as the product of factors that are irreducible over the *rationals,*

b. as the product of linear factors and quadratic factors that are irreducible over the *reals,* and

c. in completely factored form.

Solution

a. Begin by factoring the polynomial into the product of two quadratic polynomials.

$$x^4 - x^2 - 20 = (x^2 - 5)(x^2 + 4)$$

Both of these factors are irreducible over the rationals.

b. By factoring over the reals, you have

$$x^4 - x^2 - 20 = \left(x + \sqrt{5}\right)\left(x - \sqrt{5}\right)(x^2 + 4)$$

where the quadratic factor is irreducible over the reals.

c. In completely factored form, you have

$$x^4 - x^2 - 20 = \left(x + \sqrt{5}\right)\left(x - \sqrt{5}\right)(x - 2i)(x + 2i).$$

EXAMPLE 5   *Finding the Zeros of a Polynomial Function*

Find all the zeros of $f(x) = x^4 - 3x^3 + 6x^2 + 2x - 60$, given that $1 + 3i$ is a zero of f.

Solution

Because complex zeros occur in conjugate pairs, you know that $1 - 3i$ is also a zero of f. This means that both

$$[x - (1 + 3i)] \qquad \text{and} \qquad [x - (1 - 3i)]$$

are factors of $f(x)$. Multiplying these two factors produces

$$[x - (1 + 3i)][x - (1 - 3i)] = [(x - 1) - 3i][(x - 1) + 3i]$$
$$= (x - 1)^2 - 9i^2$$
$$= x^2 - 2x + 10.$$

Using long division, you can divide $x^2 - 2x + 10$ into $f(x)$ to obtain the following.

$$
\begin{array}{r}
x^2 - x - 6 \\
x^2 - 2x + 10 \overline{)\,x^4 - 3x^3 + 6x^2 + 2x - 60} \\
\underline{x^4 - 2x^3 + 10x^2} \\
-x^3 - 4x^2 + 2x \\
\underline{-x^3 + 2x^2 - 10x} \\
-6x^2 + 12x - 60 \\
\underline{-6x^2 + 12x - 60} \\
0
\end{array}
$$

Therefore, you have

$$f(x) = (x^2 - 2x + 10)(x^2 - x - 6) = (x^2 - 2x + 10)(x - 3)(x + 2)$$

and you can conclude that the zeros of f are $1 + 3i$, $1 - 3i$, 3, and -2.

Group Activities   **R e v i e w i n g   t h e   M a j o r   C o n c e p t s**

Factoring a Polynomial   Compile a list of all the various techniques for factoring a polynomial that have been covered thus far in the text. Give an example illustrating each technique, and discuss when the use of each technique is appropriate.

Warm Up The following warm-up exercises involve skills that were covered in earlier sections. You will use these skills in the exercise set for this section.

In Exercises 1–4, write the complex number in standard form and give its complex conjugate.

1. $4 - \sqrt{-29}$

2. $-5 - \sqrt{-144}$

3. $-1 + \sqrt{-32}$

4. $6 + \sqrt{-1/4}$

In Exercises 5–10, perform the indicated operations and write your answer in standard form.

5. $(-3 + 6i) - (10 - 3i)$

6. $(12 - 4i) + 20i$

7. $(4 - 2i)(3 + 7i)$

8. $(2 - 5i)(2 + 5i)$

9. $\dfrac{1 + i}{1 - i}$

10. $(3 + 2i)^3$

4.5 Exercises

In Exercises 1–26, find all the zeros of the function and write the polynomial as a product of linear factors.

1. $f(x) = x^2 + 25$

2. $f(x) = x^2 - x + 56$

3. $h(x) = x^2 - 4x + 1$

4. $g(x) = x^2 + 10x + 23$

5. $f(x) = x^4 - 81$

6. $f(y) = y^4 - 625$

7. $f(z) = z^2 - 2z + 2$

8. $h(x) = x^3 - 3x^2 + 4x - 2$

9. $g(x) = x^3 - 6x^2 + 13x - 10$

10. $f(x) = x^3 - 2x^2 - 11x + 52$

11. $f(t) = t^3 - 3t^2 - 15t + 125$

12. $f(x) = x^3 + 11x^2 + 39x + 29$

13. $f(x) = x^3 + 24x^2 + 214x + 740$

14. $f(s) = 2s^3 - 5s^2 + 12s - 5$

15. $f(x) = 16x^3 - 20x^2 - 4x + 15$

16. $f(x) = 9x^3 - 15x^2 + 11x - 5$

17. $h(x) = x^3 - x + 6$

18. $h(x) = x^3 + 9x^2 + 27x + 35$

19. $f(x) = 5x^3 - 9x^2 + 28x + 6$

20. $g(x) = 3x^3 - 4x^2 + 8x + 8$

21. $g(x) = x^4 - 4x^3 + 8x^2 - 16x + 16$

22. $h(x) = x^4 + 6x^3 + 10x^2 + 6x + 9$

23. $f(x) = x^4 + 10x^2 + 9$

24. $f(x) = x^4 + 29x^2 + 100$

25. $f(x) = 2x^4 + 5x^3 + 4x^2 + 5x + 2$

26. $g(x) = x^5 - 8x^4 + 28x^3 - 56x^2 + 64x - 32$

In Exercises 27–36, find a polynomial with integer coefficients that has the given zeros.

27. $1, 5i, -5i$

28. $4, 3i, -3i$

29. $2, 4 + i, 4 - i$

30. $6, -5 + 2i, -5 - 2i$

31. $i, -i, 6i, -6i$

32. $2, 2, 2, 4i, -4i$

33. $-5, -5, 1 + \sqrt{3}\, i$

34. $\frac{2}{3}, -1, 3 + \sqrt{2}\, i$

35. $\frac{3}{4}, -2, -\frac{1}{2} + i$

36. $0, 0, 4, 1 + i$

In Exercises 37–40, write the polynomial (a) as the product of factors that are irreducible over the *rationals*, (b) as the product of linear and quadratic factors that are irreducible over the *reals*, and (c) in completely factored form.

37. $f(x) = x^4 + 6x^2 - 27$

38. $f(x) = x^4 - 2x^3 - 3x^2 + 12x - 18$

(*Hint:* One factor is $x^2 - 6$.)

39. $f(x) = x^4 - 4x^3 + 5x^2 - 2x - 6$

(*Hint:* One factor is $x^2 - 2x - 2$.)

40. $f(x) = x^4 - 3x^3 - x^2 - 12x - 20$

(*Hint:* One factor is $x^2 + 4$.)

In Exercises 41–50, use the given zero of f to find all the zeros of f.

41. $f(x) = 2x^3 + 3x^2 + 50x + 75$, $5i$

42. $f(x) = x^3 + x^2 + 9x + 9$, $3i$

43. $f(x) = 2x^4 - x^3 + 7x^2 - 4x - 4$, $2i$

44. $f(x) = x^3 - 7x^2 - x + 87$, $5 + 2i$

45. $f(x) = 4x^3 + 23x^2 + 34x - 10$, $-3 + i$

46. $f(x) = 3x^3 - 4x^2 + 8x + 8$, $1 - \sqrt{3}\, i$

47. $f(x) = x^4 + 3x^3 - 5x^2 - 21x + 22$, $-3 + \sqrt{2}\, i$

48. $f(x) = x^3 + 4x^2 + 14x + 20$, $-1 - 3i$

49. $f(x) = 8x^3 - 14x^2 + 18x - 9$, $\frac{1}{2}\left(1 - \sqrt{5}\, i\right)$

50. $f(x) = 25x^3 - 55x^2 - 54x - 18$, $\frac{1}{5}\left(-2 + \sqrt{2}\, i\right)$

51. *Profit* The demand and cost equations for a product are $p = 140 - 0.0001x$ and $C = 80x + 150{,}000$, where p is the unit price, C is the total cost, and x is the number of units produced. The total profit obtained by producing and selling x units is given by

$$P = R - C = xp - C.$$

Determine a price p that would yield a profit of $9 million. Use a graphing utility to explain why this is not possible.

52. *Revenue* The demand equation for a certain product is given by $p = 140 - 0.0001x$, where p is the unit price (in dollars) of the product and x is the number of units produced. The total revenue obtained by producing and selling x units is given by

$$R = xp.$$

Determine a price p that would yield a revenue of $50 million. Use a graphing utility to explain why this is not possible.

53. *Reasoning* The complex number $2i$ is a zero of $f(x) = x^3 - 2ix^2 - 4x + 8i$, but the complex conjugate of $2i$ is not a zero of $f(x)$. Is this a contradiction of the conjugate pair statement on page 361? Explain.

54. *Reasoning* The complex number $1 - 2i$ is a zero of $f(x) = x^3 - (1 - 2i)x^2 - 9x + 9(1 - 2i)$, but $1 + 2i$ is not a zero of $f(x)$. Is this a contradiction of the conjugate pair statement on page 361? Explain.

55. *Think About It* Another student claims that the polynomial $f(x) = x^4 - 7x^2 + 12$ may be factored over the rational numbers as $f(x) = \left(x - \sqrt{3}\right)\left(x + \sqrt{3}\right)(x - 2)(x + 2)$. Do you agree with this claim? Explain your answer.

56. *Think About It* Another student claims that any polynomial with real coefficients can be factored over the real numbers. Give a counterexample to this claim.

CHAPTER PROJECT: Glucose Tolerance Test

In 1992, it was estimated that 11 million people in the United States had diabetes. Of those, about half were not diagnosed.

Diabetes is one of the leading causes of death in the United States. There are two basic types of diabetes. The first is caused by a lack of insulin and represents about 20% of all diabetes cases. The second type is not related to insulin production and is more common among people over the age of 40.

Diabetes affects a person's ability to use glucose (or sugar)—hence, sugar tends to build up in the blood. A test that is used to diagnose diabetes is the Glucose Tolerance Test. With this test, a person is given a dose of sugar water and the person's blood sugar level is monitored over several hours. The partial results for a person who tested positive for diabetes is shown in the table below. In the table, the time t is listed in hours and the blood-sugar level s is listed in milligrams per deciliter.

Time, t	0	0.5	1	2	3	4	5
Sugar, s	95	175	160	120	92	80	90

Use this information to investigate the questions below.

1. Fitting a Model to Data   Enter the data given in the table in a graphing utility.

(a) Use the graphing utility's cubic regression program to fit a cubic model to the data.

(b) Graphically compare the cubic model with a scatter plot of the data.

(c) Which characteristic of cubic polynomials is most responsible for making them a poor fit for this data?

2. Comparing a Model to Data   A rational model that fits the data in the table is

$$s = \frac{62.962t^2 - 366.46t - 96.06}{0.227t^3 - 1.143t^2 - 0.55t - 1}.$$

(a) Create a table that numerically compares the values of this model with the actual data. How well does the model fit the data?

(b) Graphically compare this model with the scatter plot of the actual data.

Chapter Projects provide excellent opportunities to put data analysis skills to work. For instance, this project asks the student to fit models to data, compare a model to actual data, critique several different models, and use a model for prediction purposes.

3. Using a Model   Use the model in Question 2 to estimate the times when the person's blood sugar level was 120 milligrams per deciliter. During the 5-hour test, how many times did this occur?

4. Cubic Splines   Break the given data table into two tables: the data for 0–2 hours and the data for 2–5 hours. Then use a graphing utility to fit a cubic model to each of the data sets. Compare the resulting models (which are called *cubic splines*) with the actual data. What can you conclude?

CHAPTER SUMMARY

After studying this chapter, you should have acquired the following skills. These skills are keyed to the Review Exercises that begin on page 368. Answers to odd-numbered Review Exercises are given in the back of the book.

- Divide polynomials using long division. *(Section 4.1)*   **Review Exercises 1, 2**

- Divide polynomials using synthetic division. *(Section 4.1)*   **Review Exercises 3, 4, 15, 16**

- Factor a polynomial completely using synthetic division. *(Section 4.1)*   **Review Exercises 5, 6**

- Evaluate a function using synthetic division. *(Section 4.1)*   **Review Exercises 7, 8**

- Simplify a rational expression using division. *(Section 4.1)*   **Review Exercises 9, 10**

- Determine the number of positive and negative zeros of a function using Descartes's Rule of Signs. *(Section 4.2)*   **Review Exercises 11, 12**

- List all possible rational zeros of a function using the Rational Zero Test. *(Section 4.2)*   **Review Exercises 13, 14**

- Find all real zeros of a function. *(Section 4.2)*   **Review Exercises 17–22**

- Match an equation with the number of rational and irrational zeros in its solution set. *(Section 4.2)*   **Review Exercises 23–26**

- Approximate the zeros of a function with a graphing utility. *(Section 4.3)*   **Review Exercises 27, 28**

- Approximate the zeros of a function to solve a real-life application. *(Section 4.3)*   **Review Exercises 29, 30**

- Write a complex number in standard form and find its complex conjugate. *(Section 4.4)*   **Review Exercises 31–34**

- Add, subtract, multiply, and divide complex numbers and write the results in standard form. *(Section 4.4)*   **Review Exercises 35–50**

- Solve a polynomial equation. *(Section 4.4)*   **Review Exercises 51–54**

- Plot a complex number in the complex plane. *(Section 4.4)*   **Review Exercises 55, 56**

- Write a polynomial as the product of linear factors. *(Section 4.5)*   **Review Exercises 57–62**

- Write a polynomial with integer coefficients given the zeros of the polynomial. *(Section 4.5)*   **Review Exercises 63, 64**

- Write a polynomial as the product of rational factors, as the product of real factors, and in completely factored form. *(Section 4.5)*   **Review Exercises 65, 66**

- Find all complex and real zeros of a function. *(Section 4.5)*   **Review Exercises 67–70**

REVIEW EXERCISES

In Exercises 1 and 2, divide by long division.

Dividend	*Divisor*
1. $12x^2 + 5x - 2$	$4x - 1$
2. $3x^3 + 5x^2 + 3x + 5$	$x^2 + 1$

In Exercises 3 and 4, divide by synthetic division.

Dividend	*Divisor*
3. $2x^3 - x^2 - 22x + 21$	$x - 1$
4. $x^5 - x^4 + x^3 - 13x^2 + x + 6$	$x + 2$

In Exercises 5 and 6, use synthetic division to show that x is a solution of the equation. Then factor the polynomial completely.

5. $x^3 - 4x^2 - 11x + 30 = 0, \ x = 5$

6. $3x^3 + 23x^2 + 37x - 15 = 0, \ x = \frac{1}{3}$

In Exercises 7 and 8, use synthetic division to evaluate the function.

7. $f(x) = 5x^3 - 10x + 7,$ (a) $f(1)$ (b) $f(-3)$

8. $f(x) = 2x^4 + 3x^3 + 6,$ (a) $f\left(\frac{1}{2}\right)$ (b) $f(-1)$

In Exercises 9 and 10, simplify the expression.

9. $\dfrac{x^3 + 9x^2 + 2x - 48}{x - 2}$ **10.** $\dfrac{x^4 + 5x^3 - 20x - 16}{x^2 - 4}$

In Exercises 11 and 12, use Descartes's Rule of Signs to determine the possible number of positive and negative zeros of the function. Verify using a graphing utility.

11. $g(x) = 5x^3 + 3x^2 - 6x + 9$

12. $h(x) = -2x^5 + 4x^3 - 2x^2 + 5$

In Exercises 13 and 14, use the Rational Zero Test to list all possible rational zeros of f. Verify using a graphing utility.

13. $f(x) = -4x^3 + 8x^2 - 3x + 15$

14. $f(x) = 3x^4 + 4x^3 - 5x^2 + 10x - 8$

15. *Dimensions of a Room* A rectangular room has a volume of

$$x^3 + 13x^2 + 50x + 56$$

cubic feet. The height of the room is $x + 2$. Find the number of square feet of floor space in the room.

16. *Profit* The profit for a product is given by

$$P = -130x^3 + 6500x^2 - 200,000, \qquad 0 \le x \le 50,$$

where P is the profit (in dollars) and x is the advertising expense (in tens of thousands of dollars). For this product, the advertising expense was \$400,000 ($x = 40$), and the profit was \$1,880,000. Use a graphing utility to graph the function and use the result to find another advertising expense that would have produced the same profit.

In Exercises 17–22, find all real zeros of the function.

17. $f(x) = x^3 + 2x^2 - 11x - 12$

18. $g(x) = 2x^3 - 7x^2 - 7x + 12$

19. $h(x) = 3x^4 - 27x^2 + 60$

20. $f(x) = x^5 - 4x^3 + 3x$

21. $C(x) = 3x^4 + 3x^3 - 7x^2 - x + 2$

22. $p(x) = x^4 - x^3 - 2x - 4$

In Exercises 23–26, match the equation with the number of rational and irrational zeros given in (a), (b), (c), and (d).

(a) Rational zeros: 1
   Irrational zeros: 2
(b) Rational zeros: 0
   Irrational zeros: 1
(c) Rational zeros: 3
   Irrational zeros: 0
(d) Rational zeros: 1
   Irrational zeros: 0

23. $f(x) = x^3 + 1$ **24.** $f(x) = x^3 - 3x$

25. $f(x) = x^3 - 3$ **26.** $f(x) = x^3 - 4x$

In Exercises 27 and 28, approximate the real zeros of f using a graphing utility. Use an accuracy of 0.001.

27. $f(x) = 5x^3 - 11x - 3$

28. $f(x) = 2x^4 - 9x^3 - 5x^2 + 10x + 12$

29. *Advertising Costs* A company that manufactures motorcycles estimates that the profit from selling the top-of-the-line model is given by

$$P = -35x^3 + 2000x^2 - 27,500, \quad 10 \le x \le 55$$

where P is the profit (in dollars) and x is the advertising expense (in tens of thousands of dollars). According to this model, how much money should the company spend to obtain a profit of $538,000?

30. *Age of the Groom* The average age of the groom in a marriage for a given age of the bride can be approximated by the model

$$y = -0.00428x^2 + 1.442x - 3.136, \quad 20 \le x \le 55$$

where y is the groom's age and x is the bride's age. For what age of the bride is the average age of the groom 30? (Source: U.S. National Center for Health Statistics)

In Exercises 31–34, write the complex number in standard form and find its complex conjugate.

31. $5 - \sqrt{-9}$

32. $3 + \sqrt{-12}$

33. $\sqrt{-24}$

34. 32

In Exercises 35–50, perform the indicated operation and write the result in standard form.

35. $(6 + i) + (3 - 5i)$

36. $(11 - 3i) - (5 - 8i)$

37. $\left(3 + \sqrt{-20}\right)\left(5 + \sqrt{-10}\right)$

38. $\left(5 + \sqrt{-9}\right) + \left(-3 - \sqrt{-25}\right)$

39. $(5 + 8i)(5 - 8i)$

40. $\left(\frac{1}{2} + \frac{3}{4}i\right)\left(\frac{1}{2} - \frac{3}{4}i\right)$

41. $-2i(4 - 5i)$

42. $-3(-2 + 4i)$

43. $(3 + 4i)^2$

44. $(2 - 5i)^2$

45. $\dfrac{3 + i}{3 - i}$

46. $\dfrac{6}{2 - 3i}$

47. $\dfrac{4 - 3i}{i}$

48. $\dfrac{2}{(1 + i)^2}$

49. $(3 + 2i)^2 + (3 - 2i)^2$

50. $(1 + i)^2 - (1 - i)^2$

In Exercises 51–54, solve the equation.

51. $3x^2 - 2x + 9 = 0$

52. $x^2 + 4x + 7 = 0$

53. $4x^2 + 11x + 3 = 0$

54. $9x^2 - 2x + 5 = 0$

In Exercises 55 and 56, plot the complex number.

55. $-3 + 2i$

56. $-1 - 4i$

In Exercises 57–62, write the polynomial as a product of linear factors.

57. $f(x) = x^4 - 625$

58. $h(x) = x^3 - 4x^2 + 2x - 8$

59. $f(t) = t^3 + 5t^2 + 3t + 15$

60. $h(y) = y^4 + 37y^2 + 36$

61. $g(x) = 4x^3 - 8x^2 + 9x - 18$

62. $f(x) = x^5 - 2x^4 + x^3 - x^2 + 2x - 1$

In Exercises 63 and 64, find a polynomial with integer coefficients that has the given zeros.

63. $3, 4i, -4i$

64. $2, -3, 1 - 2i, 1 + 2i$

In Exercises 65 and 66, write the polynomial (a) as the product of factors that are irreducible over the rationals, (b) as the product of linear and quadratic factors that are irreducible over the reals, and (c) in completely factored form.

65. $f(x) = x^4 + 5x^2 - 24$

66. $f(x) = x^4 - 3x^3 - 11x^2 + 15x + 30$

In Exercises 67–70, use the given zero of f to find all the zeros of f.

67. $f(x) = x^3 - 2x^2 + 9x - 18, \quad -3i$

68. $f(x) = x^3 + 3x^2 + 16x + 48, \quad 4i$

69. $f(x) = x^4 + 7x^3 + 24x^2 + 58x + 40, \quad 1 + 3i$

70. $f(x) = x^4 + 4x^3 + 8x^2 + 4x + 7, \quad 2 - \sqrt{3}\,i$

CHAPTER TEST

Take this test as you would take a test in class. After you are done, check your work against the answers given in the back of the book.

1. Use synthetic division to show that $x = \frac{3}{2}$ is a zero of $f(x) = 12x^3 + 8x^2 - 49x + 15$, and use the result to completely factor the polynomial.

2. Simplify: $\dfrac{x^4 + 4x^3 - 19x^2 - 106x - 120}{x^2 - 3x - 10}$.

3. List all possible rational zeros of $f(x) = 5x^4 - 3x^3 + 2x^2 + 11x + 12$.

4. Use Descartes's Rule of Signs to determine the possible number of positive and negative zeros of $h(x) = -3x^5 + 2x^4 - 4x^3 + 3x^2 - 7$.

In Exercises 5–12, perform the indicated operation and write the result in standard form.

5. $(12 + 3i) + (4 - 6i)$

6. $(10 - 2i) - (3 + 7i)$

7. $\left(5 + \sqrt{-12}\right)\left(3 - \sqrt{-12}\right)$

8. $(4 + 3i)(2 - 5i)$

9. $(2 - 3i)^2$

10. $(5 + 2i)^2$

11. $\dfrac{i + i}{1 - i}$

12. $\dfrac{5 - 2i}{i}$

In Exercises 13 and 14, find all real zeros of the function.

13. $f(x) = x^5 - 5x^3 + 4x$

14. $g(x) = x^4 + 2x^3 - 9x^2 - 2x + 8$

In Exercises 15 and 16, solve the quadratic equation.

15. $x^2 + 5x + 7 = 0$

16. $2x^2 - 5x + 11 = 0$

17. Find all zeros of $f(x) = x^3 + 2x^2 + 5x + 10$, given that $\sqrt{5}\,i$ is a zero.

18. Find a polynomial with integer coefficients that has $2, 5, 3i$, and $-3i$ as zeros.

19. Given that $-2 + \sqrt{3}\,i$ is a zero of $f(x) = x^4 + 4x^3 + 8x^2 + 4x + 7$, name another zero of f.

20. A company estimates that the profit from selling its product is

$$P = -11x^3 + 900x^2 - 50{,}000, \quad 0 \le x$$

where P is the profit (in dollars) and x is the advertising expense (in tens of thousands of dollars). How much should the company spend on advertising to obtain a profit of \$222,000? Explain your reasoning.

Exponential and Logarithmic Functions

5

- Exponential Functions
- Logarithmic Functions
- Properties of Logarithms
- Solving Exponential and Logarithmic Equations
- Exponential and Logarithmic Models

When a warm object is placed in a cool room, its temperature will gradually change to the temperature of the room. (The same is true for a cool object placed in a warm room.)

According to Newton's Law of Cooling, the rate at which the object's temperature T changes is proportional to the difference between its temperature and the temperature of the room L. This statement can be translated as

$$T = L + Ce^{kt}$$

where k is a constant and $L + C$ is the original temperature of the object.

The table and scatter plot at the right show the temperature T (in degrees Fahrenheit) for several times t (in seconds) for a cooling object.

t	0.2	8.4	16.6	24.8	33.0	41.1	49.3
T	155.8	133.2	117.9	107.9	100.7	94.9	90.5

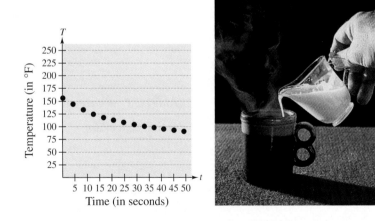

The chapter project related to this information is on page 425. The project uses actual data, collected with a *Texas Instruments CBL* unit.

5.1	**Exponential Functions**
	Exponential Functions ▪ Graphs of Exponential Functions ▪ The Natural Base e ▪ Compound Interest ▪ Another Application

Exponential Functions

So far, this text has dealt only with **algebraic functions,** which include polynomial functions and rational functions. In this chapter you will study two types of nonalgebraic functions—*exponential* functions and *logarithmic* functions. These functions are examples of **transcendental functions.**

NOTE The base $a = 1$ is excluded because it yields

$$f(x) = 1^x = 1.$$

This is a constant function, not an exponential function.

Definition of Exponential Function

The **exponential function** f **with base** a is denoted by

$$f(x) = a^x$$

where $a > 0$, $a \neq 1$, and x is any real number.

You already know how to evaluate a^x for integer and rational values of x. For example, you know that $4^3 = 64$ and $4^{1/2} = 2$. However, to evaluate 4^x for any real number x, you need to interpret forms with *irrational* exponents. For the purposes of this text, it is sufficient to think of

$$a^{\sqrt{2}} \qquad \left(\text{where } \sqrt{2} \approx 1.414214\right)$$

as that value having the successively closer approximations

$$a^{1.4}, \; a^{1.41}, \; a^{1.414}, \; a^{1.4142}, \; a^{1.41421}, \; a^{1.414214}, \ldots .$$

Example 1 shows how to use a calculator to evaluate an exponential function.

EXAMPLE 1   *Evaluating Exponential Expressions*

Scientific Calculator

Number	Keystrokes	Display
$2^{-\pi}$	2 $\boxed{y^x}$ π $\boxed{+/-}$ $\boxed{=}$	0.1133147

Graphing Calculator

Number	Keystrokes	Display
$2^{-\pi}$	2 $\boxed{\wedge}$ $\boxed{(-)}$ π $\boxed{\text{ENTER}}$	0.1133147

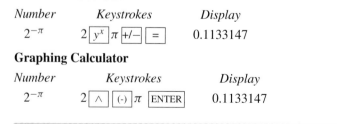

Graphs of Exponential Functions

The graphs of all exponential functions have similar characteristics, as shown in Examples 2, 3, and 4.

EXAMPLE 2   Graphs of $y = a^x$

In the same coordinate plane, sketch the graphs of the following functions.

a. $f(x) = 2^x$ **b.** $g(x) = 4^x$

Solution

The table below lists some values for each function, and Figure 5.1 shows their graphs. Note that both graphs are increasing. Moreover, the graph of $g(x) = 4^x$ is increasing more rapidly than the graph of $f(x) = 2^x$.

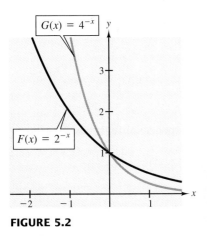

$g(x) = 4^x$

$f(x) = 2^x$

FIGURE 5.1

x	-2	-1	0	1	2	3
2^x	$\frac{1}{4}$	$\frac{1}{2}$	1	2	4	8
4^x	$\frac{1}{16}$	$\frac{1}{4}$	1	4	16	64

EXAMPLE 3   Graphs of $y = a^{-x}$

In the same coordinate plane, sketch the graphs of the following functions.

a. $F(x) = 2^{-x}$ **b.** $G(x) = 4^{-x}$

Solution

The table below lists some values for each function, and Figure 5.2 shows their graphs. Note that both graphs are decreasing. Moreover, the graph of $G(x) = 4^{-x}$ is decreasing more rapidly than the graph of $F(x) = 2^{-x}$.

$G(x) = 4^{-x}$

$F(x) = 2^{-x}$

FIGURE 5.2

x	-3	-2	-1	0	1	2
2^{-x}	8	4	2	1	$\frac{1}{2}$	$\frac{1}{4}$
4^{-x}	64	16	4	1	$\frac{1}{4}$	$\frac{1}{16}$

NOTE   The tables in Examples 2 and 3 were evaluated by hand. You could, of course, use a calculator to construct tables with even more values.

Comparing the functions in Examples 2 and 3, observe that

$$F(x) = 2^{-x} = f(-x) \qquad \text{and} \qquad G(x) = 4^{-x} = g(-x).$$

Consequently, the graph of F is a reflection (in the y-axis) of the graph of f. The graphs of G and g have the same relationship.

The graphs in Figures 5.1 and 5.2 are typical of the exponential functions a^x and a^{-x}. They have one y-intercept and one horizontal asymptote (the x-axis), and they are continuous. The basic characteristics of these exponential functions are summarized in Figure 5.3.

Graph of $y = a^x$
- Domain: $(-\infty, \infty)$
- Range: $(0, \infty)$
- Intercept: $(0, 1)$
- Increasing
- x-axis is a horizontal asymptote
  ($a^x \to 0$ as $x \to -\infty$)
- Continuous

Graph of $y = a^{-x}$
- Domain: $(-\infty, \infty)$
- Range: $(0, \infty)$
- Intercept: $(0, 1)$
- Decreasing
- x-axis is a horizontal asymptote
  ($a^{-x} \to 0$ as $x \to \infty$)
- Continuous
- Reflection of graph of
  $y = a^x$ about y-axis

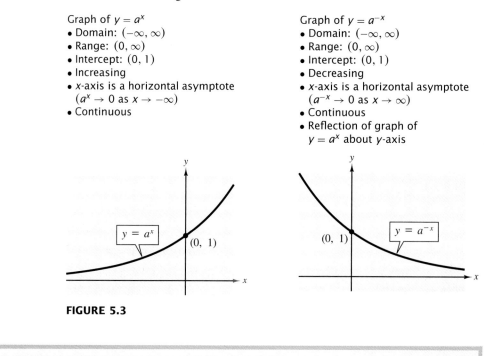

FIGURE 5.3

Technology

Try using a graphing utility to graph several exponential functions. For instance, the graph of $y = 2^x$ is shown below.

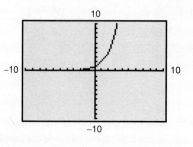

In the following example, notice how the graph of $y = a^x$ can be used to sketch the graphs of functions of the form $f(x) = b \pm a^{x+c}$.

EXAMPLE 4   Sketching Graphs of Exponential Functions

Each of the following graphs is a transformation of the graph of $f(x) = 3^x$, as shown in Figure 5.4.

a. Because $g(x) = 3^{x+1} = f(x+1)$, the graph of g can be obtained by shifting the graph of f one unit to the left.

b. Because $h(x) = 3^x - 2 = f(x) - 2$, the graph of h can be obtained by shifting the graph of f down two units.

c. Because $k(x) = -3^x = -f(x)$, the graph of k can be obtained by reflecting the graph of f in the x-axis.

d. Because $j(x) = 3^{-x} = f(-x)$, the graph of j can be obtained by reflecting the graph of f in the y-axis.

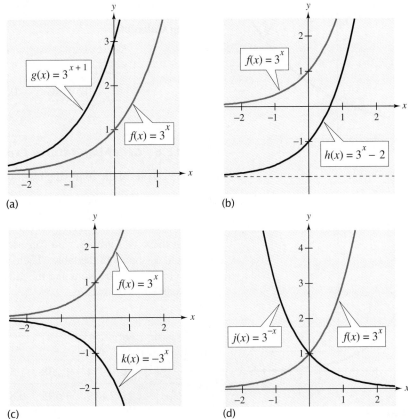

(a)   (b)   (c)   (d)

FIGURE 5.4

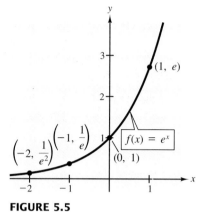

FIGURE 5.5

The Natural Base e

In many applications, the most convenient choice for a base is the irrational number

$$e \approx 2.71828\ldots$$

called the **natural base.** The function $f(x) = e^x$ is called the **natural exponential function.** Its graph is shown in Figure 5.5. Be sure you see that for the exponential function $f(x) = e^x$, e is the constant $2.71828\ldots$, whereas x is the variable.

EXAMPLE 5   Evaluating the Natural Exponential Function

Scientific Calculator

Number	Keystrokes	Display
e^2	2 $\boxed{e^x}$ $\boxed{=}$	7.3890561
e^{-1}	1 $\boxed{+/-}$ $\boxed{e^x}$ $\boxed{=}$	0.3678794

Graphing Calculator

Number	Keystrokes	Display
e^2	$\boxed{e^x}$ 2 $\boxed{\text{ENTER}}$	7.3890561
e^{-1}	$\boxed{e^x}$ $\boxed{(-)}$ 1 $\boxed{\text{ENTER}}$	0.3678794

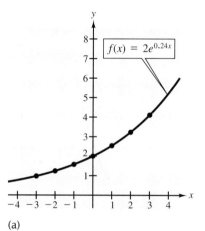

(a)

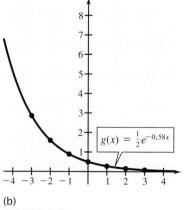

(b)

FIGURE 5.6

EXAMPLE 6   Graphing Natural Exponential Functions

Sketch the graphs of the following natural exponential functions.

a. $f(x) = 2e^{0.24x}$ **b.** $g(x) = \frac{1}{2}e^{-0.58x}$

Solution

To sketch these two graphs, you can use a calculator to plot several points on each graph, as shown in the table. Then, connect the points with smooth curves, as shown in Figure 5.6. Note that the graph in part (a) is increasing whereas the graph in part (b) is decreasing.

x	-3	-2	-1	0	1	2	3
$f(x) = 2e^{0.24x}$	0.974	1.238	1.573	2.000	2.543	3.232	4.109
$g(x) = \frac{1}{2}e^{-0.58x}$	2.849	1.595	0.893	0.500	0.280	0.157	0.088

Compound Interest

One of the most familiar examples of exponential growth is that of an investment earning *continuously compounded interest*. In Section P.3, you were introduced to the formula for the balance in an account that is compounded n times per year. Using exponential functions, you can now *develop* that formula and show how it leads to continuous compounding.

Suppose a principal P is invested at an annual percentage rate r, compounded once a year. If the interest is added to the principal at the end of the year, the balance is

$$P_1 = P + Pr = P(1 + r).$$

This pattern of multiplying the previous principal by $1 + r$ is then repeated each successive year, as shown below.

Year	Balance After Each Compounding
0	$P = P$
1	$P_1 = P(1 + r)$
2	$P_2 = P_1(1 + r) = P(1 + r)(1 + r) = P(1 + r)^2$
3	$P_3 = P_2(1 + r) = P(1 + r)^2(1 + r) = P(1 + r)^3$
$\vdots$	
t	$P_t = P(1 + r)^t$

To accommodate more frequent (quarterly, monthly, or daily) compounding of interest, let n be the number of compoundings per year and t the number of years. Then the rate per compounding is r/n and the account balance after t years is

$$A = P\left(1 + \frac{r}{n}\right)^{nt}. \qquad \text{Amount with } n \text{ compoundings}$$

If you let the number of compoundings n increase without bound, the process approaches what is called **continuous compounding.** In the formula for n compoundings per year, let $m = n/r$. This produces

$$A = P\left(1 + \frac{r}{n}\right)^{nt}$$

$$= P\left(1 + \frac{1}{m}\right)^{mrt}$$

$$= P\left[\left(1 + \frac{1}{m}\right)^m\right]^{rt}.$$

As m increases without bound, it can be shown that $[1 + (1/m)]^m$ approaches e. From this, you can conclude that the formula for continuous compounding is $A = Pe^{rt}$.

DISCOVERY

Use a calculator and the formula $A = P(1 + r/n)^{nt}$ to calculate the amount in an account when $P = \$3000$, $r = 6\%$, t is 10 years, and the number of compoundings is (1) by the day, (2) by the hour, (3) by the minute, and (4) by the second. Use these results to present an argument that increasing the number of compoundings does not mean unlimited growth of the amount in the account.

Formulas for Compound Interest

After t years, the balance A in an account with principal P and annual percentage rate r (in decimal form) is given by the following formulas.

1. For n compoundings per year: $A = P \left(1 + \dfrac{r}{n} \right)^{nt}$

2. For continuous compounding: $A = Pe^{rt}$

NOTE Be sure you see that the annual percentage rate must be expressed in decimal form. For instance, 6% should be expressed as 0.06.

Real Life

EXAMPLE 7 *Compounding n Times and Continuously*

A total of $12,000 is invested at an annual rate of 9%. Find the balance after 5 years if it is compounded

a. quarterly.   **b.** continuously.

Solution

a. For quarterly compoundings, you have $n = 4$. Thus, in 5 years at 9%, the balance is

$$A = P \left(1 + \frac{r}{n} \right)^{nt}$$ Formula for compound interest

$$= 12,000 \left(1 + \frac{0.09}{4} \right)^{4(5)}$$ Substitute for P, r, n, and t.

$$= \$18,726.11.$$ Use a calculator.

b. Compounding continuously, the balance is

$$A = Pe^{rt}$$ Formula for continuous compounding

$$= 12,000e^{0.09(5)}$$ Substitute for P, r, and t.

$$= \$18,819.75.$$ Use a calculator.

Note that continuous compounding yields

$$\$18,819.75 - \$18,726.11 = \$93.64$$

more than quarterly compounding. This is typical of the two types of compounding. That is, for a given principal, interest rate, and time, continuous compounding will always yield a larger balance than compounding n times a year.

Activities

1. Sketch the graph of the functions $f(x) = e^x$ and $g(x) = 1 + e^x$.

2. Determine the balance A at the end of 20 years if $1500 is invested at 6.5% interest and the interest is compounded (a) quarterly and (b) continuously.
   Answer: (a) $5446.73
   (b) $5503.95

3. Determine the amount of money that should be invested at 9% interest, compounded monthly, to produce a final balance of $30,000 in 15 years.
   Answer: $7816.48

Another Application

EXAMPLE 8   Radioactive Decay

In 1957, a nuclear reactor accident occurred in the Soviet Union. The explosion spread radioactive chemicals over thousands of square miles, and the government set aside a large region as a "permanent preserve." To see why this area was permanently declared off-limits to people, consider the following model.

$$P = 10e^{-0.00002845t}$$

This model represents the amount of plutonium that remains (from an initial amount of 10 pounds) after t years. Sketch the graph of this function over the interval from $t = 0$ to $t = 100,000$. How much of the 10 pounds will remain after 100,000 years?

Solution

The graph of this function is shown in Figure 5.7. Note from this graph that plutonium has a *half-life* of about 24,360 years. That is, after 24,360 years, *half* of the original amount will remain. After another 24,360 years, one-quarter of the original amount will remain, and so on. After 100,000 years, there will still be

$$P = 10e^{-0.00002845(100,000)} = 10e^{-2.845} \approx 0.58 \text{ pounds}$$

of the original amount of plutonium remaining.

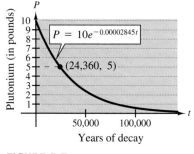

FIGURE 5.7

Group Activities

Extending the Concept

Exponential Growth   The sequence 3, 6, 9, 12, 15, . . . is given by $f(n) = 3n$ and is an example of **linear growth**. The sequence 3, 9, 27, 81, 243, . . . is given by $f(n) = 3^n$ and is an example of **exponential growth**. Explain the difference between these two types of growth. For each of the following sequences, indicate whether the sequence represents linear growth or exponential growth, and find a linear or exponential function that represents the sequence. Give several other examples of linear and exponential growth.

a. $\frac{1}{2}, \frac{1}{4}, \frac{1}{8}, \frac{1}{16}, \frac{1}{32}, \ldots$ **b.** $4, 8, 12, 16, 20, \ldots$

c. $\frac{2}{3}, \frac{4}{3}, 2, \frac{8}{3}, \frac{10}{3}, 4, \ldots$ **d.** $5, 25, 125, 625, \ldots$

Warm Up   The following warm-up exercises involve skills that were covered in earlier sections. You will use these skills in the exercise set for this section.

In Exercises 1–10, use the properties of exponents to simplify the expression.

1. $5^{2x}(5^{-x})$

2. $3^{-x}(3^{3x})$

3. $\dfrac{4^{5x}}{4^{2x}}$

4. $\dfrac{10^{2x}}{10^{x}}$

5. $(4^{x})^{2}$

6. $(4^{2x})^{5}$

7. $\left(\dfrac{2^{x}}{3^{x}}\right)^{-1}$

8. $(4^{6x})^{1/2}$

9. $(2^{3x})^{-1/3}$

10. $(16^{x})^{1/4}$

5.1 Exercises

In Exercises 1–14, use a calculator to evaluate the expression. (Round the result to three decimal places.)

1. $(3.4)^{5.6}$

2. $(1.005)^{400}$

3. $1000(1.06)^{-5}$

4. $5000(2^{-1.5})$

5. $\sqrt[4]{763}$

6. $\sqrt[3]{4395}$

7. $8^{2\pi}$

8. $5^{-\pi}$

9. $100^{\sqrt{2}}$

10. $0.6^{\sqrt{3}}$

11. e^{2}

12. $e^{1/2}$

13. $e^{-3/4}$

14. $e^{3.2}$

In Exercises 15–22, match the function with its graph. [The graphs are labeled (a), (b), (c), (d), (e), (f), (g), and (h).]

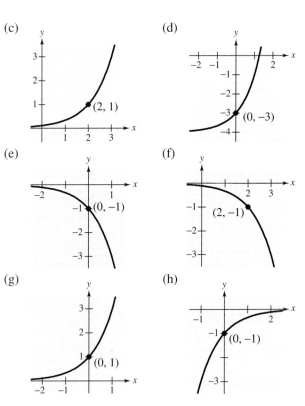

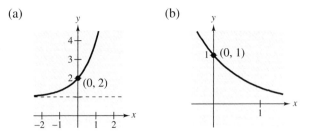

15. $f(x) = 3^x$

16. $f(x) = -3^x$

17. $f(x) = 3^{-x}$

18. $f(x) = -3^{-x}$

19. $f(x) = 3^x - 4$

20. $f(x) = 3^x + 1$

21. $f(x) = -3^{x-2}$

22. $f(x) = 3^{x-2}$

In Exercises 23–40, sketch the graph of the function. Use a graphing utility or make the sketch by hand.

23. $g(x) = 5^x$

24. $f(x) = \left(\frac{3}{2}\right)^x$

25. $f(x) = 5^{-x}$

26. $h(x) = \left(\frac{3}{2}\right)^{-x}$

27. $h(x) = 5^{x-2}$

28. $g(x) = \left(\frac{3}{2}\right)^{x+2}$

29. $g(x) = 5^{-x} - 3$

30. $f(x) = \left(\frac{3}{2}\right)^{-x} + 2$

31. $y = 2^{-x^2}$

32. $y = 3^{-x^2}$

33. $y = e^{-0.1x}$

34. $y = e^{0.2x}$

35. $f(x) = 2e^{0.12x}$

36. $f(x) = 3e^{-0.2x}$

37. $f(x) = e^{2x}$

38. $h(x) = e^{x-2}$

39. $g(x) = 1 + e^{-x}$

40. $N(t) = 1000e^{-0.2t}$

Compound Interest   In Exercises 41–44, complete the table to find the balance A for P dollars invested at rate r for t years, compounded n times per year.

n	1	2	4	12	365	Continuous
A						

41. $P = \$2500$, $r = 12\%$, $t = 10$ years

42. $P = \$1000$, $r = 10\%$, $t = 10$ years

43. $P = \$2500$, $r = 12\%$, $t = 20$ years

44. $P = \$1000$, $r = 10\%$, $t = 40$ years

Compound Interest   In Exercises 45–48, complete the table to find the amount P that must be invested at rate r to obtain a balance of $A = \$100,000$ in t years.

t	1	10	20	30	40	50
P						

45. $r = 9\%$, compounded continuously

46. $r = 12\%$, compounded continuously

47. $r = 10\%$, compounded monthly

48. $r = 7\%$, compounded daily

49. *Trust Fund*   You deposit $25,000 in a trust fund that pays 8.75% interest, compounded continuously, on the day that your grandchild is born. What will the balance of this account be on your grandchild's 25th birthday?

50. *Trust Fund*   You deposit $5000 in a trust fund that pays 7.5% interest, compounded continuously. In the trust fund, you specify that the balance of the account will be given to your college after the money has earned interest for 50 years. How much will your college receive after 50 years?

51. *Demand Function*   The demand function for a certain product is given by

$$p = 500 - 0.5e^{0.004x}.$$

(a) Find the price p for a demand of $x = 1000$ units.

(b) Find the price p for a demand of $x = 1500$ units.

(c) Use a graphing utility to graph the demand function.

(d) Does the graph indicate what a good price would be? Explain your reasoning.

52. *Demand Function*   The demand function for a certain product is given by

$$p = 5000\left(1 - \frac{4}{4 + e^{-0.002x}}\right).$$

(a) Find the price p for a demand of $x = 100$ units.

(b) Find the price p for a demand of $x = 500$ units.

(c) Use a graphing utility to graph the demand function.

(d) Does the graph indicate what a good price would be? Explain your reasoning.

53. *Bacteria Growth*   A certain type of bacteria increases according to the model

$$P(t) = 100e^{0.2197t}$$

where t is time in hours. Find (a) $P(0)$, (b) $P(5)$, and (c) $P(10)$.

54. *Population Growth*   The population of a town increases according to the model

$$P(t) = 2500e^{0.0293t}$$

where t is time in years, with $t = 0$ corresponding to 1990. Use the model to approximate the population in (a) 1995, (b) 2000, and (c) 2005.

55. *Radioactive Decay*   Five pounds of the element plutonium (Pu^{230}) is released in a nuclear accident. The amount of plutonium that is present after t years is given by

$$P = 5e^{-0.00002845t}.$$

(a) Use a graphing utility to graph this function over the interval from $t = 0$ to $t = 100,000$.

(b) How much will remain after 100,000 years?

(c) Use the graph to estimate the half-life of Pu^{230}. Explain your reasoning.

56. *Radioactive Decay*   One hundred grams of radium (Ra^{226}) is stored in a container. The amount of radium present after t years is given by

$$R = 100e^{-0.0004279t}.$$

(a) Use a graphing utility to graph this function over the interval from $t = 0$ to $t = 10,000$.

(b) How much of the 100 grams will remain after 10,000 years?

(c) Use the graph to estimate the half-life of Ra^{226}. Explain your reasoning.

57. *New York Stock Exchange*   The total number of shares of stocks *listed* on the New York Stock Exchange between 1945 and 1992 can be approximated by the function

$$y = 12.051e^{0.099327t}, \qquad 45 \leq t \leq 92$$

where y represents the number of listed shares (in millions) and $t = 45$ represents 1945. (Source: New York Stock Exchange)

(a) Use the graph at the top of the next column to *graphically* estimate the total number of shares listed in 1960, 1975, and 1992.

(b) Use the model to *algebraically* confirm the estimates obtained in part (a).

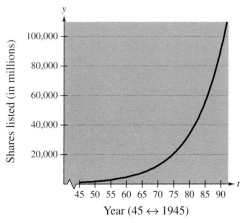

Figure for 57

58. *New York Stock Exchange*   The total number of shares of stocks *traded* on the New York Stock Exchange between 1940 and 1992 can be approximated by the function

$$y = 0.004719e^{0.11596t}, \qquad 40 \leq t \leq 92$$

where y represents the number of shares traded (in millions) and $t = 40$ represents 1940. (Source: New York Stock Exchange)

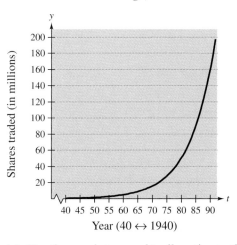

(a) Use the graph to *graphically* estimate the total number of shares traded in 1950, 1970, and 1992.

(b) Use the model to *algebraically* confirm the estimates obtained in part (a).

59. *Age at First Marriage* From 1970 to 1992, the average age of an American woman at her first marriage could be approximated by the model

$$A = 20.40 + \frac{1}{0.2163 + 2.351e^{-0.1824t}}, \quad 0 \le t \le 22$$

where A represents the average age and $t = 0$ represents 1970 (see figure). (Source: U. S. Bureau of the Census)

(a) Use the graph to estimate the average age of an American woman at her first marriage in 1970, 1980, and 1992.

(b) Algebraically confirm the results of part (a).

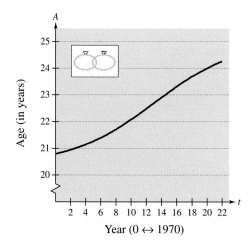

Year (0 ↔ 1970)

60. *Age at First Marriage* From 1970 to 1992, the average age of an American man at his first marriage could be approximated by the model

$$A = 21.26 + \frac{1}{0.1455 + 0.3706e^{-0.0938t}}, \quad 0 \le t \le 22$$

where A represents the average age and $t = 0$ represents 1970 (see figure). (Source: U. S. Bureau of the Census)

(a) Use the graph to estimate the average age of an American man at his first marriage in 1970, 1980, and 1992.

(b) Algebraically confirm the results of part (a).

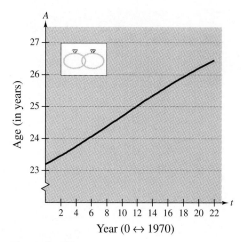

Year (0 ↔ 1970)

Figure for 60

61. *Writing* Compare the results of Exercises 59 and 60. What can you conclude about the differences in men's and women's ages at first marriages?

5.2	**Logarithmic Functions**

Logarithmic Functions ▪ Graphs of Logarithmic Functions ▪ The Natural Logarithmic Function ▪ Application

Logarithmic Functions

In Section 3.4, you studied the concept of the inverse of a function. There, you learned that if a function has the property that no horizontal line intersects the graph of a function more than once, the function must have an inverse. By looking back at the graphs of the exponential functions introduced in Section 5.1, you will see that every function of the form $f(x) = a^x$ passes the "horizontal line test," and therefore must have an inverse. This inverse function is called the **logarithmic function with base a.**

NOTE The equations

$$y = \log_a x \text{ and } x = a^y$$

are equivalent. The first equation is in logarithmic form and the second is in exponential form.

Definition of Logarithmic Function

For $x > 0$ and $0 < a \neq 1$,

$$y = \log_a x \text{ if and only if } x = a^y.$$

The function given by

$$f(x) = \log_a x$$

is called the **logarithmic function with base a.**

The logarithmic function is one of the most difficult concepts for students to learn. Capitalize on the fact—both graphically and algebraically—that the logarithmic function is the inverse of the exponential function. When explaining the logarithmic function, consider continually reminding your students that *a logarithm is an exponent*. Converting back and forth from logarithmic form to exponential form supports this concept. Also, encourage your students to become familiar with the three basic properties of logarithms and how they follow from the definition. To graph a logarithmic function, you may want your students to start with a table of values, but remind them also to use the graphing techniques learned in Sections 3.2 and 3.3.

When evaluating logarithms, remember that *a logarithm is an exponent*. This means that $\log_a x$ is the exponent to which a must be raised to obtain x. For instance, $\log_2 8 = 3$ because 2 must be raised to the third power to obtain 8.

EXAMPLE 1   Evaluating Logarithms

a. $\log_2 32 = 5$ because $2^5 = 32.$

b. $\log_3 27 = 3$ because $3^3 = 27.$

c. $\log_4 2 = \dfrac{1}{2}$ because $4^{1/2} = \sqrt{4} = 2.$

d. $\log_{10} \dfrac{1}{100} = -2$ because $10^{-2} = \dfrac{1}{10^2} = \dfrac{1}{100}.$

e. $\log_3 1 = 0$ because $3^0 = 1.$

f. $\log_2 2 = 1$ because $2^1 = 2.$

The logarithmic function with base 10 is called the **common logarithmic function.** On most calculators, this function is denoted by $\boxed{\text{log}}$.

EXAMPLE 2 *Evaluating Logarithms on a Calculator*

Scientific Calculator

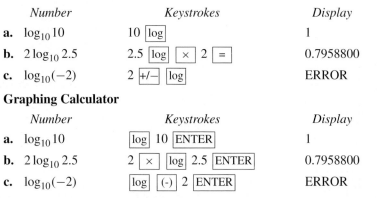

Number	Keystrokes	Display
a. $\log_{10} 10$	10 $\boxed{\text{log}}$	1
b. $2 \log_{10} 2.5$	2.5 $\boxed{\text{log}}$ $\boxed{\times}$ 2 $\boxed{=}$	0.7958800
c. $\log_{10}(-2)$	2 $\boxed{+/-}$ $\boxed{\text{log}}$	ERROR

Graphing Calculator

Number	Keystrokes	Display
a. $\log_{10} 10$	$\boxed{\text{log}}$ 10 $\boxed{\text{ENTER}}$	1
b. $2 \log_{10} 2.5$	2 $\boxed{\times}$ $\boxed{\text{log}}$ 2.5 $\boxed{\text{ENTER}}$	0.7958800
c. $\log_{10}(-2)$	$\boxed{\text{log}}$ $\boxed{(-)}$ 2 $\boxed{\text{ENTER}}$	ERROR

Note that the calculator displays an error message when you try to evaluate $\log_{10}(-2)$. The reason for this is that the domain of every logarithmic function is the set of *positive real numbers.*

The following properties follow directly from the definition of the logarithmic function with base a.

Properties of Logarithms

1. $\log_a 1 = 0$ because $a^0 = 1$.

2. $\log_a a = 1$ because $a^1 = a$.

3. $\log_a a^x = x$ because $a^x = a^x$.

4. If $\log_a x = \log_a y$, then $x = y$.

EXAMPLE 3 *Using Properties of Logarithms*

Solve the equation $\log_2 x = \log_2 3$ for x.

Solution

Using Property 4, you can conclude that $x = 3$.

STUDY TIP

Because $\log_a x$ is the inverse function of a^x, it follows that the domain of $\log_a x$ is the range of a^x, $(0, \infty)$. In other words, $\log_a x$ is defined only if x is positive.

Graphs of Logarithmic Functions

To sketch the graph of $y = \log_a x$, you can use the fact that the graphs of inverse functions are reflections of each other in the line $y = x$.

EXAMPLE 4   Graphs of Exponential and Logarithmic Functions

In the same coordinate plane, sketch the graphs of the following functions.

a. $f(x) = 2^x$ **b.** $g(x) = \log_2 x$

Solution

a. For $f(x) = 2^x$, construct a table of values, as follows.

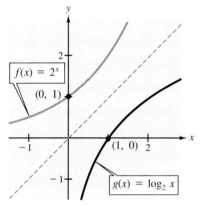

x	-2	-1	0	1	2	3
$f(x) = 2^x$	$\frac{1}{4}$	$\frac{1}{2}$	1	2	4	8

FIGURE 5.8   Inverse Functions

By plotting these points and connecting them with a smooth curve, you obtain the graph shown in Figure 5.8.

b. Because $g(x) = \log_2 x$ is the inverse of $f(x) = 2^x$, the graph of g is obtained by reflecting the graph of f in the line $y = x$, as shown in Figure 5.8.

Before you can confirm the result of Example 4 with a graphing utility, you need to know how to enter $\log_2 x$. A procedure for doing this is discussed in Section 5.3.

EXAMPLE 5   Sketching the Graph of a Logarithmic Function

Sketch the graph of the common logarithmic function $f(x) = \log_{10} x$.

Solution

Begin by constructing a table of values. Note that some of the values can be obtained without a calculator, whereas others require a calculator. Next, plot the points and connect them with a smooth curve, as shown in Figure 5.9.

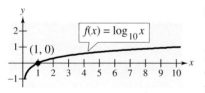

FIGURE 5.9

	Without Calculator				With Calculator		
x	$\frac{1}{100}$	$\frac{1}{10}$	1	10	2	5	8
$\log_{10} x$	-2	-1	0	1	0.301	0.699	0.903

The nature of the graph in Figure 5.9 is typical of functions of the form $f(x) = \log_a x$, $a > 1$. They have one x-intercept and one vertical asymptote. Notice how slowly the graph rises for $x > 1$. In Figure 5.9 you would need to move out to $x = 1000$ before the graph rose to $y = 3$. The basic characteristics of logarithmic graphs are summarized in Figure 5.10.

NOTE   In the graph at the right, note that the vertical asymptote occurs at $x = 0$, where $\log_a x$ is *undefined*.

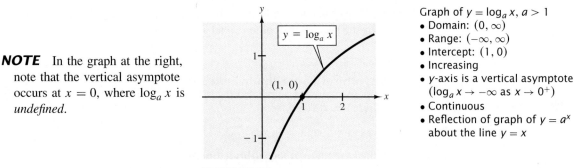

Graph of $y = \log_a x$, $a > 1$
- Domain: $(0, \infty)$
- Range: $(-\infty, \infty)$
- Intercept: $(1, 0)$
- Increasing
- y-axis is a vertical asymptote
  ($\log_a x \to -\infty$ as $x \to 0^+$)
- Continuous
- Reflection of graph of $y = a^x$
  about the line $y = x$

FIGURE 5.10

In the following example, the graph of $\log_a x$ is used to sketch the graphs of functions of the form $y = b \pm \log_a(x + c)$.

EXAMPLE 6   *Sketching the Graphs of Logarithmic Functions*

The graph of each of these functions is similar to the graph of $f(x) = \log_{10} x$, as shown in Figure 5.11.

a.   Because $g(x) = \log_{10}(x - 1) = f(x - 1)$, the graph of g can be obtained by shifting the graph of f one unit to the right.

b.   Because $h(x) = 2 + \log_{10} x = 2 + f(x)$, the graph of h can be obtained by shifting the graph of f two units up.

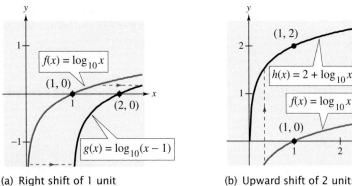
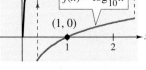

(a) Right shift of 1 unit
FIGURE 5.11

(b) Upward shift of 2 units

The Natural Logarithmic Function

As with exponential functions, the most widely used base for logarithmic functions is the number e, where

$$e \approx 2.718281828. \ldots$$

The logarithmic function with base e is the **natural logarithmic function** and is denoted by the special symbol $\ln x$, read as "el en of x."

The Natural Logarithmic Function

The function defined by

$$f(x) = \log_e x = \ln x, \qquad x > 0$$

is called the **natural logarithmic function.**

The four properties of logarithms listed on page 385 are also valid for natural logarithms.

Properties of Natural Logarithms

1. $\ln 1 = 0$ because $e^0 = 1$.

2. $\ln e = 1$ because $e^1 = e$.

3. $\ln e^x = x$ because $e^x = e^x$.

4. If $\ln x = \ln y$, then $x = y$.

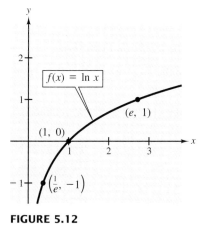

FIGURE 5.12

The graph of the natural logarithmic function is shown in Figure 5.12. Try using a graphing utility to confirm this graph. What is the domain of the natural logarithmic function?

EXAMPLE 7   Evaluating the Natural Logarithmic Function

a. $\ln \dfrac{1}{e} = \ln e^{-1} = -1$ Property 3

b. $\ln e^2 = 2$ Property 3

c. $\ln e^0 = 0$ Property 1

d. $2 \ln e = 2(1) = 2$ Property 2

On most calculators, the natural logarithm is denoted by $\boxed{\text{ln}}$, as illustrated in Example 8.

EXAMPLE 8 Evaluating the Natural Logarithmic Function

Scientific Calculator

Number	Keystrokes	Display
a. $\ln 2$	2 $\boxed{\text{ln}}$	0.6931472
b. $\ln 0.3$	$.3$ $\boxed{\text{ln}}$	-1.2039728
c. $\ln e^2$	2 $\boxed{e^x}$ $\boxed{\text{ln}}$	2
d. $\ln(-1)$	1 $\boxed{+/-}$	ERROR

Graphing Calculator

Number	Keystrokes	Display
a. $\ln 2$	$\boxed{\text{ln}}$ 2 $\boxed{\text{ENTER}}$	0.6931472
b. $\ln 0.3$	$\boxed{\text{ln}}$ $.3$ $\boxed{\text{ENTER}}$	-1.2039728
c. $\ln e^2$	$\boxed{\text{ln}}$ $\boxed{e^x}$ 2 $\boxed{\text{ENTER}}$	2
d. $\ln(-1)$	$\boxed{\text{ln}}$ $\boxed{(-)}$ 1 $\boxed{\text{ENTER}}$	ERROR

In Example 8, be sure you see that $\ln(-1)$ gives an error message. This occurs because the domain of $\ln x$ is the set of positive real numbers (see Figure 5.12). Hence, $\ln(-1)$ is undefined.

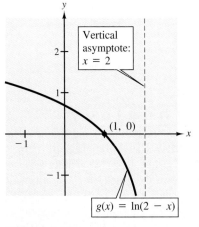

FIGURE 5.13

EXAMPLE 9 Finding the Domains of Logarithmic Functions

Find the domains of the following functions.

a. $f(x) = \ln(x - 2)$ **b.** $g(x) = \ln(2 - x)$ **c.** $h(x) = \ln x^2$

Solution

a. Because $\ln(x - 2)$ is defined only if $x - 2 > 0$, it follows that the domain of f is $(2, \infty)$.

b. Because $\ln(2 - x)$ is defined only if $2 - x > 0$, it follows that the domain of g is $(-\infty, 2)$. The graph of g is shown in Figure 5.13.

c. Because $\ln x^2$ is defined only if $x^2 > 0$, it follows that the domain of h is all real numbers except $x = 0$.

Application

EXAMPLE 10   Human Memory Model

Students participating in a psychological experiment attended several lectures on a subject. Every month for a year after that, the students were tested to see how much of the material they remembered. The average scores for the group were given by the *human memory model*

$$f(t) = 75 - 6\ln(t + 1), \qquad 0 \le t \le 12$$

where t is the time in months.

a. What was the average score on the original ($t = 0$) exam?

b. What was the average score at the end of $t = 2$ months?

c. What was the average score at the end of $t = 6$ months?

Solution

a. The original average score was

$$f(0) = 75 - 6\ln(0 + 1) = 75 - 6(0) = 75.$$

b. After 2 months, the average score was

$$f(2) = 75 - 6\ln 3 \approx 75 - 6(1.0986) \approx 68.4.$$

c. After 6 months, the average score was

$$f(6) = 75 - 6\ln 7 \approx 75 - 6(1.9459) \approx 63.3.$$

The graph of f is shown in Figure 5.14.

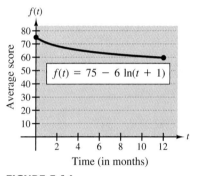

FIGURE 5.14

Group Activities   Exploring with Technology

Transforming Logarithmic Functions   Use a graphing utility to graph $f(x) = \ln x$. How will the graphs of $h(x) = \ln x + 5$, $j(x) = \ln(x + 5)$, $k(x) = \ln(x - 3)$, and $l(x) = \ln x - 4$ differ from the graph of f? Confirm your predictions with a graphing utility.

How will the basic graph of f be affected when a constant c is introduced: $g(x) = c \ln x$? Use a graphing utility to graph g with several different positive values of c, and summarize the effect of c.

Warm Up The following warm-up exercises involve skills that were covered in earlier sections. You will use these skills in the exercise set for this section.

In Exercises 1–4, solve for x.

1. $2^x = 8$

2. $4^x = 1$

3. $10^x = 0.1$

4. $e^x = e$

In Exercises 5 and 6, evaluate the expression. (Round the result to three decimal places.)

5. e^2

6. e^{-1}

In Exercises 7–10, describe how the graph of g is related to the graph of f.

7. $g(x) = f(x + 2)$

8. $g(x) = -f(x)$

9. $g(x) = -1 + f(x)$

10. $g(x) = f(-x)$

5.2 Exercises

In Exercises 1–16, evaluate the expression.

1. $\log_2 16$

2. $\log_4 64$

3. $\log_5\left(\frac{1}{25}\right)$

4. $\log_2\left(\frac{1}{8}\right)$

5. $\log_{16} 4$

6. $\log_{27} 9$

7. $\log_7 1$

8. $\log_{10} 1000$

9. $\log_{10} 0.01$

10. $\log_{10} 10$

11. $\ln e^3$

12. $\ln \dfrac{1}{e}$

13. $\ln e^{-2}$

14. $\ln 1$

15. $\log_a a^2$

16. $\log_a \dfrac{1}{a}$

In Exercises 17–26, use the definition of a logarithm to write the equation in logarithmic form. For instance, the logarithmic form of $2^3 = 8$ is $\log_2 8 = 3$.

17. $5^3 = 125$

18. $8^2 = 64$

19. $81^{1/4} = 3$

20. $9^{3/2} = 27$

21. $6^{-2} = \frac{1}{36}$

22. $10^{-3} = 0.001$

23. $e^3 = 20.0855\ldots$

24. $e^0 = 1$

25. $e^x = 4$

26. $e^{-x} = 2$

In Exercises 27–34, use a calculator to evaluate the logarithm. (Round to three decimal places.)

27. $\log_{10} 345$

28. $\log_{10}\left(\frac{4}{5}\right)$

29. $\log_{10} 145$

30. $\log_{10} 12.5$

31. $\ln 18.42$

32. $\ln 36.7$

33. $\ln\left(\sqrt{3}\right)$

34. $\ln\left(\sqrt{5}\right)$

In Exercises 35–38, sketch the graphs of f and g in the same coordinate plane.

35. $f(x) = 3^x$, $g(x) = \log_3 x$

36. $f(x) = 5^x$, $g(x) = \log_5 x$

37. $f(x) = e^x$, $g(x) = \ln x$

38. $f(x) = 10^x$, $g(x) = \log_{10} x$

In Exercises 39–44, match the function with its graph. [The graphs are labeled (a), (b), (c), (d), (e), and (f).]

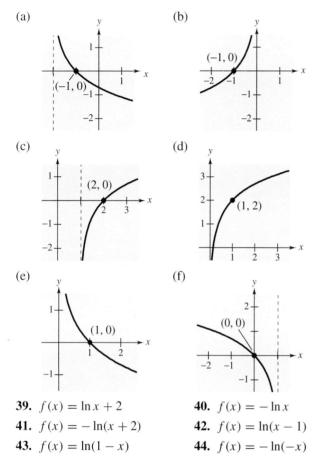

(a)

(b)

(c)

(d)

(e)

(f)

39. $f(x) = \ln x + 2$

40. $f(x) = -\ln x$

41. $f(x) = -\ln(x + 2)$

42. $f(x) = \ln(x - 1)$

43. $f(x) = \ln(1 - x)$

44. $f(x) = -\ln(-x)$

In Exercises 45–52, find the domain, vertical asymptote, and x-intercept of the logarithmic function. Then sketch its graph.

45. $f(x) = \log_4 x$

46. $g(x) = \log_6 x$

47. $h(x) = \log_4(x - 3)$

48. $f(x) = -\log_6(x + 2)$

49. $f(x) = \ln(x - 2)$

50. $h(x) = \ln(x + 1)$

51. $g(x) = \ln(-x)$

52. $f(x) = \ln(3 - x)$

53. *Human Memory Model*   Students in a mathematics class were given an exam and then retested monthly with an equivalent exam. The average score for the class was given by the human memory model

$$f(t) = 80 - 17 \log_{10}(t + 1), \qquad 0 \le t \le 12$$

where t is the time in months.

(a) What was the average score on the original exam?

(b) What was the average score after 4 months?

(c) What was the average score after 10 months?

54. *Human Memory Model*   Students in a seventh-grade class were given an exam. During the next 2 years, the same students were retested several times. The average score was given by the model

$$f(t) = 90 - 15 \log_{10}(t + 1), \qquad 0 \le t \le 24$$

where t is the time in months.

(a) What was the average score on the original exam?

(b) What was the average score after 6 months?

(c) What was the average score after 18 months?

55. *Investment Time*   A principal P, invested at $9\frac{1}{2}\%$ and compounded continuously, increases to an amount that is K times the original principal after t years, where t is given by

$$t = \frac{\ln K}{0.095}.$$

(a) Complete the table.

K	1	2	4	6	8	10	12
t							

(b) Use the table in part (a) to graph the function.

56. *Investment Time*   A principal P, invested at 7.75% and compounded continuously, increases to an amount that is K times the original principal after t years, where t is given by

$$t = \frac{\ln K}{0.0775}.$$

Use a graphing utility to graph this function.

Skill Retention Model   In Exercises 57 and 58, participants in an industrial psychology study were taught a simple mechanical task and tested monthly on this mechanical task for a period of 1 year. The average score for the class was given by the model

$$f(t) = 95 - 12 \log_{10}(t + 1), \qquad 0 \le t \le 12$$

where t is the time in months.

57. Use a graphing utility to graph this function. Sketch the graph and discuss the domain and range of this function.

58. Based on the graph of $f(t)$, do you think the study's participants practiced the simple mechanical task very often? Cite the behavior of the graph to justify your answer.

Median Age of U.S. Population   In Exercises 59–62, use the model

$$A = 18.10 - 0.016t + 5.121 \ln t, \qquad 10 \le t \le 80$$

which approximates the median age of the United States population from 1980 to 2050. In the model, A is the median age and $t = 10$ represents 1980 (see figure).   (Source: U. S. Bureau of Census)

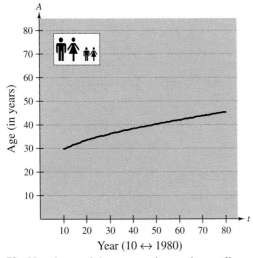

Year (10 ↔ 1980)

59. Use the model to approximate the median age in the United States in 1980.

60. Use the model to approximate the median age in the United States in 1990.

61. Use the model to approximate the change in the median age in the United States from 1980 to 2000.

62. Use the model to approximate the change in the median age in the United States from 1980 to 2050.

Monthly Payment   In Exercises 63–66, use the model

$$t = \frac{5.315}{-6.7968 + \ln x}, \qquad 1000 < x$$

which approximates the length of a home mortgage (of \$120,000 at 10%) in terms of the monthly payment. In the model, t is the length of the mortgage in years and x is the monthly payment in dollars (see figure).

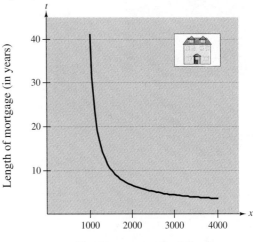

Monthly payment (in dollars)

63. Use the model to approximate the length of a home mortgage (for \$120,000 at 10%) that has a monthly payment of \$1167.41.

64. Use the model to approximate the length of a home mortgage (for \$120,000 at 10%) that has a monthly payment of \$1068.45.

65. Approximate the total amount paid over the term of the mortgage for a monthly payment of \$1167.41.

66. Approximate the total amount paid over the term of the mortgage for a monthly payment of \$1068.45.

5.3   Properties of Logarithms

Change of Base ▪ Properties of Logarithms ▪
Rewriting Logarithmic Expressions ▪ Application

Change of Base

Most calculators have only two types of log keys, one for common logarithms (base 10) and one for natural logarithms (base e). Although common logs and natural logs are the most frequently used, you may occasionally need to evaluate logarithms to other bases. To do this, you can use the following *change-of-base formula*.

John Napier, a Scottish mathematician, developed logarithms as a way to simplify some of the tedious calculations of his day. Beginning in 1594, Napier worked about 20 years on the invention of logarithms. Napier was only partially successful in his quest to simplify tedious calculations. Nonetheless, the development of logarithms was a step forward and received immediate recognition.

Change-of-Base Formula

Let a, b, and x be positive real numbers such that $a \neq 1$ and $b \neq 1$. Then $\log_a x$ is given by

$$\log_a x = \frac{\log_b x}{\log_b a}.$$

One way to look at the change-of-base formula is that logarithms to base a are simply *constant multiples* of logarithms to base b. The constant multiplier is

$$\frac{1}{\log_b a}.$$

EXAMPLE 1   *Changing Bases Using Common Logarithms*

a. $\log_4 30 = \dfrac{\log_{10} 30}{\log_{10} 4} \approx \dfrac{1.47712}{0.60206} \approx 2.4534$

b. $\log_2 14 = \dfrac{\log_{10} 14}{\log_{10} 2} \approx \dfrac{1.14613}{0.30103} \approx 3.8074$

EXAMPLE 2   *Changing Bases Using Natural Logarithms*

a. $\log_4 30 = \dfrac{\ln 30}{\ln 4} \approx \dfrac{3.40120}{1.38629} \approx 2.4534$

b. $\log_2 14 = \dfrac{\ln 14}{\ln 2} \approx \dfrac{2.63906}{0.693147} \approx 3.8074$

Properties of Logarithms

You know from the previous section that the logarithmic function with base a is the *inverse* of the exponential function with base a. Thus, it makes sense that the properties of exponents should have corresponding properties involving logarithms. For instance, the exponential property $a^0 = 1$ has the corresponding logarithmic property $\log_a 1 = 0$.

NOTE   There is no general property that can be used to rewrite $\log_a(u \pm v)$. Specifically, $\log_a(x + y)$ is not equal to $\log_a x + \log_a y$.

These properties of logarithms are very important. One way to teach them is to explain their similarities to properties of exponents. Some common errors to watch for are rewriting $\log x - \log y$ as $(\log x)/(\log y)$ instead of $\log(x/y)$, or writing $\log ax^n$ as $n\log ax$ instead of $\log a + n\log x$. Identify these errors and resolve them before proceeding to Section 5.4.

Properties of Logarithms

Let a be a positive number such that $a \neq 1$, and let n be a real number. If u and v are positive real numbers, the following properties are true.

1. $\log_a(uv) = \log_a u + \log_a v$ 1. $\ln(uv) = \ln u + \ln v$

2. $\log_a \dfrac{u}{v} = \log_a u - \log_a v$ 2. $\ln \dfrac{u}{v} = \ln u - \ln v$

3. $\log_a u^n = n \log_a u$ 3. $\ln u^n = n \ln u$

EXAMPLE 3   *Using Properties of Logarithms*

Write the logarithm in terms of $\ln 2$ and $\ln 3$.

a. $\ln 6$ **b.** $\ln \dfrac{2}{27}$

Solution

a. $\ln 6 = \ln(2 \cdot 3)$ Rewrite 6 as 2 · 3.

$= \ln 2 + \ln 3$ Property 1

b. $\ln \dfrac{2}{27} = \ln 2 - \ln 27$ Property 2

$= \ln 2 - \ln 3^3$ Rewrite 27 as 3^3.

$= \ln 2 - 3 \ln 3$ Property 3

EXAMPLE 4   *Using Properties of Logarithms*

Use the properties of logarithms to verify that $-\ln \frac{1}{2} = \ln 2$.

Solution

$$-\ln \frac{1}{2} = -\ln(2^{-1}) = -(-1)\ln 2 = \ln 2$$

Try checking this result on your calculator.

Rewriting Logarithmic Expressions

The properties of logarithms are useful for rewriting logarithmic expressions in forms that simplify the operations of algebra. This is true because they convert complicated products, quotients, and exponential forms into simpler sums, differences, and products, respectively. Examples 5 and 6 illustrate some cases.

EXAMPLE 5 Rewriting the Logarithm of a Product

$$\log_{10} 5x^3y = \log_{10} 5 + \log_{10} x^3 y$$
$$= \log_{10} 5 + \log_{10} x^3 + \log_{10} y$$
$$= \log_{10} 5 + 3\log_{10} x + \log_{10} y$$

EXAMPLE 6 Rewriting the Logarithm of a Quotient

$$\ln \frac{\sqrt{3x-5}}{7} = \ln(3x-5)^{1/2} - \ln 7 = \frac{1}{2}\ln(3x-5) - \ln 7$$

In Examples 5 and 6, the properties of logarithms were used to *expand* logarithmic expressions. In Examples 7 and 8, this procedure is reversed and the properties of logarithms are used to *condense* logarithmic expressions.

EXAMPLE 7 Condensing a Logarithmic Expression

$$\frac{1}{2}\log_{10} x - 3\log_{10}(x+1) = \log_{10} x^{1/2} - \log_{10}(x+1)^3$$
$$= \log_{10} \frac{\sqrt{x}}{(x+1)^3}$$

EXAMPLE 8 Condensing a Logarithmic Expression

$$2\ln(x+2) - \ln x = \ln(x+2)^2 - \ln x = \ln \frac{(x+2)^2}{x}$$

When applying the properties of logarithms to a logarithmic function, you should be careful to check the domain of the function. For example, the domain of $f(x) = \ln x^2$ is all real $x \neq 0$, whereas the domain of $g(x) = 2\ln x$ is all real $x > 0$.

Application

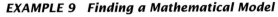

EXAMPLE 9 *Finding a Mathematical Model*

The table gives the mean distance x and the period y of the six planets that are closest to the sun. In the table, the mean distance is given in terms of astronomical units (where the earth's mean distance is defined as 1.0), and the period is given in terms of years. Find an equation that expresses y as a function of x.

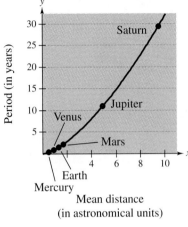

FIGURE 5.15

Planet	Mercury	Venus	Earth	Mars	Jupiter	Saturn
Period, y	0.241	0.615	1.0	1.881	11.861	29.457
Mean Distance, x	0.387	0.723	1.0	1.523	5.203	9.541

Solution

The points in the table are plotted in Figure 5.15. From this figure it is not clear how to find an equation that relates y and x. To solve this problem, take the natural log of each of the x- and y-values given in the table. This produces the following results.

Planet	Mercury	Venus	Earth	Mars	Jupiter	Saturn
$\ln y$	-1.423	-0.486	0.0	0.632	2.473	3.383
$\ln x$	-0.949	-0.324	0.0	0.421	1.649	2.256

Now, by plotting the points in the second table, you can see that all six of the points appear to lie on a line (see Figure 5.16). The slope of this line is $\frac{3}{2}$, and you can therefore conclude that

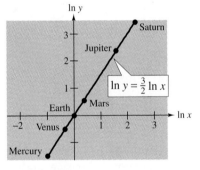

FIGURE 5.16

$$\ln y = \tfrac{3}{2} \ln x.$$

Group Activities   Extending the Concept

Kepler's Law   The relationship described in Example 9 was first discovered by Johannes Kepler. Use properties of logarithms to rewrite the relationship so that y is expressed as a function of x.

Warm Up

The following warm-up exercises involve skills that were covered in earlier sections. You will use these skills in the exercise set for this section.

In Exercises 1–4, evaluate the expression without using a calculator.

1. $\log_7 49$

2. $\log_2 \left(\frac{1}{32}\right)$

3. $\ln \frac{1}{e^2}$

4. $\log_{10} 0.001$

In Exercises 5–8, simplify the expression.

5. $e^2 e^3$

6. $\frac{e^2}{e^3}$

7. $(e^2)^3$

8. $(e^2)^0$

In Exercises 9 and 10, rewrite the expression in exponential form.

9. $\frac{1}{x^2}$

10. $\sqrt{x}$

5.3 Exercises

In Exercises 1–4, write the logarithm as a multiple of a common logarithm.

1. $\log_3 5$

2. $\log_4 10$

3. $\log_2 x$

4. $\ln 5$

In Exercises 5–8, write the logarithm as a multiple of a natural logarithm.

5. $\log_3 5$

6. $\log_4 10$

7. $\log_2 x$

8. $\log_{10} 5$

In Exercises 9–16, evaluate the logarithm. (Round to three decimal places.)

9. $\log_3 7$

10. $\log_7 4$

11. $\log_{1/2} 4$

12. $\log_4 0.55$

13. $\log_9 0.4$

14. $\log_{20} 125$

15. $\log_{15} 1250$

16. $\log_{1/3} 0.015$

In Exercises 17–36, use the properties of logarithms to write the expression as a sum, difference, and/or multiple of logarithms.

17. $\log_{10} 5x$

18. $\log_{10} 10z$

19. $\log_{10} \frac{5}{x}$

20. $\log_{10} \frac{y}{2}$

21. $\log_2 x^4$

22. $\log_2 z^{-3}$

23. $\ln \sqrt{z}$

24. $\ln \sqrt[3]{t}$

25. $\ln xyz$

26. $\ln \frac{xy}{z}$

27. $\ln \sqrt{a-1}$

28. $\ln \left(\frac{x^2-1}{x^3}\right)$

29. $\ln(z-1)^2$

30. $\ln \sqrt{\dfrac{x^2}{y^3}}$

31. $\log_b \dfrac{x^2}{y^3}$

32. $\log_b \dfrac{x^4}{z^2}$

33. $\ln \sqrt[3]{\dfrac{x}{y}}$

34. $\ln \dfrac{x}{\sqrt{x^2+1}}$

35. $\ln \dfrac{\sqrt{y}}{z^5}$

36. $\ln \sqrt{x^2(x+2)}$

In Exercises 37–52, write the expression as the logarithm of a single quantity.

37. $\ln x + \ln 2$

38. $\ln y + \ln z$

39. $\log_{10} z - \log_{10} y$

40. $\log_{10} 8 - \log_{10} t$

41. $2 \log_{10}(x+4)$

42. $-4 \log_{10} 2x$

43. $\ln x - 3 \ln(x+1)$

44. $2 \ln 8 + 5 \ln z$

45. $\frac{1}{3} \ln 5x$

46. $\frac{3}{2} \ln(z-2)$

47. $\ln(x-2) - \ln(x+2)$

48. $3 \ln x + 2 \ln y - 4 \ln z$

49. $2 \ln 3 - \frac{1}{2} \ln(x^2+1)$

50. $\frac{3}{2} \ln t^6 - \frac{3}{4} \ln t^4$

51. $\ln x - \ln(x+2) - \ln(x-2)$

52. $\ln(x+1) + 2 \ln(x-1) + 3 \ln x$

In Exercises 53–66, approximate the logarithm using the properties of logarithms, given $\log_b 2 \approx 0.3562$, $\log_b 3 \approx 0.5646$, and $\log_b 5 \approx 0.8271$.

53. $\log_b 6$

54. $\log_b 15$

55. $\log_b \left(\frac{3}{2}\right)$

56. $\log_b \left(\frac{5}{3}\right)$

57. $\log_b 25$

58. $\log_b 18 (18 = 2 \cdot 3^2)$

59. $\log_b \sqrt{2}$

60. $\log_b \left(\frac{9}{2}\right)$

61. $\log_b 40$

62. $\log_b \sqrt[3]{75}$

63. $\log_b \left(\frac{1}{4}\right)$

64. $\log_b (3b^2)$

65. $\log_b \sqrt{5b}$

66. $\log_b 1$

In Exercises 67–72, find the *exact* value of the logarithm.

67. $\log_3 9$

68. $\log_6 \sqrt[3]{6}$

69. $\log_4 16^{1.2}$

70. $\log_5 \left(\frac{1}{125}\right)$

71. $\ln e^{4.5}$

72. $\ln \sqrt[4]{e^3}$

In Exercises 73–80, use the properties of logarithms to simplify the given logarithmic expression.

73. $\log_4 8$

74. $\log_5 \left(\frac{1}{15}\right)$

75. $\log_7 \sqrt{70}$

76. $\log_2(4^2 \cdot 3^4)$

77. $\log_5 \left(\frac{1}{250}\right)$

78. $\log_{10} \left(\frac{9}{300}\right)$

79. $\ln(5e^6)$

80. $\ln \dfrac{6}{e^2}$

Curve Fitting In Exercises 81–84, find an equation that relates x and y. Explain the steps used to find the equation.

81.

x	1	2	3	4	5	6
y	1	1.189	1.316	1.414	1.495	1.565

82.

x	1	2	3	4	5	6
y	1	1.587	2.080	2.520	2.924	3.302

83.

x	1	2	3	4	5	6
y	2.5	2.102	1.9	1.768	1.672	1.597

84.

x	1	2	3	4	5	6
y	0.5	2.828	7.794	16	27.951	44.091

85. *Graphical Reasoning* Use a graphing utility to graph $f(x) = \ln \frac{x}{3}$ and $g(x) = \ln x - \ln 3$ in the same viewing rectangle. What do you observe about the two graphs? What property of logarithms is being demonstrated graphically?

86. *Graphical Reasoning* Use a graphing utility to graph $f(x) = \ln 5x$ and $g(x) = \ln 5 + \ln x$ in the same viewing rectangle. What do you observe about the two graphs? What property is demonstrated?

87. *Reasoning* You are helping another student learn the properties of logarithms. How would you use a graphing utility to demonstrate to this student the logarithm property: $\log_a u^v = v \log_a u$ (u and v are positive real numbers and a is a positive number such that $a \neq 1$). What two functions could you use? Briefly describe your explanation of this property using these functions and their graphs.

88. *Reasoning* An algebra student claims that the following is true:

$$\log_a \frac{x}{y} = \frac{\log_a x}{\log_a y} = \log_a x - \log_a y.$$

Discuss how you would use a graphing utility to demonstrate that this claim is not true. Describe how to demonstrate the actual property of logarithms that is hidden in this faulty claim.

89. *Nail Length* The approximate lengths and diameters (in inches) of common nails (see figure) are given in the table. Find a mathematical model that relates the diameter y of a common nail to its length x.

Length, x	1.00	2.00	3.00
Diameter, y	0.070	0.111	0.146

Length, x	4.00	5.00	6.00
Diameter, y	0.176	0.204	0.231

90. *Galloping Speeds of Animals* Four-legged animals run with two different types of motion: trotting and galloping. An animal that is trotting has at least one foot on the ground at all times, whereas an animal that is galloping sometimes has all four feet off the ground. The number of strides per minute at which an animal breaks from a trot to a gallop depends on the weight of the animal. Use the table at the top of the right-hand column to find a model that relates an animal's lowest galloping speed (in strides per minute) to its weight (in pounds).

Weight, x	25	35	50
Galloping Speed, y	191.5	182.7	173.8

Weight, x	75	500	1000
Galloping Speed, y	164.2	125.9	114.2

C A R E E R   I N T E R V I E W

Mary Kay Brown
Research Assistant
Dartmouth Medical School
Hanover, NH 03755

A question that is often asked in the lab is "How much drug does it take to kill 50% of a cell population?" To find out, we set up an assay using different drug concentrations to treat the same number of cells. After a period of several days, the percent of cells that survive at each drug concentration is calculated and graphed. However, there is such a wide range of concentrations (from 0.005 to 100 units in this example) that a linear-scale graph is difficult to read. So we graph this data using a *logarithmic scale*, which gives the same results as graphing $\log_{10}$ of each concentration. The logarithmic scale spreads the points out in a nice curve, and it is easy to see that a drug concentration between one and two units killed 50% of the cell population.

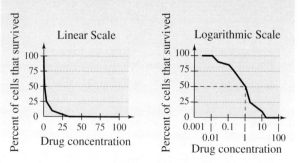

Math Matters

Archimedes and the Gold Crown

Archimedes was a citizen of Syracuse, a city on the island of Sicily. He was born in 287 B.C. and is often regarded (with Isaac Newton and Carl Gauss) as one of the three greatest mathematicians who ever lived. The following story gives some idea of Archimedes's ingenuity.

The king of Syracuse ordered a crown to be made of pure gold. When the crown was delivered, the king suspected that some silver had been mixed with the gold. He asked Archimedes to find a way of checking whether the crown was pure gold (without destroying the crown). One day, while Archimedes was at a public bath, he noticed that the level of the water rose when he stepped into the bath. He suddenly realized how he could solve the problem, and became so excited that he ran home, stark naked, shouting "Eureka!" (which means "I have found it!").

Archimedes's idea was that he could measure the volume of the crown by putting it into a bowl of water. If the crown contained any silver, the volume of the crown would be greater than the volume of an equal weight of pure gold, as shown in the accompanying figure.

1. Crown and block of pure gold balance.

2. Crown and block of pure silver balance.

3. Gold block makes water level rise.

4. Silver block makes water level rise higher.

5. Crown makes water level rise more than gold block.

MID-CHAPTER QUIZ

Take this quiz as you would take a quiz in class. After you are done, check your work against the answers given in the back of the book.

In Exercises 1–4, use a graphing utility to sketch the graph of the function.

1. $f(x) = 3^x$

2. $g(x) = 2^{-x}$

3. $h(x) = \log_2 x$

4. $h(x) = \log_3(x - 1)$

In Exercises 5 and 6, find the balance compounded monthly and continuously.

5. $P = \$2000$, $r = 8\%$, $t = 10$ years

6. $P = \$5000$, $r = 9\%$, $t = 20$ years

7. *Bacteria Growth* A bacteria population is modeled by $P(t) = 100e^{0.3012t}$, where t is the time in hours. Find (a) $P(0)$, (b) $P(6)$, and (c) $P(12)$.

8. *Demand Function* Use the demand function $p = 600 - 0.4e^{0.003x}$ to find the price for a demand of $x = 1000$ units.

In Exercises 9–12, evaluate the expression.

9. $\log_{10} 100$

10. $\ln e^4$

11. $\log_4 \frac{1}{16}$

12. $\ln 1$

13. Sketch the graph of $f(x) = 3^x$ and $g(x) = \log_3 x$ in the same coordinate plane. Discuss the special relationship between $f(x)$ and $g(x)$ that is demonstrated by their graphs.

In Exercises 14 and 15, expand the logarithmic expression.

14. $\log \sqrt[3]{\frac{xy}{z}}$

15. $\ln\left(\frac{x^2 + 3}{x^3}\right)$

In Exercises 16 and 17, condense the logarithmic expression.

16. $\ln x + \ln y - \ln 3$

17. $-3 \log_{10} 4x$

In Exercises 18 and 19, find the *exact* value of the logarithm.

18. $\log_7 \sqrt{343}$

19. $\ln \sqrt[5]{e^6}$

20. Find an equation that relates x and y.

x	1	2	3	4	6	8
y	1	1.260	1.442	1.587	1.817	2.000

5.4 Solving Exponential and Logarithmic Equations

Introduction ▪ Solving Exponential Equations ▪
Solving Logarithmic Equations ▪ Applications

Introduction

So far in this chapter, you have studied the definitions, graphs, and properties of exponential and logarithmic functions. In this section, you will study procedures for *solving equations* involving these exponential and logarithmic functions. As a simple example, consider the exponential equation

$$2^x = 32.$$

You can obtain the solution by rewriting the equation in the form

$$2^x = 2^5$$

which implies that $x = 5$. Although this method works in some cases, it does not work for an equation as simple as

$$e^x = 7.$$

To solve for x, you can take the natural logarithm of both sides to obtain

$e^x = 7$	Original equation
$\ln e^x = \ln 7$	Take log of both sides.
$x = \ln 7$	$\ln e^x = x$ because $e^x = e^x$.

Here are some guidelines for solving exponential and logarithmic equations.

Solving Exponential and Logarithmic Equations

1. *To solve an exponential equation*, first isolate the exponential expression, then take the logarithm of both sides and solve for the variable.

2. *To solve a logarithmic equation*, rewrite the equation in exponential form and solve for the variable.

These two guidelines are based on the following **inverse properties** of exponential and logarithmic functions.

	Base a	*Base e*
1.	$\log_a a^x = x$	$\ln e^x = x$
2.	$a^{\log_a x} = x$	$e^{\ln x} = x$

Solving Exponential Equations

EXAMPLE 1   Solving an Exponential Equation

Solve $e^x = 72$.

Solution

$e^x = 72$	Original equation
$\ln e^x = \ln 72$	Take log of both sides.
$x = \ln 72$	Inverse property of logs and exponents
$x \approx 4.277$	Use a calculator.

The solution is $\ln 72$. Check this in the original equation.

EXAMPLE 2   Solving an Exponential Equation

Solve $e^x + 5 = 60$.

Solution

$e^x + 5 = 60$	Original equation
$e^x = 55$	Subtract 5 from both sides.
$\ln e^x = \ln 55$	Take log of both sides.
$x = \ln 55$	Inverse property of logs and exponents
$x \approx 4.007$	Use a calculator.

The solution is $\ln 55$. Check this in the original equation.

Technology

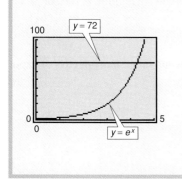

When solving an exponential or logarithmic equation, remember that you can check your solution graphically by "graphing the left and right sides separately" and estimating the x-coordinate of the point of intersection. For instance, to check the solution of the equation in Example 1, you can sketch the graphs of

$$y = e^x$$

and

$$y = 72$$

on the same screen, as shown at the left. Notice that the graphs intersect when $x \approx 4.277$, which confirms the solution found in Example 1.

EXAMPLE 3   Solving an Exponential Equation

Solve $4e^{2x} = 5$.

Solution

$$4e^{2x} = 5 \qquad\qquad \text{Original equation}$$

$$e^{2x} = \frac{5}{4} \qquad\qquad \text{Divide both sides by 4.}$$

$$\ln e^{2x} = \ln \frac{5}{4} \qquad\qquad \text{Take log of both sides.}$$

$$2x = \ln \frac{5}{4} \qquad\qquad \text{Inverse property of logs and exponents}$$

$$x = \frac{1}{2} \ln \frac{5}{4} \qquad\qquad \text{Divide both sides by 2.}$$

$$x \approx 0.112 \qquad\qquad \text{Use a calculator.}$$

The solution is $\frac{1}{2} \ln \frac{5}{4}$. Check this in the original equation.

When an equation involves two or more exponential expressions, you can still use a procedure similar to that demonstrated in the first three examples. However, the algebra is a bit more complicated. Study the next example carefully.

EXAMPLE 4   Solving an Exponential Equation

Solve $e^{2x} - 3e^x + 2 = 0$.

Solution

$$e^{2x} - 3e^x + 2 = 0 \qquad\qquad \text{Original equation}$$

$$(e^x)^2 - 3e^x + 2 = 0 \qquad\qquad \text{Quadratic form}$$

$$(e^x - 2)(e^x - 1) = 0 \qquad\qquad \text{Factor.}$$

$$e^x - 2 = 0 \implies x = \ln 2 \qquad \text{Set 1st factor equal to 0.}$$

$$e^x - 1 = 0 \implies x = 0 \qquad \text{Set 2nd factor equal to 0.}$$

The solutions are $\ln 2$ and 0. Check these in the original equation.

NOTE   In Example 4, use a graphing utility to sketch the graph of

$$y = e^{2x} - 3e^x + 2.$$

It should have two x-intercepts: one when $x = \ln 2$ and one when $x = 0$.

Solving Logarithmic Equations

To solve a logarithmic equation such as

$$\ln x = 3 \qquad \text{Logarithmic form}$$

write the equation in exponential form as follows.

$$e^{\ln x} = e^3 \qquad \text{Exponentiate both sides.}$$

$$x = e^3 \qquad \text{Exponential form}$$

This procedure is called *exponentiating* both sides of an equation.

EXAMPLE 5   Solving a Logarithmic Equation

Solve $\ln x = 2$.

Solution

$$\ln x = 2 \qquad \text{Original equation}$$

$$e^{\ln x} = e^2 \qquad \text{Exponentiate both sides.}$$

$$x = e^2 \qquad \text{Inverse property of exponents and logs}$$

$$x \approx 7.389 \qquad \text{Use a calculator.}$$

The solution is e^2. Check this in the original equation.

EXAMPLE 6   Solving a Logarithmic Equation

Solve $5 + 2\ln x = 4$.

Solution

$$5 + 2\ln x = 4 \qquad \text{Original equation}$$

$$2\ln x = -1 \qquad \text{Subtract 5 from both sides.}$$

$$\ln x = -\frac{1}{2} \qquad \text{Divide both sides by 2.}$$

$$e^{\ln x} = e^{-1/2} \qquad \text{Exponentiate both sides.}$$

$$x = e^{-1/2} \qquad \text{Inverse property of exponents and logs}$$

$$x \approx 0.607 \qquad \text{Use a calculator.}$$

The solution is $e^{-1/2}$. Check this in the original equation.

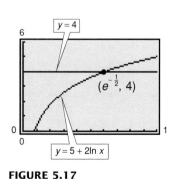

FIGURE 5.17

NOTE   To check the result of Example 6 graphically, you can sketch the graphs of $y = 5 + 2\ln x$ and $y = 4$ in the same coordinate plane, as shown in Figure 5.17.

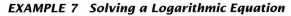

EXAMPLE 7   Solving a Logarithmic Equation

Solve $2 \ln 3x = 4$.

Solution

$2 \ln 3x = 4$	Original equation
$\ln 3x = 2$	Divide both sides by 2.
$e^{\ln 3x} = e^2$	Exponentiate both sides.
$3x = e^2$	Inverse property of exponents and logs
$x = \dfrac{1}{3}e^2$	Divide both sides by 3.
$x \approx 2.463$	Use a calculator.

The solution is $\frac{1}{3}e^2$. Check this in the original equation.

EXAMPLE 8   Solving a Logarithmic Equation

Solve $\ln x - \ln(x - 1) = 1$.

Solution

$\ln x - \ln(x - 1) = 1$	Original equation
$\ln \dfrac{x}{x - 1} = 1$	Property 2 of logarithms
$\dfrac{x}{x - 1} = e^1$	Exponentiate both sides.
$x = ex - e$	Multiply both sides by $x - 1$.
$x - ex = -e$	Subtract ex from both sides.
$x(1 - e) = -e$	Factor.
$x = \dfrac{-e}{1 - e}$	Divide both sides by $1 - e$.
$x = \dfrac{e}{e - 1}$	Simplify.

The solution is $e/(e - 1)$. Check this in the original equation.

In solving exponential or logarithmic equations, the following properties are useful. Can you see where these properties were used in the examples in this section?

1. $x = y$ if and only if $\log_a x = \log_a y$.

2. $x = y$ if and only if $a^x = a^y, \ a > 0 \neq 1$.

Activities

1. Write two or three sentences stating the general guidelines that you followed when solving (a) exponential equations and (b) logarithmic equations. *Answer:* (a) To solve exponential equations, it is useful to isolate the exponential expression first, then take the logarithm of both sides and solve for the variable. (b) To solve logarithmic equations, condense the logarithmic part into a single logarithm, then rewrite in exponential form and solve for the variable.

2. Solve for x: $7^x = 3$.
   Answer: $x = \log_7 3$
   $$= \frac{\ln 3}{\ln 7} \approx 0.5646$$

3. Solve for x:
   $\log(x + 4) + \log(x + 1) = 1$.
   Answer: $x = 1$ ($x = -6$ is not in the domain.)

Applications

EXAMPLE 9 Doubling an Investment

You have deposited $500 in an account that pays 6.75% interest, compounded continuously. How long will it take your money to double?

Solution

Using the formula for compound interest, you can find that the balance in the account is given by

$$A = Pe^{rt} = 500e^{0.0675t}.$$

To find the time required for the balance to double, let $A = 1000$, and solve the resulting equation for t.

$500e^{0.0675t} = 1000$	Original equation
$e^{0.0675t} = 2$	Divide both sides by 500.
$\ln e^{0.0675t} = \ln 2$	Take log of both sides.
$0.0675t = \ln 2$	Inverse property of logs and exponents
$t = \dfrac{1}{0.0675} \ln 2$	Divide both sides by 0.0675.
$t \approx 10.27$	Use a calculator.

The balance in the account will double after approximately 10.27 years. This result is graphically demonstrated in Figure 5.18.

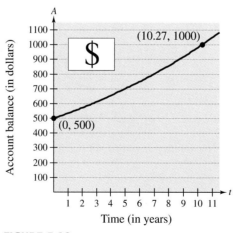

FIGURE 5.18

EXAMPLE 10 Waste Processed for Energy Recovery

From 1960 to 1986, the amount of municipal waste processed for energy recovery in the United States can be approximated by the equation

$$y = 0.00643e^{0.00533t^2}$$

where y is the amount of waste (in pounds per person) that was processed for energy recovery and $t = 0$ represents 1960 (see Figure 5.19). According to this model, during which year did the amount of waste reach 0.2 pounds? (Source: Franklin Associates)

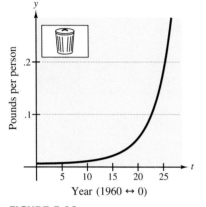

Pounds per person

.2

.1

5 10 15 20 25

Year (1960 ↔ 0)

FIGURE 5.19

Solution

$0.00643e^{0.00533t^2} = y$	Model
$0.00643e^{0.00533t^2} = 0.2$	Let y equal 0.2.
$e^{0.00533t^2} \approx 31.104$	Divide both sides by 0.00643.
$\ln e^{0.00533t^2} \approx \ln 31.104$	Take log of both sides.
$0.00533t^2 \approx 3.437$	Inverse property of logs and exponents
$t^2 \approx 644.905$	Divide both sides by 0.00533.
$t \approx 25.4$	Extract positive square root.

The solution is $t \approx 25.4$ years. Because $t = 0$ represents 1960, it follows that the amount of waste would have reached 0.2 pounds per person in 1985.

Group Activities Extending the Concept

Analyzing Relationships Numerically   Use a calculator to fill in the following table row-by-row. Discuss the resulting pattern. What can you conclude? Find two equations that summarize the relationships you've discovered.

x	$\frac{1}{2}$	1	2	10	25	50
e^x						
$\ln(e^x)$						
$\ln x$						
$e^{\ln x}$						

Warm Up The following warm-up exercises involve skills that were covered in earlier sections. You will use these skills in the exercise set for this section.

In Exercises 1–6, solve for x.

1. $x \ln 2 = \ln 3$

2. $(x - 1) \ln 4 = 2$

3. $2xe^2 = e^3$

4. $4xe^{-1} = 8$

5. $x^2 - 4x + 5 = 0$

6. $2x^2 - 3x + 1 = 0$

In Exercises 7–10, simplify the expression.

7. $\log_{10} 10^x$

8. $\log_{10} 10^{2x}$

9. $\ln e^{2x}$

10. $\ln e^{-x^2}$

5.4 Exercises

In Exercises 1–10, solve for x.

1. $4^x = 16$

2. $3^x = 243$

3. $7^x = \frac{1}{49}$

4. $8^x = 4$

5. $\left(\frac{3}{4}\right)^x = \frac{27}{64}$

6. $3^{x-1} = 27$

7. $\log_4 x = 3$

8. $\log_5 5x = 2$

9. $\log_{10} x = -1$

10. $\ln(2x - 1) = 0$

In Exercises 11–16, apply the inverse properties of $\ln x$ and e^x to simplify the expression.

11. $\ln e^{x^2}$

12. $\ln e^{2x-1}$

13. $e^{\ln(5x+2)}$

14. $-1 + \ln e^{2x}$

15. $e^{\ln x^2}$

16. $-8 + e^{\ln x^3}$

In Exercises 17–42, solve the exponential equation.

17. $e^x = 20$

18. $e^x = 5$

19. $e^x = 18$

20. $e^x = 580$

21. $2e^x = 39$

22. $4e^x = 91$

23. $e^x - 5 = 10$

24. $e^x + 6 = 38$

25. $3e^x - 8 = 23$

26. $4e^x + 10 = 605$

27. $7 - 2e^x = 5$

28. $-14 + 3e^x = 11$

29. $e^{3x} = 12$

30. $e^{2x} = 50$

31. $e^{-x} = 28$

32. $e^{-2x} = 14$

33. $500e^{-x} = 300$

34. $8e^{-5x} = 112$

35. $500e^{0.06t} = 1500$

36. $750e^{0.07t} = 2000$

37. $e^{2x} - 4e^x - 5 = 0$

38. $e^{2x} - 5e^x + 6 = 0$

39. $25e^{2x+1} = 962$

40. $80e^{-t/2} + 20 = 70$

41. $\dfrac{1}{1 - 2e^{-t}} = 1.5$

42. $\dfrac{10{,}000}{1 + 19e^{-t/5}} = 2000$

In Exercises 43–60, solve the logarithmic equation.

43. $\ln x = 5$

44. $\ln x = 2$

45. $\ln x = -10$

46. $\ln x = -2$

47. $\ln 2x = 1$

48. $\ln 4x = 2.5$

49. $2 \ln x = 7$

50. $3 \ln 5x = 10$

51. $2\ln 4x = 0$

52. $6\ln(x+1) = 2$

53. $5 + \ln 2x = 4$

54. $-6 + \ln 3x = 0$

55. $\ln 2x^2 = 5$

56. $\ln x^2 = 1$

57. $\ln x - \ln 2 = 5$

58. $\ln x + \ln(x+2) = 3$

59. $\ln x + \ln(x-2) = 1$

60. $\ln \sqrt{(x+2)} = 1$

Compound Interest   In Exercises 61 and 62, find the time required for a $1000 investment to double at interest rate r, compounded continuously.

61. $r = 0.085$

62. $r = 0.12$

Compound Interest   In Exercises 63 and 64, find the time required for a $1000 investment to triple at interest rate r, compounded continuously.

63. $r = 0.0675$

64. $r = 0.075$

65. *Demand Function*   The demand function for a product is given by
$$p = 500 - 0.5(e^{0.004x}).$$

(a) Find the demand x for a price of $p = \$350$.

(b) Find the demand x for a price of $p = \$300$.

(c) Use a graphing utility to graphically confirm the results found in parts (a) and (b).

66. *Demand Function*   The demand function for a product is given by
$$p = 5000 \left(1 - \frac{4}{4 + e^{-0.002x}}\right).$$

(a) Find the demand x for a price of $p = \$600$.

(b) Find the demand x for a price of $p = \$400$.

(c) Use a graphing utility to graphically confirm the results found in parts (a) and (b).

67. *Forest Yield*   The yield V (in millions of cubic feet per acre) for a forest at age t years is given by
$$V = 6.7e^{-48.1/t}, \qquad 0 \le t.$$

(a) Use a graphing utility to find the time necessary to have a yield of 1.3 million cubic feet.

(b) Use a graphing utility to find the time necessary to have a yield of 2 million cubic feet.

68. *Human Memory Model*   In a group project on learning theory, a mathematical model for the percent P of correct responses after n trials was found to be
$$P = \frac{0.83}{1 + e^{-0.2n}}, \qquad 0 \le n.$$

(a) After how many trials will 60% of the responses be correct? (That is, for what value of n will $P = 0.6$?)

(b) Use a graphing utility to graph the memory model and confirm the result found in part (a).

(c) Write a paragraph describing the memory model.

In Exercises 69 and 70, use a graphing utility to graph the function.

69. $y = e^{-x^2}$

70. $y = 2e^{-x^2}$

71. *Average Heights*   The percent of American males (between 18 and 24 years old) who are less than x inches tall is given by
$$p = \frac{1}{1 + e^{-0.6114(x-69.71)}}, \qquad 60 \le x \le 77$$
where p is the percent (in decimal form) and x is the height in inches (see figure).   (Source: U.S. National Center for Health Statistics)

(a) What is the median height for American males between 18 and 24 years old? (In other words, for what value of x is p equal to 0.5?)

(b) Write a paragraph describing the height model.

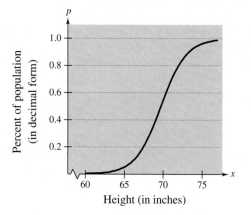

Height (in inches)

72. *Average Heights*   The percent of American females (between 18 and 24 years old) who are less than x inches tall is given by

$$p = \frac{1}{1 + e^{-0.66607(x-64.51)}}, \qquad 55 \le x \le 72$$

where p is the percent (in decimal form) and x is the height in inches (see figure).   (Source: U.S. National Center for Health Statistics)

(a) What is the median height for American females between 18 and 24 years old? (In other words, for what value of x is p equal to 0.5?)

(b) Write a paragraph describing the height model.

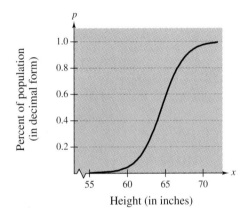

Height (in inches)

73. *Temperature in Baltimore*   The average monthly *low* temperature in Baltimore, Maryland can be approximated by the model

$$y = 21 + 46.3e^{-(t-7.2)^2/13}, \qquad 1 \le t \le 12$$

where y is the average monthly low temperature (in degrees Fahrenheit) and t represents the month of the year, with $t = 1$ corresponding to mid-January. (Source: PC U.S.A.)

(a) Use a graphing utility to graph the model.

(b) During which months of the year is the low temperature likely to fall below freezing?

(c) Write a paragraph describing the temperature model.

74. *Temperature in Baltimore*   The average monthly *high* temperature in Baltimore, Maryland can be approximated by the model

$$y = 38.5 + 50.4e^{-(t-7.1)^2/13.3}, \qquad 1 \le t \le 12$$

where y is the average monthly high temperature (in degrees Fahrenheit) and t represents the month of the year, with $t = 1$ corresponding to mid-January (see figure). During which months of the year is the high temperature likely to rise above $80°$? (Source: PC U.S.A.)

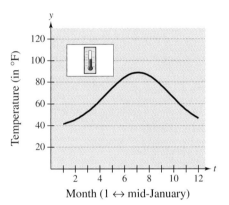

Month (1 ↔ mid-January)

75. *Native Prairie Grasses*   The number of native prairie grasses per acre A is approximated by the model

$$A = 10.5 \cdot 10^{0.04x}, \qquad 0 \le x \le 24$$

where x is the number of months since the acre was plowed. Use this model to approximate the number of months since the field was plowed in a test plot if $A = 280$.

76. *Suburban Wildlife*   The variety of suburban non-domesticated wildlife is approximated by the model

$$V = 15 \cdot 10^{0.02x}, \qquad 0 \le x \le 36$$

where x is the number of months since the development was completed. Use this model to approximate the number of months since the development was completed if $V = 50$.

5.5 **Exponential and Logarithmic Models**

Introduction

The five most common types of mathematical models involving exponential functions and logarithmic functions are as follows.

1. Exponential growth: $\qquad y = ae^{bx}, \qquad b > 0$

2. Exponential decay: $\qquad y = ae^{-bx}, \qquad b > 0$

3. Gaussian model: $\qquad y = ae^{-(x-b)^2/c}$

4. Logistics growth model: $\qquad y = \dfrac{a}{1 + be^{-(x-c)/d}}$

5. Logarithmic models: $\qquad y = \ln(ax + b), \qquad y = \log_{10}(ax + b)$

The basic shape of each of these graphs is shown in Figure 5.20.

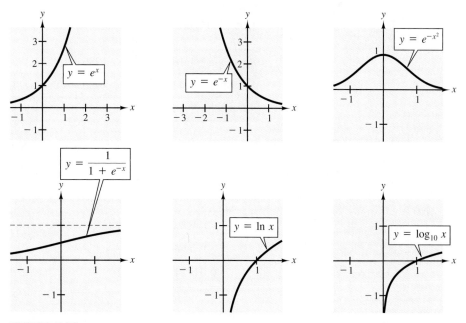

FIGURE 5.20

Exponential Growth and Decay

EXAMPLE 1   *Population Increase*

The world population (in millions) from 1980 through 1992 is shown in the table.

Year	1980	1985	1986	1987	1988	1989	1990	1991	1992
Population	4453	4850	4936	5024	5112	5202	5294	5384	5478

(Source: Statistical Office of the United Nations)

An exponential growth model that approximates this data is given by

$$P = 4453e^{0.01724t}, \qquad 0 \le t \le 12$$

where P is the population (in millions) and $t = 0$ represents 1980. Compare the values given by the model with the estimates given by the United Nations. According to this model, when will the world population reach 6 billion?

Solution

The following table compares the two sets of population figures. The model is shown in Figure 5.21.

Year	1980	1985	1986	1987	1988	1989	1990	1991	1992
Population	4453	4850	4936	5024	5112	5202	5294	5384	5478
Model	4453	4854	4938	5024	5112	5200	5291	5383	5477

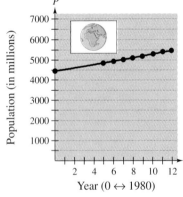

FIGURE 5.21

To find when the world population will reach 6 billion, let $P = 6000$ in the model and solve for t.

$4453e^{0.01724t} = P$	Given model
$4453e^{0.01724t} = 6000$	Let P equal 6000.
$e^{0.01724t} \approx 1.3474$	Divide both sides by 4453.
$\ln e^{0.01724t} \approx \ln 1.3474$	Take log of both sides.
$0.01724t \approx 0.2982$	Inverse property of logs and exponents
$t \approx 17.3$	Divide both sides by 0.01724.

According to the model, the world population will reach 6 billion in 1997.

NOTE   An exponential model increases (or decreases) by the same percent each year. What is the annual percent increase for the exponential model above?

In Example 1, you were given the exponential growth model. But suppose this model were not given; how could you find such a model? One technique for doing this is demonstrated in Example 2.

EXAMPLE 2   *Finding an Exponential Growth Model*

Find an exponential growth model whose graph passes through the points $(0, 4453)$ and $(7, 5024)$, as shown in Figure 5.22(a).

Solution

The general form of the model is

$$y = ae^{bx}.$$

From the fact that the graph passes through the point $(0, 4453)$, you know that $y = 4453$ when $x = 0$. By substituting these values into the general model, you have

$$4453 = ae^0 \qquad \Longrightarrow \qquad a = 4453.$$

In a similar way, from the fact that the graph passes through the point $(7, 5024)$, you know that $y = 5024$ when $x = 7$. By substituting these values into the model, you obtain

$$5024 = 4453e^{7b} \qquad \Longrightarrow \qquad b = \frac{1}{7}\ln\frac{5024}{4453} \approx 0.01724.$$

Thus, the exponential growth model is

$$y = 4453e^{0.01724x}.$$

The graph of the model is shown in Figure 5.22(b).

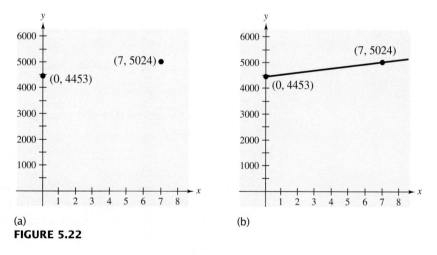

(a)

(b)

FIGURE 5.22

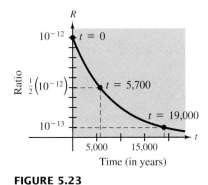

FIGURE 5.23

In living organic material, the ratio of radioactive carbon isotopes (carbon 14) to the number of nonradioactive carbon isotopes (carbon 12) is about 1 to 10^{12}. When organic material dies, its carbon 12 content remains fixed, whereas its radioactive carbon 14 begins to decay with a half-life of about 5700 years. To estimate the age of dead organic material, scientists use the following formula, which denotes the ratio of carbon 14 to carbon 12 present at any time t (in years).

$$R = \frac{1}{10^{12}} e^{-t/8223}$$

The graph of R is shown in Figure 5.23. Note that R decreases as the time t increases.

Real Life

EXAMPLE 3 Carbon Dating

The ratio of carbon 14 to carbon 12 in a newly discovered fossil is

$$R = \frac{1}{10^{13}}.$$

Estimate the age of the fossil.

Solution

In the carbon dating model, substitute the given value of R to obtain the following.

$\dfrac{1}{10^{12}} e^{-t/8223} = R$	Given model
$\dfrac{e^{-t/8223}}{10^{12}} = \dfrac{1}{10^{13}}$	Let R equal $\dfrac{1}{10^{13}}$.
$e^{-t/8223} = \dfrac{1}{10}$	Multiply both sides by 10^{12}.
$\ln e^{-t/8223} = \ln \dfrac{1}{10}$	Take log of both sides.
$-\dfrac{t}{8223} \approx -2.3026$	Inverse property of logs and exponents
$t \approx 18{,}934$	Multiply both sides by -8223.

Thus, to the nearest thousand years, you can estimate the age of the fossil to be 19,000 years.

NOTE The carbon dating model in Example 3 assumed that the carbon 14/carbon 12 ratio was one part in 10,000,000,000,000. Suppose an error in measurement occurred and the actual ratio was only one part in 8,000,000,000,000. The fossil age corresponding to the actual ratio would then be approximately 17,000 years. Try checking this result.

Gaussian Models

As mentioned at the beginning of this section, Gaussian models are of the form

$$y = ae^{-(x-b)^2/c}.$$

This type of model is commonly used in probability and statistics to represent populations that are **normally distributed.** For *standard* normal distributions, the model takes the form

$$y = \frac{1}{\sigma\sqrt{2\pi}}e^{-x^2/2\sigma^2}$$

where σ is the standard deviation (σ is the lowercase Greek letter sigma). The graph of a Gaussian model is called a **bell-shaped curve.** Try assigning a value to σ and sketching a normal distribution curve with a graphing utility. Can you see why it is called a bell-shaped curve?

Real Life

EXAMPLE 4   SAT Scores

In 1988, the Scholastic Aptitude Test (SAT) scores for males roughly followed a normal distribution given by

$$y = 0.0026e^{-(x-500)^2/48,000}, \qquad 200 \le x \le 800$$

where x is the SAT score for mathematics. Sketch the graph of this function. From the graph, estimate the average SAT score.   (Source: College Board)

Solution

The graph of the function is given in Figure 5.24. From the graph, you can see that the average mathematics score for males in 1988 was 500.

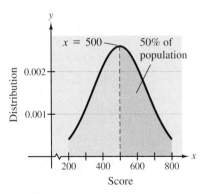

FIGURE 5.24

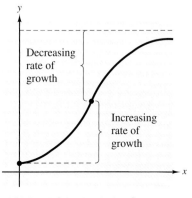

FIGURE 5.25 Logistics Curve

Logistics Growth Models

Some populations initially have rapid growth, followed by a declining rate of growth, as indicated by the graph in Figure 5.25. One model for describing this type of growth pattern is the **logistics curve** given by the function

$$y = \frac{a}{1 + be^{-(x-c)/d}}$$

where y is the population size and x is the time. An example is a bacteria culture allowed to grow initially under ideal conditions, followed by less favorable conditions that inhibit growth. A logistics growth curve is also called a **sigmoidal curve.**

Real Life

EXAMPLE 5   Spread of a Virus

On a college campus of 5000 students, one student returned from vacation with a contagious flu virus. The spread of the virus through the student body is given by

$$y = \frac{5000}{1 + 4999e^{-0.8t}}, \qquad 0 \le t$$

where y is the total number infected after t days. The college will cancel classes when 40% or more of the students are ill.

a. How many are infected after 5 days?

b. After how many days will the college cancel classes?

Solution

a. After 5 days, the number infected is

$$y = \frac{5000}{1 + 4999e^{-0.8(5)}} = \frac{5000}{1 + 4999e^{-4}} \approx 54.$$

b. In this case, the number infected is $(0.40)(5000) = 2000$. Therefore, we solve for t in the following equation.

$$2000 = \frac{5000}{1 + 4999e^{-0.8t}}$$

$$1 + 4999e^{-0.8t} = 2.5$$

$$e^{-0.8t} \approx 0.0003$$

$$\ln(e^{-0.8t}) \approx \ln 0.0003$$

$$-0.8t \approx -8.1115$$

$$t \approx 10.1$$

Hence, after 10 days, at least 40% of the students will be infected, and the college will cancel classes. The graph of the function is shown in Figure 5.26.

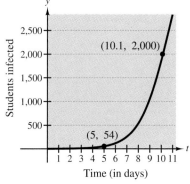

FIGURE 5.26

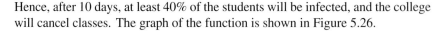

Logarithmic Models

Real Life

EXAMPLE 6   Magnitude of Earthquakes

On the Richter scale, the magnitude R of an earthquake of intensity I is given by

$$R = \log_{10} \frac{I}{I_0}$$

where $I_0 = 1$ is the minimum intensity used for comparison. Find the intensity per unit of area for the following earthquakes. (Intensity is a measure of the wave energy of an earthquake.)

a. San Francisco in 1906, $R = 8.3$

b. San Francisco Bay Area in 1989, $R = 6.9$

Solution

a. Because $I_0 = 1$ and $R = 8.3$, you have

$$8.3 = \log_{10} I$$
$$I = 10^{8.3} \approx 199{,}526{,}000.$$

b. For $R = 6.9$, you have $6.9 = \log_{10} I$, and

$$I = 10^{6.9} \approx 7{,}943{,}000.$$

Note that an increase of 1.4 units on the Richter scale (from 6.9 to 8.3) represents an intensity change by a factor of

$$\frac{199{,}526{,}000}{7{,}943{,}000} \approx 25.$$

In other words, the "great San Francisco earthquake" of 1906 had a magnitude about 25 times greater than that of the 1989 quake.

This photo was taken in Anchorage, Alaska in 1964 after a devastating earthquake that measured 9.1 on the Richter scale. This was the second largest quake registered in the world since 1900. (The largest took place in Chile in 1960, measuring 9.7 on the Richter scale.)

Group Activities   Extending the Concept

Exponential and Logarithmic Models   Use your school's library or some other reference source to find an application that fits one of the five models discussed in this section. After you have collected data for the model, plot the corresponding points and find an equation that describes the points you have plotted.

Warm Up The following warm-up exercises involve skills that were covered in earlier sections. You will use these skills in the exercise set for this section.

In Exercises 1–6, sketch the graph of the equation.

1. $y = e^{0.1x}$

2. $y = e^{-0.25x}$

3. $y = e^{-x^2/5}$

4. $y = \dfrac{2}{1 + e^{-x}}$

5. $y = \log_{10} 2x$

6. $y = \ln 4x$

In Exercises 7 and 8, solve the equation algebraically.

7. $3e^{2x} = 7$

8. $4 \ln 5x = 14$

In Exercises 9 and 10, solve the equation graphically.

9. $2e^{-0.2x} = 0.002$

10. $6 \ln 2x = 12$

5.5 Exercises

Compound Interest In Exercises 1–10, complete the table for a savings account in which interest is compounded continuously.

	Initial Investment	Annual % rate	Time to Double	Amount After 10 Years
1.	$1000	12%		
2.	$20,000	$10\frac{1}{2}$%		
3.	$750		$7\frac{3}{4}$ yr	
4.	$10,000		5 yr	
5.	$500			$1292.85
6.	$2000			$3000
7.		11%		$19,205
8.		8%		$20,000
9.	$5000			$11,127.70
10.	$250			$600

Radioactive Decay In Exercises 11–16, complete the table for the radioactive isotopes.

	Isotope	Half-Life (Years)	Initial Quantity	Amount After 1000 Years
11.	Ra^{226}	1620	10 g	
12.	Ra^{226}	1620		1.5 g
13.	C^{14}	5730		2 g
14.	C^{14}	5730	3 g	
15.	Pu^{230}	24,360		2.1 g
16.	Pu^{230}	24,360		0.52 g

In Exercises 17–20, classify the model as exponential growth or exponential decay.

17. $y = 3e^{0.5t}$

18. $y = 2e^{-0.6t}$

19. $y = 20e^{-1.5t}$

20. $y = 4e^{0.07t}$

In Exercises 21–24, find the constant k such that the exponential function $y = Ce^{kt}$ passes through the points on the graph.

21.

22.

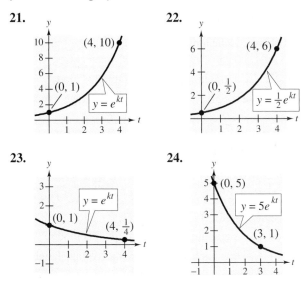

23.

24.

25. *Population* The population P of a city is given by

$$P = 105,300e^{0.015t}$$

where t is the time in years, with $t = 0$ corresponding to 1990. Sketch the graph of this equation. According to this model, in what year will the city have a population of 150,000?

26. *Population* The population P of a city is given by

$$P = 240,360e^{0.012t}$$

where t is the time in years, with $t = 0$ corresponding to 1990. Sketch the graph of this equation. According to this model, in what year will the city have a population of 250,000?

27. *Population* The population P of a city is given by

$$P = 25,000e^{kt}$$

where t is the time in years, with $t = 0$ corresponding to the year 1990. In 1945, the population was 3350. Find the value of k and use this result to predict the population in the year 2010.

28. *Population* The population P of a city is given by

$$P = 140,500e^{kt}$$

where $t = 0$ represents 1990. In 1960, the population was 100,250. Find the value of k and use this result to predict the population in the year 2000.

29. *Bacteria Growth* The number of bacteria N in a culture is given by the model

$$N = 100e^{kt}$$

where t is the time in hours, with $t = 0$ corresponding to the time when $N = 100$. When $t = 5$, there are 300 bacteria. How long does it take the population to double in size?

30. *Bacteria Growth* The number of bacteria N in a culture is given by the model

$$N = 250e^{kt}$$

where t is the time in hours, with $t = 0$ corresponding to the time when $N = 250$. When $t = 10$, there are 280 bacteria. How long does it take the population to double in size?

31. *Radioactive Decay* What percent of a present amount of radioactive radium (Ra^{226}) will remain after 100 years? Use the fact that radioactive radium has a half-life of 1620 years.

32. *Radioactive Decay* Find the half-life of a radioactive material if, after 1 year, 99.57% of the initial amount remains.

33. *Learning Curve* The management at a factory has found that the maximum number of units a worker can produce in a day is 30. The learning curve for the number of units N produced per day after a new employee has worked t days is given by

$$N = 30(1 - e^{kt}).$$

After 20 days on the job, a particular worker produced 19 units in 1 day.

(a) Find the learning curve for this worker (that is, find the value of k).

(b) How many days should pass before this worker is producing 25 units per day?

34. *Sales*   The sales S (in thousands of units) of a new product after it has been on the market for t years are given by $S = 100(1 - e^{kt})$.

(a) Find S as a function of t if 15,000 units have been sold after 1 year.

(b) How many units will have been sold after 5 years?

35. *Number of Golfers*   From 1970 to 1985, the number of golfers in the United States was increasing at the rate of 2.956% per year. Then, in 1985, the growth pattern changed. The number of golfers from 1970 to 1991 was approximately

$$y = \begin{cases} 11,245e^{0.02956t}, & 0 \le t \le 15 \\ 8677.6 + 786.8t - e^{23.03-t}, & 15 \le t \le 21 \end{cases}$$

where y is the number of golfers (in thousands) and $t = 0$ represents 1970.   (Source: National Golf Foundation)

(a) Use a graphing utility to graph this model.

(b) Describe the change in the growth pattern that occurred in 1985.

36. *Textbook Sales*   The sales of a college textbook that was published in 1988 and revised in 1992 can be approximated by the function

$$y = \begin{cases} e^{-3.04+3.38t-0.22t^2}, & 8 \le t \le 11 \\ e^{-19.61+5.13t-0.22t^2}, & 12 \le t \le 15 \end{cases}$$

where y is the number of copies sold and $t = 8$ represents 1988.

(a) Use a graphing utility to graph this model.

(b) The graph is typical of college text sales. How would you explain this sales pattern?

37. *Odometer Readings*   In 1990, the probability distribution of odometer readings for 1987 automobiles was approximately

$$p = 0.1337e^{-(x-32.2)^2/35.6}$$

where x is the odometer reading in thousands of miles. What was the average odometer reading for a 1987 automobile?   (Source: Based on Statistics from the U.S. Highway Administration)

38. *Foot Size*   A group of 4000 air force men were measured for foot size. The distribution of the various foot sizes was given by

$$p = 0.8865e^{-2.47(x-10.5)^2}$$

where x is the length of the foot in inches.   (Source: USAF Wright Air Development Center)

(a) Use a graphing utility to graph this model.

(b) Use the graph to determine the average foot size.

39. *Women's Heights*   The distribution for the heights of American women (between 25 and 34 years of age) can be approximated by the function

$$p = 0.166e^{-(x-64.5)^2/11.5}, \qquad 55 \le x \le 72$$

where x is the height in inches.   (Source: U.S. National Center for Health Statistics)

(a) Use a graphing utility to graph this model.

(b) Use the graph to determine the average height of women in this age bracket.

40. *Men's Heights*   The distribution for the heights of American men (between 25 and 34 years of age) can be approximated by the function

$$p = 0.193e^{-(x-70)^2/14.32}, \qquad 60 \le x \le 77$$

where x is the height in inches.   (Source: U.S. National Center for Health Statistics)

(a) Use a graphing utility to graph this model.

(b) Use the graph to determine the average height of men in this age bracket.

41. *Stocking a Lake with Fish*   A certain lake is stocked with 500 fish, and the fish population increases according to the logistics curve

$$P = \frac{10,000}{1 + 19e^{-t/5}}, \qquad 0 \le t$$

where t is measured in months.

(a) Use a graphing utility to graph this model.

(b) Find the fish population after 5 months.

(c) After how many months will the fish population be 2000?

42. *High School Graduates*   The percent of the American population (who are 25 years old or older) who have *not* completed high school has been decreasing according to the model

$$P = 14.9 + \frac{235.18}{1 + e^{(t+24.3)/9.87}}, \qquad 0 \le t \le 13$$

where P is the percent of the population and t is the calendar year, with $t = 0$ corresponding to 1980. (Source: *Current Population Reports*)

(a) Use a graphing utility to graph this model.

(b) According to this model, when did the percent of the population who have not completed high school fall below 25%?

(c) Use the given equation to construct a model that represents the percent of the population who *have* completed high school.

43. *Endangered Species*   A conservation organization releases 100 animals of an endangered species into a game preserve. The organization believes that the preserve has a carrying capacity of 1000 animals and that the growth of the herd will be modeled by the logistics curve

$$p = \frac{1000}{1 + 9e^{-kt}}, \qquad 0 \le t$$

where t is measured in years.

(a) Find k if the herd size is 134 after 2 years.

(b) Find the population after 5 years.

44. *Sales and Advertising*   After discontinuing all advertising for a certain product (in 1988), the manufacturer of the product found that the sales began to drop according to the model

$$S = \frac{500{,}000}{1 + 0.6e^{kt}}, \qquad 0 \le t \le 10$$

where S represents the number of units sold and t represents the calendar year, with $t = 0$ corresponding to 1988.

(a) Find k if the company sold 300,000 units in 1990.

(b) According to this model, what were the sales in 1993?

45. *Videocassette Prices*   The average price P (in dollars) of a prerecorded videocassette from 1981 to 1988 is given by

$$P = 69.52 - \frac{55.47}{1 + e^{-(t-3.83)/1.20}}, \qquad 1 \le t \le 8$$

where $t = 1$ represents 1981 (see figure). Why do you think the average price of a videocassette was falling during this time period?   (Source: Paul Kagan Associates, Inc.)

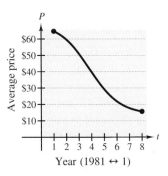

Year (1981 ↔ 1)

46. *Videocassette Sales*   From 1981 to 1988, the sales S (in thousands of cassettes) of prerecorded videocassettes in the United States was given by

$$S = 1628 + \frac{82{,}092}{1 + e^{-(t-6)/0.82}}, \qquad 1 \le t \le 8$$

where $t = 1$ represents 1981 (see figure). Use the model for the average price of a videocassette in Exercise 45 to determine a model for the total *revenue* for videocassettes from 1981 to 1988. Sketch the graph of the revenue function. Was the revenue increasing or decreasing during this period of time?   (Source: Paul Kagan Associates, Inc.)

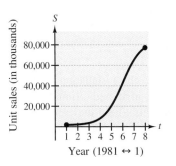

Year (1981 ↔ 1)

Earthquake Magnitudes   In Exercises 47 and 48, use the Richter scale (Example 6) for measuring the magnitude of earthquakes.

47. Find the magnitude R (on the Richter scale) of an earthquake of intensity I (let $I_0 = 1$).

   (a) $I = 80,500,000$

   (b) $I = 48,275,000$

   (c) $I = 40,000,000$

   (d) $I = 20,000,000$

48. Find the intensity I of an earthquake measuring R on the Richter scale (let $I_0 = 1$).

   (a) Colombia in 1906, $R = 8.6$

   (b) Los Angeles in 1971, $R = 6.7$

Intensity of Sound   In Exercises 49 and 50, use the information to determine the level of sound (in decibels) for the sound intensity. The level of sound β, in decibels, of a sound with an intensity of I is given by

$$\beta(I) = 10 \log_{10} \frac{I}{I_0}$$

where I_0 is an intensity of 10^{-16} watts per square centimeter, corresponding roughly to the faintest sound that can be heard.

49. (a) $I = 10^{-14}$ watts per cm^2 (faint whisper)

   (b) $I = 10^{-9}$ watts per cm^2 (busy street corner)

   (c) $I = 10^{-6.5}$ watts per cm^2 (air hammer)

   (d) $I = 10^{-4}$ watts per cm^2 (threshold of pain)

50. (a) $I = 10^{-13}$ watts per cm^2 (whisper)

   (b) $I = 10^{-7.5}$ watts per cm^2 (DC-8 4 miles away)

   (c) $I = 10^{-7}$ watts per cm^2 (diesel truck at 25 feet)

   (d) $I = 10^{-4.5}$ watts per cm^2 (automobile horn at 3 feet)

51. *Noise Level*   Due to the installation of acoustical tiling, the noise level in an auditorium was reduced from 78 to 72 decibels. Find the percent decrease in the intensity level of the noise as a result of the installation of the tiling.

52. *Noise Level*   Due to the installation of a muffler, the noise level of an automobile engine was reduced from 88 to 76. Find the percent decrease in the intensity level of the noise as a result of the installation of the muffler.

53. *Estimating the Time of Death*   At 8:30 A.M., a coroner was called to the home of a person who had died during the night. In order to estimate the time of death, the coroner took the person's temperature twice. At 9:00 A.M. the temperature was 85.7°, and at 9:30 A.M. the temperature was 82.8°. From these two temperature readings, the coroner was able to determine that the time elapsed since death and the body temperature were related by the formula

$$t = -2.5 \ln \frac{T - 70}{98.6 - 70}$$

where t is the time in hours that have elapsed since the person died and T is the temperature (in degrees Fahrenheit) of the person's body. Assume the person had a normal body temperature of 98.6° at death, and the room temperature was a constant 70°. (This formula is derived from a general cooling principle called Newton's Law of Cooling.) Use this formula to estimate the time of death of the person.

54. *Thawing a Package of Steaks*   Suppose you take a 3-pound package of steaks out of the freezer at noon. Will the steaks be thawed in time to be grilled on the barbecue at 6 P.M.? To answer this question, assume that the room temperature is 70° Fahrenheit and that the freezer temperature is 24°. Use the formula

$$t = -3.8 \ln \frac{T - 70}{24 - 70}$$

where t is the time (with $t = 0$ corresponding to noon) and T is the temperature of the package of steaks.

CHAPTER PROJECT: Newton's Law of Cooling

Time	Temperature
0.2	155.8
4.3	144.1
8.4	133.2
12.5	124.4
16.6	117.9
20.7	112.5
24.8	107.9
28.9	104
33	100.7
37.1	97.6
41.1	94.9
45.2	92.7
49.3	90.5

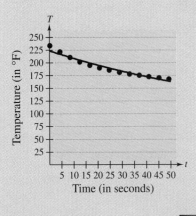

When a warm object is placed in a cool room, its temperature will gradually change to the temperature of the room. (The same is true for a cool object placed in a warm room.)

According to Newton's Law of Cooling, the rate at which the object's temperature T changes is proportional to the difference between its temperature and the temperature of the room L. This statement can be translated as

$$T = L + Ce^{kt}$$

where k is a constant and $L + C$ is the original temperature of the object.

To confirm Newton's Law of Cooling experimentally, you can place a warm object in a cool room and monitor its temperature. The table and scatter plot at the left show the temperatures (in degrees Fahrenheit) for several times (in seconds) for an object that was placed in a room whose temperature was 77.7°F. This data was collected using a *Texas Instruments CBL* unit.*

Use this information to investigate the following questions.

1. ***Fitting a Model to Data***   Enter the data given in the table in a graphing utility and use its exponential regression feature to find an exponential model for the data. The model doesn't fit the data very well—why?

2. ***Fitting a Model to Data***   Copy the table shown at the left. Then add another column to the table that shows the difference between the temperature of the object and the room temperature. Now run the graphing utility's exponential regression program on the temperature difference data. For the model

$$y = a \cdot b^x$$

what are the values of a and b?

3. ***Changing a Model's Form***   In Question 2, use the fact that $b = e^{\ln b}$ to rewrite the model in the form

$$T = Ce^{kt}.$$

4. ***Comparing a Model to Actual Data***   Using the results of Question 3, you can write a model for the temperature of the cooling object.

$$T = 77.7 + Ce^{kt}$$

Construct a table that compares the values of this model with the actual temperatures collected by the *CBL* unit.

5. ***Extended CBL Exploration***   Use the *CBL* instructions in Appendix C to collect and analyze your own data for a cooling object. Try fitting a model to a cooling object, and then to a warming object.

*Instructions for using a *CBL* to collect your own data are given in Appendix C.

CHAPTER SUMMARY

After studying this chapter, you should have acquired the following skills. These skills are keyed to the Review Exercises that begin on page 427. Answers to odd-numbered Review Exercises are given in the back of the book.

- Match an exponential function with its graph. *(Section 5.1)* — **Review Exercises 1–4**

- Sketch the graph of an exponential function. *(Section 5.1)* — **Review Exercises 9–16**

- Complete a table using the compound interest formula. *(Section 5.1)* — **Review Exercises 17–20**

- Use an exponential function in a real-life application. *(Section 5.1)* — **Review Exercises 21, 22**

- Match a logarithmic function with its graph. *(Section 5.2)* — **Review Exercises 5–8**

- Evaluate a logarithmic expression. *(Section 5.2)* — **Review Exercises 23–28**

- Write an equation in logarithmic form. *(Section 5.2)* — **Review Exercises 29, 30**

- Sketch the graphs of inverse functions. *(Section 5.2)* — **Review Exercises 31, 32**

- Find the domain, vertical asymptote, and x-intercept of a logarithmic function and sketch its graph. *(Section 5.2)* — **Review Exercises 33, 34**

- Use a logarithmic function in a real-life application. *(Section 5.2)* — **Review Exercises 35–38**

- Evaluate a logarithm using the change-of-base formula. *(Section 5.3)* — **Review Exercises 39–42**

- Write an expression as the sum, difference, and/or multiple of logarithms using the properties of logarithms. *(Section 5.3)* — **Review Exercises 43–48**

- Write an expression as the logarithm of a single quantity. *(Section 5.3)* — **Review Exercises 49–54**

- Approximate logarithms using the properties of logarithms. *(Section 5.3)* — **Review Exercises 55–58**

- Find the exact value of a logarithm. *(Section 5.3)* — **Review Exercises 59–62**

- Find an equation that relates x and y using properties of logarithms and exponentiation. *(Section 5.3)* — **Review Exercise 79**

- Solve an exponential or logarithmic equation. *(Section 5.4)* — **Review Exercises 63–74**

- Complete a table using an exponential decay model. *(Section 5.5)* — **Review Exercises 75, 76**

- Use an exponential model to solve a real-life application. *(Section 5.5)* — **Review Exercises 77, 78, 80, 81, 87**

- Use a Gaussian model to solve a real-life application. *(Section 5.5)* — **Review Exercise 82**

- Use a logistics growth model to solve a real-life application. *(Section 5.5)* — **Review Exercises 83, 86**

- Use a logarithmic model to solve a real-life application. *(Section 5.5)* — **Review Exercises 84, 85**

REVIEW EXERCISES

In Exercises 1–8, match the function with its graph. [The graphs are labeled (a), (b), (c), (d), (e), (f), (g), and (h).]

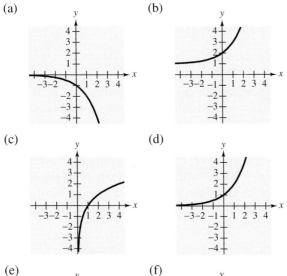

(a)

(b)

(c)

(d)

(e)

(f)

(g)

(h)

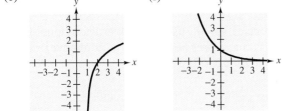

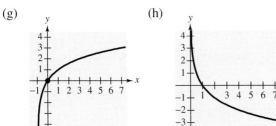

1. $f(x) = 2^x$

2. $f(x) = 2^{-x}$

3. $f(x) = -2^x$

4. $f(x) = 2^x + 1$

5. $f(x) = \log_2 x$

6. $f(x) = \log_2(x - 1)$

7. $f(x) = -\log_2 x$

8. $f(x) = \log_2(x + 1)$

In Exercises 9–16, sketch the graph of the function.

9. $y = 4^x$

10. $y = e^{2x}$

11. $y = \left(\frac{1}{2}\right)^x$

12. $f(x) = \left(\frac{1}{2}\right)^{x+1}$

13. $f(x) = 3e^{0.2x}$

14. $f(x) = 4^{x-2}$

15. $y = 4^{-x^2}$

16. $f(x) = 100e^{-0.2x}$

In Exercises 17 and 18, complete the table to determine the balance A for P dollars invested at a rate r for t years and compounded n times per year.

n	1	2	4	12	365	Continuous
A						

17. $P = \$3500$, $r = 10.5\%$, $t = 10$ years

18. $P = \$2000$, $r = 12\%$, $t = 30$ years

In Exercises 19 and 20, complete the table to determine the amount P that should be invested at a rate r to produce a final balance of $200,000 in t years.

t	1	10	20	30	40	50
P						

19. $r = 8\%$, compounded continuously

20. $r = 10\%$, compounded monthly

21. *Trust Fund*   Suppose the day your child was born you deposited $50,000 in a trust fund that paid 8.75% interest, compounded continuously. Between what birthdays would your child become a millionaire?

22. *Population*   Suppose the population of a town is increasing at the rate of 5% per year. The population of the town in 1990 is 50,000.

(a) During which year will the population of the town reach 100,000?

(b) During which year will the population of the town reach 200,000?

In Exercises 23–28, evaluate the expression without using a calculator.

23. $\log_{10} 1000$ **24.** $\log_9 3$

25. $\ln e^7$ **26.** $\log_a \dfrac{1}{a}$

27. $\ln e^{-3}$ **28.** $\ln 1$

In Exercises 29 and 30, use the definition of a logarithm to write the equation in logarithmic form.

29. $4^3 = 64$ **30.** $25^{3/2} = 125$

In Exercises 31 and 32, use the fact that f and g are inverses of each other to sketch their graphs in the same coordinate plane.

31. $f(x) = 10^x$, $g(x) = \log_{10} x$
32. $f(x) = e^x$, $g(x) = \ln x$

In Exercises 33 and 34, find the domain, vertical asymptote, and x-intercept of the logarithmic function and sketch its graph.

33. $f(x) = \ln(x + 2)$ **34.** $g(x) = \ln(3 - x)$

35. *Human Memory Model* Students in a sociology class were given an exam and then retested monthly with an equivalent exam. The average score for the class was given by the human memory model

$$f(t) = 85 - 14 \log_{10}(t + 1), \qquad 0 \le t \le 4$$

where t is the time in months. How did the average score change over the 4-month period?

36. *Investment Time* A principal P, invested at 8.75% and compounded continuously, increases to an amount that is K times the original principal after t years, where t is given by

$$t = \frac{\ln K}{0.0875}.$$

Complete the following table and describe the result.

K	1	2	3	4	6	8	10
t							

Median Age of the United States Population In Exercises 37 and 38, use the model

$$A = 18.10 - 0.016t + 5.121 \ln t, \qquad 10 \le t \le 80$$

which approximates the median age of the United States population from 1980 to 2050. In the model, assume A is the median age and t is the calendar year, with $t = 10$ corresponding to 1980. (Source: United States Bureau of Census)

37. Use the model to approximate the median ages in the United States in 1995 and 2010.

38. Based on your answers in Exercise 37, discuss the change in the median age in the United States from 1995 to 2010.

In Exercises 39–42, evaluate the logarithm using the change-of-base formula. Do each problem twice, once with common logarithms and once with natural logarithms. (Round your answer to three decimal places.)

39. $\log_4 9$ **40.** $\log_{1/2} 5$

41. $\log_{12} 200$ **42.** $\log_3 0.28$

In Exercises 43–48, use the properties of logarithms to write the expression as a sum, difference, and/or multiple of logarithms.

43. $\log_{10} \dfrac{x}{y}$ **44.** $\ln \dfrac{xy^2}{z}$

45. $\ln\left(x\sqrt{x - 3}\right)$ **46.** $\ln \sqrt[3]{\dfrac{x^3}{y^2}}$

47. $\log_{10}(y - 5)^3$ **48.** $\ln 2xyz$

In Exercises 49–54, write the expression as the logarithm of a single quantity.

49. $\ln x + \ln 5$ **50.** $\ln y + 2 \ln z$

51. $\tfrac{1}{2} \ln x$ **52.** $2 \ln x + 3 \ln y - 5 \ln z$

53. $\ln x - \ln(x - 3) - \ln(x + 1)$

54. $\ln(x + 2) + 2 \ln x - 3 \ln(x + 4)$

In Exercises 55–58, approximate the logarithm using the properties of logarithms given $\log_b 2 \approx 0.3562$, $\log_b 3 \approx 0.5646$, and $\log_b 5 \approx 0.8271$.

55. $\log_b 25$

56. $\log_b \left(\frac{25}{9}\right)$

57. $\log_b \sqrt{3}$

58. $\log_b 30$

In Exercises 59–62, find the *exact* value of the logarithm without using a calculator.

59. $\log_3 27$

60. $\log_4 \frac{1}{64}$

61. $\ln e^{3.2}$

62. $\ln \sqrt[5]{e^3}$

In Exercises 63–72, solve the equation.

63. $e^x = 12$

64. $e^{3x} = 25$

65. $3e^{-5x} = 132$

66. $e^{2x} - 7e^x + 10 = 0$

67. $\ln 3x = 8.2$

68. $2 \ln 4x = 15$

69. $-2 + \ln 5x = 0$

70. $\ln 4x^2 = 21$

71. $\ln x - \ln 3 = 2$

72. $\ln \sqrt{x + 1} = 2$

73. *Demand Function*   The demand function for a certain product is given by

$$p = 600 - 0.3(e^{0.005x}).$$

Find the demand x for a price of (a) $p = \$500$ and (b) $p = \$400$.

74. *Demand Function*   The demand function for a certain product is given by

$$p = 4000 \left(1 - \frac{3}{3 + e^{-0.004x}}\right).$$

Find the demand x for a price of (a) $p = \$700$ and (b) $p = \$400$.

Radioactive Decay   In Exercises 75 and 76, complete the table for the radioactive isotope.

Isotope	Half-Life	Initial Quantity	Amount After 1000 Years
75. C^{14}	5730	20 g	
76. Pu^{230}	24,360		2.4 g

77. *Population*   The population P of a city is given by

$$P = 270,000e^{0.019t}$$

where t is the time in years, with $t = 0$ corresponding to 1990.

(a) Use a graphing utility to graph this equation.

(b) According to this model, in what year will the city have a population of 300,000?

78. *Population*   The population P of a city is given by

$$P = 50,000e^{kt}$$

where t is the time in years, with $t = 0$ corresponding to the year 1990. In 1980, the population was 34,500.

(a) Find the value of k and use this result to predict the population in the year 2020.

(b) Use a graphing utility to confirm the result of part (a).

79. *Curve Fitting*   Use the properties of logarithms and exponentiation to find an equation that relates x and y. (*Hint:* See Example 9 in Section 5.3.)

x	1	2	3	4	5	6
y	1	2.520	4.327	6.350	8.550	10.903

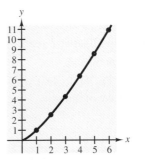

80. *Bacteria Growth*   The number of bacteria N in a culture is given by the model

$$N = 200e^{kt}$$

where t is the time in hours, with $t = 0$ corresponding to the time when $N = 200$. When $t = 6$, $N = 500$. How long does it take the population to double in size?

81. *Learning Curve* The management at a factory has found that the maximum number of units a worker can produce in a day is 50. The learning curve for the number of units N produced per day after a new employee has worked t days is given by

$$N = 50(1 - e^{kt}).$$

After 20 days on the job, a particular worker produced 31 units in 1 day.

(a) Find the learning curve for this worker.

(b) How many days should pass before this worker is producing 45 units per day?

82. *Test Scores* The scores on a general aptitude test roughly follow a normal distribution given by

$$y = 0.0028e^{-[(x-300)^2]/20,000}, \qquad 100 \le x \le 500.$$

Sketch the graph of this function. Estimate the average score on this test.

83. *Wildlife Management* A state parks and wildlife department releases 100 deer into a wilderness area. The department believes that the carrying capacity of the area is 500 deer and that the growth of the herd will be modeled by the logistics curve

$$P = \frac{500}{1 + 4e^{-kt}}, \qquad 0 \le t$$

where t is measured in years.

(a) Find k if the herd size is 170 after 2 years.

(b) Find the population after 5 years.

84. *Earthquake Magnitudes* On the Richter scale, the magnitude R of an earthquake of intensity I is given by

$$R = \log_{10} \frac{I}{I_0}$$

where $I_0 = 1$ is the minimum intensity used for comparison. Find the intensity per unit of area for the following values of R.

(a) $R = 8.4$

(b) $R = 6.85$

(c) $R = 9.1$

85. *Thawing Time* Suppose that you take a 5-pound package of steaks out of a freezer at 11 A.M. Will the steaks be thawed in time to be grilled at 6 P.M.? Assume that the room temperature is 70° Fahrenheit and the freezer temperature is 20°. Use the formula

$$t = -3.8 \ln \frac{T - 70}{20 - 70}$$

where t is the time (with $t = 0$ corresponding to 11 A.M.) and T is the temperature of the package of steaks.

86. *Sales and Advertising* In 1992, after discontinuing all advertising for a product, the manufacturer of the product found that the sales began to drop according to the model

$$S = \frac{600,000}{1 + 0.4e^{kt}}$$

where S represents the number of units sold and t represents the calendar year, with $t = 0$ corresponding to 1992.

(a) Find k if the company sold 400,000 units in 1994.

(b) According to this model, what will sales be in 1995? 1996?

87. *Bacteria Growth* The number of bacteria N in a culture is given by the model

$$N = 200e^{kt}$$

where t is the time in hours, with $t = 0$ corresponding to the time when $N = 200$. When $t = 5$, there are 350 bacteria. How long does it take for the population to triple in size?

CHAPTER TEST

Take this test as you would take a test in class. After you are done, check your work against the answers given in the back of the book.

In Exercises 1–4, sketch the graph of the function.

1. $y = 2^x$ **2.** $y = 3^{-x}$

3. $y = \ln x$ **4.** $y = \log_3(x - 1)$

5. You deposit \$20,000 into a fund that pays 8.5% interest, compounded continuously. When will the balance be greater than \$100,000?

In Exercises 6–8, expand the logarithmic expression.

6. $\ln \dfrac{x^2 y^3}{z}$ **7.** $\log_{10} 3xyz^2$ **8.** $\ln \left(x \sqrt[3]{x - 2} \right)$

In Exercises 9 and 10, condense the expression.

9. $\ln y + 2 \ln z - 3 \ln x$ **10.** $\frac{2}{3}(\log_{10} x + \log_{10} y)$

In Exercises 11–14, solve the equation.

11. $e^{4x} = 21$ **12.** $e^{2x} - 8e^x + 12 = 0$

13. $-3 + \ln 4x = 0$ **14.** $\ln \sqrt{x + 2} = 3$

In Exercises 15–17, students in a psychology class were given an exam and then retested monthly with an equivalent exam. The average score for the class is given by the human memory model $f(t) = 87 - 15 \log_{10}(t + 1)$, $0 \le t \le 4$, where t is the time in months.

15. What was the average score on the original exam?

16. What was the average score after 1 month?

17. What was the average score after 4 months?

18. The population P of a city is given by $P = 70{,}000e^{0.023t}$, where $t = 0$ represents 1990. When will the city have a population of 100,000? Explain.

19. The number of bacteria N in a culture is given by $N = 100e^{kt}$, where t is the time in hours, with $t = 0$ corresponding to the time when $N = 100$. When $t = 8$, $N = 500$. How long does it take the population to double?

20. Carbon 14 has a half-life of 5730 years. You have an initial quantity of 10 grams. How many grams will remain after 10,000 years?

CUMULATIVE TEST: CHAPTERS 3–5

Take this test as you would take a test in class. After you are done, check your work against the answers given in the back of the book.

In Exercises 1–6, given $f(x) = 2x^2 - 3$ and $g(x) = 3x + 5$, write the indicated function.

1. $(f + g)(x)$ **2.** $(f - g)(x)$ **3.** $(fg)(x)$

4. $\left(\dfrac{f}{g}\right)(x)$ **5.** $(f \circ g)(x)$ **6.** $(g \circ f)(x)$

In Exercises 7–11, sketch the graph of the function.

7. $f(x) = (x - 2)^2 + 3$ **8.** $g(x) = \dfrac{2}{x - 3}$ **9.** $h(x) = 2^{-x}$

10. $f(x) = \log_4(x - 1)$ **11.** $g(x) = \begin{cases} x + 5, & x < 0 \\ 5, & x = 0 \\ x^2 + 5, & x > 0 \end{cases}$

12. The profit for a company is given by

$$P = 500 + 20x - 0.0025x^2$$

where x is the number of units produced. What production level will yield a maximum profit?

In Exercises 13–15, perform the indicated operation and write the result in standard form.

13. $(10 + 2i)(3 - 4i)$ **14.** $(5 + 4i)^2$ **15.** $\dfrac{2 + i}{3 - i}$

16. Use the Quadratic Formula to solve $3x^2 - 5x + 7 = 0$.

17. Find all the zeros of $f(x) = x^4 + 10x^2 + 9$ given that $3i$ is a zero. Explain your reasoning.

In Exercises 18 and 19, solve the equation.

18. $e^{2x} - 11e^x + 24 = 0$ **19.** $\frac{1}{2}\ln(x - 3) = 4$

20. The population P of a city is given by

$$P = 80,000e^{0.035t}$$

where t is the time in years, with $t = 3$ corresponding to 1993. According to this model, in what year will the city have a population of 120,000?

Systems of Equations and Inequalities

6

- Systems of Equations
- Systems of Linear Equations in Two Variables
- Linear Systems in Three or More Variables
- Systems of Inequalities
- Linear Programming

The numbers N (in millions) of U.S. households that had at least one television set in selected years from 1980 to 1993 are shown in the table. A linear model for the complete set of data is given by

$$N = 1.177t + 79.0$$

where $t = 0$ represents 1980.

During that same time, the number of households n (in millions) that had at least one VCR can be modeled by

$$n = 0.185 + \frac{69.1}{1 + e^{-(t-6.5)/1.37}}.$$

Portions of the graphs of these two models are shown at the right. For the portions shown, the two graphs do not intersect. If they did intersect, it would mean that, for that year, the number of TV households was equal to the number of VCR households.

Year	0	2	4	6	8	10	12
TV Homes	77.8	81.5	83.8	85.9	88.6	92.1	92.1

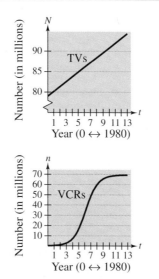

The chapter project related to this information is on page 491.

433

6.1 | Systems of Equations

The Method of Substitution ■ Graphical Approach
to Finding Solutions ■ Applications

The Method of Substitution

Up to this point in the text, most problems have involved either a function of one variable or a single equation in two variables. However, many problems in science, business, and engineering involve two or more equations in two or more variables. To solve such problems, you need to find solutions of a **system of equations.** Here is an example of a system of two equations in x and y.

$$2x + y = 5 \qquad\qquad \text{Equation 1}$$
$$3x - 2y = 4 \qquad\qquad \text{Equation 2}$$

A **solution** of this system is an ordered pair that satisfies each equation in the system. Finding the set of all solutions is called **solving the system of equations.** For instance, the ordered pair $(2, 1)$ is a solution of this system. To check this, you can substitute 2 for x and 1 for y in *each* equation.

$$2x + y = 5 \qquad\qquad \text{Equation 1}$$
$$2(2) + 1 \overset{?}{=} 5 \qquad\qquad \text{Substitute 2 for } x \text{ and 1 for } y.$$
$$4 + 1 = 5 \qquad\qquad \text{Solution checks in Equation 1.} \checkmark$$
$$3x - 2y = 4 \qquad\qquad \text{Equation 2}$$
$$3(2) - 2(1) \overset{?}{=} 4 \qquad\qquad \text{Substitute 2 for } x \text{ and 1 for } y.$$
$$6 - 2 = 4 \qquad\qquad \text{Solution checks in Equation 2.} \checkmark$$

There are several different ways to solve systems of equations. In this chapter you will study three of the most common techniques, beginning with the **method of substitution.**

Method of Substitution

1. *Solve* one of the equations for one variable in terms of the other.

2. *Substitute* the expression found in Step 1 into the other equation to obtain an equation of one variable.

3. *Solve* the equation obtained in Step 2.

4. *Back-substitute* the solution in Step 3 into the expression obtained in Step 1 to find the value of the other variable.

5. *Check* the solution to see that it satisfies *each* of the original equations.

DISCOVERY

Use a graphing utility to graph $y = -x + 4$ and $y = x - 2$ in the same viewing window. Use the TRACE feature to find the coordinates of the point of intersection. Are the coordinates the same as the solution found in Example 1? Explain.

On the *TI-82*, set the calculator to dot mode and press $\boxed{\text{ZOOM}}$ 4. What changes does this make in the coordinates of the point?

EXAMPLE 1 Solving a System of Two Equations

Solve the following system of equations.

$$x + y = 4 \qquad\qquad \text{Equation 1}$$
$$x - y = 2 \qquad\qquad \text{Equation 2}$$

Solution

Begin by solving for y in Equation 1.

$$y = 4 - x \qquad\qquad \text{Solve for } y \text{ in Equation 1.}$$

Next, substitute this expression for y in Equation 2 and solve the resulting single-variable equation for x.

$$x - y = 2 \qquad\qquad \text{Equation 2}$$
$$x - (4 - x) = 2 \qquad\qquad \text{Substitute } 4 - x \text{ for } y.$$
$$x - 4 + x = 2 \qquad\qquad \text{Simplify.}$$
$$2x = 6 \qquad\qquad \text{Combine like terms.}$$
$$x = 3 \qquad\qquad \text{Divide both sides by 2.}$$

Finally, you can solve for y by *back-substituting* $x = 3$ into the equation $y = 4 - x$, to obtain

$$y = 4 - x \qquad\qquad \text{Revised Equation 1}$$
$$y = 4 - 3 \qquad\qquad \text{Substitute 3 for } x.$$
$$y = 1 \qquad\qquad \text{Solve for } y.$$

The solution is the ordered pair $(3, 1)$. You can check this as follows.

Check

$$x + y = 4 \qquad\qquad \text{Equation 1}$$
$$3 + 1 \stackrel{?}{=} 4 \qquad\qquad \text{Substitute for } x \text{ and } y.$$
$$4 = 4 \qquad\qquad \text{Solution checks in Equation 1.} \ \checkmark$$
$$x - y = 2 \qquad\qquad \text{Equation 2}$$
$$3 - 1 \stackrel{?}{=} 2 \qquad\qquad \text{Substitute for } x \text{ and } y.$$
$$2 = 2 \qquad\qquad \text{Solution checks in Equation 2.} \ \checkmark$$

STUDY TIP

Because many steps are required to solve a system of equations, it is very easy to make errors in arithmetic. Thus, we *strongly* suggest that you always *check your solution by substituting it into each equation in the original system.*

NOTE The term *back-substitution* implies that you work *backwards*. First you solve for one of the variables, and then you substitute that value *back* into one of the equations in the system to find the value of the other variable.

Real Life

EXAMPLE 2 *Solving a System by Substitution*

A total of $12,000 is invested in two funds paying 9% and 11% simple interest. If the yearly interest is $1180, how much of the $12,000 is invested at each rate?

Solution

Verbal Model:

| 9% fund | + | 11% fund | = | Total investment |

| 9% interest | + | 11% interest | = | Total interest |

Labels:

Amount in 9% fund $= x$	(dollars)
Interest for 9% fund $= 0.09x$	(dollars)
Amount in 11% fund $= y$	(dollars)
Interest for 11% fund $= 0.11y$	(dollars)
Total investment $= \$12,000$	(dollars)
Total interest $= \$1180$	(dollars)

System:

$$x + y = 12,000 \qquad \text{Equation 1}$$
$$0.09x + 0.11y = 1180 \qquad \text{Equation 2}$$

To begin, it is convenient to multiply both sides of the second equation by 100 to obtain $9x + 11y = 118,000$. This eliminates the need to work with decimals.

$$9x + 11y = 118,000 \qquad \text{Revised Equation 2}$$

To solve this system, you can solve for x in Equation 1.

$$x = 12,000 - y \qquad \text{Solve for } x \text{ in Equation 1.}$$

Then, substitute this expression for x into Equation 2, and solve the resulting equation for y.

$$9x + 11y = 118,000 \qquad \text{Revised Equation 2}$$
$$9(12,000 - y) + 11y = 118,000 \qquad \text{Substitute } 12,000 - y \text{ for } x.$$
$$108,000 - 9y + 11y = 118,000 \qquad \text{Distributive Property}$$
$$2y = 10,000 \qquad \text{Combine like terms.}$$
$$y = 5000 \qquad \text{Divide both sides by 2.}$$

Next, back-substitute the value $y = 5000$ to solve for x.

$$x = 12,000 - y \qquad \text{Revised Equation 1}$$
$$x = 12,000 - 5000 \qquad \text{Substitute 5000 for } y.$$
$$x = 7000 \qquad \text{Solve for } x.$$

The solution is (7000, 5000). Check this in the original statement of the problem.

A graphing utility can be useful in reinforcing the solutions determined by the substitution method or by graphical solution methods.

The equations in Examples 1 and 2 are linear. The method of substitution can also be used to solve systems in which one or both of the equations are nonlinear.

EXAMPLE 3   Solving a System by Substitution: Two-Solution Case

Solve the following system of equations.

$$x^2 - x - y = 1 \qquad \text{Equation 1}$$
$$-x + y = -1 \qquad \text{Equation 2}$$

Point out that it is not practical to solve Equation 2 for x instead of y.

Solution

Begin by solving for y in Equation 2 to obtain $y = x - 1$. Next, substitute this expression for y in Equation 1 and solve for x.

$$x^2 - x - y = 1 \qquad \text{Equation 1}$$
$$x^2 - x - (x - 1) = 1 \qquad \text{Substitute } x - 1 \text{ for } y.$$
$$x^2 - 2x + 1 = 1 \qquad \text{Simplify.}$$
$$x^2 - 2x = 0 \qquad \text{Standard form}$$
$$x(x - 2) = 0 \qquad \text{Factor.}$$
$$x = 0, \ 2 \qquad \text{Solve for } x.$$

Back-substituting these values of x to solve for the corresponding values of y produces the two solutions $(0, -1)$ and $(2, 1)$. Check this in the original system.

EXAMPLE 4   Solving a System by Substitution: No-Solution Case

Solve the following system of equations.

$$-x + y = 4 \qquad \text{Equation 1}$$
$$x^2 + y = 3 \qquad \text{Equation 2}$$

Solution

Begin by solving Equation 1 for y to obtain $y = x + 4$. Next, substitute this expression for y into Equation 2, and solve for x.

$$x^2 + y = 3 \qquad \text{Equation 2}$$
$$x^2 + (x + 4) = 3 \qquad \text{Substitute } x + 4 \text{ for } y.$$
$$x^2 + x + 1 = 0 \qquad \text{Simplify.}$$
$$x = \frac{-1 \pm \sqrt{1^2 - 4(1)(1)}}{2} \qquad \text{Quadratic Formula}$$

Because the discriminant is negative, the equation $x^2 + x + 1 = 0$ has no (real) solution. Hence, this system has no (real) solution.

Graphical Approach to Finding Solutions

From Examples 2, 3, and 4, you can see that a system of two equations in two unknowns can have exactly one solution, more than one solution, or no solution. In practice, you can gain insight about the location and number of solutions of a system of equations by graphing each of the equations in the same coordinate plane. The solutions of the system correspond to the **points of intersection** of the graphs. For instance, in Figure 6.1(a) the two equations graph as two lines with a *single point* of intersection. The two equations in Example 3 graph as a parabola and a line with *two points* of intersection, as shown in Figure 6.1(b). Moreover, the two equations in Example 4 graph as a line and a parabola that happen to have *no points* of intersection, as shown in Figure 6.1(c).

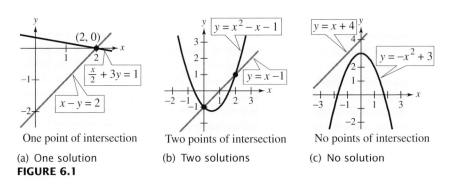

One point of intersection

(a) One solution

Two points of intersection

(b) Two solutions

No points of intersection

(c) No solution

FIGURE 6.1

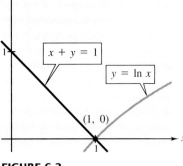

FIGURE 6.2

EXAMPLE 5   Solving a System of Equations

Solve the following system of equations.

$$y = \ln x \qquad\qquad \text{Equation 1}$$
$$x + y = 1 \qquad\qquad \text{Equation 2}$$

Solution

The graph of each equation is shown in Figure 6.2. From this, it is clear that there is only one point of intersection. Also, it appears that $(1, 0)$ is the solution point. You can confirm this by substituting in *both* equations.

Check

$$0 = \ln 1 \qquad\qquad \text{Equation 1 checks.} \checkmark$$
$$1 + 0 = 1 \qquad\qquad \text{Equation 2 checks.} \checkmark$$

Applications

The total cost C of producing x units of a product typically has two components—the initial cost and the cost per unit. When enough units have been sold so that the total revenue R equals the total cost, the sales are said to have reached the **break-even point.** You will find that the break-even point corresponds to the point of intersection of the cost and revenue curves.

Real Life

EXAMPLE 6 *An Application: Break-Even Analysis*

A small business invests $10,000 in equipment to produce a product. Each unit of the product costs $0.65 to produce and is sold for $1.20. How many items must be sold before the business breaks even?

Solution

The total cost of producing x units is

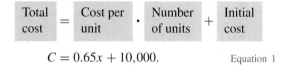

$$C = 0.65x + 10{,}000. \qquad \text{Equation 1}$$

The revenue obtained by selling x units is

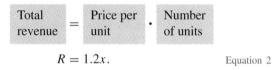

$$R = 1.2x. \qquad \text{Equation 2}$$

Because the break-even point occurs when $R = C$, you have

$1.2x = 0.65x + 10{,}000$	Equate R and C.
$0.55x = 10{,}000$	Subtract $0.65x$ from both sides.
$x = \dfrac{10{,}000}{0.55}$	Divide both sides by 0.55.
$x \approx 18{,}182$ units.	Use a calculator.

Note in Figure 6.3 that sales less than the break-even point correspond to an overall loss, whereas sales greater than the break-even point correspond to a profit.

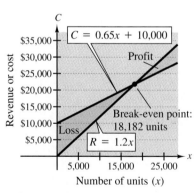

FIGURE 6.3

NOTE Another way to view the solution in Example 6 is to consider the profit function

$$P = R - C.$$

The break-even point occurs when the profit is 0, which is the same as saying that $R = C$.

Real Life

EXAMPLE 7 Compact Disc Players and Turntables

From 1983 to 1987, the number of compact disc players sold each year in the United States was *increasing* and the number of turntables was *decreasing*. Two models that approximate the sales are

$$S = -1700 + 496t \qquad \text{Compact disc players}$$
$$S = 1972 - 82t \qquad \text{Turntables}$$

where S represents the annual sales in thousands of units and t represents the calendar year, with $t = 3$ corresponding to 1983 (see Figure 6.4). According to these two models, when would you expect the sales of compact disc players to have exceeded the sales of turntables?   (Source: Dealerscope Merchandising)

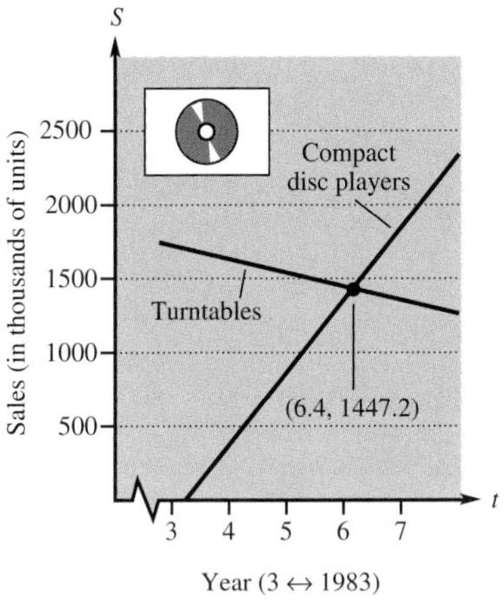

Solution

Because the first equation has already been solved for S in terms of t, we substitute this value into the second equation and solve for t, as follows.

$$-1700 + 496t = 1972 - 82t$$
$$496t + 82t = 1972 + 1700$$
$$578t = 3672$$
$$t \approx 6.4$$

Thus, from the given models, we would expect that the sales of compact disc players exceeded the sales of turntables sometime during 1986.

FIGURE 6.4

Group Activities   Extending the Concept

Interpreting Points of Intersection   You plan to rent a 14-foot truck for a 2-day local move. At truck rental agency A, you can rent a truck for $29.95 per day plus $0.49 per mile. At agency B, you can rent a truck for $50 per day plus $0.25 per mile. The total cost y (in dollars) for the truck from agency A is $y = (\$29.95 \text{ per day})(2 \text{ days}) + 0.49x = 59.90 + 0.49x$, where x is the total number of miles the truck is driven. Write a total cost equation in terms of x and y for the total cost for the truck from agency B. Use a graphing utility to graph the two equations and find the point of intersection. Interpret the meaning of the point of intersection in the context of the problem. Which agency should you choose if you plan to travel a total of 100 miles over the 2-day move? Why? How does the situation change if you plan to drive 200 miles per day instead?

Warm Up

In Exercises 1–4, sketch the graph of the equation.

1. $y = -\frac{1}{3}x + 6$

2. $y = 2(x - 3)$

3. $x^2 + y^2 = 4$

4. $y = 5 - (x - 3)^2$

In Exercises 5–8, perform the indicated operations and simplify.

5. $(3x + 2y) - 2(x + y)$

6. $(-10u + 3v) + 5(2u - 8v)$

7. $x^2 + (x - 3)^2 + 6x$

8. $y^2 - (y + 1)^2 + 2y$

In Exercises 9 and 10, solve the equation.

9. $3x + (x - 5) = 15 + 4$

10. $y^2 + (y - 2)^2 = 2$

6.1 Exercises

Exercise 39 of this set contains a system with no solution.

In Exercises 1 and 2, decide whether each ordered pair is a solution of the system.

1. $x + y = 3$
$x - y = 1$
(a) $(2, 1)$
(b) $(-2, -1)$

2. $2x - y = 2$
$x + 3y = 8$
(a) $(2, 1)$
(b) $(2, 2)$

In Exercises 3–12, use substitution to solve the system. Then use the graph to confirm your solution.

3. $2x + y = 4$
$-x + y = 1$

4. $x - y = -5$
$x + 2y = 4$

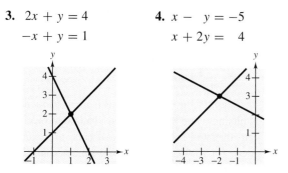

5. $x - y = -3$
$x^2 - y = -1$

6. $3x - y = -2$
$x^3 - y = 0$

7. $x + 3y = 15$
$x^2 + y^2 = 25$

8. $x - y = 0$
$x^3 - 5x + y = 0$

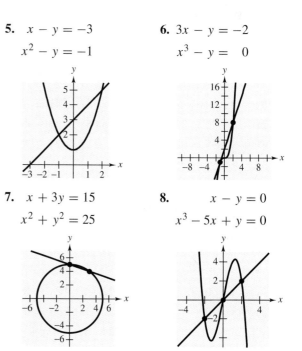

9. $x^2 - \quad y = 0$
$x^2 - 4x + y = 0$

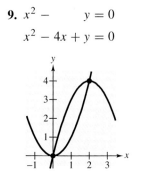

10. $y = -x^2 + 1$
$y = \quad x^2 - 1$

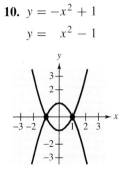

11. $x - 3y = -4$
$y = \frac{2}{3}x^2 + x - \frac{8}{3}$

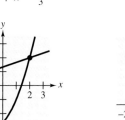

12. $y = x^3 - 3x^2 + 3$
$2x + y = 3$

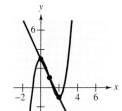

In Exercises 13–32, use substitution to solve the system.

13. $x - \quad y = \quad 0$
$5x - 3y = 10$

14. $x + 2y = \quad 1$
$5x - 4y = -23$

15. $2x - y + 2 = 0$
$4x + y - 5 = 0$

16. $6x - 3y - 4 = 0$
$x + 2y - 4 = 0$

17. $30x - 40y - 33 = 0$
$10x + 20y - 21 = 0$

18. $1.5x + 0.8y = 2.3$
$0.3x - 0.2y = 0.1$

19. $\frac{1}{5}x + \frac{1}{2}y = \quad 8$
$x + \quad y = 20$

20. $\frac{1}{2}x + \frac{3}{4}y = 10$
$\frac{3}{2}x - \quad y = \quad 4$

21. $x - y = 0$
$2x + y = 0$

22. $x - 2y = 0$
$3x - \quad y = 0$

23. $y = 2x$
$y = x^2 + 1$

24. $x + y = 4$
$x^2 - y = 2$

25. $3x - 7y + 6 = 0$
$x^2 - y^2 = 4$

26. $x^2 + y^2 = 25$
$2x + \quad y = 10$

27. $x^2 + y^2 = 5$
$x - \quad y = 1$

28. $y = x^3 - 2x^2 + \quad x - 1$
$y = \quad -x^2 + 3x - 1$

29. $y = x^4 - 2x^2 + 1$
$y = \quad 1 - \quad x^2$

30. $x^2 + y = 4$
$2x - y = 1$

31. $xy - \quad 1 = 0$
$2x - 4y + 7 = 0$

32. $x - 2y = 1$
$y = \sqrt{x - 1}$

In Exercises 33–40, use a graphing utility to graphically find the point of intersection of the graphs. Then confirm your solution algebraically.

33. $y = x^2 + 3x - 1$
$y = -x^2 - 2x + 2$

34. $y = -2x^2 + x - 1$
$y = x^2 - 2x - 1$

35. $x - y + 3 = 0$
$x^2 - 4x + 7 = y$

36. $x - y = 3$
$x - y^2 = 1$

37. $y = e^x$
$x - y + 1 = 0$

38. $y = \sqrt{x}$
$y = x$

39. $2x - y + 3 = 0$
$x^2 + y^2 - 4x = 0$

40. $x^2 + y^2 = 8$
$y = x^2$

Break-Even Analysis In Exercises 41–44, find the sales necessary to break even ($R = C$) for the given cost C of producing x units and the given revenue R obtained by selling x units. (Round your answer to the nearest whole unit.)

41. $C = 8650x + 250,000$; $R = 9950x$

42. $C = 5.5\sqrt{x} + 10,000$; $R = 3.29x$

43. $C = 2.65x + 350,000$; $R = 4.15x$

44. $C = 0.08x + 50,000$; $R = 0.25x$

45. *Break-Even Point* Suppose you are setting up a small business and have invested $16,000 to produce an item that will sell for $5.95. If each unit can be produced for $3.45, how many units must you sell to break even?

46. *Break-Even Point*   Suppose you are setting up a small business and have made an initial investment of $5000. The unit cost of the product is $21.60, and the selling price is $34.10. How many units must you sell to break even?

47. *Comparing Populations*   From 1987 to 1993, the population of the northeastern part of the United States was growing more slowly than that of the western part. Two models that represent the populations of the two regions are

$$P = 50{,}845.9 + 158.3t - 1.25t^2 \qquad \text{Northeast}$$
$$P = 52{,}922.1 + 1062.3t \qquad \text{West}$$

where P is the population in thousands and $t = 0$ represents 1990. Use a graphing utility to determine when the population of the West overtook the population of the Northeast. (Source: U.S. Bureau of Census)

48. *Comparing Populations*   In 1992, Philadelphia was the fifth most populated city in the United States and San Diego was the sixth. The populations of Philadelphia and San Diego from 1970 to 1992 can be approximated by the models

$$P = 1947.7 - 32.2t + 0.67t^2 \qquad \text{Philadelphia}$$
$$P = 696.4 + 16.0t + 0.218t^2 \qquad \text{San Diego}$$

where P is the population (within the city limits) in thousands and $t = 0$ represents 1970. Use a graphing utility to determine if the population of San Diego will overtake the population of Philadelphia. (Source: U.S. Bureau of Census)

49. *Gross National Product*   The annual gross national products G (in billions of dollars) of the United States and Japan from 1980 to 1991 can be approximated by the models

$$G = 2427.6 + 319.62t + 1.48t^2 \qquad \text{United States}$$
$$G = 1545.6 + 89.29t + 6.79t^2 \qquad \text{Japan}$$

where $t = 0$ represents 1980. Use a graphing utility to determine if the annual gross national product of Japan will exceed that of the United States. (Source: U.S. Arms Control and Disarmament Agency)

50. *Computers in the Office*   The number of business or government offices y (in thousands) with personal computers from 1984 to 1989 can be modeled by

$$y = -99.15 - 11.38t + 13.91t^2 \qquad \text{With links}$$
$$y = -2202.41 + 1934.45 \ln t \qquad \text{Without links}$$

where $t = 4$ represents 1984. In 1989, there were more offices without personal computer communications links than with personal computer communications links. Use a graphing utility to determine when this will change. (Source: Gartner Group, COMTEX data base)

51. *Investment Portfolio*   A total of $25,000 is invested in two funds paying 8% and 8.5% simple interest. If the yearly interest is $2060, how much of the $25,000 is invested at each rate?

52. *Investment Portfolio*   A total of $18,000 is invested in two funds paying 7.75% and 8.25% simple interest. If the yearly interest is $1455, how much of the $18,000 is invested at each rate?

53. *Choice of Two Jobs*   You are offered two different sales jobs. One company offers a straight commission of 6% of the sales. The other company offers a salary of $250 per week *plus* 3% of the sales. How much would you have to sell in a week in order to make the straight commission offer better?

54. *Choice of Two Jobs*   You are offered two different sales jobs. One company offers an annual salary of $20,000 *plus* a year-end bonus of 1% of your total sales. The other company offers a salary of $15,000 *plus* a year-end bonus of 2% of your total sales. How much would you have to sell in a year in order to make the second offer better?

55. *Color or Monochrome?*   The sales y (in millions) of monochrome and color computer monitors from 1981 to 1988 can be approximated by the models

$$y = -0.927 + 1.524t + 0.1176t^2 \qquad \text{Monochrome}$$
$$y = -0.010 - 0.2749t + 0.2610t^2 \qquad \text{Color}$$

where $t = 1$ represents 1981. According to these models, when will the sales of color monitors equal the sales of monochrome monitors? (Source: Future Computing/Datapro, Inc.)

6.2 Systems of Linear Equations in Two Variables

The Method of Elimination ▪ Graphical Interpretation of Solutions ▪ Applications

The Method of Elimination

Statistics Topics
Least squares regression lines are covered in this section in the Exercise Set (see 47–54). Least squares regression parabolas are covered in Section 6.3 in Exercises 47–52.

In Section 6.1, you studied two methods for solving a system of equations: substitution and graphing. In this section, you will study a third method called the **method of elimination.** The key step in the method of elimination is to obtain, for one of the variables, coefficients that differ only in sign so that *adding* the two equations eliminates this variable. The following system provides an example.

$$\begin{array}{ll} 3x + 5y = 7 & \text{Equation 1} \\ \underline{-3x - 2y = -1} & \text{Equation 2} \\ 3y = 6 & \text{Add equations.} \end{array}$$

Note that by adding the two equations, you eliminate the variable x and obtain a single equation in y. Solving this equation for y produces $y = 2$, which you can then back-substitute into one of the original equations to solve for x.

To be efficient, students should know and understand both the substitution and the elimination methods. You may wish to point out the differences between the two methods and the advantages of one over the other. For instance, you may want to point out that if each of two equations is written such that one variable is represented in terms of the other variable, substitution is a more efficient solution method.

EXAMPLE 1   The Method of Elimination

Solve the following system of linear equations.

$$\begin{array}{ll} 3x + 2y = 4 & \text{Equation 1} \\ 5x - 2y = 8 & \text{Equation 2} \end{array}$$

Solution

Because the coefficients for y differ only in sign, you can eliminate y by adding the two equations.

$$\begin{array}{ll} 3x + 2y = 4 & \text{Equation 1} \\ \underline{5x - 2y = 8} & \text{Equation 2} \\ 8x = 12 & \text{Add equations.} \end{array}$$

NOTE Try using the method of substitution to solve the system given in Example 1. Which method do you think is easier? Many people find that the method of elimination is more efficient.

Therefore, $x = \frac{3}{2}$. By back-substituting this value into the first equation, you can solve for y, as follows.

$$\begin{array}{ll} 3x + 2y = 4 & \text{Equation 1} \\ 3\left(\frac{3}{2}\right) + 2y = 4 & \text{Substitute } \frac{3}{2} \text{ for } x. \\ y = -\frac{1}{4} & \text{Solve for } y. \end{array}$$

The solution is $\left(\frac{3}{2}, -\frac{1}{4}\right)$. Check this in the original system.

EXAMPLE 2   The Method of Elimination

Solve the following system of linear equations.

$$2x - 3y = -7 \qquad \text{Equation 1}$$
$$3x + \ y = -5 \qquad \text{Equation 2}$$

Solution

For this system, you can obtain coefficients that differ only in sign by multiplying the second equation by 3.

STUDY TIP

To obtain coefficients (for one of the variables) that differ only in sign, you often need to multiply one or both of the equations by a suitable constant.

$$
\begin{array}{ll}
2x \ - \ 3y \ = \ -7 \\
3x \ + \ \ y \ = \ -5
\end{array}
\Longrightarrow
\begin{array}{ll}
2x \ - \ 3y \ = \ \ \ -7 & \text{Equation 1} \\
9x \ + \ 3y \ = \ -15 & \text{Multiply Equation 2 by 3.} \\
\hline
11x \quad\quad\ \ = \ -22 & \text{Add equations.}
\end{array}
$$

Thus, you can see that $x = -2$. By back-substituting this value of x into Equation 1, you can solve for y.

$$
\begin{array}{ll}
2x - 3y = -7 & \text{Equation 1} \\
2(-2) - 3y = -7 & \text{Substitute } -2 \text{ for } x. \\
-3y = -3 & \text{Add 4 to both sides.} \\
y = 1 & \text{Solve for } y.
\end{array}
$$

The solution is $(-2, 1)$. Check this in the original system, as follows.

Check

$$
\begin{array}{ll}
2(-2) - 3(1) \stackrel{?}{=} -7 & \text{Substitute into Equation 1.} \\
-4 - 3 = -7 & \text{Equation 1 checks.} \ \checkmark \\
3(-2) + 1 \stackrel{?}{=} -5 & \text{Substitute into Equation 2.} \\
-6 + 1 = -5 & \text{Equation 2 checks.} \ \checkmark
\end{array}
$$

If you plan to cover the least square regression line topics presented in Exercises 47–54 of this section, you may want to consider using the following activity. This activity is appropriate for making a presentation to the entire class or for group members taking turns presenting to the rest of the group.

Alternate Group Activity

Find an article from a professional journal in your field that develops or presents a regression line as an analysis tool. Explain what the variables represent and how the regression line is used. Give several examples showing the use of the regression equation and explain its value as an analysis tool.

In Example 2, the two systems of linear equations

$$
\begin{array}{lll}
2x - 3y = -7 & \text{and} & 2x - 3y = \ -7 \\
3x + \ y = -5 & & 9x + 3y = -15
\end{array}
$$

are called **equivalent** because they have precisely the same solution set. The operations that can be performed on a system of linear equations to produce an equivalent system are (1) interchanging any two equations, (2) multiplying an equation by a nonzero constant, and (3) adding a multiple of one equation to any other equation in the system.

The Method of Elimination

To use the **method of elimination** to solve a system of two linear equations in x and y, use the following steps.

1. Obtain coefficients for x (or y) that differ only in sign by multiplying all terms of one or both equations by suitably chosen constants.

2. Add the equations to eliminate one variable and solve the resulting equation.

3. Back-substitute the value obtained in Step 2 into either of the original equations and solve for the other variable.

4. Check your solution in both of the original equations.

EXAMPLE 3 The Method of Elimination

Solve the following system of linear equations.

$$5x + 3y = 9 \qquad \text{Equation 1}$$
$$2x - 4y = 14 \qquad \text{Equation 2}$$

Solution

You can obtain coefficients that differ only in sign by multiplying Equation 1 by 4 and multiplying Equation 2 by 3.

$$
\begin{array}{ll}
5x + 3y = 9 & \\
2x - 4y = 14 &
\end{array}
\quad\Longrightarrow\quad
\begin{array}{ll}
20x + 12y = 36 & \text{Multiply Equation 1 by 4.} \\
\underline{6x - 12y = 42} & \text{Multiply Equation 2 by 3.} \\
26x = 78 & \text{Add equations.}
\end{array}
$$

From this equation, you can see that $x = 3$. By back-substituting this value of x into the second equation, you can solve for y, as follows.

$$
\begin{array}{ll}
2x - 4y = 14 & \text{Equation 2} \\
2(3) - 4y = 14 & \text{Substitute 3 for } x. \\
-4y = 8 & \\
y = -2 & \text{Solve for } y.
\end{array}
$$

The solution is $(3, -2)$. Check this in the original system.

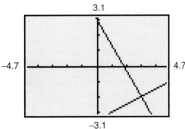

FIGURE 6.5

NOTE Remember that you can check the solution of a system of equations graphically. For instance, to check the solution found in Example 3, sketch the graphs of both equations on the same screen, as shown in Figure 6.5. Notice that the two lines intersect at $(3, -2)$.

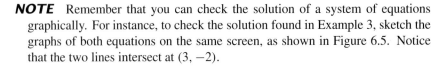

Graphical Interpretation of Solutions

It is possible for a *general* system of equations to have exactly one solution, two or more solutions, or no solution. If a system of *linear* equations has two different solutions, it must have an *infinite* number of solutions. To see why this is true, consider the following graphical interpretations of a system of two linear equations in two variables. (Remember that the graph of a linear equation in two variables is a straight line.)

STUDY TIP

Keep in mind that the terminology and methods discussed in this section and the following section apply only to systems of *linear* equations.

Graphical Interpretation of Solutions

For a system of two linear equations in two variables, the number of solutions is given by one of the following.

Number of Solutions	*Graphical Interpretation*
1. Exactly one solution	The two lines intersect at one point.
2. Infinitely many solutions	The two lines are identical.
3. No solution	The two lines are parallel.

These graphical interpretations are shown in Figure 6.6.

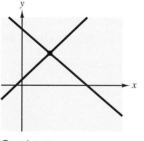

Consistent
Two lines that intersect
Single point of intersection
FIGURE 6.6

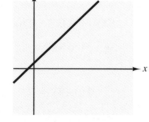

Consistent
Two lines that coincide
Infinitely many points

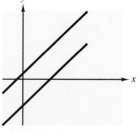

Inconsistent
Two parallel lines
No point of intersection

A system of linear equations is called **consistent** if it has at least one solution, and it is called **inconsistent** if it has no solution. In Examples 4 and 5, note how you can use the method of elimination to determine that a system of linear equations has no solution or infinitely many solutions.

EXAMPLE 4   The Method of Elimination: No-Solution Case

Solve the following system of linear equations.

$$x - 2y = 3 \qquad \text{Equation 1}$$
$$-2x + 4y = 1 \qquad \text{Equation 2}$$

Solution

To obtain coefficients that differ only in sign, multiply Equation 1 by 2.

$$
\begin{array}{ll}
x - 2y = 3 & \\
-2x + 4y = 1 &
\end{array}
\quad\Longrightarrow\quad
\begin{array}{ll}
2x - 4y = 6 & \text{Multiply Equation 1 by 2.} \\
-2x + 4y = 1 & \text{Equation 2} \\
\hline
 0 = 7 & \text{False statement}
\end{array}
$$

Because there are no values of x and y for which $0 = 7$, you can conclude that the system is inconsistent and has no solution. The lines corresponding to the two equations given in this system are shown in Figure 6.7. Note that the two lines are parallel, and therefore have no point of intersection.

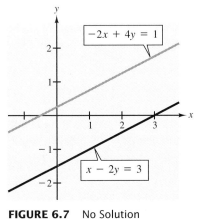

FIGURE 6.7   No Solution

In Example 4, note that the occurrence of a false statement, such as $0 = 7$, indicates that the system has no solution. In the next example, note that the occurrence of a statement that is true for all values of the variables, such as $0 = 0$, indicates that the system has infinitely many solutions.

EXAMPLE 5   The Method of Elimination: Many-Solutions Case

Solve the following system of linear equations.

$$2x - y = 1 \qquad \text{Equation 1}$$
$$4x - 2y = 2 \qquad \text{Equation 2}$$

Solution

To obtain coefficients that differ only in sign, multiply the second equation by $-\frac{1}{2}$.

$$
\begin{array}{ll}
2x - y = 1 & \\
4x - 2y = 2 &
\end{array}
\quad\Longrightarrow\quad
\begin{array}{ll}
2x - y = 1 & \text{Equation 1} \\
-2x + y = -1 & \text{Multiply Equation 2 by } -\frac{1}{2}. \\
\hline
 0 = 0 & \text{Add equations.}
\end{array}
$$

Because the two equations turn out to be equivalent (have the same solution set), you can conclude that the system has infinitely many solutions. The solution set consists of all points (x, y) lying on the line

$$2x - y = 1$$

as shown in Figure 6.8.

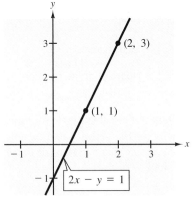

FIGURE 6.8   Infinite Number of Solutions

Example 6 illustrates a strategy for solving a system of linear equations that has decimal coefficients.

EXAMPLE 6 *A Linear System Having Decimal Coefficients*

Solve the following system of linear equations.

$$0.02x - 0.05y = -0.38 \qquad \text{Equation 1}$$
$$0.03x + 0.04y = 1.04 \qquad \text{Equation 2}$$

Solution

Because the coefficients in this system have two decimal places, you can begin by multiplying each equation by 100. (This produces a system in which the coefficients are all integers.)

$$2x - 5y = -38 \qquad \text{Revised Equation 1}$$
$$3x + 4y = 104 \qquad \text{Revised Equation 2}$$

Now, to obtain coefficients that differ only in sign, multiply the first equation by 3 and multiply the second equation by -2.

$$
\begin{array}{ll}
2x - 5y = -38 & \qquad 6x - 15y = -114 \qquad \text{Multiply Equation 1 by 3.} \\
3x + 4y = 104 & \qquad -6x - 8y = -208 \qquad \text{Multiply Equation 2 by } -2. \\
\hline
& \qquad\qquad\quad -23y = -322 \qquad \text{Add equations.}
\end{array}
$$

Thus, you can conclude that

$$y = \frac{-322}{-23} = 14.$$

Back-substituting this value into Equation 2 produces the following.

$$3x + 4y = 104 \qquad \text{Equation 2}$$
$$3x + 4(14) = 104 \qquad \text{Substitute 14 for } y.$$
$$3x = 48$$
$$x = 16 \qquad \text{Solve for } x.$$

The solution is $(16, 14)$. Check this in the original system, as follows.

Check

$$0.02(16) - 0.05(14) \overset{?}{=} -0.38 \qquad \text{Substitute into Equation 1.}$$
$$0.32 - 0.70 = -0.38 \qquad \text{Equation 1 checks.} ✔$$

$$0.03(16) + 0.04(14) \overset{?}{=} 1.04 \qquad \text{Substitute into Equation 2.}$$
$$0.48 + 0.56 = 1.04 \qquad \text{Equation 2 checks.} ✔$$

Applications

At this point, you may be asking the question "How can I tell which application problems can be solved using a system of linear equations?" The answer comes from the following considerations.

1. Does the problem involve more than one unknown quantity?

2. Are there two (or more) equations or conditions to be satisfied?

If one or both of these conditions occur, the appropriate mathematical model for the problem may be a system of linear equations. Example 7 shows how to construct such a model.

Activities

1. Solve the system by the elimination method.
   $3x - 2y = 7$
   $8x + 4y = 0$
   Answer: $(1, -2)$
2. How many milliliters of a 30% acid solution and a 55% acid solution must be mixed to obtain 100 milliliters of a 50% acid solution?
   Answer: 20 ml of 30% acid solution; 80 ml of 55% acid solution
3. Find the point of equilibrium if the demand equation is $p = 110 - 5x$ and the supply equation is $p = 60 + 20x$.
   Answer: $x = 2$, $p = \$100.00$

Real Life

EXAMPLE 7 An Application of a Linear System

An airplane flying into a head wind travels the 2000-mile flying distance between two cities in 4 hours and 24 minutes. On the return flight, the same distance is traveled in 4 hours. Find the ground speed of the plane and the speed of the wind, assuming that both remain constant.

Solution

The two unknown quantities are the speeds of the wind and the plane. If r_1 is the speed of the plane and r_2 is the speed of the wind, then

$$r_1 - r_2 = \text{speed of the plane } \textit{against} \text{ the wind}$$
$$r_1 + r_2 = \text{speed of the plane } \textit{with} \text{ the wind}$$

as shown in Figure 6.9. Using the formula

$$\text{Distance} = (\text{rate})(\text{time})$$

for these two speeds, you obtain the following equations.

$$2000 = (r_1 - r_2)\left(4 + \frac{24}{60}\right)$$
$$2000 = (r_1 + r_2)(4)$$

These two equations simplify as follows.

$$5000 = 11r_1 - 11r_2 \qquad \text{Equation 1}$$
$$500 = r_1 + r_2 \qquad \text{Equation 2}$$

By elimination, the solution is

$$r_1 = \frac{5250}{11} \approx 477.27 \text{ miles per hour}$$
$$r_2 = \frac{250}{11} \approx 22.73 \text{ miles per hour.}$$

In 1992, there were about 7 million commercial airplane flights in the United States. These flights were staffed by about 58,000 pilots and copilots and about 84,000 flight attendants.

Original flight
$r_1 - r_2$
Return flight
$r_1 + r_2$

FIGURE 6.9

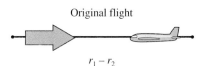

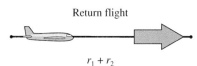

Real Life

EXAMPLE 8   Finding the Point of Equilibrium

The demand and supply functions for a certain type of calculator are given by

$$p = 150 - 0.00001x \qquad \text{Demand equation}$$
$$p = 60 + 0.00002x \qquad \text{Supply equation}$$

where p is the price in dollars and x represents the number of units. Find the point of equilibrium for this market. The point of equilibrium is the price p and number of units x that satisfy both the demand and supply equations.

Solution

Begin by substituting the value of p given in Equation 2 into Equation 1.

$$p = 150 - 0.00001x \qquad \text{Equation 1}$$
$$60 + 0.00002x = 150 - 0.00001x \qquad \text{Substitute } 60 + 0.00002x \text{ for } p.$$
$$0.00003x = 90 \qquad \text{Add } 0.00001x \text{ and subtract } 60.$$
$$x = 3,000,000 \qquad \text{Divide both sides by } 0.00003.$$

Thus, the point of equilibrium occurs when the demand and supply are each 3 million units. (See Figure 6.10.) The price that corresponds to this x-value is obtained by back-substituting $x = 3,000,000$ into either of the original equations. For instance, back-substituting into the first equation produces

$$p = 150 - 0.00001(3,000,000) = 150 - 30 = \$120.$$

Try back-substituting $x = 3,000,000$ into the second equation to see that you obtain the same price.

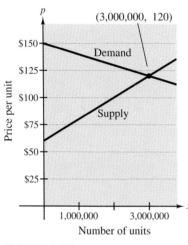

FIGURE 6.10

Group Activities   Extending the Concept

Creating Consistent and Inconsistent Systems   For each of the following systems, find the value of k so that the system has infinitely many solutions.

a. $4x + 3y = -8$ **b.** $3x - 12y = 9$
     $x + ky = -2$ $x - 4y = k$

Is it possible to find values of k such that the systems have no solutions? Explain why or why not for each system. Is it possible to find values of k such that the systems have unique solutions? Explain why or why not for each system. Summarize the differences between these two systems of equations.

Warm Up

The following warm-up exercises involve skills that were covered in earlier sections. You will use these skills in the exercise set for this section.

In Exercises 1 and 2, sketch the graph of the equation.

1. $2x + y = 4$

2. $5x - 2y = 3$

In Exercises 3 and 4, find an equation of the line passing through the two points.

3. $(-1, 3), (4, 8)$

4. $(2, 6), (5, 1)$

In Exercises 5 and 6, determine the slope of the line.

5. $3x + 6y = 4$

6. $7x - 4y = 10$

In Exercises 7–10, determine whether the lines represented by the pair of equations are parallel, perpendicular, or neither.

7. $2x - 3y = -10$
$3x + 2y = \ \ 11$

8. $\ \ 4x - 12y = 5$
$-2x + \ \ 6y = 3$

9. $5x + \ \ y = 2$
$3x + 2y = 1$

10. $\ \ x - 3y = 2$
$6x + 2y = 4$

6.2 Exercises

Exercises containing inconsistent systems (no solution): 5, 6, 19, 20, and 45. Exercises containing dependent systems (infinitely many solutions): 7, 22, and 25.

In Exercises 1–10, solve the linear system by elimination. Then use the graph to confirm your solution. Copy the graphs and label each line with the appropriate equation.

1. $2x + y = 4$
$x - y = 2$

2. $\ \ x + 3y = 2$
$-x + 2y = 3$

3. $\ \ x - \ y = \ \ 0$
$3x - 2y = -1$

4. $2x - \ y = \ 2$
$4x + 3y = 24$

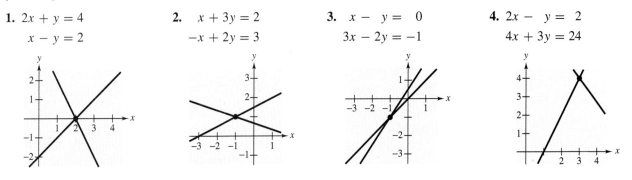

5. $x - y = 1$
$-2x + 2y = 5$

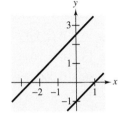

6. $3x + 2y = 2$
$6x + 4y = 14$

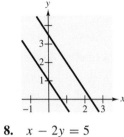

7. $3x - 2y = 6$
$-6x + 4y = -12$

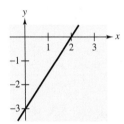

8. $x - 2y = 5$
$6x + 2y = 7$

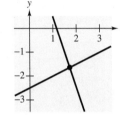

9. $9x - 3y = -1$
$3x + 6y = -5$

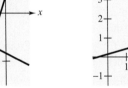

10. $5x + 3y = 18$
$2x - 7y = -1$

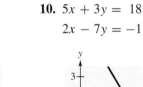

In Exercises 11–30, solve the system by elimination.

11. $x + 2y = 4$
$x - 2y = 1$

12. $3x - 5y = 2$
$2x + 5y = 13$

13. $2x + 3y = 18$
$5x - y = 11$

14. $x + 7y = 12$
$3x - 5y = 10$

15. $3x + 2y = 10$
$2x + 5y = 3$

16. $8r + 16s = 20$
$16r + 50s = 55$

17. $2u + v = 120$
$u + 2v = 120$

18. $5u + 6v = 24$
$3u + 5v = 18$

19. $6r - 5s = 3$
$-12r + 10s = 5$

20. $1.8x + 1.2y = 4$
$9x + 6y = 3$

21. $\dfrac{x}{4} + \dfrac{y}{6} = 1$
$x - y = 3$

22. $\dfrac{2}{3}x + \dfrac{1}{6}y = \dfrac{2}{3}$
$4x + y = 4$

23. $\dfrac{x+3}{4} + \dfrac{y-1}{3} = 1$
$x - y = 3$

24. $\dfrac{x-1}{2} + \dfrac{y+2}{3} = 4$
$x - 2y = 5$

25. $2.5x - 3y = 1.5$
$10x - 12y = 6$

26. $3.5x - 2y = 2.6$
$7x - 2.5y = 4$

27. $0.05x - 0.03y = 0.21$
$0.07x + 0.02y = 0.16$

28. $0.02x - 0.05y = -0.19$
$0.03x + 0.04y = 0.52$

29. $4b + 3m = 3$
$3b + 11m = 13$

30. $3b + 3m = 7$
$3b + 5m = 3$

In Exercises 31 and 32, the graphs of the two equations appear to be parallel. Are they? Explain your reasoning.

31. $200y - x = 200$
$199y - x = -198$

32. $25x - 24y = 0$
$13x - 12y = 120$

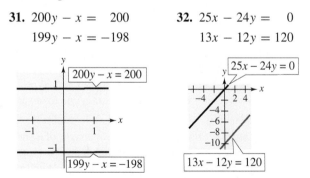

33. *Airplane Speed*   An airplane flying into a head wind travels the 1800-mile flying distance between two cities in 3 hours and 36 minutes. On the return flight, the distance is traveled in 3 hours. Find the ground speed of the plane and the speed of the wind, assuming that both remain constant.

34. *Airplane Speed* Two planes start from the same airport and fly in opposite directions. The second plane starts one-half hour after the first plane, but its speed is 50 miles per hour faster. Find the ground speed of each plane if, 2 hours after the first plane starts its flight, the planes are 2000 miles apart.

35. *Acid Mixture* Ten gallons of a 30% acid solution is obtained by mixing a 20% solution with a 50% solution. How much of each must be used?

36. *Fuel Mixture* Five hundred gallons of 82-octane gasoline is obtained by mixing 80-octane gasoline with 86-octane gasoline. How much of each must be used? (Octane ratings can be interpreted as percents. A high octane rating indicates that the gasoline is knock resistant. A low octane rating indicates that the gasoline is knock prone.)

37. *Investment Portfolio* A total of $12,000 is invested in two corporate bonds that pay 10.5% and 12% simple interest. The annual interest is $1380. How much is invested in each bond?

38. *Investment Portfolio* A total of $32,000 is invested in two municipal bonds that pay 5.75% and 6.25% simple interest. The annual interest is $1930. How much is invested in each bond?

39. *Ticket Sales* You are the manager of a theater. On Saturday morning you are going over the ticket sales for Friday evening. Five hundred tickets were sold. The tickets for adults and children sold for $7.50 and $4.00, respectively, and the receipts for the performance were $3312.50. However, your assistant manager did not record how many of each type of ticket were sold. From the information you have, can you determine how many of each type were sold?

40. *Shoe Sales* Suppose you are the manager of a shoe store. On Sunday morning you are going over the receipts for the previous week's sales. Two hundred and forty pairs of tennis shoes were sold. One style sold for $66.95 and the other sold for $84.95. The total receipts were $17,652. The cash register that was supposed to keep track of the number of each type of shoe sold malfunctioned. Can you recover the information? If so, how many of each type were sold?

Supply and Demand In Exercises 41–44, find the point of equilibrium for each pair of supply and demand equations.

	Demand	*Supply*
41.	$p = 56 - 0.0001x$	$p = 22 + 0.00001x$
42.	$p = 60 - 0.00001x$	$p = 15 + 0.00004x$
43.	$p = 140 - 0.00002x$	$p = 80 + 0.00001x$
44.	$p = 400 - 0.0002x$	$p = 225 + 0.0005x$

45. *U.S. Paper Production* The amounts of newsprint paper and tissue paper that were produced from 1985 to 1988 are as shown in the table. The numbers in the table represent millions of tons of paper. (Source: American Paper Institute)

Year	1985	1986	1987	1988
Newsprint	5.4	5.6	5.8	6.0
Tissue	4.9	5.1	5.3	5.5

(a) Find a linear equation that represents the production of each type of paper.

(b) Assuming that the amounts for the given 4 years are representative of future years, will the production of tissue ever equal the production of newsprint?

46. *U.S. Health Care Workers* From 1985 to 1990, the number of health care workers in the United States could be approximated by the models

$$y = 3861.4 + 82.9t \qquad \text{Hospital employees}$$
$$y = 2537.8 + 213.3t \qquad \text{Nonhospital employees}$$

where y represents the number of workers in thousands and $t = 5$ represents 1985. These models show an increasing trend toward nonhospital health care, such as Hospice or Visiting Nurses Associations. (Source: U.S. Bureau of Labor Statistics)

(a) Use a graphing utility to graph this system.

(b) Use the graphs to predict when the number of nonhospital health workers will equal the number of hospital workers.

Fitting a Line to Data   In Exercises 47–52, find the *least squares regression line* $y = ax + b$ for the points $(x_1, y_1), (x_2, y_2), \ldots, (x_n, y_n)$. To find the line, solve the system for a and b. (If you are unfamiliar with summation notation, look at the discussion in Section 8.1.)

$$nb + \left(\sum_{i=1}^{n} x_i\right) a = \sum_{i=1}^{n} y_i$$

$$\left(\sum_{i=1}^{n} x_i\right) b + \left(\sum_{i=1}^{n} x_i^2\right) a = \sum_{i=1}^{n} x_i y_i$$

47. $5b + 10a = 20.2$
$\quad\;\; 10b + 30a = 50.1$

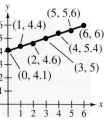

48. $5b + 10a = 11.7$
$\quad\;\; 10b + 30a = 25.6$

49. $7b + 21a = \;\; 35.1$
$\quad\;\; 21b + 91a = 114.2$

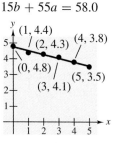

50. $6b + 15a = 23.6$
$\quad\;\; 15b + 55a = 48.8$

51. $6b + 15a = 24.9$
$\quad\;\; 15b + 55a = 58.0$

52. $7b + 21a = 20.3$
$\quad\;\; 21b + 91a = 54.8$

53. *Restaurant Sales*   The total sales (in billions of dollars) for full-service restaurants in the United States from 1990 to 1994 are shown in the table. (Source: National Restaurant Association)

(a) Use the technique demonstrated in Exercises 47–52 to find the line that best fits the data.

(b) Graph the line on a graphing utility and use the TRACE feature to predict the total restaurant sales in 1996.

Year	t	Total Sales (in billions of dollars)
1990	0	76.1
1991	1	78.4
1992	2	80.3
1993	3	83.1
1994	4	85.5

54. *U.S. Airports*   The total numbers of airports (in thousands) with paved runways in the United States for the years 1988 to 1992 are shown in the table. (Source: Federal Aviation Administration)

(a) Use the technique demonstrated in Exercises 47–52 to find the line that best fits the data.

(b) Graph the line on a graphing utility and use the TRACE feature to predict the total number of airports with paved runways in 1997.

Year	t	Number of Airports (in thousands)
1988	8	7.4
1989	9	7.6
1990	10	7.7
1991	11	7.8
1992	12	7.9

6.3 | Linear Systems in Three or More Variables

Row-Echelon Form and Back-Substitution ▪ Gaussian Elimination ▪ Nonsquare Systems ▪ Applications

Row-Echelon Form and Back-Substitution

One of the most influential Chinese mathematics books was the *Chui-chang suan-shu* or *Nine Chapters on the Mathematical Art* (written in approximately 250 B.C.). Chapter Eight of the *Nine Chapters* contained solutions of systems of linear equations using positive and negative numbers. One such system was

$$3x + 2y + z = 39$$
$$2x + 3y + z = 34$$
$$x + 2y + 3z = 26.$$

This system was solved using column operations on a matrix.

The method of elimination can be applied to a system of linear equations in more than two variables. In fact, this method easily adapts to computer use for solving linear systems with dozens of variables.

When elimination is used to solve a system of linear equations, the goal is to rewrite the system in a form to which back-substitution can be applied. To see how this works, consider the following two systems of linear equations.

$$
\begin{array}{ll}
x - 2y + 3z = 9 & \quad x - 2y + 3z = 9 \\
-x + 3y \phantom{{}+3z} = -4 & \quad y + 3z = 5 \\
2x - 5y + 5z = 17 & \quad z = 2
\end{array}
$$

The system on the right is said to be in **row-echelon form,** which means that it has a "stair-step" pattern with leading coefficients of 1. After comparing the two systems, it should be clear that it is easier to solve the system on the right.

EXAMPLE 1 Using Back-Substitution

Solve the following system of linear equations.

$$
\begin{array}{ll}
x - 2y + 3z = 9 & \qquad \text{Equation 1} \\
y + 3z = 5 & \qquad \text{Equation 2} \\
z = 2 & \qquad \text{Equation 3}
\end{array}
$$

Solution

From Equation 3, you know the value of z. To solve for y, substitute $z = 2$ into Equation 2 to obtain

$$
\begin{array}{ll}
y + 3(2) = 5 & \qquad \text{Substitute 2 for } z. \\
y = -1. & \qquad \text{Solve for } y.
\end{array}
$$

Finally, substitute $y = -1$ and $z = 2$ into Equation 1 to obtain

$$
\begin{array}{ll}
x - 2(-1) + 3(2) = 9 & \qquad \text{Substitute } -1 \text{ for } y \text{ and 2 for } z. \\
x = 1. & \qquad \text{Solve for } x.
\end{array}
$$

The solution is $x = 1$, $y = -1$, and $z = 2$, which can be written as the **ordered triple** $(1, -1, 2)$. Check this in the original system of equations.

Gaussian Elimination

Two systems of equations are **equivalent** if they have the same solution set. To solve a system that is not in row-echelon form, first convert it to an *equivalent* system that is in row-echelon form. To see how this is done, let's take another look at the method of elimination, as applied to a system of two linear equations.

EXAMPLE 2   *The Method of Elimination*

Solve the following system of linear equations.

$$3x - 2y = -1$$
$$x - y = 0$$

Solution

$$x - y = 0$$
$$3x - 2y = -1$$

You can interchange two equations in the system.

$$-3x + 3y = 0$$

Multiply the first equation by -3.

$$-3x + 3y = 0$$
$$\underline{3x - 2y = -1}$$
$$y = -1$$

You can add the multiple of the first equation to the second equation to obtain a new equation.

$$x - y = 0$$
$$y = -1$$

New system in row-echelon form

NOTE   As shown in Example 2, rewriting a system of linear equations in row-echelon form usually involves a *chain* of equivalent systems, each of which is obtained by using one of the three basic row operations. This process is called **Gaussian elimination,** after the German mathematician Carl Friedrich Gauss (1777–1855).

Now, using back-substitution, you can determine that the solution is $y = -1$ and $x = -1$, which can be written as the ordered pair $(-1, -1)$. Check this in the original system of equations.

Operations That Produce Equivalent Systems

Each of the following **row operations** on a system of linear equations produces an *equivalent* system of linear equations.

1. Interchange two equations.

2. Multiply one of the equations by a nonzero constant.

3. Add a multiple of one of the equations to another equation to replace the latter equation.

EXAMPLE 3 *Using Elimination to Solve a System*

Solve the following system of linear equations.

$$x - 2y + 3z = 9 \qquad \text{Equation 1}$$
$$-x + 3y = -4 \qquad \text{Equation 2}$$
$$2x - 5y + 5z = 17 \qquad \text{Equation 3}$$

Solution

There are many ways to begin, but we suggest working from the upper left corner, saving the x in the upper left position and eliminating the other x's from the first column.

$$x - 2y + 3z = 9$$
$$y + 3z = 5$$
$$2x - 5y + 5z = 17$$

> Adding the first equation to the second equation produces a new second equation.

$$x - 2y + 3z = 9$$
$$y + 3z = 5$$
$$-y - z = -1$$

> Adding - 2 times the first equation to the third equation produces a new third equation.

Now that all but the first x have been eliminated from the first column, go to work on the second column. (You need to eliminate y from the third equation.)

$$x - 2y + 3z = 9$$
$$y + 3z = 5$$
$$2z = 4$$

> Adding the second equation to the third equation produces a new third equation.

Finally, you need a coefficient of 1 for z in the third equation.

$$x - 2y + 3z = 9$$
$$y + 3z = 5$$
$$z = 2$$

> Multiplying the third equation by $\frac{1}{2}$ produces a new third equation.

This is the same system that was solved in Example 1, and, as in that example, you can conclude that the solution is

$$x = 1, \quad y = -1, \quad \text{and} \quad z = 2.$$

In Example 3, you can check the solution by substituting $x = 1$, $y = -1$, and $z = 2$ into each original equation, as follows.

Equation 1: $(1) - 2(-1) + 3(2) = 9$ ✓

Equation 2: $-(1) + 3(-1) = -4$ ✓

Equation 3: $2(1) - 5(-1) + 5(2) = 17$ ✓

The next example involves an inconsistent system—one that has no solution. The key to recognizing an inconsistent system is that at some stage in the elimination process, you obtain an absurdity such as $0 = -2$.

EXAMPLE 4   *An Inconsistent System*

Solve the following system of linear equations.

$$x - 3y + z = 1 \qquad \text{Equation 1}$$
$$2x - y - 2z = 2 \qquad \text{Equation 2}$$
$$x + 2y - 3z = -1 \qquad \text{Equation 3}$$

Solution

$$x - 3y + z = 1$$
$$5y - 4z = 0$$
$$x + 2y - 3z = -1$$

> Adding - 2 times the first equation to the second equation produces a new second equation.

$$x - 3y + z = 1$$
$$5y - 4z = 0$$
$$5y - 4z = -2$$

> Adding - 1 times the first equation to the third equation produces a new third equation.

$$x - 3y + z = 1$$
$$5y - 4z = 0$$
$$0 = -2$$

> Adding - 1 times the second equation to the third equation produces a new third equation.

Because the third "equation" is impossible, you can conclude that this system is inconsistent and therefore has no solution. Moreover, because this system is equivalent to the original system, you can conclude that the original system also has no solution.

As with a system of linear equations in two variables, the solution(s) of a system of linear equations in more than two variables must fall into one of three categories.

The Number of Solutions of a Linear System
For a system of linear equations, exactly one of the following is true.
1. There is exactly one solution.
2. There are infinitely many solutions.
3. There is no solution.

Arithmetic errors are often made when performing elementary row operations. Have students note the operation performed in each step so that they can go back and check their work. Because some graphing calculators perform row operations, you might consider having students use graphing calculators as a tool for checking work in this section.

You may want to point out to students that in Sections 6.1 and 6.2 graphing the system was a good way to identify the number of solutions to the system because these systems were in only two variables. With more than two variables, the row-echelon form is useful in determining the number of solutions— one solution, many solutions, or no solutions.

EXAMPLE 5   A System with Infinitely Many Solutions

Solve the following system of linear equations.

$$
\begin{aligned}
x + y - 3z &= -1 \qquad &\text{Equation 1}\\
y - z &= 0 \qquad &\text{Equation 2}\\
-x + 2y &= 1 \qquad &\text{Equation 3}
\end{aligned}
$$

Solution

$$
\begin{aligned}
x + y - 3z &= -1\\
y - z &= 0\\
3y - 3z &= 0
\end{aligned}
$$

Adding the first equation to the third equation produces a new third equation.

$$
\begin{aligned}
x + y - 3z &= -1\\
y - z &= 0\\
0 &= 0
\end{aligned}
$$

Adding - 3 times the second equation to the third equation produces a new third equation.

This means that Equation 3 depends on Equations 1 and 2 in the sense that it gives us no additional information about the variables. Thus, the original system is equivalent to the system

$$
\begin{aligned}
x + y - 3z &= -1\\
y - z &= 0.
\end{aligned}
$$

In this last equation, solve for y in terms of z to obtain $y = z$. Back-substituting for y into the previous equation produces $x = 2z - 1$. Finally, letting $z = a$, the solutions to the given system are all of the form

$$
x = 2a - 1, \quad y = a, \quad \text{and} \quad z = a
$$

where a is a real number. Thus, every ordered triple of the form

$$
(2a - 1, a, a), \qquad a \text{ is a real number}
$$

is a solution of the system.

In Example 5, there are other ways to write the same infinite set of solutions. For instance, the solutions could have been written as

$$
\left(b, \tfrac{1}{2}(b + 1), \tfrac{1}{2}(b + 1)\right), \qquad b \text{ is a real number.}
$$

Try convincing yourself of this by substituting $a = 0$, $a = 1$, $a = 2$, and $a = 3$ into the solution listed in Example 5. Then substitute $b = -1$, $b = 1$, $b = 3$, and $b = 5$ into the solution listed above. In both cases, you should obtain the same ordered triples. Thus, when comparing descriptions of an infinite solution set, keep in mind that there is more than one way to describe the set.

Nonsquare Systems

So far, each system of linear equations has been **square,** which means that the number of equations is equal to the number of variables. In a **nonsquare** system, the number of equations differs from the number of variables. A system of linear equations cannot have a unique solution unless there are at least as many equations as there are variables in the system.

EXAMPLE 6   *A System with Fewer Equations than Variables*

Solve the following system of linear equations.

$$x - 2y + z = 2 \qquad \text{Equation 1}$$
$$2x - y - z = 1 \qquad \text{Equation 2}$$

Solution

Begin by rewriting the system in row-echelon form, as follows.

$$x - 2y + z = 2$$
$$3y - 3z = -3$$

> Adding - 2 times the first equation to the second equation produces a new second equation.

$$x - 2y + z = 2$$
$$y - z = -1$$

> Multiplying the second equation by $\frac{1}{3}$ produces a new second equation.

Solving for y in terms of z, you get $y = z - 1$, and back-substitution into Equation 1 yields

$$x - 2(z - 1) + z = 2$$
$$x - 2z + 2 + z = 2$$
$$x = z.$$

Finally, by letting $z = a$, you have the solution

$$x = a, \qquad y = a - 1, \qquad \text{and} \qquad z = a$$

where a is a real number. Thus, every ordered triple of the form

$$(a, a - 1, a), \qquad a \text{ is a real number}$$

is a solution of the system.

NOTE   In Example 6, try choosing some values of a to obtain different solutions of the system, such as $(1, 0, 1)$, $(2, 1, 2)$, and $(3, 2, 3)$. Then check each of the solutions in the original system.

Applications

In 1993, Americans saved or invested about 190 billion dollars—an average of about $760 per person.

To check solutions that are not unique, encourage your students to try several values for *a*.

Activities

1. $x - y + z = 4$
    $x + 3y - 2z = -3$
    $3x + 2y + z = 5$
    Answer: $(2, -1, 1)$

2. $x - 2y - z = -5$
    $2x + y + z = 5$
    Answer: $(a, 3a, 5 - 5a)$

3. $x - 2y + z = 4$
    $3x - 6y + 3z = 7$
    $2x + y + 4z = 2$
    Answer: no solution

Real Life

EXAMPLE 7 An Investment Portfolio

You have a portfolio totaling $450,000 and want to invest in (1) certificates of deposit, (2) municipal bonds, (3) blue-chip stocks, and (4) growth or speculative stocks. The certificates pay 9% annually, and the municipal bonds pay 6% annually. Over a 5-year period, you expect the blue-chip stocks to return 10% annually and the growth stocks to return 15% annually. You want a combined annual return of 8%, and you also want to have only one-third of the portfolio invested in stocks. How much should be allocated to each type of investment?

Solution

To solve this problem, let C, M, B, and G represent the amounts in the four types of investments. Because the total investment is $450,000, you can write the following equation.

$$C + M + B + G = 450{,}000$$

A second equation can be derived from the fact that the combined annual return should be 8%.

$$0.09C + 0.06M + 0.10B + 0.15G = 0.08(450{,}000)$$

Finally, because only one-third of the total investment should be allocated to stocks, you can write

$$B + G = \tfrac{1}{3}(450{,}000).$$

These three equations make up the following system.

$$
\begin{aligned}
C + M + B + G &= 450{,}000 \\
0.09C + 0.06M + 0.10B + 0.15G &= 36{,}000 \\
B + G &= 150{,}000
\end{aligned}
$$

Using elimination, you find that the system has infinitely many solutions, which can be written as follows.

$$C = -\tfrac{5}{3}a + 100{,}000, \quad M = \tfrac{5}{3}a + 200{,}000, \quad B = -a + 150{,}000, \quad G = a$$

Thus, you have many different options. One possible solution is to choose $a = 30{,}000$, which yields the following portfolio.

1. Certificates of deposit: $50,000
2. Municipal bonds: $250,000
3. Blue-chip stocks: $120,000
4. Growth or speculative stocks: $30,000

EXAMPLE 8   Data Analysis: Curve-Fitting

Find a quadratic equation, $y = ax^2 + bx + c$, whose graph passes through the points $(-1, 3)$, $(1, 1)$, and $(2, 6)$.

Solution

Because the graph of $y = ax^2 + bx + c$ passes through the points $(-1, 3)$, $(1, 1)$, and $(2, 6)$, you can write the following.

When $x = -1$, $y = 3$: $a(-1)^2 + b(-1) + c = 3$

When $x = 1$, $y = 1$: $a(1)^2 + b(1) + c = 1$

When $x = 2$, $y = 6$: $a(2)^2 + b(2) + c = 6$

This produces the following system of linear equations.

$$
\begin{array}{ll}
a - b + c = 3 & \text{Equation 1} \\
a + b + c = 1 & \text{Equation 2} \\
4a + 2b + c = 6 & \text{Equation 3}
\end{array}
$$

The solution of this system is $a = 2$, $b = -1$, and $c = 0$. Thus, the equation of the parabola is $y = 2x^2 - x$, as shown in Figure 6.11.

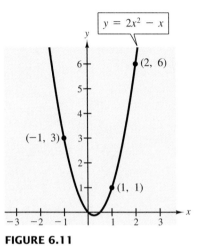

$y = 2x^2 - x$

FIGURE 6.11

Group Activities   Communicating Mathematically

Mathematical Modeling   Suppose you work for an outerwear manufacturer, and the marketing department is concerned about sales trends in Georgia. Your manager has asked you to investigate climate data in hopes of explaining sales patterns and gives you the table to the left representing the average monthly temperature y in degrees Fahrenheit for Savannah, Georgia, for the month x where $x = 1$ corresponds to November. (Source: National Climatic Data Center)

x	1	3	5
y	59	49	59

Construct a scatter plot of the data; decide what type of mathematical model might be appropriate for the data; and use the methods you have learned thus far to fit an appropriate model. Your manager would like to know the average monthly temperatures for December and February. Explain to your manager how you found your model, what it represents, and how it may be used to find the December and February average temperatures. Investigate the usefulness of this model for the rest of the year. Would you recommend using the model to predict monthly average temperatures for the whole year or just part of the year? Explain your reasoning.

The following warm-up exercises involve skills that were covered in earlier sections. You will use these skills in the exercise set for this section.

In Exercises 1–4, solve the system of linear equations.

1. $x + y = 25$
$\phantom{x + {}}y = 10$

2. $2x - 3y = 4$
$6x \phantom{{}-3y{}} = -12$

3. $x + y = 32$
$x - y = 24$

4. $2r - s = 5$
$r + 2s = 10$

In Exercises 5–8, determine whether the ordered triple is a solution of the equation.

5. $5x - 3y + 4z = 2$
$(-1, -2, 1)$

6. $x - 2y + 12z = 9$
$(6, 3, 2)$

7. $2x - 5y + 3z = -9$
$(a - 2, a + 1, a)$

8. $-5x + y + z = 21$
$(a - 4, 4a + 1, a)$

In Exercises 9 and 10, solve for x in terms of a.

9. $x + 2y - 3z = 4$
$y = 1 - a, z = a$

10. $x - 3y + 5z = 4$
$y = 2a + 3, z = a$

6.3 Exercises

Exercises containing inconsistent systems (no solution): 7, 8, 21, and 22. Exercises containing dependent systems (infinitely many solutions): 11, 12, 13, 14, 15, 16, 17, 18, 25, 26, 45, and 46.

In Exercises 1–26, solve the system of linear equations.

1. $x + y + z = 6$
$2x - y + z = 3$
$3x - \phantom{y + {}}z = 0$

2. $x + y + z = 2$
$-x + 3y + 2z = 8$
$4x + y \phantom{{}+ 2z} = 4$

3. $4x + y - 3z = 11$
$2x - 3y + 2z = 9$
$x + y + z = -3$

4. $\phantom{3x + {}}6y + 4z = -12$
$3x + 3y \phantom{{}+ 4z} = 9$
$2x \phantom{{}+ 3y} - 3z = 10$

5. $2x \phantom{{}+ 3y} + 2z = 2$
$5x + 3y \phantom{{}+ 2z} = 4$
$\phantom{5x + {}}3y - 4z = 4$

6. $2x + 4y + z = -4$
$2x - 4y + 6z = 13$
$4x - 2y + z = 6$

7. $3x - 2y + 4z = 1$
$x + y - 2z = 3$
$2x - 3y + 6z = 8$

8. $5x - 3y + 2z = 3$
$2x + 4y - z = 7$
$x - 11y + 4z = 3$

9. $3x + 3y + 5z = 1$
$3x + 5y + 9z = 0$
$5x + 9y + 17z = 0$

10. $2x + y + 3z = 1$
$2x + 6y + 8z = 3$
$6x + 8y + 18z = 5$

11. $x + 2y - 7z = -4$
$2x + y + z = 13$
$3x + 9y - 36z = -33$

12. $2x + y - 3z = 4$
$4x + 2z = 10$
$-2x + 3y - 13z = -8$

13. $x + 4z = 13$
$4x - 2y + z = 7$
$2x - 2y - 7z = -19$

14. $4x - y + 5z = 11$
$x + 2y - z = 5$
$5x - 8y + 13z = 7$

15. $x - 2y + 5z = 2$
$3x + 2y - z = -2$

16. $x - 3y + 2z = 18$
$5x - 13y + 12z = 80$

17. $2x - 3y + z = -2$
$-4x + 9y = 7$

18. $2x + 3y + 3z = 7$
$4x + 18y + 15z = 44$

19. $x + 3w = 4$
$2y - z - w = 0$
$3y - 2w = 1$
$2x - y + 4z = 5$

20. $x + y + z + w = 6$
$2x + 3y - w = 0$
$-3x + 4y + z + 2w = 4$
$x + 2y - z + w = 0$

21. $x + 4z = 1$
$x + y + 10z = 10$
$2x - y + 2z = -5$

22. $3x - 2y - 6z = -4$
$-3x + 2y + 6z = 1$
$x - y - 5z = -3$

23. $4x + 3y + 17z = 0$
$5x + 4y + 22z = 0$
$4x + 2y + 19z = 0$

24. $2x + 3y = 0$
$4x + 3y - z = 0$
$8x + 3y + 3z = 0$

25. $5x + 5y - z = 0$
$10x + 5y + 2z = 0$
$5x + 15y - 9z = 0$

26. $12x + 5y + z = 0$
$12x + 4y - z = 0$

In Exercises 27–30, find the equation of the parabola that passes through the points.

In Exercises 31–34, find the equation of the circle that passes through the points.

27.

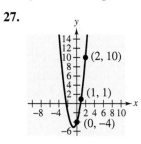

28.

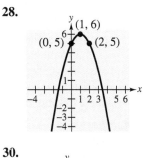

31.

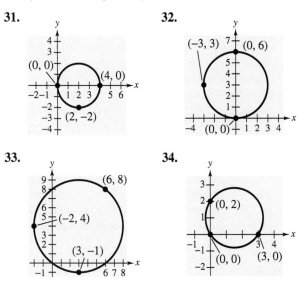

32.

29.

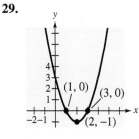

30.

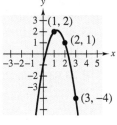

33.

34.

35. *Regular Polygons* The total numbers of sides and diagonals of regular polygons with three, four, and five sides are three, six, and ten, as shown in the accompanying figure. Find a quadratic function $y = ax^2 + bx + c$ that fits this data. Then check to see if it gives the correct answer for a polygon with six sides.

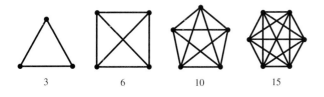

3   6   10   15

36. *Parts of a Circle* The maximum numbers of parts into which a circle can be partitioned by one, two, and three straight lines are two, four, and seven, as shown in the figure. Find a quadratic function $y = ax^2 + bx + c$ that fits this data. Then check to see if it gives the correct answer for a circle that is partitioned by four straight lines.

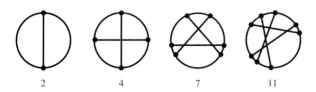

2   4   7   11

37. *Investment* An inheritance of $16,000 was divided among three investments yielding a total of $990 in interest per year. The interest rates for the three investments were 5%, 6%, and 7%. Find the amount placed in each investment if the 5% and 6% investments were $3000 and $2000 less than the 7% investment, respectively.

38. *Investment* Suppose you receive a total of $1520 a year in interest from three investments. The interest rates for the three investments are 5%, 7%, and 8%. The 5% investment is half of the 7% investment, and the 7% investment is $1500 less than the 8% investment. What is the amount of each investment?

39. *Investment* A company borrows $775,000. Some is borrowed at 8%, some at 9%, and some at 10%. How much is borrowed at each rate if the annual interest is $67,500 and the amount borrowed at 8% is four times the amount borrowed at 10%?

40. *Investment* A company borrows $800,000. Some is borrowed at 8%, some at 9%, and some at 10%. How much is borrowed at each rate if the annual interest is $67,000 and the amount borrowed at 8% is five times the amount borrowed at 10%?

41. *Grades of Paper* A manufacturer sells a 50-pound package of paper that consists of three grades of computer paper. Grade C costs $3.50 per pound, grade B costs $4.50 per pound, and grade A costs $6.00 per pound. Half of the 50-pound package consists of the two cheaper grades. The cost of the 50-pound package is $252.50. How many pounds of each grade of paper are there in the 50-pound package?

42. *Product Types* A company sells three types of products for $20, $10, and $5 per unit. In 1 year, the total revenue for the three products was $350,000, which corresponded to the sale of 32,500 units. The company sold half as many units of the $20 product as units of the $10 product. How many units of each product were sold?

43. *Crop Spraying* A mixture of 6 gallons of chemical A, 8 gallons of chemical B, and 13 gallons of chemical C is required to kill a certain destructive crop insect. Commercial spray X contains 1, 2, and 2 parts, respectively, of these chemicals. Commercial spray Y contains only chemical C. Commercial spray Z contains chemicals A, B, and C in equal amounts. How much of each type of commercial spray is needed to get the desired mixture?

44. *Chemistry* A chemist needs 10 liters of a 25% acid solution. The solution is to be mixed from three solutions whose concentrations are 10%, 20%, and 50%. How many liters of each solution should the chemist use to satisfy the following?

(a) Use as little as possible of the 50% solution.

(b) Use as much as possible of the 50% solution.

(c) Use 2 liters of the 50% solution.

Investment Portfolio   In Exercises 45 and 46, you have a total of $500,000 that is to be invested in (1) certificates of deposit, (2) municipal bonds, (3) blue-chip stocks, and (4) growth or speculative stocks. How much should be put in each type of investment?

45. The certificates of deposit pay 10% annually, and the municipal bonds pay 8% annually. Over a 5-year period, you expect the blue-chip stocks to return 12% annually and the growth stocks to return 13% annually. You want a combined annual return of 10% and you also want to have only one-fourth of the portfolio in stocks.

46. The certificates of deposit pay 9% annually, and the municipal bonds pay 5% annually. Over a 5-year period, you expect the blue-chip stocks to return 12% annually and the growth stocks to return 14% annually. You want a combined annual return of 10% and you also want to have only one-fourth of the portfolio in stocks.

Fitting a Parabola to Data   In Exercises 47–50, find the least squares regression parabola $y = ax^2 + bx + c$ for the points (x_1, y_1), (x_2, y_2), . . . , (x_n, y_n). To find the parabola, solve the system of linear equations for a, b, and c.

$$nc + \left(\sum_{i=1}^{n} x_i\right) b + \left(\sum_{i=1}^{n} x_i{}^2\right) a = \sum_{i=1}^{n} y_i$$

$$\left(\sum_{i=1}^{n} x_i\right) c + \left(\sum_{i=1}^{n} x_i{}^2\right) b + \left(\sum_{i=1}^{n} x_i{}^3\right) a = \sum_{i=1}^{n} x_i y_i$$

$$\left(\sum_{i=1}^{n} x_i{}^2\right) c + \left(\sum_{i=1}^{n} x_i{}^3\right) b + \left(\sum_{i=1}^{n} x_i{}^4\right) a = \sum_{i=1}^{n} x_i{}^2 y_i$$

47. $5c \quad\;\; + 10a = 15.5$
$\quad\quad\;\; 10b \quad\quad = 6.3$
$\quad\;\; 10c \quad\; + 34a = 32.1$

48. $5c \quad\;\; + 10a = 15.0$
$\quad\quad\;\; 10b \quad\quad = 17.3$
$\quad\;\; 10c \quad\; + 34a = 34.5$

49. $6c + \;\; 3b + \;\; 19a = 23.9$
$\quad 3c + 19b + \;\; 27a = -7.2$
$\quad 19c + 27b + 115a = 48.8$

50. $6c + \;\; 3b + \;\; 19a = 13.1$
$\quad 3c + 19b + \;\; 27a = -2.6$
$\quad 19c + 27b + 115a = 29.0$

51. *Computer Use in Schools*   The total numbers of personal computers used in grades K through 12 in the United States from 1982 to 1988 are shown in the accompanying table. In the table, y represents the number of personal computers in millions and $x = 0$ represents 1985. Find the least squares regression parabola $y = ax^2 + bx + c$ that best fits this data by solving the following system. Then use the model to predict the number of personal computers in use in 1992. (Source: Future Computing/Datapro, Inc.)

$7a \quad\quad\; + 28c = \;\; 6.48$
$\quad 28b \quad\quad\;\; = \;\; 7.39$
$\; 28a \quad\; + 196c = 27.01$

x	-3	-2	-1	0	1	2	3
y	0.21	0.40	0.59	0.87	1.17	1.48	1.76

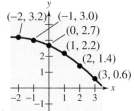

52. *Prices of Mobile Homes*   The average prices of new mobile homes sold in the United States from 1989 to 1993 are shown in the accompanying table. In the table, y represents the average price (in thousands of dollars) and $x = 0$ represents 1990. Find the least squares regression parabola

$$y = ax^2 + bx + c$$

that best fits this data by solving the system. Then use the model to predict the average price of a new mobile home in 1996. (Source: National Association of Realtors)

$$a + c = 27.6$$
$$70b = 12.4$$
$$6a + 20c = 548.2$$

x	-1	0	1	2	3
y	27.2	27.8	27.7	28.4	30.5

Math Matters   Regular Polygons and Regular Polyhedra

A regular polygon is a polygon that has n sides of equal length and n equal angles. For instance, a regular three-sided polygon is called an equilateral triangle, a regular four-sided polygon is called a square, and so on. There are infinitely many different types of regular polygons. For instance, it is possible to construct a regular polygon with 100, 1000, or even 1,000,000 sides (they would look very much like circles, but it is still possible to do).

For solid figures, the story is quite different. A regular polyhedron is a solid figure each of whose sides is a regular polygon (of the same size) and each of whose angles is formed by the same number of sides. At first, one might think that there are infinitely many different types of regular polyhedra, but in fact it can be shown that there are only five. The five different types are tetrahedron (4 triangular sides), cube (6 square sides), octahedron (8 triangular sides), dodecahedron (12 pentagonal sides), and icosahedron (20 triangular sides), as shown in the figure.

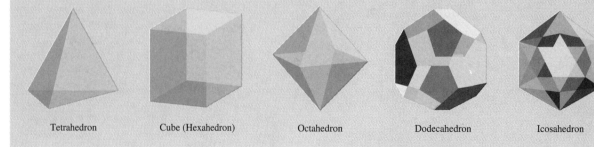

Tetrahedron   Cube (Hexahedron)   Octahedron   Dodecahedron   Icosahedron

MID-CHAPTER QUIZ

Take this quiz as you would take a quiz in class. After you are done, check your work against the answers given in the back of the book.

In Exercises 1 and 2, write a system that has the given solution.

1. $(3, 2)$

2. $(5, 3)$

In Exercises 3 and 4, use a graphing utility to graphically solve the system.

3. $y = 2\sqrt{x} + 1$, $y = x - 2$

4. $x^2 + y^2 = 9$, $y = 2x + 1$

In Exercises 5 and 6, find the number of sales necessary to break even.

5. $C = 12.50x + 10,000$, $R = 19.95x$

6. $C = 3.79x + 400,000$, $R = 4.59x$

In Exercises 7 and 8, solve the system by elimination. Verify the solution graphically.

7. $2.5x - y = 6$
$3x + 4y = 2$

8. $\frac{1}{2}x + \frac{1}{3}y = 1$
$x - 2y = -2$

In Exercises 9 and 10, find the point of equilibrium.

9. Demand: $p = 45 - 0.001x$;   supply: $p = 23 + 0.0002x$

10. Demand: $p = 95 - 0.0002x$;   supply: $p = 80 + 0.00001x$

In Exercises 11–13, solve the system of equations.

11. $2x + 3y - z = -7$
$x + 3z = 10$
$2y + z = -1$

12. $x + y + 3z = 11$
$2x - y - z = -11$
$2y + 3z = 17$

13. $2x + 3y = 12$
$-x + 2y + z = 0$
$y - z = 7$

In Exercises 14 and 15, write three ordered triples of the given form.

14. $(a, a - 2, 3a)$

15. $(2a, a + 5, a)$

16. When solving a square system of equations with three variables, you arrive at an equation $0 = -3$. What conclusion can you make?

17. How many solutions does the system $2x - 3y + z = 4$, $y - z = 7$ have?

18. *Acid Mixture*   Ten gallons of a 35% acid solution is obtained by mixing a 25% solution with a 40% solution. How much of each must be used?

19. *True or False?*   A system of equations may have infinitely many solutions.

20. *True or False?*   A square linear system must have exactly one solution.

6.4	**Systems of Inequalities**
	The Graph of an Inequality ■ Systems of Inequalities ■ Applications

The Graph of an Inequality

The following statements are inequalities in two variables:

$$3x - 2y < 6 \qquad \text{and} \qquad 2x^2 + 3y^2 \geq 6.$$

An ordered pair (a, b) is a **solution of an inequality** in x and y if the inequality is true when a and b are substituted for x and y, respectively. The **graph** of an inequality is the collection of all solutions of the inequality. To sketch the graph of an inequality, begin by sketching the graph of the *corresponding equation*. The graph of the equation will normally separate the plane into two or more regions. In each such region, one of the following must be true.

1. *All* points in the region are solutions of the inequality.

2. *No* point in the region is a solution of the inequality.

Thus, you can determine whether the points in an entire region satisfy the inequality by simply testing *one* point in the region.

Sketching the Graph of an Inequality in Two Variables

1. Replace the inequality sign by an equal sign, and sketch the graph of the resulting equation. (Use a dashed line for $<$ or $>$ and a solid line for $\leq$ or $\geq$.)

2. Test one point in each of the regions formed by the graph in Step 1. If the point satisfies the inequality, shade the entire region to denote that every point in the region satisfies the inequality.

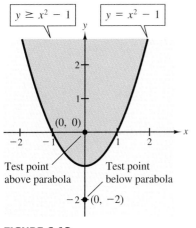

FIGURE 6.12

EXAMPLE 1 *Sketching the Graph of an Inequality*

Sketch the graph of the inequality $y \geq x^2 - 1$.

Solution

The graph of the corresponding *equation* $y = x^2 - 1$ is a parabola, as shown in Figure 6.12. By testing a point *above* the parabola $(0, 0)$ and a point *below* the parabola $(0, -2)$, you can see that the points that satisfy the inequality are those lying above (or on) the parabola.

The inequality given in Example 1 is a nonlinear inequality in two variables. Most of the following examples involve **linear inequalities** such as $ax + by < c$. The graph of a linear inequality is a half-plane lying on one side of the line $ax + by = c$. The simplest linear inequalities are those corresponding to horizontal or vertical lines, as shown in Example 2.

EXAMPLE 2   *Sketching the Graph of a Linear Inequality*

Sketch the graphs of the following linear inequalities.

a. $x > -2$ **b.** $y \leq 3$

Solution

a. The graph of the corresponding equation $x = -2$ is a vertical line. The points that satisfy the inequality $x > -2$ are those lying to the right of this line, as shown in Figure 6.13.

b. The graph of the corresponding equation $y = 3$ is a horizontal line. The points that satisfy the inequality $y \leq 3$ are those lying below (or on) this line, as shown in Figure 6.14.

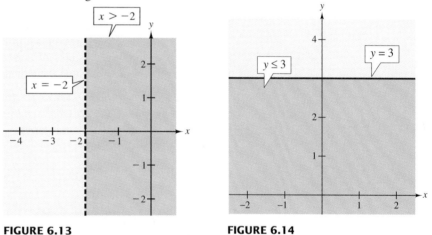

FIGURE 6.13   **FIGURE 6.14**

EXAMPLE 3   *Sketching the Graph of a Linear Inequality*

Sketch the graph of $x - y < 2$.

Solution

The graph of the corresponding equation $x - y = 2$ is a line, as shown in Figure 6.15. Because the origin $(0, 0)$ satisfies the inequality, the graph consists of the half-plane lying above the line. (Try checking a point below the line. Regardless of which point you choose, you will see that it does not satisfy the inequality.)

STUDY TIP

To graph a linear inequality, it can help to write the inequality in slope-intercept form. For instance, by writing $x - y < 2$ in the form

$$y > x - 2$$

you can see that the solution points lie *above* the line $x - y = 2$ (or $y = x - 2$), as shown in Figure 6.15.

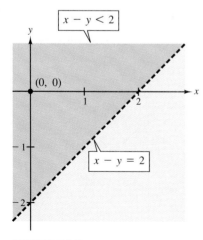

FIGURE 6.15

Systems of Inequalities

Many practical problems in business, science, and engineering involve systems of linear inequalities. A **solution** of a system of inequalities in x and y is a point (x, y) that satisfies each inequality in the system.

To sketch the graph of a system of inequalities in two variables, first sketch the graph of each individual inequality (on the same coordinate system) and then find the region that is *common* to every graph in the system. For systems of *linear* inequalities, it is helpful to find the vertices of the solution region.

EXAMPLE 4 Solving a System of Inequalities

Sketch the graph (and label the vertices) of the solution set of the following system.

$$x - y < 2$$
$$x > -2$$
$$y \leq 3$$

Solution

The graphs of these inequalities are shown in Figures 6.13 to 6.15. The triangular region common to all three graphs can be found by superimposing the graphs on the same coordinate plane, as shown in Figure 6.16. To find the vertices of the region, solve the three systems of corresponding equations obtained by taking *pairs* of equations representing the boundaries of the individual regions.

Vertex A: $(-2, -4)$	*Vertex B:* $(5, 3)$	*Vertex C:* $(-2, 3)$
Obtained by solving the system	*Obtained by solving the system*	*Obtained by solving the system*
$x - y = 2$ $x = -2$	$x - y = 2$ $y = 3$	$x = -2$ $y = 3$

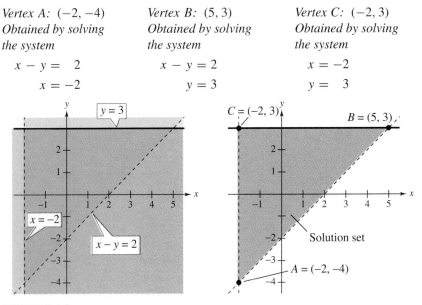

FIGURE 6.16

For the triangular region shown in Figure 6.16, each point of intersection of a pair of boundary lines corresponds to a vertex. With more complicated regions, two border lines can sometimes intersect at a point that is not a vertex of the region, as shown in Figure 6.17. In order to keep track of which points of intersection are actually vertices of the region, we suggest that you make a careful sketch of the region and refer to your sketch as you find each point of intersection.

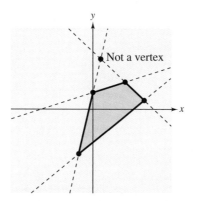

FIGURE 6.17   Boundary lines can intersect at a point that is not a vertex.

EXAMPLE 5   *Solving a System of Inequalities*

Sketch the region containing all points that satisfy the following system.

$$x^2 - y \le 1$$
$$-x + y \le 1$$

Solution

As shown in Figure 6.18, the points that satisfy the inequality $x^2 - y \le 1$ are the points lying above (or on) the parabola given by

$$y = x^2 - 1. \qquad \text{Parabola}$$

The points satisfying the inequality $-x + y \le 1$ are the points lying on or below the line given by

$$y = x + 1. \qquad \text{Line}$$

To find the points of intersection of the parabola and the line, solve the system of corresponding equations.

$$x^2 - y = 1$$
$$-x + y = 1$$

Using the method of substitution, you can find the solutions to be $(-1, 0)$ and $(2, 3)$, as shown in Figure 6.18.

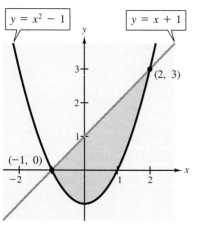

FIGURE 6.18

When solving a system of inequalities, you should be aware that the system might have no solution. For instance, the system

$$x + y > 3$$
$$x + y < -1$$

has no solution points, because the quantity $(x + y)$ cannot be both less than -1 and greater than 3, as shown in Figure 6.19.

Activities

1. Sketch the graph of the inequality.
   $2x - 3y \leq 6$.
   Answer: the region above the solid line $2x - 3y = 6$

2. Sketch the graph of the solution of the system of inequalities.
   $x^2 + y^2 < 16$
   $\quad y > \quad x$
   Answer: the region above the line $y = x$, but in the interior of the circle

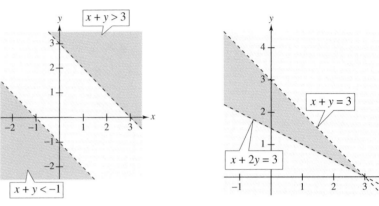

FIGURE 6.19   No Solution

FIGURE 6.20   Unbounded Region

Another possibility is that the solution set of a system of inequalities can be unbounded. For instance, the solution set of

$$x + \quad y < 3$$
$$x + 2y > 3$$

forms an *infinite wedge*, as shown in Figure 6.20.

Technology

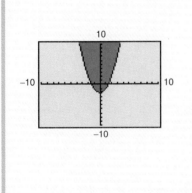

Inequalities and Graphing Utilities

A graphing utility can be used to sketch the graph of an inequality. For instance, to sketch the graph of $y \geq x^2 - 2$ on a *TI-82*, you can use the following steps.

1. Clear the $\boxed{Y=}$ screen.

2. Press $\boxed{ZOOM}$ 6 to clear the graphing screen.

3. $\boxed{DRAW}$ Shade($X^2-2,12$) $\boxed{ENTER}$

Note that $x^2 - 2$ is the lower boundary of the shaded region and 12 is the upper boundary. The resulting graph is shown at the left. Try using a graphing utility to graph the following inequalities.

a. $y \leq 2x + 2$ **b.** $y \geq \dfrac{1}{2}x^2 - 4$

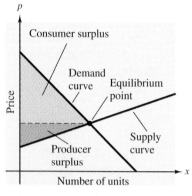

FIGURE 6.21

Applications

Example 8 in Section 6.2 discussed the *point of equilibrium* for a demand and supply function. The next example discusses two related concepts that economists call **consumer surplus** and **producer surplus.** As shown in Figure 6.21, the consumer surplus is defined as the area of the region that lies *below* the demand curve, *above* the horizontal line passing through the equilibrium point, and to the right of the *y*-axis. Similarly, the producer surplus is defined as the area of the region that lies *above* the supply curve, *below* the horizontal line passing through the equilibrium point, and to the right of the *y*-axis. The consumer surplus is a measure of the amount that consumers would have been willing to pay *above what they actually paid*, whereas the producer surplus is a measure of the amount that producers would have been willing to receive *below what they actually received*.

Real Life

EXAMPLE 6   *Consumer and Producer Surplus*

The demand and supply functions for a certain type of calculator are given by

$$p = 150 - 0.00001x \qquad \text{Demand equation}$$
$$p = 60 + 0.00002x \qquad \text{Supply equation}$$

where p is the price in dollars and x represents the number of units. Find the consumer surplus and producer surplus for these two equations.

Solution

Begin by finding the point of equilibrium by solving the equation

$$60 + 0.00002x = 150 - 0.00001x.$$

In Example 8 in Section 6.2, you saw that the solution is $x = 3{,}000{,}000$, which corresponds to an equilibrium price of $p = \$120$. Thus, the consumer surplus and producer surplus are the areas of the triangular regions given by the system of inequalities.

Consumer Surplus	Producer Surplus
$p \le 150 - 0.00001x$	$p \ge 60 + 0.00002x$
$p \ge 120$	$p \le 120$
$x \ge 0$	$x \ge 0$

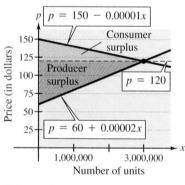

FIGURE 6.22   No Solution

In Figure 6.22, you can see that the consumer surplus is

$$\text{Consumer surplus} = \tfrac{1}{2}(\text{base})(\text{height}) = \tfrac{1}{2}(30)(3{,}000{,}000) = \$45{,}000{,}000$$

and the producer surplus is

$$\text{Producer surplus} = \tfrac{1}{2}(\text{base})(\text{height}) = \tfrac{1}{2}(60)(3{,}000{,}000) = \$90{,}000{,}000.$$

Real Life

EXAMPLE 7   Nutrition

The liquid portion of a diet is to provide at least 300 calories, 36 units of vitamin A, and 90 units of vitamin C daily. A cup of dietary drink X provides 60 calories, 12 units of vitamin A, and 10 units of vitamin C. A cup of dietary drink Y provides 60 calories, 6 units of vitamin A, and 30 units of vitamin C. Set up a system of linear inequalities that describes the minimum daily requirements for calories and vitamins.

Solution

Begin by letting x and y represent the following.

x = number of cups of dietary drink X

y = number of cups of dietary drink Y

To meet the minimum daily requirements, the following inequalities must be satisfied.

For calories:	$60x + 60y \geq 300$
For vitamin A:	$12x + 6y \geq 36$
For vitamin C:	$10x + 30y \geq 90$
	$x \geq 0$
	$y \geq 0$

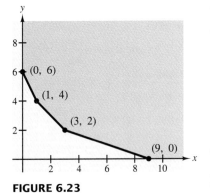

FIGURE 6.23

The last two inequalities are included because x and y cannot be negative. The graph of this system of inequalities is shown in Figure 6.23. (More is said about this application in Example 7 in Section 6.5.)

Group Activities   Exploring with Technology

Graphing Systems of Inequalities   The Technology feature on page 474 describes how to sketch the graph of a single inequality with a graphing utility. With the user's guide for your graphing utility, find how to graph a system of inequalities. Then use a graphing utility to graph the following systems.

a. $y \leq 4 - x^2$
　　　$y \geq 2x - 3$

b. $y \leq 4 - x^2$
　　　$y \geq x^2 - 4$

Warm Up The following warm-up exercises involve skills that were covered in earlier sections. You will use these skills in the exercise set for this section.

In Exercises 1–6, identify the graph of the given equation.

1. $x + y = 3$

2. $4x - y = 8$

3. $y = x^2 - 4$

4. $y = -x^2 + 1$

5. $x^2 + y^2 = 9$

6. $\dfrac{x^2}{4} + \dfrac{y^2}{9} = 1$

In Exercises 7–10, solve the system of equations.

7. $x + 2y = 3$
$4x - 7y = -3$

8. $2x - 3y = 4$
$x + 5y = 2$

9. $x^2 + y = 5$
$2x - 4y = 0$

10. $x^2 + y^2 = 13$
$x + y = 5$

6.4 Exercises

Exercise containing an inconsistent system (no solution): 27

In Exercises 1–6, match the inequality with its graph. [The graphs are labeled (a), (b), (c), (d), (e), and (f).]

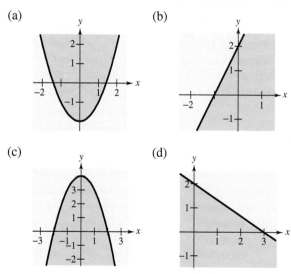

(a) (b) (c) (d)

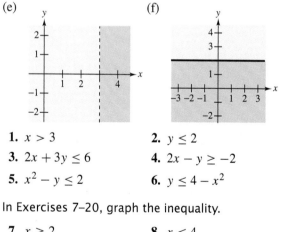

(e) (f)

1. $x > 3$

2. $y \leq 2$

3. $2x + 3y \leq 6$

4. $2x - y \geq -2$

5. $x^2 - y \leq 2$

6. $y \leq 4 - x^2$

In Exercises 7–20, graph the inequality.

7. $x \geq 2$

8. $x \leq 4$

9. $y \geq -1$

10. $y \leq 3$

11. $y < 2 - x$

12. $y > 2x - 4$

13. $2y - x \geq 4$

14. $5x + 3y \geq -15$

15. $y^2 + 2x > 0$

16. $y^2 - x < 0$

17. $y \le \dfrac{1}{1 + x^2}$

18. $y < \ln x$

19. $x^2 + y^2 \le 4$

20. $x^2 + y^2 \ge 4$

In Exercises 21–40, graph the solution set of the system of inequalities.

21. $\begin{aligned} x + y &\le 1 \\ -x + y &\le 1 \\ y &\ge 0 \end{aligned}$

22. $\begin{aligned} 3x + 2y &< 6 \\ x &> 0 \\ y &> 0 \end{aligned}$

23. $\begin{aligned} x + y &\le 5 \\ x &\ge 2 \\ y &\ge 0 \end{aligned}$

24. $\begin{aligned} 2x + y &\ge 2 \\ x &\le 2 \\ y &\le 1 \end{aligned}$

25. $\begin{aligned} -3x + 2y &< 6 \\ x + 4y &> -2 \\ 2x + y &< 3 \end{aligned}$

26. $\begin{aligned} x - 7y &> -36 \\ 5x + 2y &> 5 \\ 6x - 5y &> 6 \end{aligned}$

27. $\begin{aligned} 2x + y &< 2 \\ 6x + 3y &> 2 \end{aligned}$

28. $\begin{aligned} x - 2y &< -6 \\ 5x - 3y &> -9 \end{aligned}$

29. $\begin{aligned} x &\ge 1 \\ x - 2y &\le 3 \\ 3x + 2y &\ge 9 \\ x + y &\le 6 \end{aligned}$

30. $\begin{aligned} x - y^2 &> 0 \\ x - y &< 2 \end{aligned}$

31. $\begin{aligned} x^2 + y^2 &\le 9 \\ x^2 + y^2 &\ge 1 \end{aligned}$

32. $\begin{aligned} x^2 + y^2 &\le 25 \\ 4x - 3y &\le 0 \end{aligned}$

33. $\begin{aligned} x &> y^2 \\ x &< y + 2 \end{aligned}$

34. $\begin{aligned} x &< 2y - y^2 \\ 0 &< x + y \end{aligned}$

35. $\begin{aligned} y &\le \sqrt{3x} + 1 \\ y &\ge x + 1 \end{aligned}$

36. $\begin{aligned} y &< -x^2 + 2x + 3 \\ y &> -3x - 9 \end{aligned}$

37. $\begin{aligned} y &< x^3 - 2x + 1 \\ y &> -2x \\ x &\le 1 \end{aligned}$

38. $\begin{aligned} y &\ge -3 \\ y &\le 1 - x^2 \end{aligned}$

39. $\begin{aligned} y &\le e^x \\ y &\ge 0 \\ x &\ge -1 \\ x &\le 1 \end{aligned}$

40. $\begin{aligned} y &\le e^{-x^2/2} \\ y &\ge 0 \\ -2 &\le x \le 2 \end{aligned}$

In Exercises 41–46, write a system of inequalities that describes the region.

41. Rectangular region with vertices at $(2, 1)$, $(5, 1)$, $(5, 7)$, and $(2, 7)$

42. Parallelogram region with vertices at $(0, 0)$, $(4, 0)$, $(1, 4)$, and $(5, 4)$

43. Triangular region with vertices at $(0, 0)$, $(5, 0)$, and $(2, 3)$

44. Triangular region with vertices at $(-1, 0)$, $(1, 0)$, and $(0, 1)$

45. Sector of a circle

46. Sector of a circle

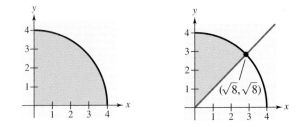

47. *Furniture Production* A furniture company can sell all the tables and chairs it produces. Each table requires 1 hour in the assembly center and $1\frac{1}{3}$ hours in the finishing center. Each chair requires $1\frac{1}{2}$ hours in the assembly center and $1\frac{1}{2}$ hours in the finishing center. The company's assembly center is available 12 hours per day, and its finishing center is available 15 hours per day. If x is the number of tables produced per day and y is the number of chairs, find a system of inequalities describing all possible production levels. Sketch the graph of the system.

48. *Computer Inventory* A store sells two models of a certain brand of computer. Because of the demand, it is necessary to stock twice as many units of model A as units of model B. The costs to the store for the two models are $800 and $1200, respectively. The management does not want more than $20,000 in computer inventory at any one time, and it wants at least four model A computers and two model B computers in inventory at all times. Devise a system of inequalities describing all possible inventory levels, and sketch the graph of the system.

49. *Investment* A person plans to invest $20,000 in two different interest-bearing accounts. Each account is to contain at least $5000. Moreover, one account should have at least twice the amount that is in the other account.

(a) Find a system of inequalities that describes the amounts that can be deposited in each account.

(b) Sketch the graph of the system.

50. *Concert Ticket Sales* Two types of tickets are to be sold for a concert. One type costs $15 and the other type costs $25. The promoter of the concert must sell at least 15,000 tickets, including at least 8000 of the $15 tickets and at least 4000 of the $25 tickets. Moreover, the gross receipts must total at least $275,000 in order for the concert to be held.

(a) Find a system of inequalities describing the different numbers of tickets that can be sold.

(b) Sketch the graph of the system.

51. *Diet Supplement* A dietitian is asked to design a special diet supplement using two different foods. Each ounce of food X contains 20 units of calcium, 15 units of iron, and 10 units of vitamin B. Each ounce of food Y contains 10 units of calcium, 10 units of iron, and 20 units of vitamin B. The minimum daily requirements in the diet are 300 units of calcium, 150 units of iron, and 200 units of vitamin B.

(a) Find a system of inequalities describing the different amounts of food X and food Y that can be used in the diet.

(b) Sketch the graph of the system.

52. *Diet Supplement* A dietitian is asked to design a special diet supplement using two different foods. Each ounce of food X contains 20 units of calcium, 15 units of iron, and 10 units of vitamin B. Each ounce of food Y contains 10 units of calcium, 10 units of iron, and 20 units of vitamin B. The minimum daily requirements in the diet are 280 units of calcium, 160 units of iron, and 180 units of vitamin B.

(a) Find a system of inequalities describing the different amounts of food X and food Y that can be used in the diet.

(b) Sketch the graph of the system.

Consumer and Producer Surpluses In Exercises 53–56, find the consumer surplus and producer surplus for the pair of supply and demand equations.

	Demand	Supply
53.	$p = 56 - 0.0001x$	$p = 22 + 0.00001x$
54.	$p = 60 - 0.00001x$	$p = 15 + 0.00004x$
55.	$p = 140 - 0.00002x$	$p = 80 + 0.00001x$
56.	$p = 400 - 0.0002x$	$p = 225 + 0.0005x$

57. *Births in Nevada* The number of babies y (in thousands) born to Nevada residents each year from 1987 to 1991 can be approximated by the linear model

$$y = 7.5 + 1.4t$$

where $t = 7$ represents 1987. The *total* number of babies born to Nevada residents during this 5-year period can be approximated by finding the area of the trapezoid represented by the system

$$y \le 7.5 + 1.4t, \ y \ge 0, \ t \ge 6.5, \ t \le 11.5.$$

(Source: U.S. Bureau of Census)

(a) Sketch this region using a graphing utility.

(b) Use the formula for the area of a trapezoid to approximate the total number of births.

58. *Audio and Video Sales* Annual retail sales of household audio and video equipment in the United States from 1988 to 1992 can be approximated by the linear model

$$y = 9.52 + 0.59t$$

where y is the retail sales (in billions of dollars) and $t = 0$ represents 1990. The *total* amount of retail sales during this 5-year period can be approximated by finding the area of the trapezoid represented by the system

$$y \le 9.52 + 0.59t, \ y \ge 0, \ t \ge -2.5, \ t \le 2.5.$$

(Source: Dealerscope Merchandising)

(a) Sketch this region using a graphing utility.

(b) Use the formula for the area of a trapezoid to approximate the total retail sales.

6.5 Linear Programming

Linear Programming: A Graphical Approach ▪ Applications

Linear Programming: A Graphical Approach

It is important that students become familiar with the terminology of this section. Emphasize the fact that this technique is powerful in that it quickly identifies those few points out of many that can be easily tested to find the point that gives the maximum or minimum value.

Many applications in business and economics involve a process called **optimization,** in which you are asked to find the minimum cost, the maximum profit, or the minimum use of resources. In this section you will study an optimization strategy called **linear programming.**

A two-dimensional linear programming problem consists of a linear **objective function** and a system of linear inequalities called **constraints.** The objective function gives the quantity that is to be maximized (or minimized), and the constraints determine the set of **feasible solutions.** For example, consider a linear programming problem in which you are asked to maximize the value of

$$z = ax + by \qquad \text{Objective function}$$

subject to a set of constraints that determines the region in Figure 6.24. Because every point in the region satisfies each constraint, it is not clear how you should go about finding the point that yields a maximum value of z. Fortunately, it can be shown that if there is an optimal solution, it must occur at one of the vertices. This means that *you can find the maximum value by testing z at each of the vertices.*

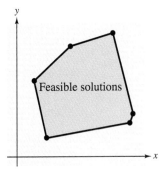

FIGURE 6.24

Optimal Solution of a Linear Programming Problem

If a linear programming problem has a solution, it must occur at a vertex of the set of feasible solutions. If the problem has more than one solution, then at least one of them must occur at a vertex of the set of feasible solutions. In either case, the value of the objective function is unique.

EXAMPLE 1 Solving a Linear Programming Problem

Find the maximum value of

$$z = 3x + 2y \qquad \text{Objective function}$$

subject to the following constraints.

$$\left.\begin{array}{r} x \geq 0 \\ y \geq 0 \\ x + 2y \leq 4 \\ x - y \leq 1 \end{array}\right\} \quad \text{Constraints}$$

Solution

The constraints form the region shown in Figure 6.25. At the four vertices of this region, the objective function has the following values.

$$\begin{array}{ll} \text{At } (0, 0): & z = 3(0) + 2(0) = 0 \\ \text{At } (1, 0): & z = 3(1) + 2(0) = 3 \\ \text{At } (2, 1): & z = 3(2) + 2(1) = 8 \qquad \text{Maximum value of } z \\ \text{At } (0, 2): & z = 3(0) + 2(2) = 4 \end{array}$$

Thus, the maximum value of z is 8, and this occurs when $x = 2$ and $y = 1$.

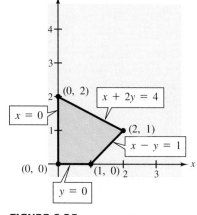

FIGURE 6.25

NOTE In Example 1, try testing some of the *interior* points in the region. You will see that the corresponding values of z are less than 8. Here are some examples.

$$\begin{array}{ll} \text{At } (1, 1): & z = 3(1) + 2(1) = 5 \\ \text{At } \left(1, \frac{1}{2}\right): & z = 3(1) + 2\left(\frac{1}{2}\right) = 4 \\ \text{At } \left(\frac{1}{2}, \frac{3}{2}\right): & z = 3\left(\frac{1}{2}\right) + 2\left(\frac{3}{2}\right) = \frac{9}{2} \end{array}$$

To see why the maximum value of the objective function in Example 1 must occur at a vertex, consider writing the objective function in the form

$$y = -\frac{3}{2}x + \frac{z}{2} \qquad \text{Family of lines}$$

where $z/2$ is the y-intercept of the objective function. This equation represents a family of lines, each of slope $-\frac{3}{2}$. Of these infinitely many lines, you want the one that has the largest z-value while still intersecting the region determined by the constraints. In other words, of all the lines whose slope is $-\frac{3}{2}$, you want the one that has the largest y-intercept *and* intersects the given region, as shown in Figure 6.26. It should be clear that such a line will pass through one (or more) of the vertices of the region.

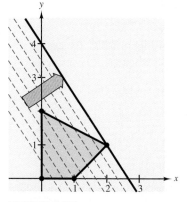

FIGURE 6.26

Solving a Linear Programming Problem

To solve a linear programming problem involving two variables by the graphical method, use the following steps.

1. Sketch the region corresponding to the system of constraints. (The points inside or on the boundary of the region are called *feasible solutions*.)

2. Find the vertices of the region.

3. Test the objective function at each of the vertices and select the values of the variables that optimize the objective function. For a bounded region, both a minimum and a maximum value will exist. (For an unbounded region, *if* an optimal solution exists, it will occur at a vertex.)

STUDY TIP

Remember that a vertex of the region can be found using a system of linear equations. The system will consist of the equations of the lines passing through the vertex.

These guidelines will work whether the objective function is to be maximized or minimized. For instance, the same test used in Example 1 to find the maximum value of z can be used to conclude that the minimum value of z is 0 and that this value occurs at the vertex $(0, 0)$.

EXAMPLE 2 Solving a Linear Programming Problem

Find the maximum value of the objective function

$$z = 4x + 6y \qquad \text{Objective function}$$

where $x \geq 0$ and $y \geq 0$, subject to the following constraints.

$$\left. \begin{array}{r} -x + y \leq 11 \\ x + y \leq 27 \\ 2x + 5y \leq 90 \end{array} \right\} \quad \text{Constraints}$$

Solution

The region bounded by the constraints is shown in Figure 6.27. By testing the objective function at each vertex, you obtain the following.

At $(0, 0)$:	$z = 4(0) + 6(0) = 0$	
At $(0, 11)$:	$z = 4(0) + 6(11) = 66$	
At $(5, 16)$:	$z = 4(5) + 6(16) = 116$	
At $(15, 12)$:	$z = 4(15) + 6(12) = 132$	Maximum value of z
At $(27, 0)$:	$z = 4(27) + 6(0) = 108$	

Thus, the maximum value of z is 132, and this occurs when $x = 15$ and $y = 12$.

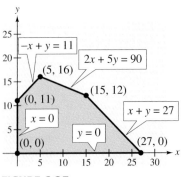

FIGURE 6.27

The next example shows that the same basic procedure can be used to solve a linear programming problem in which the objective function is to be *minimized*.

EXAMPLE 3 Minimizing an Objective Function

Find the minimum value of the objective function

$$z = 5x + 7y \qquad \text{Objective function}$$

where $x \geq 0$ and $y \geq 0$, subject to the following constraints.

$$\left.\begin{array}{rcrcl} 2x &+& 3y &\geq& 6 \\ 3x &-& y &\leq& 15 \\ -x &+& y &\leq& 4 \\ 2x &+& 5y &\leq& 27 \end{array}\right\} \quad \text{Constraints}$$

Solution

The region bounded by the constraints is shown in Figure 6.28. By testing the objective function at each vertex, you obtain the following.

At $(0, 2)$: $\quad z = 5(0) + 7(2) = 14 \qquad$ Minimum value of z

At $(0, 4)$: $\quad z = 5(0) + 7(4) = 28$

At $(1, 5)$: $\quad z = 5(1) + 7(5) = 40$

At $(6, 3)$: $\quad z = 5(6) + 7(3) = 51$

At $(5, 0)$: $\quad z = 5(5) + 7(0) = 25$

At $(3, 0)$: $\quad z = 5(3) + 7(0) = 15$

Thus, the minimum value of z is 14, and this occurs when $x = 0$ and $y = 2$.

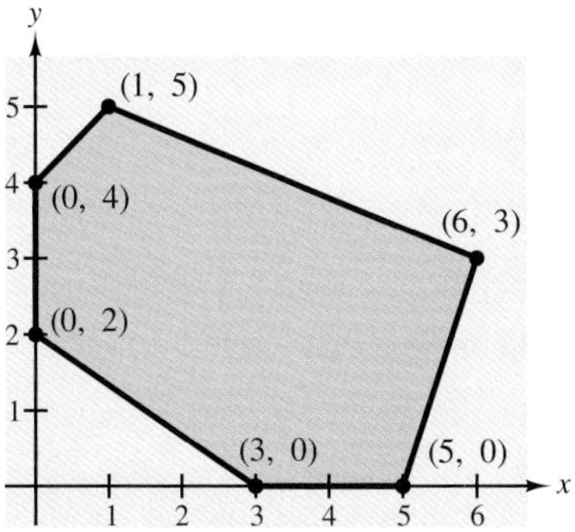

FIGURE 6.28

EXAMPLE 4 Maximizing an Objective Function

Find the maximum value of the objective function

$$z = 5x + 7y \qquad \text{Objective function}$$

where $x \geq 0$ and $y \geq 0$, subject to the following constraints.

$$\left.\begin{array}{rcrcl} 2x &+& 3y &\geq& 6 \\ 3x &-& y &\leq& 15 \\ -x &+& y &\leq& 4 \\ 2x &+& 5y &\leq& 27 \end{array}\right\} \quad \text{Constraints}$$

NOTE In Examples 3 and 4, note that the steps used to find the minimum and maximum values are precisely the same. In other words, once you have evaluated the objective function at the vertices of the feasible region, you simply choose the largest value as the maximum and the smallest value as the minimum.

Solution

This linear programming problem is identical to that given in Example 3 above, *except* that the objective function is maximized instead of minimized. Using the values of z at the vertices shown above, you can conclude that the maximum value of z is 51, and that this value occurs when $x = 6$ and $y = 3$.

It is possible for the maximum (or minimum) value in a linear programming problem to occur at *two* different vertices. For instance, at the vertices of the region shown in Figure 6.29, the objective function

$$z = 2x + 2y \qquad \text{Objective function}$$

has the following values.

At $(0, 0)$: $\quad z = 2(0) + 2(0) = \quad 0$

At $(0, 4)$: $\quad z = 2(0) + 2(4) = \quad 8$

At $(2, 4)$: $\quad z = 2(2) + 2(4) = 12 \qquad$ Maximum value of z

At $(5, 1)$: $\quad z = 2(5) + 2(1) = 12 \qquad$ Maximum value of z

At $(5, 0)$: $\quad z = 2(5) + 2(0) = 10$

In this case, you can conclude that the objective function has a maximum value not only at the vertices $(2, 4)$ and $(5, 1)$; it also has a maximum value (of 12) at *any point on the line segment connecting these two vertices*. Note that the objective function

$$y = -x + \tfrac{1}{2}z$$

has the same slope as the line through the vertices $(2, 4)$ and $(5, 1)$.

Some linear programming problems have no optimal solution. This can occur if the region determined by the constraints is *unbounded*. Example 5 illustrates such a problem.

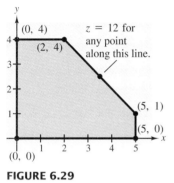

FIGURE 6.29

EXAMPLE 5   An Unbounded Region

Find the maximum value of

$$z = 4x + 2y \qquad \text{Objective function}$$

where $x \geq 0$ and $y \geq 0$, subject to the following constraints.

$$\left.\begin{array}{r} x + 2y \geq 4 \\ 3x + y \geq 7 \\ -x + 2y \leq 7 \end{array}\right\} \quad \text{Constraints}$$

Solution

The region determined by the constraints is shown in Figure 6.30. For this unbounded region, there is no maximum value of z. To see this, note that the point $(x, 0)$ lies in the region for all values of $x \geq 4$. By choosing x to be large, you can obtain values of

$$z = 4(x) + 2(0) = 4x$$

that are as large as you want. Thus, there is no maximum value of z. For this problem, there *is* a minimum value of $z = 10$, which occurs at the vertex $(2, 1)$.

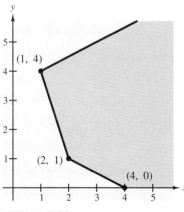

FIGURE 6.30

Applications

Example 6 shows how linear programming can be used to find the maximum profit in a business application.

EXAMPLE 6   *Maximum Profit*

A manufacturer wants to maximize the profit for two products. The first product yields a profit of $1.50 per unit, and the second product yields a profit of $2.00 per unit. Market tests and available resources have indicated the following constraints.

1. The combined production level should not exceed 1200 units per month.
2. The demand for product II is no more than half the demand for product I.
3. The production level of product I is less than or equal to 600 units plus three times the production level of product II.

Solution

If you let x be the number of units of product I and y be the number of units of product II, the objective function (for the combined profit) is given by

$$P = 1.5x + 2y. \qquad \text{Objective function}$$

The three constraints translate into the following linear inequalities.

1.	$x + y \le 1200$		$x + y \le 1200$
2.	$y \le \frac{1}{2}x$		$-x + 2y \le 0$
3.	$x \le 3y + 600$		$x - 3y \le 600$

Because neither x nor y can be negative, you also have the two additional constraints of $x \ge 0$ and $y \ge 0$. Figure 6.31 shows the region determined by the constraints. To find the maximum profit, test the value of P at the vertices of the region.

$$\begin{aligned}
\text{At } (0, 0): \quad & P = 1.5(0) &+\; 2(0) &= 0 \\
\text{At } (800, 400): \quad & P = 1.5(800) &+\; 2(400) &= 2000 \qquad \text{Maximum profit} \\
\text{At } (1050, 150): \quad & P = 1.5(1050) &+\; 2(150) &= 1875 \\
\text{At } (600, 0): \quad & P = 1.5(600) &+\; 2(0) &= 900
\end{aligned}$$

Thus, the maximum profit is $2000, and it occurs when the monthly production consists of 800 units of product I and 400 units of product II.

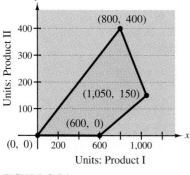

FIGURE 6.31

NOTE   In Example 6, suppose the manufacturer improved the production of product I so that it yielded a profit of $2.50 per unit. How would this affect the number of units the manufacturer should sell to obtain a maximum profit?

Real Life

EXAMPLE 7 Minimum Cost

The liquid part of a diet is to provide at least 300 calories, 36 units of vitamin A, and 90 units of vitamin C daily. A cup of dietary drink X costs $0.12 and provides 60 calories, 12 units of vitamin A, and 10 units of vitamin C. A cup of dietary drink Y costs $0.15 and provides 60 calories, 6 units of vitamin A, and 30 units of vitamin C. How many cups of each drink should be consumed each day to minimize the cost and still meet the daily requirements?

Solution

As in Example 7 on page 476, let x be the number of cups of dietary drink X and let y be the number of cups of dietary drink Y.

$$\left.\begin{array}{llrcr}\text{For calories:} & 60x + 60y &\geq& 300 \\ \text{For vitamin A:} & 12x + 6y &\geq& 36 \\ \text{For vitamin C:} & 10x + 30y &\geq& 90 \\ & x &\geq& 0 \\ & y &\geq& 0 \end{array}\right\} \text{Constraints}$$

The cost C is given by

$$C = 0.12x + 0.15y. \qquad \text{Objective function}$$

The graph of the region corresponding to the constraints is shown in Figure 6.32. To determine the minimum cost, test C at each vertex of the region, as follows.

$$\begin{array}{lll}\text{At } (0, 6): & C = 0.12(0) + 0.15(6) = 0.90 \\ \text{At } (1, 4): & C = 0.12(1) + 0.15(4) = 0.72 \\ \text{At } (3, 2): & C = 0.12(3) + 0.15(2) = 0.66 & \text{Minimum value of } C \\ \text{At } (9, 0): & C = 0.12(9) + 0.15(0) = 1.08 \end{array}$$

Thus, the minimum cost is $0.66 per day, and this occurs when three cups of drink X and two cups of drink Y are consumed each day.

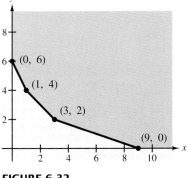

FIGURE 6.32

Group Activities Problem Solving

Analysis of Constraints
Explain what difficulties you might encounter with the following sets of linear programming constraints.

$$\textbf{a.} \quad \begin{array}{rcr} x - y &<& 0 \\ 3x + y &>& 9 \\ -4x + y &>& -2 \end{array} \qquad \textbf{b.} \quad \begin{array}{rcr} 2x + y &>& 11 \\ x - y &>& 0 \\ x &<& 4 \\ y &<& 0 \end{array}$$

Warm Up The following warm-up exercises involve skills that were covered in earlier sections. You will use these skills in the exercise set for this section.

In Exercises 1–4, sketch the graph of the linear equation.

1. $y + x = 3$
3. $x = 0$

2. $y - x = 12$
4. $y = 4$

In Exercises 5–8, find the point of intersection of the two lines.

5. $x + y = 4, x = 0$
7. $x + y = 4, 2x + 3y = 9$

6. $x + 2y = 12, y = 0$
8. $x + 2y = 12, 2x + y = 9$

In Exercises 9 and 10, sketch the graph of the inequality.

9. $2x + 3y \geq 18$
10. $4x + 3y \geq 12$

6.5 **Exercises**

In Exercises 1–12, find the minimum and maximum values of the objective function, subject to the indicated constraints. (For each exercise, the graph of the region determined by the constraints is provided.)

1. *Objective function:*
$$z = 4x + 5y$$
Constraints:
$$x \geq 0$$
$$y \geq 0$$
$$x + y \leq 6$$

2. *Objective function:*
$$z = 2x + 8y$$
Constraints:
$$x \geq 0$$
$$y \geq 0$$
$$2x + y \leq 4$$

3. *Objective function:*
$$z = 10x + 6y$$
Constraints:
(See Exercise 1.)

4. *Objective function:*
$$z = 7x + 3y$$
Constraints:
(See Exercise 2.)

5. *Objective function:*
$$z = 3x + 2y$$
Constraints:
$$x \geq 0$$
$$y \geq 0$$
$$x + 3y \leq 15$$
$$4x + y \leq 16$$

6. *Objective function:*
$$z = 4x + 3y$$
Constraints:
$$x \geq 0$$
$$2x + 3y \geq 6$$
$$3x - 2y \leq 9$$
$$x + 5y \leq 20$$

7. *Objective function:*
$z = 5x + 0.5y$
Constraints:
(See Exercise 5.)

8. *Objective function:*
$z = x + 6y$
Constraints:
(See Exercise 6.)

9. *Objective function:*
$z = 10x + 7y$
Constraints:
$$0 \le x \le 60$$
$$0 \le y \le 45$$
$$5x + 6y \le 420$$

10. *Objective function:*
$z = 50x + 35y$
Constraints:
$$x \ge 0$$
$$y \ge 0$$
$$8x + 9y \le 7200$$
$$8x + 9y \ge 5400$$

11. *Objective function:*
$z = 25x + 30y$
Constraints:
(See Exercise 9.)

12. *Objective function:*
$z = 16x + 18y$
Constraints:
(See Exercise 10.)

In Exercises 13–24, sketch the region determined by the constraints. Then find the minimum and maximum values of the objective function, subject to the indicated constraints.

13. *Objective function:*
$z = 6x + 10y$
Constraints:
$$x \ge 0$$
$$y \ge 0$$
$$2x + 5y \le 10$$

14. *Objective function:*
$z = 7x + 8y$
Constraints:
$$x \ge 0$$
$$y \ge 0$$
$$x + \tfrac{1}{2}y \le 4$$

15. *Objective function:*
$z = 9x + 4y$
Constraints:
$$x \ge 0$$
$$y \ge 0$$
$$2x + 5y \le 10$$

16. *Objective function:*
$z = 7x + 2y$
Constraints:
$$x \ge 0$$
$$y \ge 0$$
$$x + \tfrac{1}{2}y \le 4$$

17. *Objective function:*
$z = 4x + 5y$
Constraints:
$$x \ge 0$$
$$y \ge 0$$
$$x + y \ge 8$$
$$3x + 5y \ge 30$$

18. *Objective function:*
$z = 4x + 5y$
Constraints:
$$x \ge 0$$
$$y \ge 0$$
$$2x + 2y \le 10$$
$$x + 2y \le 6$$

19. *Objective function:*
$z = 2x + 7y$
Constraints:
(See Exercise 17.)

20. *Objective function:*
$z = 2x - y$
Constraints:
(See Exercise 18.)

21. *Objective function:*
$z = 4x + y$
Constraints:
$$x \ge 0$$
$$y \ge 0$$
$$x + 2y \le 40$$
$$x + y \ge 30$$
$$2x + 3y \ge 72$$

22. *Objective function:*
$z = x$
Constraints:
$$x \ge 0$$
$$y \ge 0$$
$$2x + 3y \le 60$$
$$2x + y \le 28$$
$$4x + y \le 48$$

23. *Objective function:*
$z = x + 4y$
Constraints:
(See Exercise 21.)

24. *Objective function:*
$z = y$
Constraints:
(See Exercise 22.)

In Exercises 25–28, maximize the objective function subject to the constraints

$3x + y \le 15$ and $4x + 3y \le 30$

where $x \ge 0$ and $y \ge 0$.

25. $z = 2x + y$

26. $z = 5x + y$

27. $z = x + y$

28. $z = 3x + y$

In Exercises 29–32, maximize the objective function subject to the constraints

$x + 4y \le 20$, $x + y \le 8$,   and $3x + 2y \le 21$

where $x \ge 0$ and $y \ge 0$.

29. $z = x + 5y$

30. $z = 2x + 4y$

31. $z = 4x + 5y$

32. $z = 4x + y$

33. *Maximum Profit*   A store plans to sell two models of computers at costs of $250 and $400. The $250 model yields a profit of $45 and the $400 model yields a profit of $50. How many units of each model should be stocked to maximize profit, subject to the following constraints?

- The merchant estimates that the total monthly demand will not exceed 250 units.

- The merchant does not want to invest more than $70,000 in computer inventory.

34. *Maximum Profit*   A fruit grower has 150 acres of land available to raise two crops, A and B. The profit is $140 per acre for crop A and $235 per acre for crop B. Find the number of acres of each fruit that should be planted to maximize profit, subject to the following constraints.

- It takes 1 day to trim an acre of crop A and 2 days to trim an acre of crop B, and there are 240 days per year available for trimming.

- It takes 0.3 day to pick an acre of crop A and 0.1 day to pick an acre of crop B, and there are 30 days per year available for picking.

35. *Minimum Cost*   A farming cooperative mixes two brands of cattle feed. Brand X costs $25 per bag, and brand Y costs $20 per bag. Find the number of bags of each brand that should be mixed to produce a mixture having a minimum cost per bag, subject to the following constraints.

- Brand X contains 2 units of nutritional element A, 2 units of element B, and 2 units of element C.

- Brand Y contains 1 unit of nutritional element A, 9 units of element B, and 3 units of element C.

- The minimum requirements of nutrients A, B, and C are 12 units, 36 units, and 24 units, respectively.

36. *Minimum Cost*   Two gasolines, type A and type B, have octane ratings of 80 and 92, respectively. Type A costs $0.83 per gallon and type B costs $0.98 per gallon. Determine the blend of minimum cost with an octane rating of at least 90. (*Hint:* Let x be the fraction of each gallon that is type A and y be the fraction that is type B.)

37. *Maximum Profit*   A manufacturer produces two models of bicycles. The times (in hours) required for assembling, painting, and packaging each model are as follows.

Process	Model A	Model B
Assembling	2	2.5
Painting	4	1
Packaging	1	0.75

The total times available for assembling, painting, and packaging are 4000 hours, 4800 hours, and 1500 hours, respectively. The profits per unit are $45 for model A and $50 for model B. How many of each should be made to obtain a maximum profit?

38. *Maximum Profit*   A company makes two models of a product. The times (in hours) required for assembling, painting, and packaging are as follows.

Process	Model A	Model B
Assembling	2.5	3
Painting	2	1
Packaging	0.75	1.25

The total times available for assembling, painting, and packaging are 4000 hours, 2500 hours, and 1500 hours, respectively. The profits per unit are $50 for model A and $52 for model B. How many of each should be made to obtain a maximum profit?

39. *Maximum Revenue*   An accounting firm has 900 hours of staff time and 100 hours of review time available each week. The firm charges $2000 for an audit and $300 for a tax return. What numbers of audits and tax returns will bring in a maximum revenue, subject to the following constraints?

- Each audit requires 100 hours of staff time and 10 hours of review time.

- Each tax return requires 12.5 hours of staff time and 2.5 hours of review time.

40. *Maximum Revenue* The accounting firm in Exercise 39 lowers its charge for an audit to $1800. What numbers of audits and tax returns will bring in a maximum revenue?

41. *Media Selection* A company has budgeted a maximum of $600,000 for advertising a product nationally. Each minute of television time costs $60,000 and each one-page newspaper ad costs $15,000. Each television ad is expected to be viewed by 15 million viewers, and each newspaper ad is expected to be seen by 3 million readers. The company's market research department recommends that at most 90% of the advertising budget be spent on television ads. How should the advertising budget be spent to maximize the total audience?

42. *Maximum Profit* A fruit juice company makes two drinks by blending apple and pineapple juices. The first uses 30% apple juice and 70% pineapple, and the second uses 60% apple and 40% pineapple. There are 1000 liters of apple juice and 1500 liters of pineapple juice available. The profit for the first drink is $0.60 per liter and the profit for the second drink is $0.50. Find the number of liters of each drink that should be produced in order to maximize profit.

43. *Investments* An investor has up to $250,000 to invest in two types of investments. Type A pays 8% annually and type B pays 10% annually. To have a well-balanced portfolio, the investor imposes the following conditions. At least one-fourth of the total portfolio is to be allocated to type A investments and at least one-fourth of the portfolio is to be allocated to type B investments. How much should be allocated to each type of investment to obtain a maximum return?

44. *Investments* An investor has up to $450,000 to invest in two types of investments. Type A pays 6% annually and type B pays 10% annually. To have a well-balanced portfolio, the investor imposes the following conditions. At least one-half of the total portfolio is to be allocated to type A investments and at least one-fourth of the portfolio is to be allocated to type B investments. How much should be allocated to each type of investment to obtain a maximum return?

In Exercises 45–50, the given linear programming problem has an unusual characteristic. Sketch a graph of the solution region for the problem and describe the unusual characteristic. (In each problem, the objective function is to be maximized.)

45. *Objective function:*
$$z = 2.5x + y$$
Constraints:
$$x \geq 0$$
$$y \geq 0$$
$$3x + 5y \leq 15$$
$$5x + 2y \leq 10$$

46. *Objective function:*
$$z = x + y$$
Constraints:
$$x \geq 0$$
$$y \geq 0$$
$$-x + y \leq 1$$
$$-x + 2y \leq 4$$

47. *Objective function:*
$$z = -x + 2y$$
Constraints:
$$x \geq 0$$
$$y \geq 0$$
$$x \leq 10$$
$$x + y \leq 7$$

48. *Objective function:*
$$z = x + y$$
Constraints:
$$x \geq 0$$
$$y \geq 0$$
$$-x + y \leq 1$$
$$-3x + y \geq 3$$

49. *Objective function:*
$$z = 3x + 4y$$
Constraints:
$$x \geq 0$$
$$y \geq 0$$
$$x + y \leq 1$$
$$2x + y \leq 4$$

50. *Objective function:*
$$z = x + 2y$$
Constraints:
$$x \geq 0$$
$$y \geq 0$$
$$x + 2y \leq 4$$
$$2x + y \leq 4$$

51. *Maximum Profit* A company makes two models of a patio furniture set. The times for assembling, finishing, and packaging model A are 3.0 hours, 2.5 hours, and 0.60 hour, respectively. The times for model B are 2.75 hours, 1.0 hour, and 1.25 hours. The total times available for assembling, finishing, and packaging are 3000 hours, 2400 hours, and 1200 hours, respectively. The profit per unit for model A is $80 and the profit per unit for model B is $65. How many of each model should be produced to obtain a maximum profit?

CHAPTER PROJECT: Televisions and Video Recorders

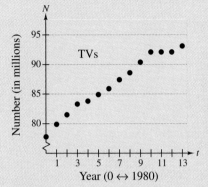

Figure A

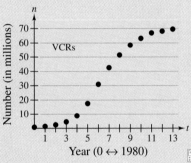

Figure B

The following table gives the numbers N (in millions) of households in the United States that had at least one television set in the years 1980 to 1993.

t	0	1	2	3	4	5	6
N	77.8	79.9	81.5	83.3	83.8	84.9	85.9

t	7	8	9	10	11	12	13
N	87.4	88.6	90.4	92.1	92.1	92.1	93.1

In the table, $t = 0$ represents 1980. From the scatter plot in Figure A, you can see that this data follows a pattern that is nearly linear. During the same years, the numbers n (in millions) of households with VCRs followed a very different pattern (see table below). A scatter plot for this data is shown in Figure B.

t	0	1	2	3	4	5	6
n	0.84	1.44	2.53	4.58	8.88	17.60	30.92

t	7	8	9	10	11	12	13
n	42.58	51.39	58.40	63.18	66.94	68.10	69.60

Use this information to investigate the following questions.

1. ***Fitting a Model to Data***   Use a graphing utility to find the linear model that best fits the data for television households.

2. ***Comparing a Model to Actual Data***   Compare the values given by the model found in Question 1 with the actual data given in the table above. Make your comparison in two ways—numerically by constructing a table and graphically by comparing the graph of the model with Figure A.

3. ***Comparing a Model to Actual Data***   A model for the number of VCR households is given by

$$n = 0.185 + \frac{69.1}{1 + e^{-(t-6.5)/1.37}}.$$

   Compare the values given by this model with the actual data given in the table above. Make your comparison in two ways—numerically by constructing a table and graphically by comparing the graph of the model with Figure B.

4. ***Comparing Two Models***   Use a graphing utility to graph both models on the same viewing rectangle. Do the models intersect? If they did intersect, what could you conclude?

CHAPTER SUMMARY

After studying this chapter, you should have acquired the following skills. These skills are keyed to the Review Exercises that begin on page 493. Answers to odd-numbered Review Exercises are given in the back of the book.

- Solve a system of equations by the method of substitution. *(Section 6.1)* — **Review Exercises 1–6**

- Find points of intersection using a graphing utility. *(Section 6.1)* — **Review Exercises 7, 8, 10**

- Solve a real-life application with a system of equations. *(Section 6.1)* — **Review Exercises 9, 11, 12, 19, 20, 48, 49**

- Solve a system of equations by the method of elimination. *(Section 6.2)* — **Review Exercises 13–18**

- Find the point of equilibrium for a pair of supply and demand equations. *(Section 6.2)* — **Review Exercises 21, 22**

- Find the least squares regression line for the given points. *(Section 6.2)* — **Review Exercise 36**

- Solve a system of three or more variables. *(Section 6.3)* — **Review Exercises 23–28, 33–35**

- Find the equation of a parabola. *(Section 6.3)* — **Review Exercises 29, 30**

- Find the equation of a circle. *(Section 6.3)* — **Review Exercises 31, 32**

- Find the least squares regression parabola for the given points. *(Section 6.3)* — **Review Exercise 37**

- Sketch the graph of the solution set of a system of inequalities. *(Section 6.4)* — **Review Exercises 38–43**

- Derive a set of inequalities that describes a region. *(Section 6.4)* — **Review Exercises 44–47**

- Find the consumer surplus and producer surplus for a given pair of supply and demand equations. *(Section 6.4)* — **Review Exercises 50, 51**

- Find the minimum and maximum values of an objective function, subject to the given constraints. *(Section 6.5)* — **Review Exercises 52–55**

- Sketch the region determined by a given set of constraints, and find the minimum and maximum values of the objective function, subject to the constraints. *(Section 6.5)* — **Review Exercises 56–59**

- Use linear programming to solve a real-life application. *(Section 6.5)* — **Review Exercises 60–67**

Exercises containing inconsistent systems (no solution):
17. Exercises containing dependent systems (infinitely many solutions): 15, 25, and 27

REVIEW EXERCISES

In Exercises 1–6, solve the system by the method of substitution.

1. $x + 3y = 10$
$4x - 5y = -28$

2. $3x - y - 13 = 0$
$4x + 3y - 26 = 0$

3. $\frac{1}{2}x + \frac{3}{5}y = -2$
$2x + y = 6$

4. $1.3x + 0.9y = 7.5$
$0.4x - 0.5y = -0.8$

5. $x^2 + y^2 = 100$
$x + 2y = 20$

6. $y = x^3 - 2x^2 - 2x - 3$
$y = -x^2 + 4x - 3$

In Exercises 7 and 8, use a graphing utility to find all points of intersection of the graphs of the equations.

7. $y = x^2 - 3x + 11$
$y = -x^2 + 2x + 8$

8. $x^2 + y^2 = 16$
$y = e^x$

9. *Break-Even Point*   You are setting up a business and have made an initial investment of $10,000. The unit cost of the product is $2.85 and the selling price is $4.95. How many units must you sell to break even? (Round to the nearest whole unit.)

10. *Comparing Product Sales*   Your company produces three types of CD players. The research and development department has designed a new CD player that is predicted to sell better than the three players combined. The sales equations for the CD players are

$S = 1725.18 - 49.6t + 0.45t^2$ Three players combined

$S = 792.95 + 115.8t + 3.98t^2$ New player

where S is the sales in thousands and $t = 0$ represents 1990. Use a graphing utility to determine when the sales of the new CD player will overtake the combined sales of the other players.

11. *Investment Portfolio*   A total of $34,000 is invested in two funds paying 7.0% and 7.5% simple interest. If the yearly interest is $2480, how much of the $34,000 is invested at each rate?

12. *Choice of Two Jobs*   You are offered two different jobs selling computers. One company offers an annual salary of $22,500 plus a year-end bonus of 1.5% of your total sales. The other company offers a salary of $20,000 plus a year-end bonus of 2% of your total sales. How much would you have to sell in order to make the second offer better?

In Exercises 13–18, solve the system by elimination.

13. $2x - 3y = 21$
$3x + y = 4$

14. $3u + 5v = 9$
$12u + 10v = 22$

15. $1.25x - 2y = 3.5$
$5x - 8y = 14$

16. $\dfrac{x-2}{3} + \dfrac{y+3}{4} = 5$
$2x - y = 7$

17. $1.5x + 2.5y = 8.5$
$6x + 10y = 24$

18. $\frac{3}{5}x + \frac{2}{7}y = 10$
$x + 2y = 38$

19. *Acid Mixture*   Twelve gallons of a 25% acid solution are obtained by mixing a 10% solution with a 50% solution.

(a) Write a system of equations that represents the problem and use a graphing utility to graph both equations.

(b) How much of each solution must be used?

20. *Cassette Tape Sales*   Suppose you are the manager of a music store. At the end of the week you are going over receipts for the previous week's sales. Six hundred and fifty cassette tapes were sold. One type of cassette sold for $9.95 and a second type sold for $14.95. The total cassette receipts were $7717.50. The cash register that was supposed to keep track of the number of each type of cassette sold malfunctioned. Can you recover the information? If so, how many of each type of cassette were sold?

Supply and Demand   In Exercises 21 and 22, find the point of equilibrium for the given pair of demand and supply equations.

	Demand	Supply
21.	$p = 37 - 0.0002x$	$p = 22 + 0.00001x$
22.	$p = 120 - 0.0001x$	$p = 45 + 0.0002x$

In Exercises 23–28, solve the system of equations.

23. $2x + y + z = 6$
$x - 4y - z = 3$
$x + y + z = 4$

24. $x + 3y - z = 13$
$2x \quad\quad - 5z = 23$
$4x - y - 2z = 14$

25. $x + y + z = 10$
$-2x + 3y + 4z = 22$

26. $x + y + z = 1$
$2x - 4y + 3z = -20$
$2x + 3y + 2z = 5$

27. $2x + y - z = 3$
$x - 3y + z = 7$
$3x + 5y - 3z = -1$

28. $x + y + z + w = 2$
$4y + 5z + 2w = 7$
$2x + 3y - z = -2$
$3x + 2y \quad\quad - 4w = 20$

In Exercises 29 and 30, find the equation of the parabola $y = ax^2 + bx + c$ that passes through the points.

29. $(0, -6), (1, -3), (2, 4)$

30. $(-5, 0), (1, -6), (2, 14)$

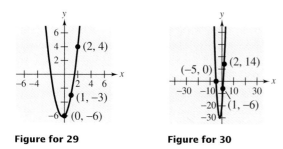

Figure for 29   **Figure for 30**

In Exercises 31 and 32, find the equation of the circle $x^2 + y^2 + Dx + Ey + F = 0$ that passes through the points.

31. $(2, 2), (5, -1), (-1, -1)$

32. $(4, 2), (1, 3), (-2, -6)$

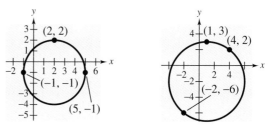

Figure for 31   **Figure for 32**

33. *Investment*   Suppose you receive $1060 a year in interest from three investments. The interest rates for the three investments are 6%, 7%, and 8%. The 7% investment is five times the 6% investment, and the 8% investment is $1000 more than the 6% investment. What is the amount of each investment?

34. *Product Types*   A company sells three types of products for $20, $15, and $10 per unit. In 1 year, the total revenue for the three products was $1,500,000, which corresponded to the sale of 110,000 units. The company sold half as many units of the $20 product as units of the $15 product. How many units of each product were sold?

35. *Investment Portfolio*   Suppose an investor has a portfolio totaling $500,000 that is to be allocated among the following types of investments: (1) certificates of deposit, (2) municipal bonds, (3) blue-chip stocks, and (4) growth or speculative stocks. The certificates of deposit pay 8% annually, and the municipal bonds pay 6% annually. Over a 5-year period, the investor expects the blue-chip stocks to return 10% annually and the growth stocks to return 15% annually. The investor wishes a combined return of $8\frac{1}{2}\%$ and also wants to have only two-fifths of the portfolio invested in stocks. How much should be allocated to each type of investment if the amount invested in certificates of deposit is twice that invested in municipal bonds?

36. *Fitting a Line to Data*   Find the least squares regression line $y = ax + b$ for the points

(0, 1.6), (1, 2.4), (2, 3.6), (3, 4.7), (4, 5.5).

To find the equation of the line, solve the system of linear equations for a and b.

$$5b + 10a = 17.8$$
$$10b + 30a = 45.7$$

37. *Fitting a Parabola to Data*   Find the least squares regression parabola $y = ax^2 + bx + c$ for the points

(−2, 0.4), (−1, 0.9), (0, 1.9), (1, 2.1), (2, 3.8).

To find the parabola, solve the system of linear equations for a, b, and c.

$$5c \quad + 10a = \quad 9.1$$
$$10b \qquad = \quad 8.0$$
$$10c \quad + 34a = 19.8$$

In Exercises 38–43, sketch the graph of the solution set of the system of inequalities.

38. $2x + 3y < 9$
$\qquad x > 0$
$\qquad y > 0$

39. $2x - y > 6$
$\qquad x < 5$
$\qquad y \le 8$

40. $3x - y > -4$
$\quad 2x + y > -1$
$\quad 7x + y < \quad 4$

41. $\quad x + y > \quad 4$
$\quad 3x + y < 10$

42. $\quad x^2 + y^2 \le 9$
$\quad x^2 - x - 2 \le y$

43. $\ln x < y$
$\qquad y > -1$
$\qquad x < 4$

In Exercises 44 and 45, derive a set of inequalities that describes the region.

44. Parallelogram with the following vertices:

(1, 5), (3, 1), (6, 10), (8, 6)

45. Triangle with the following vertices:

(1, 2), (6, 7), (8, 1)

46. *Video Cassette Recorder Inventory*   A store sells two models of a certain brand of VCR. Because of the demand, it is necessary to stock twice as many units of model A as units of model B. The costs to the store for the two models are $200 and $300, respectively. The management does not want more than $4000 in VCR inventory at any one time, and it wants at least four model A VCRs and two model B VCRs in inventory at all times. Derive a system of inequalities that describes all possible inventory levels, and sketch the graph of the system.

47. *Concert Ticket Sales*   Two types of tickets are to be sold for a concert. One type costs $20 per ticket and the other type costs $30 per ticket. The promoter of the concert must sell at least 16,000 tickets, including at least 9000 of the $20 tickets and at least 4000 of the $30 tickets. Moreover, the gross receipts must total at least $400,000 in order for the concert to be held. Find a system of inequalities describing the different numbers of tickets that can be sold, and sketch the graph of the system.

48. *Morning or Evening Paper?*   The total circulations of morning and evening newspapers in the U.S. from 1980 to 1992 can be approximated by the models

$$y = 30.63 + 1.08t \qquad \text{Morning}$$
$$y = 32.62 - 1.23t \qquad \text{Evening}$$

where y is the circulation (in millions) and t represents the year, with $t = 0$ corresponding to 1980. During which year did the circulation of morning papers overtake the circulation of evening papers? (Source: Editor and Publisher Company)

49. *News Programs*   A local television station tracked the numbers of viewers of its two prime news shows for 1 year. The numbers of viewers for the two news shows can be approximated by the models

$$y = 9500 + 350t \qquad \text{6 P.M. news}$$
$$y = 15,000 - 275t \qquad \text{10 P.M. news}$$

where y is the number of viewers and $t = 1$ corresponds to the month of January. During what month did the number of viewers of the 6 P.M. news program surpass the number of viewers of the 10 P.M. news program?

In Exercises 50 and 51, find the consumer surplus and producer surplus for the given pair of demand and supply equations.

	Demand	*Supply*
50.	$p = 160 - 0.0001x$	$p = 70 + 0.0002x$
51.	$p = 130 - 0.0002x$	$p = 30 + 0.0003x$

In Exercises 52–55, find the minimum and maximum values of the objective function, subject to the indicated constraints. (For each exercise, the graph of the constraints is provided.)

52. *Objective function:*

$$z = 3x + 4y$$

Constraints:

$$x \geq 0$$
$$y \geq 0$$
$$x + y \leq 8$$

53. *Objective function:*

$$z = 15x + 12y$$

Constraints:

$$x \geq 0$$
$$y \geq 0$$
$$x + 3y \leq 12$$
$$3x + 2y \leq 15$$

54. *Objective function:*

$$z = 10x + 6y$$

Constraints:

$$0 \leq x \leq 50$$
$$0 \leq y \leq 35$$
$$4x + 5y \leq 275$$

55. *Objective function:*

$$z = 50x + 60y$$

Constraints:

$$x \geq 0$$
$$y \geq 0$$
$$3x + 4y \geq 1200$$
$$5x + 6y \leq 3000$$

In Exercises 56–59, sketch the region determined by the constraints. Find the minimum and maximum values of the objective function, subject to the indicated constraints.

56. *Objective function:*

$$z = 6x + 8y$$

Constraints:

$$x \geq 0$$
$$y \geq 0$$
$$x + 4y \leq 16$$
$$3x + 2y \leq 18$$

57. *Objective function:*

$$z = 9x + 7y$$

Constraints:

$$0 \leq x \leq 5$$
$$y \geq 0$$
$$x + 2y \leq 12$$
$$2x + 3y \leq 19$$

58. *Objective function:*

$$z = 8x + 3y$$

Constraints:

$$0 \leq x \leq 5$$
$$0 \leq y \leq 7$$
$$x + y \leq 9$$
$$3x + y \leq 17$$

59. *Objective function:*

$$z = 4x + 5y$$

Constraints:

$$x \geq 0$$
$$y \geq 0$$
$$2x + 5y \leq 30$$
$$x + y \geq 3$$
$$2x + y \leq 14$$

60. *Maximum Profit* A merchant plans to sell two models of camcorders at prices of $800 and $1000. The $800 model yields a profit of $75 and the $1000 model yields a profit of $125. The merchant estimates that the total monthly demand will not exceed 300 units. Find the number of units of each model that should be stocked in order to maximize profit. Assume that the merchant does not want to invest more than $280,000 in camcorder inventory.

61. *Maximum Profit* A merchant plans to sell two models of color televisions at prices of $300 and $500. The $300 model yields a profit of $30 and the $500 model yields a profit of $45. The merchant estimates that the total monthly demand will not exceed 100 units. Find the number of units of each model that should be stocked in order to maximize profit. Assume that the merchant does not want to invest more than $40,000 in color television inventory.

62. *Maximum Profit*   A factory manufactures two color television set models: a basic model that yields $100 profit and a deluxe model that yields a profit of $140. The times (in hours) required for assembling, finishing, and packaging are as follows.

Process	Basic Model	Deluxe Model
Assembling	2	5
Finishing	1	2
Packaging	1	1

The total times available for assembling, finishing, and packaging are 3000 hours, 1400 hours, and 1000 hours, respectively. How many of each model should be produced for maximum profit?

63. *Maximum Profit*   A company makes two models of a product. The times (in hours) required for assembling, finishing, and packaging each model are as follows.

Process	Model A	Model B
Assembling	3.5	8
Finishing	2.5	2
Packaging	1.3	0.7

The total times available for assembling, finishing, and packaging are 5600 hours, 2000 hours, and 910 hours, respectively. The profits per unit are $80 for model A and $100 for model B. How many of each model should be produced to obtain a maximum profit?

64. *Minimum Cost*   A pet supply company mixes two brands of dry dog food. Brand X costs $15 per bag and contains 8 units of nutritional element A, 1 unit of nutritional element B, and 2 units of nutritional element C. Brand Y costs $30 per bag and contains 2 units of nutritional element A, 1 unit of nutritional element B, and 7 units of nutritional element C. Each bag of mixed dog food must contain at least 16 units of element A, 5 units of element B, and 20 units of element C. Find the number of bags of each brand that should be mixed to produce a mixture having a minimum cost per bag.

65. *Minimum Cost*   Two gasolines, type A and type B, have octane ratings of 80 and 92, respectively. Type A costs $1.25 per gallon and type B costs $1.55 per gallon. Determine the blend of minimum cost that has an octane rating of at least 88. (*Hint:* Let x be the fraction of each gallon that is type A and let y be the fraction of each gallon that is type B.)

66. *Maximum Revenue*   An accounting firm has 800 hours of staff time and 90 hours of review time available each week. The firm charges $2400 for an audit and $300 for a tax return. Each audit requires 100 hours of staff time and 10 hours of review time. Each tax return requires 10 hours of staff time and 2 hours of review time. What number of audits and tax returns will bring in maximum revenue?

67. *Maximum Profit*   Suppose the accounting firm in Exercise 66 realizes a profit of $800 for each audit and $120 for each tax return. What combination of audits and tax returns will yield a maximum profit?

CHAPTER TEST

Take this test as you would take a test in class. After you are done, check your work against the answers given in the back of the book.

In Exercises 1–8, solve the system of equations using the indicated method.

1. *Substitution*
$$3x - 2y = -2$$
$$4x + 3y = 20$$

2. *Substitution*
$$x + y = 3$$
$$x^2 + y = 9$$

3. *Substitution*
$$2x - y = 11$$
$$3x + 5y = 8$$

4. *Graphing*
$$5x - y = 6$$
$$2x^2 + y = 8$$

5. *Graphing*
$$1.5x - 2.25y = 8$$
$$2.5x + 2y = 5.75$$

6. *Elimination*
$$2x - 4y + z = 11$$
$$x + 2y + 3z = 9$$
$$3y + 5z = 12$$

7. *Elimination*
$$3x - 2y + z = 16$$
$$5x - z = 6$$
$$2x - y - z = 3$$

8. *Elimination*
$$-x + y - 2z = 3$$
$$x - 4y - 2z = 1$$
$$x + 2y + 6z = 5$$

9. Find the point of equilibrium for a system that has a demand function of $p = 45 - 0.0003x$ and a supply function of $p = 29 + 0.00002x$.

10. A system of linear equations reduces to $-23 = 0$. What can you conclude?

11. A system of linear equations reduces to $0 = 0$. What can you conclude?

12. Find an equation of the parabola passing through $(1, 10)$, $(-1, 4)$, and $(0, 5)$.

13. A total of $35,000 is invested in two funds paying 8% and 8.5% simple interest. The annual interest is $2890. How much is invested in each fund?

In Exercises 14–17, sketch the inequality.

14. $x \geq 0$ **15.** $y \geq 0$ **16.** $x + 3y \leq 12$ **17.** $3x + 2y \leq 15$

18. Sketch the solution of the system of inequalities comprised of the inequalities in Exercises 14–17.

19. Find the minimum and maximum values of the objective function $z = 5x + 11y$, subject to the constraints given in Exercises 14–17.

20. A manufacturer produces two models of stair climbers. The times (in hours) required for assembling, painting, and packaging each model are as follows.

- Assembling: 3.5 hours for model A; 8 hours for model B
- Painting: 2.5 hours for model A; 2 hours for model B
- Packaging: 1.3 hours for model A; 0.9 hour for model B

The total times available for assembling, painting, and packaging are 5600 hours, 2000 hours, and 900 hours, respectively. The profits per unit are $100 for model A and $150 for model B. How many of each type should be produced to obtain a maximum profit? Explain your reasoning.

Matrices and Determinants

7

- Matrices and Systems of Linear Equations
- Operations with Matrices
- The Inverse of a Square Matrix
- The Determinant of a Square Matrix
- Applications of Determinants and Matrices

Your company, Brand X, is introducing a new type of film in a market that is dominated by Brand A and Brand B. The current percents of the market are given by the matrix

$$X_0 = \begin{bmatrix} 0.4 \\ 0.6 \\ 0.0 \end{bmatrix}. \quad \begin{matrix} \text{Brand A} \\ \text{Brand B} \\ \text{Brand X} \end{matrix}$$

You believe that each month after you introduce your product, the following changes will occur. (1) Brand A will retain 91%, and acquire 3% of Brand B's share and 2% of yours. (2) Brand B will retain 94%, and acquire 4% of Brand A's share and 1% of yours. (3) Brand X will retain 97%, and acquire 5% of Brand A's share and 3% of Brand B's. If these assumptions prove to be true, your market share during the first 12 months will change as shown in the table and bar graph at the right.

Month	0	2	4	6	8	10	12
Share	0	0.073	0.137	0.192	0.240	0.282	0.318

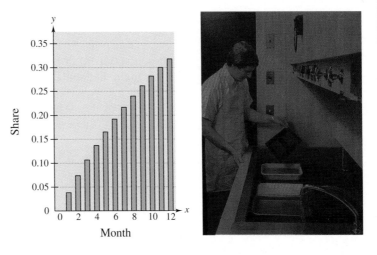

The chapter project related to this information is on page 558.

7.1 Matrices and Systems of Linear Equations

Matrices ▪ Elementary Row Operations ▪ Gaussian Elimination with Back-Substitution ▪ Gauss-Jordan Elimination

Matrices

In this section you will study a streamlined technique for solving systems of linear equations. This technique involves the use of a rectangular array of real numbers called a **matrix.**

Definition of a Matrix

If m and n are positive integers, an $m \times n$ **matrix** (read "m by n") is a rectangular array

$$\begin{bmatrix} a_{11} & a_{12} & a_{13} & \cdots & a_{1n} \\ a_{21} & a_{22} & a_{23} & \cdots & a_{2n} \\ a_{31} & a_{32} & a_{33} & \cdots & a_{3n} \\ \vdots & \vdots & \vdots & & \vdots \\ a_{m1} & a_{m2} & a_{m3} & \cdots & a_{mn} \end{bmatrix} \Bigg\} m \text{ rows}$$

$$\underbrace{\phantom{a_{11} \quad a_{12} \quad a_{13} \quad \cdots \quad a_{1n}}}_{n \text{ columns}}$$

NOTE The plural of matrix is *matrices.*

in which each **entry,** a_{ij}, of the matrix is a number. An $m \times n$ matrix has m **rows** (horizontal lines) and n **columns** (vertical lines).

The entry in the ith row and jth column is denoted by the *double subscript* notation a_{ij}. A matrix having m rows is said to be of **order** $m \times n$. If $m = n$, the matrix is **square** of order n. For a square matrix, the entries a_{11}, a_{22}, a_{33}, . . . are the **main diagonal** entries.

EXAMPLE 1   Examples of Matrices

The following matrices have the indicated orders.

NOTE A matrix that has only one row is called a **row matrix,** and a matrix that has only one column is called a **column matrix.**

a. *Order:* 1×1

$[2]$

b. *Order:* 1×4

$\begin{bmatrix} 1 & -3 & 0 & \frac{1}{2} \end{bmatrix}$

c. *Order:* 2×2

$\begin{bmatrix} 0 & 0 \\ 0 & 0 \end{bmatrix}$

d. *Order:* 3×2

$\begin{bmatrix} 5 & 0 \\ 2 & -2 \\ -7 & 4 \end{bmatrix}$

A matrix derived from a system of linear equations (each written in standard form with the constant term on the right) is the **augmented matrix** of the system. Moreover, the matrix derived from the coefficients of the system (but that does not include the constant terms) is the **coefficient matrix** of the system.

System	*Augmented Matrix*	*Coefficient Matrix*

$$
\begin{aligned}
x - 4y + 3z &= 5 \\
-x + 3y - z &= -3 \\
2x \quad\; - 4z &= 6
\end{aligned}
\qquad
\left[\begin{array}{ccc:c}
1 & -4 & 3 & 5 \\
-1 & 3 & -1 & -3 \\
2 & 0 & -4 & 6
\end{array}\right]
\qquad
\left[\begin{array}{ccc}
1 & -4 & 3 \\
-1 & 3 & -1 \\
2 & 0 & -4
\end{array}\right]
$$

NOTE   Note the use of 0 for the missing y-variable in the third equation, and also note the fourth column of constant terms in the augmented matrix.

When forming either the coefficient matrix or the augmented matrix of a system, you should begin by vertically aligning the variables in the equations and using 0's for the missing variables.

Given System	*Line Up Variables*	*Form Augmented Matrix*

$$
\begin{aligned}
x + 3y &= 9 \\
-y + 4z &= -2 \\
x - 5z &= 0
\end{aligned}
\qquad
\begin{aligned}
x + 3y \quad\;\; &= 9 \\
-y + 4z &= -2 \\
x \quad\;\; - 5z &= 0
\end{aligned}
\qquad
\left[\begin{array}{ccc:c}
1 & 3 & 0 & 9 \\
0 & -1 & 4 & -2 \\
1 & 0 & -5 & 0
\end{array}\right]
$$

Elementary Row Operations

In Section 6.3, you studied three operations that can be used on a system of linear equations to produce an equivalent system.

1. Interchange two equations.

2. Multiply an equation by a nonzero constant.

3. Add a multiple of an equation to another equation.

In matrix terminology these three operations correspond to **elementary row operations.** An elementary row operation on an augmented matrix of a given system of linear equations produces a new augmented matrix corresponding to a new (but equivalent) system of linear equations. Two matrices are **row-equivalent** if one can be obtained from the other by a sequence of elementary row operations.

Elementary Row Operations

1. Interchange two rows.

2. Multiply a row by a nonzero constant.

3. Add a multiple of a row to another row.

Although elementary row operations are simple to perform, they involve a lot of arithmetic. Because it is easy to make a mistake, we suggest that you get in the habit of noting the elementary row operations performed in each step so that you can go back and check your work.

EXAMPLE 2 Elementary Row Operations

a. Interchange the first and second rows.

Original Matrix

$$\begin{bmatrix} 0 & 1 & 3 & 4 \\ -1 & 2 & 0 & 3 \\ 2 & -3 & 4 & 1 \end{bmatrix}$$

New Row-Equivalent Matrix

$$\begin{matrix} R_2 \\ R_1 \end{matrix} \begin{bmatrix} -1 & 2 & 0 & 3 \\ 0 & 1 & 3 & 4 \\ 2 & -3 & 4 & 1 \end{bmatrix}$$

b. Multiply the first row by $\frac{1}{2}$.

Original Matrix

$$\begin{bmatrix} 2 & -4 & 6 & -2 \\ 1 & 3 & -3 & 0 \\ 5 & -2 & 1 & 2 \end{bmatrix}$$

New Row-Equivalent Matrix

$$\frac{1}{2}R_1 \rightarrow \begin{bmatrix} 1 & -2 & 3 & -1 \\ 1 & 3 & -3 & 0 \\ 5 & -2 & 1 & 2 \end{bmatrix}$$

c. Add -2 times the first row to the third row.

Original Matrix

$$\begin{bmatrix} 1 & 2 & -4 & 3 \\ 0 & 3 & -2 & -1 \\ 2 & 1 & 5 & -2 \end{bmatrix}$$

New Row-Equivalent Matrix

$$-2R_1 + R_3 \rightarrow \begin{bmatrix} 1 & 2 & -4 & 3 \\ 0 & 3 & -2 & -1 \\ 0 & -3 & 13 & -8 \end{bmatrix}$$

Note that we write the elementary row operation beside the row that we are *changing*.

Technology

Most graphing utilities can perform elementary row operations on matrices. For instance, on a *TI-82* you can perform the elementary row operation shown in Example 2(c) as follows.

1. Use the matrix edit feature to enter the matrix as [A].

2. Choose the "* row + (" feature in the matrix math menu.

$$* \text{ row} + (-2, [A], 1, 3) \boxed{\text{ENTER}}$$

The new row-equivalent matrix will be displayed. To do a sequence of row operations, use $\boxed{\text{ANS}}$ in place of [A] in each operation. If you want to save this new matrix, you must do this with separate steps.

In Example 3 of Section 6.3, you used Gaussian elimination with back-substitution to solve a system of linear equations. The next example demonstrates the matrix version of Gaussian elimination. The two methods are essentially the same. The basic difference is that with matrices you do not need to keep writing the variables.

EXAMPLE 3 Using Elementary Row Operations

Linear System   *Associated Augmented Matrix*

$$x - 2y + 3z = 9$$
$$-x + 3y \qquad = -4$$
$$2x - 5y + 5z = 17$$

$$\begin{bmatrix} 1 & -2 & 3 & \vdots & 9 \\ -1 & 3 & 0 & \vdots & -4 \\ 2 & -5 & 5 & \vdots & 17 \end{bmatrix}$$

Add the first equation to the second equation.

$$x - 2y + 3z = 9$$
$$y + 3z = 5$$
$$2x - 5y + 5z = 17$$

Add the first row to the second row $(R_1 + R_2)$.

$$R_1 + R_2 \rightarrow \begin{bmatrix} 1 & -2 & 3 & \vdots & 9 \\ 0 & 1 & 3 & \vdots & 5 \\ 2 & -5 & 5 & \vdots & 17 \end{bmatrix}$$

Add -2 times the first equation to the third equation.

$$x - 2y + 3z = 9$$
$$y + 3z = 5$$
$$-y - z = -1$$

Add -2 times the first row to the third row $(-2R_1 + R_3)$.

$$-2R_1 + R_3 \rightarrow \begin{bmatrix} 1 & -2 & 3 & \vdots & 9 \\ 0 & 1 & 3 & \vdots & 5 \\ 0 & -1 & -1 & \vdots & -1 \end{bmatrix}$$

Add the second equation to the third equation.

$$x - 2y + 3z = 9$$
$$y + 3z = 5$$
$$2z = 4$$

Add the second row to the third row $(R_2 + R_3)$.

$$R_2 + R_3 \rightarrow \begin{bmatrix} 1 & -2 & 3 & \vdots & 9 \\ 0 & 1 & 3 & \vdots & 5 \\ 0 & 0 & 2 & \vdots & 4 \end{bmatrix}$$

Multiply the third equation by $\frac{1}{2}$.

$$x - 2y + 3z = 9$$
$$y + 3z = 5$$
$$z = 2$$

Multiply the third row by $\frac{1}{2}$.

$$\tfrac{1}{2}R_3 \rightarrow \begin{bmatrix} 1 & -2 & 3 & \vdots & 9 \\ 0 & 1 & 3 & \vdots & 5 \\ 0 & 0 & 1 & \vdots & 2 \end{bmatrix}$$

At this point, you can use back-substitution to find that the solution is $x = 1$, $y = -1$, and $z = 2$, as was done in Example 3 of Section 6.3.

NOTE Remember that you can check a solution by substituting the values of x, y, and z into each equation in the original system.

The last matrix in Example 3 is said to be in **row-echelon form.** The term *echelon* refers to the stair-step pattern formed by the nonzero elements of the matrix. To be in this form, a matrix must have the following properties.

Row-Echelon Form and Reduced Row-Echelon Form

A matrix in **row-echelon form** has the following properties.

1. All rows consisting entirely of zeros occur at the bottom of the matrix.

2. For each row that does not consist entirely of zeros, the first nonzero entry is 1 (called a **leading 1**).

3. For two successive (nonzero) rows, the leading 1 in the higher row is farther to the left than the leading 1 in the lower row.

A matrix in *row-echelon form* is in **reduced row-echelon form** if every column that has a leading 1 has zeros in every position above and below its leading 1.

EXAMPLE 4   Row-Echelon Form

The following matrices are in row-echelon form.

a. $\begin{bmatrix} 1 & 2 & -1 & 4 \\ 0 & 1 & 0 & 3 \\ 0 & 0 & 1 & -2 \end{bmatrix}$
b. $\begin{bmatrix} 0 & 1 & 0 & 5 \\ 0 & 0 & 1 & 3 \\ 0 & 0 & 0 & 0 \end{bmatrix}$

c. $\begin{bmatrix} 1 & -5 & 2 & -1 & 3 \\ 0 & 0 & 1 & 3 & -2 \\ 0 & 0 & 0 & 1 & 4 \\ 0 & 0 & 0 & 0 & 1 \end{bmatrix}$
d. $\begin{bmatrix} 1 & 0 & 0 & -1 \\ 0 & 1 & 0 & 2 \\ 0 & 0 & 1 & 3 \\ 0 & 0 & 0 & 0 \end{bmatrix}$

The matrices in (b) and (d) also happen to be in *reduced* row-echelon form. The following matrices are not in row-echelon form.

e. $\begin{bmatrix} 1 & 2 & -3 & 4 \\ 0 & 2 & 1 & -1 \\ 0 & 0 & 1 & -3 \end{bmatrix}$
f. $\begin{bmatrix} 1 & 2 & -1 & 2 \\ 0 & 0 & 0 & 0 \\ 0 & 1 & 2 & -4 \end{bmatrix}$

Every matrix is row-equivalent to a matrix in row-echelon form. For instance, in Example 4, you can change the matrix in part (e) to row-echelon form by multiplying its second row by $\frac{1}{2}$. What elementary row operation could you perform on the matrix in part (f) so that it would be in row-echelon form?

Gaussian Elimination with Back-Substitution

EXAMPLE 5 Gaussian Elimination with Back-Substitution

Solve the following system.

$$\begin{aligned} y + z - 2w &= -3 \\ x + 2y - z &= 2 \\ 2x + 4y + z - 3w &= -2 \\ x - 4y - 7z - w &= -19 \end{aligned}$$

STUDY TIP

Gaussian elimination with back-substitution works well for solving systems of linear equations by hand or with a computer. For this algorithm, the order in which the elementary row operations are performed is important. We suggest operating from *left to right by columns,* using elementary row operations to obtain zeros in all entries directly below the leading 1's.

Solution

$$\begin{matrix} \curvearrowright R_2 \\ R_1 \end{matrix} \left[\begin{array}{cccc:c} 1 & 2 & -1 & 0 & 2 \\ 0 & 1 & 1 & -2 & -3 \\ 2 & 4 & 1 & -3 & -2 \\ 1 & -4 & -7 & -1 & -19 \end{array}\right]$$

First column has leading 1 in upper left corner.

$$\begin{matrix} \\ \\ -2R_1 + R_3 \rightarrow \\ -R_1 + R_4 \rightarrow \end{matrix} \left[\begin{array}{cccc:c} 1 & 2 & -1 & 0 & 2 \\ 0 & 1 & 1 & -2 & -3 \\ 0 & 0 & 3 & -3 & -6 \\ 0 & -6 & -6 & -1 & -21 \end{array}\right]$$

First column has zeros below its leading 1.

$$\begin{matrix} \\ \\ \\ 6R_2 + R_4 \rightarrow \end{matrix} \left[\begin{array}{cccc:c} 1 & 2 & -1 & 0 & 2 \\ 0 & 1 & 1 & -2 & -3 \\ 0 & 0 & 3 & -3 & -6 \\ 0 & 0 & 0 & -13 & -39 \end{array}\right]$$

Second column has zeros below its leading 1.

$$\begin{matrix} \\ \\ \tfrac{1}{3}R_3 \rightarrow \\ \\ \end{matrix} \left[\begin{array}{cccc:c} 1 & 2 & -1 & 0 & 2 \\ 0 & 1 & 1 & -2 & -3 \\ 0 & 0 & 1 & -1 & -2 \\ 0 & 0 & 0 & -13 & -39 \end{array}\right]$$

Third column has zeros below its leading 1.

$$\begin{matrix} \\ \\ \\ -\tfrac{1}{13}R_4 \rightarrow \end{matrix} \left[\begin{array}{cccc:c} 1 & 2 & -1 & 0 & 2 \\ 0 & 1 & 1 & -2 & -3 \\ 0 & 0 & 1 & -1 & -2 \\ 0 & 0 & 0 & 1 & 3 \end{array}\right]$$

Fourth column has a leading 1.

The matrix is now in row-echelon form, and the corresponding system is

$$\begin{aligned} x + 2y - z &= 2 \\ y + z - 2w &= -3 \\ z - w &= -2 \\ w &= 3 \end{aligned}$$

Using back-substitution, you can determine that the solution is $x = -1$, $y = 2$, $z = 1$, and $w = 3$. Check this in the original system of equations.

> ### Gaussian Elimination with Back-Substitution
>
> **1.** Write the augmented matrix of the system of linear equations.
>
> **2.** Use elementary row operations to rewrite the augmented matrix in row-echelon form.
>
> **3.** Write the system of linear equations corresponding to the matrix in row-echelon form, and use back-substitution to find the solution.

When solving a system of linear equations, remember that it is possible for the system to have no solution. If, in the elimination process, you obtain a row with zeros except for the last entry, it is unnecessary to continue the elimination process. You can simply conclude that the system is inconsistent.

EXAMPLE 6   A System with No Solution

Solve the following system.

$$\begin{aligned}
x - y + 2z &= 4 \\
x \quad\;\; + z &= 6 \\
2x - 3y + 5z &= 4 \\
3x + 2y - z &= 1
\end{aligned}$$

Solution

$$\begin{bmatrix}
1 & -1 & 2 & \vdots & 4 \\
1 & 0 & 1 & \vdots & 6 \\
2 & -3 & 5 & \vdots & 4 \\
3 & 2 & -1 & \vdots & 1
\end{bmatrix}
\begin{matrix}
\\ -R_1 + R_2 \to \\ -2R_1 + R_3 \to \\ -3R_1 + R_4 \to
\end{matrix}
\begin{bmatrix}
1 & -1 & 2 & \vdots & 4 \\
0 & 1 & -1 & \vdots & 2 \\
0 & -1 & 1 & \vdots & -4 \\
0 & 5 & -7 & \vdots & -11
\end{bmatrix}$$

$$\begin{matrix}
\\ \\ R_2 + R_3 \to \\ \;
\end{matrix}
\begin{bmatrix}
1 & -1 & 2 & \vdots & 4 \\
0 & 1 & -1 & \vdots & 2 \\
0 & 0 & 0 & \vdots & -2 \\
0 & 5 & -7 & \vdots & -11
\end{bmatrix}$$

Note that the third row of this matrix consists of zeros except for the last entry. This means that the original system of linear equations is *inconsistent*. You can see why this is true by converting back to a system of linear equations.

$$\begin{aligned}
x - y + 2z &= 4 \\
y - z &= 2 \\
0 &= -2 \\
5y - 7z &= -11
\end{aligned}$$

Because the third equation is not possible, the system has no solution.

Gauss-Jordan Elimination

With Gaussian elimination, we apply elementary row operations to a matrix to obtain a (row-equivalent) row-echelon form. A second method of elimination called **Gauss-Jordan elimination,** after Carl Friedrich Gauss and Wilhelm Jordan (1842–1899), continues the reduction process until a *reduced* row-echelon form is obtained. We demonstrate this procedure in the following example.

EXAMPLE 7   *Gauss-Jordan Elimination*

Use Gauss-Jordan elimination to solve the following system.

$$\begin{aligned} x - 2y + 3z &= 9 \\ -x + 3y &= -4 \\ 2x - 5y + 5z &= 17 \end{aligned}$$

Solution

In Example 3 we used Gaussian elimination to obtain the row-echelon form

$$\begin{bmatrix} 1 & -2 & 3 & \vdots & 9 \\ 0 & 1 & 3 & \vdots & 5 \\ 0 & 0 & 1 & \vdots & 2 \end{bmatrix}.$$

Now, rather than using back-substitution, apply elementary row operations until you obtain a matrix in reduced row-echelon form. To do this, you must produce zeros above each of the leading 1's, as follows.

$$2R_2 + R_1 \rightarrow \begin{bmatrix} 1 & 0 & 9 & \vdots & 19 \\ 0 & 1 & 3 & \vdots & 5 \\ 0 & 0 & 1 & \vdots & 2 \end{bmatrix}$$

Second column has zeros above its leading 1.

$$\begin{aligned} -9R_3 + R_1 \rightarrow \\ -3R_3 + R_2 \rightarrow \end{aligned} \begin{bmatrix} 1 & 0 & 0 & \vdots & 1 \\ 0 & 1 & 0 & \vdots & -1 \\ 0 & 0 & 1 & \vdots & 2 \end{bmatrix}$$

Third column has zeros above its leading 1.

Now, converting back to a system of linear equations, you have

$$\begin{aligned} x &= 1 \\ y &= -1 \\ z &= 2 \ . \end{aligned}$$

The beauty of Gauss-Jordan elimination is that, from the reduced row-echelon form, you can simply read the solution.

NOTE   Which technique do you prefer: Gaussian elimination or Gauss-Jordan elimination?

The elimination procedures described in this section employ an algorithmic approach that is easily adapted to computer use. However, the procedure makes no effort to avoid fractional coefficients. For instance, if the system given in Example 7 had been listed as

$$2x - 5y + 5z = 17$$
$$x - 2y + 3z = 9$$
$$-x + 3y = -4$$

the procedure would have required multiplication of the first row by $\frac{1}{2}$, which would have introduced fractions in the first row. For hand computations, fractions can sometimes be avoided by judiciously choosing the order in which the elementary row operations are applied.

EXAMPLE 8   A System with an Infinite Number of Solutions

Solve the following system.

$$2x + 4y - 2z = 0$$
$$3x + 5y = 1$$

Solution

$$\begin{bmatrix} 2 & 4 & -2 & \vdots & 0 \\ 3 & 5 & 0 & \vdots & 1 \end{bmatrix} \qquad \frac{1}{2}R_1 \rightarrow \begin{bmatrix} 1 & 2 & -1 & \vdots & 0 \\ 3 & 5 & 0 & \vdots & 1 \end{bmatrix}$$

$$-3R_1 + R_2 \rightarrow \begin{bmatrix} 1 & 2 & -1 & \vdots & 0 \\ 0 & -1 & 3 & \vdots & 1 \end{bmatrix}$$

$$-R_2 \rightarrow \begin{bmatrix} 1 & 2 & -1 & \vdots & 0 \\ 0 & 1 & -3 & \vdots & -1 \end{bmatrix}$$

$$-2R_2 + R_1 \rightarrow \begin{bmatrix} 1 & 0 & 5 & \vdots & 2 \\ 0 & 1 & -3 & \vdots & -1 \end{bmatrix}$$

The corresponding system of equations is

$$x + 5z = 2$$
$$y - 3z = -1 \ .$$

Solving for x and y in terms of z, you have $x = -5z + 2$ and $y = 3z - 1$. Then, letting $z = a$, the solution set has the form

$$(-5a + 2, \ 3a - 1, \ a)$$

where a is a real number. Try substituting values for a to obtain a few solutions. Then check each solution in the original system of equations.

It is worth noting that the row-echelon form of a matrix is not unique. That is, two different sequences of elementary row operations may yield different row-echelon forms. For instance, the following sequence of elementary row operations on the matrix in Example 3 produces a slightly different row-echelon form.

$$\begin{bmatrix} 1 & -2 & 3 & \vdots & 9 \\ -1 & 3 & 0 & \vdots & -4 \\ 2 & -5 & 5 & \vdots & 17 \end{bmatrix} \quad \begin{matrix} R_2 \\ R_1 \end{matrix} \begin{bmatrix} -1 & 3 & 0 & \vdots & -4 \\ 1 & -2 & 3 & \vdots & 9 \\ 2 & -5 & 5 & \vdots & 17 \end{bmatrix}$$

$$-R_1 \rightarrow \begin{bmatrix} 1 & -3 & 0 & \vdots & 4 \\ 1 & -2 & 3 & \vdots & 9 \\ 2 & -5 & 5 & \vdots & 17 \end{bmatrix}$$

$$\begin{matrix} -R_1 + R_2 \rightarrow \\ -2R_1 + R_3 \rightarrow \end{matrix} \begin{bmatrix} 1 & -3 & 0 & \vdots & 4 \\ 0 & 1 & 3 & \vdots & 5 \\ 0 & 1 & 5 & \vdots & 9 \end{bmatrix}$$

$$-R_2 + R_3 \rightarrow \begin{bmatrix} 1 & -3 & 0 & \vdots & 4 \\ 0 & 1 & 3 & \vdots & 5 \\ 0 & 0 & 2 & \vdots & 4 \end{bmatrix}$$

$$\tfrac{1}{2} R_3 \rightarrow \begin{bmatrix} 1 & -3 & 0 & \vdots & 4 \\ 0 & 1 & 3 & \vdots & 5 \\ 0 & 0 & 1 & \vdots & 2 \end{bmatrix}$$

The corresponding system of linear equations is

$$\begin{aligned} x - 3y \quad\quad &= 4 \\ y + 3z &= 5 \\ z &= 2 \end{aligned}$$

Try using back-substitution on this system to see that you obtain the same solution that was obtained in Example 3.

Group Activities   You Be the Instructor

Error Analysis   One of your students has handed in the following steps for solution of a system by Gauss-Jordan elimination. Find the error(s) in the solution and discuss how to explain the error(s) to your student.

$$\begin{bmatrix} 1 & 1 & \vdots & 4 \\ 2 & 3 & \vdots & 5 \end{bmatrix} \xrightarrow{} \begin{matrix} \\ -2R_1 + R_2 \rightarrow \end{matrix} \begin{bmatrix} 1 & 1 & \vdots & 4 \\ 0 & 1 & \vdots & 5 \end{bmatrix} \xrightarrow{-1R_2 + R_1 \rightarrow} \begin{bmatrix} 1 & 0 & \vdots & 4 \\ 0 & 1 & \vdots & 5 \end{bmatrix}$$

Warm Up The following warm-up exercises involve skills that were covered in earlier sections. You will use these skills in the exercise set for this section.

In Exercises 1–4, evaluate the expression.

1. $2(-1) - 3(5) + 7(2)$

2. $-4(-3) + 6(7) + 8(-3)$

3. $11\left(\frac{1}{2}\right) - 7\left(-\frac{3}{2}\right) - 5(2)$

4. $\frac{2}{3}\left(\frac{1}{2}\right) + \frac{4}{3}\left(-\frac{1}{3}\right)$

In Exercises 5 and 6, decide whether $x = 1$, $y = 3$, and $z = -1$ is a solution of the system.

5.
$$4x - 2y + 3z = -5$$
$$x + 3y - z = 11$$
$$-x + 2y = 5$$

6.
$$-x + 2y + z = 4$$
$$2x - 3z = 5$$
$$3x + 5y - 2z = 21$$

In Exercises 7–10, use back-substitution to solve the system of linear equations.

7.
$$2x - 3y = 4$$
$$y = 2$$

8.
$$5x + 4y = 0$$
$$y = -3$$

9.
$$x - 3y + z = 0$$
$$y - 3z = 8$$
$$z = 2$$

10.
$$2x - 5y + 3z = -2$$
$$y - 4z = 0$$
$$z = 1$$

7.1 Exercises

Exercises containing inconsistent systems (no solution): 40 and 43. Exercises containing dependent systems (infinitely many solutions): 44, 45, 47, 49, 50, 51, 52, 54, 56, 57, and 58.

In Exercises 1–6, determine the order of the matrix.

1. $\begin{bmatrix} 4 & -2 \\ 7 & 0 \\ 0 & 8 \end{bmatrix}$

2. $\begin{bmatrix} 5 & -3 & 8 & 7 \end{bmatrix}$

3. $\begin{bmatrix} -9 \\ 2 \\ 36 \\ 11 \\ 3 \end{bmatrix}$

4. $\begin{bmatrix} 11 & 0 & 8 & 5 \\ -3 & 7 & 15 & 0 \\ 0 & 6 & 3 & 3 \\ 12 & 4 & 16 & 9 \\ 1 & 1 & 6 & 7 \end{bmatrix}$

5. $\begin{bmatrix} 33 & 45 \\ -9 & 20 \\ 12 & 15 \\ 16 & -2 \end{bmatrix}$

6. $\begin{bmatrix} 12 & -2 & 4 \\ -3 & 4 & 0 \\ -8 & 12 & 2 \end{bmatrix}$

In Exercises 7–10, determine whether the matrix is in row-echelon form. If it is, determine if it is also in reduced row-echelon form.

7. $\begin{bmatrix} 1 & 0 & 0 & 0 \\ 0 & 1 & 1 & 5 \\ 0 & 0 & 0 & 0 \end{bmatrix}$

8. $\begin{bmatrix} 1 & 0 & 2 & 1 \\ 0 & 1 & -3 & 10 \\ 0 & 0 & 1 & 0 \end{bmatrix}$

9. $\begin{bmatrix} 2 & 0 & 4 & 0 \\ 0 & -1 & 3 & 6 \\ 0 & 0 & 1 & 5 \end{bmatrix}$

10. $\begin{bmatrix} 1 & 3 & 0 & 0 & 0 & 0 \\ 0 & 0 & 1 & 8 & 1 & 0 \\ 0 & 0 & 0 & 0 & 1 & 1 \\ 0 & 0 & 0 & 0 & 1 & 1 \end{bmatrix}$

11. Use a graphing utility to perform the row operations to reduce the matrix to row-echelon form.

$$\begin{bmatrix} 1 & 2 & 3 \\ 2 & -1 & -4 \\ 3 & 1 & -1 \end{bmatrix}$$

(a) Add -2 times Row 1 to Row 2.

(b) Add -3 times Row 1 to Row 3.

(c) Add -1 times Row 2 to Row 3.

(d) Multiply Row 2 by $-\frac{1}{5}$.

(e) Add -2 times Row 2 to Row 1.

12. Use a graphing utility to perform the operations to reduce the matrix to *reduced* row-echelon form.

$$\begin{bmatrix} 7 & 1 \\ 0 & 2 \\ -3 & 4 \\ 4 & 1 \end{bmatrix}$$

(a) Add Row 3 to Row 4.

(b) Interchange Rows 1 and 4.

(c) Add 3 times Row 1 to Row 3.

(d) Add -7 times Row 1 to Row 4.

(e) Multiply Row 2 by $\frac{1}{2}$.

(f) Add the appropriate multiple of Row 2 to Rows 1, 3, and 4.

In Exercises 13–16, write the matrix in row-echelon form. (Note: row-echelon forms are not unique.)

13. $\begin{bmatrix} 1 & 1 & 0 & 5 \\ -2 & -1 & 2 & -10 \\ 3 & 6 & 7 & 14 \end{bmatrix}$

14. $\begin{bmatrix} 1 & 2 & -1 & 3 \\ 3 & 7 & -5 & 14 \\ -2 & -1 & -3 & 8 \end{bmatrix}$

15. $\begin{bmatrix} 1 & -1 & -1 & 1 \\ 5 & -4 & 1 & 8 \\ -6 & 8 & 18 & 0 \end{bmatrix}$

16. $\begin{bmatrix} 1 & -3 & 0 & -7 \\ -3 & 10 & 1 & 23 \\ 1 & 0 & 1 & 12 \\ 4 & -10 & 2 & -24 \end{bmatrix}$

In Exercises 17–20, write the matrix in *reduced* row-echelon form.

17. $\begin{bmatrix} 3 & 3 & 3 \\ -1 & 0 & -4 \\ 2 & 4 & -2 \end{bmatrix}$

18. $\begin{bmatrix} 1 & 3 & 2 \\ 5 & 15 & 9 \\ 2 & 6 & 10 \end{bmatrix}$

19. $\begin{bmatrix} 1 & 2 & 3 & -5 \\ 1 & 2 & 4 & -9 \\ -2 & -4 & -4 & 3 \\ 4 & 8 & 11 & -14 \end{bmatrix}$

20. $\begin{bmatrix} 1 & -3 \\ -1 & 8 \\ 0 & 4 \\ -2 & 10 \end{bmatrix}$

In Exercises 21–24, write the system of linear equations represented by the augmented matrix. (Use variables w, x, y, and z.)

21. $\begin{bmatrix} 4 & 3 & \vdots & 8 \\ 1 & -2 & \vdots & 3 \end{bmatrix}$

22. $\begin{bmatrix} 9 & -4 & \vdots & 0 \\ 6 & 1 & \vdots & -4 \end{bmatrix}$

23. $\begin{bmatrix} 1 & 0 & 2 & \vdots & -10 \\ 0 & 3 & -1 & \vdots & 5 \\ 4 & 2 & 0 & \vdots & 3 \end{bmatrix}$

24. $\begin{bmatrix} 5 & 8 & 2 & 0 & \vdots & -1 \\ -2 & 15 & 5 & 1 & \vdots & 9 \\ 1 & 6 & -7 & 0 & \vdots & -3 \end{bmatrix}$

In Exercises 25–28, write the system of equations represented by the augmented matrix. Use back-substitution to find the solution. (Use w, x, y, and z.)

25. $\begin{bmatrix} 1 & -2 & \vdots & 4 \\ 0 & 1 & \vdots & -3 \end{bmatrix}$

26. $\begin{bmatrix} 1 & -1 & 2 & \vdots & 4 \\ 0 & 1 & -1 & \vdots & 2 \\ 0 & 0 & 1 & \vdots & -2 \end{bmatrix}$

27. $\begin{bmatrix} 1 & 2 & -2 & \vdots & -1 \\ 0 & 1 & 1 & \vdots & 9 \\ 0 & 0 & 1 & \vdots & -3 \end{bmatrix}$

28. $\begin{bmatrix} 1 & 2 & -2 & 0 & \vdots & -1 \\ 0 & 1 & 1 & 2 & \vdots & 9 \\ 0 & 0 & 1 & 0 & \vdots & 2 \\ 0 & 0 & 0 & 1 & \vdots & -3 \end{bmatrix}$

In Exercises 29–32, an augmented matrix that represents a system of linear equations (in variables x, y, and z) has been reduced using Gauss-Jordan elimination. Write the solution represented by the augmented matrix.

29. $\begin{bmatrix} 1 & 0 & \vdots & 7 \\ 0 & 1 & \vdots & -5 \end{bmatrix}$ **30.** $\begin{bmatrix} 1 & 0 & \vdots & -2 \\ 0 & 1 & \vdots & 4 \end{bmatrix}$

31. $\begin{bmatrix} 1 & 0 & 0 & \vdots & -4 \\ 0 & 1 & 0 & \vdots & -8 \\ 0 & 0 & 1 & \vdots & 2 \end{bmatrix}$

32. $\begin{bmatrix} 1 & 0 & 0 & \vdots & 3 \\ 0 & 1 & 0 & \vdots & -1 \\ 0 & 0 & 1 & \vdots & 0 \end{bmatrix}$

In Exercises 33–36, write the augmented matrix for the system of linear equations.

33. $4x - 5y = -2$
$-x + 8y = 10$

34. $8x + 3y = 25$
$3x - 9y = 12$

35. $x + 10y - 3z = 2$
$5x - 3y + 4z = 0$
$2x + 4y = 6$

36. $9w - 3x + 20y + z = 13$
$12w - 8y = 5$
$w + 2x + 3y - 4z = -2$
$-w - x + y + z = 1$

In Exercises 37–58, solve the system of equations. Use Gaussian elimination with back-substitution or Gauss-Jordan elimination.

37. $x + 2y = 7$
$2x + y = 8$

38. $2x + 6y = 16$
$2x + 3y = 7$

39. $-3x + 5y = -22$
$3x + 4y = 4$
$4x - 8y = 32$

40. $x + 2y = 0$
$x + y = 6$
$3x - 2y = 8$

41. $8x - 4y = 7$
$5x + 2y = 1$

42. $2x - y = -0.1$
$3x + 2y = 1.6$

43. $-x + 2y = 1.5$
$2x - 4y = 3$

44. $x - 3y = 5$
$-2x + 6y = -10$

45. $2x + 3z = 3$
$4x - 3y + 7z = 5$
$8x - 9y + 15z = 9$

46. $2x - y + 3z = 24$
$2y - z = 14$
$7x - 5y = 6$

47. $x + y - 5z = 3$
$x - 2z = 1$
$2x - y - z = 0$

48. $x - 3z = -2$
$3x + y - 2z = 5$
$2x + 2y + z = 4$

49. $x + 2y + z = 8$
$3x + 7y + 6z = 26$

50. $4x + 12y - 7z - 20w = 22$
$3x + 9y - 5z - 28w = 30$

51. $3x + 3y + 12z = 6$
$x + y + 4z = 2$
$2x + 5y + 20z = 10$
$-x + 2y + 8z = 4$

52. $2x + 10y + 2z = 6$
$x + 5y + 2z = 6$
$x + 5y + z = 3$
$-3x - 15y - 3z = -9$

53. $2x + y - z + 2w = -16$
$3x + 4y + w = 1$
$x + 5y + 2z + 6w = -3$
$5x + 2y - z - w = 3$

54. $x + 2y + 2z + 4w = 11$
$3x + 6y + 5z + 12w = 30$

55. $x + 2y = 0$
$-x - y = 0$

56. $x + 2y = 0$
$2x + 4y = 0$

57. $x + y + z = 0$
$2x + 3y + z = 0$
$3x + 5y + z = 0$

58. $x + 2y + z + 3w = 0$
$x - y + w = 0$
$y - z + 2w = 0$

59. *Borrowing Money* A small corporation borrowed $500,000 to expand its product line. Some of the money was borrowed at 9%, some at 10%, and some at 12%. How much was borrowed at each rate if the annual interest was $52,000 and the amount borrowed at 10% was two and one-half times the amount borrowed at 9%?

60. *Borrowing Money* A company borrowed $800,000 to expand its product line. Some of the money was borrowed at 10%, some at 10.5%, and some at 11.5%. How much was borrowed at each rate if the annual interest was $83,500 and the amount borrowed at 10% was four times the amount borrowed at 11.5%?

Curve Fitting In Exercises 61 and 62, find the parabola $y = ax^2 + bx + c$ passing through the points.

61.   **62.**

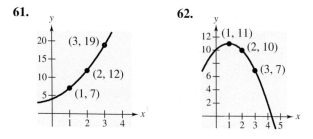

Curve Fitting In Exercises 63 and 64, find the cubic $y = ax^3 + bx^2 + cx + d$ passing through the points.

63.   **64.**

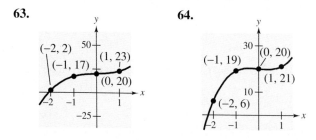

65. *Average Salary* From 1984 to 1993, the average annual salary for assistant professors at public colleges and universities in the United States increased in a pattern that was approximately linear (see figure). Find the least squares regression line $y = a + bt$ for the data shown in the figure by solving the following system using a graphing utility. (y represents the average salary in thousands of dollars and $t = 0$ represents 1988.)

$$10a + 5b = 307$$
$$5a + 85b = 283.6$$

Use the result to predict the average salary for an assistant professor at a public college in 1996. (Source: Maryse Eymonerie Associates)

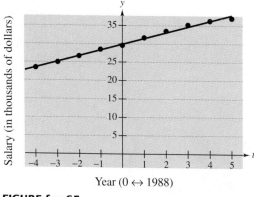

Year (0 ↔ 1988)

FIGURE for 65

66. *Average Salary* From 1984 to 1993, the average annual salary for assistant professors at private (non-church related) colleges and universities in the United States increased in a pattern that was approximately linear (see figure). Find the least squares regression line $y = a + bt$ for the data shown in the figure by solving the following system using a graphing utility. (y represents the average salary in thousands of dollars and $t = 0$ represents 1988.)

$$10a + 5b = 311.7$$
$$5a + 85b = 298.9$$

Use the result to predict the average salary for an assistant professor at a private college in 1996. (Source: Maryse Eymonerie Associates)

Year (0 ↔ 1988)

Equality of Matrices

A British mathematician, Arthur Cayley, invented matrices around 1858. Cayley was a Cambridge University graduate and a lawyer by profession. His ground-breaking work on matrices was begun as he studied the theory of transformations. Cayley also was instrumental in the development of determinants. Cayley and two American mathematicians, Benjamin Peirce (1809–1880) and his son Charles S. Peirce (1839–1914), are credited with developing "matrix algebra."

In Section 7.1, you used matrices to solve systems of linear equations. Matrices, however, can do much more than this. There is a rich mathematical theory of matrices, and its applications are numerous. This section and the next introduce some fundamentals of matrix theory. It is standard mathematical convention to represent matrices in any of the following three ways.

1. A matrix can be denoted by an uppercase letter such as A, B, or C.

2. A matrix can be denoted by a representative element enclosed in brackets, such as $[a_{ij}]$, $[b_{ij}]$, or $[c_{ij}]$.

3. A matrix can be denoted by a rectangular array of numbers such as

$$A = [a_{ij}] = \begin{bmatrix} a_{11} & a_{12} & a_{13} & \cdots & a_{1n} \\ a_{21} & a_{22} & a_{23} & \cdots & a_{2n} \\ a_{31} & a_{32} & a_{33} & \cdots & a_{3n} \\ \vdots & \vdots & \vdots & & \vdots \\ a_{m1} & a_{m2} & a_{m3} & \cdots & a_{mn} \end{bmatrix}.$$

Two matrices $A = [a_{ij}]$ and $B = [b_{ij}]$ are **equal** if they have the same order $(m \times n)$ and

$$a_{ij} = b_{ij}$$

for $1 \leq i \leq m$ and $1 \leq j \leq n$. In other words, two matrices are equal if their corresponding entries are equal.

EXAMPLE 1   Equality of Matrices

Solve for a_{11}, a_{12}, a_{21}, and a_{22} in the following matrix equation.

$$\begin{bmatrix} a_{11} & a_{12} \\ a_{21} & a_{22} \end{bmatrix} = \begin{bmatrix} 2 & -1 \\ -3 & 0 \end{bmatrix}$$

Solution

Because two matrices are equal only if their corresponding entries are equal, you can conclude that

$$a_{11} = 2, \quad a_{12} = -1, \quad a_{21} = -3, \quad \text{and} \quad a_{22} = 0.$$

Matrix Addition and Scalar Multiplication

You can **add** two matrices (of the same order) by adding their corresponding entries.

Definition of Matrix Addition

If $A = [a_{ij}]$ and $B = [b_{ij}]$ are matrices of order $m \times n$, their **sum** is the $m \times n$ matrix given by

$$A + B = [a_{ij} + b_{ij}].$$

The sum of two matrices of different orders is undefined.

EXAMPLE 2 Addition of Matrices

a. $\begin{bmatrix} -1 & 2 \\ 0 & 1 \end{bmatrix} + \begin{bmatrix} 1 & 3 \\ -1 & 2 \end{bmatrix} = \begin{bmatrix} -1+1 & 2+3 \\ 0-1 & 1+2 \end{bmatrix} = \begin{bmatrix} 0 & 5 \\ -1 & 3 \end{bmatrix}$

b. $\begin{bmatrix} 0 & 1 & -2 \\ 1 & 2 & 3 \end{bmatrix} + \begin{bmatrix} 0 & 0 & 0 \\ 0 & 0 & 0 \end{bmatrix} = \begin{bmatrix} 0 & 1 & -2 \\ 1 & 2 & 3 \end{bmatrix}$

c. $\begin{bmatrix} 1 \\ -3 \\ -2 \end{bmatrix} + \begin{bmatrix} -1 \\ 3 \\ 2 \end{bmatrix} = \begin{bmatrix} 0 \\ 0 \\ 0 \end{bmatrix}$

d. The sum of

$$A = \begin{bmatrix} 2 & 1 & 0 \\ 4 & 0 & -1 \\ 3 & -2 & 2 \end{bmatrix} \quad \text{and} \quad B = \begin{bmatrix} 0 & 1 \\ -1 & 3 \\ 2 & 4 \end{bmatrix}$$

is undefined.

When working with matrices, numbers are usually referred to as **scalars**. In this text, scalars will always be real numbers. You can multiply a matrix A by a scalar c by multiplying each entry in A by c.

Definition of Scalar Multiplication

If $A = [a_{ij}]$ is an $m \times n$ matrix and c is a scalar, the **scalar multiple** of A by c is the $m \times n$ matrix given by

$$cA = [ca_{ij}].$$

The symbol $-A$ represents the scalar product $(-1)A$. Moreover, if A and B are of the same order, then $A - B$ represents the sum of A and $(-1)B$. That is,

$$A - B = A + (-1)B.$$ Subtraction of matrices

EXAMPLE 3   Scalar Multiplication and Matrix Subtraction

For the following matrices find (a) $3A$, (b) $-B$, and (c) $3A - B$.

$$A = \begin{bmatrix} 2 & 2 & 4 \\ -3 & 0 & -1 \\ 2 & 1 & 2 \end{bmatrix} \quad \text{and} \quad B = \begin{bmatrix} 2 & 0 & 0 \\ 1 & -4 & 3 \\ -1 & 3 & 2 \end{bmatrix}$$

Solution

a. $3A = 3 \begin{bmatrix} 2 & 2 & 4 \\ -3 & 0 & -1 \\ 2 & 1 & 2 \end{bmatrix}$ Scalar multiplication

$\quad = \begin{bmatrix} 3(2) & 3(2) & 3(4) \\ 3(-3) & 3(0) & 3(-1) \\ 3(2) & 3(1) & 3(2) \end{bmatrix}$ Multiply each entry by 3.

$\quad = \begin{bmatrix} 6 & 6 & 12 \\ -9 & 0 & -3 \\ 6 & 3 & 6 \end{bmatrix}$ Simplify.

b. $-B = (-1) \begin{bmatrix} 2 & 0 & 0 \\ 1 & -4 & 3 \\ -1 & 3 & 2 \end{bmatrix}$ Definition of negation

$\quad = \begin{bmatrix} -2 & 0 & 0 \\ -1 & 4 & -3 \\ 1 & -3 & -2 \end{bmatrix}$ Multiply each entry by -1.

c. $3A - B = \begin{bmatrix} 6 & 6 & 12 \\ -9 & 0 & -3 \\ 6 & 3 & 6 \end{bmatrix} - \begin{bmatrix} 2 & 0 & 0 \\ 1 & -4 & 3 \\ -1 & 3 & 2 \end{bmatrix}$ Matrix subtraction

$\quad = \begin{bmatrix} 4 & 6 & 12 \\ -10 & 4 & -6 \\ 7 & 0 & 4 \end{bmatrix}$ Subtract corresponding entries.

It is often convenient to rewrite the scalar multiple cA by factoring c out of every entry in the matrix. For instance, in the following example, the scalar $\frac{1}{2}$ has been factored out of the matrix.

$$\begin{bmatrix} \frac{1}{2} & -\frac{3}{2} \\ \frac{5}{2} & \frac{1}{2} \end{bmatrix} = \frac{1}{2} \begin{bmatrix} 1 & -3 \\ 5 & 1 \end{bmatrix}$$

The properties of matrix addition and scalar multiplication are similar to those of addition and multiplication of real numbers.

Properties of Matrix Addition and Scalar Multiplication

If A, B, and C are $m \times n$ matrices and c and d are scalars, the following properties are true.

1. $A + B = B + A$ Commutative Property of Matrix Addition

2. $A + (B + C) = (A + B) + C$ Associative Property of Matrix Addition

3. $(cd)A = c(dA)$ Associative Property of Scalar Multiplication

4. $1A = A$ Scalar Identity

5. $c(A + B) = cA + cB$ Distributive Property

6. $(c + d)A = cA + dA$ Distributive Property

Note that the Associative Property of Matrix Addition allows you to write expressions such as $A + B + C$ without ambiguity because the same sum occurs no matter how the matrices are grouped. In other words, you obtain the same sum whether you group $A + B + C$ as $(A + B) + C$ or as $A + (B + C)$. This same reasoning applies to sums of four or more matrices.

EXAMPLE 4   Addition of More than Two Matrices

By adding corresponding entries, you obtain the following sum of four matrices.

$$
\begin{bmatrix} 1 \\ 2 \\ -3 \end{bmatrix} + \begin{bmatrix} -1 \\ -1 \\ 2 \end{bmatrix} + \begin{bmatrix} 0 \\ 1 \\ 4 \end{bmatrix} + \begin{bmatrix} 2 \\ -3 \\ -2 \end{bmatrix} = \begin{bmatrix} 2 \\ -1 \\ 1 \end{bmatrix}
$$

Technology

Most graphing utilities can add and subtract matrices and multiply matrices by scalars. For instance, on a *TI-82*, you can find the sum of the matrices

$$
A = \begin{bmatrix} 2 & -3 \\ -1 & 0 \end{bmatrix} \quad \text{and} \quad B = \begin{bmatrix} -1 & 4 \\ 2 & -5 \end{bmatrix}
$$

by entering the matrices and then using the following keystrokes.

[A] $\boxed{+}$ [B] $\boxed{\text{ENTER}}$

One important property of addition of real numbers is that the number 0 is the additive identity. That is, $c + 0 = c$ for any real number c. For matrices, a similar property holds. That is, if A is an $m \times n$ matrix and O is the $m \times n$ **zero matrix** consisting entirely of zeros, then $A + O = A$.

In other words, O is the **additive identity** for the set of all $m \times n$ matrices. For example, the following matrices are the additive identities for the set of all 2×3 and 2×2 matrices.

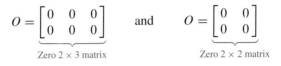

$$O = \begin{bmatrix} 0 & 0 & 0 \\ 0 & 0 & 0 \end{bmatrix} \quad \text{and} \quad O = \begin{bmatrix} 0 & 0 \\ 0 & 0 \end{bmatrix}.$$

Zero 2×3 matrix   Zero 2×2 matrix

The algebra of real numbers and the algebra of matrices have many similarities. For example, compare the following solutions.

Real Numbers *(Solve for x.)*	$m \times n$ *Matrices* *(Solve for X.)*
$x + a = b$	$X + A = B$
$x + a + (-a) = b + (-a)$	$X + A + (-A) = B + (-A)$
$x + 0 = b - a$	$X + O = B - A$
$x = b - a$	$X = B - A$

EXAMPLE 5 Solving a Matrix Equation

Solve for X in the equation $3X + A = B$, where

$$A = \begin{bmatrix} 1 & -2 \\ 0 & 3 \end{bmatrix} \quad \text{and} \quad B = \begin{bmatrix} -3 & 4 \\ 2 & 1 \end{bmatrix}.$$

Solution

Begin by solving the equation for X to obtain

$$3X = B - A \quad \Longrightarrow \quad X = \frac{1}{3}(B - A).$$

Now, using the matrices A and B, you have

$$X = \frac{1}{3}\left(\begin{bmatrix} -3 & 4 \\ 2 & 1 \end{bmatrix} - \begin{bmatrix} 1 & -2 \\ 0 & 3 \end{bmatrix} \right)$$

$$= \frac{1}{3}\begin{bmatrix} -4 & 6 \\ 2 & -2 \end{bmatrix}$$

$$= \begin{bmatrix} -\frac{4}{3} & 2 \\ \frac{2}{3} & -\frac{2}{3} \end{bmatrix}.$$

Matrix Multiplication

NOTE The definition of matrix multiplication indicates a *row-by-column* multiplication, where the entry in the ith row and jth column of the product AB is obtained by multiplying the entries in the ith row of A by the corresponding entries in the jth column of B and then adding the results. Example 6 illustrates this process.

The third basic matrix operation is **matrix multiplication.** At first glance the definition may seem unusual. You will see later, however, that this definition of the product of two matrices has many practical applications.

Definition of Matrix Multiplication

If $A = [a_{ij}]$ is an $m \times n$ matrix and $B = [b_{ij}]$ is an $n \times p$ matrix, the **product** AB is an $m \times p$ matrix

$$AB = [c_{ij}]$$

where $c_{ij} = a_{i1}b_{1j} + a_{i2}b_{2j} + a_{i3}b_{3j} + \cdots + a_{in}b_{nj}.$

EXAMPLE 6   *Finding the Product of Two Matrices*

Find the product AB where

$$A = \begin{bmatrix} -1 & 3 \\ 4 & -2 \\ 5 & 0 \end{bmatrix} \quad \text{and} \quad B = \begin{bmatrix} -3 & 2 \\ -4 & 1 \end{bmatrix}.$$

Solution

First, note that the product AB is defined because the number of columns of A is equal to the number of rows of B. Moreover, the product AB has order 3×2, and is of the form

$$\begin{bmatrix} -1 & 3 \\ 4 & -2 \\ 5 & 0 \end{bmatrix} \begin{bmatrix} -3 & 2 \\ -4 & 1 \end{bmatrix} = \begin{bmatrix} c_{11} & c_{12} \\ c_{21} & c_{22} \\ c_{31} & c_{32} \end{bmatrix}.$$

To find the entries of the product, multiply each row of A by each column of B, as follows. Use a graphing utility to check this result.

$$\begin{aligned} AB &= \begin{bmatrix} -1 & 3 \\ 4 & -2 \\ 5 & 0 \end{bmatrix} \begin{bmatrix} -3 & 2 \\ -4 & 1 \end{bmatrix} \\[2mm] &= \begin{bmatrix} (-1)(-3) + (3)(-4) & (-1)(2) + (3)(1) \\ (4)(-3) + (-2)(-4) & (4)(2) + (-2)(1) \\ (5)(-3) + (0)(-4) & (5)(2) + (0)(1) \end{bmatrix} \\[2mm] &= \begin{bmatrix} -9 & 1 \\ -4 & 6 \\ -15 & 10 \end{bmatrix} \end{aligned}$$

Technology

Some graphing utilities, such as the *TI-82*, are able to add, subtract, and multiply matrices. If you have such a graphing utility, try entering the matrices

$$A = \begin{bmatrix} 1 & 2 & 3 \\ 2 & -5 & 1 \end{bmatrix} \text{ and}$$

$$B = \begin{bmatrix} -3 & 2 & 1 \\ 4 & -2 & 0 \\ 1 & 2 & 3 \end{bmatrix}$$

and use the following keystrokes to find the product of the matrices.

[A] $\boxed{\times}$ [B] $\boxed{\text{ENTER}}$

You should get:

$$\begin{bmatrix} 8 & 4 & 10 \\ -25 & 16 & 5 \end{bmatrix}$$

Be sure you understand that for the product of two matrices to be defined, the number of columns of the first matrix must equal the number of rows of the second matrix. That is, the middle two indices must be the same and the outside two indices give the order of the product, as shown in the following diagram.

$$\underset{m \times n}{A} \quad \underset{n \times p}{B} \quad = \quad \underset{m \times p}{AB}$$

equal

order of AB

EXAMPLE 7 Matrix Multiplication

a. $\begin{bmatrix} 1 & 0 & 3 \\ 2 & -1 & -2 \end{bmatrix} \begin{bmatrix} -2 & 4 & 2 \\ 1 & 0 & 0 \\ -1 & 1 & -1 \end{bmatrix} = \begin{bmatrix} -5 & 7 & -1 \\ -3 & 6 & 6 \end{bmatrix}$

   $\quad\quad 2 \times 3 \quad\quad\quad 3 \times 3 \quad\quad\quad\quad 2 \times 3$

b. $\begin{bmatrix} 3 & 4 \\ -2 & 5 \end{bmatrix} \begin{bmatrix} 1 & 0 \\ 0 & 1 \end{bmatrix} = \begin{bmatrix} 3 & 4 \\ -2 & 5 \end{bmatrix}$

   $\quad\quad 2 \times 2 \quad 2 \times 2 \quad\quad 2 \times 2$

c. $\begin{bmatrix} 1 & 2 \\ 1 & 1 \end{bmatrix} \begin{bmatrix} -1 & 2 \\ 1 & -1 \end{bmatrix} = \begin{bmatrix} 1 & 0 \\ 0 & 1 \end{bmatrix}$

   $\quad\quad 2 \times 2 \quad\quad 2 \times 2 \quad\quad 2 \times 2$

d. $\begin{bmatrix} 1 & -2 & -3 \end{bmatrix} \begin{bmatrix} 2 \\ -1 \\ 1 \end{bmatrix} = \begin{bmatrix} 1 \end{bmatrix}$

   $\quad\quad 1 \times 3 \quad\quad 3 \times 1 \quad 1 \times 1$

e. $\begin{bmatrix} 2 \\ -1 \\ 1 \end{bmatrix} \begin{bmatrix} 1 & -2 & -3 \end{bmatrix} = \begin{bmatrix} 2 & -4 & -6 \\ -1 & 2 & 3 \\ 1 & -2 & -3 \end{bmatrix}$

   $\quad\quad 3 \times 1 \quad\quad 1 \times 3 \quad\quad\quad\quad 3 \times 3$

f. The product AB for the following matrices is not defined.

$$A = \begin{bmatrix} -2 & 1 \\ 1 & -3 \\ 1 & 4 \end{bmatrix} \quad \text{and} \quad B = \begin{bmatrix} -2 & 3 & 1 & 4 \\ 0 & 1 & -1 & 2 \\ 2 & -1 & 0 & 1 \end{bmatrix}$$

   $\quad\quad\quad 3 \times 2 \quad\quad\quad\quad\quad\quad\quad 3 \times 4$

NOTE   In parts (d) and (e) of Example 7, note that the two products are different. Matrix multiplication is not, in general, commutative. That is, for most matrices, $AB \neq BA$.

The general pattern for matrix multiplication is as follows. To obtain the entry in the ith row and the jth column of the product AB, use the ith row of A and the jth column of B.

$$\begin{bmatrix} a_{11} & a_{12} & a_{13} & \cdots & a_{1n} \\ a_{21} & a_{22} & a_{23} & \cdots & a_{2n} \\ a_{31} & a_{32} & a_{33} & \cdots & a_{3n} \\ \vdots & \vdots & \vdots & & \vdots \\ a_{i1} & a_{i2} & a_{i3} & \cdots & a_{in} \\ \vdots & \vdots & \vdots & & \vdots \\ a_{m1} & a_{m2} & a_{m3} & \cdots & a_{mn} \end{bmatrix} \begin{bmatrix} b_{12} & b_{11} & \cdots & b_{1j} & \cdots & b_{1p} \\ b_{22} & b_{21} & \cdots & b_{2j} & \cdots & b_{2p} \\ b_{32} & b_{31} & \cdots & b_{3j} & \cdots & b_{3p} \\ \vdots & \vdots & & \vdots & & \vdots \\ b_{n1} & b_{n2} & \cdots & b_{nj} & \cdots & b_{np} \end{bmatrix} = \begin{bmatrix} c_{11} & c_{12} & \cdots & c_{1j} & \cdots & c_{1p} \\ c_{21} & c_{22} & \cdots & c_{2j} & \cdots & c_{2p} \\ \vdots & \vdots & & & & \\ c_{i1} & c_{i2} & \cdots & c_{ij} & \cdots & c_{ip} \\ \vdots & \vdots & & & & \\ c_{m1} & c_{m2} & \cdots & c_{mj} & \cdots & c_{mp} \end{bmatrix}$$

$$a_{i1}b_{1j} + a_{i2}b_{2j} + a_{i3}b_{3j} + \cdots + a_{in}b_{nj} = c_{ij}$$

Properties of Matrix Multiplication

Let A, B, and C be matrices and let c be a scalar.

1. $A(BC) = (AB)C$ Associative Property of Multiplication

2. $A(B + C) = AB + AC$ Left Distributive Property

3. $(A + B)C = AC + BC$ Right Distributive Property

4. $c(AB) = (cA)B = A(cB)$

The $n \times n$ matrix that consists of 1's on its main diagonal and 0's elsewhere is called the **identity matrix of order n** and is denoted by

$$I_n = \begin{bmatrix} 1 & 0 & 0 & \cdots & 0 \\ 0 & 1 & 0 & \cdots & 0 \\ 0 & 0 & 1 & \cdots & 0 \\ \vdots & \vdots & \vdots & & \vdots \\ 0 & 0 & 0 & \cdots & 1 \end{bmatrix}. \qquad \text{Identity matrix}$$

Note that an identity matrix must be *square*. When the order is understood to be n, you can denote I_n simply by I. If A is an $n \times n$ matrix, the identity matrix has the property that $AI_n = A$ and $I_n A = A$. For example,

$$\begin{bmatrix} 3 & -2 & 5 \\ 1 & 0 & 4 \\ -1 & 2 & -3 \end{bmatrix} \begin{bmatrix} 1 & 0 & 0 \\ 0 & 1 & 0 \\ 0 & 0 & 1 \end{bmatrix} = \begin{bmatrix} 3 & -2 & 5 \\ 1 & 0 & 4 \\ -1 & 2 & -3 \end{bmatrix}$$

and

$$\begin{bmatrix} 1 & 0 & 0 \\ 0 & 1 & 0 \\ 0 & 0 & 1 \end{bmatrix} \begin{bmatrix} 3 & -2 & 5 \\ 1 & 0 & 4 \\ -1 & 2 & -3 \end{bmatrix} = \begin{bmatrix} 3 & -2 & 5 \\ 1 & 0 & 4 \\ -1 & 2 & -3 \end{bmatrix}.$$

Applications

One application of matrix multiplication is representation of a system of linear equations. Note how the system

$$a_{11}x_1 + a_{12}x_2 + a_{13}x_3 = b_1$$
$$a_{21}x_1 + a_{22}x_2 + a_{23}x_3 = b_2$$
$$a_{31}x_1 + a_{32}x_2 + a_{33}x_3 = b_3$$

can be written as the matrix equation $AX = B$, where A is the *coefficient matrix* of the system, and X and B are column matrices.

$$\underset{A}{\begin{bmatrix} a_{11} & a_{12} & a_{13} \\ a_{21} & a_{22} & a_{23} \\ a_{31} & a_{32} & a_{33} \end{bmatrix}} \quad \underset{\times \quad X}{\begin{bmatrix} x_1 \\ x_2 \\ x_3 \end{bmatrix}} \underset{=}{=} \underset{B}{\begin{bmatrix} b_1 \\ b_2 \\ b_3 \end{bmatrix}}$$

EXAMPLE 8   Solving a System of Linear Equations

Solve the matrix equation $AX = B$ for X, where

Coefficient matrix   Constant matrix

$$A = \begin{bmatrix} 1 & -2 & 1 \\ 0 & 1 & 2 \\ 2 & 3 & -2 \end{bmatrix} \quad \text{and} \quad B = \begin{bmatrix} -4 \\ 4 \\ 2 \end{bmatrix}.$$

Solution

As a system of linear equations, $AX = B$ is as follows.

$$x_1 - 2x_2 + x_3 = -4$$
$$x_2 + 2x_3 = 4$$
$$2x_1 + 3x_2 - 2x_3 = 2$$

Using Gauss-Jordan elimination on the augmented matrix of this system, you obtain the following reduced row-echelon matrix.

$$\begin{bmatrix} 1 & 0 & 0 & \vdots & -1 \\ 0 & 1 & 0 & \vdots & 2 \\ 0 & 0 & 1 & \vdots & 1 \end{bmatrix}$$

Thus, the solution of the system of linear equations is $x_1 = -1$, $x_2 = 2$, and $x_3 = 1$, and the solution of the matrix equation is

$$X = \begin{bmatrix} x_1 \\ x_2 \\ x_3 \end{bmatrix} = \begin{bmatrix} -1 \\ 2 \\ 1 \end{bmatrix}.$$

Check this solution in the original system of equations.

Real Life

EXAMPLE 9 *Softball Team Expenses*

In 1992, there were about 200,000 adult softball teams in the United States—almost four times the number of youth softball teams.

Two softball teams submit equipment lists to their sponsors.

	Women's Team	Men's Team
Bats	12	15
Balls	45	38
Gloves	15	17

Each bat costs $21, each ball costs $4, and each glove costs $30. Use matrices to find the total cost of equipment for each team.

Solution

The equipment lists and the costs per item can be written in matrix form as

$$E = \begin{bmatrix} 12 & 15 \\ 45 & 38 \\ 15 & 17 \end{bmatrix} \quad \text{and} \quad C = \begin{bmatrix} 21 & 4 & 30 \end{bmatrix}.$$

The total cost of equipment for each team is given by the product

$$CE = \begin{bmatrix} 21 & 4 & 30 \end{bmatrix} \begin{bmatrix} 12 & 15 \\ 45 & 38 \\ 15 & 17 \end{bmatrix} = \begin{bmatrix} 882 & 977 \end{bmatrix}.$$

Thus, the total cost of equipment for the women's team is $822, and the total cost of equipment for the men's team is $977.

Activities

1. Find $2A - 3B$.

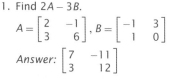

Answer: $\begin{bmatrix} 7 & -11 \\ 3 & 12 \end{bmatrix}$

2. Find AB, if possible.

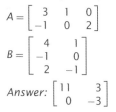

Answer: $\begin{bmatrix} 11 & 3 \\ 0 & -3 \end{bmatrix}$

Group Activities

Extending the Concept

Group Activity Suggestion
You may want to consider approaching this group activity as a class activity after discussing the section. You could start the activity with a brief discussion of commutativity in general.

Matrix Multiplication Discuss the requirements for matrix order $m \times n$ for multiplication of two matrices. Determine which of the following matrix multiplications AB is (are) defined. For each case in which AB is defined, what is the order of the resulting matrix?

a. A is of order 1×3
B is of order 2×1

b. A is of order 2×3
B is of order 2×3

c. A is of order 3×4
B is of order 4×2

d. A is of order 3×1
B is of order 3×3

Discuss why matrix multiplication is not, in general, commutative. Give an example of two 2×2 matrices such that $AB \neq BA$. Find an example of two 2×2 matrices such that $AB = BA$.

Warm Up
The following warm-up exercises involve skills that were covered in earlier sections. You will use these skills in the exercise set for this section.

In Exercises 1 and 2, evaluate the expression.

1. $-3\left(-\frac{5}{6}\right) + 10\left(-\frac{3}{4}\right)$

2. $-22\left(\frac{5}{2}\right) + 6(8)$

In Exercises 3 and 4, determine whether the matrix is in *reduced* row-echelon form.

3. $\begin{bmatrix} 0 & 1 & 0 & -5 \\ 1 & 0 & 3 & 2 \\ 0 & 0 & 1 & 0 \end{bmatrix}$

4. $\begin{bmatrix} 1 & 0 & 0 & 2 & 3 \\ 0 & 0 & 0 & 0 & 0 \\ 0 & 1 & 1 & 3 & 10 \end{bmatrix}$

In Exercises 5 and 6, write the augmented matrix for the system of linear equations.

5. $\begin{aligned} -5x + 10y &= 12 \\ 7x - 3y &= 0 \\ -x + 7y &= 25 \end{aligned}$

6. $\begin{aligned} 10x + 15y - 9z &= 42 \\ 6x - 5y &= 0 \end{aligned}$

In Exercises 7–10, solve the system of linear equations represented by the augmented matrix.

7. $\left[\begin{array}{cc:c} 1 & 0 & 0 \\ 0 & 1 & 2 \end{array}\right]$

8. $\left[\begin{array}{ccc:c} 1 & 0 & -1 & 2 \\ 0 & 1 & 1 & 3 \end{array}\right]$

9. $\left[\begin{array}{ccc:c} 1 & 2 & 1 & 0 \\ 0 & 0 & 1 & -1 \\ 0 & 0 & 0 & 0 \end{array}\right]$

10. $\left[\begin{array}{ccc:c} 1 & -1 & 0 & 3 \\ 0 & 1 & -2 & 1 \\ 0 & 0 & 1 & -1 \end{array}\right]$

7.2 Exercises

In Exercises 1–4, find x and y.

1. $\begin{bmatrix} x & -2 \\ 7 & y \end{bmatrix} = \begin{bmatrix} -4 & -2 \\ 7 & 22 \end{bmatrix}$

2. $\begin{bmatrix} -5 & x \\ y & 8 \end{bmatrix} = \begin{bmatrix} -5 & 13 \\ 12 & 8 \end{bmatrix}$

3. $\begin{bmatrix} 16 & 4 & 5 & 4 \\ -3 & 13 & 15 & 6 \\ 0 & 2 & 4 & 0 \end{bmatrix} = \begin{bmatrix} 16 & 4 & 2x+1 & 4 \\ -3 & 13 & 15 & 3x \\ 0 & 2 & 3y-5 & 0 \end{bmatrix}$

4. $\begin{bmatrix} x+2 & 8 & -3 \\ 1 & 2y & 2x \\ 7 & -2 & y+2 \end{bmatrix} = \begin{bmatrix} 2x+6 & 8 & -3 \\ 1 & 18 & -8 \\ 7 & -2 & 11 \end{bmatrix}$

In Exercises 5–10, find (a) $A + B$, (b) $A - B$, (c) $3A$, and (d) $3A - 2B$.

5. $A = \begin{bmatrix} 1 & -1 \\ 2 & -1 \end{bmatrix}$, $B = \begin{bmatrix} 2 & -1 \\ -1 & 8 \end{bmatrix}$

6. $A = \begin{bmatrix} 1 & 2 \\ 2 & 1 \end{bmatrix}$, $B = \begin{bmatrix} -3 & -2 \\ 4 & 2 \end{bmatrix}$

7. $A = \begin{bmatrix} 6 & -1 \\ 2 & 4 \\ -3 & 5 \end{bmatrix}$, $B = \begin{bmatrix} 1 & 4 \\ -1 & 5 \\ 1 & 10 \end{bmatrix}$

8. $A = \begin{bmatrix} 2 & 1 & 1 \\ -1 & -1 & 4 \end{bmatrix}$, $B = \begin{bmatrix} 2 & -3 & 4 \\ -3 & 1 & -2 \end{bmatrix}$

9. $A = \begin{bmatrix} 2 & 2 & -1 & 0 & 1 \\ 1 & 1 & -2 & 0 & -1 \end{bmatrix}$

$B = \begin{bmatrix} 1 & 1 & -1 & 1 & 0 \\ -3 & 4 & 9 & -6 & -7 \end{bmatrix}$

10. $A = \begin{bmatrix} 3 \\ 2 \\ -1 \end{bmatrix}$, $B = \begin{bmatrix} -4 \\ 6 \\ 2 \end{bmatrix}$

In Exercises 11–16, find (a) AB, (b) BA, and, if possible, (c) A^2. (*Note*: $A^2 = AA$.)

11. $A = \begin{bmatrix} 1 & 2 \\ 4 & 2 \end{bmatrix}$, $B = \begin{bmatrix} 2 & -1 \\ -1 & 8 \end{bmatrix}$

12. $A = \begin{bmatrix} 2 & -1 \\ 1 & 4 \end{bmatrix}$, $B = \begin{bmatrix} 0 & 0 \\ 3 & -3 \end{bmatrix}$

13. $A = \begin{bmatrix} 3 & -1 \\ 1 & 3 \end{bmatrix}$, $B = \begin{bmatrix} 1 & -3 \\ 3 & 1 \end{bmatrix}$

14. $A = \begin{bmatrix} 1 & -1 \\ 1 & 1 \end{bmatrix}$, $B = \begin{bmatrix} 1 & 3 \\ -3 & 1 \end{bmatrix}$

15. $A = \begin{bmatrix} 1 & -1 & 7 \\ 2 & -1 & 8 \\ 3 & 1 & -1 \end{bmatrix}$, $B = \begin{bmatrix} 1 & 1 & 2 \\ 2 & 1 & 1 \\ 1 & -3 & 2 \end{bmatrix}$

16. $A = \begin{bmatrix} 3 & 2 & 1 & 0 \end{bmatrix}$, $B = \begin{bmatrix} 2 \\ 3 \\ 1 \\ 0 \end{bmatrix}$

In Exercises 17–24, find AB, if possible.

17. $A = \begin{bmatrix} 2 & 1 \\ -3 & 4 \\ 1 & 6 \end{bmatrix}$, $B = \begin{bmatrix} 0 & -1 & 0 \\ 4 & 0 & 2 \\ 8 & -1 & 7 \end{bmatrix}$

18. $A = \begin{bmatrix} 0 & -1 & 0 \\ 4 & 0 & 2 \\ 8 & -1 & 7 \end{bmatrix}$, $B = \begin{bmatrix} 2 & 1 \\ -3 & 4 \\ 1 & 6 \end{bmatrix}$

19. $A = \begin{bmatrix} -1 & 3 \\ 4 & -5 \\ 0 & 2 \end{bmatrix}$, $B = \begin{bmatrix} 1 & 2 \\ 0 & 7 \end{bmatrix}$

20. $A = \begin{bmatrix} 1 & 0 & 0 \\ 0 & 4 & 0 \\ 0 & 0 & -2 \end{bmatrix}$, $B = \begin{bmatrix} 3 & 0 & 0 \\ 0 & -1 & 0 \\ 0 & 0 & 5 \end{bmatrix}$

21. $A = \begin{bmatrix} 5 & 0 & 0 \\ 0 & -8 & 0 \\ 0 & 0 & 7 \end{bmatrix}$, $B = \begin{bmatrix} \frac{1}{5} & 0 & 0 \\ 0 & -\frac{1}{8} & 0 \\ 0 & 0 & \frac{1}{2} \end{bmatrix}$

22. $A = \begin{bmatrix} 0 & 0 & 5 \\ 0 & 0 & -3 \\ 0 & 0 & 4 \end{bmatrix}$, $B = \begin{bmatrix} 6 & -11 & 4 \\ 8 & 16 & 4 \\ 0 & 0 & 0 \end{bmatrix}$

23. $A = \begin{bmatrix} 6 \\ -2 \\ 1 \\ 6 \end{bmatrix}$, $B = \begin{bmatrix} 10 & 12 \end{bmatrix}$

24. $A = \begin{bmatrix} 1 & 0 & 3 & -2 & 4 \\ 6 & 13 & 8 & -17 & 10 \end{bmatrix}$, $B = \begin{bmatrix} 1 & 6 \\ 4 & 2 \end{bmatrix}$

In Exercises 25–28, solve for X given

$$A = \begin{bmatrix} -2 & -1 \\ 1 & 0 \\ 3 & -4 \end{bmatrix} \quad \text{and} \quad B = \begin{bmatrix} 0 & 3 \\ 2 & 0 \\ -4 & -1 \end{bmatrix}.$$

25. $X = 3A - 2B$ **26.** $2X = 2A - B$

27. $2X + 3A = B$ **28.** $2A + 4B = -2X$

In Exercises 29–34, find matrices A, X, and B such that the system of linear equations can be written as the matrix equation $AX = B$. Solve the system of equations.

29. $\begin{aligned} -x + y &= 4 \\ -2x + y &= 0 \end{aligned}$ **30.** $\begin{aligned} 2x + 3y &= 5 \\ x + 4y &= 10 \end{aligned}$

31. $\begin{aligned} x + 2y &= 3 \\ 3x - y &= 2 \end{aligned}$ **32.** $\begin{aligned} 2x - 4y + z &= 0 \\ -x + 3y + z &= 1 \\ x + y &= 3 \end{aligned}$

33. $\begin{aligned} x - 2y + 3z &= 9 \\ -x + 3y - z &= -6 \\ 2x - 5y + 5z &= 17 \end{aligned}$

34. $\begin{aligned} x + y - 3z &= -1 \\ -x + 2y &= 1 \\ -y + z &= 0 \end{aligned}$

35. If a, b, and c are real numbers such that $c \neq 0$ and $ac = bc$, then $a = b$. However, if A, B, and C are matrices such that $AC = BC$, then A is *not* necessarily equal to B. Illustrate this using the following matrices.

$$A = \begin{bmatrix} 1 & 2 & 3 \\ 0 & 5 & 4 \\ 3 & -2 & 1 \end{bmatrix}, \quad B = \begin{bmatrix} 4 & -6 & 3 \\ 5 & 4 & 4 \\ -1 & 0 & 1 \end{bmatrix},$$

and $\quad C = \begin{bmatrix} 0 & 0 & 0 \\ 0 & 0 & 0 \\ 4 & -2 & 3 \end{bmatrix}$

36. If a and b are real numbers such that $ab = 0$, then $a = 0$ or $b = 0$. However, if A and B are matrices such that $AB = O$, then it is *not* necessarily true that $A = O$ or $B = O$. Illustrate this using the following matrices.

$$A = \begin{bmatrix} 3 & 3 \\ 4 & 4 \end{bmatrix} \quad \text{and} \quad B = \begin{bmatrix} 1 & -1 \\ -1 & 1 \end{bmatrix}$$

Find another example of two nonzero matrices whose product is the zero matrix.

37. *Factory Production* A certain corporation has four factories, each of which manufactures two products. The number of units of product i produced at factory j in one day is represented by a_{ij} in the matrix

$$A = \begin{bmatrix} 100 & 90 & 70 & 30 \\ 40 & 20 & 60 & 60 \end{bmatrix}.$$

Find the output if production is increased by 10%. (*Hint:* Because an increase of 10% corresponds to $100\% + 10\%$, multiply the matrix by 1.10.)

38. *Factory Production* A certain corporation has three factories, each of which manufactures two products. The number of units of product i produced at factory j in one day is represented by a_{ij} in the matrix

$$A = \begin{bmatrix} 60 & 40 & 20 \\ 30 & 90 & 60 \end{bmatrix}.$$

Find the output if production is increased by 20%. (*Hint:* Because an increase of 20% corresponds to $100\% + 20\%$, multiply the matrix by 1.20.)

39. *Crop Production* A fruit grower raises two crops, which are shipped to three outlets. The number of units of product i that are shipped to outlet j is represented by a_{ij} in the matrix

$$A = \begin{bmatrix} 100 & 75 & 75 \\ 125 & 150 & 100 \end{bmatrix}.$$

The profit per unit is represented by the matrix

$$B = \begin{bmatrix} \$3.75 & \$7.00 \end{bmatrix}.$$

(a) Find the product BA.

(b) Use the context of the problem to interpret what each entry of the product represents.

40. *Inventory Levels* A company sells five different models of computers through three retail outlets. The inventories of the three models at the three outlets are given by the matrix S.

	Model					
	A	B	C	D	E	
$S =$	3	2	2	3	0	1
	0	2	3	4	3	2 } Outlet
	4	2	1	3	2	3

The wholesale and retail prices for each model are given by matrix T.

	Price		
	Wholesale	Retail	
	\$840	\$1100	A
	\$1200	\$1350	B
$T =$	\$1450	\$1650	C } Model
	\$2650	\$3000	D
	\$3050	\$3200	E

(a) What is the total retail price of the inventory at Outlet 1?

(b) What is the total wholesale price of the inventory at Outlet 3?

(c) Compute the product ST and use the context of the problem to interpret the result.

41. *Total Revenue*   A manufacturer produces three different models of a given product, which are shipped to two different warehouses. The number of units of model i that are shipped to warehouse j is represented by a_{ij} in the matrix

$$A = \begin{bmatrix} 5000 & 4000 \\ 6000 & 10{,}000 \\ 8000 & 5000 \end{bmatrix}.$$

The price per unit is represented by the matrix

$$B = \begin{bmatrix} \$20.50 & \$26.50 & \$29.50 \end{bmatrix}.$$

Find the product BA, and state what each entry of the product represents.

42. *Labor/Wage Requirements*   A company that manufactures boats has the following labor-hour and wage requirements.

Labor-Hour Requirements (per boat)

Department

	Cutting	Assembly	Packaging	
$S =$	1.0 hour	0.5 hour	0.2 hour	Small
	1.6 hours	1.0 hour	0.2 hour	Medium
	2.5 hours	2.0 hours	0.4 hour	Large

Boat size

Wage Requirements (per hour)

Plant

	A	B	
$T =$	\$12	\$10	Cutting
	\$9	\$8	Assembly
	\$6	\$5	Packaging

Department

(a) What is the labor cost for a medium-sized boat at Plant B?

(b) What is the labor cost for a large-sized boat at Plant A?

(c) Compute ST and interpret the result.

Think About It   In Exercises 43 and 44, find a matrix B such that AB is the identity matrix. Is there more than one correct result?

43. $A = \begin{bmatrix} 1 & 3 \\ 1 & 2 \end{bmatrix}$ **44.** $A = \begin{bmatrix} 2 & 1 \\ 5 & 2 \end{bmatrix}$

45. *Voting Preference*   The matrix

From

$$P = \begin{bmatrix} 0.6 & 0.1 & 0.1 \\ 0.2 & 0.7 & 0.1 \\ 0.2 & 0.2 & 0.8 \end{bmatrix} \begin{matrix} R \\ D \\ I \end{matrix} \Bigg\} \text{To}$$

is called a stochastic matrix. Each entry p_{ij} ($i \neq j$) represents the proportion of the voting population that changes from party i to party j, and p_{ii} represents the proportion that remains loyal to the party from one election to the next. Use a graphing utility to find P^2. (This matrix gives the transition probabilities from the first election to the third.)

46. *Voting Preference*   Use a graphing utility to find P^3, P^4, P^5, P^6, P^7, and P^8 for the matrix given in Exercise 45. Can you detect a pattern as P is raised to higher and higher powers?

47. *Contract Bonuses*   Professional athletes frequently have bonus or incentive clauses in their contracts. For example, a defensive football player might receive a bonus for a sack, interception, and/or key tackle. Suppose in a contract a sack is worth \$1000, an interception is worth \$750, and a key tackle is worth \$500. Use matrices to calculate the bonuses for defensive players A, B, and C if the following matrix describes the numbers of sacks, interceptions, and key tackles in a game.

Player	Sacks	Interceptions	Key Tackles
Player A	2	0	4
Player B	0	1	5
Player C	1	3	2

7.3	**The Inverse of a Square Matrix**

The Inverse of a Matrix ▪ Finding Inverse Matrices ▪ The Inverse of a 2 × 2 Matrix (Quick Method) ▪ Systems of Linear Equations

The Inverse of a Matrix

This section further develops the algebra of matrices. To begin, consider the real number equation $ax = b$. To solve this equation for x, multiply both sides of the equation by a^{-1} (provided $a \neq 0$).

$$ax = b$$
$$(a^{-1}a)x = a^{-1}b$$
$$(1)x = a^{-1}b$$
$$x = a^{-1}b$$

The number a^{-1} is called the *multiplicative inverse of a* because it has the property that $a^{-1}a = 1$. The definition of a multiplicative inverse of a matrix is similar.

NOTE The symbol A^{-1} is read "A inverse."

Definition of an Inverse of a Square Matrix

Let A be an $n \times n$ matrix. If there exists a matrix A^{-1} such that

$$AA^{-1} = I_n = A^{-1}A$$

A^{-1} is called the **inverse** of A.

NOTE Recall that it is not always true that $AB = BA$, even if both products are defined. However, if A and B are both square matrices and $AB = I_n$, it can be shown that $BA = I_n$. Hence, in Example 1, you need only to check that $AB = I_2$.

EXAMPLE 1 The Inverse of a Matrix

Show that B is the inverse of A, where

$$A = \begin{bmatrix} -1 & 2 \\ -1 & 1 \end{bmatrix} \quad \text{and} \quad B = \begin{bmatrix} 1 & -2 \\ 1 & -1 \end{bmatrix}.$$

Solution

To show that B is the inverse of A, show that $AB = I = BA$, as follows.

$$AB = \begin{bmatrix} -1 & 2 \\ -1 & 1 \end{bmatrix} \begin{bmatrix} 1 & -2 \\ 1 & -1 \end{bmatrix} = \begin{bmatrix} -1+2 & 2-2 \\ -1+1 & 2-1 \end{bmatrix} = \begin{bmatrix} 1 & 0 \\ 0 & 1 \end{bmatrix}$$

$$BA = \begin{bmatrix} 1 & -2 \\ 1 & -1 \end{bmatrix} \begin{bmatrix} -1 & 2 \\ -1 & 1 \end{bmatrix} = \begin{bmatrix} -1+2 & 2-2 \\ -1+1 & 2-1 \end{bmatrix} = \begin{bmatrix} 1 & 0 \\ 0 & 1 \end{bmatrix}$$

If a matrix A has an inverse, A is called **invertible** (or **nonsingular**); otherwise, A is called **singular.** A nonsquare matrix cannot have an inverse. To see this, note that if A is of order $m \times n$ and B is of order $n \times m$ (where $m \neq n$), the products AB and BA are of different orders and therefore cannot be equal to each other. Not all square matrices possess inverses (see the matrix at the bottom of page 531). If, however, a matrix does possess an inverse, that inverse is unique. The following example shows how to use a system of equations to find the inverse.

EXAMPLE 2   *Finding the Inverse of a Matrix*

Find the inverse of the matrix

$$A = \begin{bmatrix} 1 & 4 \\ -1 & -3 \end{bmatrix}.$$

Solution

To find the inverse of A, try to solve the matrix equation $AX = I$ for X.

$$\overset{A}{\begin{bmatrix} 1 & 4 \\ -1 & -3 \end{bmatrix}} \overset{X}{\begin{bmatrix} x_{11} & x_{12} \\ x_{21} & x_{22} \end{bmatrix}} = \overset{I}{\begin{bmatrix} 1 & 0 \\ 0 & 1 \end{bmatrix}}$$

$$\begin{bmatrix} x_{11} + 4x_{21} & x_{12} + 4x_{22} \\ -x_{11} - 3x_{21} & -x_{12} - 3x_{22} \end{bmatrix} = \begin{bmatrix} 1 & 0 \\ 0 & 1 \end{bmatrix}$$

Equating corresponding entries, you obtain the following two systems of linear equations.

$$\begin{aligned} x_{11} + 4x_{21} &= 1 & x_{12} + 4x_{22} &= 0 \\ -x_{11} - 3x_{21} &= 0 & -x_{12} - 3x_{22} &= 1 \end{aligned}$$

From the first system you can determine that $x_{11} = -3$ and $x_{21} = 1$, and from the second system you can determine that $x_{12} = -4$ and $x_{22} = 1$. Therefore, the inverse of A is

$$X = A^{-1} = \begin{bmatrix} -3 & -4 \\ 1 & 1 \end{bmatrix}.$$

You can use matrix multiplication to check this result.

Check

$$AA^{-1} = \begin{bmatrix} 1 & 4 \\ -1 & -3 \end{bmatrix} \begin{bmatrix} -3 & -4 \\ 1 & 1 \end{bmatrix} = \begin{bmatrix} 1 & 0 \\ 0 & 1 \end{bmatrix} \quad ✓$$

$$A^{-1}A = \begin{bmatrix} -3 & -4 \\ 1 & 1 \end{bmatrix} \begin{bmatrix} 1 & 4 \\ -1 & -3 \end{bmatrix} = \begin{bmatrix} 1 & 0 \\ 0 & 1 \end{bmatrix} \quad ✓$$

Finding Inverse Matrices

In Example 2, note that the two systems of linear equations have the *same coefficient matrix A*. Rather than solve the two systems represented by

$$\begin{bmatrix} 1 & 4 & \vdots & 1 \\ -1 & -3 & \vdots & 0 \end{bmatrix} \quad \text{and} \quad \begin{bmatrix} 1 & 4 & \vdots & 0 \\ -1 & -3 & \vdots & 1 \end{bmatrix}$$

separately, you can solve them *simultaneously* by **adjoining** the identity matrix to the coefficient matrix to obtain

$$\begin{matrix} A & & I \\ \begin{bmatrix} 1 & 4 & \vdots & 1 & 0 \\ -1 & -3 & \vdots & 0 & 1 \end{bmatrix} \end{matrix}.$$

Then, applying Gauss-Jordan elimination to this matrix, you can solve *both* systems with a single elimination process, as follows.

$$\begin{bmatrix} 1 & 4 & \vdots & 1 & 0 \\ -1 & -3 & \vdots & 0 & 1 \end{bmatrix}$$

$$R_1 + R_2 \rightarrow \begin{bmatrix} 1 & 4 & \vdots & 1 & 0 \\ 0 & 1 & \vdots & 1 & 1 \end{bmatrix}$$

$$-4R_2 + R_1 \rightarrow \begin{bmatrix} 1 & 0 & \vdots & -3 & -4 \\ 0 & 1 & \vdots & 1 & 1 \end{bmatrix}$$

Thus, from the "doubly augmented" matrix $[A \,:\, I]$, you obtained the matrix $[I \,:\, A^{-1}]$.

$$\begin{matrix} A & & I \\ \begin{bmatrix} 1 & 4 & \vdots & 1 & 0 \\ -1 & -3 & \vdots & 0 & 1 \end{bmatrix} \end{matrix} \quad \Longrightarrow \quad \begin{matrix} I & & A^{-1} \\ \begin{bmatrix} 1 & 0 & \vdots & -3 & -4 \\ 0 & 1 & \vdots & 1 & 1 \end{bmatrix} \end{matrix}$$

This procedure (or algorithm) works for an arbitrary square matrix that has an inverse.

Finding an Inverse Matrix

Let A be a square matrix of order n.

1. Write the $n \times 2n$ matrix that consists of the given matrix A on the left and the $n \times n$ identity matrix I on the right to obtain $[A \,:\, I]$. Note that we separate the matrices A and I by a dotted line. We call this process **adjoining** the matrices A and I.

2. If possible, row reduce A to I using elementary row operations on the *entire* matrix $[A \,:\, I]$. The result will be the matrix $[I \,:\, A^{-1}]$. If this is not possible, A is not invertible.

3. Check your work by multiplying to see that $AA^{-1} = I = A^{-1}A$.

EXAMPLE 3   *Finding the Inverse of a Matrix*

Find the inverse of the matrix

$$A = \begin{bmatrix} 1 & -1 & 0 \\ 1 & 0 & -1 \\ 6 & -2 & -3 \end{bmatrix}.$$

Solution

Begin by adjoining the identity matrix to A to form the matrix

$$\begin{bmatrix} A & \vdots & I \end{bmatrix} = \begin{bmatrix} 1 & -1 & 0 & \vdots & 1 & 0 & 0 \\ 1 & 0 & -1 & \vdots & 0 & 1 & 0 \\ 6 & -2 & -3 & \vdots & 0 & 0 & 1 \end{bmatrix}.$$

Use elementary row operations to obtain the form $[I \ \vdots \ A^{-1}]$, as follows.

$$\begin{bmatrix} 1 & -1 & 0 & \vdots & 1 & 0 & 0 \\ 1 & 0 & -1 & \vdots & 0 & 1 & 0 \\ 6 & -2 & -3 & \vdots & 0 & 0 & 1 \end{bmatrix}$$

$$\begin{matrix} -R_1 + R_2 \rightarrow \\ -6R_1 + R_3 \rightarrow \end{matrix} \begin{bmatrix} 1 & -1 & 0 & \vdots & 1 & 0 & 0 \\ 0 & 1 & -1 & \vdots & -1 & 1 & 0 \\ 0 & 4 & -3 & \vdots & -6 & 0 & 1 \end{bmatrix}$$

$$\begin{matrix} R_2 + R_1 \rightarrow \\ \\ -4R_2 + R_3 \rightarrow \end{matrix} \begin{bmatrix} 1 & 0 & -1 & \vdots & 0 & 1 & 0 \\ 0 & 1 & -1 & \vdots & -1 & 1 & 0 \\ 0 & 0 & 1 & \vdots & -2 & -4 & 1 \end{bmatrix}$$

$$\begin{matrix} R_3 + R_1 \rightarrow \\ R_3 + R_2 \rightarrow \end{matrix} \begin{bmatrix} 1 & 0 & 0 & \vdots & -2 & -3 & 1 \\ 0 & 1 & 0 & \vdots & -3 & -3 & 1 \\ 0 & 0 & 1 & \vdots & -2 & -4 & 1 \end{bmatrix}$$

Therefore, the matrix A is invertible and its inverse is

$$A^{-1} = \begin{bmatrix} -2 & -3 & 1 \\ -3 & -3 & 1 \\ -2 & -4 & 1 \end{bmatrix}.$$

Try confirming this result by multiplying A and A^{-1} to obtain I.

You may want to show what happens when A^{-1} or B^{-1} does not exist. For example,

$$A = \begin{bmatrix} 3 & 6 \\ 1 & 2 \end{bmatrix} \text{ or }$$

$$B = \begin{bmatrix} 3 & 2 & 1 \\ 1 & 0 & -1 \\ 0 & 1 & 2 \end{bmatrix}.$$

The process shown in Example 3 applies to any $n \times n$ matrix A. If A has an inverse, this process will find it. On the other hand, if A does not have an inverse, the process will tell us so. For instance, the following matrix has no inverse.

$$A = \begin{bmatrix} 1 & 2 & 0 \\ 3 & -1 & 2 \\ -2 & 3 & -2 \end{bmatrix}$$

Explain how the elimination process shows that this matrix is singular.

The Inverse of a 2 x 2 Matrix (Quick Method)

Using Gauss-Jordan elimination to find the inverse of a matrix works well (even as a computer technique) for matrices of order 3×3 or greater. For 2×2 matrices, however, many people prefer to use a formula for the inverse rather than Gauss-Jordan elimination. This simple formula, which works *only* for 2×2 matrices, is explained as follows. If A is a 2×2 matrix given by

$$A = \begin{bmatrix} a & b \\ c & d \end{bmatrix}$$

then A is invertible if and only if $ad - bc \neq 0$. Moreover, if $ad - bc \neq 0$, the inverse is given by

$$A^{-1} = \frac{1}{ad - bc} \begin{bmatrix} d & -b \\ -c & a \end{bmatrix}.$$

Try verifying this inverse by multiplication.

NOTE   The denominator $ad - bc$ is called the **determinant** of the 2×2 matrix A. You will study determinants in the next section.

EXAMPLE 4   Finding the Inverse of a 2 x 2 Matrix

If possible, find the inverses of the following matrices.

a. $A = \begin{bmatrix} 3 & -1 \\ -2 & 2 \end{bmatrix}$ **b.** $B = \begin{bmatrix} 3 & -1 \\ -6 & 2 \end{bmatrix}$

Solution

a. For the matrix A, apply the formula for the inverse of a 2×2 matrix to obtain

$$ad - bc = (3)(2) - (-1)(-2) = 4.$$

Because this quantity is not zero, the inverse is formed by interchanging the entries on the main diagonal and changing the sign of the other two entries, as follows.

$$A^{-1} = \frac{1}{4} \begin{bmatrix} 2 & 1 \\ 2 & 3 \end{bmatrix} = \begin{bmatrix} \frac{1}{2} & \frac{1}{4} \\ \frac{1}{2} & \frac{3}{4} \end{bmatrix}$$

b. For the matrix B, you have

$$ad - bc = (3)(2) - (-1)(-6) = 0$$

which means that B is not invertible.

Systems of Linear Equations

We know that a system of linear equations can have exactly one solution, infinitely many solutions, or no solution. If the coefficient matrix A of a *square* system (a system that has the same number of equations as variables) is invertible, the system has a unique solution, which is given as follows.

A System of Equations with a Unique Solution

If A is an invertible matrix, the system of linear equations represented by $AX = B$ has a unique solution given by

$$X = A^{-1}B.$$

EXAMPLE 5   Solving a System of Equations Using an Inverse

Use an inverse matrix to solve the following system.

$$2x + 3y + z = -1$$
$$3x + 3y + z = 1$$
$$2x + 4y + z = -2$$

Solution

$$X = A^{-1}B = \begin{bmatrix} -1 & 1 & 0 \\ -1 & 0 & 1 \\ 6 & -2 & -3 \end{bmatrix} \begin{bmatrix} -1 \\ 1 \\ -2 \end{bmatrix} = \begin{bmatrix} 2 \\ -1 \\ -2 \end{bmatrix}$$

Thus, the solution is $x = 2$, $y = -1$, and $z = -2$.

Group Activities   Exploring with Technology

Finding an Inverse Matrix   Explain how to use a graphing utility to find the inverse matrix for each of the following.

a. $A = \begin{bmatrix} -3 & 2 \\ 7 & 4 \end{bmatrix}$ **b.** $B = \begin{bmatrix} 1 & -4 & 2 \\ 2 & -9 & 5 \\ 1 & -5 & 4 \end{bmatrix}$ **c.** $C = \begin{bmatrix} 3 & 1 & 0 \\ 1 & 1 & 1 \\ 1 & -1 & 2 \end{bmatrix}$

Warm Up The following warm-up exercises involve skills that were covered in earlier sections. You will use these skills in the exercise set for this section.

In Exercises 1–8, perform the indicated matrix operations.

1. $4\begin{bmatrix} 1 & 6 \\ 0 & -4 \\ 12 & 2 \end{bmatrix}$

2. $\dfrac{1}{2}\begin{bmatrix} 11 & 10 & 48 \\ 1 & 0 & 16 \\ 0 & 2 & 8 \end{bmatrix}$

3. $\begin{bmatrix} 1 & -10 & 3 \\ 4 & 1 & 0 \end{bmatrix} - 2\begin{bmatrix} 3 & -4 & 8 \\ 0 & 7 & 1 \end{bmatrix}$

4. $\begin{bmatrix} 5 & 20 \\ -7 & 15 \end{bmatrix} - 3\begin{bmatrix} 6 & 3 \\ 4 & -2 \end{bmatrix}$

5. $\begin{bmatrix} 1 & -2 \\ -1 & 3 \end{bmatrix}\begin{bmatrix} 3 & 2 \\ 1 & 1 \end{bmatrix}$

6. $\begin{bmatrix} 1 & 0 \\ 0 & 1 \end{bmatrix}\begin{bmatrix} 6 & 5 \\ 3 & -2 \end{bmatrix}$

7. $\begin{bmatrix} 2 & 0 & 0 \\ 0 & -1 & 0 \\ 0 & 0 & 3 \end{bmatrix}\begin{bmatrix} \frac{1}{2} & 0 & 0 \\ 0 & -1 & 0 \\ 0 & 0 & \frac{1}{3} \end{bmatrix}$

8. $\begin{bmatrix} 1 & -1 & 0 \\ 1 & 0 & -1 \\ 6 & -2 & -3 \end{bmatrix}\begin{bmatrix} -2 & -3 & 1 \\ -3 & -3 & 1 \\ -2 & -4 & 1 \end{bmatrix}$

In Exercises 9 and 10, rewrite the matrix in reduced row-echelon form.

9. $\begin{bmatrix} 3 & -2 & 1 & 0 \\ 4 & -3 & 0 & 1 \end{bmatrix}$

10. $\begin{bmatrix} 1 & 1 & 2 & 1 & 0 & 0 \\ -1 & 0 & 3 & 0 & 1 & 0 \\ 1 & 2 & 8 & 0 & 0 & 1 \end{bmatrix}$

7.3 Exercises

In Exercises 1–10, show that B is the inverse of A.

1. $A = \begin{bmatrix} 2 & 1 \\ 5 & 3 \end{bmatrix}$, $B = \begin{bmatrix} 3 & -1 \\ -5 & 2 \end{bmatrix}$

2. $A = \begin{bmatrix} 1 & -1 \\ -1 & 2 \end{bmatrix}$, $B = \begin{bmatrix} 2 & 1 \\ 1 & 1 \end{bmatrix}$

3. $A = \begin{bmatrix} 1 & 2 \\ 3 & 4 \end{bmatrix}$, $B = \begin{bmatrix} -2 & 1 \\ \frac{3}{2} & -\frac{1}{2} \end{bmatrix}$

4. $A = \begin{bmatrix} 1 & -1 \\ 2 & 3 \end{bmatrix}$, $B = \begin{bmatrix} \frac{3}{5} & \frac{1}{5} \\ -\frac{2}{5} & \frac{1}{5} \end{bmatrix}$

5. $A = \begin{bmatrix} -2 & 2 & 3 \\ 1 & -1 & 0 \\ 0 & 1 & 4 \end{bmatrix}$, $B = \dfrac{1}{3}\begin{bmatrix} -4 & -5 & 3 \\ -4 & -8 & 3 \\ 1 & 2 & 0 \end{bmatrix}$

6. $A = \begin{bmatrix} 2 & -17 & 11 \\ -1 & 11 & -7 \\ 0 & 3 & -2 \end{bmatrix}$, $B = \begin{bmatrix} 1 & 1 & 2 \\ 2 & 4 & -3 \\ 3 & 6 & -5 \end{bmatrix}$

7. $A = \begin{bmatrix} -1 & 0 & 2 \\ 1 & -2 & 0 \\ 1 & 0 & 3 \end{bmatrix}$, $B = \dfrac{1}{10}\begin{bmatrix} -6 & 0 & 4 \\ -3 & -5 & 2 \\ 2 & 0 & 2 \end{bmatrix}$

8. $A = \begin{bmatrix} -1 & 1 & -3 \\ 2 & -1 & 4 \\ -1 & 1 & -2 \end{bmatrix}$, $B = \begin{bmatrix} 2 & 1 & -1 \\ 0 & 1 & 2 \\ -1 & 0 & 1 \end{bmatrix}$

9. $A = \begin{bmatrix} 2 & 0 & 1 & 1 \\ 3 & 0 & 0 & 1 \\ -1 & 1 & -2 & 1 \\ 4 & -1 & 1 & 0 \end{bmatrix}$, $B = \begin{bmatrix} -1 & 2 & -1 & -1 \\ -4 & 9 & -5 & -6 \\ 0 & 1 & -1 & -1 \\ 3 & -5 & 3 & 3 \end{bmatrix}$

10. $A = \begin{bmatrix} -1 & 1 & 0 & -1 \\ 1 & -1 & 2 & 0 \\ -1 & 1 & 2 & 0 \\ 0 & -1 & 1 & 1 \end{bmatrix}$

$B = \dfrac{1}{4} \begin{bmatrix} -4 & 1 & 1 & -4 \\ -4 & -1 & 3 & -4 \\ 0 & 1 & 1 & 0 \\ -4 & -2 & 2 & 0 \end{bmatrix}$

In Exercises 11–36, find the inverse of the matrix (if it exists).

11. $\begin{bmatrix} 2 & 0 \\ 0 & 3 \end{bmatrix}$ 　　　**12.** $\begin{bmatrix} 1 & 2 \\ 3 & 7 \end{bmatrix}$

13. $\begin{bmatrix} 1 & -2 \\ 2 & -3 \end{bmatrix}$ 　　**14.** $\begin{bmatrix} -7 & 33 \\ 4 & -19 \end{bmatrix}$

15. $\begin{bmatrix} -1 & 1 \\ -2 & 1 \end{bmatrix}$ 　　**16.** $\begin{bmatrix} 11 & 1 \\ -1 & 0 \end{bmatrix}$

17. $\begin{bmatrix} 2 & 4 \\ 4 & 8 \end{bmatrix}$ 　　　**18.** $\begin{bmatrix} 2 & 3 \\ 1 & 4 \end{bmatrix}$

19. $\begin{bmatrix} 2 & 7 & 1 \\ -3 & -9 & 2 \end{bmatrix}$ 　　**20.** $\begin{bmatrix} -2 & 5 \\ 6 & -15 \\ 0 & 1 \end{bmatrix}$

21. $\begin{bmatrix} 1 & 1 & 1 \\ 3 & 5 & 4 \\ 3 & 6 & 5 \end{bmatrix}$ 　　**22.** $\begin{bmatrix} 1 & 2 & 2 \\ 3 & 7 & 9 \\ -1 & -4 & -7 \end{bmatrix}$

23. $\begin{bmatrix} 1 & 2 & -1 \\ 3 & 7 & -10 \\ -5 & -7 & -15 \end{bmatrix}$ 　**24.** $\begin{bmatrix} 10 & 5 & -7 \\ -5 & 1 & 4 \\ 3 & 2 & -2 \end{bmatrix}$

25. $\begin{bmatrix} 1 & 1 & 2 \\ 3 & 1 & 0 \\ -2 & 0 & 3 \end{bmatrix}$ 　**26.** $\begin{bmatrix} 3 & 2 & 2 \\ 2 & 2 & 2 \\ -4 & 4 & 3 \end{bmatrix}$

27. $\begin{bmatrix} 0.1 & 0.2 & 0.3 \\ -0.3 & 0.2 & 0.2 \\ 0.5 & 0.4 & 0.4 \end{bmatrix}$ 　**28.** $\begin{bmatrix} 2 & 0 & 0 \\ 0 & 3 & 0 \\ 0 & 0 & 5 \end{bmatrix}$

29. $\begin{bmatrix} 1 & 0 & 0 \\ 3 & 4 & 0 \\ 2 & 5 & 5 \end{bmatrix}$ 　**30.** $\begin{bmatrix} 1 & 0 & 0 \\ 3 & 0 & 0 \\ 2 & 5 & 5 \end{bmatrix}$

31. $\begin{bmatrix} 1 & 0 & 3 & 0 \\ 0 & 2 & 0 & 4 \\ 1 & 0 & 3 & 0 \\ 0 & 2 & 0 & 4 \end{bmatrix}$ 　**32.** $\begin{bmatrix} 1 & 3 & -2 & 0 \\ 0 & 2 & 4 & 6 \\ 0 & 0 & -2 & 1 \\ 0 & 0 & 0 & 5 \end{bmatrix}$

33. $\begin{bmatrix} -8 & 0 & 0 & 0 \\ 0 & 1 & 0 & 0 \\ 0 & 0 & 4 & 0 \\ 0 & 0 & 0 & -5 \end{bmatrix}$

34. $\begin{bmatrix} -1 & 0 & 1 & 0 \\ 0 & 2 & 0 & -1 \\ 2 & 0 & -1 & 0 \\ 0 & -1 & 0 & 1 \end{bmatrix}$

35. $\begin{bmatrix} 1 & -2 & -1 & -2 \\ 3 & -5 & -2 & -3 \\ 2 & -5 & -2 & -5 \\ -1 & 4 & 4 & 11 \end{bmatrix}$

36. $\begin{bmatrix} 4 & 8 & -7 & 14 \\ 2 & 5 & -4 & 6 \\ 0 & 2 & 1 & -7 \\ 3 & 6 & -5 & 10 \end{bmatrix}$

In Exercises 37–40, use an inverse matrix to solve the system of linear equations. (Use the inverse matrix found in Exercise 15.)

37. $-x + y = 4$
$\quad -2x + y = 0$

38. $-x + y = -3$
$\quad -2x + y = 5$

39. $-x + y = 20$
$\quad -2x + y = 10$

40. $-x + y = 0$
$\quad -2x + y = 7$

In Exercises 41–44, use an inverse matrix to solve the system of linear equations. (Use the inverse matrix found in Exercise 18.)

41. $2x + 3y = 5$
$\quad x + 4y = 10$

42. $2x + 3y = 0$
$\quad x + 4y = 3$

43. $2x + 3y = 4$
$\quad x + 4y = 2$

44. $2x + 3y = 1$
$\quad x + 4y = -2$

In Exercises 45 and 46, use an inverse matrix to solve the system of linear equations. (Use the inverse matrix found in Exercise 26.)

45. $3x + 2y + 2z = 0$
$\quad 2x + 2y + 2z = 5$
$\quad -4x + 4y + 3z = 2$

46. $3x + 2y + 2z = -1$
$\quad 2x + 2y + 2z = 2$
$\quad -4x + 4y + 3z = 0$

In Exercises 47 and 48, use an inverse matrix to solve the system of linear equations. (Use the inverse matrix found in Exercise 35.)

47.
$$\begin{aligned} x_1 - 2x_2 - x_3 - 2x_4 &= 0 \\ 3x_1 - 5x_2 - 2x_3 - 3x_4 &= 1 \\ 2x_1 - 5x_2 - 2x_3 - 5x_4 &= -1 \\ -x_1 + 4x_2 + 4x_3 + 11x_4 &= 2 \end{aligned}$$

48.
$$\begin{aligned} x_1 - 2x_2 - x_3 - 2x_4 &= 1 \\ 3x_1 - 5x_2 - 2x_3 - 3x_4 &= -2 \\ 2x_1 - 5x_2 - 2x_3 - 5x_4 &= 0 \\ -x_1 + 4x_2 + 4x_3 + 11x_4 &= -3 \end{aligned}$$

49. *Starting Salaries for Engineers*   The average starting salaries for new 4-year graduates in engineering for the years 1980 to 1993 are shown in the figure. The least squares regression line $y = a + bt$ for this data is found by solving the system

$$14a + 91b = 402.2$$
$$91a + 819b = 2853.7$$

where y is the average salary (in thousands of dollars) and $t = 0$ represents 1980.   (Source: Northwestern University Placement Center)

(a) Use a graphing utility to find an inverse matrix to solve this system and find the equation of the least squares regression line.

(b) Use the result of part (a) to approximate the average salary in 1996.

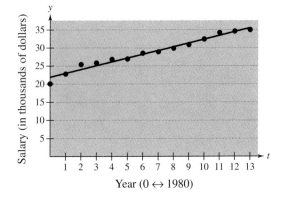

Year (0 ↔ 1980)

50. *Service Industry in U.S.*   The total numbers of people employed in the "service industry" (excluding self-employed people) from 1988 to 1992 are shown in the figure. The least squares regression parabola $y = at^2 + bt + c$ for this data is found by solving the system

$$\begin{aligned} 4c + 6b + 4a &= 110.8 \\ 6c + 14b + 36a &= 173.7 \\ 14c + 36b + 98a &= 408.9 \end{aligned}$$

where y is the number of workers (in millions) and $t = 0$ represents 1988.   (Source: U.S. Bureau of Labor Statistics)

(a) Use a graphing utility to find an inverse matrix to solve this system and write the equation of the least squares regression line.

(b) Use the result of part (a) to approximate the number of workers in the service industry in 1992.

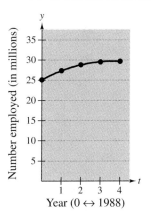

Year (0 ↔ 1988)

In Exercises 51 and 52, develop for the given matrix a system of equations that will have the given solution. Use an inverse matrix to verify that the system of equations gives the desired solution.

51. $\begin{bmatrix} 2 & 1 & 3 \\ 4 & 0 & -2 \\ 0 & 3 & 2 \end{bmatrix}$ $\begin{aligned} x &= 2 \\ y &= -3 \\ z &= 5 \end{aligned}$

52. $\begin{bmatrix} 1 & 0 & 2 \\ 1 & 1 & 1 \\ 2 & -1 & 0 \end{bmatrix}$ $\begin{aligned} x &= 5 \\ y &= -2 \\ z &= 1 \end{aligned}$

Bond Investment In Exercises 53–56, you are investing in AAA-rated bonds, A-rated bonds, and B-rated
bonds. Your average yield is 8% on AAA bonds, 6%
on A bonds, and 7% on B bonds. Twice as much is invested in B bonds as in A bonds. Moreover, the total
annual return for all three types of bonds is $2800.
The desired system of linear equations (where x, y,
and z represent the amounts invested in AAA, A, and
B bonds) is as follows.

$$\begin{aligned}
x + \quad y + \quad z &= \text{(total investment)} \\
0.08x + 0.06y + 0.07z &= 2800 \\
2y - \quad z &= \quad 0
\end{aligned}$$

Use the inverse of the coefficient matrix of this system to find the amount invested in each type of
bond, with the given total investment.

53. Total investment = $37,500

54. Total investment = $40,000

55. Total investment = $38,000

56. Total investment = $42,000

Acquisition of Raw Materials In Exercises 57–60, consider a company that produces computer chips, resistors, and transistors. Each computer chip requires
2 units of copper, 2 units of zinc, and 1 unit of glass.
Each resistor requires 1 unit of copper, 3 units of
zinc, and 2 units of glass. Each transistor requires 3
units of copper, 2 units of zinc, and 2 units of glass.
The desired system of linear equations (where x, y,
and z represent the numbers of computer chips, resistors, and transistors) is as follows.

$$\begin{aligned}
2x + \quad y + 3z &= \text{(units of copper)} \\
2x + 3y + 2z &= \text{(units of zinc)} \\
x + 2y + 2z &= \text{(units of glass)}
\end{aligned}$$

Use the inverse of the coefficient matrix of this system to find the numbers of computer chips, resistors,
and transistors that the company can produce with
the given amounts of raw materials.

57. 70 units of copper
80 units of zinc
55 units of glass

58. 85 units of copper
105 units of zinc
70 units of glass

59. 200 units of copper
250 units of zinc
150 units of glass

60. 300 units of copper
400 units of zinc
250 units of glass

Raw Materials In Exercises 61 and 62, consider a
company that specializes in gourmet chocolate
baked goods—chocolate muffins, cookies, and
brownies. Each chocolate muffin requires 2 units of
chocolate, 3 units of flour, and 2 units of sugar. Each
chocolate cookie requires 1 unit of chocolate, 1 unit
of flour, and 1 unit of sugar. Each chocolate brownie
requires 2 units of chocolate, 1 unit of flour, and 1.5
units of sugar. Find the numbers of muffins, cookies,
and brownies that the company can produce with the
given amounts of raw materials.

61. 550 units of chocolate
525 units of flour
500 units of sugar

62. 800 units of chocolate
750 units of flour
700 units of sugar

MID-CHAPTER QUIZ

Take this quiz as you would take a quiz in class. After you are done, check your work against the answers given in the back of the book.

In Exercises 1 and 2, write a matrix of the given order.

1. 3×4 **2.** 1×3

In Exercises 3 and 4, write the system of linear equations represented by the augmented matrix. (Use variables x, y, and z.)

3. $\begin{bmatrix} 3 & 2 & \vdots & -2 \\ 5 & -1 & \vdots & 19 \end{bmatrix}$ **4.** $\begin{bmatrix} 1 & 0 & 3 & \vdots & -5 \\ 1 & 2 & -1 & \vdots & 3 \\ 3 & 0 & 4 & \vdots & 0 \end{bmatrix}$

5. Use Gauss-Jordan elimination to solve the system in Exercise 3.

6. Use Gauss-Jordan elimination to solve the system in Exercise 4.

In Exercises 7–12, use the following matrices to find the indicated matrix.

$$A = \begin{bmatrix} 1 & -2 \\ 3 & 4 \end{bmatrix}, \quad B = \begin{bmatrix} -1 & 2 & -3 \\ 2 & 0 & 5 \end{bmatrix}, \quad C = \begin{bmatrix} 0 & -2 \\ 3 & 1 \end{bmatrix}$$

7. $2A + 3C$ **8.** AB **9.** $A - 2C$

10. C^2 **11.** A^{-1} **12.** B^{-1}

In Exercises 13 and 14, solve for X using matrices A and C from Exercises 7–12.

13. $X = 2A - 3C$ **14.** $2X = 3A - C$

In Exercises 15 and 16, find matrices A, X, and B such that the system can be written as $AX = B$. Then solve for X.

15. $\begin{aligned} x - 3y &= 10 \\ -2x + y &= -10 \end{aligned}$ **16.** $\begin{aligned} 2x - y + z &= 3 \\ 3x \quad\;\; - z &= 15 \\ 4y + 3z &= -1 \end{aligned}$

Labor/Wage Requirements   In Exercises 17–20, a hang-glider manufacturer has the labor-hour and wage requirements indicated at the right.

17. What is the labor cost for model A?

18. What is the labor cost for model B?

19. What is the labor cost for model C?

20. Find LW and interpret the result.

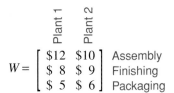

$$L = \begin{bmatrix} 1.0 & 0.6 & 0.2 \\ 2.4 & 1.0 & 0.2 \\ 2.8 & 2.0 & 0.5 \end{bmatrix} \begin{matrix} \text{Model A} \\ \text{Model B} \\ \text{Model C} \end{matrix}$$

$$\begin{matrix} \text{Assembly} & \text{Finishing} & \text{Packaging} \end{matrix}$$

Labor Requirements
(in hours)

$$W = \begin{bmatrix} \$12 & \$10 \\ \$\,8 & \$\,9 \\ \$\,5 & \$\,6 \end{bmatrix} \begin{matrix} \text{Assembly} \\ \text{Finishing} \\ \text{Packaging} \end{matrix}$$

$$\begin{matrix} \text{Plant 1} & \text{Plant 2} \end{matrix}$$

Wage Requirements
(in dollars per hour)

7.4   The Determinant of a Square Matrix

The Determinant of a 2 x 2 Matrix ▪ Minors and Cofactors ▪
The Determinant of a Square Matrix ▪ Triangular Matrices

The Determinant of a 2 x 2 Matrix

Every *square* matrix can be associated with a real number called its **determinant.**
Determinants have many uses, and several will be discussed in this and the next
section. Historically, the use of determinants arose from special number patterns
that occur when systems of linear equations are solved. For instance, the system

$$a_1 x + b_1 y = c_1$$
$$a_2 x + b_2 y = c_2$$

has a solution given by

$$x = \frac{c_1 b_2 - c_2 b_1}{a_1 b_2 - a_2 b_1} \quad \text{and} \quad y = \frac{a_1 c_2 - a_2 c_1}{a_1 b_2 - a_2 b_1}$$

provided $a_1 b_2 - a_2 b_1 \neq 0$. Note that the denominator of each fraction is the
same. This denominator is called the **determinant** of the coefficient matrix of the
system.

$$\begin{array}{cc} \textit{Coefficient Matrix} & \textit{Determinant} \\ A = \begin{bmatrix} a_1 & b_1 \\ a_2 & b_2 \end{bmatrix} & \det(A) = a_1 b_2 - a_2 b_1 \end{array}$$

The determinant of the matrix A can also be denoted by vertical bars on both sides
of the matrix, as indicated in the following definition.

Definition of the Determinant of a 2 x 2 Matrix

The **determinant** of the matrix

$$A = \begin{bmatrix} a_1 & b_1 \\ a_2 & b_2 \end{bmatrix}$$

is given by

$$\det(A) = |A| = \begin{vmatrix} a_1 & b_1 \\ a_2 & b_2 \end{vmatrix} = a_1 b_2 - a_2 b_1.$$

NOTE   In this text, $\det(A)$ and $|A|$ are used interchangeably to represent the
determinant of A. Although vertical bars are also used to denote the absolute
value of a real number, the context will show which use is intended.

A convenient method for remembering the formula for the determinant of a 2×2 matrix is shown in the following diagram.

$$\det(A) = \begin{vmatrix} a_1 & b_1 \\ a_2 & b_2 \end{vmatrix} = a_1 b_2 - a_2 b_1$$

Note that the determinant is given by the difference of the products of the two diagonals of the matrix.

EXAMPLE 1   The Determinant of a 2 x 2 Matrix

Find the determinant of each matrix.

a. $A = \begin{bmatrix} 2 & -3 \\ 1 & 2 \end{bmatrix}$ **b.** $B = \begin{bmatrix} 2 & 1 \\ 4 & 2 \end{bmatrix}$ **c.** $C = \begin{bmatrix} 0 & 3 \\ 2 & 4 \end{bmatrix}$

Solution

NOTE   Notice in Example 1 that the determinant of a matrix can be positive, zero, or negative.

a. $\det(A) = \begin{vmatrix} 2 & -3 \\ 1 & 2 \end{vmatrix} = 2(2) - 1(-3) = 4 + 3 = 7$

b. $\det(B) = \begin{vmatrix} 2 & 1 \\ 4 & 2 \end{vmatrix} = 2(2) - 4(1) = 4 - 4 = 0$

c. $\det(C) = \begin{vmatrix} 0 & 3 \\ 2 & 4 \end{vmatrix} = 0(4) - 2(3) = 0 - 6 = -6$

The determinant of a matrix of order 1×1 is defined simply as the entry of the matrix. For instance, if $A = [-2]$, then $\det(A) = -2$.

Technology

Most graphing utilities can evaluate the determinant of a matrix. For instance, on a *TI-82*, you can evaluate the determinant of

$$A = \begin{bmatrix} 2 & -3 \\ 1 & 2 \end{bmatrix}$$

by entering the matrix as "A" and then using the keystrokes

det [A] ENTER .

The result should be 7, as in Example 1(a). Try evaluating determinants of other matrices. What happens when you try to evaluate the determinant of a nonsquare matrix?

Minors and Cofactors

To define the determinant of a square matrix of order 3×3 or higher, it is convenient to introduce the concepts of **minors** and **cofactors.**

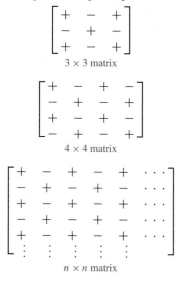

Sign Pattern for Cofactors

3×3 matrix

4×4 matrix

$n \times n$ matrix

Minors and Cofactors of a Square Matrix

If A is a square matrix, the **minor** M_{ij} of the entry a_{ij} is the determinant of the matrix obtained by deleting the ith row and jth column of A. The **cofactor** C_{ij} of the entry a_{ij} is given by

$$C_{ij} = (-1)^{i+j} M_{ij}.$$

EXAMPLE 2   *Finding the Minors and Cofactors of a Matrix*

Find all the minors and cofactors of $A = \begin{bmatrix} 0 & 2 & 1 \\ 3 & -1 & 2 \\ 4 & 0 & 1 \end{bmatrix}$.

Solution

To find the minor M_{11}, delete the first row and first column of A and evaluate the determinant of the resulting matrix.

$$\begin{bmatrix} 0 & 2 & 1 \\ 3 & -1 & 2 \\ 4 & 0 & 1 \end{bmatrix}, \quad M_{11} = \begin{vmatrix} -1 & 2 \\ 0 & 1 \end{vmatrix} = -1(1) - 0(2) = -1$$

Similarly, to find M_{12}, delete the first row and second column.

$$\begin{bmatrix} 0 & 2 & 1 \\ 3 & -1 & 2 \\ 4 & 0 & 1 \end{bmatrix}, \quad M_{12} = \begin{vmatrix} 3 & 2 \\ 4 & 1 \end{vmatrix} = 3(1) - 4(2) = -5$$

Continuing this pattern, you obtain the following minors.

$$\begin{array}{lll} M_{11} = -1 & M_{12} = -5 & M_{13} = 4 \\ M_{21} = 2 & M_{22} = -4 & M_{23} = -8 \\ M_{31} = 5 & M_{32} = -3 & M_{33} = -6 \end{array}$$

Now, to find the cofactors, combine the checkerboard pattern of signs with these minors to obtain the following.

$$\begin{array}{lll} C_{11} = -1 & C_{12} = 5 & C_{13} = 4 \\ C_{21} = -2 & C_{22} = -4 & C_{23} = 8 \\ C_{31} = 5 & C_{32} = 3 & C_{33} = -6 \end{array}$$

The Determinant of a Square Matrix

The definition given below is called **inductive** because it uses determinants of matrices of order $n - 1$ to define the determinant of a matrix of order n.

Determinant of a Square Matrix

If A is a square matrix (of order 2×2 or greater), the determinant of A is the sum of the entries in any row (or column) of A multiplied by their respective cofactors. For instance, expanding along the first row yields

$$|A| = a_{11}C_{11} + a_{12}C_{12} + \cdots + a_{1n}C_{1n}.$$

Applying this definition to find a determinant is called **expanding by cofactors.**

NOTE Try checking that for a 2×2 matrix this definition yields $|A| = a_{11}a_{22} - a_{12}a_{21}$, as previously defined.

EXAMPLE 3   The Determinant of a Matrix of Order 3 x 3

Find the determinant of

$$A = \begin{bmatrix} 0 & 2 & 1 \\ 3 & -1 & 2 \\ 4 & 0 & 1 \end{bmatrix}.$$

Solution

Note that this is the same matrix that was given in Example 2. There we found the cofactors of the entries in the first row to be

$$C_{11} = -1, \quad C_{12} = 5, \quad \text{and} \quad C_{13} = 4.$$

Therefore, by the definition of a determinant, you have the following.

$$\begin{aligned} |A| &= a_{11}C_{11} + a_{12}C_{12} + a_{13}C_{13} \qquad \text{First-row expansion} \\ &= 0(-1) + 2(5) + 1(4) \\ &= 14 \end{aligned}$$

In Example 3 the determinant was found by expanding by the cofactors in the first row. You could have used any row or column. For instance, you could have expanded along the second row to obtain

$$\begin{aligned} |A| &= a_{21}C_{21} + a_{22}C_{22} + a_{23}C_{23} \qquad \text{Second-row expansion} \\ &= 3(-2) + (-1)(-4) + 2(8) \\ &= 14. \end{aligned}$$

When expanding by cofactors, you do not need to find cofactors of zero entries, because zero times its cofactor is zero.

$$a_{ij}C_{ij} = (0)C_{ij} = 0$$

Thus, the row (or column) containing the most zeros is usually the best choice for expansion by cofactors. This is demonstrated in the next example.

EXAMPLE 4   *The Determinant of a Matrix of Order 4 x 4*

Find the determinant of

$$A = \begin{bmatrix} 1 & -2 & 3 & 0 \\ -1 & 1 & 0 & 2 \\ 0 & 2 & 0 & 3 \\ 3 & 4 & 0 & 2 \end{bmatrix}.$$

Solution

Students may have difficulty choosing the best row or column about which to expand a determinant.

After inspecting this matrix, you can see that three of the entries in the third column are zeros. Thus, you can eliminate some of the work in the expansion by using the third column.

$$|A| = 3(C_{13}) + 0(C_{23}) + 0(C_{33}) + 0(C_{43})$$

Because C_{23}, C_{33}, and C_{43} have zero coefficients, you need only find the cofactor C_{13}. To do this, delete the first row and third column of A and evaluate the determinant of the resulting matrix.

$$C_{13} = (-1)^{1+3} \begin{vmatrix} -1 & 1 & 2 \\ 0 & 2 & 3 \\ 3 & 4 & 2 \end{vmatrix}$$

$$= \begin{vmatrix} -1 & 1 & 2 \\ 0 & 2 & 3 \\ 3 & 4 & 2 \end{vmatrix}$$

When expanding a determinant by cofactors, you may need to remind students to use the appropriate sign.

Expanding by cofactors in the second row yields the following.

$$C_{13} = 0(-1)^3 \begin{vmatrix} 1 & 2 \\ 4 & 2 \end{vmatrix} + 2(-1)^4 \begin{vmatrix} -1 & 2 \\ 3 & 2 \end{vmatrix} + 3(-1)^5 \begin{vmatrix} -1 & 1 \\ 3 & 4 \end{vmatrix}$$

$$= 0 + 2(1)(-8) + 3(-1)(-7)$$

$$= 5$$

Thus, you obtain

$$|A| = 3C_{13} = 3(5) = 15.$$

NOTE   Try using a graphing utility to confirm the result of Example 4.

There is an alternative method that is commonly used for evaluating the determinant of a 3×3 matrix A. (This method works *only* for 3×3 matrices.) To apply this method, copy the first and second columns of A to form fourth and fifth columns. The determinant of A is then obtained by adding the products of the three "downward diagonals" and subtracting the products of the three "upward diagonals," as shown in the following diagram.

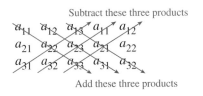

Thus, the determinant of the 3×3 matrix A is given by

$$|A| = a_{11}a_{22}a_{33} + a_{12}a_{23}a_{31} + a_{13}a_{21}a_{32}$$
$$- a_{31}a_{22}a_{13} - a_{32}a_{23}a_{11} - a_{33}a_{21}a_{12}.$$

EXAMPLE 5   The Determinant of a 3 x 3 Matrix

Find the determinant of

$$A = \begin{bmatrix} 0 & 2 & 1 \\ 3 & -1 & 2 \\ 4 & -4 & 1 \end{bmatrix}.$$

Solution

Because A is a 3×3 matrix, you can use the alternative procedure for finding $|A|$. Begin by recopying the first two columns and then computing the six diagonal products, as follows.

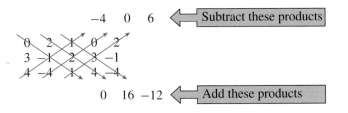

NOTE Be sure you understand that the diagonal process illustrated in Example 5 is valid *only* for matrices of order 3×3. For matrices of higher orders, another method must be used.

Now, by adding the lower three products and subtracting the upper three products, you find the determinant of A to be

$$|A| = 0 + 16 - 12 - (-4) - 0 - 6 = 2.$$

Triangular Matrices

Evaluating determinants of matrices of order 4 or higher can be tedious. There is, however, an important exception: the determinant of a **triangular** matrix. A square matrix is **upper triangular** if it has all zero entries below its main diagonal and **lower triangular** if it has all zero entries above its main diagonal. A matrix that is both upper and lower triangular is called **diagonal.** That is, a diagonal matrix is one in which all entries above and below the main diagonal are zero.

$$
\begin{array}{cc}
\textit{Upper Triangular Matrix} & \textit{Lower Triangular Matrix} \\
\begin{bmatrix}
a_{11} & a_{12} & a_{13} & \cdots & a_{1n} \\
0 & a_{22} & a_{23} & \cdots & a_{2n} \\
0 & 0 & a_{33} & \cdots & a_{3n} \\
\vdots & \vdots & \vdots & & \vdots \\
0 & 0 & 0 & \cdots & a_{nn}
\end{bmatrix}
&
\begin{bmatrix}
a_{11} & 0 & 0 & \cdots & 0 \\
a_{21} & a_{22} & 0 & \cdots & 0 \\
a_{31} & a_{32} & a_{33} & \cdots & 0 \\
\vdots & \vdots & \vdots & & \vdots \\
a_{n1} & a_{n2} & a_{n3} & \cdots & a_{nn}
\end{bmatrix}
\end{array}
$$

To find the determinant of a triangular matrix of any order, simply form the product of the entries on the main diagonal.

EXAMPLE 6 The Determinant of a Triangular Matrix

a.
$$
\begin{vmatrix}
2 & 0 & 0 & 0 \\
4 & -2 & 0 & 0 \\
-5 & 6 & 1 & 0 \\
1 & 5 & 3 & 3
\end{vmatrix} = (2)(-2)(1)(3) = -12
$$

b.
$$
\begin{vmatrix}
-1 & 0 & 0 & 0 & 0 \\
0 & 3 & 0 & 0 & 0 \\
0 & 0 & 2 & 0 & 0 \\
0 & 0 & 0 & 4 & 0 \\
0 & 0 & 0 & 0 & -2
\end{vmatrix} = (-1)(3)(2)(4)(-2) = 48
$$

Group Activities · Extending the Concept

The Determinant of a Triangular Matrix   Write an argument that explains why the determinant of a 3×3 triangular matrix is the product of its main-diagonal entries.

$$
\begin{bmatrix}
a_{11} & a_{12} & a_{13} \\
0 & a_{22} & a_{23} \\
0 & 0 & a_{33}
\end{bmatrix} = a_{11}\,a_{22}\,a_{33}
$$

In Exercises 1–4, perform the indicated matrix operations.

1. $\begin{bmatrix} 1 & -2 \\ 0 & 3 \end{bmatrix} + \begin{bmatrix} 2 & 7 \\ 4 & -3 \end{bmatrix}$

2. $\begin{bmatrix} -2 & 5 \\ 3 & -2 \end{bmatrix} - \begin{bmatrix} 0 & -3 \\ 1 & 2 \end{bmatrix}$

3. $3 \begin{bmatrix} 3 & -4 & 2 \\ 1 & 0 & -1 \\ 0 & 1 & -2 \end{bmatrix}$

4. $4 \begin{bmatrix} 0 & 2 & 3 \\ -1 & 2 & 3 \\ -2 & 1 & -2 \end{bmatrix}$

In Exercises 5–10, perform the indicated arithmetic operations.

5. $[(1)(3) + (-3)(2)] - [(1)(4) + (3)(5)]$

6. $[(4)(4) + (-1)(-3)] - [(-1)(2) + (-2)(7)]$

7. $\dfrac{4(7) - 1(-2)}{(-5)(-2) - 3(4)}$

8. $\dfrac{3(6) - 2(7)}{6(-5) - 2(1)}$

9. $-5(-1)^2[6(-2) - 7(-3)]$

10. $4(-1)^3[3(6) - 2(7)]$

7.4 Exercises

In Exercises 1–24, find the determinant of the matrix.

1. $[5]$

2. $[-8]$

3. $\begin{bmatrix} 2 & 1 \\ 3 & 4 \end{bmatrix}$

4. $\begin{bmatrix} -3 & 1 \\ 5 & 2 \end{bmatrix}$

5. $\begin{bmatrix} 5 & 2 \\ -6 & 3 \end{bmatrix}$

6. $\begin{bmatrix} 2 & -2 \\ 4 & 3 \end{bmatrix}$

7. $\begin{bmatrix} -7 & 6 \\ \frac{1}{2} & 3 \end{bmatrix}$

8. $\begin{bmatrix} 4 & -3 \\ 0 & 0 \end{bmatrix}$

9. $\begin{bmatrix} 2 & 6 \\ 0 & 3 \end{bmatrix}$

10. $\begin{bmatrix} 2 & -3 \\ -6 & 9 \end{bmatrix}$

11. $\begin{bmatrix} 2 & -1 & 0 \\ 4 & 2 & 1 \\ 4 & 2 & 1 \end{bmatrix}$

12. $\begin{bmatrix} -2 & 2 & 3 \\ 1 & -1 & 0 \\ 0 & 1 & 4 \end{bmatrix}$

13. $\begin{bmatrix} 0.3 & 0.2 & 0.2 \\ 0.2 & 0.2 & 0.2 \\ -0.4 & 0.4 & 0.3 \end{bmatrix}$

14. $\begin{bmatrix} 0.1 & 0.2 & 0.3 \\ -0.3 & 0.2 & 0.2 \\ 0.5 & 0.4 & 0.4 \end{bmatrix}$

15. $\begin{bmatrix} 1 & 4 & -2 \\ 3 & 6 & -6 \\ -2 & 1 & 4 \end{bmatrix}$

16. $\begin{bmatrix} 2 & 3 & 1 \\ 0 & 5 & -2 \\ 0 & 0 & -2 \end{bmatrix}$

17. $\begin{bmatrix} 6 & 3 & -7 \\ 0 & 0 & 0 \\ 4 & -6 & 3 \end{bmatrix}$

18. $\begin{bmatrix} 1 & 1 & 2 \\ 3 & 1 & 0 \\ -2 & 0 & 3 \end{bmatrix}$

19. $\begin{bmatrix} -1 & 2 & 5 \\ 0 & 3 & 4 \\ 0 & 0 & 3 \end{bmatrix}$

20. $\begin{bmatrix} 1 & 0 & 0 \\ -4 & -1 & 0 \\ 5 & 1 & 5 \end{bmatrix}$

21. $\begin{bmatrix} -1 & 0 & 0 & 0 \\ 2 & 3 & 0 & 0 \\ -4 & 5 & 3 & 0 \\ 1 & 0 & 2 & 2 \end{bmatrix}$

22. $\begin{bmatrix} 4 & 0 & 0 & 0 \\ 1 & -4 & 0 & 0 \\ 2 & 1 & -1 & 0 \\ 6 & -2 & 3 & -1 \end{bmatrix}$

23. $\begin{bmatrix} 1 & 0 & 0 & 0 & 0 \\ 0 & 2 & 0 & 0 & 0 \\ 0 & 0 & 3 & 0 & 0 \\ 0 & 0 & 0 & 4 & 0 \\ 0 & 0 & 0 & 0 & 5 \end{bmatrix}$

24. $\begin{bmatrix} -2 & 0 & 0 & 0 & 0 \\ 0 & 3 & 0 & 0 & 0 \\ 0 & 0 & -1 & 0 & 0 \\ 0 & 0 & 0 & 2 & 0 \\ 0 & 0 & 0 & 0 & -4 \end{bmatrix}$

In Exercises 25–28, find (a) all minors and (b) all cofactors of the given matrix.

25. $\begin{bmatrix} 3 & 4 \\ 2 & -5 \end{bmatrix}$ **26.** $\begin{bmatrix} 11 & 0 \\ -3 & 2 \end{bmatrix}$

27. $\begin{bmatrix} 3 & -2 & 8 \\ 3 & 2 & -6 \\ -1 & 3 & 6 \end{bmatrix}$ **28.** $\begin{bmatrix} -2 & 9 & 4 \\ 7 & -6 & 0 \\ 6 & 7 & -6 \end{bmatrix}$

In Exercises 29–34, find the determinant of the matrix by the method of expansion by cofactors. Expand using the indicated row or column.

29. $\begin{bmatrix} -3 & 2 & 1 \\ 4 & 5 & 6 \\ 2 & -3 & 1 \end{bmatrix}$ **30.** $\begin{bmatrix} -3 & 4 & 2 \\ 6 & 3 & 1 \\ 4 & -7 & -8 \end{bmatrix}$
   (a) Row 1   (a) Row 2
   (b) Column 2   (b) Column 3

31. $\begin{bmatrix} 5 & 0 & -3 \\ 0 & 12 & 4 \\ 1 & 6 & 3 \end{bmatrix}$ **32.** $\begin{bmatrix} 10 & -5 & 5 \\ 30 & 0 & 10 \\ 0 & 10 & 1 \end{bmatrix}$
   (a) Row 2   (a) Row 3
   (b) Column 2   (b) Column 1

33. $\begin{bmatrix} 6 & 0 & -3 & 5 \\ 4 & 13 & 6 & -8 \\ -1 & 0 & 7 & 4 \\ 8 & 6 & 0 & 2 \end{bmatrix}$
   (a) Row 2
   (b) Column 2

34. $\begin{bmatrix} 10 & 8 & 3 & -7 \\ 4 & 0 & 5 & -6 \\ 0 & 3 & 2 & 7 \\ 1 & 0 & -3 & 2 \end{bmatrix}$
   (a) Row 3
   (b) Column 1

⊞ In Exercises 35–44, find the determinant of the matrix. Use a graphing utility to confirm your result.

35. $\begin{bmatrix} 1 & 4 & -2 \\ 3 & 2 & 0 \\ -1 & 4 & 3 \end{bmatrix}$ **36.** $\begin{bmatrix} 2 & -1 & 3 \\ 1 & 4 & 4 \\ 1 & 0 & 2 \end{bmatrix}$

37. $\begin{bmatrix} 2 & 4 & 6 \\ 0 & 3 & 1 \\ 0 & 0 & -5 \end{bmatrix}$ **38.** $\begin{bmatrix} -3 & 0 & 0 \\ 7 & 11 & 0 \\ 1 & 2 & 2 \end{bmatrix}$

39. $\begin{bmatrix} 3 & 6 & -5 & 4 \\ -2 & 0 & 6 & 0 \\ 1 & 1 & 2 & 2 \\ 0 & 3 & -1 & -1 \end{bmatrix}$

40. $\begin{bmatrix} 2 & 6 & 6 & 2 \\ 2 & 7 & 3 & 6 \\ 1 & 5 & 0 & 1 \\ 3 & 7 & 0 & 7 \end{bmatrix}$

41. $\begin{bmatrix} 5 & 3 & 0 & 6 \\ 4 & 6 & 4 & 12 \\ 0 & 2 & -3 & 4 \\ 0 & 1 & -2 & 2 \end{bmatrix}$ **42.** $\begin{bmatrix} 1 & 4 & 3 & 2 \\ -5 & 6 & 2 & 1 \\ 0 & 0 & 0 & 0 \\ 3 & -2 & 1 & 5 \end{bmatrix}$

43. $\begin{bmatrix} 3 & 2 & 4 & -1 & 5 \\ -2 & 0 & 1 & 3 & 2 \\ 1 & 0 & 0 & 4 & 0 \\ 6 & 0 & 2 & -1 & 0 \\ 3 & 0 & 5 & 1 & 0 \end{bmatrix}$

44. $\begin{bmatrix} 5 & 2 & 0 & 0 & -2 \\ 0 & 1 & 4 & 3 & 2 \\ 0 & 0 & 2 & 6 & 3 \\ 0 & 0 & 3 & 4 & 1 \\ 0 & 0 & 0 & 0 & 2 \end{bmatrix}$

⊞ In Exercises 45 and 46, find a 4×4 *upper* triangular matrix whose determinant is equal to the given value and a 4×4 *lower* triangular matrix whose determinant is equal to the given value. Use a graphing utility to confirm your results.

45. -24 **46.** 32

In Exercises 47 and 48, determine why the determinant of the matrix is equal to zero.

47. $\begin{bmatrix} 3 & 4 & -2 & 7 \\ 1 & 3 & -1 & 2 \\ 0 & 5 & 7 & 1 \\ 1 & 3 & -1 & 2 \end{bmatrix}$ **48.** $\begin{bmatrix} 3 & 2 & -1 \\ 0 & 0 & 0 \\ 5 & -7 & 9 \end{bmatrix}$

Math Matters   Guess the Number

Here is a guessing game that can be made using the four cards shown in the figure (note that the fourth card has numbers on the front *and* back.) To play the game, ask someone to think of a number between 1 and 15. Ask the person if the number is on the first card. If it is, place the card face up with the "YES" on top. If it isn't, place the card face up with the "NO" on top. Repeat this with each of the four cards (using the same number), stacking the cards one on top of another. Be sure that the fourth card is used last. After all four cards are in a stack, turn the stack over. The person's number will be the number that is showing through a window in the cards. Can you explain why this card game works?

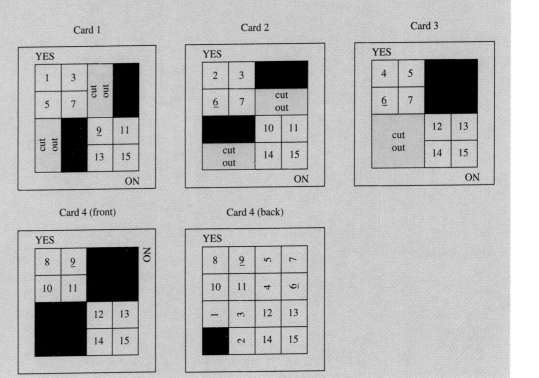

7.5	**Applications of Determinants and Matrices**
	Area of a Triangle ▪ Lines in the Plane ▪ Cryptography

Area of a Triangle

In this section, you will study some additional applications of matrices and determinants. The first involves a formula for finding the area of a triangle whose vertices are given by three points on a rectangular coordinate system.

Area of a Triangle

The area of a triangle with vertices (x_1, y_1), (x_2, y_2), and (x_3, y_3) is given by

$$\text{Area} = \pm\frac{1}{2}\begin{vmatrix} x_1 & y_1 & 1 \\ x_2 & y_2 & 1 \\ x_3 & y_3 & 1 \end{vmatrix}$$

where the symbol $(\pm)$ indicates that the appropriate sign should be chosen to yield a positive area.

EXAMPLE 1   *Finding the Area of a Triangle*

Find the area of the triangle whose vertices are $(1, 0)$, $(2, 2)$, and $(4, 3)$, as shown in Figure 7.1.

Solution

Let $(x_1, y_1) = (1, 0)$, $(x_2, y_2) = (2, 2)$, and $(x_3, y_3) = (4, 3)$. Then, to find the area of the triangle, evaluate the determinant

$$\begin{vmatrix} x_1 & y_1 & 1 \\ x_2 & y_2 & 1 \\ x_3 & y_3 & 1 \end{vmatrix} = \begin{vmatrix} 1 & 0 & 1 \\ 2 & 2 & 1 \\ 4 & 3 & 1 \end{vmatrix} = 1\begin{vmatrix} 2 & 1 \\ 3 & 1 \end{vmatrix} - 0\begin{vmatrix} 2 & 1 \\ 4 & 1 \end{vmatrix} + 1\begin{vmatrix} 2 & 2 \\ 4 & 3 \end{vmatrix}$$

$$= 1(-1) - 0(-2) + 1(-2)$$

$$= -3.$$

Using this value, you can conclude that the area of the triangle is

$$\text{Area} = -\frac{1}{2}\begin{vmatrix} 1 & 0 & 1 \\ 2 & 2 & 1 \\ 4 & 3 & 1 \end{vmatrix} = -\frac{1}{2}(-3) = \frac{3}{2}.$$

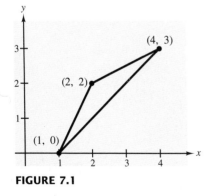

FIGURE 7.1

Lines in the Plane

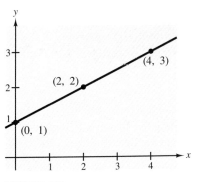

FIGURE 7.2

Suppose the three points in Example 1 had been on the same line. What would have happened had the area formula been applied to three such points? The answer is that the determinant would have been zero. Consider, for instance, the three collinear points $(0, 1)$, $(2, 2)$, and $(4, 3)$, as shown in Figure 7.2. The area of the "triangle" that has these three points as vertices is

$$\frac{1}{2}\begin{vmatrix} 0 & 1 & 1 \\ 2 & 2 & 1 \\ 4 & 3 & 1 \end{vmatrix} = \frac{1}{2}\left(0\begin{vmatrix} 2 & 1 \\ 3 & 1 \end{vmatrix} - 1\begin{vmatrix} 2 & 1 \\ 4 & 1 \end{vmatrix} + 1\begin{vmatrix} 2 & 2 \\ 4 & 3 \end{vmatrix}\right)$$

$$= \frac{1}{2}[0(-1) - 1(-2) + 1(-2)] = 0.$$

This result is generalized as follows.

Test for Collinear Points

Three points (x_1, y_1), (x_2, y_2), and (x_3, y_3) are collinear (lie on the same line) if and only if

$$\begin{vmatrix} x_1 & y_1 & 1 \\ x_2 & y_2 & 1 \\ x_3 & y_3 & 1 \end{vmatrix} = 0.$$

EXAMPLE 2   Testing for Collinear Points

Determine whether the points $(-2, -2)$, $(1, 1)$, and $(7, 5)$ lie on the same line. (See Figure 7.3.)

Solution

Letting $(x_1, y_1) = (-2, -2)$, $(x_2, y_2) = (1, 1)$, and $(x_3, y_3) = (7, 5)$, you have

$$\begin{vmatrix} x_1 & y_1 & 1 \\ x_2 & y_2 & 1 \\ x_3 & y_3 & 1 \end{vmatrix} = \begin{vmatrix} -2 & -2 & 1 \\ 1 & 1 & 1 \\ 7 & 5 & 1 \end{vmatrix}$$

$$= -2\begin{vmatrix} 1 & 1 \\ 5 & 1 \end{vmatrix} - (-2)\begin{vmatrix} 1 & 1 \\ 7 & 1 \end{vmatrix} + 1\begin{vmatrix} 1 & 1 \\ 7 & 5 \end{vmatrix}$$

$$= -2(-4) - (-2)(-6) + 1(-2)$$

$$= -6.$$

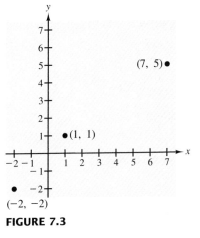

FIGURE 7.3

Because the value of this determinant is *not* zero, you can conclude that the three points do not lie on the same line.

The test for collinear points can be adapted to another use. That is, if you are given two points on a rectangular coordinate system, you can find an equation of the line passing through the two points, as follows.

Two-Point Form of the Equation of a Line

An equation of the line passing through the distinct points (x_1, y_1) and (x_2, y_2) is given by

$$\begin{vmatrix} x & y & 1 \\ x_1 & y_1 & 1 \\ x_2 & y_2 & 1 \end{vmatrix} = 0.$$

EXAMPLE 3   *Finding an Equation of a Line*

Find an equation of the line passing through the two points $(2, 4)$ and $(-1, 3)$, as shown in Figure 7.4.

Solution

Applying the determinant formula for the equation of a line produces

$$\begin{vmatrix} x & y & 1 \\ 2 & 4 & 1 \\ -1 & 3 & 1 \end{vmatrix} = 0.$$

To evaluate this determinant, you can expand by cofactors along the first row to obtain the following.

$$x \begin{vmatrix} 4 & 1 \\ 3 & 1 \end{vmatrix} - y \begin{vmatrix} 2 & 1 \\ -1 & 1 \end{vmatrix} + 1 \begin{vmatrix} 2 & 4 \\ -1 & 3 \end{vmatrix} = x - 3y + 10 = 0$$

Therefore, an equation of the line is

$$x - 3y + 10 = 0.$$

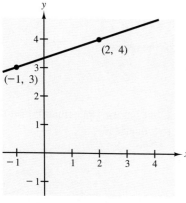

FIGURE 7.4

Note that this method of finding the equation of a line works for all lines, including horizontal and vertical lines. For instance, the equation of the vertical line through $(2, 0)$ and $(2, 2)$ is

$$\begin{vmatrix} x & y & 1 \\ 2 & 0 & 1 \\ 2 & 2 & 1 \end{vmatrix} = 0$$

$$4 - 2x = 0$$

$$x = 2.$$

Cryptography

A **cryptogram** is a message written according to a secret code. (The Greek word *kryptos* means "hidden.") Matrix multiplication can be used to **encode** and **decode** messages. To begin, you need to assign a number to each letter in the alphabet (with 0 assigned to a blank space), as follows.

Both governments and private industry use codes to transmit messages. Many different systems are used. The first mechanical device for coding and decoding was invented in 1917 by an American engineer, Gilbert Vernam.

0 = _	9 = I	18 = R
1 = A	10 = J	19 = S
2 = B	11 = K	20 = T
3 = C	12 = L	21 = U
4 = D	13 = M	22 = V
5 = E	14 = N	23 = W
6 = F	15 = O	24 = X
7 = G	16 = P	25 = Y
8 = H	17 = Q	26 = Z

Then the message is converted to numbers and partitioned into **uncoded row matrices,** each having n entries, as demonstrated in Example 4.

EXAMPLE 4   *Forming Uncoded Row Matrices*

Write the uncoded row matrices of order 1×3 for the message

MEET ME MONDAY.

Solution

Partitioning the message (including blank spaces, but ignoring punctuation) into groups of three produces the following uncoded row matrices.

$$[13 \quad 5 \quad 5] \quad [20 \quad 0 \quad 13] \quad [5 \quad 0 \quad 13] \quad [15 \quad 14 \quad 4] \quad [1 \quad 25 \quad 0]$$
$$\ \ \text{M} \quad \text{E} \quad \text{E} \quad \ \ \text{T} \quad \quad \text{M} \quad \text{E} \quad \quad \ \ \text{M} \quad \text{O} \quad \text{N} \quad \ \ \text{D} \quad \text{A} \quad \text{Y}$$

Note that a blank space is used to fill out the last uncoded row matrix.

Activities
1. Use a determinant to find the area of the triangle with vertices $(1, -3)$, $(2, 3)$, and $(3, 1)$.
   Answer: 4 square units
2. Use a determinant to find an equation of the line through the points $(-1, 2)$ and $(3, 4)$.
   Answer: $x - 2y + 5 = 0$

To **encode** a message, choose an $n \times n$ invertible matrix A and multiply the uncoded row matrices (on the right) by A to obtain **coded row matrices.** Here is an example.

Uncoded Matrix   *Encoding Matrix A*   *Coded Matrix*

$$\begin{bmatrix} 13 & 5 & 5 \end{bmatrix} \begin{bmatrix} 1 & -2 & 2 \\ -1 & 1 & 3 \\ 1 & -1 & -4 \end{bmatrix} = \begin{bmatrix} 13 & -26 & 21 \end{bmatrix}$$

This technique is further illustrated in Example 5.

EXAMPLE 5   Encoding a Message

Use the following matrix to encode the message MEET ME MONDAY.

$$A = \begin{bmatrix} 1 & -2 & 2 \\ -1 & 1 & 3 \\ 1 & -1 & -4 \end{bmatrix}$$

Solution

The coded row matrices are obtained by multiplying each of the uncoded row matrices found in Example 4 by the matrix A, as follows.

Uncoded Matrix	*Encoding Matrix A*	*Coded Matrix*

$$\begin{bmatrix} 13 & 5 & 5 \end{bmatrix} \begin{bmatrix} 1 & -2 & 2 \\ -1 & 1 & 3 \\ 1 & -1 & -4 \end{bmatrix} = \begin{bmatrix} 13 & -26 & 21 \end{bmatrix}$$

$$\begin{bmatrix} 20 & 0 & 13 \end{bmatrix} \begin{bmatrix} 1 & -2 & 2 \\ -1 & 1 & 3 \\ 1 & -1 & -4 \end{bmatrix} = \begin{bmatrix} 33 & -53 & -12 \end{bmatrix}$$

$$\begin{bmatrix} 5 & 0 & 13 \end{bmatrix} \begin{bmatrix} 1 & -2 & 2 \\ -1 & 1 & 3 \\ 1 & -1 & -4 \end{bmatrix} = \begin{bmatrix} 18 & -23 & -42 \end{bmatrix}$$

$$\begin{bmatrix} 15 & 14 & 4 \end{bmatrix} \begin{bmatrix} 1 & -2 & 2 \\ -1 & 1 & 3 \\ 1 & -1 & -4 \end{bmatrix} = \begin{bmatrix} 5 & -20 & 56 \end{bmatrix}$$

$$\begin{bmatrix} 1 & 25 & 0 \end{bmatrix} \begin{bmatrix} 1 & -2 & 2 \\ -1 & 1 & 3 \\ 1 & -1 & -4 \end{bmatrix} = \begin{bmatrix} -24 & 23 & 77 \end{bmatrix}$$

Thus, the sequence of coded row matrices is

$$[13 \ -26 \ 21] \ [33 \ -53 \ -12] \ [18 \ -23 \ -42] \ [5 \ -20 \ 56] \ [-24 \ 23 \ 77].$$

Finally, removing the matrix notation produces the following cryptogram.

$$13 \ -26 \ 21 \ 33 \ -53 \ -12 \ 18 \ -23 \ -42 \ 5 \ -20 \ 56 \ -24 \ 23 \ 77$$

For those who do not know the matrix A, decoding the cryptogram found in Example 5 is difficult. But for an authorized receiver who knows the matrix A, decoding is simple. The receiver need only multiply the coded row matrices by A^{-1} (on the right) to retrieve the uncoded row matrices. Here is an example.

$$\underbrace{\begin{bmatrix} 13 & -26 & 21 \end{bmatrix}}_{\text{Coded}} A^{-1} = \underbrace{\begin{bmatrix} 13 & 5 & 5 \end{bmatrix}}_{\text{Uncoded}}$$

EXAMPLE 6 Decoding a Message

Use the inverse of the matrix $A = \begin{bmatrix} 1 & -2 & 2 \\ -1 & 1 & 3 \\ 1 & -1 & -4 \end{bmatrix}$ to decode the cryptogram

13 −26 21 33 −53 −12 18 −23 −42 5 −20 56 −24 23 77.

Solution

Partition the message into groups of three to form the coded row matrices. Multiply each coded row matrix on the right by A^{-1} to obtain the decoded row matrices.

Coded Matrix   *Decoding Matrix A^{-1}*   *Decoded Matrix*

$$\begin{bmatrix} 13 & -26 & 21 \end{bmatrix} \begin{bmatrix} -1 & -10 & -8 \\ -1 & -6 & -5 \\ 0 & -1 & -1 \end{bmatrix} = \begin{bmatrix} 13 & 5 & 5 \end{bmatrix}$$

$$\begin{bmatrix} 33 & -53 & -12 \end{bmatrix} \begin{bmatrix} -1 & -10 & -8 \\ -1 & -6 & -5 \\ 0 & -1 & -1 \end{bmatrix} = \begin{bmatrix} 20 & 0 & 13 \end{bmatrix}$$

$$\begin{bmatrix} 18 & -23 & -42 \end{bmatrix} \begin{bmatrix} -1 & -10 & -8 \\ -1 & -6 & -5 \\ 0 & -1 & -1 \end{bmatrix} = \begin{bmatrix} 5 & 0 & 13 \end{bmatrix}$$

$$\begin{bmatrix} 5 & -20 & 56 \end{bmatrix} \begin{bmatrix} -1 & -10 & -8 \\ -1 & -6 & -5 \\ 0 & -1 & -1 \end{bmatrix} = \begin{bmatrix} 15 & 14 & 4 \end{bmatrix}$$

$$\begin{bmatrix} -24 & 23 & 77 \end{bmatrix} \begin{bmatrix} -1 & -10 & -8 \\ -1 & -6 & -5 \\ 0 & -1 & -1 \end{bmatrix} = \begin{bmatrix} 1 & 25 & 0 \end{bmatrix}$$

Thus, the message is as follows.

[13 5 5]   [20 0 13]   [5 0 13]   [15 14 4]   [1 25 0]
 M E E   T   M E   M   O N   D A Y

Group Activities

Exploring with Technology

Decoding   Show how to use a graphing utility with matrix operations to decode the following cryptogram. (Use the matrix in Example 6.)

12 −25 15 28 −32 −89 10 −10 −49 12 −12 −51 17 −31 10 10 −28 55 4 −8 8

Warm Up

The following warm-up exercises involve skills that were covered in earlier sections. You will use these skills in the exercise set for this section.

In Exercises 1–6, evaluate the determinant.

1. $\begin{vmatrix} 4 & 3 \\ -3 & -2 \end{vmatrix}$

2. $\begin{vmatrix} 10 & -20 \\ -1 & 2 \end{vmatrix}$

3. $\begin{vmatrix} 4 & 0 \\ -3 & -2 \end{vmatrix}$

4. $\begin{vmatrix} x & x^2 \\ 1 & 2x \end{vmatrix}$

5. $\begin{vmatrix} 4 & 0 & -2 \\ 3 & 1 & 2 \\ -8 & 0 & 6 \end{vmatrix}$

6. $\begin{vmatrix} 3 & 2 & 5 \\ 0 & 0 & -4 \\ -6 & 1 & 1 \end{vmatrix}$

In Exercises 7 and 8, find the inverse of the matrix.

7. $A = \begin{bmatrix} 1 & 3 \\ 2 & 7 \end{bmatrix}$

8. $A = \begin{bmatrix} 10 & 5 & -2 \\ -4 & -2 & 1 \\ 1 & 1 & 0 \end{bmatrix}$

In Exercises 9 and 10, perform the indicated matrix multiplication.

9. $\begin{bmatrix} 0.1 & 0.2 & 0.2 \\ 0.4 & 0.3 & 0.5 \\ 0.5 & 0.5 & 0.3 \end{bmatrix} \begin{bmatrix} 0.4 \\ 0.5 \\ 0.1 \end{bmatrix}$

10. $\begin{bmatrix} 2 & 5 & 8 \end{bmatrix} \begin{bmatrix} 1 & 2 & -1 \\ 1 & 2 & 2 \\ 2 & 5 & 0 \end{bmatrix}$

7.5 Exercises

In Exercises 1–10, use a determinant to find the area of the triangle with the given vertices.

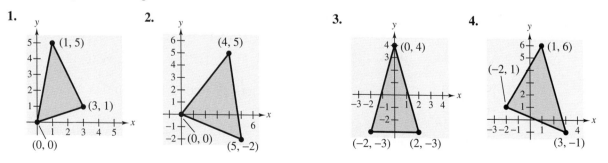

1. (1, 5) (3, 1) (0, 0)

2. (4, 5) (0, 0) (5, −2)

3. (0, 4) (−2, −3) (2, −3)

4. (1, 6) (−2, 1) (3, −1)

5.

6.

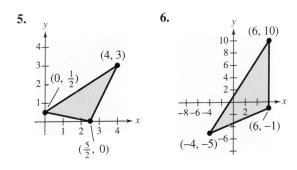

7. $(-2, 4), (2, 3), (-1, 5)$

8. $(0, -2), (-1, 4), (3, 5)$

9. $(-3, 5), (2, 6), (3, -5)$

10. $(-2, 4), (1, 5), (3, -2)$

In Exercises 11 and 12, find a value of x so that the triangle has an area of 4.

11. $(-5, 1), (0, 2), (-2, x)$

12. $(-4, 2), (-3, 5), (-1, x)$

13. *Area of a Region*   A large region of forest has been infected with gypsy moths. The region is roughly triangular, as shown in the figure. From the northern-most vertex A of the region, the distances to the other vertices are 25 miles south and 10 miles east (for vertex B), and 20 miles south and 28 miles east (for vertex C). Use a graphing utility to approximate the number of square miles in this region.

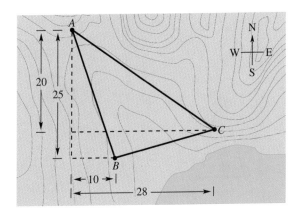

14. *Area of a Region*   You own a triangular tract of land, as shown in the figure. To estimate the number of square feet in the tract, you start at one vertex, walk 65 feet east and 50 feet north to the second vertex, and then walk 85 feet west and 30 feet north to the third vertex. Use a graphing utility to determine how many square feet there are in the tract of land.

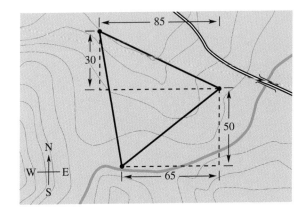

In Exercises 15–20, use a determinant to determine whether the points are collinear.

15. $(3, -1), (0, -3), (12, 5)$

16. $(-3, -5), (6, 1), (10, 2)$

17. $\left(2, -\frac{1}{2}\right), (-4, 4), (6, -3)$

18. $(0, 1), (4, -2), (-8, 7)$

19. $(0, 2), (1, 2.4), (-1, 1.6)$

20. $(2, 3), (3, 3.5), (-1, 2)$

In Exercises 21–26, use a determinant to find an equation of the line through the points.

21. $(0, 0), (5, 3)$ **22.** $(0, 0), (-2, 2)$

23. $(-4, 3), (2, 1)$ **24.** $(10, 7), (-2, -7)$

25. $\left(-\frac{1}{2}, 3\right), \left(\frac{5}{2}, 1\right)$ **26.** $\left(\frac{2}{3}, 4\right), (6, 12)$

In Exercises 27 and 28, find x so that the points are collinear.

27. $(2, -5), (4, x), (5, -2)$

28. $(-6, 2), (-5, x), (-3, 5)$

In Exercises 29 and 30, find the uncoded 1×2 row matrices for the message. Then encode the message using the matrix.

Message	Matrix
29. COME HOME SOON	$\begin{bmatrix} 1 & 2 \\ 3 & 5 \end{bmatrix}$
30. HELP IS ON THE WAY	$\begin{bmatrix} -2 & 3 \\ -1 & 1 \end{bmatrix}$

In Exercises 31 and 32, find the uncoded 1×3 row matrices for the message. Then encode the message using the matrix.

Message	Matrix
31. TROUBLE IN RIVER CITY	$\begin{bmatrix} 1 & -1 & 0 \\ 1 & 0 & -1 \\ -6 & 2 & 3 \end{bmatrix}$
32. PLEASE SEND MONEY	$\begin{bmatrix} 4 & 2 & 1 \\ -3 & -3 & -1 \\ 3 & 2 & 1 \end{bmatrix}$

In Exercises 33–36, write a cryptogram for the message using the matrix

$$A = \begin{bmatrix} 1 & 2 & 2 \\ 3 & 7 & 9 \\ -1 & -4 & -7 \end{bmatrix}.$$

33. LANDING SUCCESSFUL

34. BEAM ME UP SCOTTY

35. HAPPY BIRTHDAY

36. OPERATION OVERLORD

In Exercises 37–40, use A^{-1} to decode the cryptogram.

37. $A = \begin{bmatrix} 1 & 2 \\ 3 & 5 \end{bmatrix}$

11, 21, 64, 112, 25, 50, 29, 53, 23, 46, 40, 75, 55, 92

38. $A = \begin{bmatrix} 2 & 3 \\ 3 & 4 \end{bmatrix}$

85, 120, 6, 8, 10, 15, 84, 117, 42, 56, 90, 125, 60, 80, 30, 45, 19, 26

39. $A = \begin{bmatrix} 4 & 2 & 1 \\ -3 & -3 & -1 \\ 3 & 2 & 1 \end{bmatrix}$

33, 9, 9, 55, 28, 14, 95, 50, 25, 99, 53, 29, -22, -32, -9

40. $A = \begin{bmatrix} 1 & -1 & 0 \\ 1 & 0 & -1 \\ -6 & 2 & 3 \end{bmatrix}$

9, -1, -9, 38, -19, -19, 28, -9, -19, -80, 25, 41, -64, 21, 31, -7, -4, 7

In Exercises 41 and 42, decode the cryptogram by using the inverse of the matrix

$$A = \begin{bmatrix} 1 & 2 & 2 \\ 3 & 7 & 9 \\ -1 & -4 & -7 \end{bmatrix}.$$

41. 20, 17, -15, -12, -56, -104, 1, -25, -65, 62, 143, 181

42. 13, -9, -59, 61, 112, 106, -17, -73, -131, 11, 24, 29, 65, 144, 172

43. The following cryptogram was encoded with a 2×2 matrix.

8, 21, -15, -10, -13, -13, 5, 10, 5, 25,

5, 19, -1, 6, 20, 40, -18, -18, 1, 16

The last word of the message is _RON. What is the message?

44. The following cryptogram was encoded with a 2×2 matrix.

5, 2, 25, 11, -2, -7, -15, -15, 32,

14, -8, -13, 38, 19, -19, -19, 37, 16

The last word of the message is _SUE. What is the message?

CHAPTER PROJECT: Marketing a New Product

Your company, Brand X, is introducing a new type of film in a market that is dominated by two companies: Brand A and Brand B. The current percents of the market held by the three companies are given by the matrix

$$X_0 = \begin{bmatrix} 0.4 \\ 0.6 \\ 0.0 \end{bmatrix} . \quad \begin{matrix} \text{Brand A} \\ \text{Brand B} \\ \text{Brand X} \end{matrix}$$

After conducting market research, you believe that each month after you introduce your product, the following changes will occur. (1) Brand A will retain 91% of its share, acquire 3% of Brand B's share, and acquire 2% of your share. (2) Brand B will retain 94% of its share, acquire 4% of Brand A's share, and acquire 1% of your share. (3) Brand X will retain 97% of its share, acquire 5% of Brand A's share, and acquire 3% of Brand B's share. After one month, the shares of the three companies will be given by

$$X_1 = P X_0 = \begin{bmatrix} 0.91 & 0.03 & 0.02 \\ 0.04 & 0.94 & 0.01 \\ 0.05 & 0.03 & 0.97 \end{bmatrix} \begin{bmatrix} 0.4 \\ 0.6 \\ 0.0 \end{bmatrix} = \begin{bmatrix} 0.382 \\ 0.58 \\ 0.038 \end{bmatrix} . \quad \begin{matrix} \text{Brand A} \\ \text{Brand B} \\ \text{Brand X} \end{matrix}$$

Your market share during the first 12 months is shown in the table below.

Month	0	2	4	6	8	10	12
Share	0	0.073	0.137	0.192	0.240	0.282	0.318

Use this information to investigate the following questions.

1. **Making a Table**   The table above shows Brand X's share for 6 months during the first year. Construct a complete table that shows the shares of the three brands during each of the first 12 months. The matrix operations on your graphing utility can help with the calculations.

2. **Fitting a Model to Data**   Enter the data for Brand X's share into a graphing utility.

   (a) Sketch a scatter plot of the data.

   (b) What type of model do you think best fits the data? Will any of the curve-fitting programs on your graphing utility find the model that you think fits best?

3. **Fitting a Model to Data**   Construct a new table that shows the portion of the market that Brand X *does not* have. Enter this data into a graphing utility, plot the data, and use one of the utility's curve-fitting programs to find a model. One program will produce a very good fit. Which one? What model?

4. **Think About It**   What is your brand's future? According to your assumptions, when will your brand have over 50% of the market? Over 75%?

CHAPTER SUMMARY

After studying this chapter, you should have acquired the following skills. These skills are keyed to the Review Exercises that begin on page 560. Answers to odd-numbered Review Exercises are given in the back of the book.

- Write a matrix in row-echelon form. *(Section 7.1)* **Review Exercises 1, 2**

- Write a matrix in reduced row-echelon form. *(Section 7.1)* **Review Exercises 3, 4**

- Write a system of linear equations represented by an augmented matrix. *(Section 7.1)* **Review Exercises 5, 6**

- Use matrices to solve a linear system of equations. *(Section 7.1)* **Review Exercises 7–16**

- Use matrix addition to add two matrices. *(Section 7.2)* **Review Exercises 17–20**

- Use matrix subtraction to subtract two matrices. *(Section 7.2)* **Review Exercises 17–20**

- Multiply a matrix by a scalar. *(Section 7.2)* **Review Exercises 17–20**

- Multiply two matrices. *(Section 7.2)* **Review Exercises 21–30**

- Solve a matrix equation. *(Section 7.2)* **Review Exercises 31, 32**

- Use the properties of scalar multiplication to solve a real-life application. *(Section 7.2)* **Review Exercises 33, 34**

- Use the properties of matrix multiplication to solve a real-life application. *(Section 7.2)* **Review Exercises 35, 36**

- Show that a matrix is an inverse of another matrix. *(Section 7.3)* **Review Exercises 37, 38**

- Find the inverse of a matrix. *(Section 7.3)* **Review Exercises 39–42**

- Use an inverse matrix to solve a system of linear equations. *(Section 7.3)* **Review Exercises 43–50**

- Find the determinant of a matrix. *(Section 7.4)* **Review Exercises 51–62**

- Find the area of a triangle using a determinant. *(Section 7.5)* **Review Exercises 63, 64**

- Use a determinant to ascertain whether three points are collinear. *(Section 7.5)* **Review Exercises 65, 66**

- Find the equation of a line passing through two points using a determinant. *(Section 7.5)* **Review Exercises 67, 68**

- Use a matrix to encode a message. *(Section 7.5)* **Review Exercises 69, 70**

- Use a matrix to decode a cryptogram. *(Section 7.5)* **Review Exercises 71, 72**

REVIEW EXERCISES

In Exercises 1 and 2, write the matrix in row-echelon form.

1. $\begin{bmatrix} 1 & 3 & 0 & 2 \\ 3 & 10 & 1 & 8 \\ 2 & 3 & 3 & 10 \end{bmatrix}$ **2.** $\begin{bmatrix} 1 & 2 & -1 & 0 \\ -2 & -3 & 3 & 4 \\ 4 & 0 & 1 & 3 \end{bmatrix}$

In Exercises 3 and 4, write the matrix in reduced row-echelon form.

3. $\begin{bmatrix} 1 & 2 & 3 \\ -2 & 0 & 2 \\ 2 & 1 & 2 \end{bmatrix}$ **4.** $\begin{bmatrix} 2 & 3 & 1 & -5 \\ 1 & 0 & 5 & 2 \\ -1 & 4 & 3 & 6 \\ 0 & -2 & 6 & -8 \end{bmatrix}$

In Exercises 5 and 6, write the system of linear equations represented by the augmented matrix. (Use variables x, y, and z.)

5. $\begin{bmatrix} 3 & -2 & 4 & \vdots & 9 \\ 1 & 0 & 5 & \vdots & 11 \\ -2 & 1 & 3 & \vdots & -5 \end{bmatrix}$

6. $\begin{bmatrix} 1 & 0 & 2 & \vdots & 4 \\ 0 & 1 & 3 & \vdots & 9 \\ 2 & -3 & -1 & \vdots & -6 \end{bmatrix}$

In Exercises 7–14, use matrices to solve the system.

7. $\begin{aligned} 4x - 3y &= 18 \\ x + y &= 1 \end{aligned}$ **8.** $\begin{aligned} 2x + 4y &= 16 \\ -x + 3y &= 17 \end{aligned}$

9. $\begin{aligned} 2x + 3y - z &= 13 \\ 3x \qquad + z &= 8 \\ x - 2y + 3z &= -4 \end{aligned}$

10. $\begin{aligned} 3x + 4y + 2z &= 5 \\ 2x + 3y \qquad &= 7 \\ 2y - 3z &= 12 \end{aligned}$

11. $\begin{aligned} x + 2y + 2z &= 10 \\ 2x + 3y + 5z &= 20 \end{aligned}$

12. $\begin{aligned} 3x + 10y + 4z &= 20 \\ x + 3y - 2z &= 8 \end{aligned}$

13. $\begin{aligned} 2x + y - 3z &= 4 \\ x + 2y + 2z &= 10 \\ x \qquad - 2z &= 12 \\ x + y + z &= 6 \end{aligned}$

14. $\begin{aligned} 2x + 4y + 2z &= 10 \\ x \qquad + 3z &= 9 \\ 3x - 2y \qquad &= 4 \\ x + y + z &= 8 \end{aligned}$

15. *Borrowing Money* A company borrowed $600,000 to expand its product line. Some of the money was borrowed at 9%, some at 10%, and some at 12%. How much was borrowed at each rate if the annual interest was $63,000 and the amount borrowed at 10% was three times the amount borrowed at 9%?

16. *Borrowing Money* A company borrowed $700,000 to expand its product line. Some of the money was borrowed at 9.5%, some at 10.5%, and some at 11%. How much was borrowed at each rate if the annual interest was $70,000 and the amount borrowed at 9.5% was four times the amount borrowed at 11%?

In Exercises 17–20, find (a) $A + B$, (b) $A - B$, (c) $4A$, and (d) $4A - 3B$.

17. $A = \begin{bmatrix} 1 & 3 \\ 1 & -2 \end{bmatrix}$, $B = \begin{bmatrix} 2 & -2 \\ 4 & 7 \end{bmatrix}$

18. $A = \begin{bmatrix} 1 & 0 & 2 \\ -1 & 3 & 5 \\ 2 & -2 & 3 \end{bmatrix}$, $B = \begin{bmatrix} 2 & 0 & 1 \\ 3 & -4 & 6 \\ 1 & 2 & -3 \end{bmatrix}$

19. $A = \begin{bmatrix} 1 & 3 & -2 & 6 \\ 0 & 1 & 3 & 2 \end{bmatrix}$

$B = \begin{bmatrix} 2 & 1 & 4 & -5 \\ 3 & -6 & 3 & -2 \end{bmatrix}$

20. $A = \begin{bmatrix} 3 \\ -2 \\ 3 \end{bmatrix}$, $B = \begin{bmatrix} -1 \\ 4 \\ 5 \end{bmatrix}$

In Exercises 21–24, find (a) AB, (b) BA, and, if possible, (c) A^2. (*Note:* $A^2 = AA$.)

21. $A = \begin{bmatrix} 1 & 3 \\ 2 & 4 \end{bmatrix}$, $B = \begin{bmatrix} 2 & -1 \\ 3 & 3 \end{bmatrix}$

22. $A = \begin{bmatrix} 2 & -3 \\ 4 & 5 \end{bmatrix}$, $B = \begin{bmatrix} -1 & 0 \\ 2 & -1 \end{bmatrix}$

23. $A = \begin{bmatrix} 1 & 0 & 2 \\ 3 & 1 & -2 \\ 1 & 1 & 1 \end{bmatrix}$, $B = \begin{bmatrix} 2 & 0 & 0 \\ 1 & -2 & 1 \\ 5 & 4 & -2 \end{bmatrix}$

24. $A = \begin{bmatrix} 0 & 2 & 3 \\ 1 & -1 & 0 \\ 1 & 2 & 1 \end{bmatrix}$, $B = \begin{bmatrix} 1 & 1 & 4 \\ 2 & 3 & 0 \\ 4 & 3 & 2 \end{bmatrix}$

In Exercises 25–30, find AB, if possible.

25. $A = \begin{bmatrix} 1 & 0 & 2 \\ 1 & 3 & 4 \\ -2 & 1 & 1 \end{bmatrix}$, $B = \begin{bmatrix} 2 & 1 \\ 3 & 0 \\ -1 & 5 \end{bmatrix}$

26. $A = \begin{bmatrix} 3 & 1 \\ 4 & 7 \\ 1 & 1 \end{bmatrix}$, $B = \begin{bmatrix} 1 & 2 & -2 \\ 3 & 4 & 0 \\ 0 & 1 & 0 \end{bmatrix}$

27. $A = \begin{bmatrix} 4 & 0 & 0 \\ 0 & 3 & 0 \\ 0 & 0 & -2 \end{bmatrix}$, $B = \begin{bmatrix} \frac{1}{4} & 0 & 0 \\ 0 & \frac{1}{3} & 0 \\ 0 & 0 & -\frac{1}{2} \end{bmatrix}$

28. $A = \begin{bmatrix} 3 \\ 2 \\ 4 \\ 6 \end{bmatrix}$, $B = \begin{bmatrix} 2 & 0 & -1 \end{bmatrix}$

29. $A = \begin{bmatrix} 1 & 2 & 3 & 6 & -1 \\ 2 & 8 & 0 & 0 & 2 \end{bmatrix}$

$B = \begin{bmatrix} 3 & 2 \\ 4 & -1 \end{bmatrix}$

30. $A = \begin{bmatrix} 0 & 0 & 2 \\ 1 & 0 & 6 \\ 0 & 2 & 2 \end{bmatrix}$, $B = \begin{bmatrix} 3 & 4 & 0 & 1 \\ 2 & 1 & 0 & 0 \\ 0 & 0 & 1 & 1 \end{bmatrix}$

In Exercises 31 and 32, solve for X given

$A = \begin{bmatrix} 1 & -2 \\ 0 & 1 \\ 2 & 3 \end{bmatrix}$ and $B = \begin{bmatrix} 0 & 1 \\ 1 & 1 \\ 3 & 5 \end{bmatrix}$.

31. $X = 5A - 3B$ **32.** $3X = 2A + 3B$

33. *Factory Production*   A corporation has four factories, each of which manufactures three products. The number of units of product i produced at factory j in one day is represented by a_{ij} in the matrix

$$A = \begin{bmatrix} 80 & 70 & 90 & 40 \\ 50 & 30 & 80 & 20 \\ 90 & 60 & 100 & 50 \end{bmatrix}.$$

Find the production levels if production is increased by 20%.

34. *Factory Production*   A corporation has three factories, each of which manufactures four products. The number of units of product j produced at factory i in one day is represented by a_{ij} in the matrix

$$A = \begin{bmatrix} 30 & 80 & 70 & 20 \\ 50 & 100 & 90 & 90 \\ 60 & 70 & 80 & 100 \end{bmatrix}.$$

Find the production levels if production is decreased by 10%. (*Hint:* Because a 10% decrease corresponds to $100\% - 10\%$, multiply the given matrix by 0.90.)

35. *Inventory Levels*   A company sells four different models of car sound systems through three retail outlets. The inventories of the four models at the three outlets are given by matrix S.

Model

$$S = \begin{array}{c} \\ \\ \\ \end{array} \begin{matrix} A & B & C & D \\ \end{matrix}$$

$$S = \begin{bmatrix} 2 & 3 & 2 & 1 \\ 0 & 4 & 3 & 3 \\ 4 & 0 & 1 & 2 \end{bmatrix} \begin{matrix} 1 \\ 2 \\ 3 \end{matrix} \Bigg\} \text{Outlet}$$

The wholesale and retail prices of the four models are given by matrix T.

Price

Wholesale   Retail

$$T = \begin{bmatrix} 300 & 500 \\ 400 & 650 \\ 200 & 350 \\ 800 & 1200 \end{bmatrix} \begin{matrix} A \\ B \\ C \\ D \end{matrix} \Bigg\} \text{Model}$$

Use a graphing utility to compute ST and interpret the result.

36. *Labor/Wage Requirements* A company that manufactures racing bicycles has the following labor-hour and wage requirements.

Labor-Hour Requirements (per bicycle)

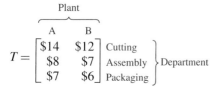

$$S = \begin{bmatrix} 0.9 \text{ hour} & 0.8 \text{ hour} & 0.2 \text{ hour} \\ 1.5 \text{ hours} & 1.0 \text{ hour} & 0.4 \text{ hour} \\ 3.5 \text{ hours} & 3.0 \text{ hours} & 0.5 \text{ hour} \end{bmatrix} \begin{matrix} \text{Basic} \\ \text{Light} \\ \text{Ultra-} \\ \text{light} \end{matrix} \text{Models}$$

where the columns are Cutting, Assembly, Packaging under Department.

Wage Requirements (per hour)

$$T = \begin{bmatrix} \$14 & \$12 \\ \$8 & \$7 \\ \$7 & \$6 \end{bmatrix} \begin{matrix} \text{Cutting} \\ \text{Assembly} \\ \text{Packaging} \end{matrix} \text{Department}$$

where the columns A and B are Plant.

(a) What is the labor cost for a light racing bicycle at Plant A?

(b) What is the labor cost for an ultra-light racing bicycle at Plant B?

(c) Use a graphing utility to compute ST and interpret the result.

In Exercises 37 and 38, show that B is the inverse of A.

37. $A = \begin{bmatrix} 1 & 2 & 1 \\ 3 & 6 & 4 \\ 0 & 1 & 3 \end{bmatrix}$, $B = \begin{bmatrix} -14 & 5 & -2 \\ 9 & -3 & 1 \\ -3 & 1 & 0 \end{bmatrix}$

38. $A = \begin{bmatrix} 2 & 0 & 1 & 2 \\ 3 & 0 & 0 & 1 \\ -1 & 1 & 2 & 0 \\ 0 & -1 & 2 & 2 \end{bmatrix}$

$B = \dfrac{1}{9} \begin{bmatrix} -4 & 6 & 1 & 1 \\ 10 & -6 & 2 & -7 \\ -7 & 6 & 4 & 4 \\ 12 & -9 & -3 & -3 \end{bmatrix}$

In Exercises 39–42, find the inverse of the matrix (if it exists).

39. $\begin{bmatrix} 1 & 3 \\ 2 & 5 \end{bmatrix}$ **40.** $\begin{bmatrix} -2 & 1 \\ 4 & 3 \end{bmatrix}$

41. $\begin{bmatrix} -1 & 0 & 0 & 0 \\ 0 & 2 & 0 & 0 \\ 0 & 0 & 4 & 0 \\ 0 & 0 & 0 & 6 \end{bmatrix}$ **42.** $\begin{bmatrix} 3 & 2 & 2 \\ 0 & 2 & 1 \\ 1 & 0 & 1 \end{bmatrix}$

In Exercises 43 and 44, use an inverse matrix to solve the system of linear equations. (Use the inverse matrix found in Exercise 39.)

43. $x + 3y = 15$
$2x + 5y = 26$

44. $x + 3y = 7$
$2x + 5y = 11$

In Exercises 45 and 46, use an inverse matrix to solve the system of linear equations. (Use the inverse matrix found in Exercise 42.)

45. $3x + 2y + 2z = 13$
$2y + z = 4$
$x + z = 5$

46. $3x + 2y + 2z = 12$
$2y + z = 13$
$x + z = 3$

Computer Models In Exercises 47 and 48, consider a company that produces three computer models. Model A requires 2 units of plastic, 2 units of computer chips, and 1 unit of computer "cards." Model B requires 1 unit of plastic, 3 units of computer chips, and 2 units of computer "cards." Model C requires 3 units of plastic, 2 units of computer chips, and 2 units of computer "cards." A system of linear equations (where x, y, and z represent models A, B, and C) is as follows.

$2x + y + 3z = \text{(units of plastic)}$
$2x + 3y + 2z = \text{(units of computer chips)}$
$x + 2y + 2z = \text{(units of computer "cards")}$

Use the inverse of the coefficient matrix of this system to find the numbers of models A, B, and C that the company can produce with the given amounts of components.

47. 85 units of plastic
110 units of computer chips
70 units of computer "cards"

48. 600 units of plastic
800 units of computer chips
500 units of computer "cards"

Investment Portfolio In Exercises 49 and 50, consider a person who invests in blue-chip stocks, common stocks, and municipal bonds. The average yields are 8% on blue-chip stocks, 6% on common stocks, and 7% on municipal bonds. Twice as much is invested in municipal bonds as in common stocks. Moreover, the total annual return for all three types of investments (stocks and bonds) is $2400. A system of linear equations (where x, y, and z represent the amounts invested in blue-chip stocks, common stocks, and municipal bonds) is as follows.

$$x + y + z = \text{(total investment)}$$
$$0.08x + 0.06y + 0.07z = 2400$$
$$2y - z = 0$$

Use the inverse of the coefficient matrix of this system to find the amount invested in each type of stock or bond for the indicated total investment.

49. Total investment = $32,000

50. Total investment = $34,000

In Exercises 51–62, find the determinant of the matrix.

51. $\begin{bmatrix} 3 & -2 \\ 6 & 5 \end{bmatrix}$ **52.** $\begin{bmatrix} 2 & 0 \\ 3 & 6 \end{bmatrix}$

53. $\begin{bmatrix} 5 & 2 \\ 0 & 0 \end{bmatrix}$ **54.** $\begin{bmatrix} 3 & 0 \\ 0 & -7 \end{bmatrix}$

55. $\begin{bmatrix} 1 & 2 & 3 \\ 8 & 6 & 7 \\ 0 & 2 & -1 \end{bmatrix}$ **56.** $\begin{bmatrix} -2 & 3 & 3 \\ -1 & 0 & 5 \\ 1 & 2 & -1 \end{bmatrix}$

57. $\begin{bmatrix} 2 & 0 & 0 & 0 \\ 0 & 5 & 0 & 0 \\ 0 & 0 & 4 & 0 \\ 0 & 0 & 0 & 3 \end{bmatrix}$ **58.** $\begin{bmatrix} 2 & 0 & 0 & 0 \\ 3 & 4 & 0 & 0 \\ 5 & 1 & -2 & 0 \\ 6 & 3 & 1 & 1 \end{bmatrix}$

59. $\begin{bmatrix} 3 & 0 & 0 & 0 \\ 0 & 2 & 0 & 0 \\ 0 & 0 & -1 & 0 \\ 0 & 0 & 0 & -10 \end{bmatrix}$

60. $\begin{bmatrix} 1 & 3 & 2 & 4 \\ 0 & -1 & 2 & 2 \\ 0 & 0 & 3 & 0 \\ 0 & 0 & 0 & 4 \end{bmatrix}$

61. $\begin{bmatrix} 1 & 3 & 0 & 7 \\ 5 & 2 & 4 & 2 \\ 0 & 1 & 0 & 1 \\ 2 & 1 & 0 & 0 \end{bmatrix}$ **62.** $\begin{bmatrix} 1 & 0 & 0 & 0 & 0 \\ 0 & 2 & 0 & 0 & 0 \\ 0 & 0 & 0 & 8 & 0 \\ 0 & 0 & 0 & 0 & \frac{1}{2} \end{bmatrix}$

In Exercises 63 and 64, use a determinant to find the area of the triangle with the given vertices.

63. $(-1, 1)$, $(2, 3)$, $(4, -1)$

64. $(-4, -2)$, $(2, 4)$, $(8, -4)$

In Exercises 65 and 66, use a determinant to ascertain whether the points are collinear.

65. $(0, 3)$, $(1, 5)$, $(2, 8)$ **66.** $(2, 6)$, $(-2, 3)$, $(0, 5)$

In Exercises 67 and 68, use a determinant to find an equation of the line through the points (x_1, y_1) and (x_2, y_2).

67. $(-5, 2)$, $(3, 0)$ **68.** $(4, 1)$, $(-2, 2)$

In Exercises 69 and 70, use the matrix to encode the message.

Message	*Matrix*

69. HAPPY ANNIVERSARY $\begin{bmatrix} 2 & 3 \\ 3 & 4 \end{bmatrix}$

70. MEET ME IN ST LOUIS $\begin{bmatrix} 1 & 2 & 2 \\ 3 & 7 & 9 \\ -1 & -4 & -7 \end{bmatrix}$

In Exercises 71 and 72, use A^{-1} to decode the cryptogram.

71. $A = \begin{bmatrix} 1 & 2 & 2 \\ 3 & 7 & 9 \\ -1 & -4 & -7 \end{bmatrix}$

$-5, -41, -87, 91, 207, 257, 41, 73, 67, 20, 42, 47, 82, 183, 221$

72. $A = \begin{bmatrix} 1 & -1 & 0 \\ 1 & 0 & -1 \\ -6 & 2 & 3 \end{bmatrix}$

$-14, -1, 10, -38, 2, 27, -94, 18, 57, 7, -11, -1, -96, 20, 57, -74, 23, 35, 27, -12, -5$

CHAPTER TEST

Take this test as you would take a test in class. After you are done, check your work against the answers given in the back of the book.

In Exercises 1 and 2, write an augmented matrix that represents the system.

1. $2x - y = -3$
$\quad x + y = 2$

2. $3x + 2y + 2z = 17$
$\quad\quad\quad 2y + z = -1$
$\quad x + z = 8$

In Exercises 3–5, use matrices to solve the system.

3. $3x + 4y = -6$
$\quad -x + 3y = -11$

4. $x - 2y + z = 14$
$\quad\quad y - 3z = 2$
$\quad\quad\quad\quad z = -6$

5. $2x - 3y + z = 14$
$\quadx + 2y = -4$
$\quad\quad y - z = -4$

In Exercises 6–9, use the matrices to find the indicated matrix.

$$A = \begin{bmatrix} 1 & 3 \\ 2 & 4 \end{bmatrix}, \quad B = \begin{bmatrix} 2 & -1 & 3 \\ 4 & 0 & 1 \end{bmatrix}, \quad C = \begin{bmatrix} 0 & -2 \\ 3 & 5 \end{bmatrix}, \quad D = \begin{bmatrix} 3 \\ 2 \\ -1 \end{bmatrix}$$

6. $2A + C$ **7.** AB **8.** BD **9.** A^2

In Exercises 10–12, find the inverse of the matrix.

10. $A = \begin{bmatrix} 2 & -1 \\ -3 & 4 \end{bmatrix}$ **11.** $A = \begin{bmatrix} 1 & 0 \\ 0 & 1 \end{bmatrix}$ **12.** $A = \begin{bmatrix} 3 & 2 & 2 \\ 0 & 2 & 1 \\ 1 & 0 & 1 \end{bmatrix}$

In Exercises 13–15, find the determinant of the matrix.

13. $\begin{bmatrix} 3 & -1 \\ 4 & 7 \end{bmatrix}$ **14.** $\begin{bmatrix} 3 & 2 & -1 \\ 1 & 0 & 2 \\ 4 & 5 & 2 \end{bmatrix}$ **15.** $\begin{bmatrix} 2 & 0 & 0 \\ 0 & 5 & 0 \\ 0 & 0 & -2 \end{bmatrix}$

16. Use the inverse found in Exercise 12 to solve the system in Exercise 2.

17. Find two nonzero matrices whose product is a zero matrix.

18. Use a determinant to decide whether $(2, 9)$, $(-2, 1)$, and $(3, 11)$ are collinear.

19. Find the area of the triangle whose vertices are $(-2, 4)$, $(0, 5)$, and $(3, -1)$.

20. A manufacturer produces three models of a product, which are shipped to two warehouses. The number of units i that are shipped to warehouse j is represented by a_{ij} in matrix A at the right. The price per unit is represented by matrix B. Find the product BA and interpret the result.

$$A = \begin{bmatrix} 1000 & 3000 \\ 2000 & 4000 \\ 5000 & 8000 \end{bmatrix}$$

$$B = \begin{bmatrix} \$25 & \$20 & \$32 \end{bmatrix}$$

Matrices for 20

Sequences and Probability

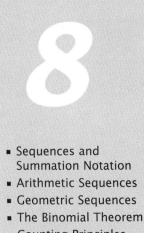

- Sequences and Summation Notation
- Arithmetic Sequences
- Geometric Sequences
- The Binomial Theorem
- Counting Principles
- Probability

When a rubber ball is dropped and bounces on a hard, level surface, each rebound height is a percent of the previous rebound height. This percent *p* is called the *rebound rate* and is a function of the ball's composition.

To model the rebound heights *y* of a bouncing ball, you can use the exponential model

$$y = y_0 p^n$$

where y_0 is the initial height and *n* is the number of bounces.

The table and graph at the right show the heights of several bounces for a ball with a rebound rate of $p = 0.8$ and an initial height of 10 feet. The total distance traveled by the ball, up and down, during its eight rebounds can be found using the formula for the sum of a geometric sequence.

n	0	1	2	3	4	5	6	7	8
y	10.00	8.00	6.40	5.12	4.10	3.28	2.62	2.10	1.68

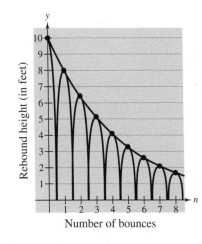

Number of bounces

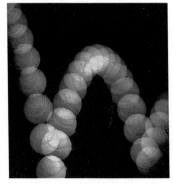

The chapter project related to this information is on page 627. This project uses actual data, collected with a *Texas Instruments CBL* unit.

565

8.1	**Sequences and Summation Notation**
	Sequences ▪ Factorial Notation ▪ Summation Notation ▪ Application

Sequences

In mathematics, the word *sequence* is used in much the same way as in ordinary English. Saying that a collection is listed *in sequence* means that it is ordered so that it has a first member, a second member, a third member, and so on.

Mathematically, you can think of a sequence as a *function* whose domain is the set of positive integers. Rather than using function notation, however, sequences are usually written using subscript notation, as indicated in the definition below.

NOTE On occasion it is convenient to begin subscripting a sequence with 0 instead of 1 so that the terms of the sequence become

$$a_0, \ a_1, \ a_2, \ a_3, \ \ldots .$$

Definition of a Sequence

An **infinite sequence** is a function whose domain is the set of positive integers. The function values

$$a_1, \ a_2, \ a_3, \ a_4, \ \ldots, \ a_n, \ \ldots$$

are the **terms** of the sequence. If the domain of the function consists of the first n positive integers only, the sequence is called a **finite sequence.**

EXAMPLE 1   Finding Terms of a Sequence

a. The first four terms of the sequence given by $a_n = 3n - 2$ are

$a_1 = 3(1) - 2 = 1$	1st term
$a_2 = 3(2) - 2 = 4$	2nd term
$a_3 = 3(3) - 2 = 7$	3rd term
$a_4 = 3(4) - 2 = 10.$	4th term

b. The first four terms of the sequence given by $a_n = 3 + (-1)^n$ are

$a_1 = 3 + (-1)^1 = 3 - 1 = 2$	1st term
$a_2 = 3 + (-1)^2 = 3 + 1 = 4$	2nd term
$a_3 = 3 + (-1)^3 = 3 - 1 = 2$	3rd term
$a_4 = 3 + (-1)^4 = 3 + 1 = 4.$	4th term

The terms of a sequence need not all be positive, as shown in Example 2.

EXAMPLE 2 Finding Terms of a Sequence

The first four terms of the sequence given by $a_n = (-1)^n/(2n-1)$ are

$$a_1 = \frac{(-1)^1}{2(1)-1} = \frac{-1}{2-1} = -1 \qquad \text{1st term}$$

$$a_2 = \frac{(-1)^2}{2(2)-1} = \frac{1}{4-1} = \frac{1}{3} \qquad \text{2nd term}$$

$$a_3 = \frac{(-1)^3}{2(3)-1} = \frac{-1}{6-1} = -\frac{1}{5} \qquad \text{3rd term}$$

$$a_4 = \frac{(-1)^4}{2(4)-1} = \frac{1}{8-1} = \frac{1}{7}. \qquad \text{4th term}$$

NOTE Try finding the first four terms of the sequence whose nth term is

$$a_n = \frac{(-1)^{n+1}}{2n-1}.$$

How do they differ from the first four terms of the sequence in Example 2?

It is important to realize that simply listing the first few terms is not sufficient to define a sequence—the nth term *must be given*. To see this, consider the following sequences, both of which have the same first three terms.

$$\frac{1}{2}, \frac{1}{4}, \frac{1}{8}, \frac{1}{16}, \cdots, \frac{1}{2^n}, \cdots$$

$$\frac{1}{2}, \frac{1}{4}, \frac{1}{8}, \frac{1}{15}, \cdots, \frac{6}{(n+1)(n^2-n+6)}, \cdots$$

Technology

Some graphing utilities, such as the *TI-82*, have spreadsheet capabilities that create tables. For instance, you can use the following steps to create a table that shows the terms of the sequence given by $a_n = 2^n + 1$.

1. Enter the sequence in $\boxed{\text{Y=}}$ as $Y_1 = 2 \wedge X + 1$.

2. With the TblSet Menu, set TblMin = 1 and $\triangle$Tbl = 1.

3. View the terms of the sequence using $\boxed{\text{TABLE}}$.

When viewing the table, you can use the cursor to look at additional terms.

Factorial Notation

Some very important sequences in mathematics involve terms that are defined with special types of products called **factorials.**

Definition of Factorial

If n is a positive integer, **n factorial** is defined by

$$n! = 1 \cdot 2 \cdot 3 \cdot 4 \cdots (n-1) \cdot n.$$

As a special case, zero factorial is defined as $0! = 1$.

Here are some values of $n!$ for the first several nonnegative integers. Notice that 0! is 1 by definition.

$$0! = 1$$
$$1! = 1$$
$$2! = 1 \cdot 2 = 2$$
$$3! = 1 \cdot 2 \cdot 3 = 6$$
$$4! = 1 \cdot 2 \cdot 3 \cdot 4 = 24$$
$$5! = 1 \cdot 2 \cdot 3 \cdot 4 \cdot 5 = 120$$

The value of n does not have to be very large before the value of $n!$ becomes huge. For instance, $10! = 3,628,800$.

EXAMPLE 3   *Finding Terms of a Sequence Involving Factorials*

List the first five terms of the sequence whose nth term is $a_n = 2/n!$. Begin with $n = 0$.

Solution

$$a_0 = \frac{2}{0!} = \frac{2}{1} = 2 \qquad\qquad \text{0th term}$$

$$a_1 = \frac{2}{1!} = \frac{2}{1} = 2 \qquad\qquad \text{1st term}$$

$$a_2 = \frac{2}{2!} = \frac{2}{2} = 1 \qquad\qquad \text{2nd term}$$

$$a_3 = \frac{2}{3!} = \frac{2}{6} = \frac{1}{3} \qquad\qquad \text{3rd term}$$

$$a_4 = \frac{2}{4!} = \frac{2}{24} = \frac{1}{12} \qquad\qquad \text{4th term}$$

Factorials follow the same conventions for order of operations as do exponents. For instance,

$$2n! = 2(n!) = 2(1 \cdot 2 \cdot 3 \cdot 4 \cdots n)$$

whereas $(2n)! = 1 \cdot 2 \cdot 3 \cdot 4 \cdots 2n$.

EXAMPLE 4   Finding Terms of a Sequence Involving Factorials

List the first five terms of the sequence whose nth term is $a_n = 2^n/n!$. Begin with $n = 0$.

Solution

$$a_0 = \frac{2^0}{0!} = \frac{1}{1} = 1 \qquad\qquad \text{0th term}$$

$$a_1 = \frac{2^1}{1!} = \frac{2}{1} = 2 \qquad\qquad \text{1st term}$$

$$a_2 = \frac{2^2}{2!} = \frac{4}{2} = 2 \qquad\qquad \text{2nd term}$$

$$a_3 = \frac{2^3}{3!} = \frac{8}{6} = \frac{4}{3} \qquad\qquad \text{3rd term}$$

$$a_4 = \frac{2^4}{4!} = \frac{16}{24} = \frac{2}{3} \qquad\qquad \text{4th term}$$

When working with fractions involving factorials, you will often find that the fractions can be reduced.

EXAMPLE 5   Evaluating Factorial Expressions

Evaluate the following factorial expressions.

a. $\dfrac{n!}{(n-1)!}$ **b.** $\dfrac{8!}{2! \cdot 6!}$ **c.** $\dfrac{2! \cdot 6!}{3! \cdot 5!}$

Solution

a. $\dfrac{n!}{(n-1)!} = \dfrac{1 \cdot 2 \cdot 3 \cdots (n-1) \cdot n}{1 \cdot 2 \cdot 3 \cdots (n-1)} = n$

b. $\dfrac{8!}{2! \cdot 6!} = \dfrac{1 \cdot 2 \cdot 3 \cdot 4 \cdot 5 \cdot 6 \cdot 7 \cdot 8}{1 \cdot 2 \cdot 1 \cdot 2 \cdot 3 \cdot 4 \cdot 5 \cdot 6} = \dfrac{7 \cdot 8}{2} = 28$

c. $\dfrac{2! \cdot 6!}{3! \cdot 5!} = \dfrac{1 \cdot 2 \cdot 1 \cdot 2 \cdot 3 \cdot 4 \cdot 5 \cdot 6}{1 \cdot 2 \cdot 3 \cdot 1 \cdot 2 \cdot 3 \cdot 4 \cdot 5} = \dfrac{6}{3} = 2$

Summation Notation

There is a convenient notation for the sum of the terms of a finite sequence. It is called **summation notation** or **sigma notation** because it involves the use of the uppercase Greek letter sigma, written as Σ.

Definition of Summation Notation

The sum of the first n terms of a sequence is represented by

$$\sum_{i=1}^{n} a_i = a_1 + a_2 + a_3 + a_4 + \cdots + a_n$$

where i is called the **index of summation,** n is the **upper limit of summation,** and 1 is the **lower limit of summation.**

EXAMPLE 6   Summation Notation for Sums

a. $\displaystyle\sum_{i=1}^{5} 3i = 3(1) + 3(2) + 3(3) + 3(4) + 3(5)$

$\qquad\qquad = 3(1 + 2 + 3 + 4 + 5)$

$\qquad\qquad = 3(15) = 45$

b. $\displaystyle\sum_{k=3}^{6} (1 + k^2) = (1 + 3^2) + (1 + 4^2) + (1 + 5^2) + (1 + 6^2)$

$\qquad\qquad\qquad = 10 + 17 + 26 + 37 = 90$

c. $\displaystyle\sum_{i=0}^{8} \frac{1}{i!} = \frac{1}{0!} + \frac{1}{1!} + \frac{1}{2!} + \frac{1}{3!} + \frac{1}{4!} + \frac{1}{5!} + \frac{1}{6!} + \frac{1}{7!} + \frac{1}{8!}$

$\qquad\qquad = 1 + 1 + \frac{1}{2} + \frac{1}{6} + \frac{1}{24} + \frac{1}{120} + \frac{1}{720} + \frac{1}{5040} + \frac{1}{40,320}$

$\qquad\qquad \approx 2.71828$

For this summation, note that the sum is very close to the irrational number $e \approx 2.718281828$. It can be shown that as more terms of the sequence whose nth term is $1/n!$ are added, the sum becomes closer and closer to e.

NOTE   In Example 6, note that the lower index of a summation does not have to be 1. Also note that the index does not have to be the letter i. For instance, in part (b), the letter k is the index.

Properties of Sums

1. $\displaystyle\sum_{i=1}^{n} ca_i = c\sum_{i=1}^{n} a_i,$ c is any constant

2. $\displaystyle\sum_{i=1}^{n} (a_i + b_i) = \sum_{i=1}^{n} a_i + \sum_{i=1}^{n} b_i$

3. $\displaystyle\sum_{i=1}^{n} (a_i - b_i) = \sum_{i=1}^{n} a_i - \sum_{i=1}^{n} b_i$

Variations in the upper and lower limits of summation can produce quite different-looking summation notations for *the same sum.* For example, consider the following two sums.

$$\sum_{i=1}^{5} 3(2^i) = 3\sum_{i=1}^{5} 2^i = 3(2^1 + 2^2 + 2^3 + 2^4 + 2^5)$$

$$\sum_{i=0}^{4} 3(2^{i+1}) = 3\sum_{i=0}^{4} 2^{i+1} = 3(2^1 + 2^2 + 2^3 + 2^4 + 2^5)$$

The sum of the terms of any *finite* sequence must be a finite number. An important discovery in mathematics was that for some special types of *infinite* sequences, the sum of *all* the terms is a finite number. For instance, it can be shown that the sum of all of the terms of the sequence whose nth term is $1/2^n$ (with n beginning at 0) is

$$\sum_{n=0}^{\infty} \frac{1}{2^n} = 1 + \frac{1}{2} + \frac{1}{4} + \frac{1}{8} + \frac{1}{16} + \frac{1}{32} + \cdots$$

$$= 2.$$

EXAMPLE 7 *Finding the Sum of an Infinite Sequence*

$$\sum_{n=1}^{\infty} \frac{3}{10^n} = \frac{3}{10^1} + \frac{3}{10^2} + \frac{3}{10^3} + \frac{3}{10^4} + \frac{3}{10^5} + \cdots$$

$$= 0.3 + 0.03 + 0.003 + 0.0003 + 0.00003 + \cdots$$

$$= 0.33333\ldots$$

$$= \frac{1}{3}$$

Application

Sequences have many applications in business and science. One is illustrated in Example 8.

EXAMPLE 8   Population of the United States

From 1950 to 1993, the resident population of the United States can be approximated by the model

$$a_n = \sqrt{23{,}107 + 874.7n + 2.74n^2}, \qquad n = 0, 1, \ldots, 43$$

where a_n is the population in millions and n represents the calendar year, with $n = 0$ corresponding to 1950. Find the last five terms of this finite sequence. (Source: U.S. Bureau of Census)

Solution

The last five terms of this finite sequence are as follows.

$$a_{39} = \sqrt{23{,}107 + 874.7(39) + 2.74(39)^2} \approx 247.8 \qquad \text{1989 population}$$

$$a_{40} = \sqrt{23{,}107 + 874.7(40) + 2.74(40)^2} \approx 250.0 \qquad \text{1990 population}$$

$$a_{41} = \sqrt{23{,}107 + 874.7(41) + 2.74(41)^2} \approx 252.1 \qquad \text{1991 population}$$

$$a_{42} = \sqrt{23{,}107 + 874.7(42) + 2.74(42)^2} \approx 254.3 \qquad \text{1992 population}$$

$$a_{43} = \sqrt{23{,}107 + 874.7(43) + 2.74(43)^2} \approx 256.5 \qquad \text{1993 population}$$

The bar graph in Figure 8.1 graphically represents the population given by this sequence for the entire 44-year period from 1950 to 1993.

The United States Constitution specifies that a census occur every 10 years. The first census took place in 1790. Although the country had fewer than 4 million people, the census took 18 months to complete.

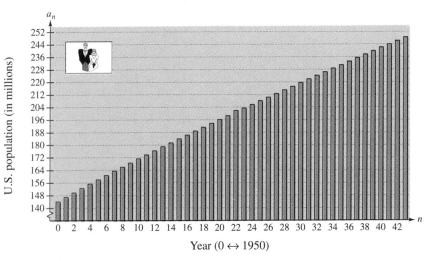

FIGURE 8.1

Activities

1. Write the first five terms of the sequence. (Assume n begins with 1.)

   $$a_n = \frac{2n-1}{2n}$$

   Answer: $\frac{1}{2}, \frac{3}{4}, \frac{5}{6}, \frac{7}{8}, \frac{9}{10}$

2. Write an expression for the most apparent nth term of the sequence $0, \frac{1}{2}, \frac{2}{6}, \frac{3}{24}, \frac{4}{120}$.

   Answer: $\dfrac{n-1}{n}$

3. Find the sum.

   $$\sum_{k=1}^{4} (-1)^k 2k$$

   Answer: 4

Group Activities   **Exploring with Technology**

A Summation Program   A graphing calculator may be programmed to calculate the sum of the first n terms of a sequence. For instance, the program below lists the program steps for the *TI–82*. See the Appendix for programs for other calculator models.

PROGRAM:SUM	Title of program
:Prompt M	M is the lower limit of summation.
:Prompt N	N is the upper limit of summation.
:0→S	Initialize S for use as stored sum.
:For (X, M, N)	Program for loop summing from lower limit to upper limit.
:S + Y$_1$→S	Evaluate function stored in Y$_1$ and add to sum.
:Disp S	Display partial sum.
:End	End For loop.

To use this program to find the sum of a finite sequence, store the nth term of the sequence as Y_1 and run the program. For instance, to find the sum

$$\sum_{i=1}^{10} 2^i$$

you should press $\boxed{\text{Y=}}$, enter $2 \wedge X$, and then use the keystroke sequence

$\boxed{\text{PRGM}}$, cursor to program SUM, $\boxed{\text{ENTER}}$

1 $\boxed{\text{ENTER}}$ 10 $\boxed{\text{ENTER}}$.

The result should be 2046. (Note that the calculator displayed 10 numbers—the last number displayed is the final sum.) To pause the program after each partial sum, insert :Pause after :Disp S. Then press $\boxed{\text{ENTER}}$ to display the next partial sum. Try using this program to find the following sums.

a. $\sum_{i=1}^{15} (i - 2)^2$ **b.** $\sum_{i=1}^{20} 3^{(i-1)}$ **c.** $\sum_{i=1}^{10} \sqrt{e^i}$

Another way to calculate the sum of the first n terms of a sequence using the *TI–82* is to combine the **sum** and **seq(** commands under the LIST menu. However, this method does not display any of the partial sums leading up to the final sum. Try using this method to find the sums in (a), (b), and (c).

Warm Up The following warm-up exercises involve skills that were covered in earlier sections. You will use these skills in the exercise set for this section.

1. Find $f(2)$ for $f(n) = \dfrac{2n}{n^2 + 1}$.

2. Find $f(3)$ for $f(n) = \dfrac{4}{3(n + 1)}$.

In Exercises 3–6, factor the expression.

3. $4n^2 - 1$

4. $4n^2 - 8n + 3$

5. $n^2 - 3n + 2$

6. $n^2 + 3n + 2$

In Exercises 7–10, perform the indicated operations and/or simplify.

7. $\left(\dfrac{2}{3}\right)\left(\dfrac{3}{4}\right)\left(\dfrac{4}{5}\right)\left(\dfrac{5}{6}\right)$

8. $\dfrac{2 \cdot 4 \cdot 6 \cdot 8}{2^4}$

9. $\dfrac{1}{2 \cdot 2} + \dfrac{1}{2 \cdot 3} + \dfrac{1}{2 \cdot 4}$

10. $\dfrac{1}{1 \cdot 2} + \dfrac{1}{2 \cdot 3} + \dfrac{1}{3 \cdot 4}$

8.1 Exercises

In Exercises 1–22, write the first five terms of the sequence. (Assume n begins with 1.)

1. $a_n = 2n + 1$

2. $a_n = 4n - 3$

3. $a_n = 2^n$

4. $a_n = \left(\dfrac{1}{2}\right)^n$

5. $a_n = (-2)^n$

6. $a_n = \left(-\dfrac{1}{2}\right)^n$

7. $a_n = \dfrac{n}{n + 1}$

8. $a_n = \dfrac{n + 1}{n}$

9. $a_n = 1 + (-1)^n$

10. $a_n = \dfrac{1 + (-1)^n}{n}$

11. $a_n = 3 - \dfrac{1}{2^n}$

12. $a_n = \dfrac{3^n}{4^n}$

13. $a_n = \dfrac{3^n}{n!}$

14. $a_n = \dfrac{n!}{n}$

15. $a_n = 5 - \dfrac{1}{n} + \dfrac{1}{n^2}$

16. $a_n = \dfrac{n^2 - 1}{n^2 + 2}$

17. $a_n = \dfrac{(-1)^n}{n^2}$

18. $a_n = \dfrac{(-1)^n n}{n + 1}$

19. $a_n = \dfrac{3n!}{(n - 1)!}$

20. $a_n = \dfrac{1}{n^{3/2}}$

21. $a_n = \dfrac{n^2 - 1}{n + 1}$

22. $a_n = \dfrac{3n^2 - n + 4}{2n^2 + 1}$

In Exercises 23–26, evaluate the expression.

23. $\dfrac{6!}{4!}$

24. $\dfrac{7!}{4!}$

25. $\dfrac{10!}{8!}$

26. $\dfrac{25!}{23!}$

In Exercises 27–30, simplify the expression. Then evaluate the expression when $n = 0$ and $n = 1$.

27. $\dfrac{(n + 1)!}{n!}$

28. $\dfrac{(n + 2)!}{n!}$

29. $\dfrac{(2n-1)!}{(2n+1)!}$ **30.** $\dfrac{(2n+2)!}{(2n)!}$

In Exercises 31–44, write an expression for the most apparent nth term of the sequence. (Assume that n begins with 1.)

31. $1, 4, 7, 10, 13, \ldots$ **32.** $3, 7, 11, 15, 19, \ldots$

33. $0, 3, 8, 15, 24, \ldots$ **34.** $1, \frac{1}{4}, \frac{1}{9}, \frac{1}{16}, \frac{1}{25}, \ldots$

35. $\frac{2}{3}, \frac{3}{4}, \frac{4}{5}, \frac{5}{6}, \frac{6}{7}, \ldots$ **36.** $\frac{2}{1}, \frac{3}{3}, \frac{4}{5}, \frac{5}{7}, \frac{6}{9}, \ldots$

37. $\frac{1}{2}, -\frac{1}{4}, \frac{1}{8}, -\frac{1}{16}, \frac{1}{32}, \ldots$

38. $\frac{1}{3}, \frac{2}{9}, \frac{4}{27}, \frac{8}{81}, \frac{16}{243}, \ldots$

39. $1+\frac{1}{1}, 1+\frac{1}{2}, 1+\frac{1}{3}, 1+\frac{1}{4}, 1+\frac{1}{5}, \ldots$

40. $1+\frac{1}{2}, 1+\frac{3}{4}, 1+\frac{7}{8}, 1+\frac{15}{16}, 1+\frac{31}{32}, \ldots$

41. $1, \frac{1}{2}, \frac{1}{6}, \frac{1}{24}, \frac{1}{120}, \ldots$

42. $2, -4, 6, -8, 10, \ldots$

43. $1, -1, 1, -1, 1, \ldots$

44. $1, 2, \dfrac{2^2}{2}, \dfrac{2^3}{6}, \dfrac{2^4}{24}, \dfrac{2^5}{120}, \ldots$

In Exercises 45–58, find the sum.

45. $\displaystyle\sum_{i=1}^{4} (4i+2)$ **46.** $\displaystyle\sum_{i=1}^{6} (3i-1)$

47. $\displaystyle\sum_{i=1}^{5} (2i+1)$ **48.** $\displaystyle\sum_{i=1}^{6} 2i$

49. $\displaystyle\sum_{k=1}^{4} 10$ **50.** $\displaystyle\sum_{k=1}^{5} 4$

51. $\displaystyle\sum_{i=0}^{4} i^2$ **52.** $\displaystyle\sum_{i=0}^{5} 3i^2$

53. $\displaystyle\sum_{k=0}^{3} \frac{1}{k^2+1}$ **54.** $\displaystyle\sum_{j=3}^{5} \frac{1}{j}$

55. $\displaystyle\sum_{i=1}^{4} (i-1)^2$ **56.** $\displaystyle\sum_{k=2}^{5} (k+1)(k-3)$

57. $\displaystyle\sum_{i=1}^{4} (-1)^i (2i+4)$ **58.** $\displaystyle\sum_{i=1}^{4} (-2)^i$

In Exercises 59–68, use summation notation to write the sum.

59. $\dfrac{1}{3(1)} + \dfrac{1}{3(2)} + \dfrac{1}{3(3)} + \cdots + \dfrac{1}{3(9)}$

60. $\dfrac{5}{1+1} + \dfrac{5}{1+2} + \dfrac{5}{1+3} + \cdots + \dfrac{5}{1+15}$

61. $\left[2\left(\frac{1}{8}\right)+3\right] + \left[2\left(\frac{2}{8}\right)+3\right] + \cdots + \left[2\left(\frac{8}{8}\right)+3\right]$

62. $\left[1-\left(\frac{1}{6}\right)^2\right] + \left[1-\left(\frac{2}{6}\right)^2\right] + \cdots + \left[1-\left(\frac{6}{6}\right)^2\right]$

63. $3 - 9 + 27 - 81 + 243 - 729$

64. $1 - \frac{1}{2} + \frac{1}{4} - \frac{1}{8} + \cdots - \frac{1}{128}$

65. $\dfrac{1}{1^2} - \dfrac{1}{2^2} + \dfrac{1}{3^2} - \dfrac{1}{4^2} + \cdots - \dfrac{1}{20^2}$

66. $\dfrac{1}{1\cdot 3} + \dfrac{1}{2\cdot 4} + \dfrac{1}{3\cdot 5} + \cdots + \dfrac{1}{10\cdot 12}$

67. $\frac{1}{4} + \frac{3}{8} + \frac{7}{16} + \frac{15}{32} + \frac{31}{64}$

68. $\frac{1}{2} + \frac{2}{4} + \frac{6}{8} + \frac{24}{16} + \frac{120}{32} + \frac{720}{64}$

69. *Compound Interest*   A deposit of \$5000 is made in an account that earns 8% interest compounded quarterly. The balance in the account after n quarters is given by

$$A_n = 5000\left(1 + \frac{0.08}{4}\right)^n, \quad n = 1, 2, 3, \ldots.$$

(a) Compute the first eight terms of this sequence.

(b) Find the balance in this account after 10 years by computing the 40th term of the sequence.

(c) Is the balance after 20 years twice the balance after 10 years? Explain.

70. *Compound Interest*   A deposit of \$100 is made *each* month in an account that earns 12% interest compounded monthly. The balance in the account after n months is given by

$$A_n = 100(101)[(1.01)^n - 1], \quad n = 1, 2, 3, \ldots.$$

(a) Compute the first six terms of this sequence.

(b) Find the balance after 5 years by computing the 60th term of the sequence.

(c) Find the balance after 20 years by computing the 240th term of the sequence.

71. *Ratio of Men to Women*   Until the mid-1940s, the population of the United States had more men than women. After that, there were more women than men. The ratio of men to women is approximately given by the model

$$a_n = 1.09 - 0.027n + 0.0012n^2,$$

$$n = 1, 2, \ldots, 9$$

where a_n is the ratio of men to women and n is the year, with $n = 1, 2, 3, \ldots, 9$ corresponding to 1910, 1920, 1930, ..., 1990. (Source: U.S. Bureau of Census)

(a) Use a graphing utility to find the terms of this finite sequence.

(b) Construct a bar graph that represents the sequence.

(c) In 1990, the population of the United States was 250 million. How many of these were women? How many were men?

72. *Total Sales*   The total annual sales for Toys R Us (a chain of stores selling children's toys) from 1988 to 1993 can be approximated by the model

$$a_n = 4032.0 + 674.0n + 22.08n^2,$$

$$n = 0, 1, 2, 3, 4, 5$$

where a_n is the annual sales (in millions of dollars) and $n = 0$ represents 1988. Find the total sales from 1988 to 1993 by evaluating the sum.

$$\sum_{n=0}^{5} (4032.0 + 674.0n + 22.08n^2)$$

(Source: Toys R Us)

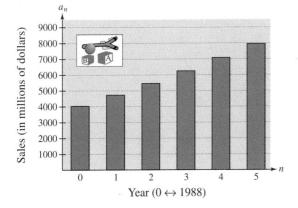
Year (0 ↔ 1988)

73. *Hospital Costs*   The average cost of a day in a hospital from 1985 to 1992 is given by the model

$$a_n = 462.3 + 33.33n + 2.50n^2, \quad n = 0, 1, \ldots, 7$$

where a_n is the average cost in dollars and $n = 0$ represents 1985. Use a graphing utility to find the terms of this finite sequence and construct a bar graph that represents the sequence. (Source: American Hospital Association)

74. *Total Revenue*   The total annual sales for MCI Communications from 1984 to 1993 can be approximated by the model

$$a_n = 741.8 + 64.56n + 61.28n^2,$$

$$n = 4, 5, \ldots, 13$$

where a_n is the annual sales (in millions of dollars) and $n = 4$ represents 1984. Find the total revenue from 1984 to 1993 by evaluating the sum. (Source: MCI Communications)

$$\sum_{n=4}^{13} (741.8 + 64.56n + 61.28n^2)$$

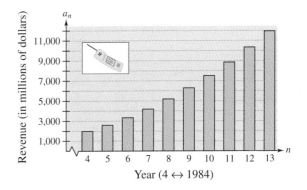
Year (4 ↔ 1984)

75. *Dividends*   A company declares dividends per share of common stock that can be approximated by the model $a_n = 0.35n + 1.25$, $n = 0, 1, 2, 3, 4$, where a_n is the dividend in dollars and $n = 0$ represents 1990. Approximate the sum of the dividends per share of common stock for the years 1990 through 1994 by evaluating

$$\sum_{n=0}^{4} (0.35n + 1.25).$$

8.2	**Arithmetic Sequences**
	Arithmetic Sequences ▪ The Sum of an Arithmetic Sequence ▪ Applications

Arithmetic Sequences

A sequence whose consecutive terms have a common difference is called an **arithmetic sequence.**

> ### Definition of an Arithmetic Sequence
>
> A sequence is **arithmetic** if the differences between consecutive terms are the same. Thus, the sequence
>
> $$a_1, a_2, a_3, a_4, \ldots, a_n, \ldots$$
>
> is arithmetic if there is a number d such that
>
> $$a_2 - a_1 = d, \quad a_3 - a_2 = d, \quad a_4 - a_3 = d,$$
>
> and so on. The number d is the **common difference** of the arithmetic sequence.

EXAMPLE 1   Examples of Arithmetic Sequences

a. The sequence whose nth term is $4n + 3$ is arithmetic. For this sequence, the common difference between consecutive terms is 4.

$$\underbrace{7, \; 11,}_{11 - 7 = 4} \; 15, \; 19, \; \ldots, \; 4n + 3, \; \ldots$$

b. The sequence whose nth term is $7 - 5n$ is arithmetic. For this sequence, the common difference between consecutive terms is -5.

$$\underbrace{2, \; -3,}_{-3 - 2 = -5} \; -8, \; -13, \; \ldots, \; 7 - 5n, \; \ldots$$

c. The sequence whose nth term is $\frac{1}{4}(n + 3)$ is arithmetic. For this sequence, the common difference between consecutive terms is $\frac{1}{4}$.

$$\underbrace{1, \; \frac{5}{4},}_{\frac{5}{4} - 1 = \frac{1}{4}} \; \frac{3}{2}, \; \frac{7}{4}, \; \ldots, \; \frac{n + 3}{4}, \; \ldots$$

In Example 1, notice that each of the arithmetic sequences has an nth term that is of the form $dn + c$, where the common difference of the sequence is d. This result is summarized as follows.

The nth Term of an Arithmetic Sequence

The nth term of an arithmetic sequence has the form

$$a_n = dn + c$$

where d is the common difference between consecutive terms of the sequence and $c = a_1 - d$.

EXAMPLE 2   Finding the nth Term of an Arithmetic Sequence

Find a formula for the nth term of the arithmetic sequence whose common difference is 3 and whose first term is 2.

Solution

Because the sequence is arithmetic, you know that the formula for the nth term is of the form $a_n = dn + c$. Moreover, because the common difference is $d = 3$, the formula must have the form

$$a_n = 3n + c.$$

Because $a_1 = 2$, it follows that

$$c = a_1 - d = 2 - 3 = -1.$$

Thus, the formula for the nth term is

$$a_n = 3n - 1.$$

The sequence therefore has the following form.

$$2, \ 5, \ 8, \ 11, \ 14, \ \ldots, \ 3n - 1, \ \ldots$$

Another way to find a formula for the nth term of the sequence in Example 2 is to begin by writing the terms of the sequence.

a_1	a_2	a_3	a_4	a_5	a_6	a_7	
2	$2+3$	$5+3$	$8+3$	$11+3$	$14+3$	$17+3$	$\ldots$
2	5	8	11	14	17	20	$\ldots$

From these terms, you can reason that the nth term is of the form

$$a_n = dn + c = 3n - 1.$$

EXAMPLE 3   *Finding the nth Term of an Arithmetic Sequence*

The fourth term of an arithmetic sequence is 20, and the 13th term is 65. Write the first several terms of this sequence.

Solution

The fourth and 13th terms of the sequence are related by

$$a_{13} = a_4 + 9d.$$

Using $a_4 = 20$ and $a_{13} = 65$, you can conclude that $d = 5$, which implies that the sequence is as follows.

$$
\begin{array}{ccccccccccc}
a_1 & a_2 & a_3 & a_4 & a_5 & a_6 & a_7 & a_8 & a_9 & a_{10} & a_{11} \\
5, & 10, & 15, & 20, & 25, & 30, & 35, & 40, & 45, & 50, & 55, \ldots
\end{array}
$$

If you know the nth term of an arithmetic sequence *and* you know the common difference of the sequence, you can find the $(n+1)$th term by using the **recursion formula**

$$a_{n+1} = a_n + d.$$

With this formula, you can find any term of an arithmetic sequence, *provided* that you know the previous term. For instance, if you know the first term, you can find the second term. Then, knowing the second term, you can find the third term, and so on.

EXAMPLE 4   *Using a Recursion Formula*

Find the ninth term of the arithmetic sequence whose first two terms are 2 and 9.

Solution

For this sequence, the common difference is $d = 9 - 2 = 7$. There are two ways to find the ninth term. One way is to simply write out the first nine terms (by repeatedly adding 7).

$$2, 9, 16, 23, 30, 37, 44, 51, 58$$

Another way to find the ninth term is to first find a formula for the nth term. Because the first term is 2, it follows that

$$c = a_1 - d = 2 - 7 = -5.$$

Therefore, a formula for the nth term is $a_n = 7n - 5$, which implies that the ninth term is $a_9 = 7(9) - 5 = 58$.

STUDY TIP

If you substitute $a_1 - d$ for c in the formula $a_n = dn + c$, then the nth term of an arithmetic sequence has the alternative recursion formula

$$a_n = a_1 + (n-1)d.$$

Use this formula to solve Example 4 and Example 9.

The Sum of an Arithmetic Sequence

There is a simple formula for the *sum* of a finite arithmetic sequence.

The Sum of a Finite Arithmetic Sequence

The sum of a finite arithmetic sequence with n terms is given by

$$S = \frac{n}{2}(a_1 + a_n).$$

EXAMPLE 5 *Finding the Sum of an Arithmetic Sequence*

Find the sum: $1 + 3 + 5 + 7 + 9 + 11 + 13 + 15 + 17 + 19$.

Solution

To begin, notice that the sequence is arithmetic (with a common difference of 2). Moreover, the sequence has 10 terms. Thus, the sum of the sequence is

$$S = 1 + 3 + 5 + 7 + 9 + 11 + 13 + 15 + 17 + 19$$
$$= \frac{n}{2}(a_1 + a_n)$$
$$= \frac{10}{2}(1 + 19)$$
$$= 5(20)$$
$$= 100.$$

DISCOVERY

To *develop* the formula for the sum of a finite arithmetic sequence, consider writing the sum in two different ways. One way is to write the sum as

$$S = a_1 + (a_1 + d) + (a_1 + 2d)$$
$$+ \cdots + [a_1 + (n - 1)d].$$

In the second way, you repeatedly subtract d from the nth term to obtain

$$S = a_n + (a_n - d) + (a_n - 2d)$$
$$+ \cdots + [a_n - (n - 1)d].$$

Can you discover a way to combine these two versions of S to obtain the following formula?

$$S = \frac{n}{2}(a_1 + a_n)$$

EXAMPLE 6 *Finding the Sum of an Arithmetic Sequence*

Find the sum of the integers from 1 to 100.

Solution

The integers from 1 to 100 form an arithmetic sequence that has 100 terms. Thus, you can use the formula for the sum of an arithmetic sequence, as follows.

$$S = 1 + 2 + 3 + 4 + 5 + 6 + \cdots + 99 + 100$$
$$= \frac{n}{2}(a_1 + a_n)$$
$$= \frac{100}{2}(1 + 100)$$
$$= 50(101)$$
$$= 5050$$

EXAMPLE 7   *Finding the Sum of an Arithmetic Sequence*

Find the sum of the first 150 terms of the arithmetic sequence

$$5, 16, 27, 38, 49, \ldots .$$

Solution

For this arithmetic sequence, you have $a_1 = 5$ and $d = 16 - 5 = 11$. Thus, $c = a_1 - d = 5 - 11 = -6$, and the nth term is

$$a_n = 11n - 6.$$

Therefore, $a_{150} = 11(150) - 6 = 1644$, and the sum of the first 150 terms is as follows.

$$\begin{aligned} S &= \frac{n}{2}(a_1 + a_n) \\ &= \frac{150}{2}(5 + 1644) \\ &= 75(1649) \\ &= 123{,}675 \end{aligned}$$

Applications

EXAMPLE 8   *Seating Capacity*

An auditorium has 20 rows of seats. There are 20 seats in the first row, 21 seats in the second row, 22 seats in the third row, and so on (see figure). How many seats are there in all 20 rows?

Solution

The number of seats in the rows forms an arithmetic sequence in which the common difference is $d = 1$. Because $c = a_1 - d = 20 - 1 = 19$, you can determine that the formula for the nth term in the sequence is $a_n = n + 19$. Therefore, the 20th term in the sequence is $a_{20} = 20 + 19 = 39$, and the total number of seats is

$$\begin{aligned} S &= 20 + 21 + 22 + \cdots + 39 \\ &= \frac{n}{2}(a_1 + a_{20}) \\ &= \frac{20}{2}(20 + 39) \\ &= 10(59) \\ &= 590. \end{aligned}$$

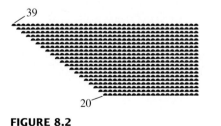

39

20

FIGURE 8.2

EXAMPLE 9 Total Sales

A small business sells $10,000 worth of products during its first year. The owner of the business has set a goal of increasing annual sales by $7500 each year for 9 years. Assuming that this goal is met, find the total sales during the first 10 years this business is in operation.

Solution

The annual sales form an arithmetic sequence in which $a_1 = 10,000$ and $d = 7500$. Thus, $c = a_1 - d = 10,000 - 7500 = 2500$, and the nth term of the sequence is

$$a_n = 7500n + 2500.$$

This implies that the 10th term of the sequence is $a_{10} = 77,500$. Therefore, the total sales for the first 10 years are as follows.

$$S = \frac{n}{2}(a_1 + a_{10})$$
$$= \frac{10}{2}(10,000 + 77,500)$$
$$= 5(87,500)$$
$$= \$437,500$$

Activities

1. Determine which of the following are arithmetic sequences.

   (a) $3, 5, 7, 9, 11, \ldots$

   (b) $3, 6, 12, 24, 48, \ldots$

   (c) $-3, 6, -9, 12, -15, \ldots$

   (d) $5, 0, -5, -10, -15, \ldots$

   (e) $1, 3, 6, 10, 15, 21, \ldots$

   Answer: (a) and (d)

2. Find the first five terms of the arithmetic sequence $a_1 = 13$, $d = -4$.
   Answer: $13, 9, 5, 1, -3$

3. Find the sum.
   $$\sum_{n=1}^{100} (2 + 3n)$$
   Answer: $15,350$

Group Activities   Problem Solving

Numerical Relationships   Decide whether it is possible to fill in the blanks in each of the following such that the resulting sequence is arithmetic. If so, find a recursion formula for the sequence.

a. $-7, \underline{\quad}, \underline{\quad}, \underline{\quad}, \underline{\quad}, \underline{\quad}, 11$

b. $17, \underline{\quad}, \underline{\quad}, \underline{\quad}, \underline{\quad}, \underline{\quad}, \underline{\quad}, \underline{\quad}, \underline{\quad}, 71$

c. $2, 6, \underline{\quad}, \underline{\quad}, 162$

d. $4, 7.5, \underline{\quad}, \underline{\quad}, \underline{\quad}, \underline{\quad}, \underline{\quad}, \underline{\quad}, \underline{\quad}, 39$

e. $8, 12, \underline{\quad}, \underline{\quad}, \underline{\quad}, 60.75$

Warm Up   The following warm-up exercises involve skills that were covered in earlier sections. You will use these skills in the exercise set for this section.

In Exercises 1 and 2, find the sum.

1. $\sum_{i=1}^{6} (2i - 1)$

2. $\sum_{i=1}^{10} (4i + 2)$

In Exercises 3 and 4, find the distance between the two real numbers.

3. $\frac{5}{2}$, 8

4. $\frac{4}{3}$, $\frac{14}{3}$

In Exercises 5 and 6, evaluate the function as indicated.

5. Find $f(3)$ for $f(n) = 10 + (n - 1)4$.

6. Find $f(10)$ for $f(n) = 1 + (n - 1)\frac{1}{3}$.

In Exercises 7–10, evaluate the expression.

7. $\frac{11}{2}(1 + 25)$

8. $\frac{16}{2}(4 + 16)$

9. $\frac{20}{2}[2(5) + (12 - 1)3]$

10. $\frac{8}{2}[2(-3) + (15 - 1)5]$

8.2 **Exercises**

In Exercises 1–10, determine whether the sequence is arithmetic. If it is, find the common difference.

1. 4, 7, 10, 13, 16, 19, . . .

2. 10, 8, 6, 4, 2, 0, . . .

3. 1, 2, 4, 8, 16, 32, . . .

4. 3, $\frac{5}{2}$, 2, $\frac{3}{2}$, 1, $\frac{1}{2}$, . . .

5. $\frac{9}{4}$, 2, $\frac{7}{4}$, $\frac{3}{2}$, $\frac{5}{4}$, 1, . . .

6. −12, −8, −4, 0, 4, 8, . . .

7. $\frac{1}{3}$, $\frac{2}{3}$, $\frac{4}{3}$, $\frac{8}{3}$, $\frac{16}{3}$, $\frac{32}{3}$, . . .

8. ln 1, ln 2, ln 3, ln 4, ln 5, . . .

9. 5.3, 5.7, 6.1, 6.5, 6.9, . . .

10. 1^2, 2^2, 3^2, 4^2, 5^2, . . .

In Exercises 11–18, write the first five terms of the specified sequence. Determine whether the sequence is arithmetic. If it is, find the common difference.

11. $a_n = 5 + 3n$

12. $a_n = (2^n)n$

13. $a_n = \dfrac{1}{n + 1}$

14. $a_n = 1 + (n - 1)4$

15. $a_n = 100 - 3n$

16. $a_n = 2^{n-1}$

17. $a_n = (2 + n) - (1 + n)$

18. $a_n = (-1)^n$

In Exercises 19–28, find a formula for a_n for the arithmetic sequence.

19. $a_1 = 1$, $d = 3$

20. $a_1 = 15$, $d = 4$

21. $a_1 = 100$, $d = -8$

22. $a_1 = 0$, $d = -\frac{2}{3}$

23. $a_1 = 5$, $a_4 = 14$ **24.** $a_1 = -4$, $a_5 = 16$

25. $a_4 = 12$, $a_{10} = 30$ **26.** $a_5 = 30$, $a_{15} = 145$

27. 2, 8, 14, 20, 26, . . .   **28.** 3, 13, 23, 33, 43, . . .

In Exercises 29–36, write the first five terms of the arithmetic sequence.

29. $a_1 = 5$, $d = 6$ **30.** $a_1 = 5$, $d = -\frac{3}{4}$

31. $a_1 = 4$, $a_{n+1} = a_n + 6$

32. $a_1 = 6$, $a_{n+1} = a_n + 12$

33. $a_1 = 2$, $a_{12} = 46$ **34.** $a_5 = 28$, $a_{10} = 53$

35. $a_8 = 26$, $a_{12} = 42$ **36.** $a_4 = 16$, $a_{10} = 46$

In Exercises 37–44, find the sum of the first n terms of the arithmetic sequence.

37. 8, 20, 32, 44, . . . , $n = 10$

38. 2, 8, 14, 20, . . . , $n = 25$

39. -6, -2, 2, 6, . . . , $n = 50$

40. 0.5, 0.9, 1.3, 1.7, . . . , $n = 10$

41. 40, 37, 34, 31, . . . , $n = 10$

42. 1.50, 1.45, 1.40, 1.35, . . . , $n = 20$

43. $a_1 = 100$, $a_{25} = 220$, $n = 25$

44. $a_1 = 15$, $a_{100} = 307$, $n = 100$

In Exercises 45–54, find the indicated sum.

45. $\displaystyle\sum_{n=1}^{50} n$ **46.** $\displaystyle\sum_{n=1}^{100} 2n$

47. $\displaystyle\sum_{n=1}^{100} 5n$ **48.** $\displaystyle\sum_{n=51}^{100} 7n$

49. $\displaystyle\sum_{n=1}^{500} (n + 3)$ **50.** $\displaystyle\sum_{n=1}^{250} (1000 - n)$

51. $\displaystyle\sum_{n=1}^{20} (2n + 5)$ **52.** $\displaystyle\sum_{n=1}^{100} \frac{n+4}{2}$

53. $\displaystyle\sum_{n=0}^{50} (1000 - 5n)$ **54.** $\displaystyle\sum_{n=0}^{100} \frac{8 - 3n}{16}$

55. Find the sum of the first 100 odd integers.

56. Find the sum of the integers from -10 to 50.

57. *Job Offer*   A person accepts a position with a company and will receive a salary of $27,500 for the first year. The person is guaranteed a raise of $1500 per year for the first 5 years.

(a) What will the salary be during the sixth year of employment?

(b) Use a graphing utility to find how much the company will have paid the person by the end of the sixth year.

58. *Job Offer*   A person accepts a position with a company and will receive a salary of $26,800 for the first year. The person is guaranteed a raise of $1750 per year for the first 5 years.

(a) What will the salary be during the sixth year of employment?

(b) Use a graphing utility to find how much the company will have paid the person by the end of the sixth year.

59. *Seating Capacity*   Determine the seating capacity of an auditorium with 30 rows of seats if there are 20 seats in the first row, 24 seats in the second row, 28 seats in the third row, and so on.

60. *Baling Hay*   As a farmer bales a field of hay, each trip around the field gets shorter. Suppose that on the first round there are 267 bales and on the second round there are 253 bales. Assume that the decrease will be the same on each round and that there will be 11 more trips. How many bales of hay will the farmer get from the field?

61. *Total Sales*   The annual sales for Campbell Soup Company from 1984 to 1994 can be approximated by the model

$$a_n = 327.0n + 2421.1, \quad n = 4, 5, 6, 7, \ldots, 14$$

where a_n is the total annual sales (in millions of dollars) and $n = 4$ represents 1984. (Source: Campbell Soup Company)

(a) Sketch a bar graph showing the annual sales for Campbell Soup Company from 1984 to 1994.

(b) Find the total sales from 1984 to 1994.

62. *Total Sales*   The annual sales for the BIC Corporation from 1988 to 1993 can be modeled by

$$a_n = 56.4 + 29.23n, \ n = 8, 9, 10, 11, 12, 13$$

where a_n is the annual sales (in millions of dollars) and $n = 8$ represents 1988 (see figure). Find the total sales from 1988 to 1993.   (Source: BIC Corporation)

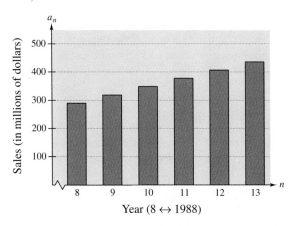

Year (8 ↔ 1988)

63. *Falling Object*   A heavy object (with negligible air resistance) is dropped from a plane. During the first second of fall, the object falls 16 feet; during the second second, it falls 48 feet; during the third second, it falls 80 feet; during the fourth second, it falls 112 feet. If this (arithmetic) pattern continues, how many feet will the object fall in 8 seconds?

64. *Falling Object*   A heavy object (with negligible air resistance) is dropped from a plane. During the first second of fall, the object falls 4.9 meters; during the second second, it falls 14.7 meters; during the third second, it falls 24.5 meters; during the fourth second, it falls 34.3 meters. How many meters will the object fall in 10 seconds?

65. *Number of Logs*   Logs are stacked in a pile, as shown in the figure. The top row has 15 logs and the bottom row has 21 logs. How many logs are in the stack?

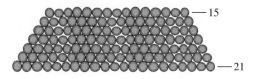

— 15

— 21

66. *Brick Pattern*   A brick patio is roughly the shape of a trapezoid (see figure). The patio has 20 rows of bricks. The first row has 14 bricks and the 20th row has 33 bricks. How many bricks are in the patio?

33 —

14 —

67. *Total Profit*   A small company makes a profit of $20,000 during its first year. The company president sets a goal of increasing profit by $5000 each year for 4 years. Assuming this goal is met, find the total profit during the first 5 years of business.

68. *Total Sales*   An entrepreneur sells $16,000 worth of gumballs in the first year of operation. The entrepreneur has set a goal of increasing annual sales by $4000 each year for 9 years. Assuming this goal is met, find the total sales during the first 10 years of this gumball business.

8.3	**Geometric Sequences**
	Geometric Sequences ▪ The Sum of a Geometric Sequence ▪ Application

Geometric Sequences

In Section 8.2, you learned that a sequence whose consecutive terms have a common *difference* is an arithmetic sequence. In this section, you will study another important type of sequence called a **geometric sequence.** Consecutive terms of a geometric sequence have a common *ratio*, as indicated in the following definition.

Definition of a Geometric Sequence

A sequence is **geometric** if the ratios of consecutive terms are the same. Thus, a sequence is geometric if there is a number r, $r \neq 0$, such that

$$\frac{a_2}{a_1} = r, \quad \frac{a_3}{a_2} = r, \quad \frac{a_4}{a_3} = r,$$

and so on. The number r is the **common ratio** of the geometric sequence.

EXAMPLE 1 *Examples of Geometric Sequences*

a. The sequence whose nth term is 2^n is geometric. For this sequence, the common ratio between consecutive terms is 2.

$$\underbrace{2, \ 4,}_{\frac{4}{2} = 2} \ 8, \ 16, \ \ldots, \ 2^n, \ \ldots$$

b. The sequence whose nth term is $4(3^n)$ is geometric. For this sequence, the common ratio between consecutive terms is 3.

$$\underbrace{12, \ 36,}_{\frac{36}{12} = 3} \ 108, \ 324, \ \ldots, \ 4(3^n), \ \ldots$$

c. The sequence whose nth term is $\left(-\frac{1}{3}\right)^n$ is geometric. For this sequence, the common ratio between consecutive terms is $-\frac{1}{3}$.

$$\underbrace{-\frac{1}{3}, \ \frac{1}{9},}_{\frac{1/9}{-1/3} = -\frac{1}{3}} \ -\frac{1}{27}, \ \frac{1}{81}, \ \ldots, \ \left(-\frac{1}{3}\right)^n, \ \ldots$$

In Example 1, notice that each of the geometric sequences has an nth term that is of the form ar^n, where the common ratio of the sequence is r.

The nth Term of a Geometric Sequence

The nth term of a geometric sequence has the form

$$a_n = a_1 r^{n-1}$$

where r is the common ratio of consecutive terms of the sequence. Thus, every geometric sequence can be written in the following form.

$$a_1, \quad a_2, \quad a_3, \quad a_4, \quad a_5, \ldots \ldots, \quad a_n, \ldots \ldots$$
$$\downarrow \quad \downarrow \quad \downarrow \quad \downarrow \quad \downarrow \cdots \cdots, \quad \downarrow \cdots \cdots$$
$$a_1, a_1 r, a_1 r^2, a_1 r^3, a_1 r^4, \ldots, a_1 r^{n-1}, \ldots$$

NOTE If you know the nth term of a geometric sequence, you can find the $(n+1)$th term by multiplying by r. That is, $a_{n+1} = r a_n$.

Technology

You can generate the geometric sequence in Example 2 with a graphing utility using the following steps.

3 ENTER

2 X ANS

Now press ENTER repeatedly to generate the terms of the sequence.

EXAMPLE 2 *Finding the Terms of a Geometric Sequence*

Write the first five terms of the geometric sequence whose first term is $a_1 = 3$ and whose common ratio is $r = 2$.

Solution

Starting with 3, repeatedly multiply by 2 to obtain the following.

$a_1 = 3$	1st term
$a_2 = 3(2^1) = 6$	2nd term
$a_3 = 3(2^2) = 12$	3rd term
$a_4 = 3(2^3) = 24$	4th term
$a_5 = 3(2^4) = 48$	5th term

EXAMPLE 3 *Finding a Term of a Geometric Sequence*

Find the 15th term of the geometric sequence whose first term is 20 and whose common ratio is 1.05.

Solution

$a_{15} = a_1 r^{n-1}$	Formula for geometric sequence
$\quad = 20(1.05)^{15-1}$	Substitute for a_1, r, and n.
$\quad \approx 39.599$	Use a calculator.

EXAMPLE 4 *Finding a Term of a Geometric Sequence*

Find the 12th term of the geometric sequence

$$5, \ 15, \ 45, \ \ldots .$$

Solution

The common ratio of this sequence is $r = 15/5 = 3$. Because the first term is $a_1 = 5$, you can determine the 12th term ($n = 12$) to be

$a_{12} = a_1 r^{n-1}$	Formula for geometric sequence
$= 5(3)^{12-1}$	Substitute 12 for n.
$= 5(177,147)$	Use a calculator.
$= 885,735.$	Simplify.

STUDY TIP

Remember that r is the common ratio of consecutive terms of a sequence. So in Example 5

$$a_{10} = a_1 r^9$$
$$= a_1 \cdot r \cdot r \cdot r \cdot r^6$$
$$= a_1 \cdot \frac{a_2}{a_1} \cdot \frac{a_3}{a_2} \cdot \frac{a_4}{a_3} \cdot r^6$$
$$= a_4 r^6.$$

If you know any two terms of a geometric sequence, you can use that information to find a formula for the nth term of the sequence.

EXAMPLE 5 *Finding a Term of a Geometric Sequence*

The fourth term of a geometric sequence is 125, and the 10th term is 125/64. Find the 14th term. (Assume that the terms of the sequence are positive.)

Solution

The tenth term is related to the fourth term by the equation

$$a_{10} = a_4 r^6.$$

Because $a_{10} = 125/64$ and $a_4 = 125$, you can solve for r as follows.

$$\frac{125}{64} = 125r^6$$
$$\frac{1}{64} = r^6$$
$$\frac{1}{2} = r$$

You can obtain the 14th term by multiplying the tenth term by r^4.

$$a_{14} = a_{10} r^4$$
$$= \frac{125}{64} \left(\frac{1}{2}\right)^4 = \frac{125}{1024}$$

The Sum of a Geometric Sequence

The formula for the sum of a *finite* geometric sequence is as follows.

The Sum of a Finite Geometric Sequence

The sum of the geometric sequence

$$a_1, \ a_1 r, \ a_1 r^2, \ a_1 r^3, \ a_1 r^4, \ \dots, \ a_1 r^{n-1}$$

with common ratio $r \neq 1$ is given by

$$S = a_1 \left(\frac{1 - r^n}{1 - r} \right).$$

DISCOVERY

To *develop* the formula for the sum of a finite geometric sequence, consider the following two equations.

$$S = a_1 + a_1 r + a_1 r^2$$
$$+ \cdots + a_1 r^{n-2} + a_1 r^{n-1}$$
$$rS = a_1 r + a_1 r^2 + a_1 r^3$$
$$+ \cdots + a_1 r^{n-1} + a_1 r^n$$

Can you discover a way to simplify the difference of these two equations to obtain the formula for the sum of a geometric sequence?

EXAMPLE 6   Finding the Sum of a Finite Geometric Sequence

Find the sum $\displaystyle\sum_{n=1}^{12} 4(0.3)^n$.

Solution

By writing out a few terms, you have

$$\sum_{n=1}^{12} 4(0.3)^n = 4(0.3) + 4(0.3)^2 + 4(0.3)^3 + \cdots + 4(0.3)^{12}.$$

Now, because $a_1 = 4(0.3)$, $r = 0.3$, and $n = 12$, you can apply the formula for the sum of a finite geometric sequence to obtain

$$\sum_{n=1}^{12} 4(0.3)^n = a_1 \left(\frac{1 - r^n}{1 - r} \right)$$

$$= 4(0.3) \left[\frac{1 - (0.3)^{12}}{1 - 0.3} \right]$$

$$\approx 1.714.$$

When using the formula for the sum of a geometric sequence, be careful to check that the index begins at $i = 1$. If the index begins at $i = 0$, you must adjust the formula for the nth partial sum. For instance, if the index in Example 6 had begun with $n = 0$, the sum would have been

$$\sum_{n=0}^{12} 4(0.3)^n = 4 + \sum_{n=1}^{12} 4(0.3)^n \approx 4 + 1.714 = 5.714.$$

The formula for the sum of a *finite* geometric sequence can, depending on the value of r, be extended to produce a formula for the sum of an *infinite* geometric sequence. Specifically, if the common ratio r has the property that $|r| < 1$, it can be shown that r^n becomes arbitrarily close to zero as n increases without bound. Consequently,

$$a_1 \left(\frac{1 - r^n}{1 - r} \right) \longrightarrow a_1 \left(\frac{1 - 0}{1 - r} \right) \quad \text{as} \quad n \longrightarrow \infty.$$

This result is summarized as follows.

The Sum of an Infinite Geometric Sequence

If $|r| < 1$, the infinite geometric sequence

$$a_1, \; a_1 r, \; a_1 r^2, \; a_1 r^3, \; \ldots, \; a_1 r^{n-1}, \; \ldots$$

has the sum

$$S = \frac{a_1}{1 - r}.$$

EXAMPLE 7   *Finding the Sum of an Infinite Geometric Sequence*

Find the sum of the infinite geometric sequence.

a. $\displaystyle\sum_{n=1}^{\infty} 4(0.6)^{n-1}$ **b.** $\displaystyle\sum_{n=1}^{\infty} 3(0.1)^{n-1}$

Solution

a. $\displaystyle\sum_{n=1}^{\infty} 4(0.6)^{n-1} = 4 + 4(0.6) + 4(0.6)^2 + 4(0.6)^3 + \cdots + 4(0.6)^{n-1} + \cdots$

$$= \frac{4}{1 - (0.6)} \qquad \frac{a_1}{1 - r}$$

$$= 10$$

b. $\displaystyle\sum_{n=1}^{\infty} 3(0.1)^{n-1} = 3 + 3(0.1) + 3(0.1)^2 + 3(0.1)^3 + \cdots + 3(0.1)^{n-1} + \cdots$

$$= \frac{3}{1 - (0.1)} \qquad \frac{a_1}{1 - r}$$

$$= \frac{10}{3}$$

$$\approx 3.33$$

Application

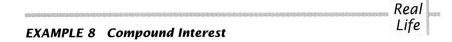

EXAMPLE 8   *Compound Interest*

A deposit of $50 is made on the first day of each month in a savings account that pays 6% compounded monthly. What is the balance at the end of 2 years?

Solution

The first deposit will gain interest for 24 months, and its balance will be

$$A_{24} = 50 \left(1 + \frac{0.06}{12} \right)^{24} = 50(1.005)^{24}.$$

The second deposit will gain interest for 23 months, and its balance will be

$$A_{23} = 50 \left(1 + \frac{0.06}{12} \right)^{23} = 50(1.005)^{23}.$$

The last deposit will gain interest for only 1 month, and its balance will be

$$A_1 = 50 \left(1 + \frac{0.06}{12} \right)^{1} = 50(1.005).$$

The total balance in the account will be the sum of the balances of the 24 deposits. Using the formula for the sum of a geometric sequence, with $A_1 = 50(1.005)$ and $r = 1.005$, you have

$$S = 50(1.005) \left[\frac{1 - (1.005)^{24}}{1 - 1.005} \right] = \$1277.96.$$

Group Activities   Extending the Concept

An Experiment   You will need a piece of string or yarn, a pair of scissors, and a tape measure. Measure out any length of string at least 5 feet long. Double over the string and cut it in half. Take one of the resulting halves, double it over, and cut it in half. Continue this process until you are no longer able to cut a length of string in half. How many cuts were you able to make? Construct a sequence of the resulting string lengths after each cut, starting with the original length of the string. Find a formula for the nth term of this sequence. How many cuts could you theoretically make? Discuss why you were not able to make that many cuts.

Warm Up The following warm-up exercises involve skills that were covered in earlier sections. You will use these skills in the exercise set for this section.

In Exercises 1–4, evaluate the expression.

1. $\left(\frac{4}{5}\right)^3$

2. $\left(\frac{3}{4}\right)^2$

3. 2^{-4}

4. $5(3^4)$

In Exercises 5–8, simplify the expression.

5. $(2n)(3n^2)$

6. $n(3n)^3$

7. $\dfrac{4n^5}{n^2}$

8. $\dfrac{(2n)^3}{8n}$

In Exercises 9 and 10, use sigma notation to write the sum.

9. $2 + 2(3) + 2(3^2) + 2(3^3)$

10. $3 + 3(2) + 3(2^2) + 3(2^3)$

8.3 Exercises

In Exercises 1–10, determine whether the sequence is geometric. If it is, find its common ratio and write a formula for a_n.

1. 5, 15, 45, 135, . . .

2. 3, 12, 48, 192, . . .

3. 3, 12, 21, 30, . . .

4. 1, −2, 4, −8, . . .

5. $1, -\frac{1}{2}, \frac{1}{4}, -\frac{1}{8}, \ldots$

6. 5, 1, 0.2, 0.04, . . .

7. $\frac{1}{2}, \frac{2}{3}, \frac{3}{4}, \frac{4}{5}, \ldots$

8. $9, -6, 4, -\frac{8}{3}, \ldots$

9. $1, \frac{1}{2}, \frac{1}{3}, \frac{1}{4}, \ldots$

10. $\frac{1}{5}, \frac{2}{3}, \frac{3}{9}, \frac{4}{11}, \ldots$

In Exercises 11–18, write the first five terms of the geometric sequence.

11. $a_1 = 2, \ r = 3$

12. $a_1 = 6, \ r = 2$

13. $a_1 = 1, \ r = \frac{1}{2}$

14. $a_1 = 1, \ r = \frac{1}{3}$

15. $a_1 = 5, \ r = -\frac{1}{10}$

16. $a_1 = 1, \ r = -\frac{1}{4}$

17. $a_1 = 1, \ r = e$

18. $a_1 = 2, \ r = e^{0.1}$

In Exercises 19–30, find the indicated term of the geometric sequence.

19. $a_1 = 4, \ r = \frac{1}{2}$, 10th term

20. $a_1 = 5, \ r = \frac{3}{2}$, 8th term

21. $a_1 = 6, \ r = -\frac{1}{3}$, 12th term

22. $a_1 = 1, \ r = -\frac{2}{3}$, 7th term

23. $a_1 = 100, \ r = e$, 9th term

24. $a_1 = 8, \ r = \sqrt{5}$, 9th term

25. $a_1 = 500, \ r = 1.02$, 40th term

26. $a_1 = 1000, \ r = 1.005$, 60th term

27. $a_1 = 16, \ a_4 = \frac{27}{4}$, 3rd term

28. $a_2 = 3, \ a_5 = \frac{3}{64}$, 1st term

29. $a_2 = -18, \ a_5 = \frac{2}{3}$, 6th term

30. $a_3 = \frac{16}{3}, \ a_5 = \frac{64}{27}$, 7th term

In Exercises 31–38, find the indicated sum.

31. $\displaystyle\sum_{n=1}^{10} 8(2^n)$

32. $\displaystyle\sum_{n=1}^{6} 3(4^n)$

33. $\displaystyle\sum_{i=1}^{10} 8\left(\tfrac{1}{4}\right)^{i-1}$

34. $\displaystyle\sum_{i=1}^{10} 5\left(\tfrac{1}{3}\right)^{i-1}$

35. $\displaystyle\sum_{n=0}^{8} 2^n$

36. $\displaystyle\sum_{n=0}^{8} (-2)^n$

37. $\displaystyle\sum_{n=0}^{20} 3\left(\tfrac{3}{2}\right)^n$

38. $\displaystyle\sum_{n=0}^{15} 2\left(\tfrac{4}{3}\right)^n$

In Exercises 39–44, find the sum.

39. $\displaystyle\sum_{n=0}^{\infty} \left(\tfrac{1}{2}\right)^n = 1 + \tfrac{1}{2} + \tfrac{1}{4} + \tfrac{1}{8} + \cdots$

40. $\displaystyle\sum_{n=0}^{\infty} 2\left(\tfrac{2}{3}\right)^n = 2 + \tfrac{4}{3} + \tfrac{8}{9} + \tfrac{16}{27} + \cdots$

41. $\displaystyle\sum_{n=1}^{\infty} \left(-\tfrac{1}{2}\right)^{n-1} = 1 - \tfrac{1}{2} + \tfrac{1}{4} - \tfrac{1}{8} + \cdots$

42. $\displaystyle\sum_{n=1}^{\infty} 2\left(-\tfrac{2}{3}\right)^{n-1} = 2 - \tfrac{4}{3} + \tfrac{8}{9} - \tfrac{16}{27} + \cdots$

43. $\displaystyle\sum_{n=0}^{\infty} 4\left(\tfrac{1}{4}\right)^n = 4 + 1 + \tfrac{1}{4} + \tfrac{1}{16} + \cdots$

44. $\displaystyle\sum_{n=0}^{\infty} \left(\tfrac{1}{10}\right)^n = 1 + 0.1 + 0.01 + 0.001 + \cdots$

45. *Compound Interest*   A deposit of $100 is made at the beginning of each month for 5 years in an account that pays 6%, compounded monthly. The balance A in the account at the end of 5 years is

$$A = 100\left(1 + \frac{0.06}{12}\right)^1 + \cdots + 100\left(1 + \frac{0.06}{12}\right)^{60}.$$

(a) Find the balance after 5 years.

(b) How much would the balance increase if the interest rate were raised to 7%?

(c) At 7% interest, find the balance after 10 years.

46. *Compound Interest*   A deposit of $50 is made at the beginning of each month for 5 years in an account that pays 5%, compounded monthly. Use a graphing utility to find the balance A in the account at the end of 5 years.

$$A = 50\left(1 + \frac{0.05}{12}\right)^1 + \cdots + 50\left(1 + \frac{0.05}{12}\right)^{60}$$

47. *Compound Interest*   Suppose you deposited $75 in an account at the beginning of each month for 10 years. The account pays 8% compounded monthly. Use a graphing utility to find your balance at the end of 10 years. If the interest were compounded continuously, what would the balance be?

48. *Compound Interest*   Suppose you deposited $200 in an account at the beginning of each month for 40 years. The account pays 6% compounded monthly. What would your balance be at the end of 40 years? If the interest were compounded continuously, what would the balance be?

49. *Profit*   The annual profit for the Hershey Foods Corp. from 1984 to 1993 can be approximated by the model

$$a_n = 106.9 e^{0.10n}, \quad n = 0, 1, 2, 3, \ldots, 9$$

where a_n is the annual profit (in millions of dollars) and n represents the year, with $n = 0$ corresponding to 1984 (see figure). Use the formula for the sum of a geometric sequence to approximate the total profit earned during this 10-year period. (Source: Hershey Foods Corp.)

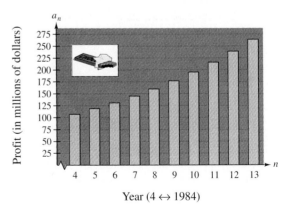

Year (4 $\leftrightarrow$ 1984)

50. *Profit* The annual profit for the Bristol Myers Squibb Company from 1980 to 1988 can be approximated by the model

$$a_n = 267.11e^{0.14n}, \quad n = 0, 1, 2, 3, \ldots, 8$$

where a_n is the annual profit (in millions of dollars) and n represents the year, with $n = 0$ corresponding to 1980 (see figure). Use the formula for the sum of a geometric sequence to approximate the total profit earned during this 9-year period. (Source: Bristol Myers Squibb Company)

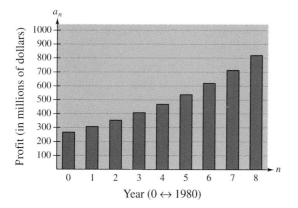

51. *Sales* The annual sales for the La-Z-Boy Chair Company from 1980 to 1989 can be modeled by

$$a_n = 153e^{0.159n}, \quad n = 0, 1, 2, 3, \ldots, 9$$

where a_n is the annual sales (in millions of dollars) and n represents the year, with $n = 0$ corresponding to 1980. Use the formula for the sum of a geometric sequence to approximate the total sales earned during this 10-year period. (Source: La-Z-Boy Chair Company)

52. *Sales* The annual sales for Rubbermaid, Inc. from 1990 to 1993 can be approximated by the model

$$a_n = 1534.96e^{0.0815n}, \quad n = 0, 1, 2, 3$$

where a_n is the annual sales (in millions of dollars) and n represents the year, with $n = 0$ corresponding to 1990. Use the formula for the sum of a geometric sequence to approximate the total sales earned during this 4-year period. (Source: Rubbermaid, Inc.)

53. *Would You Take This Job?* Suppose you went to work at a company that pays $0.01 for the first day, $0.02 for the second day, $0.04 for the third day, and so on. If the daily wage keeps doubling, what would your total income be for working 29 days? 30 days? 31 days?

54. *Gambler's Ruin* Suppose you were planning to visit a gambling casino and a friend suggested the following gambling strategy. Find a game that pays $2 for each dollar bet (such as the pass line at craps). Start by betting $2, and every time you win, return the $2 bet. If you lose the $2 bet, then bet $4. If you lose the $4 bet, then bet $8. For instance, if you lose four times in a row and then win, you will have bet $2 + 4 + 8 + 16 + 32$ and you would have won $64 on the last bet (which puts you $2 ahead). What is wrong with this gambling strategy? How much money would you have to take to cover the possibility of losing 12 times in a row?

55. *Area* The sides of a square are 16 inches in length. A new square is formed by connecting the midpoints of the sides of the original square, and two of the triangles are shaded (see figure). If this process is repeated five more times, what will the total area of the shaded region be?

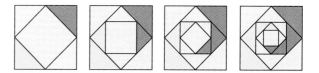

56. *Area* The sides of a square are 16 inches in length. The square is divided into nine smaller squares and the center square is shaded (see figure). Each of the eight unshaded squares is then divided into nine smaller squares and the center square of each is shaded. If this process is repeated four more times, what will the total area of the shaded region be?

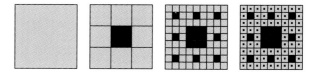

Math Matters   The Snowflake Curve

To create a **snowflake curve,** begin with an equilateral triangle (called Stage 1).   To form Stage 2, trisect each side of the triangle and at each of the middle points construct an equilateral triangle pointing outward. To form each additional stage, repeat this process. The following diagram shows the first six stages of the snowflake curve, which is the curve obtained by continuing this process infinitely many times.

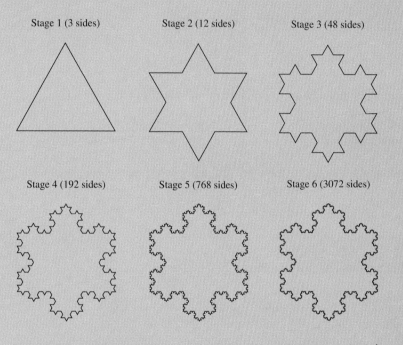

Stage 1 (3 sides)   Stage 2 (12 sides)   Stage 3 (48 sides)

Stage 4 (192 sides)   Stage 5 (768 sides)   Stage 6 (3072 sides)

In this figure, note that the nth stage of the snowflake curve has $3(4^{n-1})$ sides. If each side in Stage 1 has a length of 1, then each side in Stage 2 will have a length of $\frac{1}{3}$, each side in Stage 3 will have a length of $\left(\frac{1}{3}\right)^2 = \frac{1}{9}$, and so on. Thus, the total perimeter of the nth stage is

$$3(4^{n-1}) \cdot \left(\frac{1}{3}\right)^{n-1} = 3\left(\frac{4}{3}\right)^{n-1}.$$

Because the value of this expression is unbounded and n approaches infinity, you can conclude that the perimeter of the snowflake curve is infinite! This seems to be a paradox because it is difficult to conceive of a curve that is infinitely long being drawn on a piece of paper that has a finite area.

MID-CHAPTER QUIZ

Take this quiz as you would take a quiz in class. After you are done, check your work against the answers given in the back of the book.

In Exercises 1 and 2, write the first five terms of the sequence. (Begin with $n = 1$.)

1. $a_n = 3n - 1$

2. $a_n = \dfrac{n}{n+1}$

In Exercises 3 and 4, evaluate the expression.

3. $\dfrac{7!}{4!}$

4. $\dfrac{27!}{24!}$

In Exercises 5-8, write an expression for a_n. (Begin with $n = 1$.)

5. $2, \frac{1}{2}, \frac{2}{9}, \frac{1}{8}, \frac{2}{25}, \ldots$

6. $\frac{1}{2}, \frac{1}{3}, \frac{1}{4}, \frac{1}{5}, \frac{1}{6}, \ldots$

7. *Arithmetic*: $a_1 = 2, d = 5$

8. *Arithmetic*: $a_1 = 7, d = 3$

In Exercises 9-12, decide whether the sequence is arithmetic, geometric, or neither. Explain your reasoning.

9. $a_n = 2 + 3(n - 1)$

10. $1, -2, 4, -8, \ldots$

11. $1, \frac{1}{2}, \frac{1}{4}, \frac{1}{8}, \ldots$

12. $a_n = 2^n$

In Exercises 13-18, evaluate the sum.

13. $\displaystyle\sum_{i=1}^{5} (2i - 1)$

14. $\displaystyle\sum_{i=1}^{4} 2i^2$

15. $\displaystyle\sum_{n=1}^{25} 2n$

16. $\displaystyle\sum_{n=1}^{20} (3n + 1)$

17. $\displaystyle\sum_{n=0}^{\infty} \left(\tfrac{1}{3}\right)^n$

18. $\displaystyle\sum_{n=0}^{\infty} 5 \left(\tfrac{1}{5}\right)^n$

19. *Job Offer*   A person accepts a position with a company and will receive a salary of $28,500 for the first year. The person is guaranteed a raise of $1500 per year for the first 5 years.

(a) What will the salary be during the sixth year of employment?

(b) How much will the company have paid the person by the end of the sixth year?

20. *Compound Interest*   Suppose that you deposit $50 in an account at the beginning of each month for 10 years. The account pays 6% compounded monthly. What will your balance be at the end of the 10 years?

8.4 The Binomial Theorem

Binomial Coefficients ■ Pascal's Triangle ■ Binomial Expansions

Binomial Coefficients

Recall that a **binomial** is a polynomial that has two terms. In this section, you will study a formula that gives a quick method of raising a binomial to a power. To begin, let's look at the expansion of $(x + y)^n$ for several values of n.

$$(x + y)^0 = 1$$
$$(x + y)^1 = x + y$$
$$(x + y)^2 = x^2 + 2xy + y^2$$
$$(x + y)^3 = x^3 + 3x^2y + 3xy^2 + y^3$$
$$(x + y)^4 = x^4 + 4x^3y + 6x^2y^2 + 4xy^3 + y^4$$
$$(x + y)^5 = x^5 + 5x^4y + 10x^3y^2 + 10x^2y^3 + 5xy^4 + y^5$$

There are several observations you can make about these expansions.

1. In each expansion, there are $n + 1$ terms.

2. In each expansion, x and y have symmetrical roles. The powers of x decrease by 1 in successive terms, whereas the powers of y increase by 1.

3. The sum of the powers of each term is n. For instance, in the expansion of $(x + y)^5$, the sum of the powers of each term is 5.

$$\overset{4+1=5}{}\quad\overset{3+2=5}{}$$
$$(x + y)^5 = x^5 + 5\overset{\frown}{x^4y^1} + 10\overset{\frown}{x^3y^2} + 10x^2y^3 + 5xy^4 + y^5$$

4. The coefficients increase and then decrease in a symmetric pattern.

The coefficients of a binomial expansion are called **binomial coefficients.** To find them, you can use the following theorem.

The Binomial Theorem

In the expansion of $(x + y)^n$

$$(x + y)^n = x^n + nx^{n-1}y + \cdots + {}_nC_r x^{n-r}y^r + \cdots + nxy^{n-1} + y^n$$

the coefficient of $x^{n-r}y^r$ is given by

$$ {}_nC_r = \frac{n!}{(n - r)!r!}. $$

EXAMPLE 1 Finding Binomial Coefficients

Find the binomial coefficients.

a. $_8C_2$ **b.** $_{10}C_3$ **c.** $_7C_0$ **d.** $_8C_8$

Solution

a. $_8C_2 = \dfrac{8!}{6! \cdot 2!} = \dfrac{(8 \cdot 7) \cdot 6!}{6! \cdot 2!} = \dfrac{8 \cdot 7}{2 \cdot 1} = 28$

b. $_{10}C_3 = \dfrac{10!}{7! \cdot 3!} = \dfrac{(10 \cdot 9 \cdot 8) \cdot 7!}{7! \cdot 3!} = \dfrac{10 \cdot 9 \cdot 8}{3 \cdot 2 \cdot 1} = 120$

c. $_7C_0 = \dfrac{7!}{7! \cdot 0!} = 1$

d. $_8C_8 = \dfrac{8!}{0! \cdot 8!} = 1$

NOTE When $r \neq 0$ and $r \neq n$, as in parts (a) and (b) above, there is a simple pattern for evaluating binomial coefficients.

$$_8C_2 = \overset{\text{2 factors}}{\dfrac{8 \cdot 7}{\underset{\text{2 factorial}}{2 \cdot 1}}} \quad \text{and} \quad _{10}C_3 = \overset{\text{3 factors}}{\dfrac{10 \cdot 9 \cdot 8}{\underset{\text{3 factorial}}{3 \cdot 2 \cdot 1}}}$$

EXAMPLE 2 Finding Binomial Coefficients

Find the binomial coefficients.

a. $_7C_3$ **b.** $_7C_4$ **c.** $_{12}C_1$ **d.** $_{12}C_{11}$

Solution

a. $_7C_3 = \dfrac{7 \cdot 6 \cdot 5}{3 \cdot 2 \cdot 1} = 35$

b. $_7C_4 = \dfrac{7 \cdot 6 \cdot 5 \cdot 4}{4 \cdot 3 \cdot 2 \cdot 1} = 35$

c. $_{12}C_1 = \dfrac{12}{1} = 12$

d. $_{12}C_{11} = \dfrac{12!}{1! \cdot 11!} = \dfrac{(12) \cdot 11!}{1! \cdot 11!} = \dfrac{12}{1} = 12$

NOTE It is not a coincidence that the results in parts (a) and (b) of Example 2 are the same and that the results in parts (c) and (d) are the same. In general, it is true that

$$_nC_r = {}_nC_{n-r}.$$

This shows the symmetric property of binomial coefficients that was identified earlier.

Pascal's Triangle

There is a convenient way to remember a pattern for binomial coefficients. By arranging the coefficients in a triangular pattern, you obtain the following array, which is called **Pascal's Triangle.** This triangle is named after the famous French mathematician Blaise Pascal (1623–1662).

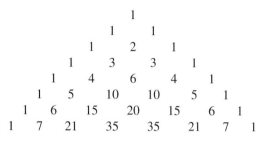

NOTE The top row in Pascal's Triangle is called the *zero row* because it corresponds to the binomial expansion

$$(x + y)^0 = 1.$$

Similarly, the next row is called the *first row* because it corresponds to the binomial expansion

$$(x + y)^1 = 1(x) + 1(y).$$

In general, the *nth row* in Pascal's Triangle gives the coefficients of $(x + y)^n$.

The first and last number in each row of Pascal's Triangle is 1. Every other number in each row is formed by adding the two numbers immediately above the number. Pascal noticed that numbers in this triangle are precisely the same numbers that are the coefficients of binomial expansions, as follows.

$$(x + y)^0 = 1$$
$$(x + y)^1 = 1x + 1y$$
$$(x + y)^2 = 1x^2 + 2xy + 1y^2$$
$$(x + y)^3 = 1x^3 + 3x^2y + 3xy^2 + 1y^3$$
$$(x + y)^4 = 1x^4 + 4x^3y + 6x^2y^2 + 4xy^3 + 1y^4$$
$$(x + y)^5 = 1x^5 + 5x^4y + 10x^3y^2 + 10x^2y^3 + 5xy^4 + 1y^5$$
$$(x + y)^6 = 1x^6 + 6x^5y + 15x^4y^2 + 20x^3y^3 + 15x^2y^4 + 6xy^5 + 1y^6$$
$$(x + y)^7 = 1x^7 + 7x^6y + 21x^5y^2 + 35x^4y^3 + 35x^3y^4 + 21x^2y^5 + 7xy^6 + 1y^7$$

EXAMPLE 3   Using Pascal's Triangle

Use the seventh row of Pascal's Triangle to find the binomial coefficients.

$$_8C_0, \ _8C_1, \ _8C_2, \ _8C_3, \ _8C_4, \ _8C_5, \ _8C_6, \ _8C_7, \ _8C_8$$

Solution

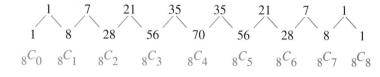

Binomial Expansions

As mentioned at the beginning of this section, when you write out the coefficients for a binomial that is raised to a power, you are **expanding a binomial.** The formulas for binomial coefficients give you an easy way to expand binomials, as demonstrated in the next four examples.

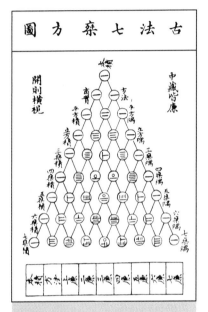

圖 方 蔡 七 法 古

開則橫視

中藏皆廉

"Pascal's" Triangle and forms of the Binomial Theorem were known in Eastern cultures prior to the Western "discovery" of the theorem. A Chinese text *Precious Mirror* contains a triangle of binomial expansions through the eighth power.

EXAMPLE 4 Expanding a Binomial

Write the expansion for the expression

$(x + 1)^3$.

Solution

The binomial coefficients from the third row of Pascal's Triangle are

1, 3, 3, 1.

Therefore, the expansion is as follows.

$$(x + 1)^3 = (1)x^3 + (3)x^2(1) + (3)x(1^2) + (1)(1^3)$$
$$= x^3 + 3x^2 + 3x + 1$$

To expand binomials representing *differences*, rather than sums, you alternate signs. Here are two examples.

$$(x - 1)^3 = x^3 - 3x^2 + 3x - 1$$
$$(x - 1)^4 = x^4 - 4x^3 + 6x^2 - 4x + 1$$

EXAMPLE 5 Expanding a Binomial

Write the expansion for the expression

$(x - 3)^4$.

Solution

The binomial coefficients from the fourth row of Pascal's Triangle are

1, 4, 6, 4, 1.

Therefore, the expansion is as follows.

$$(x - 3)^4 = (1)x^4 - (4)x^3(3) + (6)x^2(3^2) - (4)x(3^3) + (1)(3^4)$$
$$= x^4 - 12x^3 + 54x^2 - 108x + 81$$

EXAMPLE 6   Expanding a Binomial

Write the expansion for $(x - 2y)^4$.

Solution

Use the fourth row of Pascal's Triangle, as follows.

$$(x - 2y)^4 = (1)x^4 - (4)x^3(2y) + (6)x^2(2y)^2 - (4)x(2y)^3 + (1)(2y)^4$$
$$= x^4 - 8x^3y + 24x^2y^2 - 32xy^3 + 16y^4$$

EXAMPLE 7   Finding a Term in a Binomial Expansion

Find the sixth term of $(a + 2b)^8$.

Solution

From the Binomial Theorem, you can see that the $r + 1$ term is $_nC_r x^{n-r} y^r$. So in this case, $6 = r + 1$ means that $r = 5$. Because $n = 8$, $x = a$, and $y = 2b$, the sixth term in the binomial expansion is

$$_8C_5 a^{8-5}(2b)^5 = 56 \cdot a^3 \cdot (2b)^5$$
$$= 56(2^5)a^3b^5$$
$$= 1792a^3b^5$$

Group Activities   Extending the Concept

Finding a Pattern   By adding the terms in each of the rows of Pascal's Triangle, you obtain the following.

Row 0: $1 = 1$

Row 1: $1 + 1 = 2$

Row 2: $1 + 2 + 1 = 4$

Row 3: $1 + 3 + 3 + 1 = 8$

Row 4: $1 + 4 + 6 + 4 + 1 = 16$

Find a pattern for this sequence. Then use the pattern to find the sum of the terms in the 10th row of Pascal's Triangle. Check your answer by actually adding the terms of the 10th row.

Warm Up The following warm-up exercises involve skills that were covered in earlier sections. You will use these skills in the exercise set for this section.

In Exercises 1–6, perform the indicated operations and/or simplify.

1. $5x^2(x^3 + 3)$ **2.** $(x + 5)(x^2 - 3)$

3. $(x + 4)^2$ **4.** $(2x - 3)^2$

5. $x^2 y(3xy^{-2})$ **6.** $(-2z)^5$

In Exercises 7–10, evaluate the expression.

7. $5!$ **8.** $\dfrac{8!}{5!}$

9. $\dfrac{10!}{7!}$ **10.** $\dfrac{6!}{3!3!}$

8.4 Exercises

In Exercises 1–10, evaluate $_nC_r$.

1. $_5C_3$ **2.** $_8C_6$

3. $_{12}C_0$ **4.** $_{20}C_{20}$

5. $_{20}C_{15}$ **6.** $_{12}C_5$

7. $_{100}C_{98}$ **8.** $_{10}C_4$

9. $_{100}C_2$ **10.** $_{10}C_6$

In Exercises 11–30, use the Binomial Theorem to expand the expression. Simplify your answer.

11. $(x + 1)^4$ **12.** $(x + 1)^6$

13. $(x + 2)^3$ **14.** $(x + 3)^4$

15. $(x - 2)^4$ **16.** $(x - 2)^5$

17. $(x + y)^5$ **18.** $(x + y)^6$

19. $(a + 2)^4$ **20.** $(s + 3)^5$

21. $(2 + 3s)^6$ **22.** $(5 + 2y)^4$

23. $(x - y)^5$ **24.** $(2x - y)^5$

25. $(1 - 2x)^3$ **26.** $\left(\dfrac{x}{2} - 3y\right)^3$

27. $(x^2 + 5)^4$ **28.** $(x^2 + y^2)^6$

29. $\left(\dfrac{1}{x} + y\right)^5$ **30.** $\left(\dfrac{1}{x} + 2y\right)^6$

In Exercises 31–36, use the Binomial Theorem to expand the complex number. Simplify your answer by using the fact that $i^2 = -1$.

31. $(1 + i)^4$ **32.** $(2 - i)^5$

33. $(2 - 3i)^6$ **34.** $\left(5 + \sqrt{-9}\right)^3$

35. $\left(\dfrac{-1}{2} + \dfrac{\sqrt{3}}{2}i\right)^3$ **36.** $\left(5 - \sqrt{3}\,i\right)^4$

In Exercises 37–40, expand the expression using Pascal's Triangle to determine the coefficients.

37. $(2t - 1)^5$ **38.** $(x - 4)^5$

39. $(3 - 2z)^4$ **40.** $(3y + 2)^5$

In Exercises 41–48, find the indicated term in the expansion of the binomial expression.

41. $(x + 3)^{12}$ $\qquad$ ax^5

42. $(x^2 + 3)^{12}$ $\qquad$ ax^8

43. $(x - 2y)^{10}$ $\qquad$ ax^8y^2

44. $(4x - y)^{10}$ $\qquad$ ax^2y^8

45. $(3x - 2y)^9$ $\qquad$ ax^4y^5

46. $(2x - 3y)^{11}$ $\qquad$ ax^3y^8

47. $(x^2 - 1)^8$ $\qquad$ ax^8

48. $(x^2 + 1)^{12}$ $\qquad$ ax^{10}

In Exercises 49–54, use the Binomial Theorem to expand the expression. [In the study of probability, it is sometimes necessary to use the expansion of $(p + q)^n$, where $p + q = 1$.]

49. $\left(\frac{1}{2} + \frac{1}{2}\right)^7$

50. $\left(\frac{1}{4} + \frac{3}{4}\right)^{10}$

51. $\left(\frac{1}{3} + \frac{2}{3}\right)^8$

52. $(0.3 + 0.7)^{12}$

53. $(0.6 + 0.4)^5$

54. $(0.35 + 0.65)^6$

55. *Finding a Pattern* Describe the pattern formed by the sums of the numbers along the diagonal segments of Pascal's Triangle (see figure).

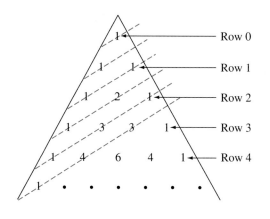

56. *Finding a Pattern* Use each of the encircled groups of numbers in the figure to form a 2×2 matrix. Find the determinant of the matrix. Then describe the pattern.

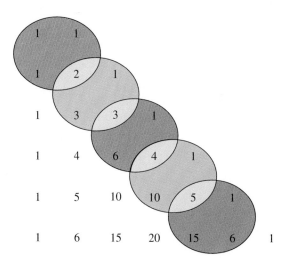

57. *Finding a Pattern* Each of the "flowers" in the figure is taken from Pascal's Triangle. Copy the flowers and fill in the missing numbers. Which rows did the flowers come from?

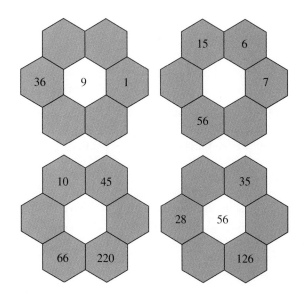

8.5 Counting Principles

Simple Counting Problems ▪ Counting Principles ▪ Permutations ▪ Combinations

Simple Counting Problems

You may want to consider opening class by asking students if they can predict which offers more choices for license plates: (a) a plate with three different letters of the alphabet in any order or (b) a plate with four different nonzero digits in any order. You may want to end class by verifying the answer: (a) offers $26 \cdot 25 \cdot 24 = 15{,}600$ choices and (b) offers $9 \cdot 8 \cdot 7 \cdot 6 = 3024$ choices.

The last two sections of this chapter contain a brief introduction to some of the basic counting principles and their application to probability. In the next section, you will see that much of probability has to do with counting the number of ways an event can occur.

EXAMPLE 1   *Selecting Pairs of Numbers at Random*

Eight pieces of paper are numbered from 1 to 8 and placed in a box. One piece of paper is drawn from the box, its number is written down, and the piece of paper is replaced in the box. Then, a second piece of paper is drawn from the box, and its number is written down. Finally, the two numbers are added together. How many different ways can a total of 12 be obtained?

Solution

To solve this problem, count the different ways that a total of 12 can be obtained using two numbers from 1 to 8.

First number	4	5	6	7	8
Second number	8	7	6	5	4

From this list, you can see that a total of 12 can occur in five different ways.

EXAMPLE 2   *Selecting Pairs of Numbers at Random*

NOTE The difference between the counting problems in Examples 1 and 2 can be distinguished by saying that the random selection in Example 1 occurs **with replacement,** whereas the random selection in Example 2 occurs **without replacement,** which eliminates the possibility of choosing two 6's.

Eight pieces of paper are numbered from 1 to 8 and placed in a box. Two pieces of paper are drawn from the box, and the numbers on the pieces of paper are written down and totaled. How many different ways can a total of 12 be obtained?

Solution

To solve this problem, you can count the different ways that a total of 12 can be obtained *using two different numbers* from 1 to 8.

First number	4	5	7	8
Second number	8	7	5	4

Thus, a total of 12 can be obtained in four different ways.

Counting Principles

Examples 1 and 2 describe simple counting problems in which you can *list* each possible way that an event can occur. When it is possible, this is always the best way to solve a counting problem. However, some events can occur in so many different ways that it is not feasible to write out the entire list. In such cases, you must rely on formulas and counting principles. The most important of these is the **Fundamental Counting Principle.**

NOTE   The Fundamental Counting Principle can be extended to three or more events. For instance, the number of ways that three events E_1, E_2, and E_3 can occur is $m_1 \cdot m_2 \cdot m_3$.

Fundamental Counting Principle

Let E_1 and E_2 be two events. The first event E_1 can occur in m_1 different ways. After E_1 has occurred, E_2 can occur in m_2 different ways. The number of ways that the two events can occur is $m_1 \cdot m_2$.

EXAMPLE 3   *Using the Fundamental Counting Principle*

How many different pairs of letters from the English alphabet are possible?

Solution

This experiment has two events. The first event is the choice of the first letter, and the second event is the choice of the second letter. Because the English alphabet contains 26 letters, it follows that the number of two-letter words is $26 \cdot 26 = 676$.

EXAMPLE 4   *Using the Fundamental Counting Principle*   *Real Life*

Telephone numbers in the United States have ten digits. The first three are the *area code* and the next seven are the *local telephone number*. How many different telephone numbers are possible within each area code? (Note that a local telephone number cannot begin with 0 or 1.)

Solution

Because the first digit cannot be 0 or 1, there are only 8 choices for the first digit. For each of the other six digits, there are 10 choices.

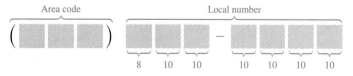

Thus, the number of local telephone numbers that are possible within each area code is $8 \cdot 10 \cdot 10 \cdot 10 \cdot 10 \cdot 10 \cdot 10 = 8{,}000{,}000$.

Permutations

One important application of the Fundamental Counting Principle is in determining the number of ways that n elements can be arranged (in order). An ordering of n elements is called a **permutation** of the elements.

Definition of Permutation

A **permutation** of n different elements is an ordering of the elements such that one element is first, one is second, one is third, and so on.

EXAMPLE 5   Finding the Number of Permutations of n Elements

How many permutations are possible for the letters A, B, C, D, E, and F?

Solution

Consider the following reasoning.

First position:	Any of the *six* letters.
Second position:	Any of the remaining *five* letters.
Third position:	Any of the remaining *four* letters.
Fourth position:	Any of the remaining *three* letters.
Fifth position:	Any of the remaining *two* letters.
Sixth position:	The *one* remaining letter.

Thus, the number of choices for the six positions are as follows.

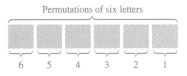

Permutations of six letters

6 5 4 3 2 1

The total number of permutations of the six letters is $6! = 720$.

Number of Permutations of n Elements

The number of permutations of n elements is given by

$$n \cdot (n - 1) \cdots 4 \cdot 3 \cdot 2 \cdot 1 = n!.$$

In other words, there are $n!$ different ways that n elements can be ordered.

Occasionally, you are interested in ordering a *subset* of a collection of elements rather than the entire collection. For example, you might want to choose (and order) r elements out of a collection of n elements. Such an ordering is called a **permutation of n elements taken r at a time.**

EXAMPLE 6   *Counting Horse Race Finishes*

Eight horses are running in a race. In how many different ways can these horses come in first, second, and third? (Assume that there are no ties.)

Solution

Here are the different possibilities.

Win (first position):	*Eight* choices
Place (second position):	*Seven* choices
Show (third position):	*Six* choices

Using the Fundamental Counting Principle, multiply these three numbers together to obtain the following.

Different orders of horses

$$8 \quad 7 \quad 6$$

Thus, there are $8 \cdot 7 \cdot 6 = 336$ different orders.

Permutations of n Elements Taken r at a Time

The number of permutations of n elements taken r at a time is

$$_nP_r = \frac{n!}{(n-r)!} = n(n-1)(n-2)\cdots(n-r+1).$$

Using this formula, you can rework Example 6 to find that the number of permutations of eight horses taken three at a time is

$$_8P_3 = \frac{8!}{5!}$$
$$= \frac{8 \cdot 7 \cdot 6 \cdot \cancel{5!}}{\cancel{5!}}$$
$$= 336$$

which is the same answer obtained in Example 6.

Remember that for permutations, order is important. Thus, if you are looking at the possible permutations of the letters A, B, C, and D taken three at a time, the permutations (A, B, D) and (B, A, D) would be different because the *order* of the elements is different.

Suppose, however, that you are asked to find the possible permutations of the letters A, A, B, and C. The total number of permutations of the four letters would be $_4P_4 = 4!$. However, not all of these arrangements would be *distinguishable* because there are two A's in the list. To find the number of distinguishable permutations, you can use the following formula.

Distinguishable Permutations

Suppose a set of n objects has n_1 of one kind of object, n_2 of a second kind, n_3 of a third kind, and so on, with $n = n_1 + n_2 + n_3 + \cdots + n_k$. Then the number of **distinguishable permutations** of the n objects is

$$\frac{n!}{n_1! \cdot n_2! \cdot n_3! \cdots n_k!}.$$

EXAMPLE 7 *Distinguishable Permutations*

In how many distinguishable ways can the letters in BANANA be written?

Solution

This word has six letters, of which three are A's, two are N's, and one is a B. Thus, the number of distinguishable ways the letters can be written is

$$\frac{6!}{3! \cdot 2! \cdot 1!} = \frac{6 \cdot 5 \cdot 4 \cdot \cancel{3!}}{\cancel{3!} \cdot 2!} = 60.$$

The 60 different "words" are as follows.

AAABNN	AAANBN	AAANNB	AABANN	AABNAN	AABNNA
AANABN	AANANB	AANBAN	AANBNA	AANNAB	AANNBA
ABAANN	ABANAN	ABANNA	ABNAAN	ABNANA	ABNNAA
ANAABN	ANAANB	ANABAN	ANABNA	ANANAB	ANANBA
ANBAAN	ANBANA	ANBNAA	ANNAAB	ANNABA	ANNBAA
BAAANN	BAANAN	BAANNA	BANAAN	BANANA	BANNAA
BNAAAN	BNAANA	BNANAA	BNNAAA	NAAABN	NAAANB
NAABAN	NAABNA	NAANAB	NAANBA	NABAAN	NABANA
NABNAA	NANAAB	NANABA	NANBAA	NBAAAN	NBAANA
NBANAA	NBNAAA	NNAAAB	NNAABA	NNABAA	NNBAAA

Combinations

When one counts the number of possible permutations of a set of elements, *order* is important. As a final topic in this section, we look at a method of selecting subsets of a larger set in which order is *not important*. Such subsets are called **combinations of *n* elements taken *r* at a time.** For instance, the combinations

$$\{A, \ B, \ C\} \qquad \text{and} \qquad \{B, \ A, \ C\}$$

are equivalent because both sets contain the same three elements, and the order in which the elements are listed is *not important*. Hence, you would count only one of the two sets. A common example of how a combination occurs is a card game in which the player is free to reorder the cards after they have been dealt.

EXAMPLE 8   Combinations of n Elements Taken r at a Time

In how many different ways can three letters be chosen from the letters A, B, C, D, and E? (The order of the three letters is not important.)

Solution

The following subsets represent the different combinations of three letters that can be chosen from five letters.

$\{A, B, C\}$	$\{A, B, D\}$
$\{A, B, E\}$	$\{A, C, D\}$
$\{A, C, E\}$	$\{A, D, E\}$
$\{B, C, D\}$	$\{B, C, E\}$
$\{B, D, E\}$	$\{C, D, E\}$

From this list, you can conclude that there are 10 different ways that three letters can be chosen from five letters.

Technology

Most graphing utilities have keys that will evaluate the formulas for the number of permutations or combinations of *m* elements taken *n* at a time. For instance, on a *TI-82*, you can evaluate $_8C_5$ as follows.

The display should be 56. You can evaluate $_8P_5$ (the number of permutations of eight elements taken five at a time) in a similar way.

Combinations of *n* Elements Taken *r* at a Time

The number of combinations of *n* elements taken *r* at a time is

$$_nC_r = \frac{n!}{(n-r)!\,r!}.$$

Note that the formula for $_nC_r$ is the same one given for binomial coefficients. To see how this formula is used, let's solve the counting problem in Example 8. In that problem, you are asked to find the number of combinations of five elements taken three at a time. Thus, $n = 5$, $r = 3$, and the number of combinations is

$$_5C_3 = \frac{5!}{2!3!} = \frac{5 \cdot \overset{2}{\cancel{4}} \cdot \cancel{3}}{\cancel{3} \cdot \cancel{2} \cdot 1} = 10$$

which is the same answer obtained in Example 8.

Real Life

EXAMPLE 9   Counting Card Hands

A standard poker hand consists of five cards dealt from a deck of 52. How many different poker hands are possible? (After the cards are dealt, the player may reorder them, and therefore order is not important.)

Solution

You can find the number of different poker hands by using the formula for the number of combinations of 52 elements taken five at a time, as follows.

$$_{52}C_5 = \frac{52!}{47!5!} = \frac{52 \cdot 51 \cdot 50 \cdot 49 \cdot 48}{5 \cdot 4 \cdot 3 \cdot 2 \cdot 1} = 2{,}598{,}960 \text{ different hands}$$

Group Activities   Problem Solving

Problem Posing   According to NASA, each space shuttle astronaut consumes an average of 3000 calories per day. An evening meal normally consists of a main dish, a vegetable dish, and two different desserts. The space shuttle food and beverage list contains 10 items classified as main dishes, 8 vegetable dishes, and 13 desserts. How many different evening meal menus are possible? Create two other problems that could be asked about the evening meal menus (*hint:* you may want to think of variety versus repetition and length of mission) and solve them.   (Source: NASA)

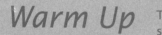

The following warm-up exercises involve skills that were covered in earlier sections. You will use these skills in the exercise set for this section.

In Exercises 1–4, evaluate the expression.

1. $13 \cdot 8^2 \cdot 2^3$

2. $10^2 \cdot 9^3 \cdot 4$

3. $\dfrac{12!}{2!(7!)(3!)}$

4. $\dfrac{25!}{22!}$

In Exercises 5 and 6, find the binomial coefficient.

5. $_{12}C_7$

6. $_{25}C_{22}$

In Exercises 7–10, simplify the expression.

7. $\dfrac{n!}{(n-4)!}$

8. $\dfrac{(2n)!}{4(2n-3)!}$

9. $\dfrac{2 \cdot 4 \cdot 6 \cdot 8 \cdots (2n)}{2^n}$

10. $\dfrac{3 \cdot 6 \cdot 9 \cdot 12 \cdots (3n)}{3^n}$

8.5 Exercises

1. *Job Applicants*   A small college needs two additional faculty members: a chemist and a statistician. In how many ways can these positions be filled if there are three applicants for the chemistry position and four for the position in statistics?

2. *Computer Systems*   A customer in a computer store can choose one of three monitors, one of two keyboards, and one of four computers. If all the choices are compatible, how many different systems could be chosen?

3. *Toboggan Ride*   Four people are lining up for a ride on a toboggan, but only two of the four are willing to take the first position. With that constraint, in how many ways can the four people be seated on the toboggan? Draw a diagram to illustrate the number of ways that the people can sit.

4. *Course Schedule*   A college student is preparing her course schedule for the next semester. She may select one of two mathematics courses, one of three science courses, and one of five courses from the social sciences and humanities. In how many ways can she select her schedule?

5. *License Plate Numbers*   In a certain state, each automobile license plate number consists of two letters followed by a four-digit number. How many distinct license plate numbers can be formed?

6. *License Plate Numbers*   In a certain state, each automobile license plate number consists of two letters followed by a four-digit number. To avoid confusion between "O" and "Zero" and "I" and "One," the letters "O" and "I" are not used. How many distinct license plate numbers can be formed?

7. *True-False Exam* In how many ways can a six-question true-false exam be answered? (Assume that no questions are omitted.)

8. *True-False Exam* In how many ways can a 10-question true-false exam be answered? (Assume that no questions are omitted.)

9. *Three-Digit Numbers* How many three-digit numbers can be formed under the following conditions?

   (a) Leading digits cannot be zero.

   (b) Leading digits cannot be zero and no repetition of digits is allowed.

   (c) Leading digits cannot be zero and the number must be a multiple of 5.

10. *Four-Digit Numbers* How many four-digit numbers can be formed under the following conditions?

    (a) Leading digits cannot be zero.

    (b) Leading digits cannot be zero and no repetition of digits is allowed.

    (c) Leading digits cannot be zero and the number must be a multiple of 5.

11. *Combination Lock* A combination lock will open when the right choice of three numbers (from 1 to 40, inclusive) is selected. How many different lock combinations are possible?

12. *Combination Lock* A combination lock will open when the right choice of three numbers (from 1 to 50, inclusive) is selected. How many different lock combinations are possible?

13. *Concert Seats* Three couples have reserved seats in a given row for a concert. In how many different ways can they be seated, given the following conditions?

    (a) There are no restrictions.

    (b) The two members of each couple wish to sit together.

14. *Single File* In how many orders can three girls and two boys walk through a doorway single-file, given the following conditions?

    (a) There are no restrictions.

    (b) The boys go before the girls.

    (c) The girls go before the boys.

In Exercises 15–24, evaluate $_nP_r$.

15. $_4P_4$ 16. $_5P_5$

17. $_8P_3$ 18. $_{20}P_2$

19. $_{20}P_5$ 20. $_{100}P_1$

21. $_{100}P_2$ 22. $_{10}P_2$

23. $_5P_4$ 24. $_7P_4$

25. *Posing for a Photograph* In how many ways can five children line up in one row to have their picture taken?

26. *Riding in a Car* In how many ways can six people sit in a six-passenger car?

27. *Choosing Officers* From a pool of 12 candidates, the offices of president, vice-president, secretary, and treasurer will be filled. In how many different ways can the offices be filled if each of the 12 candidates can hold any office?

28. *Assembly-Line Production* There are four processes involved in assembling a certain product, and these can be performed in any order. The management wants to test each order to determine which is the least time consuming. How many different orders will have to be tested?

29. *Forming an Experimental Group* In order to conduct a certain experiment, four students are randomly selected from a class of 20. How many different groups of four students are possible?

30. *Test Questions* A student may answer any 10 questions from a total of 12 questions on an exam. How many different ways can the student select the questions?

31. *Lottery Choices* There are 40 numbers in a particular state lottery. In how many ways can a player select six of the numbers? (The order of selection is not important.)

32. *Lottery Choices* There are 50 numbers in a particular state lottery. In how many ways can a player select eight of the numbers? (The order of selection is not important.)

33. *Number of Subsets* How many subsets of four elements can be formed from a set of 100 elements?

34. *Number of Subsets*   How many subsets of five elements can be formed from a set of 80 elements?

35. *Forming a Committee*   A committee composed of three graduate students and two undergraduate students is to be selected from a group of eight graduates and five undergraduates. How many different committees can be formed?

36. *Defective Units*   A shipment of 12 microwave ovens contains three defective units. In how many ways can a vending company purchase four of these units and receive (a) all good units, (b) two good units, and (c) at least two good units?

37. *Job Applicants*   An employer interviews eight people for four openings in the company. Three of the eight people are women. If all eight are qualified, in how many ways can the employer fill the four positions if (a) the selection is random and (b) exactly two are women?

38. *Forming a Committee*   Four people are randomly chosen from a group of four couples. In how many ways can this happen, given the following conditions?

(a) There are no restrictions.

(b) The selection must include one member from each couple.

39. *Poker Hand*   Five cards are selected from an ordinary deck of 52 playing cards. In how many ways can you get a full house? (A full house consists of three of one kind and two of another. For example, A-A-A-5-5 and K-K-K-10-10 are full houses.)

40. *Poker Hand*   Five cards are selected from an ordinary deck of 52 playing cards. In how many ways can you get a straight flush? (A straight flush consists of five cards that are in order and of the same suit. For example, A♡, 2♡, 3♡, 4♡, 5♡ and 10♠, J♠, Q♠, K♠, A♠ are straight flushes.)

Diagonals of a Polygon   In Exercises 41–44, find the number of diagonals of the polygon. (A line segment connecting any two nonadjacent vertices is called a *diagonal* of the polygon.)

41. Pentagon

42. Hexagon

43. Octagon

44. Decagon (10 sides)

In Exercises 45–50, find the number of distinguishable permutations of the group of letters.

45. A, A, G, E, E, E, M

46. B, B, B, T, T, T, T, T

47. A, A, Y, Y, Y, Y, X, X, X

48. K, K, M, M, M, L, L, N, N

49. A, L, G, E, B, R, A

50. M, I, S, S, I, S, S, I, P, P, I

In Exercise 51 and 52, use the example of the tree diagram to determine the possible arrangements needed to answer the question.

A college theater company is planning a tour with performances in Phoenix, Dallas, and New Orleans. If there are no restrictions on the order of the performances, how many ways may the tour be arranged?

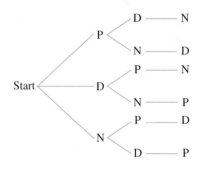

To determine the possible arrangements, we count the paths. A path moves from the start to the end of the last branch. In this case, there are six paths so there are six possible ways to arrange the itinerary of the tour. The tree diagram verifies visually what we know by using counting principles. That is, the number of possible arrangements would be $3 \cdot 2 \cdot 1 = 6$.

51. Use a tree diagram to show the possible arrangements for the college theater company if the city of Houston is added to their tour.

52. Use a tree diagram to describe visually the possible arrangements for the genders of three children within a family.

| 8.6 | **Probability** |

The Probability of an Event ▪ Mutually Exclusive Events ▪
Independent Events ▪ The Complement of an Event

The Probability of an Event

In Section P.7, you were introduced to the concept of the **probability of a simple event.** Recall that any happening whose result is uncertain is called an **experiment.** The possible results of the experiment are **outcomes,** the set of all possible outcomes of an experiment is the **sample space** of the experiment, and any subcollection of a sample space is an **event.**

For instance, when a six-sided die is tossed, the sample space can be represented by the numbers from 1 through 6. For this experiment, each of the outcomes is *equally likely*.

To describe sample spaces in such a way that each outcome is equally likely, you must sometimes distinguish between various outcomes in ways that appear artificial. Example 1 illustrates such a situation.

Real Life

EXAMPLE 1   Finding the Sample Space

Find the sample spaces for the following.

a. One coin is tossed.   **b.** Two coins are tossed.   **c.** Three coins are tossed.

Solution

a. Because the coin will land either heads up (denoted by H) or tails up (denoted by T), the sample space is $S = \{H, T\}$.

b. Because either coin can land heads up or tails up, the possible outcomes are as follows.

HH = heads up on both coins

HT = heads up on first coin and tails up on second coin

TH = tails up on first coin and heads up on second coin

TT = tails up on both coins

Thus, the sample space is $S = \{HH, HT, TH, TT\}$. Note that this list distinguishes between the two cases HT and TH, even though these two outcomes appear to be similar.

c. Following the notation of part (b), the sample space is

$$S = \{HHH, HHT, HTH, HTT, THH, THT, TTH, TTT\}.$$

To calculate the probability of an event, count the number of outcomes in the event and in the sample space. The *number of outcomes* in event E is denoted by $n(E)$ and the number of outcomes in the sample space S is denoted by $n(S)$. The probability that event E will occur is given by $n(E)/n(S)$.

The Probability of an Event

If an event E has $n(E)$ equally likely outcomes and its sample space S has $n(S)$ equally likely outcomes, the **probability** of event E is

$$P(E) = \frac{n(E)}{n(S)}.$$

Because the number of outcomes in an event must be less than or equal to the number of outcomes in the sample space, the probability of an event must be a number between 0 and 1. That is,

$$0 \le P(E) \le 1.$$

If $P(E) = 0$, event E *cannot occur*, and E is called an **impossible event.** If $P(E) = 1$, event E *must occur*, and E is called a **certain event.**

Real Life

EXAMPLE 2   *Finding the Probability of an Event*

a. Two coins are tossed. What is the probability that both land heads up?

b. A card is drawn from a standard deck of playing cards. What is the probability that it is an ace?

Solution

a. Following the procedure in Example 1(b), let

$$E = \{HH\}$$

and

$$S = \{HH,\ HT,\ TH,\ TT\}.$$

The probability of getting two heads is

$$P(E) = \frac{n(E)}{n(S)} = \frac{1}{4}.$$

b. Because there are 52 cards in a standard deck of playing cards and there are four aces (one in each suit), the probability of drawing an ace is

$$P(E) = \frac{n(E)}{n(S)} = \frac{4}{52} = \frac{1}{13}.$$

FIGURE 8.3

Real Life

EXAMPLE 3   *Finding the Probability of an Event*

Two six-sided dice are tossed. What is the probability that the total of the two dice is 7? (See Figure 8.3.)

Solution

Because there are six possible outcomes on each die, you can use the Fundamental Counting Principle to conclude that there are 6 • 6 or 36 different outcomes when two dice are tossed. To find the probability of rolling a total of 7, you must first count the number of ways this can occur.

First Die	1	2	3	4	5	6
Second Die	6	5	4	3	2	1

Thus, a total of 7 can be rolled in six ways, which means that the probability of rolling a 7 is

$$P(E) = \frac{n(E)}{n(S)} = \frac{6}{36} = \frac{1}{6}.$$

You could have written out each sample space in Examples 2 and 3 and simply counted the outcomes in the desired events. For larger sample spaces, however, you must make more use of the counting principles discussed in Section 8.5.

Real Life

EXAMPLE 4   *Finding the Probability of an Event*

Twelve-sided dice can be constructed (in the shape of regular dodecahedrons) so that each of the numbers from 1 to 6 appears twice on each die, as shown in Figure 8.4. Prove that these dice can be used in any game requiring ordinary six-sided dice without changing the probabilities of different outcomes.

Solution .

For an ordinary six-sided die, each of the numbers 1, 2, 3, 4, 5, and 6 occurs only once, so the probability of any particular number coming up is

$$P(E) = \frac{n(E)}{n(S)} = \frac{1}{6}.$$

For one of the twelve-sided dice, each number occurs twice, so the probability of any particular number coming up is

$$P(E) = \frac{n(E)}{n(S)} = \frac{2}{12} = \frac{1}{6}.$$

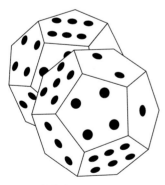

FIGURE 8.4

Although popular in the early 1800's, lotteries were banned in more and more states until, by 1894, no state allowed lotteries. In 1964, New Hampshire became the first state to reinstitute a state lottery. Today, lotteries are conducted by almost all states.

Real Life

EXAMPLE 5 *The Probability of Winning a Lottery*

In a state lottery, a player chooses six numbers from 1 to 40. If these six numbers match the six numbers drawn by the lottery commission, the player wins (or shares) the top prize. What is the probability of winning the top prize?

Solution

To find the number of elements in the sample space, use the formula for the number of combinations of 40 elements taken six at a time.

$$n(S) = {}_{40}C_6 = \frac{40 \cdot 39 \cdot 38 \cdot 37 \cdot 36 \cdot 35}{6 \cdot 5 \cdot 4 \cdot 3 \cdot 2 \cdot 1} = 3{,}838{,}380.$$

If a person buys only one ticket, the probability of winning is

$$P(E) = \frac{n(E)}{n(S)} = \frac{1}{3{,}838{,}380}.$$

Real Life

EXAMPLE 6 *Random Selection*

The total number of colleges and universities in the United States in 1992 is shown in Figure 8.5. One institution is selected at random. What is the probability that the institution is in one of the three southern regions?   (Source: U.S. National Center for Education Statistics)

Solution

From the figure, the total number of colleges and universities is 3628. Because there are $600 + 272 + 289 = 1161$ colleges and universities in the three southern regions, the probability that the institution is in one of these regions is

$$P(E) = \frac{n(E)}{n(S)} = \frac{1161}{3628} \approx 0.320.$$

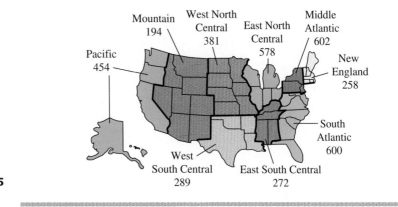

FIGURE 8.5

Mutually Exclusive Events

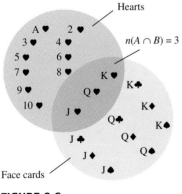

Hearts

$n(A \cap B) = 3$

Face cards

FIGURE 8.6

Two events A and B (from the same sample space) are **mutually exclusive** if A and B have no outcomes in common. In the terminology of sets, the **intersection of A and B** is the empty set and

$$P(A \cap B) = 0.$$

For instance, if two dice are tossed, the event A of rolling a total of 6 and the event B of rolling a total of 9 are mutually exclusive. To find the probability that one or the other of two mutually exclusive events will occur, you can *add* their individual probabilities.

Probability of the Union of Two Events

If A and B are events in the same sample space, the probability of A *or* B occurring is given by

$$P(A \cup B) = P(A) + P(B) - P(A \cap B).$$

If A and B are mutually exclusive, then

$$P(A \cup B) = P(A) + P(B).$$

Real Life

EXAMPLE 7 *The Probability of a Union*

One card is selected from a standard deck of 52 playing cards. What is the probability that the card is either a heart or a face card?

Solution

Because the deck has 13 hearts, the probability of selecting a heart (event A) is $P(A) = \frac{13}{52}$. Similarly, because the deck has 12 face cards, the probability of selecting a face card (event B) is $P(B) = \frac{12}{52}$. Because three of the cards are hearts and face cards (see Figure 8.6), it follows that $P(A \cap B) = \frac{3}{52}$. Finally, applying the formula for the probability of the union of two events, you can conclude that the probability of selecting a heart or a face card is

$$P(A \cup B) = P(A) + P(B) - P(A \cap B)$$
$$= \frac{13}{52} + \frac{12}{52} - \frac{3}{52}$$
$$= \frac{22}{52}$$
$$\approx 0.423.$$

EXAMPLE 8 Probability of Mutually Exclusive Events

The personnel department of a company has compiled data showing the number of years of service for each employee. The results are shown in the table.

Years of Service	Number of Employees
0–4	157
5–9	89
10–14	74
15–19	63
20–24	42
25–29	38
30–34	37
35–39	21
40–44	8

If an employee is chosen at random, what is the probability that the employee has had nine or fewer years of service?

Solution

To begin, add the number of employees and find that the total is 529. Next, let event A represent choosing an employee with 0 to 4 years of service and event B represent choosing an employee with 5 to 9 years of service. Then

$$P(A) = \frac{157}{529} \quad \text{and} \quad P(B) = \frac{89}{529}.$$

Because A and B have no outcomes in common, you can conclude that these two events are mutually exclusive and that

$$P(A \cup B) = P(A) + P(B) = \frac{157}{529} + \frac{89}{529}$$
$$= \frac{246}{529}$$
$$\approx 0.465.$$

Thus, the probability of choosing an employee who has nine or fewer years of service is about 0.465.

Independent Events

Two events are **independent** if the occurrence of one has no effect on the occurrence of the other. To find the probability that two independent events will occur, *multiply* the probabilities of each. For instance, rolling a total of 12 with two six-sided dice has no effect on the outcome of future rolls of the dice.

> **Probability of Independent Events**
>
> If A and B are independent events, the probability that both A and B will occur is
>
> $$P(A \text{ and } B) = P(A) \cdot P(B).$$

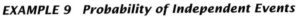

Real Life

EXAMPLE 9   *Probability of Independent Events*

A random number generator on a computer selects three integers from 1 to 20. What is the probability that all three numbers are less than or equal to 5?

Solution

The probability of selecting a number from 1 to 5 is

$$P(A) = \frac{5}{20} = \frac{1}{4}.$$

Thus, the probability that all three numbers are less than or equal to 5 is

$$P(A) \cdot P(A) \cdot P(A) = \left(\frac{1}{4}\right)\left(\frac{1}{4}\right)\left(\frac{1}{4}\right)$$
$$= \frac{1}{64}.$$

Real Life

EXAMPLE 10   *Probability of Independent Events*

In 1992, 60.0% of the population of the United States were 30 years old or older. Suppose that in a survey, 10 people were chosen at random from the population. What is the probability that all 10 were 30 years old or older?

Solution

Let A represent choosing a person who was 30 years old or older. Because the probability of choosing a person who was 30 years old or older was 0.600, you can conclude that the probability that all 10 people were 30 years old or older is

$$[P(A)]^{10} = (0.600)^{10}$$
$$\approx 0.0060.$$

The Complement of an Event

The **complement of an event** A is the collection of all outcomes in the sample space that are *not* in A. The complement of event A is denoted by A'. Because $P(A \text{ or } A') = 1$ and because A and A' are mutually exclusive, it follows that $P(A) + P(A') = 1$. Therefore, the probability of A' is given by

$$P(A') = 1 - P(A).$$

For instance, if the probability of *winning* a certain game is

$$P(A) = \frac{1}{4}$$

the probability of *losing* the game is

$$P(A') = 1 - \frac{1}{4} = \frac{3}{4}.$$

Probability of a Complement

Let A be an event and let A' be its complement. If the probability of A is $P(A)$, then the probability of the complement is

$$P(A') = 1 - P(A).$$

Real Life

EXAMPLE 11   Finding the Probability of a Complement

A manufacturer has determined that a certain machine averages one faulty unit for every 1000 it produces. What is the probability that an order of 200 units will have one or more faulty units?

Solution

To solve this problem as stated, you would need to find the probability of having exactly one faulty unit, exactly two faulty units, exactly three faulty units, and so on. However, using complements, you can simply find the probability that all units are perfect and then subtract this value from 1. Because the probability that any given unit is perfect is 999/1000, the probability that all 200 units are perfect is

$$P(A) = \left(\frac{999}{1000} \right)^{200} \approx 0.8186.$$

Therefore, the probability that at least one unit is faulty is

$$P(A') = 1 - P(A) \approx 0.1814.$$

DISCOVERY

You are in a class with 22 other people. What is the probability that at least two out of the 23 people will have a birthday on the same day of the year?

The complement of the probability that at least two people have the same birthday is the probability that all 23 birthdays are different. So, first find the probability that all 23 people have different birthdays and then find the complement.

Now, determine the probability that in a room with 50 people at least two people have the same birthday.

Group Activities

Exploring with Technology

An Experiment in Probability   In this section you have been finding probabilities from a *theoretical* point of view. Another way to find probabilities is from an *experimental* point of view. For instance, suppose you wanted to find the probability of obtaining a given total when two six-sided dice are tossed. The following BASIC program simulates the tossing of a pair of dice 5000 times.

BASIC Program:

```
10  RANDOMIZE
20  DIM TALLY(12)
30  FOR I=1 TO 5000
40  ROLLONE=INT(6*RND)+1
50  ROLLTWO=INT(6*RND)+1
60  DICETOTAL=ROLLONE+ROLLTWO
70  TALLY(DICETOTAL)=TALLY(DICETOTAL)+1
80  NEXT
90  FOR I=2 TO 12
100 PRINT ''TOTAL OF '',I,'' OCCURRED '',TALLY(I),'' TIMES''
110 NEXT
120 END
```

When we ran this program, the printout was as follows.

```
TOTAL OF   2  OCCURRED  139  TIMES
TOTAL OF   3  OCCURRED  264  TIMES
TOTAL OF   4  OCCURRED  443  TIMES
TOTAL OF   5  OCCURRED  553  TIMES
TOTAL OF   6  OCCURRED  691  TIMES
TOTAL OF   7  OCCURRED  810  TIMES
TOTAL OF   8  OCCURRED  715  TIMES
TOTAL OF   9  OCCURRED  557  TIMES
TOTAL OF  10  OCCURRED  398  TIMES
TOTAL OF  11  OCCURRED  270  TIMES
TOTAL OF  12  OCCURRED  160  TIMES
```

In Example 3 you found that the theoretical probability of tossing a total of 7 on a pair of dice is $\frac{1}{6} \approx 0.167$. From this experiment, the experimental probability of tossing a total of 7 on a pair of dice is $810/5000 \approx 0.162$. Try this experiment on a computer that has the BASIC language. When you increase the number of trials from 5000 to 10,000, does your experimental result get closer to the theoretical result?

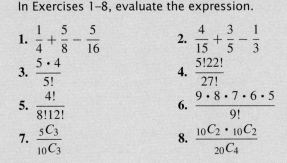

Warm Up The following warm-up exercises involve skills that were covered in earlier sections. You will use these skills in the exercise set for this section.

In Exercises 1–8, evaluate the expression.

1. $\dfrac{1}{4} + \dfrac{5}{8} - \dfrac{5}{16}$

2. $\dfrac{4}{15} + \dfrac{3}{5} - \dfrac{1}{3}$

3. $\dfrac{5 \cdot 4}{5!}$

4. $\dfrac{5!22!}{27!}$

5. $\dfrac{4!}{8!12!}$

6. $\dfrac{9 \cdot 8 \cdot 7 \cdot 6 \cdot 5}{9!}$

7. $\dfrac{{}_5C_3}{{}_{10}C_3}$

8. $\dfrac{{}_{10}C_2 \cdot {}_{10}C_2}{{}_{20}C_4}$

In Exercises 9 and 10, evaluate the expression. (Round your answer to three decimal places.)

9. $\left(\dfrac{99}{100}\right)^{100}$

10. $1 - \left(\dfrac{89}{100}\right)^{50}$

8.6 Exercises

Heads or Tails? In Exercises 1–4, a coin is tossed three times. Find the probability of the event.

1. Getting exactly one tail

2. Getting a head on the first toss

3. Getting at least one head

4. Getting at least two heads

Tossing a Die In Exercises 5–10, two six-sided dice are tossed. Find the probability of the event.

5. The sum is 4

6. The sum is less than 11

7. The sum is at least 7

8. The total is 2, 3, or 12

9. The sum is odd and no more than 7

10. The sum is odd or a prime

Drawing a Card In Exercises 11–14, a card is selected from a standard deck of 52 cards. Find the probability of the event.

11. Getting a face card

12. Not getting a face card

13. Getting a black card that is not a face card

14. Getting a card that is a 6 or less

Drawing Marbles In Exercises 15–18, two marbles are drawn from a bag containing one green, two yellow, and three red marbles. Find the probability of the event.

15. Drawing two red marbles

16. Drawing two yellow marbles

17. Drawing neither of the yellow marbles

18. Drawing marbles of different colors

In Exercises 19 and 20, you are given the probability that an event *will* happen. Find the probability that the event *will not* happen.

19. $p = 0.7$ **20.** $p = 0.36$

In Exercises 21 and 22, you are given the probability that an event *will not* happen. Find the probability that the event *will* happen.

21. $p = 0.15$ **22.** $p = 0.84$

23. *Winning an Election*   Three people have been nominated for president of a college class. From a small poll, it is estimated that the probability of Jane winning the election is 0.37, and the probability of Larry winning the election is 0.44. What is the probability of the third candidate winning the election?

24. *Winning an Election*   Taylor, Moore, and Jenkins are candidates for public office. It is estimated that Moore and Jenkins have about the same probability of winning, and Taylor is believed to be twice as likely to win as either of the others. Find each candidate's probability of winning the election.

25. *College Bound*   In a high school graduating class of 72 students, 28 are on the honor roll. Of these, 18 are going on to college, and of the other 44 students, 12 are going on to college. If a student is selected at random from the class, what are the probabilities that the person chosen is

(a) going to college?

(b) not going to college?

(c) on the honor roll, but not going to college?

26. *Alumni Association*   The alumni office of a college is sending a survey to selected members of the class of 1990. Of the 1254 people who graduated that year, 672 are women, and of those, 124 went to graduate school. Of the 582 male graduates, 198 went to graduate school. If an alumnus is selected at random, what are the probabilities that the person is

(a) female?

(b) male?

(c) female and did not attend graduate school?

27. *Random Number Generator*   Two integers (from 1 to 30, inclusive) are chosen by a random number generator on a computer. What are the probabilities that (a) the numbers are both even, (b) one number is even and one is odd, (c) both numbers are less than 10, and (d) the same number is chosen twice?

28. *Random Number Generator*   Two integers (from 1 to 40, inclusive) are chosen by a random number generator on a computer. What are the probabilities that (a) the numbers are both even, (b) one number is even and one is odd, (c) both numbers are less than 30, and (d) the same number is chosen twice?

29. *Preparing for a Test*   An instructor gives her class a list of eight study problems, from which she will select five to be answered on an exam. If a given student knows how to solve six of the problems, find the probability that the student will be able to answer all five questions on the exam.

30. *Preparing for a Test*   An instructor gives his class a list of 20 study problems, from which he will select 10 to be answered on an exam. If a given student knows how to solve 15 of the problems, find the probability that the student will be able to answer all 10 questions on the exam.

31. *Drawing Cards from a Deck*   Two cards are selected at random from an ordinary deck of 52 playing cards. Find the probabilities that two aces are selected under the following conditions.

(a) The cards are drawn in sequence, with the first card being replaced and the deck reshuffled prior to the second drawing.

(b) The two cards are drawn consecutively, without replacement.

32. *Drawing Cards from a Deck*   Two cards are selected at random from an ordinary deck of 52 playing cards. Find the probabilities that two hearts are selected under the following conditions.

(a) The cards are drawn in sequence, with the first card being replaced and the deck reshuffled prior to the second drawing.

(b) The two cards are drawn consecutively, without replacement.

33. *Game Show*   On a game show, you are given four digits to arrange in the proper order to represent the price of a car. If you are correct, you win the car. Find the probabilities of winning if

(a) You guess the position of each digit.

(b) You know the first digit, but must guess the remaining three.

34. *Game Show*   On a game show, you are given five digits to arrange in the proper order to represent the price of a car. If you are correct, you win the car. Find the probabilities of winning if

(a) You guess the position of each digit.

(b) You know the first digit, but must guess the remaining four.

35. *Letter Mix-Up*   Four letters and envelopes are addressed to four different people. If the letters are randomly inserted into the envelopes, what are the probabilities that (a) exactly one will be inserted in the correct envelope and (b) at least one will be inserted in the correct envelope?

36. *Payroll Mix-Up*   Five paychecks and envelopes are addressed to five different people. If the paychecks get mixed up and are randomly inserted into the envelopes, what are the probabilities that (a) exactly one will be inserted in the correct envelope and (b) at least one will be inserted in the correct envelope?

37. *Poker Hand*   Five cards are drawn from a standard deck of 52 cards. What is the probability of getting a full house? (See Exercise 39 in Section 8.5.)

38. *Poker Hand*   Five cards are drawn from a standard deck of 52 cards. What is the probability of getting a straight flush? (See Exercise 40 in Section 8.5.)

39. *Defective Units*   A shipment of 1000 compact disc players contains four defective units. A retail outlet has ordered 20 units. (a) What is the probability that all 20 units are good? (b) What is the probability that at least one unit is defective?

40. *Defective Units*   A shipment of 12 microwave ovens contains three defective units. Four of the units are shipped to a vending company. What are the probabilities that (a) all four units are good, (b) exactly two units are good, and (c) at least two units are good?

41. *Backup System*   A space vehicle has an independent backup system for one of its communication networks. The probability that either system will function satisfactorily for the duration of a flight is 0.985. What are the probabilities that during a given flight (a) both systems function satisfactorily, (b) at least one system functions satisfactorily, and (c) both systems fail?

42. *Backup Vehicle*   A fire company keeps two rescue vehicles to serve the community. Because of the demand on the company's time and the chance of mechanical failure, the probability that a specific vehicle is available when needed is 90%. If the availability of one vehicle is *independent* of the other, find the probabilities that (a) both vehicles are available at a given time, (b) neither vehicle is available at a given time, and (c) at least one is available at a given time.

43. *Making a Sale*   A sales representative makes a sale at approximately one-third of the offices she calls on. If, on a given day, she goes to four offices, what are the probabilities that she will make a sale at (a) all four offices, (b) none of the offices, and (c) at least one office?

44. *Making a Sale*   A sales representative makes a sale at approximately one-fourth of the businesses he calls on. If, on a given day, he goes to five businesses, what are the probabilities that he will make a sale at (a) all five businesses, (b) none of the businesses, and (c) at least one of the businesses?

45. *A Boy or a Girl?*   Assume that the probability of the birth of a child of a particular sex is 50%. In a family with six children, what are the probabilities that

(a) all six children are girls?

(b) all six children are of the same sex?

(c) there is at least one girl?

46. *A Boy or a Girl?*   Assume that the probability of the birth of a child of a particular sex is 50%. In a family with four children, what are the probabilities that

(a) all four children are boys?

(b) all four children are of the same sex?

(c) there is at least one boy?

47. *Is That Cash or Charge?* According to a survey by *USA Today*, the methods used by Christmas shoppers to pay for gifts are as shown in the pie graph. Suppose two Christmas shoppers are chosen at random. What is the probability that both shoppers paid for their gifts in cash only?

How Shoppers Pay for Gifts

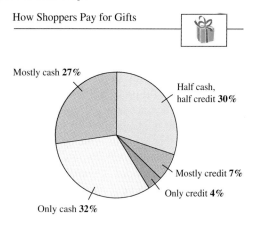

49. *Number of Telephones* According to a survey by Maritz Ameripoll, the numbers of telephones in American households are as shown in the figure. Suppose three households are chosen at random. What is the probability that all three households have at least four telephones?

Telephones in the Home

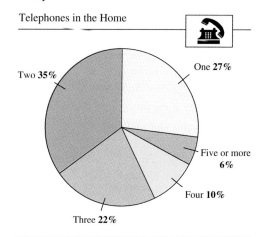

48. *Watching the Ads* According to a survey by Video Storyboard Tests, Inc., television viewers pay varying amounts of attention to television advertisements as compared with the television programs (see figure). Suppose two television viewers are chosen at random. What is the probability that both viewers pay no attention to the television ads they see?

Watching the Ads

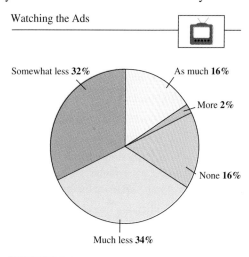

50. *Flexible Work Hours* In a survey by Robert Hall International, people were asked if they would prefer to work flexible hours—even if it meant slower career advancement—so that they could spend more time with their families. The results of the survey are shown in the figure. Suppose three people are chosen at random. What is the probability that all three people prefer flexible work hours?

Flexible Work Hours

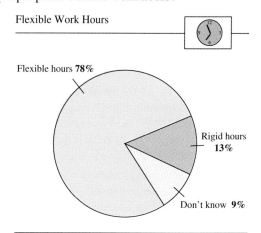

CHAPTER PROJECT: Analyzing a Bouncing Ball

The figure at the left shows the graph of data that was collected with a *Texas Instruments CBL* unit.* Once the data had been collected and stored on a *TI-82* calculator, the TRACE feature was used to approximate the maximum height of each rebound. The heights (in meters) are given in the following table.

Bounce, n	0	1	2	3	4	5	6	7	8
Height, h	0.82	0.64	0.50	0.39	0.30	0.24	0.18	0.14	0.11

Height (in meters) vs. Bounce graph shown at left, with Height axis from 0 to 1 and Bounce axis from 0 to 8.

Use this information to investigate the following questions.

1. *Writing a Model*   Consider the heights given in the table above as terms of a sequence. What type of sequence does the data most closely resemble? Write a formula that approximates the nth term of the sequence. Then construct a table that compares the actual data with the data given by your model.

2. *Fitting a Model to Data*   Enter the data given in the table above in a graphing calculator. Then run the calculator's exponential regression program. Compare the results with those obtained in Question 1.

3. *Total Distance Traveled*   Use the formula for the sum of a sequence to find the total vertical distance, up and down, traveled by the ball.

    (a) Assume that the ball bounces 40 times before coming to rest.

    (b) Assume that the ball bounces infinitely many times.

    (c) Do parts (a) and (b) give the same answer? Explain.

4. *Comparing Data Sets*   The data in the table above depends on the type of ball used and the initial height of the ball. The following two tables represent experimental data that was collected with a *CBL* unit. Did either experiment use the same ball that was used to collect the data above? Explain your reasoning.

Bounce, n	0	1	2	3	4	5	6	7	8
Height, h	1.13	0.88	0.69	0.54	0.42	0.33	0.25	0.20	0.15

Bounce, n	0	1	2	3	4	5	6	7	8
Height, h	1.13	0.86	0.65	0.50	0.38	0.22	0.17	0.13	0.10

5. *Extended CBL Exploration*   The *rebound rate* of a bouncing ball is the ratio of each rebound height to the previous rebound height. Use the *CBL* instructions given in Appendix C to collect and analyze data for four different types of balls. For instance, you might use a ping-pong ball, a tennis ball, a racquetball, and a "super ball." Do all of the balls have the same rebound rate?

*Instructions for using a *CBL* to collect your own data are given in Appendix C.

CHAPTER SUMMARY

After studying this chapter, you should have acquired the following skills. These skills are keyed to the Review Exercises that begin on page 629. Answers to odd-numbered Review Exercises are given in the back of the book.

- Write the first several terms of a sequence. *(Section 8.1)* — **Review Exercises 1–6**
- Evaluate an expression involving factorials. *(Section 8.1)* — **Review Exercises 7, 8**
- Write an expression for the nth term of a sequence. *(Section 8.1)* — **Review Exercises 9–14**
- Find the sum of the terms of a finite sequence. *(Section 8.1)* — **Review Exercises 15–18**
- Use summation notation to write the sum of a sequence. *(Section 8.1)* — **Review Exercises 19–22**
- Find the balance in an account described by a sequence. *(Section 8.1)* — **Review Exercises 23, 24**
- Determine if a sequence is arithmetic, and, if it is, find its common difference. *(Section 8.2)* — **Review Exercises 25–30**
- Find the formula for a_n for an arithmetic sequence. *(Section 8.2)* — **Review Exercises 31, 32**
- Write the first several terms of an arithmetic sequence. *(Section 8.2)* — **Review Exercises 33, 34**
- Find the sum of the first n terms of an arithmetic sequence. *(Section 8.2)* — **Review Exercises 35–40**
- Determine if a sequence is geometric, and, if it is, find its common ratio. *(Section 8.3)* — **Review Exercises 41–44**
- Write the first several terms of a geometric sequence. *(Section 8.3)* — **Review Exercises 45–48**
- Find the indicated term of a geometric sequence. *(Section 8.3)* — **Review Exercises 49–52**
- Find the sum of a geometric sequence. *(Section 8.3)* — **Review Exercises 53–60**
- Evaluate a binomial coefficient. *(Section 8.4)* — **Review Exercises 61–64**
- Expand a binomial using the Binomial Theorem. *(Section 8.4)* — **Review Exercises 65–70, 75, 76**
- Expand a binomial using Pascal's Triangle to determine the coefficients. *(Section 8.4)* — **Review Exercises 71, 72**
- Find a term in the expansion of a binomial. *(Section 8.4)* — **Review Exercises 73, 74**
- Use the Fundamental Counting Principle. *(Section 8.5)* — **Review Exercises 77–83**
- Determine the number of permutations of n elements taken r at a time. *(Section 8.5)* — **Review Exercises 84–87, 91**
- Determine the number of combinations of n elements taken r at a time. *(Section 8.5)* — **Review Exercises 88–90**
- Find the probability of an event. *(Section 8.6)* — **Review Exercises 92–95, 98–104**
- Find the probability that an event will not happen. *(Section 8.6)* — **Review Exercises 96, 97**
- Find the probability of the complement of an event. *(Section 8.6)* — **Review Exercises 100, 102–104**

REVIEW EXERCISES

In Exercises 1–6, write the first five terms of the sequence. (Begin with $n = 1$.)

1. $a_n = 3n + 2$

2. $a_n = 2n - 3$

3. $a_n = \dfrac{2n}{2n + 1}$

4. $a_n = 5 - \dfrac{1}{n}$

5. $a_n = (-1)^n 2^n$

6. $a_n = (-1)^n \dfrac{3^n - 1}{3^n + 1}$

In Exercises 7 and 8, evaluate the fraction.

7. $\dfrac{15!}{13!}$

8. $\dfrac{22!}{20!}$

In Exercises 9–14, write an expression for the apparent nth term of the sequence. (Begin with $n = 1$.)

9. $7, 9, 11, 13, 15, \ldots$

10. $1 + \frac{1}{2}, 1 + \frac{2}{3}, 1 + \frac{3}{4}, 1 + \frac{4}{5}, 1 + \frac{5}{6}, \ldots$

11. $1, \frac{1}{2}, \frac{1}{3}, \frac{1}{4}, \frac{1}{5}, \ldots$

12. $3, -9, 27, -81, 243, \ldots$

13. $-\frac{2}{3}, \frac{3}{4}, -\frac{4}{5}, \frac{5}{6}, -\frac{6}{7}, \ldots$

14. $-2, 4, -8, 16, -32, \ldots$

In Exercises 15–18, find the sum.

15. $\displaystyle\sum_{i=1}^{5} (2i + 3)$

16. $\displaystyle\sum_{i=1}^{4} (5i - 2)$

17. $\displaystyle\sum_{k=1}^{6} (k - 1)(k + 2)$

18. $\displaystyle\sum_{k=1}^{4} k^2 + 1$

In Exercises 19–22, use summation notation to write the sum.

19. $\dfrac{3}{1+1} + \dfrac{3}{1+2} + \dfrac{3}{1+3} + \dfrac{3}{1+4} + \cdots + \dfrac{3}{1+12}$

20. $1 + \frac{1}{2} + \frac{1}{3} + \frac{1}{4} + \frac{1}{5} + \cdots + \frac{1}{60}$

21. $2 - 4 + 8 - 16 + 32 - 64 + \cdots + 8192$

22. $2 + 8 + 18 + 32 + \cdots + 98$

23. *Compound Interest*   Suppose on your next birthday you deposit $2000 in an account that earns 8% compounded quarterly. The balance in the account after n quarters is given by

$$A_n = 2000 \left(1 + \frac{0.08}{4} \right)^n, \quad n = 1, 2, 3, \ldots.$$

(a) Compute the first eight terms of the sequence.

(b) Find the balance in this account 10 years from the date of deposit by computing the 40th term of the sequence.

(c) Assuming you do not withdraw money from the account, find the balance 30 years from the date of deposit by computing the 120th term of the sequence.

24. *Compound Deposit*   A deposit of $20,000 is made in an account that earns 6% compounded monthly. The balance in the account after n months is given by

$$A_n = 20{,}000 \left(1 + \frac{0.06}{12} \right)^n, \quad n = 1, 2, 3, \ldots.$$

(a) Compute the first 10 terms of the sequence.

(b) Find the balance in this account after 10 years by computing the 120th term of the sequence.

In Exercises 25–28, determine whether the sequence is arithmetic. If it is, find the common difference.

25. $5, 9, 13, 17, 21, \ldots$

26. $1, 3, 9, 27, \ldots$

27. $4, \frac{7}{2}, 3, \frac{5}{2}, 2, \ldots$

28. $3.2, 4.0, 4.8, 5.6, 6.4, \ldots$

In Exercises 29 and 30, write the first five terms of the specified sequence. Determine whether the sequence is arithmetic, and, if it is, find the common difference.

29. $a_n = 3 + 4n$

30. $a_n = 2 + (n - 2)5$

In Exercises 31 and 32, find a formula for a_n for the arithmetic sequence.

31. $a_1 = 3, d = 5$ **32.** $a_1 = 29, d = -2$

In Exercises 33 and 34, write the first five terms of the arithmetic sequence.

33. $a_1 = 7, a_{10} = 52$ **34.** $a_6 = 9, a_{12} = 21$

In Exercises 35–38, find the sum of the first n terms of the arithmetic sequence.

35. $3, 9, 15, 21, 27, \ldots, n = 10$

36. $-8, -4, 0, 4, 8, \ldots, n = 20$

37. $a_1 = 27, a_{40} = 300, n = 40$

38. $a_1 = 17, a_{20} = 492, n = 20$

39. *Job Offers*   A person is offered a job by two companies. The position with Company A has a salary of $24,500 for the first year with a guaranteed annual raise of $1500 per year for the first 5 years. The position with Company B has a salary of $22,000 for the first year with a guaranteed annual raise of $2400 per year for the first 5 years.

(a) What will the salary be during the sixth year of employment at Company A? Company B?

(b) How much will Company A have paid the person at the end of 6 years?

(c) How much will Company B have paid the person at the end of 6 years?

(d) Which job should the person accept, and why?

40. *Seating Capacity*   Determine the seating capacity of an auditorium with 20 rows of seats if there are 25 seats in the first row, 28 seats in the second row, and so on, in arithmetic progression.

In Exercises 41–44, determine whether the sequence is geometric. If it is, find its common ratio.

41. $1, -3, 9, -27, \ldots$ **42.** $3, 9, 15, 21, \ldots$

43. $16, 8, 4, 2, 1, \frac{1}{2}, \frac{1}{4}, \frac{1}{8}, \ldots$

44. $1, -\frac{1}{3}, \frac{1}{9}, -\frac{1}{27}, \frac{1}{81}, -\frac{1}{243}, \ldots$

In Exercises 45–48, write the first five terms of the geometric sequence.

45. $a_1 = 2, r = 3$ **46.** $a_1 = 1, r = \frac{3}{4}$

47. $a_1 = 10, r = -\frac{1}{5}$ **48.** $a_1 = 2, r = e$

In Exercises 49–52, find the indicated term of the geometric sequence.

49. $a_1 = 8, r = \frac{1}{2}$, 10th term

50. $a_1 = 100, r = -\frac{1}{10}$, 5th term

51. $a_1 = 200, r = -\frac{1}{2}$, 10th term

52. $a_1 = 10, r = 1.05$, 50th term

In Exercises 53 and 54, find the sum.

53. $\displaystyle\sum_{n=1}^{10} 4(2^n)$ **54.** $\displaystyle\sum_{n=0}^{20} 2\left(\frac{2}{3}\right)^n$

In Exercises 55 and 56, find the sum of the infinite geometric sequence.

55. $\displaystyle\sum_{n=0}^{\infty} \left(\frac{1}{3}\right)^n = 1 + \frac{1}{3} + \frac{1}{9} + \frac{1}{27} + \cdots$

56. $\displaystyle\sum_{n=0}^{\infty} 2\left(\frac{1}{10}\right)^n = 2 + 0.2 + 0.02 + 0.002 + \cdots$

57. *Compound Interest*   A deposit of $100 is made at the beginning of each month for 10 years in an account that pays 8% compounded monthly. The balance A at the end of 10 years is as follows.

$$A = 100\left(1 + \frac{0.08}{12}\right)^1 + \cdots + 100\left(1 + \frac{0.08}{12}\right)^{120}$$

(a) Find the balance.

(b) Is the balance after 20 years twice what it is after 10 years? Explain your reasoning.

58. *Compound Interest*   Suppose you deposit $125 in an account at the beginning of each month for 10 years. The account pays 6% compounded monthly. What will your balance be at the end of 10 years? If the interest were compounded continuously, what would the balance be?

59. *Profit* The annual profit for a company from 1980 to 1994 can be approximated by the model

$$a_n = 156.4e^{0.15n}$$

where a_n is the annual profit (in millions of dollars) and n represents the year, with $n = 0$ corresponding to 1980.

(a) Sketch a bar graph that represents the company's profit during the 15-year period.

(b) Find the total profit during the 15-year period.

60. *Sales* The annual sales for a company from 1980 to 1994 can be approximated by the model

$$a_n = 221e^{0.139n}$$

where a_n is the annual sales (in millions of dollars) and n represents the year, with $n = 0$ corresponding to 1980. Use the formula for the sum of a geometric sequence to approximate the total sales earned during the 15-year period.

In Exercises 61–64, evaluate $_nC_r$.

61. $_6C_3$

62. $_{10}C_5$

63. $_{30}C_{30}$

64. $_{20}C_{12}$

In Exercises 65–70, use the Binomial Theorem to expand the binomial. Simplify your answer.

65. $(x + 2)^6$

66. $(x - 1)^7$

67. $(x + y)^{10}$

68. $(2x + 3)^8$

69. $(x - 2y)^8$

70. $(x^2 + 4)^5$

In Exercises 71 and 72, expand the binomial using Pascal's Triangle to determine the coefficients.

71. $(3 - 2y)^4$

72. $(3x + 4y)^5$

In Exercises 73 and 74, find the required term in the expansion of the binomial.

Binomial	Term
73. $(x + 2)^{12}$	ax^7
74. $(2x - 5y)^9$	ax^4y^5

In Exercises 75 and 76, use the Binomial Theorem to expand the expression. [In the study of probability, it is sometimes necessary to use the expansion of $(p + q)^n$, where $p + q = 1$.]

75. $\left(\frac{1}{3} + \frac{2}{3}\right)^5$

76. $(0.2 + 0.8)^{10}$

77. *Computer Systems* A customer in a computer store can choose one of four monitors, one of three keyboards, and one of four computers. If all of the choices are compatible, how many different systems can be chosen?

78. *Sound System* A customer in an electronics store can choose one of four CD players, one of five speaker systems, and one of three radio/cassette players to design a sound system. How many different systems can be designed?

79. *Roller Coaster Ride* Six people are lining up for a ride on a roller coaster, but only three of the six are willing to sit in the front two seats of the coaster car. With that constraint, in how many ways can the six people be seated in the roller coaster car?

80. *True-False Exam* In how many ways can a 12-question true-false exam be answered? (Assume that no questions are omitted.)

81. *True-False Exam* In how many ways can a 20-question true-false exam be answered? (Assume that no questions are omitted.)

82. *Four-Digit Numbers* How many four-digit numbers can be formed under the following conditions?

(a) Leading digits cannot be zero.

(b) Leading digits cannot be zero and no repetition of digits is allowed.

(c) Leading digits cannot be zero and the number must be divisible by two.

83. *Five-Digit Numbers* How many five-digit numbers can be formed under the following conditions?

(a) Leading digits cannot be zero.

(b) Leading digits cannot be zero and no repetition of digits is allowed.

(c) Leading digits cannot be zero and the number must be odd.

In Exercises 84–87, evaluate $_nP_r$.

84. $_5P_5$

85. $_{30}P_1$

86. $_8P_5$

87. $_{10}P_3$

88. *Test Questions* A student may answer any 15 questions from a total of 20 questions on an exam. In how many different ways can the student select the questions?

89. *Starting Lineup* In how many ways can the starting lineup for a coed, six-member flag football team be selected from a team of eight men and eight women? League rules require that every team have three men and three women on the field.

90. *Defective Units* A shipment of 20 VCRs contains three defective units. In how many ways can a vending company purchase five of these units and receive (a) all good units, (b) two good units, and (c) at least two good units?

91. *Job Applicants* An employer interviews 10 people for five openings in the company. Six of the 10 people are women. If all 10 are qualified, in how many ways can the employer fill the five positions if (a) the selection is random, (b) exactly three are women, and (c) all five are women?

Drawing a Card In Exercises 92 and 93, find the indicated probability in the experiment of selecting one card from a standard deck of 52 playing cards.

92. The probability of getting a ten, jack, or queen

93. The probability of getting a red card that is not a face card

Tossing a Die In Exercises 94 and 95, find the indicated probability in the experiment of tossing a six-sided die twice.

94. The probability that the sum is 7

95. The probability that the sum is 7 or 11

In Exercises 96 and 97, you are given the probability that an event will happen. Find the probability that the event will not happen.

96. $p = 0.75$

97. $p = 0.48$

98. *Preparing for a Test* An instructor gives a class a list of 12 study problems, from which 8 will be selected to construct an exam. If a given student knows how to solve 10 of the problems, find the probabilities that the student will be able to answer (a) all 8 questions on the exam, (b) exactly 7 questions on the exam, and (c) at least 7 questions on the exam.

99. *Game Show* On a game show, you are given five digits to arrange in the proper order to represent the price of a car. If you are correct, you win the car. What are the probabilities of winning under the following conditions?

(a) You guess the position of each digit.

(b) You know the first two digits, but must guess the remaining three.

100. *Payroll Mix-Up* Six paychecks and envelopes are addressed to six different people. If the paychecks get mixed up and are randomly inserted into the envelopes, what are the probabilities that (a) exactly one will be inserted in the correct envelope, and (b) at least one will be inserted in the correct envelope?

101. *Poker Hand* Five cards are drawn from an ordinary deck of 52 playing cards. What is the probability of getting a full house? (A full house consists of three of one kind and two of another. For example, Q-Q-Q-2-2 and A-A-A-6-6 are full houses.)

102. *A Boy or a Girl?* Assume that the probability of the birth of a child of a particular sex is 50%. In a family with five children, what are the probabilities that (a) all the children are boys, (b) all the children are of the same sex, and (c) there is at least one boy?

103. *A Boy or a Girl?* Assume that the probability of the birth of a child of a particular sex is 50%. In a family with 10 children, what are the probabilities that (a) all the children are girls, (b) all the children are of the same sex, and (c) there is at least one girl?

104. *Defective Units* A shipment of 1000 radar detectors contains five defective units. A retail outlet has ordered 50 units.

(a) What is the probability that all 50 units are good?

(b) What is the probability that at least one unit is defective?

CHAPTER TEST

Take this test as you would take a test in class. After you are done, check your work against the answers given in the back of the book.

In Exercises 1–4, write the first five terms of the sequence. (Begin with $n = 1$.)

1. $a_n = 3n + 1$ **2.** $a_n = (-1)^n n^2$ **3.** $a_n = n!$ **4.** $a_n = \left(\frac{1}{2}\right)^n$

In Exercises 5–8, decide whether the sequence is arithmetic, geometric, or neither. If possible, find its common difference or common ratio.

5. $a_n = 5n - 2$ **6.** $a_n = 2^n + 2$ **7.** $a_n = 5(2^n)$ **8.** $a_n = 5n^2 - 2$

In Exercises 9–11, evaluate the sum.

9. $\displaystyle\sum_{n=1}^{50}(2n + 1)$ **10.** $\displaystyle\sum_{n=1}^{20}3\left(\frac{3}{2}\right)^n$ **11.** $\displaystyle\sum_{n=1}^{\infty}\left(\frac{1}{3}\right)^n$

12. A deposit of \$10,000 is made in an account that pays 8% compounded monthly. The balance in the account after n months is given by

$$A = 10{,}000\left(1 + \frac{0.08}{12}\right)^n, \quad n = 1, 2, 3, \ldots$$

Find the balance in the account after 10 years.

13. You deposit \$100 in an account at the beginning of each month for 10 years. The account pays 6% compounded monthly. What is the balance at the end of 10 years?

14. Use the Binomial Theorem to expand $(x + 2)^7$.

15. Determine the numerical coefficient of the x^7 term of $(x - 3)^{12}$.

16. A customer in an electronics store can choose one of five CD players, one of six speaker systems, and one of four radio/cassette players to design a sound system. How many different systems can be designed?

17. In how many ways can a 15-question true-false exam be answered? (Assume that no question is omitted.) Explain your reasoning.

18. How many five-digit numbers can be formed if the leading digit cannot be zero and the number must be odd?

19. On a game show, you are given five different digits to arrange in the proper order to represent the price of a car. You know only the first digit. What is the probability that you will arrange the other four digits correctly?

20. Five coins are tossed. What is the probability that all are heads?

CUMULATIVE TEST: CHAPTERS 6–8

Take this test as you would take a test in class. After you are done, check your work against the answers given in the back of the book.

In Exercises 1–4, solve the system by the indicated method.

1. *Substitution*

$$x + 2y = 22$$
$$-x + 4y = 20$$

2. *Graphing*

$$2x - 3y = 0$$
$$4x + y = 14$$

3. *Elimination*

$$2x - 3y + z = 18$$
$$3x \quad\quad - 2z = -4$$
$$x - y + 3z = 20$$

4. *Matrices*

$$3x - 4y + 2z = -32$$
$$2x + 3y \quad\quad = 8$$
$$y - 3z = 19$$

In Exercises 5–8, use the matrices to find the indicated matrix.

$$A = \begin{bmatrix} 3 & 2 & 2 \\ 1 & 2 & 2 \\ 1 & 0 & 1 \end{bmatrix}, \quad B = \begin{bmatrix} 4 & 1 \\ -1 & 2 \\ 3 & 1 \end{bmatrix}, \quad C = \begin{bmatrix} 5 & 0 & -4 \\ 3 & 0 & 1 \\ 2 & -1 & -3 \end{bmatrix}, \quad D = \begin{bmatrix} 1 & 3 \\ -2 & 0 \end{bmatrix}$$

5. $2A - C$

6. AB

7. BD

8. A^{-1}

In Exercises 9 and 10, evaluate the determinant of the matrix.

9. $\begin{bmatrix} 1 & -2 & 4 \\ 3 & 7 & -5 \\ 6 & 1 & 4 \end{bmatrix}$

10. $\begin{bmatrix} 1 & 0 & 0 \\ 0 & 7 & 0 \\ 0 & 0 & -1 \end{bmatrix}$

11. Find the point of equilibrium for a system with demand function $p = 75 - 0.0005x$ and supply function $p = 30 + 0.002x$.

12. A total of $45,000 is invested at 7.5% and 8.5% simple interest. If the yearly interest is $3625, how much of the $45,000 is invested at each rate?

13. A factory produces three different models of a product, which are shipped to three different warehouses. The number of units i that are shipped to warehouse j is represented by a_{ij} in matrix A at the right. The price per unit is represented by matrix B. Find the product BA and state what each entry of the product represents.

$$A = \begin{bmatrix} 1000 & 3000 & 2000 \\ 2000 & 4000 & 5000 \\ 3000 & 1000 & 1000 \end{bmatrix}$$

$$B = \begin{bmatrix} \$20 & \$30 & \$25 \end{bmatrix}$$

Matrices for 13

14. Write the first five terms of the sequence given by $a_n = 3 + 4n$.

15. Write the first five terms of the geometric sequence with $a_1 = 3$ and $r = \frac{1}{2}$.

16. Use the Binomial Theorem to expand $(y - 4)^5$.

17. In how many ways can the letters A, B, C, D, and E be arranged?

18. How many five-digit numbers are multiples of 5?

19. A shipment of 2000 microwave ovens contains 10 defective units. You order three units from the shipment. What is the probability that all three are good?

20. In how many ways can a 10-question true-false exam be answered? (Assume that no questions are omitted.)

Conic Sections

Conic Sections

Introduction to Conic Sections • Parabolas • Ellipses • Hyperbolas

Introduction to Conic Sections

Conic sections were discovered during the classical Greek period, which lasted from 600 to 300 B.C. By the beginning of the Alexandrian period, enough was known of conics for Apollonius (262–190 B.C.) to produce an eight-volume work on the subject.

This early Greek study was largely concerned with the geometrical properties of conics. It was not until the early seventeenth century that the broad applicability of conics became apparent.

A **conic section** (or simply **conic**) can be described as the intersection of a plane and a double-napped cone. Notice from Figure A.1 that in the formation of the four basic cones, the intersecting plane does not pass through the vertex of the cone. When the plane does pass through the vertex, we call the resulting figure a **degenerate conic**, as shown in Figure A.2.

FIGURE A.1 Conic Sections

Circle   Parabola   Ellipse   Hyperbola

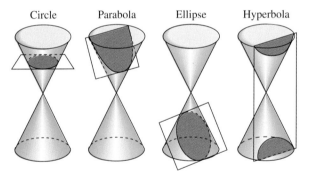

FIGURE A.2 Degenerate Conics

Point   Line   Two intersecting lines

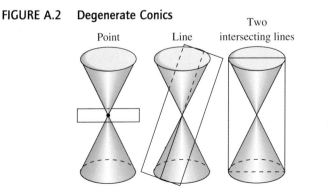

There are several ways to approach the study of conics. We could begin by defining conics in terms of the intersections of planes and cones, as the Greeks did, or we could define them algebraically, in terms of the general second-degree equation

$$Ax^2 + Bxy + Cy^2 + Dx + Ey + F = 0.$$

However, we will use a third approach, in which each of the conics is defined as a *locus*, or collection of points satisfying a certain geometric property. For example, in Section 3.2 we saw how the definition of a circle as *the collection of all points (x, y) that are equidistant from a fixed point (h, k)* led easily to the standard equation of a circle, $(x - h)^2 + (y - k)^2 = r^2$.

We will restrict our study of conics in this section to parabolas with vertices at the origin and ellipses and hyperbolas with centers at the origin. In the following section, we will look at the more general cases.

Parabolas

In Section 4.5 we determined that the graph of the quadratic function $f(x) = ax^2 + bx + c$ is a parabola that opens upward or downward. The following definition of a parabola is more general in the sense that it is independent of the orientation of the parabola.

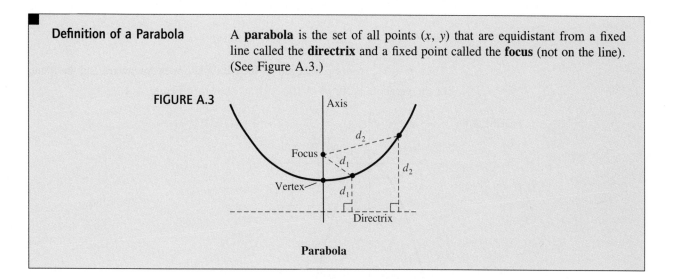

Definition of a Parabola

A **parabola** is the set of all points (x, y) that are equidistant from a fixed line called the **directrix** and a fixed point called the **focus** (not on the line). (See Figure A.3.)

FIGURE A.3

Parabola

The midpoint between the focus and the directrix is called the **vertex**, and the line passing through the focus and the vertex is called the **axis** of the parabola.

Using this definition, we can derive the following standard form of the equation of a parabola.

Standard Equation of a Parabola (Vertex at Origin)	The **standard form of the equation of a parabola** with a vertex at $(0, 0)$ and directrix $y = -p$ is

$$x^2 = 4py, \quad p \neq 0.$$ Vertical axis

For directrix $x = -p$, the equation is

$$y^2 = 4px, \quad p \neq 0.$$ Horizontal axis

The focus is on the axis p units (directed distance) from the vertex.

EXAMPLE 1 ■ Finding the Focus of a Parabola

Find the focus of the parabola whose equation is $y = -2x^2$.

Solution

Since the squared term in the equation involves x, we know that the axis is vertical, and we should use the standard form

$$x^2 = 4py.$$ Standard form, vertical axis

Writing the given equation in this form, we have

$$-2x^2 = y$$ Given equation

$$x^2 = -\frac{1}{2}y$$ Divide both sides by -2

$$x^2 = 4\left(-\frac{1}{8}\right)y.$$ Standard form

Thus, $p = -\frac{1}{8}$. Since p is negative, the parabola opens downward and the focus of the parabola is $(0, p) = \left(0, -\frac{1}{8}\right)$, as shown in Figure A.4.

FIGURE A.4

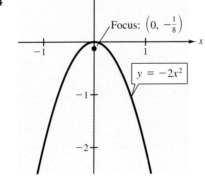

EXAMPLE 2 ■ A Parabola with a Horizontal Axis

Write the standard form of the equation of the parabola with the vertex at the origin and the focus at $(2, 0)$.

Solution

The axis of the parabola is horizontal, passing through $(0, 0)$ and $(2, 0)$, as shown in Figure A.5. Thus, we consider the standard form

$$y^2 = 4px. \qquad \text{Standard form, horizontal axis}$$

Since the focus is $p = 2$ units from the vertex, the equation is

$$y^2 = 4(2)x$$
$$y^2 = 8x. \qquad \text{Standard form} \qquad \blacksquare$$

FIGURE A.5

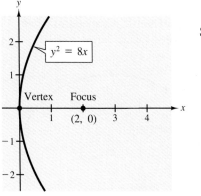

NOTE: Be sure you understand that the term *parabola* is a technical term used in mathematics and does not simply refer to *any* U-shaped curve.

Parabolas occur in a wide variety of applications. For instance, a parabolic reflector can be formed by revolving a parabola about its axis. The resulting surface has the property that all incoming rays parallel to the axis are reflected through the focus of the parabola—this is the principle behind the construction of the parabolic mirrors used in reflecting telescopes. Conversely, the light rays emanating from the focus of a parabolic reflector used in a flashlight are all parallel to one another, as shown in Figure A.6.

FIGURE A.6

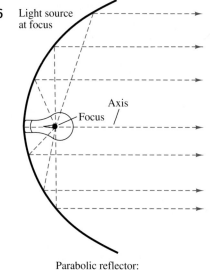

Parabolic reflector:
Light is reflected in parallel rays.

Ellipses

The second basic type of conic is called an **ellipse**, and it is defined as follows.

Definition of an Ellipse

An **ellipse** is the set of all points (x, y) the sum of whose distances from two distinct fixed points, called **foci**, is constant. (See Figure A.7.)

FIGURE A.7

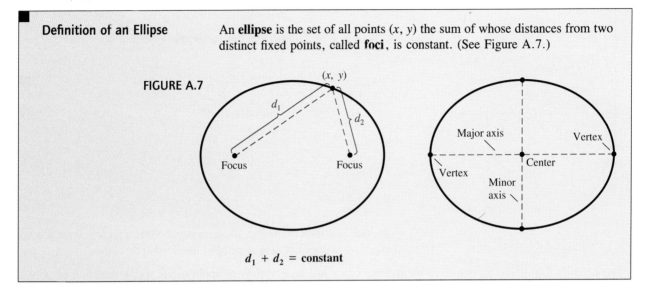

$d_1 + d_2 = \text{constant}$

The line through the foci intersects the ellipse at two points, called the **vertices**. The chord joining the vertices is called the **major axis**, and its midpoint is called the **center** of the ellipse. The chord perpendicular to the major axis at the center is called the **minor axis** of the ellipse.

You can visualize the definition of an ellipse by imagining two thumbtacks placed at the foci, as shown in Figure A.8. If the ends of a fixed length of string are fastened to the thumbtacks and the string is drawn taut with a pencil, the path traced by the pencil will be an ellipse.

FIGURE A.8

The standard form of the equation of an ellipse takes one of two forms, depending upon whether the major axis is horizontal or vertical.

Standard Equation of an Ellipse (Center at Origin)

The **standard form of the equation of an ellipse** with the center at the origin and major and minor axes of lengths $2a$ and $2b$, where $0 < b < a$, is

$$\frac{x^2}{a^2} + \frac{y^2}{b^2} = 1 \quad \text{or} \quad \frac{x^2}{b^2} + \frac{y^2}{a^2} = 1, \quad 0 < b < a.$$

FIGURE A.9

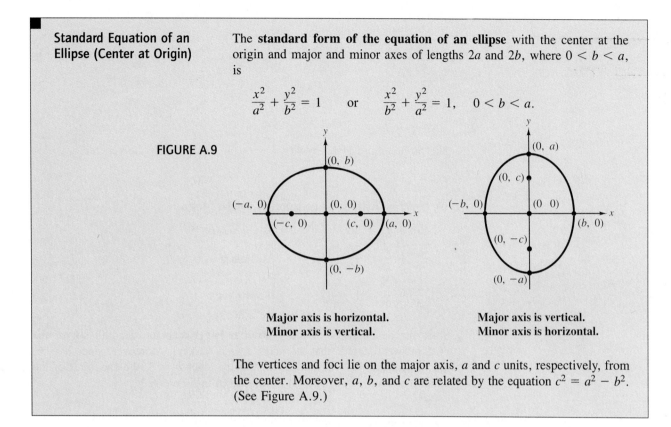

Major axis is horizontal.
Minor axis is vertical.

Major axis is vertical.
Minor axis is horizontal.

The vertices and foci lie on the major axis, a and c units, respectively, from the center. Moreover, a, b, and c are related by the equation $c^2 = a^2 - b^2$. (See Figure A.9.)

EXAMPLE 3   ■   Finding the Standard Equation of an Ellipse

Find the standard form of the equation of the ellipse that has a major axis of length 6 and foci at $(-2, 0)$ and $(2, 0)$, as shown in Figure A.10.

FIGURE A.10

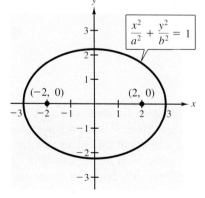

Solution

Since the foci occur at $(-2, 0)$ and $(2, 0)$, the center of the ellipse is $(0, 0)$, and the major axis is horizontal. Thus, the ellipse has an equation of the form

$$\frac{x^2}{a^2} + \frac{y^2}{b^2} = 1. \qquad \text{Standard form, vertical major axis}$$

Since the length of the major axis is 6, we have

$$2a = 6 \qquad \text{Length of major axis}$$

which implies that $a = 3$. Moreover, the distance from the center to either focus is $c = 2$. Finally, we have

$$b^2 = a^2 - c^2 = 3^2 - 2^2 = 9 - 4 = 5$$

which yields the equation

$$\frac{x^2}{9} + \frac{y^2}{5} = 1. \qquad \text{Standard form}$$ ■

EXAMPLE 4 ■ Sketching an Ellipse

Sketch the ellipse given by $4x^2 + y^2 = 36$, and identify the vertices.

Solution

We begin by writing the equation in standard form.

$$4x^2 + y^2 = 36 \qquad \text{Given equation}$$

$$\frac{4x^2}{36} + \frac{y^2}{36} = \frac{36}{36} \qquad \text{Divide both sides by 36}$$

$$\frac{x^2}{3^2} + \frac{y^2}{6^2} = 1 \qquad \text{Standard form}$$

Since the denominator of the y^2-term is larger than the denominator of the x^2-term, we conclude that the major axis is vertical. Moreover, since $a = 6$, the vertices are $(0, -6)$ and $(0, 6)$. Finally, since $b = 3$, the endpoints of the minor axis are $(-3, 0)$ and $(3, 0)$, as shown in Figure A.11.

FIGURE A.11

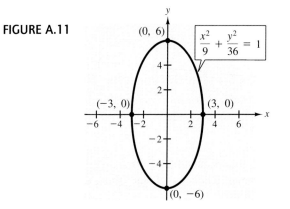

■

In Example 4 note that from the standard form of the equation we can sketch the ellipse by locating the endpoints of the two axes. Since 3^2 is the denominator of the x^2-term, we move three units to the *right and left* of the center to locate the endpoints of the horizontal axis. Similarly, since 6^2 is the denominator of the y^2-term, we move six units *up and down* from the center to locate the endpoints of the vertical axis.

Hyperbolas

The definition of a **hyperbola** is similar to that of an ellipse. The distinction is that, for an ellipse, the *sum* of the distances between the foci and a point on the ellipse is constant, while, for a hyperbola, the *difference* of these distances is constant.

Definition of a Hyperbola

A **hyperbola** is the set of all points (x, y), the difference of whose distances from two distinct fixed points, called **foci**, is constant. (See Figure A.12.)

FIGURE A.12

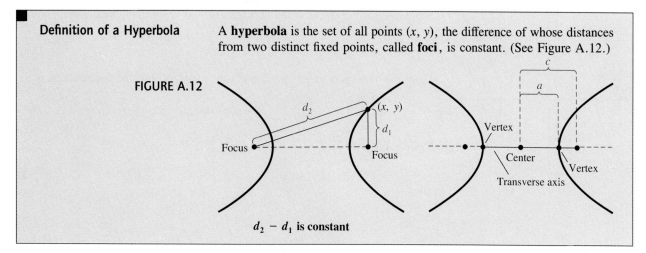

$d_2 - d_1$ **is constant**

The graph of a hyperbola has two disconnected parts, called **branches**. The line through the two foci intersects the hyperbola at two points, called **vertices**. The line segment connecting the vertices is called the **transverse axis**, and the midpoint of the transverse axis is called the **center** of the hyperbola.

Standard Equation of a Hyperbola (Center at Origin)

The **standard form of the equation of a hyperbola** with the center at $(0, 0)$ is $\dfrac{x^2}{a^2} - \dfrac{y^2}{b^2} = 1$ or $\dfrac{y^2}{a^2} - \dfrac{x^2}{b^2} = 1$.

FIGURE A.13

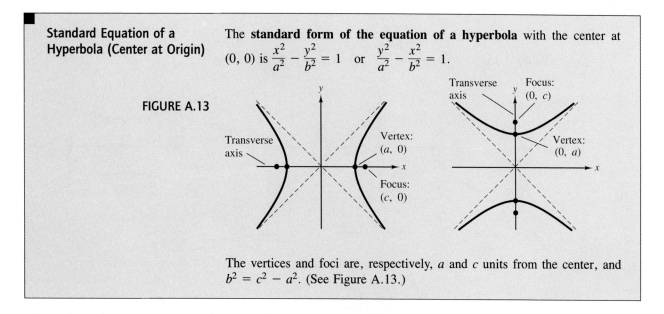

The vertices and foci are, respectively, a and c units from the center, and $b^2 = c^2 - a^2$. (See Figure A.13.)

EXAMPLE 5   ■   Finding the Standard Equation of a Hyperbola

Find the standard form of the equation of the hyperbola with foci at $(-3, 0)$ and $(3, 0)$ and vertices at $(-2, 0)$ and $(2, 0)$, as shown in Figure A.14.

FIGURE A.14

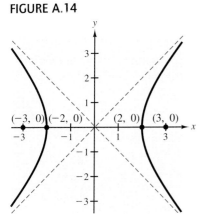

Solution

It can be seen that $c = 3$ because the foci are three units from the center. Moreover, $a = 2$ because the vertices are two units from the center. Thus, it follows that

$$b^2 = c^2 - a^2 = 3^2 - 2^2 = 9 - 4 = 5.$$

Since the transverse axis is horizontal, the standard form of the equation is

$$\frac{x^2}{a^2} - \frac{y^2}{b^2} = 1. \qquad \text{Standard form, horizontal transverse axis}$$

Finally, substituting $a^2 = 4$ and $b^2 = 5$, we have

$$\frac{x^2}{4} - \frac{y^2}{5} = 1. \qquad \text{Standard form} \qquad ■$$

An important aid in sketching the graph of a hyperbola is the determination of its **asymptotes**, as shown in Figure A.15. Each hyperbola has two asymptotes that intersect at the center of the hyperbola. Furthermore, the asymptotes pass through the corners of a rectangle of dimensions $2a$ by $2b$. The line segment of length $2b$, joining $(0, b)$ and $(0, -b)$ [or $(-b, 0)$ and $(b, 0)$], is referred to as the **conjugate axis** of the hyperbola.

FIGURE A.15

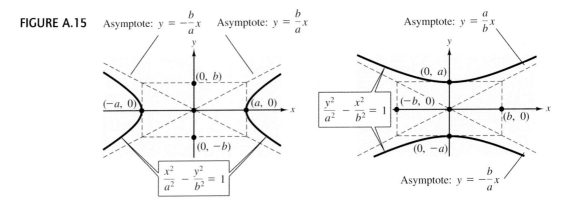

Transverse axis is horizontal.   **Transverse axis is vertical.**

Asymptotes of a Hyperbola (Center at Origin)

The asymptotes of a hyperbola with the center at $(0, 0)$ are

$$y = \frac{b}{a}x \quad \text{and} \quad y = -\frac{b}{a}x \qquad \text{Transverse axis is horizontal}$$

or

$$y = \frac{a}{b}x \quad \text{and} \quad y = -\frac{a}{b}x. \qquad \text{Transverse axis is vertical}$$

EXAMPLE 6   ■   Sketching the Graph of a Hyperbola

Sketch the graph of the hyperbola whose equation is $4x^2 - y^2 = 16$.

Solution

We begin by rewriting the equation in standard form.

$$4x^2 - y^2 = 16 \qquad \text{Given equation}$$

$$\frac{4x^2}{16} - \frac{y^2}{16} = \frac{16}{16} \qquad \text{Divide both sides by 16}$$

$$\frac{x^2}{2^2} - \frac{y^2}{4^2} = 1 \qquad \text{Standard form}$$

Because the x^2-term is positive, we conclude that the transverse axis is horizontal and the vertices occur at $(-2, 0)$ and $(2, 0)$. Moreover, the endpoints of the conjugate axis occur at $(0, -4)$ and $(0, 4)$, and we are able to sketch the rectangle shown in Figure A.16. Finally, by drawing the asymptotes through the corners of this rectangle, we complete the sketch shown in Figure A.17.

FIGURE A.16

FIGURE A.17

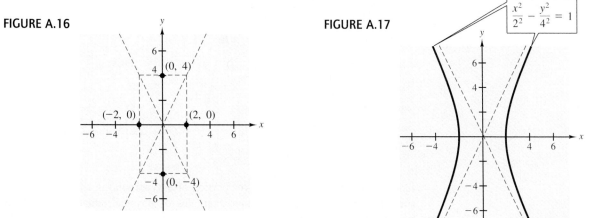

EXAMPLE 7 ■   Finding the Standard Equation of a Hyperbola

Find the standard form of the equation of the hyperbola having vertices at $(0, -3)$ and $(0, 3)$ and with asymptotes $y = -2x$ and $y = 2x$, as shown in Figure A.18.

Solution

Since the transverse axis is vertical, we have asymptotes of the form

$$y = \frac{a}{b}x \quad \text{and} \quad y = -\frac{a}{b}x. \qquad \text{Transverse axis is vertical}$$

Using the given equations for the asymptotes, it follows that

$$\frac{a}{b} = 2$$

and since $a = 3$, we can determine that $b = \frac{3}{2}$. Finally, we conclude that the hyperbola has the following equation.

$$\frac{y^2}{3^2} - \frac{x^2}{(3/2)^2} = 1 \qquad \text{Standard form} \qquad ■$$

FIGURE A.18

DISCUSSION PROBLEM ■   Hyperbolas in Applications

At the beginning of this section, we mentioned that each type of conic section can be formed by the intersection of a plane and a double-napped cone. Figure A.19 shows three examples of how such an intersection can occur in physical situations.

FIGURE A.19

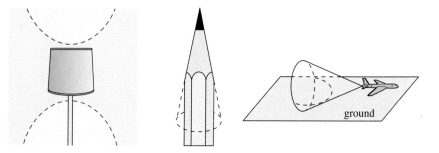

Identify the cone and hyperbola (or portion of a hyperbola) in each of the three situations. Can you think of other examples of physical situations in which hyperbolas are formed? ■

Warm-Up

The following warm-up exercises involve skills that were covered in earlier sections. You will use these skills in the exercise set for this section.

In Exercises 1–4, rewrite the equations so that they have no fractions.

1. $\dfrac{x^2}{16} + \dfrac{y^2}{9} = 1$

2. $\dfrac{x^2}{32} + \dfrac{4y^2}{32} = \dfrac{32}{32}$

3. $\dfrac{x^2}{1/4} - \dfrac{y^2}{4} = 1$

4. $\dfrac{3x^2}{1/9} + \dfrac{4y^2}{9} = 1$

In Exercises 5–8, solve for c. (Assume $c > 0$.)

5. $c^2 = 3^2 - 1^2$

6. $c^2 = 2^2 + 3^2$

7. $c^2 + 2^2 = 4^2$

8. $c^2 - 1^2 = 2^2$

In Exercises 9 and 10, find the distance between the point and the origin.

9. $(0, -4)$

10. $(-2, 0)$

A.1 EXERCISES

In Exercises 1–8, match the equation with its graph. [The graphs are labeled (a), (b), (c), (d), (e), (f), (g), and (h).]

1. $x^2 = 4y$

2. $x^2 = -4y$

3. $y^2 = 4x$

4. $y^2 = -4x$

5. $\dfrac{x^2}{1} + \dfrac{y^2}{4} = 1$

6. $\dfrac{x^2}{4} + \dfrac{y^2}{1} = 1$

7. $\dfrac{x^2}{1} - \dfrac{y^2}{4} = 1$

8. $\dfrac{y^2}{4} - \dfrac{x^2}{1} = 1$

(a)   (b)   (c)   (d)

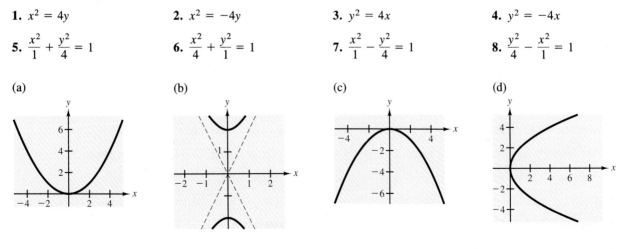

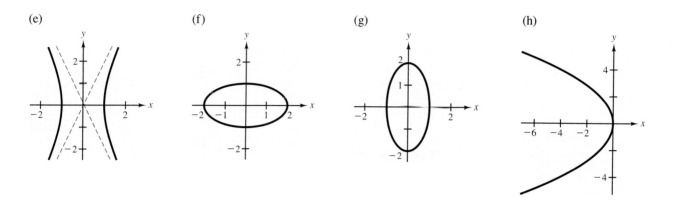

(e)   (f)   (g)   (h)

In Exercises 9–16, find the vertex and focus of the parabola and sketch its graph.

9. $y = 4x^2$ **10.** $y = 2x^2$ **11.** $y^2 = -6x$ **12.** $y^2 = 3x$

13. $x^2 + 8y = 0$ **14.** $x + y^2 = 0$ **15.** $y^2 - 8x = 0$ **16.** $x^2 + 12y = 0$

In Exercises 17–26, find an equation of the specified parabola with a vertex at the origin.

17. Focus: $\left(0, -\dfrac{3}{2}\right)$ **18.** Focus: $(2, 0)$ **19.** Focus: $(-2, 0)$ **20.** Focus: $(0, -2)$

21. Directrix: $y = -1$ **22.** Directrix: $x = 3$ **23.** Directrix: $y = 2$ **24.** Directrix: $x = -2$

25. Horizontal axis; Passes through the point $(4, 6)$ **26.** Vertical axis; Passes through the point $(-2, -2)$

In Exercises 27–34, find the center and vertices of the ellipse and sketch its graph.

27. $\dfrac{x^2}{25} + \dfrac{y^2}{16} = 1$ **28.** $\dfrac{x^2}{144} + \dfrac{y^2}{169} = 1$ **29.** $\dfrac{x^2}{16} + \dfrac{y^2}{25} = 1$ **30.** $\dfrac{x^2}{169} + \dfrac{y^2}{144} = 1$

31. $\dfrac{x^2}{9} + \dfrac{y^2}{5} = 1$ **32.** $\dfrac{x^2}{28} + \dfrac{y^2}{64} = 1$ **33.** $5x^2 + 3y^2 = 15$ **34.** $x^2 + 4y^2 = 4$

In Exercises 35–42, find an equation of the specified ellipse with the center at the origin.

35. Vertices: $(0, \pm 2)$; Minor axis of length 2   **36.** Vertices: $(\pm 2, 0)$; Minor axis of length 3

37. Vertices: $(\pm 5, 0)$; Foci $(\pm 2, 0)$ **38.** Vertices: $(0, \pm 8)$; Foci: $(0, \pm 4)$

39. Foci: $(\pm 5, 0)$; Major axis of length 12   **40.** Foci: $(\pm 2, 0)$; Major axis of length 8

41. Vertices: $(0, \pm 5)$; Passes through the point $(4, 2)$ **42.** Major axis vertical; Passes through the points $(0, 4)$ and $(2, 0)$

In Exercises 43–50, find the center and vertices of the hyperbola and sketch its graph, using asymptotes as an aid.

43. $x^2 - y^2 = 1$

44. $\dfrac{x^2}{9} - \dfrac{y^2}{16} = 1$

45. $\dfrac{y^2}{1} - \dfrac{x^2}{4} = 1$

46. $\dfrac{y^2}{9} - \dfrac{x^2}{1} = 1$

47. $\dfrac{y^2}{25} - \dfrac{x^2}{144} = 1$

48. $\dfrac{x^2}{36} - \dfrac{y^2}{4} = 1$

49. $2x^2 - 3y^2 = 6$

50. $3y^2 - 5x^2 = 15$

In Exercises 51–58, find an equation of the specified hyperbola with the center at the origin.

51. Vertices: $(0, \pm 2)$; Foci: $(0, \pm 4)$

52. Vertices: $(\pm 3, 0)$; Foci: $(\pm 5, 0)$

53. Vertices: $(\pm 1, 0)$; Asymptotes: $y = \pm 3x$

54. Vertices: $(0, \pm 3)$; Asymptotes: $y = \pm 3x$

55. Foci: $(0, \pm 8)$; Asymptotes: $y = \pm 4x$

56. Foci: $(\pm 10, 0)$; Asymptotes: $y = \pm \dfrac{3}{4}x$

57. Vertices: $(0, \pm 3)$; Passes through the point $(-2, 5)$

58. Vertices: $(\pm 2, 0)$; Passes through the point $\left(3, \sqrt{3}\right)$

59. *Satellite Antenna* The receiver in a parabolic television dish antenna is three feet from the vertex and is located at the focus (see figure). Find an equation of a cross section of the reflector. (Assume that the dish is directed upward and the vertex is at the origin.)

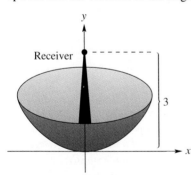

60. *Suspension Bridge* Each cable of a suspension bridge is suspended (in the shape of a parabola) between two towers that are 400 feet apart and 50 feet above the roadway (see figure). The cables touch the roadway midway between the towers. Find an equation for the parabolic shape of each cable.

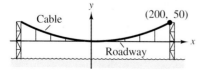

61. *Fireplace Arch* A fireplace arch is to be constructed in the shape of a semi-ellipse. The opening is to have a height of two feet at the center and a width of five feet along the base (see figure). The contractor draws the outline of the ellipse by the method shown in Figure A.8. Where should the tacks be placed and what should be the length of the piece of string?

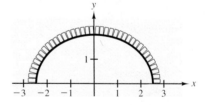

62. *Mountain Tunnel* A semi-elliptical arch over a tunnel for a road through a mountain has a major axis of 100 feet, and its height at the center is 30 feet (see figure). Determine the height of the arch 5 feet from the edge of the tunnel.

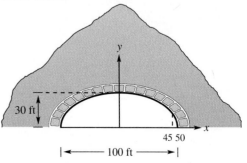

63. Sketch a graph of the ellipse that consists of all points (x, y) such that the sum of the distances between (x, y) and two fixed points is 15 units and the foci are located at the centers of the two sets of concentric circles, as shown in the accompanying figure.

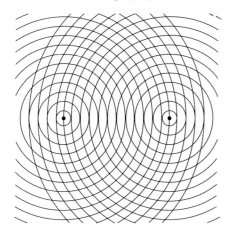

64. A line segment through a focus with endpoints on the ellipse and perpendicular to the major axis is called a **latus rectum** of the ellipse. Therefore, an ellipse has two latus recta. Knowing the length of the latus recta is helpful in sketching an ellipse because it yields other points on the curve (see figure). Show that the length of each latus rectum is $2b^2/a$.

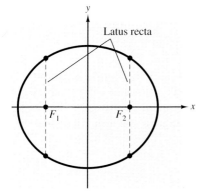

In Exercises 65–68, sketch the graph of the ellipse, making use of the latus recta (see Exercise 64).

65. $\dfrac{x^2}{4} + \dfrac{y^2}{1} = 1$ **66.** $\dfrac{x^2}{9} + \dfrac{y^2}{16} = 1$

67. $9x^2 + 4y^2 = 36$ **68.** $5x^2 + 3y^2 = 15$

69. *Loran* Long-range navigation for aircraft and ships is accomplished by synchronized pulses transmitted by widely separated transmitting stations. These pulses travel at the speed of light (186,000 miles per second). The difference in the arrival-times of these pulses at an aircraft or ship is constant on a hyperbola having the transmitting stations as foci. Assume that two stations 300 miles apart are positioned on the rectangular coordinate system at points with coordinates $(-150, 0)$ and $(150, 0)$ and that a ship is traveling on a path with coordinates $(x, 75)$ (see figure). Find the x-coordinate of the position of the ship if the time-difference between the pulses from the transmitting stations is 1,000 microseconds (0.001 second).

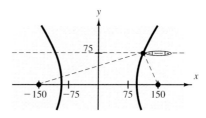

70. *Hyperbolic Mirror* A hyperbolic mirror (used in some telescopes) has the property that a light ray directed at one focus will be reflected to the other focus (see figure). The focus of a hyperbolic mirror has coordinates $(12, 0)$. Find the vertex of the mirror if its **mount** has coordinates $(12, 12)$.

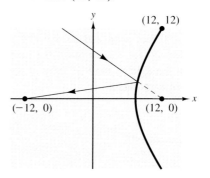

71. Use the definition of an ellipse to derive the standard form of the equation of an ellipse.

72. Use the definition of a hyperbola to derive the standard form of the equation of a hyperbola.

Conic Sections and Translations
Vertical and Horizontal Shifts of Conics   •   *Writing Equations of Conics in Standard Form*

Vertical and Horizontal Shifts of Conics

In Section A.1 we looked at conic sections whose graphs were in *standard position*. In this section we will study the equations of conic sections that have been shifted vertically or horizontally in the plane. The following summary lists the standard forms of the equations of the four basic conics.

Standard Forms of Equations of Conics

Circle (r = radius)

$$(x - h)^2 + (y - k)^2 = r^2$$

Ellipse ($2a$ = major axis length,
                $2b$ = minor axis length)

Center: (h, k)

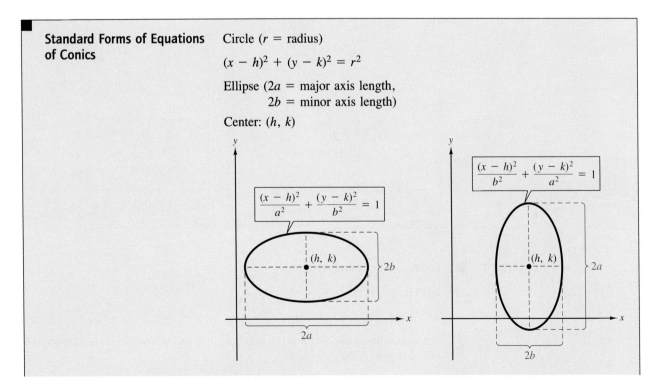

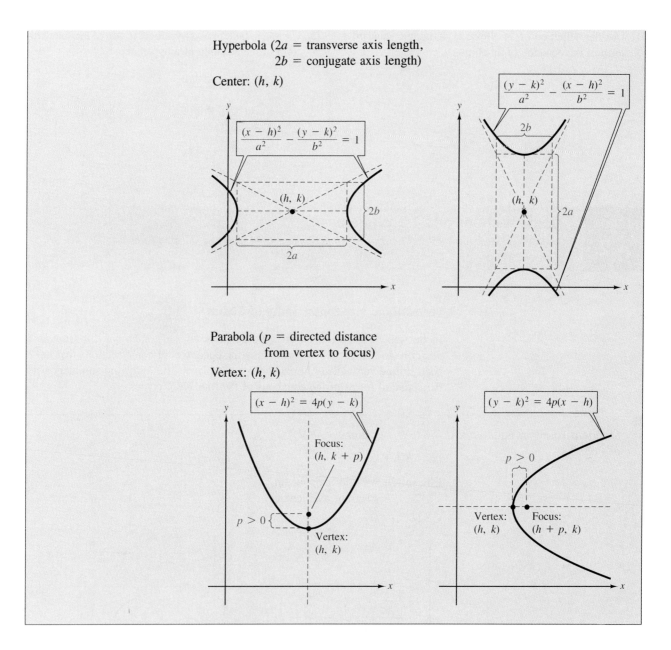

Hyperbola ($2a$ = transverse axis length,
$2b$ = conjugate axis length)

Center: (h, k)

$$\frac{(x - h)^2}{a^2} - \frac{(y - k)^2}{b^2} = 1$$

$$\frac{(y - k)^2}{a^2} - \frac{(x - h)^2}{b^2} = 1$$

Parabola (p = directed distance
from vertex to focus)

Vertex: (h, k)

$$(x - h)^2 = 4p(y - k)$$

Focus:
$(h, k + p)$

$p > 0$

Vertex:
(h, k)

$$(y - k)^2 = 4p(x - h)$$

$p > 0$

Vertex: Focus:
(h, k) $(h + p, k)$

EXAMPLE 1 ■ **Equations of Conic Sections**

(a) The graph of

$$(x - 1)^2 + (y + 2)^2 = 3^2 \qquad \text{Circle}$$

is a circle whose center is the point $(1, -2)$ and whose radius is 3, as shown in Figure A.20(a).

(b) The graph of

$$\frac{(x - 2)^2}{3^2} + \frac{(y - 1)^2}{2^2} = 1 \qquad \text{Ellipse}$$

is an ellipse whose center is the point (2, 1). The major axis of the ellipse is horizontal with a length of 2(3) = 6. The minor axis of the ellipse is vertical with a length of 2(2) = 4, as shown in Figure A.20(b).

(c) The graph of

$$\frac{(x - 3)^2}{1^2} - \frac{(y - 2)^2}{3^2} = 1 \qquad \text{Hyperbola}$$

is a hyperbola whose center is the point (3, 2). The transverse axis is horizontal with a length of 2(1) = 2. The conjugate axis is vertical with a length of 2(3) = 6, as shown in Figure A.20(c).

(d) The graph of

$$(x - 2)^2 = 4(-1)(y - 3) \qquad \text{Parabola}$$

is a parabola whose vertex is the point (2, 3). The axis of the parabola is vertical. The focus of the parabola is 1 unit above or below the vertex. Moreover, since $p = -1$, it follows that the focus lies *below* the vertex, as shown in Figure A.20(d).

FIGURE A.20

(a)
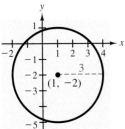
(b)

(c)

(d)

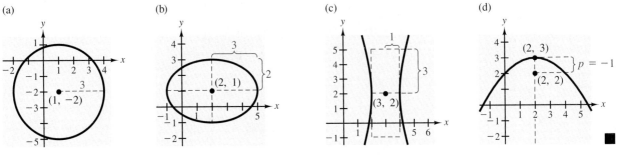

Writing Equations of Conics in Standard Form

To write the equation of a conic in standard form, we complete the square as demonstrated in Examples 2, 3, and 4.

EXAMPLE 2   ■   Finding the Standard Form of a Parabola

Find the vertex and focus of the parabola given by

$$x^2 - 2x + 4y - 3 = 0.$$

FIGURE A.21

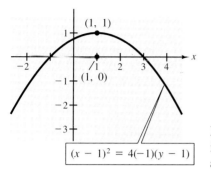

$$(x - 1)^2 = 4(-1)(y - 1)$$

Solution

$x^2 - 2x + 4y - 3 = 0$	Given equation
$x^2 - 2x = -4y + 3$	Group terms
$x^2 - 2x + 1 = -4y + 3 + 1$	Add 1 to both sides
$(x - 1)^2 = -4y + 4$	Completed square form
$(x - 1)^2 = 4(-1)(y - 1)$	Standard form
$(x - h)^2 = 4p(y - k)$	

From this standard form, we see that $h = 1$, $k = 1$, and $p = -1$. Since the axis is vertical and p is negative, the parabola opens downward. The vertex and focus are

Vertex: $(h, k) = (1, 1)$

Focus: $(h, k + p) = (1, 0)$.

The graph of this parabola is shown in Figure A.21.   ■

NOTE: In Example 1 p is the *directed distance* from the vertex to the focus. Because the axis of the parabola is vertical and $p = -1$, the focus is one unit *below* the vertex, and the parabola opens downward.

EXAMPLE 3   ■   **Sketching an Ellipse**

Sketch the graph of the ellipse whose equation is

$$x^2 + 4y^2 + 6x - 8y + 9 = 0.$$

Solution

$x^2 + 4y^2 + 6x - 8y + 9 = 0$	Given equation
$(x^2 + 6x + \blacksquare) + (4y^2 - 8y + \blacksquare) = -9$	Group terms
$(x^2 + 6x + \blacksquare) + 4(y^2 - 2y + \blacksquare) = -9$	Factor 4 out of y-terms
$(x^2 + 6x + 9) + 4(y^2 - 2y + 1) = -9 + 9 + 4(1)$	Add 9 and 4 to both sides
$(x + 3)^2 + 4(y - 1)^2 = 4$	Completed square form
$\dfrac{(x + 3)^2}{4} + \dfrac{(y - 1)^2}{1} = 1$	Standard form
$\dfrac{(x - h)^2}{a^2} + \dfrac{(y - k)^2}{b^2} = 1$	

FIGURE A.22

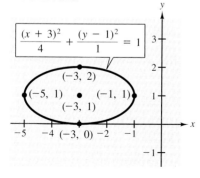

From this standard form, we see that the center is $(h, k) = (-3, 1)$. Since the denominator of the x-term is $4 = a^2 = 2^2$, the endpoints of the major axis lie two units to the right and left of the center. Similarly, since the denominator of the y-term is $1 = b^2 = 1^2$, the endpoints of the minor axis lie one unit up and down from the center. The graph of this ellipse is shown in Figure A.22.

■

EXAMPLE 4   ■   Sketching a Hyperbola

Sketch the graph of the hyperbola given by the equation

$$y^2 - 4x^2 + 4y + 24x - 41 = 0.$$

Solution

$y^2 - 4x^2 + 4y + 24x - 41 = 0$	Given equation
$(y^2 + 4y + \blacksquare) - (4x^2 - 24x + \blacksquare) = 41$	Group terms
$(y^2 + 4y + \blacksquare) - 4(x^2 - 6x + \blacksquare) = 41$	Factor 4 out of x-terms
$(y^2 + 4y + 4) - 4(x^2 - 6x + 9) = 41 + 4 - 4(9)$	Add 4, subtract 36
$(y + 2)^2 - 4(x - 3)^2 = 9$	Completed square form
$\dfrac{(y + 2)^2}{9} - \dfrac{4(x - 3)^2}{9} = 1$	Divide both sides by 9
$\dfrac{(y + 2)^2}{9} - \dfrac{(x - 3)^2}{9/4} = 1$	Change 4 to $1/(1/4)$
$\dfrac{(y + 2)^2}{3^2} - \dfrac{(x - 3)^2}{(3/2)^2} = 1$	Standard form
$\dfrac{(y - h)^2}{a^2} - \dfrac{(x - h)^2}{b^2} = 1$	

FIGURE A.23

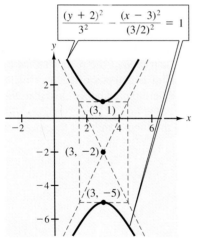

$$\frac{(y + 2)^2}{3^2} - \frac{(x - 3)^2}{(3/2)^2} = 1$$

From the standard form, we see that the transverse axis is vertical and the center lies at $(h, k) = (3, -2)$. Since the denominator of the y-term is $a^2 = 3^2$, we know that the vertices occur three units above and below the center.

Vertices: $(3, -5)$ and $(3, 1)$

To sketch the hyperbola, we draw a rectangle whose top and bottom pass through the vertices. Since the denominator of the x-term is $b^2 = (3/2)^2$, we locate the sides of the rectangle $3/2$ units to the right and left of the center, as shown in Figure A.23. Finally, we sketch the asymptotes by drawing lines through the opposite corners of the rectangle. Using these asymptotes, we complete the graph of the hyperbola, as shown in Figure A.23.   ■

To find the foci in Example 4, we first find c.

$$c^2 = a^2 + b^2 = 9 + \frac{9}{4} = \frac{45}{4} \qquad \Longrightarrow \qquad c = \frac{3\sqrt{5}}{2}$$

Because the transverse axis is vertical, the foci lie c units above and below the center.

Foci: $\left(3, -2 + \dfrac{3\sqrt{5}}{2}\right)$ and $\left(3, -2 - \dfrac{3\sqrt{5}}{2}\right)$

EXAMPLE 5 ■ Writing the Equation of an Ellipse

Write the standard form of the equation of the ellipse whose vertices are $(2, -2)$ and $(2, 4)$. The length of the minor axis of the ellipse is 4, as shown in Figure A.24.

FIGURE A.24

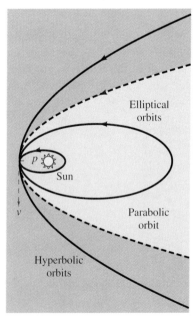

Solution

The center of the ellipse lies at the midpoint of its vertices. Thus, the center is

$$(h, k) = (2, 1).$$ Center

Since the vertices lie on a vertical line and are 6 units apart, it follows that the major axis is vertical and has a length of $2a = 6$. Thus, $a = 3$. Moreover, since the minor axis has a length of 4, it follows that $2b = 4$, which implies that $b = 2$. Therefore, we can conclude that the standard form of the equation of the ellipse is as follows.

$$\frac{(x - h)^2}{a^2} + \frac{(y - k)^2}{b^2} = 1$$ Major axis is vertical

$$\frac{(x - 2)^2}{2^2} + \frac{(y - 1)^2}{3^2} = 1$$ Standard form   ■

FIGURE A.25

An interesting application of conic sections involves the orbits of comets in our solar system. Of the 610 comets identified prior to 1970, 245 have elliptical orbits, 295 have parabolic orbits, and 70 have hyperbolic orbits. For example, Halley's Comet has an elliptical orbit, and we can predict the reappearance of this comet every 75 years. The center of the Sun is a focus of each of these orbits, and each orbit has a vertex at the point where the comet is closest to the Sun, as shown in Figure A.25.

If p is the distance between the vertex and the focus, and v is the speed of the comet at the vertex, then the orbit is:

an *ellipse* if $v < \sqrt{\dfrac{2GM}{p}}$

a *parabola* if $v = \sqrt{\dfrac{2GM}{p}}$

a *hyperbola* if $v > \sqrt{\dfrac{2GM}{p}}$

where M is the mass of the Sun and $G \approx 6.67(10^{-8})$ cm³/(gm · sec²).

DISCUSSION PROBLEM   ■   Asymptotes of a Hyperbola

In Section A.2 you learned how to find equations for the asymptotes of a hyperbola whose center is the origin. Write a paragraph describing how to find equations for the asymptotes of a hyperbola whose center is the point (h, k). Then use your procedure to find the asymptotes of the hyperbola given by

$$\frac{(x-1)^2}{3^2} - \frac{(y-2)^2}{2^2} = 1.$$

■

Warm-Up

The following warm-up exercises involve skills that were covered in earlier sections. You will use these skills in the exercise set for this section.

In Exercises 1–10, identify the conic represented by each equation.

1. $\dfrac{x^2}{4} - \dfrac{y^2}{4} = 1$

2. $\dfrac{x^2}{9} + \dfrac{y^2}{1} = 1$

3. $2x + y^2 = 0$

4. $\dfrac{x^2}{9} - \dfrac{y^2}{4} = 1$

5. $\dfrac{x^2}{4} + \dfrac{y^2}{16} = 1$

6. $4x^2 + 4y^2 = 25$

7. $\dfrac{y^2}{4} - \dfrac{x^2}{2} = 1$

8. $x^2 - 6y = 0$

9. $3x - y^2 = 0$

10. $\dfrac{x^2}{9/4} + \dfrac{y^2}{4} = 1$

A.2   EXERCISES

In Exercises 1–12, find the vertex, focus, and directrix of the parabola and sketch its graph.

1. $(x-1)^2 + 8(y+2) = 0$

2. $(x+3) + (y-2)^2 = 0$

3. $\left(y + \dfrac{1}{2}\right)^2 = 2(x-5)$

4. $\left(x + \dfrac{1}{2}\right)^2 = 4(y-3)$

5. $y = \dfrac{1}{4}(x^2 - 2x + 5)$

6. $y = -\dfrac{1}{6}(x^2 + 4x - 2)$

7. $4x - y^2 - 2y - 33 = 0$

8. $y^2 + x + y = 0$

9. $y^2 + 6y + 8x + 25 = 0$

10. $x^2 - 2x + 8y + 9 = 0$

11. $y^2 - 4y - 4x = 0$

12. $y^2 - 4x - 4 = 0$

In Exercises 13–20, find an equation of the specified parabola.

13. Vertex: (3, 2); Focus: (1, 2)

14. Vertex: (−1, 2); Focus: (−1, 0)

15. Vertex: (0, 4); Directrix: $y = 2$

16. Vertex: (−2, 1); Directrix: $x = 1$

17. Focus: (2, 2); Directrix: $x = -2$

18. Focus: (0, 0); Directrix: $y = 4$

19.

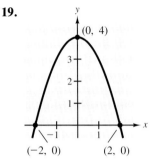

20.

In Exercises 21–28, find the center, foci, and vertices of the ellipse and sketch its graph.

21. $\dfrac{(x - 1)^2}{9} + \dfrac{(y - 5)^2}{25} = 1$

22. $(x + 2)^2 + \dfrac{(y + 4)^2}{1/4} = 1$

23. $9x^2 + 4y^2 + 36x - 24y + 36 = 0$

24. $9x^2 + 4y^2 - 36x + 8y + 31 = 0$

25. $16x^2 + 25y^2 - 32x + 50y + 16 = 0$

26. $9x^2 + 25y^2 - 36x - 50y + 61 = 0$

27. $12x^2 + 20y^2 - 12x + 40y - 37 = 0$

28. $36x^2 + 9y^2 + 48x - 36y + 43 = 0$

In Exercises 29–36, find an equation for the specified ellipse.

29. Vertices: (0, 2), (4, 2); Minor axis of length 2

30. Foci: (0, 0), (4, 0); Major axis of length 8

31. Foci: (0, 0), (0, 8); Major axis of length 16

32. Center: (2, −1); Vertex: $\left(2, \dfrac{1}{2}\right)$; Minor axis of length 2

33. Vertices: (3, 1), (3, 9); Minor axis of length 6

34. Center: (3, 2); $a = 3c$; Foci: (1, 2), (5, 2)

35. Center: (0, 4); $a = 2c$; Vertices: (−4, 4), (4, 4)

36. Vertices: (5, 0), (5, 12); Endpoints of the minor axis: (0, 6), (10, 6)

In Exercises 37–46, find the center, vertices, and foci of the hyperbola and sketch its graph, using asymptotes as an aid.

37. $\dfrac{(x - 1)^2}{4} - \dfrac{(y + 2)^2}{1} = 1$

38. $\dfrac{(x + 1)^2}{144} - \dfrac{(y - 4)^2}{25} = 1$

39. $(y + 6)^2 - (x - 2)^2 = 1$

40. $\dfrac{(y - 1)^2}{1/4} - \dfrac{(x + 3)^2}{1/9} = 1$

41. $9x^2 - y^2 - 36x - 6y + 18 = 0$

42. $x^2 - 9y^2 + 36y - 72 = 0$

43. $9y^2 - x^2 + 2x + 54y + 62 = 0$

44. $16y^2 - x^2 + 2x + 64y + 63 = 0$

45. $x^2 - 9y^2 + 2x - 54y - 107 = 0$

46. $9x^2 - y^2 + 54x + 10y + 55 = 0$

In Exercises 47–54, find an equation for the specified hyperbola.

47. Vertices: (2, 0), (6, 0); Foci: (0, 0), (8, 0)

48. Vertices: (2, 3), (2, −3); Foci: (2, 5), (2, −5)

49. Vertices: (4, 1), (4, 9); Foci: (4, 0), (4, 10)

50. Vertices: (−2, 1), (2, 1); Foci: (−3, 1), (3, 1)

51. Vertices: (2, 3), (2, −3); Passes through the point (0, 5)

52. Vertices: (−2, 1), (2, 1); Passes through the point (4, 3)

53. Vertices: (0, 2), (6, 2); Asymptotes: $y = \frac{2}{3}x$, $y = 4 - \frac{2}{3}x$

54. Vertices: (3, 0), (3, 4); Asymptotes: $y = \frac{2}{3}x$, $y = 4 - \frac{2}{3}x$

In Exercises 55–62, classify the graph of each equation as a circle, a parabola, an ellipse, or a hyperbola.

55. $x^2 + y^2 - 6x + 4y + 9 = 0$

56. $x^2 + 4y^2 - 6x + 16y + 21 = 0$

57. $4x^2 - y^2 - 4x - 3 = 0$

58. $y^2 - 4y - 4x = 0$

59. $4x^2 + 3y^2 + 8x - 24y + 51 = 0$

60. $4y^2 - 2x^2 - 4y - 8x - 15 = 0$

61. $25x^2 - 10x - 200y - 119 = 0$

62. $4x^2 + 4y^2 - 16y + 15 = 0$

63. *Satellite Orbit* A satellite in a 100-mile-high circular orbit around the Earth has a velocity of approximately 17,500 miles per hour. If this velocity is multiplied by $\sqrt{2}$, then the satellite will have the minimum velocity necessary to escape the Earth's gravity, and it will follow a parabolic path with the center of the Earth as the focus (see figure).

(a) Find the escape velocity of the satellite.

(b) Find an equation of its path (assume the radius of the Earth is 4,000 miles).

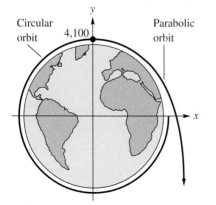

64. *Fluid Flow* Water is flowing from a horizontal pipe 48 feet above the ground. The falling stream of water has the shape of a parabola whose vertex (0, 48) is at the end of the pipe (see figure). The stream of water strikes the ground at the point $\left(10\sqrt{3},\ 0\right)$. Find the equation of the path taken by the water.

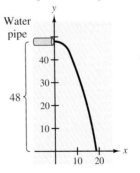

In Exercises 65–71, e is called the **eccentricity** of the ellipse and is defined by $e = c/a$. It measures the flatness of the ellipse.

65. Find an equation of the ellipse with vertices $(\pm 5, 0)$ and eccentricity $e = \frac{3}{5}$.

66. Find an equation of the ellipse with vertices $(0, \pm 8)$ and eccentricity $e = \frac{1}{2}$.

67. *Orbit of the Earth* The Earth moves in an elliptical orbit with the Sun at one of the foci (see figure). The length of half of the major axis is 92.957×10^6 miles and the eccentricity is 0.017. Find the shortest distance (*perihelion*) and the greatest distance (*aphelion*) that the Earth can get from the Sun.

Figure for 67–69

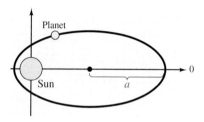

68. *Orbit of Pluto* The planet Pluto moves in an elliptical orbit with the Sun at one of the foci (see figure). The length of half of the major axis is 3.666×10^9 miles and the eccentricity is 0.248. Find the shortest distance (*perihelion*) and the greatest distance (*aphelion*) that Pluto can get from the Sun.

69. *Orbit of Saturn* The planet Saturn moves in an elliptical orbit with the Sun at one of the foci (see figure). The shortest distance and the greatest distance that the planet can get from the Sun are 1.3495×10^9 kilometers and 1.5045×10^9 kilometers, respectively. Find the eccentricity of the orbit.

70. *Satellite Orbit* The first artificial satellite to orbit the Earth was Sputnik I (launched by Russia in 1957). Its highest point above the Earth's surface was 583 miles, and its lowest point was 132 miles (see figure). Assume that the center of the Earth is the focus of the elliptical orbit and the radius of the Earth is 4,000 miles. Find the eccentricity of the orbit.

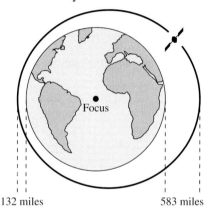

132 miles   583 miles

71. Show that the equation of an ellipse can be written as

$$\frac{(x - h)^2}{a^2} + \frac{(y - k)^2}{a^2(1 - e^2)} = 1.$$

Note that as e approaches zero, the ellipse approaches a circle of radius a.

Further Concepts in Statistics

The material in this appendix may be taught as a stand-alone chapter or used as desired in conjunction with topics as they occur in the text. For convenience, marginal notes point out earlier occurrences of topics in the text.

Representing Data and Linear Modeling

Stem-and-Leaf Plots • *Histograms and Frequency Distributions* • *Scatter Plots* • *Fitting a Line to Data*

Stem-and-Leaf Plots

Statistics is the branch of mathematics that studies techniques for collecting, organizing, and interpreting data. In this section, you will study several ways to organize and interpret data.

One type of plot that can be used to organize sets of numbers by hand is a **stem-and-leaf plot.** A set of test scores and the corresponding stem-and-leaf plot are shown below.

Test Scores	*Stems*	*Leaves*
93, 70, 76, 58, 86, 93, 82, 78, 83, 86,	5	8
64, 78, 76, 66, 83, 83, 96, 74, 69, 76,	6	4 4 6 9
64, 74, 79, 76, 88, 76, 81, 82, 74, 70	7	0 0 4 4 4 6 6 6 6 6 8 8 9
	8	1 2 2 3 3 3 6 6 8
	9	3 3 6

Note that the *leaves* represent the units digits of the numbers and the *stems* represent the tens digits. Stem-and-leaf plots can also be used to compare two sets of data, as shown in the following example.

EXAMPLE 1 ■ Comparing Two Sets of Data

Use a stem-and-leaf plot to compare the test scores given above with the following test scores. Which set of test scores is better?

90, 81, 70, 62, 64, 73, 81, 92, 73, 81, 92, 93, 83, 75, 76,
83, 94, 96, 86, 77, 77, 86, 96, 86, 77, 86, 87, 87, 79, 88

Solution

Begin by ordering the second set of scores.

62, 64, 70, 73, 73, 75, 76, 77, 77, 77, 79, 81, 81, 81, 83,
83, 86, 86, 86, 86, 87, 87, 88, 90, 92, 92, 93, 94, 96, 96

Now that the data has been ordered, you can construct a *double* stem-and-leaf plot by letting the leaves to the right of the stems represent the units digits for the first group of test scores and letting the leaves to the left of the stems represent the units digits for the second group of test scores.

Leaves (2nd Group)	Stems	Leaves (1st Group)
	5	8
4 2	6	4 4 6 9
9 7 7 7 6 5 3 3 0	7	0 0 4 4 4 6 6 6 6 6 8 8 9
8 7 7 6 6 6 6 3 3 1 1 1	8	1 2 2 3 3 3 6 6 8
6 6 4 3 2 2 0	9	3 3 6

By comparing the two sets of leaves, you can see that the second group of test scores is better than the first group.   ■

EXAMPLE 2   ■   Using a Stem-and-Leaf Plot

Table B.1 shows the percent of the population of each state and the District of Columbia that was at least 65 years old in 1989. Use a stem-and-leaf plot to organize the data. (*Source:* U.S. Bureau of Census)

TABLE B.1

AK	4.1	AL	12.7	AR	14.8	AZ	13.1	CA	10.6
CO	9.8	CT	13.6	DC	12.5	DE	11.8	FL	18.0
GA	10.1	HI	10.7	IA	15.1	ID	11.9	IL	12.3
IN	12.4	KS	13.7	KY	12.7	LA	11.1	MA	13.8
MD	10.8	ME	13.4	MI	11.9	MN	12.6	MO	13.9
MS	12.4	MT	13.2	NC	12.1	ND	13.9	NE	13.9
NH	11.4	NJ	13.2	NM	10.5	NV	10.9	NY	13.0
OH	12.8	OK	13.3	OR	13.9	PA	15.1	RI	14.8
SC	11.1	SD	14.4	TN	12.6	TX	10.1	UT	8.6
VA	10.8	VT	11.9	WA	11.9	WI	13.4	WV	14.6
WY	9.8								

Solution

Begin by ordering the numbers, as shown below.

4.1, 8.6, 9.8, 9.8, 10.1, 10.1, 10.5, 10.6, 10.7, 10.8, 10.8
10.9, 11.1, 11.1, 11.4, 11.8, 11.9, 11.9, 11.9, 11.9, 12.1, 12.3
12.4, 12.4, 12.5, 12.6, 12.6, 12.7, 12.7, 12.8, 13.0, 13.1, 13.2
13.2, 13.3, 13.4, 13.4, 13.6, 13.7, 13.8, 13.9, 13.9, 13.9, 13.9
14.4, 14.6, 14.8, 14.8, 15.1, 15.1, 18.0

Next construct the stem-and-leaf plot using the leaves to represent the digits to the right of the decimal points.

Stems	Leaves	
4.	1	Alaska has the lowest percent.
5.		
6.		
7.		
8.	6	
9.	8 8	
10.	1 1 5 6 7 8 8 9	
11.	1 1 4 8 9 9 9 9	
12.	1 3 4 4 5 6 6 7 7 8	
13.	0 1 2 2 3 4 4 6 7 8 9 9 9 9	
14.	4 6 8 8	
15.	1 1	
16.		
17.		
18.	0	Florida has the highest percent.

∎

Technology Note

Try using a computer or graphing calculator to create a histogram for the data at the right. How does the histogram change when the intervals change?

NOTE Bar graphs are used in exercises throughout this book. For example, see Exercises 85 and 86 on page 32 and Exercise 49 on page 593. The material presented here allows for more in-depth discussion.

Histograms and Frequency Distributions

With data such as that given in Example 2, it is useful to group the numbers into intervals and plot the frequency of the data in each interval. For instance, the **frequency distribution** and **histogram** shown in Figure B.1 represent the data given in Example 2.

Frequency Distribution

Interval	Tally
[4, 6)	I
[6, 8)	
[8, 10)	III
[10, 12)	ЖЖ ЖЖ ЖЖ I
[12, 14)	ЖЖ ЖЖ ЖЖ ЖЖ IIII
[14, 16)	ЖЖ I
[16, 18)	
[18, 20)	I

Histogram

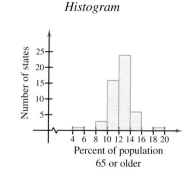

FIGURE B.1

A histogram has a portion of a real number line as its horizontal axis. A **bar graph** is similar to a histogram, except that the rectangles (bars) can be either horizontal or vertical and the labels of the bars are not necessarily numbers.

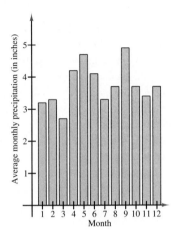

FIGURE B.2

Another difference between a bar graph and a histogram is that the bars in a bar graph are usually separated by spaces, whereas the bars in a histogram are not separated by spaces.

EXAMPLE 3   ■   Constructing a Bar Graph

The data below shows the average monthly precipitation (in inches) in Houston, Texas. Construct a bar graph for this data. What can you conclude? (*Source:* PC USA)

January	3.2	February	3.3	March	2.7
April	4.2	May	4.7	June	4.1
July	3.3	August	3.7	September	4.9
October	3.7	November	3.4	December	3.7

Solution

To create a bar graph, begin by drawing a vertical axis to represent the precipitation and a horizontal axis to represent the months. The bar graph is shown in Figure B.2. From the graph, you can see that Houston receives a fairly consistent amount of rain throughout the year–the driest month tends to be March and the wettest month tends to be September.   ■

Scatter Plots

Many real-life situations involve finding relationships between two variables, such as the year and the number of people in the labor force. In a typical situation, data is collected and written as a set of ordered pairs. The graph of such a set is called a **scatter plot.**

From the scatter plot in Figure B.3, it appears that the points describe a relationship that is nearly linear. (The relationship is not *exactly* linear because the labor force did not increase by precisely the same amount each year.) A mathematical equation that approximates the relationship between t and P is called a *mathematical model.* When developing a mathematical model, you strive for two (often conflicting) goals—accuracy and simplicity.

Consider a collection of ordered pairs of the form (x, y). If y tends to increase as x increases, the collection is said to have a **positive correlation.** If y tends to decrease as x increases, the collection is said to have a **negative correlation.** Figure B.4 shows three examples: one with a positive correlation, one with a negative correlation, and one with no (discernible) correlation.

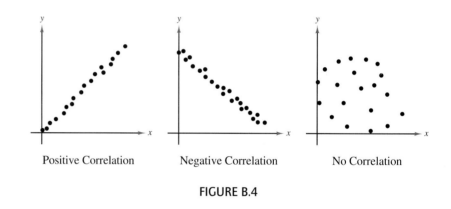

Positive Correlation   Negative Correlation   No Correlation

FIGURE B.4

> *Technology Note*
>
> Most graphing utilities have built-in statistical programs that can create scatter plots. Use your graphing utility to plot the points given in the table below. The data shows the number of people P (in millions) in the United States who were part of the labor force from 1980 through 1990. In the table, t represents the year, with $t = 0$ corresponding to 1980. (*Source:* U.S. Bureau of Labor Statistics)

t	P
0	109
1	110
2	112
3	113
4	115
5	117

t	P
6	120
7	122
8	123
9	126
10	126

NOTE   Scatter plots are discussed in Section 2.5. Exercises 35–42 on page 219 are related to the idea of correlation. The material presented here allows a more in-depth discussion.

FIGURE B.3

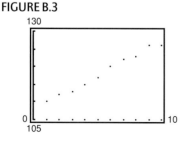

Fitting a Line to Data

Finding a linear model that represents the relationship described by a scatter plot is called **fitting a line to data.** You can do this graphically by simply sketching the line that appears to fit the points, finding two points on the line, and then finding the equation of the line that passes through the two points.

EXAMPLE 4   ■   Fitting a Line to Data

Find a linear model that relates the year with the number of people P (in millions) who were part of the United States labor force from 1980 through 1990. In Table B.2, t represents the year, with $t = 0$ corresponding to 1980. (*Source:* U.S. Bureau of Labor Statistics)

TABLE B.2

t	0	1	2	3	4	5	6	7	8	9	10
P	109	110	112	113	115	117	120	122	123	126	126

Solution

After plotting the data from Table B.2, draw the line that you think best represents the data, as shown in Figure B.5. Two points that lie on this line are (0, 109) and (9, 126). Using the point-slope form, you can find the equation of the line to be

$$P = \frac{17}{9}t + 109. \qquad \text{Linear model}$$

NOTE   Fitting a line to data is discussed in Section 2.5. The material presented here allows for a more in-depth discussion.

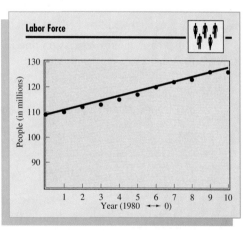

FIGURE B.5

Once you have found a model, you can measure how well the model fits the data by comparing the actual values with the values given by the model, as shown in Table B.3.

TABLE B.3

	t	0	1	2	3	4	5	6	7	8	9	10
Actual →	P	109	110	112	113	115	117	120	122	123	126	126
Model →	P	109	110.9	112.8	114.7	116.6	118.4	120.3	122.2	124.1	126	127.9

The sum of the squares of the differences between the actual values and the model's values is the **sum of the squared differences.** The model that has the least sum is called the **least squares regression line** for the data. For the model in Example 4, the sum of the squared differences is 13.81. The least squares regression line for the data is

$$P = 1.864t + 108.2. \qquad \text{Best-fitting linear model}$$

Its sum of squared differences is 4.7.

Least Squares Regression Line

The least squares regression line, $y = ax + b$, for the points (x_1, y_1), (x_2, y_2), $(x_3, y_3), \ldots, (x_n, y_n)$ is given by

$$a = \frac{n\sum_{i=1}^{n} x_i y_i - \sum_{i=1}^{n} x_i \sum_{i=1}^{n} y_i}{n\sum_{i=1}^{n} x_i^2 - \left(\sum_{i=1}^{n} x_i\right)^2} \quad \text{and} \quad b = \frac{1}{n}\left(\sum_{i=1}^{n} y_i - a\sum_{i=1}^{n} x_i\right).$$

EXAMPLE 5 ■ Finding a Least Squares Regression Line

Find the least squares regression line for the points $(-3, 0)$, $(-1, 1)$, $(0, 2)$, and $(2, 3)$.

Solution

Begin by constructing a table like that shown in Table B.4.

TABLE B.4

x	y	xy	x^2
-3	0	0	9
-1	1	-1	1
0	2	0	0
2	3	6	4
$\sum_{i=1}^{n} x_i = -2$	$\sum_{i=1}^{n} y_i = 6$	$\sum_{i=1}^{n} x_i y_i = 5$	$\sum_{i=1}^{n} x_i^2 = 14$

NOTE Least squares regression lines are mentioned in Exercises 47–54 on page 455. The material presented here allows a more in-depth discussion.

Applying the formulas for the least squares regression line with $n = 4$ produces

$$a = \frac{n\sum_{i=1}^{n} x_i y_i - \sum_{i=1}^{n} x_i \sum_{i=1}^{n} y_1}{n\sum_{i=1}^{n} x_i^2 - \left(\sum_{i=1}^{n} x_i\right)^2} = \frac{4(5) - (-2)(6)}{4(14) - (-2)^2} = \frac{8}{13}$$

and

$$b = \frac{1}{n} \left(\sum_{i=1}^{n} y_i - a \sum_{i=1}^{n} x_i \right) = \frac{1}{4} \left[6 - \frac{8}{13}(-2) \right] = \frac{47}{26}.$$

Thus, the least squares regression line is $y = \frac{8}{13}x + \frac{47}{26}$, as shown in Figure B.6.

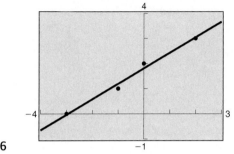

FIGURE B.6   ■

Many calculators have "built-in" least squares regression programs. If your calculator has such a program, try using it to duplicate the results shown in the following example.

EXAMPLE 6   ■   Finding a Least Squares Regression Line

The following ordered paris (w, h) represent the shoe sizes w, and the heights h (in inches), of 25 men. Use a computer program or a statistical calculator to find the least squares regression line for this data.

(10.0, 70.0), (10.5, 71.0), (9.5, 70.0), (11.0, 72.0), (12.0, 74.0),
(8.5, 66.0), (9.0, 68.5), (13.0, 76.0), (10.5, 71.5), (10.5, 70.5),
(10.0, 72.0), (9.5, 70.0), (10.0, 71.0), (10.5, 69.5), (11.0, 71.5)
(12.0, 73.5), (12.5, 74.0), (11.0, 71.5), (9.0, 67.5), (10.0, 70.0),
(13.0, 73.5), (10.5, 72.5), (10.5, 71.0), (11.0, 73.0), (8.5, 68.0)

Solution

A scatter plot for the data is shown in Figure B.7. Note that the plot does not have 25 points because some of the ordered pairs graph as the same point. After entering the data into a statistical calculator, you can obtain

$$a = 1.67 \quad \text{and} \quad b = 53.57.$$

Thus, the least squares regression line for the data is

$$h = 1.67w + 53.57.$$

NOTE Built-in regression features of calculators are mentioned in Chapter Projects throughout the book. For example, see the Chapter 2 Project on page 221 and the Chapter 6 Project on page 491.

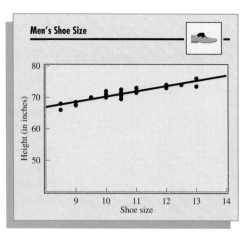

FIGURE B.7

If you use a statistical calculator or computer program to duplicate the results of Example 6, you will notice that the program also outputs a value of $r \approx 0.918$. This number is called the **correlation coefficient** of the data. Correlation coefficients vary between -1 and 1. Basically, the closer $|r|$ is to 1, the better the points can be described by a line. Three examples are shown in Figure B.8.

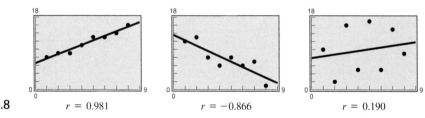

FIGURE B.8 $r = 0.981$ $r = -0.866$ $r = 0.190$

B.1   EXERCISES

Exam Scores   In Exercises 1 and 2, use the following scores from a math class with 30 students. The scores are given for two 100-point exams.

Exam #1:   77, 100, 77, 70, 83, 89, 87, 85, 81, 84, 81, 78, 89, 78, 88, 85, 90, 92, 75, 81, 85, 100, 98, 81, 78, 75, 85, 89, 82, 75

Exam #2:   76, 78, 73, 59, 70, 81, 71, 66, 66, 73, 68, 67, 63, 67, 77, 84, 87, 71, 78, 78, 90, 80, 77, 70, 80, 64, 74, 68, 68, 68

1. Construct a stem-and-leaf plot for Exam #1.

2. Construct a double stem-and-leaf plot to compare the scores for Exam #1 and Exam #2. Which set of test scores is higher?

3. *Education Expenses*   The following table shows the per capita expenditures for public elementary and secondary education in the 50 states and the District of Columbia in 1991. Use a stem-and-leaf plot to organize the data. (*Source:* National Education Association)

AK	1626	AL	694	AR	668	AZ	892	CA	918
CO	841	CT	1151	DC	1010	DE	891	FL	862
GA	859	HI	784	IA	846	ID	725	IL	788
IN	925	KS	906	KY	725	LA	758	MA	866
MD	944	ME	1062	MI	926	MN	990	MO	742
MS	671	MT	983	NC	813	ND	719	NE	757
NH	881	NJ	1223	NM	915	NV	1004	NY	1186
OH	861	OK	776	OR	925	PA	889	RI	892
SC	835	SD	716	TN	618	TX	905	UT	828
VA	941	VT	992	WA	1095	WI	928	WV	883
WY	1178								

4. *Snowfall*   The data below shows the seasonal snowfall (in inches) at Erie, Pennsylvania for the years 1960 through 1989 (the amounts are listed in order by year). How would you organize this data? Explain your reasoning. (*Source:* National Oceanic and Atmospheric Administration)

69.6, 42.5, 75.9, 115.9, 92.9, 84.8, 68.6, 107.9, 79.7, 85.6, 120.0, 92.3, 53.7, 68.6, 66.7, 66.0, 111.5, 142.8, 76.5, 55.2, 89.4, 71.3, 41.2, 110.0, 106.3, 124.9, 68.2, 103.5, 76.5, 114.9

5. *Fruit Crops*   The data below shows the cash receipts (in millions of dollars) from fruit crops for farmers in 1990. Construct a bar graph for this data. (*Source:* U.S. Department of Agriculture)

Apples	1159	Peaches	365
Grapefruit	317	Pears	266
Grapes	1668	Plums and Prunes	293
Lemons	278	Strawberries	560
Oranges	1707		

6. *Travel to the United States*   The data below gives the places of origin and the numbers of travelers (in millions) to the United States in 1991. Construct a horizontal bar graph for this data. (*Source:* U.S. Travel and Tourism Administration)

Canada	18.9	Mexico	7.0
Europe	7.4	Latin America	2.0
Other	6.8		

Crop Yield   In Exercises 7–10, use the data in the table, where x is the number of units of fertilizer applied to sample plots and y is the yield (in bushels) of a crop.

x	0	1	2	3	4	5	6	7	8
y	58	60	59	61	63	66	65	67	70

7. Sketch a scatter plot of the data.

8. Determine whether the points are positively correlated, are negatively correlated, or have no discernible correlation.

9. Sketch a linear model that you think best represents the data. Find and equation of the line you sketched. Use the line to predict the yield if 10 units of fertilizer are used.

10. Can the model found in Exercise 9 be used to predict yields for arbitrarily large values of x? Explain.

Speed of Sound   In Exercises 11–14, use the data in the table, where h is altitude in thousands of feet and v is the speed of sound in feet per second.

h	0	5	10	15	20	25	30	35
v	1116	1097	1077	1057	1036	1015	995	973

11. Sketch a scatter plot of the data.

12. Determine whether the points are positively correlated, are negatively correlated, or have no discernible correlation.

13. Sketch a linear model that you think best represents the data. Find an equation of the line you sketched. Use the line to predict the speed of sound at an altitude of 27,000 feet.

14. The speed of sound at an altitude of 70,000 feet is approximately 971 feet per second. What does this suggest about the validity of using the model in Exercise 13 to extrapolate beyond the data given in the table?

In Exercises 15 and 16, (a) sketch a scatter plot of the points, (b) find an equation of the linear model you think best represents the data and find the sum of the squared differences, and (c) use the formulas of this section to find the least squares regression line and the sum of the squared differences.

15. $(-1, 0)$, $(0, 1)$, $(1, 3)$, $(2, 3)$

16. $(0, 4)$, $(1, 3)$, $(2, 2)$, $(4, 1)$

In Exercises 17–20, (a) Sketch a scatter plot of the points, (b) use the formulas of this section to find the least squares regression line, and (c) sketch the graph of the line.

17. $(-2, 0)$, $(-1, 1)$, $(0, 1)$, $(2, 2)$

18. $(-3, 1)$, $(-1, 2)$, $(0, 2)$, $(1, 3)$, $(3, 5)$

19. $(1, 5)$, $(2, 8)$, $(3, 13)$, $(4, 16)$, $(5, 22)$, $(6, 26)$

20. $(1, 10)$, $(2, 8)$, $(3, 8)$, $(4, 6)$, $(5, 5)$, $(6, 3)$

In Exercises 21–24, use a graphing utility to find the least squares regression line for the data. Sketch a scatter plot and the regression line.

21. $(0, 23)$, $(1, 20)$, $(2, 19)$, $(3, 17)$, $(4, 15)$, $(5, 11)$, $(6, 10)$

22. $(4, 52.8)$, $(5, 54.7)$, $(6, 55.7)$, $(7, 57.8)$, $(8, 60.2)$, $(9, 63.1)$, $(10, 66.5)$

23. $(-10, 5.1)$, $(-5, 9.8)$, $(0, 17.5)$, $(2, 25.4)$, $(4, 32.8)$, $(6, 38.7)$, $(8, 44.2)$, $(10, 50.5)$

24. $(-10, 213.5)$, $(-5, 174.9)$, $(0, 141.7)$, $(5, 119.7)$, $(8, 102.4)$, $(10, 87.6)$

25. *Advertising* The management of a department store ran an experiment to determine if a relationship existed between sales S (in thousands of dollars) and the amount spent on advertising x (in thousands of dollars). The following data were collected.

x	1	2	3	4	5	6	7	8
S	405	423	455	466	492	510	525	559

(a) Use a graphing utility to find the least squares regression line. Use the equation to estimate sales if $4500 is spent on advertising.

(b) Make a scatter plot of the data and sketch the graph of the regression line.

(c) Use a computer or calculator to determine the correlation coefficient.

26. *School Enrollment* The table gives the preprimary school enrollments y (in millions) for the years 1985 through 1991 where $t = 5$ corresponds to 1985. (*Source:* U.S. Bureau of the Census)

t	5	6	7	8	9	10	11
y	10.73	10.87	10.87	11.00	11.04	11.21	11.37

(a) Use a computer or calculator to find the least squares regression line. Use the equation to estimate enrollment in 1992.

(b) Make a scatter plot of the data and sketch the graph of the regression line.

(c) Use the computer or calculator to determine the correlation coefficient.

Measures of Central Tendency and Dispersion

Mean, Median, and Mode • *Choosing a Measure of Central Tendency* • *Variance and Standard Deviation*

Mean, Median, and Mode

NOTE   Mean is discussed on page 40. The material presented here allows a more in-depth discussion.

In many real-life situations, it is helpful to describe data by a single number that is most representative of the entire collection of numbers. Such a number is called a **measure of central tendency.** The most commonly used measures are as follows.

1. The **mean,** or **average,** of n numbers is the sum of the numbers divided by n.

2. The **median** of n numbers is the middle number when the numbers are written in order. If n is even, the median is the average of the two middle numbers.

3. The **mode** of n numbers is the number that occurs most frequently. If two numbers tie for most frequent occurrence, the collection has two modes and is called **bimodal.**

EXAMPLE 1   ■   Comparing Measures of Central Tendency

You are interviewing for a job. The interviewer tells you that the average income of the company's 25 employees is $60,849. The actual annual incomes of the 25 employees are shown below. What are the mean, median, and mode of the incomes? Was the person telling you the truth?

$17,305,	$478,320,	$45,678,	$18,980,	$17,408,
$25,676,	$28,906,	$12,500,	$24,540,	$33,450,
$12,500,	$33,855,	$37,450,	$20,432,	$28,956,
$34,983,	$36,540,	$250,921,	$36,853,	$16,430,
$32,654,	$98,213,	$48,980,	$94,024,	$35,671

Technology Note

Statistical calculators have built-in programs for calculating the mean of a collection of numbers. Try using your calculator to find the mean shown in Example 1. Then use your calculator to sort the incomes shown in Example 1.

Solution

The mean of the incomes is

$$\text{Mean} = \frac{17,305 + 478,320 + 45,678 + 18,980 + \cdots + 35,671}{25}$$

$$= \frac{1,521,225}{25}$$

$$= \$60,849.$$

To find the median, order the incomes as follows.

$12,500,	$12,500,	$16,430,	$17,305,	$17,408,
$18,980,	$20,432,	$24,540,	$25,676,	$28,906,
$28,956,	$32,654,	$33,450,	$33,855,	$34,983,
$35,671,	$36,540,	$36,853,	$37,450,	$45,678,
$48,980,	$94,024,	$98,213,	$250,921,	$478,320

NOTE Median is mentioned in exercises throughout the book. For example, see Exercise 60 on page 276 and Exercises 59–62 on page 393.

From this list, you can see that the median (the middle number) is $33,450. From the same list, you can see that $12,500 is the only income that occurs more than once. Thus, the mode is $12,500. Technically, the person was telling the truth because the average is (generally) defined to be the mean. However, of the three measures of central tendency

Mean: $60,849   *Median:* $33,450   *Mode:* $12,500

it seems clear that the median is the most representative. The mean is inflated by the two highest salaries. ■

Choosing a Measure of Central Tendency

Which of the three measures of central tendency is the most representative? The answer is that it depends on the distribution of the data *and* the way in which you plan to use the data.

For instance, in Example 1, the mean salary of $60,849 does not seem very representative to a potential employee. To a city income tax collector who wants to estimate 1% of the total income of the 25 employees, however, the mean is precisely the right measure.

EXAMPLE 2 ■ Choosing a Measure of Central Tendency

Which measure of central tendency is the most representative of the data given in each of the following frequency distributions?

a. *Number*	*Tally*		b. *Number*	*Tally*		c. *Number*	*Tally*
1	7		1	9		1	6
2	20		2	8		2	1
3	15		3	7		3	2
4	11		4	6		4	3
5	8		5	5		5	5
6	3		6	6		6	5
7	2		7	7		7	4
8	0		8	8		8	3
9	15		9	9		9	0

Solution

a. For these data, the mean is 4.23, the median is 3, and the mode is 2. Of these, the mode is probably the most representative.

b. For these data, the mean and median are each 5 and the modes are 1 and 9 (the distribution is bimodal). Of these, the mean or median is the most representative.

c. For these data, the mean is 4.59, the median is 5, and the mode is 1. Of these, the mean or median is the most representative.   ■

Variance and Standard Deviation

Very different sets of numbers can have the same mean. You will now study two **measures of dispersion,** which give you an idea of how much the numbers in the set differ from the mean of the set. These two measures are called the *variance* of the set and the *standard deviation* of the set.

Definitions of Variance and Standard Deviation	Consider a set of numbers $\{x_1, x_2, \ldots, x_n\}$ with a mean of $\overline{x}$. The **variance** of the set is

$$v = \frac{(x_1 - \overline{x})^2 + (x_2 - \overline{x})^2 + \cdots + (x_n - \overline{x})^2}{n}$$

and the **standard deviation** of the set is

$$\sigma = \sqrt{v}$$

(σ is the lowercase Greek letter *sigma*).

The standard deviation of a set is a measure of how much a typical number in the set differs from the mean. The greater the standard deviation, the more the numbers in the set *vary* from the mean. For instance, each of the following sets has a mean of 5.

NOTE Standard deviation is discussed on page 40. The material here allows a more in-depth discussion.

$$\{5, 5, 5, 5\}, \quad \{4, 4, 6, 6\}, \quad \text{and} \quad \{3, 3, 7, 7\}$$

The standard deviations of the sets are 0, 1, and 2.

$$\sigma_1 = \sqrt{\frac{(5-5)^2 + (5-5)^2 + (5-5)^2 + (5-5)^2}{4}} = 0$$

$$\sigma_2 = \sqrt{\frac{(4-5)^2 + (4-5)^2 + (6-5)^2 + (6-5)^2}{4}} = 1$$

$$\sigma_3 = \sqrt{\frac{(3-5)^2 + (3-5)^2 + (7-5)^2 + (7-5)^2}{4}} = 2$$

EXAMPLE 3 ■ Estimations of Standard Deviation

Consider the three sets of data represented by the following bar graphs. Which set has the smallest standard deviation? Which has the largest?

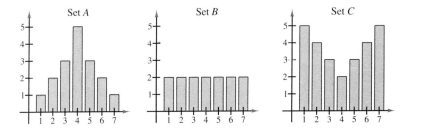

Set A   Set B   Set C

Solution

Of the three sets, the numbers in set A are grouped most closely to the center and the numbers in set C are the most dispersed. Thus, set A has the smallest standard deviation and set C has the largest standard deviation. ■

EXAMPLE 4 ■ Finding Standard Deviation

Find the standard deviation of each set shown in Example 3.

Solution

Because of the symmetry of each bar graph, you can conclude that each has a mean of $\bar{x} = 4$. The standard deviation of set A is

$$\sigma = \sqrt{\frac{(-3)^2 + 2(-2)^2 + 3(-1)^2 + 5(0)^2 + 3(1)^2 + 2(2)^2 + (3)^2}{17}}$$

$$\approx 1.53.$$

The standard deviation of set B is

$$\sigma = \sqrt{\frac{2(-3)^2 + 2(-2)^2 + 2(-1)^2 + 2(0)^2 + 2(1)^2 + 2(2)^2 + 2(3)^2}{14}}$$

$$= 2.$$

The standard deviation of set C is

$$\sigma = \sqrt{\frac{5(-3)^2 + 4(-2)^2 + 3(-1)^2 + 2(0)^2 + 3(1)^2 + 4(2)^2 + 5(3)^2}{26}}$$

$$\approx 2.22.$$

These values confirm the results of Example 3. That is, set A has the smallest standard deviation and set C has the largest. ■

Technology Note

If you have access to a computer or calculator with a standard deviation program, try using it to obtain the result given in Example 4. If you do this, the program will probably output two versions of the standard deviation. In one, the sum of the squared differences is divided by n, and in the other it is divided by $n - 1$. In this text, we always divide by n.

The following alternative formula provides a more efficient way to compute the standard deviation.

Alternative Formula for Standard Deviation	The standard deviation of $\{x_1, x_2, \ldots, x_n\}$ is $$\sigma = \sqrt{\frac{x_1{}^2 + x_2{}^2 + \cdots + x_n{}^2}{n} - \overline{x}^2}.$$

Because of messy computations, this formula is difficult to verify. Conceptually, however, the process is straightforward. It consists of showing that the expressions

$$\sqrt{\frac{(x_1 - \overline{x})^2 + (x_2 - \overline{x})^2 + \cdots (x_n - \overline{x})^2}{n}}$$

and

$$\sqrt{\frac{x_1{}^2 + x_2{}^2 + \cdots + x_n{}^2}{n} - \overline{x}^2}$$

are equivalent. Try verifying this equivalence for the set $\{x_1, x_2, x_3\}$ with $\overline{x} = (x_1 + x_2 + x_3)/3$.

EXAMPLE 5 ■   Using the Alternative Formula

Use the alternative formula for standard deviation to find the standard deviation of the following set of numbers.

5, 6, 6, 7, 7, 8, 8, 8, 9, 10

Solution

Begin by finding the mean of the set, which is 7.4. Thus, the standard deviation is

$$\sigma = \sqrt{\frac{5^2 + 2(6^2) + 2(7^2) + 3(8^2) + 9^2 + 10^2}{10} - (7.4)^2}$$

$$= \sqrt{\frac{568}{10} - 54.76}$$

$$= \sqrt{2.04}$$

$$\approx 1.43.$$

You can use the statistical features of a graphing utility to check this result.   ■

A well-known theorem in statistics, called *Chebychev's Theorem,* states that at least

$$1 - \frac{1}{k^2}$$

of the numbers in a distribution must lie within k standard deviations of the mean. Thus, 75% of the numbers in a collection must lie within two standard deviations of the mean, and at least 88.9% of the numbers must lie within three standard deviations of the mean. For most distributions, these percentages are low. For instance, in all three distributions shown in Example 3, 100% of the numbers lie within two standard deviations of the mean.

EXAMPLE 6   ■   Describing a Distribution

The following table shows the number of dentists (per 100,000 people) in each state and the District of Columbia. Find the mean and standard deviation of the numbers. What percent of the numbers lies within two standard deviations of the mean? (*Source:* American Dental Association)

AK	66	AL	40	AR	39	AZ	51	CA	62
CO	69	CT	80	DC	94	DE	44	FL	50
GA	46	HI	80	IA	55	ID	53	IL	61
IN	47	KS	51	KY	53	LA	45	MA	74
MD	68	ME	47	MI	62	MN	67	MO	53
MS	37	MT	62	NC	42	ND	47	NE	63
NH	59	NJ	77	NM	45	NV	49	NY	73
OH	55	OK	47	OR	70	PA	61	RI	56
SC	41	SD	49	TN	53	TX	47	UT	66
VA	54	VT	57	WA	68	WI	65	WV	43
WY	52								

FIGURE B.9

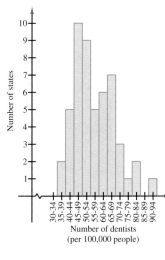

Number of dentists
(per 100,000 people)

Solution

Begin by entering the numbers into a computer or calculator that has a standard deviation program. After running the program, you should obtain

$$\bar{x} \approx 56.76 \quad \text{and} \quad \sigma \approx 12.14.$$

The interval that contains all numbers that lie within two standard deviations of the mean is

$$[56.76 - 2(12.14),\ 56.76 + 2(12.14)] \quad \text{or} \quad [32.48, 81.04].$$

From the histogram in Figure B.9, you can see that all but one of the numbers (98%) lie in this interval—all but the number that corresponds to the number of dentists (per 100,000 people) in Washington, D.C.   ■

| **B.2** | **EXERCISES** |

In Exercises 1–6, find the mean, median, and mode of the set of measurements.

1. 5, 12, 7, 14, 8, 9, 7

2. 30, 37, 32, 39, 33, 34, 32

3. 5, 12, 7, 24, 8, 9, 7

4. 20, 37, 32, 39, 33, 34, 32

5. 5, 12, 7, 14, 9, 7

6. 30, 37, 32, 39, 34, 32

7. Compare your answers for Exercises 1 and 3 with those for Exercises 2 and 4. Which of the measures of central tendency is sensitive to extreme measurements? Explain your reasoning.

8. (a) Add 6 to each measurement in Exercise 1 and calculate the mean, median, and mode of the revised measurements. How are the measures of central tendency changed?

(b) If a constant k is added to each measurement in a set of data, how will the measures of central tendency change?

9. *Electric Bills*   A person had the following monthly bills for electricity. What are the mean and median of this collection of bills?

January	$67.92	February	$59.84
March	$52.00	April	$52.50
May	$57.99	June	$65.35
July	$81.76	August	$74.98
September	$87.82	October	$83.18
November	$65.35	December	$57.00

10. *Car Rental*   A car rental company kept the following record of the number of miles driven by a car that was rented. What are the mean, median, and mode of these data?

Monday	410	Tuesday	260
Wednesday	320	Thursday	320
Friday	460	Saturday	150

11. *Six-child Families*   A study was done on families having six children. The table gives the numbers of families in the study with the indicated numbers of girls. Determine the mean, median, and mode of this set of data.

Number of Girls	0	1	2	3	4	5	6
Frequency	1	24	45	54	50	19	7

12. *Baseball*   A baseball fan examined the records of a favorite baseball player's performance during his last 50 games. The numbers of games in which the player had 0, 1, 2, 3, and 4 hits are recorded in the table.

Number of Hits	0	1	2	3	4
Frequency	14	26	7	2	1

(a) Determine the average number of hits per game.

(b) Determine the player's batting average if he had 200 at bats during the 50-game series.

13. Construct a collection of numbers that has the following properties. If this is not possible, explain why it is not.

Mean = 6, Median = 4, Mode = 4

14. Construct a collection of numbers that has the following properties. If this is not possible, explain why it is not.

Mean = 6, Median = 6, Mode = 4

15. *Test Scores*   A professor records the following scores for a 100-point exam.

99, 64, 80, 77, 59, 72, 87, 79, 92, 88,
90, 42, 20, 89, 42, 100, 98, 84, 78, 91

Which measure of central tendency best describes these test scores?

16. *Shoe Sales*   A salesman sold eight pairs of a certain style of men's shoes. The sizes of the eight pairs were as follows: $10\frac{1}{2}$, 8, 12, $10\frac{1}{2}$, 10, $9\frac{1}{2}$, 11, and $10\frac{1}{2}$. Which measure (or measures) of central tendency best describes the typical shoe size for this data?

In Exercises 17–24, find the mean, variance, and standard deviation of the numbers.

17. 4, 10, 8, 2

18. 3, 15, 6, 9, 2

19. 0, 1, 1, 2, 2, 2, 3, 3, 4

20. 2, 2, 2, 2, 2, 2

21. 1, 2, 3, 4, 5, 6, 7

22. 1, 1, 1, 5, 5, 5

23. 49, 62, 40, 29, 32, 70

24. 1.5, 0.4, 2.1, 0.7, 0.8

In Exercises 25–30, use the alternative formula to find the standard deviation of the numbers.

25. 2, 4, 6, 6, 13, 5

26. 10, 25, 50, 26, 15, 33, 29, 4

27. 246, 336, 473, 167, 219, 359

28. 6.0, 9.1, 4.4, 8.7, 10.4

29. 8.1, 6.9, 3.7, 4.2, 6.1

30. 9.0, 7.5, 3.3, 7.4, 6.0

31. Without calculating the standard deviation, explain why the set {4, 4, 20, 20} has a standard deviation of 8.

32. If the standard deviation of a set of numbers is 0, what does this imply about the set?

33. *Test Scores* An instructor adds five points to each student's exam score. Will this change the mean or standard deviation of the exam scores? Explain.

34. Consider the four sets of data represented by the histograms. Order the sets from the smallest to the largest variance.

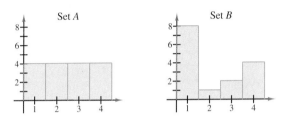

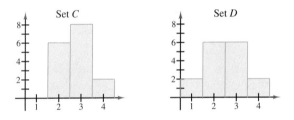

35. *Test Scores* The scores of a mathematics exam given to 600 science and engineering students at a college had a mean and standard deviation equal to 235 and 28, respectively. Use Chebychev's Theorem to determine the intervals containing at least $\frac{3}{4}$ and at least $\frac{8}{9}$ of the scores. How would the intervals change if the standard deviation were 16?

36. *Precipitation* The following data represents the annual precipitation (in inches) at Erie, Pennsylvania for the years 1960 through 1989. Use a computer or calculator to find the mean, variance, and standard deviation of the data. What percent of the data lies within two standard deviations of the mean? (*Source:* National Oceanic and Atmospheric Administration)

27.41,	36.50,	36.90,	28.11,	36.47,
38.41,	37.74,	37.78,	34.33,	36.58,
41.50,	34.06,	43.55,	38.04,	41.83,
43.03,	43.85,	61.70,	35.04,	55.31,
47.04,	41.97,	41.56,	46.25,	37.79,
45.87,	47.30,	44.86,	38.87,	41.88

Technology

C.1   CBL Experiments

C.2   Programs

The material in Section C.1 explains how to use the *Texas Instruments Calculator-Based Laboratory (CBL) System* in conjunction with the Chapter Projects in Chapters 2, 3, 5, and 8. A description of the necessary equipment, experiment set-up, and experiment execution steps (including the necessary *TI-82* and *TI-85* programs) is included for each Chapter Project.

Programs for the *TI-82* are given in several sections of the text. Section C.2 contains additional programming hints for the *TI-82* and translations of the text programs for the following graphics calculators from Texas Instruments, Casio, and Sharp.

- *TI-80*
- *TI-81*
- *TI-82*
- *TI-85*
- *Sharp EL-9200*
- *Sharp EL-9300*
- *Casio 6300*
- *Casio 7700*
- *Casio fx-9700GE*
- *Casio CFX-9800G.*

Similar programs can be written for other brands and models of graphics calculators.

Enter a program in your calculator, then refer to the text discussion and apply the program as appropriate. Section references are provided to help you locate the text discussion of the programs and their uses.

To illustrate the power and versatility of programmable calculators, a variety of programs is presented, including a simulation program (Graph Reflection Program) and a tutorial program (Reflections and Shifts Program).

CBL Experiments

Chapter 2

Equipment Needed

- *CBL* unit (power adapter AC–9201 recommended)
- *TI–82* or *TI–85* graphing calculator with unit-to-unit link cable
- *Vernier CBL Ultrasonic Motion Detector (MD–CBL)*
- Small kitchen garbage bag (parachute)
- Small binder clip (weight)

Experiment Set-Up

Set up the experiment as described below.

1. Connect the *CBL* unit to the *TI–82* or *TI–85* graphing calculator using the unit-to-unit link cable and the I/O ports located on the bottom edge of each unit.
2. Connect the *Ultrasonic Motion Detector* to the SONIC port located on the left side of the *CBL* unit. Then set the motion detector on the floor so that it is facing upward.
3. Turn on the *CBL* unit and the graphing calculator.
4. Construct a parachute from the garbage bag and use the binder clip to gather and hold together the edges.

Note: Two people are needed to perform the experiment: one to operate the *CBL* unit and the graphing calculator and one to drop the parachute.

TI–82 Program

```
PROGRAM:PARACHUT
:Lbl 2
:PlotsOff
:FnOff
:AxesOn
:ClrDraw
:ClrHome
:ClrList L1,L2,L3,L4,L5
:{1,0}→L1
:Send(L1)
:{1,11,3}→L1
:Send(L1)
:{3,.02,99,1,0,0,0,0,1}→L1
:Send(L1)
:Get(L3):Get(L2)
:Plot1(Scatter,L2,L3,●)
```

```
:1→Xscl:1→Yscl
:ZoomStat
:Text(25,1,"HEIGHT")
:Text(57,53,"TIME")
:Pause
:Menu("GRAPH OK?",
    "YES",1,"NO",2)
:Lbl 1
:ClrHome
:ClrDraw
:Text(45,2,"LOWER BOUND?")
:Input:X→A
:Vertical A
:Text(45,2,"UPPER BOUND?")
:Input:X→B
:Vertical B
```

```
:dim L2→N
:1→C
:Text(28,36,"ANALYZING...")
:For(I,1,N)
:If L2(I)≥A and L2(I)≤B
:Then
:L2(I)→L4(C)
:L3(I)→L5(C)
:C+1→C
:End:End
:L4-L4(1)→L4
:Plot1(Scatter,L4,L5,●)
:ZoomStat
:Stop
```

Source: *CBL Explorations in Precalculus for the TI–82*™ and *CBL Explorations in Precalculus for the TI–85*™. Copyright 1995 by Meridian Creative Group, A Division of Larson Texts, Inc., 5178 Station Road, Erie, Pennsylvania 16510.

TI–85 Program

```
PROGRAM:PARACHUT            :Lbl M
:Lbl N                      :ClLCD
:ClDrw                      :Input
:FnOff                      :x→A
:AxesOn                     :Vert A
:ClLCD                      :Input
:1→dimL L1:1→dimL L2        :x→B
:1→dimL L3:1→dimL L4        :Vert B
:1→dimL L5                  :1→C
:{1,0}→L1                   :ClDrw
:Outpt("CBLSEND",L1)        :Disp "ANALYZING..."
:{1,11,3}→L1                :For(I,1,dimL L2)
:Outpt("CBLSEND",L1)        :If L2(I)≥A and L2(I)≤B
:{3,.02,80,1,0,0,0,0,1}→L1  :Then
:Outpt("CBLSEND",L1)        :L2(I)→L4(C)
:Input "CBLGET",L3          :L3(I)→L5(C)
:Input "CBLGET",L2          :C+1→C
:min(L2)→xMin               :End:End
:max(L2)→xMax               :L4-L4(1)→L4
:min(L3)→yMin               :min(L4)→xMin
:max(L3)→yMax               :max(L4)→xMax
:1→xScl:1→yScl              :min(L5)→yMin
:Scatter L2,L3              :max(L5)→yMax
:Pause                      :Scatter L4,L5
:Menu(1,"CONT",M,2,"REDO",N) :Stop
```

Performing the Experiment

1. With the *CBL* unit and the graphing calculator on, start program PARACHUT on the *TI–82* or *TI–85*. The motion detector will begin clicking, signaling that it is ready to collect data.

2. With the second person holding the parachute so that the weight is at least 4.5 feet above the motion detector, press TRIGGER on the *CBL*.

3. Because the *CBL* unit is programmed to gather data for 1.98 seconds, release the parachute immediately after TRIGGER is pressed.

4. If the parachute falls straight down, press ENTER 1 on the *TI–82* or ENTER F1 on the *TI–85*. If it does not fall straight down, press ENTER 2 on the *TI–82* or ENTER F2 on the *TI–85* to repeat the experiment. (*Note:* If the motion detector has trouble detecting the binder clip, attach a 2-inch × 2-inch section of paper towel to it.)

Chapter 3

Equipment Needed

- *CBL* unit (power adapter AC–9201 recommended)
- *TI–82* or *TI–85* graphing calculator with unit-to-unit link cable
- *Vernier CBL Ultrasonic Motion Detector (MD–CBL)*
- A large, heavy object that can be dropped, such as a basketball

Experiment Set-Up

Set up the experiment as described below.

1. Connect the *CBL* unit to the *TI–82* or *TI–85* graphing calculator using the unit-to-unit link cable and the I/O ports located on the bottom edge of each unit.
2. Connect the *Ultrasonic Motion Detector* to the SONIC port located on the left side of the *CBL* unit.
3. Set the motion detector on the floor so that it is facing upward.
4. Turn on the *CBL* unit and the graphing calculator.

Note: Three people are needed to perform the experiment: one to operate the *CBL* unit and the graphing calculator, one to drop the object, and one to catch the object. Be careful when dropping and catching the object. Do not let it fall on the motion detector.

TI–82 Program

```
PROGRAM:FALLOBJ
:Lbl 2
:PlotsOff
:FnOff
:AxesOn
:ClrDraw
:ClrHome
:ClrList L1,L2,L3,L4
:{1,0}→L1
:Send(L1)
:{1,11,3}→L1
:Send(L1)
:{3,.02,80,1,0,0,0,0,1}→L1
:Send(L1)
:Get(L2)
:Get(L1)
:Plot1(Scatter,L1,L2,●)
:1→Xscl :1→Yscl
:ZoomStat
:Text(25,1,"HEIGHT")
:Text(57,53,"TIME")
:Pause
:Menu("GRAPH OK?",
    "YES",1,"NO",2)
```

```
:Lbl 1
:ClrHome :ClrDraw
:Text(45,2,"LOWER BOUND?")
:Input
:X→A
:Vertical A
:Text(45,2,"UPPER BOUND?")
:Input
:X→B
:Vertical B
:dim L1→N
:1→C
:Text(2,2,"ANALYZING... ")
:For(I,1,N)
:If L1(I)≥A and L1(I)≤B
:Then
:L1(I)→L3(C)
:L2(I)→L4(C)
:C+1→C
:End:End
:L3-L3(1)→L3
:Plot1(Scatter,L3,L4,●)
:ZoomStat
:Stop
```

TI–85 Program

```
PROGRAM:FALLOBJ
:Lbl N
:ClDrw
:FnOff
:AxesOn
:ClLCD
:1→dimL L1:1→dimL L2
:1→dimL L3:1→dimL L4
:{1,0}→L1
:Outpt("CBLSEND",L1)
:{1,11,3}→L1
:Outpt("CBLSEND",L1)
:{3,.02,80,1,0,0,0,0,1}→L1
:Outpt("CBLSEND",L1)
:Input "CBLGET",L2
:Input "CBLGET",L1
:min(L1)→xMin
:max(L1)→xMax
:min(L2)→yMin
:max(L2)→yMax
:1→xScl:1→yScl
:Scatter L1,L2
:Pause
:Menu(1,"CONT",M,2,"REDO",N)
:Lbl M
:ClLCD
:Input
:x→A
:Vert A
:Input
:x→B
:Vert B
:1→C
:ClDrw
:Disp "ANALYZING..."
:For(I,1,dimL L1)
:If L1(I)≥A and L1(I)≤B
:Then
:L1(I)→L3(C)
:L2(I)→L4(C)
:C+1→C
:End:End
:L3-L3(1)→L3
:min(L3)→xMin
:max(L3)→xMax
:min(L4)→yMin
:max(L4)→yMax
:Scatter L3,L4
:Stop
```

Performing the Experiment

1. With the *CBL* unit and the graphing calculator on, start program FALLOBJ on the *TI–82* or *TI–85*. Once the program is started, the motion detector will begin clicking, meaning it is ready to collect data.

2. With a second person holding the object five feet above the motion detector and a third person in place to catch the object, press $\boxed{\text{TRIGGER}}$ on the *CBL* unit.

3. Because the *CBL* unit is programmed to gather data for just 1.6 seconds, the person holding the object should drop it (with no initial velocity) as soon as $\boxed{\text{TRIGGER}}$ is pressed. The person catching the object should do so when it is about 1.75 feet above the motion detector.

4. If the display shows reasonable data, press $\boxed{\text{ENTER}}$ $\boxed{1}$ on the *TI–82* or $\boxed{\text{ENTER}}$ $\boxed{\text{F1}}$ on the *TI-85*. If it does not give reasonable data, press $\boxed{\text{ENTER}}$ $\boxed{2}$ on the *TI–82* or $\boxed{\text{ENTER}}$ $\boxed{\text{F2}}$ on the *TI–85* to repeat the experiment.

Chapter 5

Equipment Needed

- *CBL* unit (power adapter AC–9201 recommended)
- *TI–82* or *TI–85* graphing calculator with unit-to-unit link cable
- *TI Temperature Probe* (provided with the *CBL*)
- A cup or beaker
- Hot water

Experiment Set-Up

Set up the experiment as described below.

1. Connect the *CBL* unit to the *TI–82* or *TI–85* graphing calculator using the unit-to-unit link cable and I/O ports located on the bottom edge of each unit.
2. Connect the *TI Temperature Probe* to the Channel 1 port (CH1) located on the right of the top edge of the *CBL* unit.
3. Turn on the *CBL* unit and the graphing calculator.

TI–82 Program

```
PROGRAM:NEWTCOOL                    :Pause
:PlotsOff                           :ClrHome
:Func:FnOff                         :N+1→N
:ClrDraw:ClrHome                    :Goto 1
:AxesOn                             :End
:1→N                                :Lbl 2
:Lbl 1                              :ClrList L1,L2,L3
:If N=3:Then                        :{1,0}→L1
:Goto 2:Else                        :Send(L1)
:ClrList L1,L2,L3                   :{1,1,11}→L1
:{1,0}→L1                           :Send(L1)
:Send(L1)                           :Disp "PRESS ENTER TO"
:{1,1,11}→L1                        :Disp "COLLECT DATA"
:Send(L1)                           :Pause
:Disp "PRESS ENTER TO"              :{3,.5,99,0,0,0,0,0,1}→L1
:Disp "DETERMINE"                   :Send(L1)
:Disp "TEMPERATURE"                 :Get(L3)
:Pause                              :Get(L2)
:{3,1,5,0,0,0,0,0,1}→L1             :L3-L4(1)→L3
:Send(L1)                           :Plot1(Scatter,L2,L3, • )
:Get(L3)                            :5→Xscl:10→Yscl
:(sum L3-max(L3)-min(L3))           :ZoomStat
      /3→L4(N)                      :Text(0,2,"TEMP")
:Disp "TEMPERATURE"                 :Text(54,60,"TIME")
:Disp L4(N)                         :Stop
```

TI–85 Program

```
PROGRAM:NEWTCOOL
:ClDrw
:Func:FnOff
:AxesOn
:ClLCD
:1→dimL L1:1→dimL L2
:1→dimL L3:1→dimL L4
:1→N
:Lbl A
:If N==3:Then
:Goto B:Else
:ClLCD
:{1,0}→L1
:Outpt("CBLSEND",L1)
:{1,1,11}→L1
:Outpt("CBLSEND",L1)
:Disp "PRESS ENTER TO"
:Disp "DETERMINE"
:Disp "TEMPERATURE"
:Pause
:{3,1,5,0,0,0,0,0,1}→L1
:Outpt("CBLSEND",L1)
:Input "CBLGET",L3
:(sum L3-max(L3)-min(L3))
      /3→L4(N)
:Disp "TEMPERATURE"
```

```
:Disp L4(N)
:Pause
:ClLCD
:N+1→N
:Goto A
:End
:Lbl B
:{1,0}→L1
:Outpt("CBLSEND",L1)
:{1,1,11}→L1
:Outpt("CBLSEND",L1)
:Disp "PRESS ENTER TO"
:Disp "COLLECT DATA"
:Pause
:{3,.5,99,0,0,0,0,0,1}→L1
:Outpt("CBLSEND",L1)
:Input "CBLGET",L3
:Input "CBLGET",L2
:L3-L4(1)→L3
:min(L2)→xMin
:max(L2)→xMax
:min(L3)→yMin
:max(L3)→yMax
:5→xScl:10→yScl
:Scatter L2,L3
:Stop
```

Performing the Experiment

1. Fill the cup or beaker with hot water. Then, with the *CBL* unit and the graphing calculator on, start program NEWTCOOL on the *TI–82* or *TI–85*.

2. Hold the probe so that the thermometer is exposed to the air. Press ENTER on the *TI–82* or *TI–85*. It will display the room temperature.

3. Place the thermometer in the hot water for at least one minute. Then press ENTER twice. The displayed temperature is the water temperature.

4. Press ENTER and remove the thermometer from the water exposing it to the air. As you remove the thermometer from the water, press ENTER again. The *CBL* unit will now record the temperatures of the cooling thermometer for approximately fifty seconds. After the data has been collected, a scatter plot should be displayed on the viewing screen.

Note: While the *CBL* unit is collecting the temperatures of the cooling thermometer, make sure the thermometer is exposed to the air. Do not place it on a table top or expose it to any drafts, as this may allow conduction and evaporation effects to occur.

Chapter 8

Equipment Needed

- *CBL* unit (power adapter AC–9201 recommended)
- *TI–82* or *TI–85* graphing calculator with unit-to-unit link cable
- *Vernier CBL Ultrasonic Motion Detector (MD-CBL)*
- Rubber ball (racquetballs work well)
- 3-foot ring stand and clamp
- Hard, level surface

Experiment Set-up

Set up the experiment as described below.

1. Connect the *CBL* unit to the *TI–82* or *TI–85* graphing calculator using the unit-to-unit link cable and the I/O ports located on the bottom edge of each unit.
2. Connect the *Ultrasonic Motion Detector* to the SONIC port located on the left side of the *CBL* unit.
3. Attach the motion detector to the top of the ring stand using the clamp. Make sure the motion detector is level and facing a hard, level surface.
4. Turn on the *CBL* unit and the graphing calculator.

TI–82 Program

```
PROGRAM:BOUNCE                :Send(L1)
:PlotsOff                     :Get(L4)
:FnOff                        :Get(L2)
:AxesOn                       :max(L4)-L4→L4
:ClrDraw                      :0→Xmin
:ClrHome                      :max(L2)→Xmax
:ClrList L1,L2,L3,L4          :.5→Xscl
:{1,0}→L1                     :max(L4)*1.4→Ymax
:Send(L1)                     :min(0,min(L4)*1.4)→Ymin
:{1,11,2}→L1                  :.5→Yscl
:Send(L1)                     :Plot1(xyLine,L2,L4, ●)
:Disp "PRESS ENTER TO"        :DispGraph
:Disp "COLLECT DATA"          :Text(4,1,"HEIGHT(M)")
:Pause                        :Text(47,81,"T(S)")
:{3,.04,99,0,0,0,0,0,1}→L1    :Stop
```

TI–85 Program

PROGRAM:BOUNCE
:ClDrw
:FnOff
:AxesOn
:ClLCD
:1→dimL L1:1→dimL L2
:1→dimL L3:1→dimL L4
:{1,0}→L1
:Outpt("CBLSEND",L1)
:{1,11,2}→L1
:Outpt("CBLSEND",L1)
:Disp "PRESS ENTER TO"
:Disp "COLLECT DATA"

:Pause
:{3,.04,99,0,0,0,0,0,1}→L1
:Outpt("CBLSEND",L1)
:Input "CBLGET",L4
:Input "CBLGET",L2
:max(L4)-L4→L4
:0→xMin
:max(L2)→xMax
:min(0,min(L4)*1.4)→yMin
:max(L4)*1.4→yMax
:.5→xScl:.5→yScl
:xyline L2,L4
:Stop

Performing the Experiment

1. With the *CBL* unit and the graphing calculator on, run program BOUNCE on the *TI–82* or *TI–85* and stop when "PRESS ENTER TO COLLECT DATA" is displayed on the screen.

2. Position the ball at least 0.5 meter directly below the motion detector and press ENTER on the *TI–82* or *TI–85*.

3. As soon as the motion detector starts clicking, drop the ball so that it bounces repeatedly directly below the motion detector. Make sure that when you drop the ball, it has no initial velocity.

Note: After the *CBL* collects the data, a graph should be displayed on the viewing screen of the *TI–82* or *TI–85*. Repeat the experiment if necessary. Notice that the heights are in meters and the time is in seconds.

SECTION C.2	Programs

Simple Interest Program (Section 1.2)

The program, shown in the marginal Programming note on page 94, can be used to find the amount of simple interest earned by a given principal at a given annual interest rate for a certain amount of time.

Note: The program requires a line of code that restricts the displayed result to two decimal places. For instance, the *TI-82* program requires "Fix 2" as the first line. To do this, press the MODE key, cursoring to "2" on the line beginning with "Float," and press ENTER. Similarly, the last line of the program may be entered by pressing the MODE key, cursoring to "Float," and pressing ENTER. On the *TI-82,* other commands such as "Disp" and "Input" may be entered through the "I/O" menu accessed by pressing the PRGM key. The $\rightarrow$ symbol represents pressing the STO▷ key. Keystroke sequences required for similar commands on other calculators will vary. Consult the user's manual for your calculator.

TI-80
TI-81
TI-82

```
SIMPINT
:Fix 2
:Disp "PRINCIPAL"
:Input P
:Disp "INTEREST RATE"
:Disp "IN DECIMAL FORM"
:Input R
:Disp "NO. OF YEARS"
:Input T
:PRT→I
:Disp "THE INTEREST IS"
:Disp I
:Float
```

TI-85

```
SimpInt
:Disp "Principal"
:Input P
:Disp "Interest rate"
:Disp "in decimal form"
:Input R
:Disp "No. of years"
:Input T
:P*R*T→I
:Disp "The interest is"
:Disp I
:Float
```

Sharp EL 9200
Sharp EL 9300

```
SimpInt
Input principal
Print "Interest rate
Print "in decimal form
Input rate
Print "No. of years
Input time
interest=principal*rate*time
Print interest
```

Casio 6300

```
SIMPINT:Fix 2:"PRINCIPAL"?→P:
"INTEREST"◢"RATE AS"◢"DECIMAL"?→R:
"NO. YEARS"?→T:PRT→I:
"INTEREST IS"◢I:Norm
```

Casio 7700
Casio fx-9700GE
Casio CFX-9800G

```
SIMPINT↵
Fix 2↵
"PRINCIPAL"?→P↵
"INTEREST RATE"↵
"IN DECIMAL FORM"?→R↵
"NO. OF YEARS"?→T↵
PRT→I↵
"THE INTEREST IS":I↵
Norm
```

Quadratic Formula Program (Section 1.4)

The program, shown in the marginal Programming note on page 118, will display the solutions to quadratic equiations or the words "No Real Solution." To use the program, write the quadratic equation in standard form and enter the values of *a, b,* and *c.*

Note: On the *TI-82,* the "Prompt" command may be entered through the "I/O" menu accessed by pressing the PRGM key. The "=" and "≥" symbols may be entered through the "TEST" menu accessed by pressing the TEST key. Other commands such as "If," "Then," "Else," and "End" may be entered through the "CTL" menu accessed by pressing the PRGM key. For additional keystroke instructions, see previous programs in this appendix. Keystroke sequences required for similar commands on other calculators will vary. Consult the user's manual for your calculator.

TI-80

```
PROGRAM:QUADRAT
:Disp "AX²+BX+C=0"
:Input "ENTER A ", A
:Input "ENTER B ", B
:Input "ENTER C ", C
:B²-4AC→D
:If D≥0
:Then
:(-B+√D)/(2A)→M
:Disp M
:(-B-√D)/(2A)→N
:Disp N
:Else
:Disp "NO REAL SOLUTION"
:End
```

TI-81

```
Prgm4: QUADRAT
:Disp "ENTER A"
:Input A
:Disp "ENTER B"
:Input B
:Disp "ENTER C"
:Input C
:B²-4AC → D
:If D<0
:Goto 1
:((-B+√D)/(2A)) → S
:Disp S
:((-B-√D)/(2A)) → S
:Disp S
:End
:Lbl 1
:Disp "NO REAL"
:Disp "SOLUTION"
:End
```

TI-82

```
PROGRAM:QUADRAT
:Disp "AX²+BX+C=0"
:Prompt A
:Prompt B
:Prompt C
:B²-4AC → D
:If D≥0
:Then
:(-B+√D)/(2A) → M
:Disp M
:(-B-√D)/(2A) → N
:Disp N
:Else
:Disp "NO REAL SOLUTION"
:End
```

TI-85

```
PROGRAM:
:Input "ENTER A ",A
:Input "ENTER B ",B
:Input "ENTER C ",C
:B²-4*A*C→D
:Disp (-B+√D)/(2A)
:Disp (-B-√D)/(2A)
```

Solutions to quadratic equations are also available directly by using the TI-85 POLY function.

Sharp EL 9200
Sharp EL 9300

```
Quadratic
----------COMPLEX
Input A
Input B
Input C
D=B²-4AC
x1=(-B+√D)/(2A)
x2=(-B-√D)/(2A)
Print x1
Print x2
X=x1
Y=x2
End
```

This program is written in the program's complex mode, so both real and complex answers are given. The answers are also stored under variables X and Y so they can be used in the calculator mode.

Casio 6300

QUADRATICS:
"AX²+BX+C=0"◢
"A="?→A:
"B="?→B:
"C="?→C:
B²-4AC→D:
D<0⇒Goto 1:
"X="◢(-B+√D)÷2A◢
"OR X="◢(-B-√D)÷2A:
Goto 2:
Lbl 1:
"NO REAL SOLUTION"
Lbl 2

Casio 7700

QUADRATICS
"AX²+BX+C=0"
"A="?→A
"B="?→B
"C="?→C
B²-4AC→D
D<0⇒Goto 1
"X=":(-B+√D)÷2A ◢
"OR X=":(-B-√D)÷2A
Goto 2
Lbl 1
"NO REAL SOLUTION"
Lbl 2

Casio fx-9700GE
Casio CFX-9800G

QUADRATIC↵
"AX²+BX+C=0"↵
"A="?→A↵
"B="?→B↵
"C="?→C↵
B²-4AC→D↵
(-B+√D)÷2A◢
(-B-√D)÷2A

Both real and complex answers are given. Solutions to quadratic equations are also available directly from the Casio calculator's EQUATION MENU.

Two-Point Form of a Line (Section 2.4)

The program, referenced in the marginal Programming note on page 199, will display the slope and y-intercept of the line that passes through two points, (x_1, y_1) and (x_2, y_2), entered by the user.

Note: For help with keystrokes, see previous programs in this appendix.

TI-80
TI-81
TI-82

```
TWOPTFM
:Disp "ENTER X1,Y1"
:Input X
:Input Y
:Disp "ENTER X2,Y2"
:Input C
:Input D
:(D-Y)/(C-X)→M
:M*(-X)+Y→B
:Disp "SLOPE ="
:Disp M
:Disp "Y-INT ="
:Disp B
```

TI-85

```
TwoPtFrm
:Disp "Enter X1,Y1"
:Input X
:Input Y
:Disp "Enter X2,Y2"
:Input C
:Input D
:(D-Y)/(C-X)→M
:M*(-X+Y→B
:Disp "Slope ="
:Disp M
:Disp "Y-int ="
:Disp B
```

Sharp EL 9200
Sharp EL 9300

```
TwoPtForm
Print "Enter X1,Y1
Input x
c=x
Input y
d=y
Print "Enter X2,Y2
Input x
Input y
m=(d-y)÷(c-x)
b=m*(-x)+y
Print "Slope
Print m
Print "Y-int
Print b
```

Casio 6300

```
TWOPTFORM:"ENTER X1,Y1"?→X:?→Y:
"ENTER X2,Y2"?→C:?→D:
(D-Y)÷(C-X)→M:M×(-X)+Y→B:
"SLOPE ="◢M◢"Y-INT ="◢B
```

Casio 7700
Casio fx-9700GE
Casio CFX-9800G

```
TWOPTFORM↵
"ENTER X1,Y1"?→X:?→Y↵
"ENTER X2,Y2"?→C:?→D↵
(D-Y)÷(C-X)→M↵
M×(-X)+Y→B↵
"SLOPE =":M◢
"Y-INT =":B
```

Reflections and Shifts Program (Section 3.3)

The program, referenced in the marginal Programming note on page 256, will sketch a graph of the function $y = R(x + H)^2 + V$, where $R = \pm 1$, H is an integer between -6 and 6, and V is an integer between -3 and 3. This program gives you practice working with reflections, horizontal shifts, and vertical shifts.

Note: On a *TI-82*, the "int" and "rand" commands may be entered through the "NUM" and "PRB" menus, respectively, accessed by pressing the MATH key. The symbol "Y_1" may be entered through the "Function" menu accessed by pressing the Y-VARS key. The commands "Xmin," "Xmax," "Xscl," "Ymin," "Ymax," and "Yscl" may be entered through the "Window" menu accessed by pressing the VARS key. The commands "DispGraph" and "Pause" may be entered through the "I/O" and "CTL" menus, respectively, accessed by pressing the PRGM key. For additional keystroke instructions, see previous programs in this appendix. Keystroke sequences required for similar commands on other calculators will vary. Consult the user's manual for your calculator.

TI-80

```
PROGRAM:PARABOLA
:-6+int (12rand)→H
:-3+int (6rand)→V
:rand→R
:If R<.5
:Then
:-1→R
:Else
:1→R
:End
:"R(X+H)²+V"→Y₁
:-9→Xmin
:9→Xmax
:1→Xscl
:-6→Ymin
:6→Ymax
:1→Yscl
:DispGraph
:Pause
:Disp "R=",R
:Disp "H=",H
:Disp "V=",V
```

Press ENTER after the graph to display the coordinates.

TI-81

```
Prgm2: PARABOLA
:Rand→H
:-6+Int (12H)→H
:Rand→V
:-3+Int (6V)→V
:Rand→R
:If R<.5
:-1→R
:If R>.49
:1→R
:"R(X+H)²+V"→Y₁
:-9→Xmin
:9→Xmax
:1→Xscl
:-6→Ymin
:6→Ymax
:1→Yscl
:DispGraph
:Pause
:Disp "Y=R(X+H)²+V"
:Disp "R="
:Disp R
:Disp "H="
:Disp H
:Disp "V="
:Disp V
:End
```

Press ENTER after the graph to display the coordinates.

TI-82

```
PROGRAM:PARABOLA
:-6+int (12rand)→H
:-3+int (6rand)→V
:rand→R
:If R<.5
:Then
:-1→R
:Else
:1→R
:End
:"R(X+H)²+V"→Y₁
:-9→Xmin
:9→Xmax
:1→Xscl
:-6→Ymin
:6→Ymax
:1→Yscl
:DispGraph
:Pause
:Disp "R=",R
:Disp "H=",H
:Disp "V=",V
```

Press ENTER after the graph to display the coordinates.

TI-85

```
 PROGRAM:
:rand→H
:-6+int(12H)→H
:rand→V
:-3+int(6V)→V
:rand→R
:If R<.5
:-1→R
:If R>.49
:1→R
:y1=R(x+H)²+V
:-9→xMin
:9→xMax
:1→xScl
:-6→yMin
:6→yMax
:1→yScl
:DispG
:Pause
:Disp "Y=R(X+H)²+V"
:Disp "R=",R
:Disp "H=",H
:Disp "V=",V
```

Sharp EL 9200
Sharp EL 9300

```
Parabola
------------REAL
H=int (random*12)-6
V=int (random*6)-3
S=(random*2)-1
R=S/abs S
Range -9,9,1,-6,6,1
Graph R(X+H)²+V
Wait
Print "Y=R(X+H)²+V
Print R
Print H
Print V
End
```

Pressing Enter after the graph will display the coefficients.

Casio 6300

```
R(X+H)²+V:
-6+Int (12Ran#)→H:
-3+Int (6Ran#)→V:
Ran#<.5⇒-1→R:1→R:
Range -9,9,1,-6,6,1:
Graph Y=R(X+H)²+V◢
"Y=R(X+H)²+V"◢
"R="◢R◢
"H="◢H◢
"V="◢V
```

Casio 7700

```
R(X+H)²+V
-6+Int (12Ran#)→H
-3+Int (6Ran#)→V
Ran#<.5⇒-1→R:1→R
Range -9,9,1,-6,6,1
Graph Y=R(X+H)²+V◢
"Y=R(X+H)²+V"
"R=":R◢
"H=":H◢
"V=":V
```

Casio fx-9700GE
Casio CFX-9800G

```
R(X+H)²+V↵
-6+Int (12Ran#)→H↵
-3+Int (6Ran#)→V↵
Ran#→R↵
R<.5⇒-1→R↵
R≥.5⇒1→R↵
Range -9,9,1,-6,6,1↵
Graph Y=R(X+H)²+V◢
"Y=R(X+H)²+V"↵
"R=":R◢
"H=":H◢
"V=":V
```

Press ENTER after the graph to display the coefficients.

Graph Reflection Program (Section 3.4)

The program, given in the marginal Programming note on page 271, will graph a function f and its reflection in the line $y = x$.

Note: On a *TI-82*, the "While" command may be entered through the "CTL" menu accessed by pressing the PRGM key. The "Pt-On(" command may be entered through the "POINTS" menu accessed by pressing the DRAW key. For additional keystroke instructions, see previous programs in this appendix. Keystroke sequences required for similar commands on other calculations will vary. Consult the user's manual for your calculator.

TI-80

```
PROGRAM:REFLECT
:47Xmin/63→Ymin
:47Xmax/63→Ymax
:Xscl→Yscl
:"X"→Y₂
:DispGraph
:(Xmax−Xmin)/62→I
:Xmin→X
:Lbl A
:Pt-On(Y₁,X)
:X + I→X
:If X > Xmax
:Stop
:Goto A
```

To use this program, enter the function in Y_1 and set a viewing rectangle.

TI-81

```
Prgm3: REFLECT
:2Xmin/3 → Ymin
:2Xmax/3 → Ymax
:Xscl → Yscl
:"X" → Y₂
:DispGraph
:(Xmax-Xmin)/95 → I
:Xmin → X
:Lbl 1
:PT-On(Y₁,X)
:X+I → X
:If X>Xmax
:End
:Goto 1
```

To use this program, enter the function in Y_1 and set a viewing rectangle.

TI-82

```
PROGRAM:REFLECT
:63Xmin/95 → Ymin
:63Xmax/95 → Ymax
:Xscl → Yscl
:"X" → Y₂
:DispGraph
:(Xmax-Xmin)/94 → N
:Xmin → X
:While X≤Xmax
:Pt-On(Y₁,X)
:X+N → X
:End
```

To use this program, enter the function in Y_1 and set a viewing rectangle.

TI-85

```
 PROGRAM:
:63xMin/127→yMin
:63xMax/127→yMax
:xScl→yScl
:y2=x
:DispG
:(xMax-xMin)/126→I
:xMin→x
:Lbl A
:PtOn(y1,x)
:x+I→x
:If x>xMax
:Stop
:Goto A
```

Sharp EL 9200
Sharp EL 9300

```
Reflection
-----------REAL
Goto top
Label eqtn
Y=X^3+X+1
Return
Label rng
xmin=-10
xmax=10
xstp=(xmax-xmin)/10
ymin=2xmin/3
ymax=2xmax/3
ystp=xstp
Range xmin,xmax,xstp,ymin,ymax,ystp
Return
Label top
Gosub rng
Graph X
step=(xmax-xmin)/(94*2)
X=xmin
Label 1
Gosub eqtn
Plot X,Y
Plot Y,X
X=X+step
If X<=xmax Goto 1
End
```

To use this program, enter a function
in X in the third line.

Casio 7700

```
REFLECTION
"GRAPH -A TO A"
"A="?→A
Range -A,A,1,-2A÷3,2A÷3,1
Graph Y=f₁
-A→B
Lbl 1
B→X
Plot f₁,B
B+A÷32→B
B≤A⇒Goto 1:Graph Y=X
```

To use this program, enter the
function in f_1 and set a viewing
rectangle.

Casio CFX-9800G

```
REFLECTION↵
63Xmin÷95→A↵
63Xmax÷95→B↵
Xscl→C↵
Range ,,,A,B,C↵
(Xmax–Xmin)÷94→I↵
Xmax→M↵
Xmin→D↵
Graph Y=f₁↵
Lbl 1↵
D→X↵
Plot f₁,D↵
D+I→D↵
D≤M⇒Goto 1:Graph Y=X
```

Casio 6300

```
REFLECTION:
"-A TO A"◢
"A="?→A:
Range -A,A,1,-2A÷3,2A÷3,1:
-A→B:
Lbl 1:
B→X:
Prog 0:
Ans→Y:
Plot B,Y:
B+A÷24→B:
B≤A⇒Goto 1:
-A→B:
Lbl 2:
B→X:
Prog 0:
Ans→Y:
Plot Y,B:
B+A÷24→B:
B≤A⇒Goto 2:Graph Y=X
```

To use this program, write the
function as Prog 0 and set a
viewing rectangle.

Casio fx-9700GE

```
REFLECTION↵
63Xmin÷127→A↵
63Xmax÷127→B↵
Xscl→C↵
Range ,,,A,B,C↵
(Xmax–Xmin)÷126→I↵
Xmax→M↵
Xmin→D↵
Graph Y=f₁↵
Lbl 1↵
D→X↵
Plot f₁,D↵
D+I→D↵
D≤M⇒Goto 1:Graph Y=X
```

To use either program, enter a
function in f_1 and set a viewing
rectangle.

Systems of Linear Equations (Section 6.2)

The program, given in the marginal Programming note on page 449, will display the solution of a system of two linear equations in two variables of the form

$ax + by = c$

$dx + ey = f$

if a unique solution exists.

Note: For help with keystrokes, see previous programs in this appendix.

TI-80

```
PROGRAM:SOLVE
:Disp "AX + BY = C"
:Input "ENTER A ", A
:Input "ENTER B ", B
:Input "ENTER C ", C
:Disp "DX + EY = F"
:Input "ENTER D ", D
:Input "ENTER E ", E
:Input "ENTER F ", F
:If AE−DB = 0
:Then
:Disp "NO SOLUTION"
:Else
:(CE−BF)/(AE−DB)→X
:(AF−CD)/(AE−DB)→Y
:Disp X
:Disp Y
:End
```

TI-81

```
Prgm5: SOLVE
:Disp "ENTER A, B, C, D, E, F"
:Input A
:Input B
:Input C
:Input D
:Input E
:Input F
:If AE−DB=0
:Goto 1
:(CE−BF)/(AE−DB) →X
:(AF−CD)/(AE−DB) →Y
:Disp X
:Disp Y
:Lbl 1
:End
```

TI-82

```
PROGRAM:SOLVE
:Disp "AX + BY = C"
:Prompt A
:Prompt B
:Prompt C
:Disp "DX + EY = F"
:Prompt D
:Prompt E
:Prompt F
:If AE−DB = 0
:Then
:Disp "NO SOLUTION"
:Else
:(CE−BF)/(AE−DB)→X
:(AF−CD)/(AE−DB)→Y
:Disp X
:Disp Y
:End
```

TI-85

```
 PROGRAM:
:Disp "ax+by=c"
:Input "Enter a ",A
:Input "Enter b ",B
:Input "Enter c ",C
:Disp "dx+ey=f"
:Input "Enter d ",D
:Input "Enter e ",E
:Input "Enter f ",F
:If A*E−D*B=0
:Goto A
:(C*E−B*F)/(A*E−D*B)→X
:(A*F−C*D)/(A*E−D*B)→Y
:Disp X
:Disp Y
:Lbl A
```

Sharp EL 9200
Sharp El 9300

```
Solve
----------REAL
Input A
Input B
Input C
Input D
Input E
Input F
If A*E-D*B=0 Goto 1
X=(C*E-B*F)/(A*E-D*B)
Y=(A*F-C*D)/(A*E-D*B)
Print X
Print Y
End
Label 1
Print "No solution
End
```

Equations must be entered in the
form: $Ax + By = C$; $Dx + Ey = F$.
Uppercase letters are used so that the
values can be accessed in the
calculation mode of the calculator.

Casio 6300

```
SOLVE:
"A="?→A:
"B="?→B:
"C="?→C:
"D="?→D:
"E="?→E:
"F="?→F:
AE-DB=0⇒Goto 1:
"X="◢(CE-BF)÷(AE-DB)◢
"Y="◢(AF-CD)÷(AE-DB):
Goto 2:
Lbl 1:
"NO UNIQUE SOLUTION":
Lbl 2
```

Casio 7700

```
SOLVE
"ENTER A,B,C,D,E,F"
"A="?→A
"B="?→B
"C="?→C
"D="?→D
"E="?→E
"F="?→F
AE-DB=0⇒Goto 1
"X=":(CE-BF)÷(AE-DB)◢
"Y=":(AF-CD)÷(AE-DB)
Goto 2
Lbl 1
"NO UNIQUE SOLUTION"
Lbl 2
```

Casio fx-9700GE
Casio CFX-9800G

```
SOLVE↲
"ENTER A,B,C,D,E,F"↲
"A":?→A↲
"B":?→B↲
"C":?→C↲
"D":?→D↲
"E":?→E↲
"F":?→F↲
AE-DB=0⇒Goto 1↲
"X=":(CE-BF)÷(AE-DB)◢
"Y=":(AF-CD)÷(AE-DB)↲
Goto 2↲
Lbl 1↲
"NO UNIQUE SOLUTION"↲
Lbl 2
```

Solutions to systems of linear
equations are also available directly
from the Casio calculator's
EQUATION MENU.

Summation Program (Section 8.1)

The program, given in the Group Activity on page 573, finds the sum of a finite sequence.

Note: On a *TI-82*, the "sum" and "seq(" commands may be entered through the "MATH" and "OPS" menus, respectively, accessed by pressing the LIST key. For additional keystroke instructions, see previous programs in this appendix. Keystroke sequences required for similar commands on other calculators will vary. Consult the user's manual for your calculator.

TI-80

```
PROGRAM:SUM
:Input "LOWER LIMIT ", M
:Input "UPPER LIMIT ", N
:sum seq(Y₁,X,M,N,1)→S
:Disp "SUM = ",S
```

To use this program, store the formula for the nth term as Y_1. You may also find the sum of a sequence directly on the TI-80's home screen.

TI-81

```
Prgm6: SUM
:Disp "ENTER M"
:Input M
:Disp "ENTER N"
:Input N
:0 → S
:M → X
:Lbl 1
:S+Y₁ → S
:If X=N
:Goto 2
:X+1 → X
:Goto 1
:Lbl 2
:Disp S
```

To use this program, store the formula for the nth term as Y_1.

TI-82

```
PROGRAM:SUM
:Input "LOWER LIMIT ", M
:Input "UPPER LIMIT ", N
:sum seq(Y₁,X,M,N,1)→S
:Disp "SUM = ",S
```

To use this program, store the formula for the nth term as Y_1. You may also find the sum of a sequence directly on the TI-82's home screen.

TI-85

```
PROGRAM:
:Input "ENTER M ",M
:Input "ENTER N ",N
:0→S
:M→x
:Lbl A
:S+y1→S
:If x=N
:Goto B
:x+1→x
:Goto A
:Lbl B
:Disp S
```

To use this program, store the formula for the nth term as $y1$.

Sharp EL 9200
Sharp EL 9300

```
Sum
---------REAL
Goto 1
Label eqtn
y=√ (22926¡902.5x¡2.01x²)
Return
Label 1
Print "Enter start
Input m
Print "Enter end
Input n
If n<m Goto 1
sum=0
x=m
Label 2
Gosub eqtn
sum=sum+y
x=x+1
If x<=n Goto 2
Print sum
End
```

To use this program, enter the formula in *x* for the *n*th term into the *y*-line at the beginning of the program.

Casio 7700

```
SUM
"LOWER LIMIT"?→M
"UPPER LIMIT"?→N
0→S
M→X
Lbl 1
S+f₁→S
X+1→X
X≤N⇒Goto 1
"SUM=":S
```

To use this program, store the formula for the *n*th term in f_1.

Casio 6300

```
SUM:
"LOWER LIMIT"?→M:
"UPPER LIMIT"?→N:
0→S:
M→X:
Lbl 1:
Prog 0:
Ans+S→S:
X+1→X:
X≤N⇒Goto 1:
"SUM="◢S
```

To use this program, store the formula for the *n*th term as Prog 0.

Casio fx-9700GE
Casio CFX-9800G

```
SUM↵
"LOWER LIMIT":?→M↵
"UPPER LIMIT":?→N↵
0→S↵
M→X↵
Lbl 1↵
S+f₁→S↵
X+1→X↵
X≤N⇒Goto 1↵
"SUM=":S
```

To use this program, store the formula for the *n*th term in f_1. The sum of a series is also available directly from the Casio calculator's TABLE&GRAPH MENU.

PHOTO CREDITS

CHAPTER P

Section P.1 *(page 9)*

1. (a) Natural: $\{5\}$
   (b) Integer: $\{-9, 5\}$
   (c) Rational: $\left\{-9, -\frac{7}{2}, 5, \frac{2}{3}, 0.1\right\}$
   (d) Irrational: $\{\sqrt{2}\}$

3. (a) Natural: $\{1, 2, 12\}$ *(Note:* $\sqrt{4} = 2$*)*
   (b) Integer: $\{-13, 1, 2, 12\}$
   (c) Rational: $\left\{-13, 1, \frac{3}{2}, 2, 12\right\}$
   (d) Irrational: $\{\sqrt{6}\}$

5. (a) Natural: $\{3, 2\}$ *(Note:* $\frac{6}{3} = 2$*)*
   (b) Integer: $\{-1, 2, 3\}$
   (c) Rational: $\left\{-7.5, -1, \frac{1}{3}, 2, 3\right\}$
   (d) Irrational: $\left\{-\frac{1}{2}\sqrt{2}\right\}$

7. $\frac{3}{2} < 7$

9. $-4 > -8$

11. $\frac{5}{6} > \frac{2}{3}$

13. $x \le 5$ denotes all real numbers less than or equal to 5.

15. $x < 0$ denotes all real numbers less than 0.

17. $x \ge 4$ denotes all real numbers greater than or equal to 4.

19. $-2 < x < 2$ denotes all real numbers greater than -2 and less than 2.

21. $-1 \le x < 0$ denotes all real numbers greater than or equal to -1 and less than 0.

23. $x < 0$ **25.** $A \ge 30$ **27.** $3.5\% \le r \le 6\%$

29. 10   **31.** $\pi - 3$ **33.** -1 **35.** -9

37. $|-4| = |4|$ **39.** $-5 = -|5|$ **41.** $-|-2| = -|2|$

43. 4   **45.** $\frac{5}{2}$ **47.** $\frac{7}{2}$ **49.** 51   **51.** $\frac{128}{75}$

53. $|x - 5| \le 3$ **55.** $\left|z - \frac{3}{2}\right| > 1$ **57.** $|y| \ge 6$

59. $\frac{127}{90}, \frac{584}{413}, \frac{7071}{5000}, \sqrt{2}, \frac{47}{33}$ **61.** 0.625   **63.** $0.\overline{123}$

	$|a - b|$	$0.05b$	Passes Budget Variance Test
65.	\$127.88	\$1250	Yes
67.	\$572.59	\$470	No
69.	\$671.75	\$1882	No

	Median Income, y	Household Income, x	$|y - x|$
71.	\$52,725	\$65,400	\$12,675
73.	\$52,984	\$37,300	\$15,684
75.	\$36,538	\$21,300	\$15,238

Section P.2 *(page 20)*

Warm Up *(page 20)*

1. $-4 < -2$ **2.** $0 > -3$ **3.** $\sqrt{3} > 1.73$

4. $-\pi < -3$ **5.** $|6 - 4| = 2$ **6.** $|2 - (-2)| = 4$

7. $|0 - (-5)| = 5$ **8.** $|3 - (-1)| = 4$

9. $|-7| + |7| = 7 + 7 = 14$

10. $-|8 - 10| = -|-2| = -2$

1. $7x, 4$ **3.** $x^2, -4x, 8$ **5.** $4x^3, x, -5$

7. (a) -10 (b) -6 **9.** (a) 14   (b) 2

11. (a) 0   (b) 0   **13.** (a) Undefined   (b) 0

15. Commutative (addition)   **17.** Inverse (addition)

19. Distributive Property   **21.** Distributive Property

23. Identity (addition)   **25.** Identity (addition)

27. Associative (addition)

29. Associative and Commutative (multiplication)

31. $2^2 \cdot 3^2$ **33.** $2 \cdot 7$ **35.** -6 **37.** 2

39. 6   **41.** -14 **43.** 96   **45.** $\frac{2}{7}$ **47.** $\frac{7}{20}$

49. $\frac{3}{10}$ **51.** $\frac{1}{12}$ **53.** 48   **55.** 4.45   **57.** 0.27

59. -0.13 **61.** 1.56   **63.** 0.22   **65.** 18.81

67. 198.65   **69.** 17.88   **71.** 16.6%

73. General Motors had 35.8% of the market.
General Motors: 5.4 million vehicles
Ford: 3.6 million vehicles
Chrysler: 1.8 million vehicles
Toyota: 1.1 million vehicles
Honda: 0.9 million vehicles
Nissan: 0.7 million vehicles
Other: 1.5 million vehicles

75. $5(2.7 - 9.4)$

77. Scientific:

5 $\boxed{\times}$ $\boxed{(}$ 18 $\boxed{-}$ 2 $\boxed{y^x}$ 3 $\boxed{)}$ $\boxed{\div}$ 10 $\boxed{=}$

Graphing:

5 $\boxed{(}$ 18 $\boxed{-}$ 2 $\boxed{\wedge}$ 3 $\boxed{)}$ $\boxed{\div}$ 10 $\boxed{\text{ENTER}}$

79. 7402

Section P.3 *(page 30)*

Warm Up *(page 30)*

1. 1   **2.** 5   **3.** 4   **4.** $\frac{1}{4}$ **5.** 0

6. 1   **7.** $\frac{3}{7}$ **8.** 4   **9.** $-\frac{1}{8}$ **10.** 1

---

1. 64   **3.** 125   **5.** 729   **7.** -81 **9.** 8

11. $\frac{7}{10}$ **13.** $\frac{9}{16}$ **15.** 5184   **17.** $-\frac{3}{5}$ **19.** 36

21. 1   **23.** -24 **25.** $\frac{1}{2}$ **27.** 5   **29.** $9x^2$

31. $5x^6$ **33.** $10x^4$ **35.** $24y^{10}$ **37.** $3x^2$

39. $\frac{7}{x}$ **41.** $\frac{4}{3}(x+y)^2$ **43.** 1

45. $\frac{1}{4x^4}$ **47.** $-2x^3$ **49.** $\frac{10}{x}$ **51.** $\frac{x^2}{9z^4}$

53. 3^{3n} **55.** $\frac{x^{3n}}{x^{3n}} = 1$, $x \neq 0$

57. 5.75×10^7 square miles   **59.** 9.461×10^{15} kilometers

61. 8.99×10^{-5} **63.** 3.937×10^{-5} inch

65. 524,000,000 servings   **67.** $13,000,000\,°C$

69. 0.00000000048 electrostatic unit

71. 0.000000001 meter   **73.** (a) 7697.125   (b) 954.448

75. (a) 3,070,500   (b) 3.077×10^{10}

77. (a) 4.907×10^{17} (b) 1.479

79.

Number of years	Balance
5	$1489.85
10	$2219.64
15	$3306.92
20	$4926.80
25	$7340.18

81. (a) $19,055.59   (b) $19,121.84

83. Insert parentheses around the exponent.   **85.** 9.31%

$$A = 2000\left(1 + \frac{0.06}{12}\right)^{(12\cdot5)}$$

87. $(5.1 - 3.6)^5$ **89.** No. Let $a = -1$ and $n = 2$.

Section P.4 *(page 41)*

Warm Up *(page 41)*

1. $\frac{4}{27}$ **2.** 48   **3.** $-8x^3$ **4.** $6x^7$ **5.** $28x^6$

6. $\frac{1}{5}x^2$ **7.** $3z^4$ **8.** $\frac{25}{4x^2}$ **9.** 1

10. $(x+2)^{10}$

---

1. $9^{1/2} = 3$ **3.** $\sqrt[5]{32} = 2$ **5.** $\sqrt{196} = 14$

7. $(-216)^{1/3} = -6$ **9.** $\sqrt[3]{27^2} = 9$

11. $81^{3/4} = 27$ **13.** 3   **15.** 2   **17.** 6

19. 3   **21.** $\frac{1}{2}$ **23.** -125 **25.** 4   **27.** 216

29. $\frac{1}{8}$ **31.** $\frac{27}{8}$ **33.** -4 **35.** $2\sqrt{2}$ **37.** $2x\sqrt[3]{2x^2}$

39. $\dfrac{5|x|\sqrt{3}}{y^2}$　**41.** $\dfrac{\sqrt{3}}{3}$　**43.** $4\sqrt[3]{4}$　**45.** $\dfrac{x(5+\sqrt{3})}{11}$

47. $3(\sqrt{6}-\sqrt{5})$　**49.** $2\sqrt{x}$　**51.** $34\sqrt{2}$

53. $-2\sqrt{y}$　**55.** $\sqrt{6}$　**57.** 25　**59.** $6^{1/6}$

61. $x^{3/2},\, x\neq 0$　**63.** $2\sqrt[4]{2}$　**65.** $3^{1/2}=\sqrt{3}$

67. $(x+1)^{2/3}=\sqrt[3]{(x+1)^2}$　**69.** 7.550

71. 2.006　**73.** 2.938　**75.** 0.382

77. $\sqrt{5}+\sqrt{3}>\sqrt{5+3}$　**79.** $5>\sqrt{3^2+2^2}$

81. $\sqrt{3}\cdot\sqrt[4]{3}>\sqrt[8]{3}$　**83.** 24 in. $\times\,24$ in. $\times\,24$ in.

85. 12.8%　**87.** $\pi/2$ or 1.57 seconds

89. 494 vibrations/second　**91.** 1

Mid-Chapter Quiz　*(page 44)*

1. $<$　**2.** $=$　**3.** $x\geq 0$　**4.** $0.965\leq r\leq 1$

5. All real numbers that are greater than or equal to 0 and are also less than 3.

6. $3x^2,\,-7x,\,2$　**7.** 2　**8.** -4　**9.** $\dfrac{5}{14}$

10. $\dfrac{11}{9}$　**11.** $-2x^7$　**12.** $\dfrac{y^4}{3}$　**13.** $18t^3$

14. $\dfrac{27x^6}{y^6}$　**15.** $\$2219.64$　**16.** -1　**17.** -64

18. 9　**19.** $-\sqrt[3]{3}$　**20.** 25 cm $\times\,25$ cm $\times\,25$ cm

Section P.5　*(page 51)*

> ### Warm Up　*(page 51)*
>
> **1.** $42x^3$　**2.** $-20z^2$　**3.** $-27x^6$　**4.** $-3x^6$
>
> **5.** $\dfrac{9}{4}z^3,\, z\neq 0$　**6.** $4\sqrt{3}$　**7.** $\dfrac{9}{4x^2}$　**8.** 8
>
> **9.** $\sqrt{2}$　**10.** $-3x$

	Degree	Leading Coefficient
1.	2	2
3.	5	1
5.	5	4

7. Polynomial, $-3x^3+2x+8$, degree 3

9. Not a polynomial

11. Polynomial, $-y^4+y^3+y^2$, degree 4

13. (a) -1　(b) 3　(c) 7　(d) 11

15. (a) 11　(b) 6　(c) 3　(d) 2

17. (a) -6　(b) 0　(c) -2　(d) -6　**19.** $-2x-10$

21. $3x^3-2x+2$　**23.** $8x^3+29x^2+11$　**25.** $12z+8$

27. $3x^3-6x^2+3x$　**29.** $-15z^2+5z$　**31.** $30x^3+12x^2$

33. $x^2+7x+12$　**35.** $6x^2-7x-5$

37. $x^2+12x+36$　**39.** $4x^2-20xy+25y^2$

41. $x^2+2xy+y^2-6x-6y+9$

43. x^2-100　**45.** x^2-4y^2　**47.** m^2-n^2-6m+9

49. $4r^4-25$　**51.** x^3+3x^2+3x+1

53. $8x^3-12x^2y+6xy^2-y^3$　**55.** $x-y$

57. $16x^6-24x^3+9$　**59.** $64x^2+48x+9$

61. x^4+x^2+1　**63.** $2x^2+2x$

65. $\$9419.27$　**67.** $500r^2+1000r+500$

69. $\begin{array}{ll}x=1\text{ inch}, & V=384\text{ cubic inches}\\ x=2\text{ inches}, & V=616\text{ cubic inches}\\ x=3\text{ inches}, & V=720\text{ cubic inches}\end{array}$

71. $3x^2+7x$　**73.** $x^2+42x+360$　**75.** $1,\,81,\,36$

Section P.6　*(page 59)*

> ### Warm Up　*(page 59)*
>
> **1.** $15x^2-6x$　**2.** $-2y^2-2y$　**3.** $4x^2+12x+9$
>
> **4.** $9x^2-48x+64$　**5.** $2x^2+13x-24$
>
> **6.** $-5z^2-z+4$　**7.** $4y^2-1$　**8.** x^2-a^2
>
> **9.** $x^3+12x^2+48x+64$
>
> **10.** $8x^3-36x^2+54x-27$

1. $3(x+2)$　**3.** $2x(x^2-3)$　**5.** $(x-1)(x+5)$

7. $(x+6)(x-6)$　**9.** $(4y+3)(4y-3)$

11. $(x+1)(x-3)$　**13.** $(x-2)^2$　**15.** $(2t+1)^2$

17. $(5y-1)^2$　**19.** $(x+2)(x-1)$　**21.** $(s-3)(s-2)$

23. $(y+5)(y-4)$　**25.** $(x-20)(x-10)$

27. $(3x-2)(x-1)$　**29.** $(3z+1)(3z-2)$

31. $(5x+1)(x+5)$　**33.** $(x-2)(x^2+2x+4)$

35. $(y + 4)(y^2 - 4y + 16)$ **37.** $(2t - 1)(4t^2 + 2t + 1)$

39. $(x - 1)(x^2 + 2)$ **41.** $(2x - 1)(x^2 - 3)$

43. $(3 + x)(2 - x^3)$ **45.** $4x(x - 2)$

47. $x(x + 3)(x - 3)$ **49.** $x^2(x - 4)$ **51.** $(x - 1)^2$

53. $(1 - 2x)^2$ **55.** $2x(x + 1)(x - 2)$

57. $(9x + 1)(x + 1)$ **59.** $(3x + 1)(x^2 + 5)$

61. $x(x - 4)(x^2 + 1)$ **63.** $-x(x + 10)$

65. $(x + 1)^2(x - 1)^2$ **67.** $2(t - 2)(t^2 + 2t + 4)$

69. $(2x - 1)(6x - 1)$ **71.** $(x + 2)^3$

73.

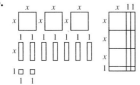

75.

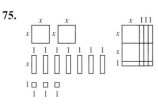

77. (b) **79.** (a) **81.** $(2x + 1)$ feet

Math Matters *(page 61)*

Crystal, freshman, economics; Helene, sophomore, history; Juan, junior, math; Tito, senior, science.

Section P.7 *(page 70)*

Warm Up *(page 70)*

1. $5x^2(1 - 3x)$ **2.** $(4x + 3)(4x - 3)$

3. $(3x - 1)^2$ **4.** $(2y + 3)^2$ **5.** $(z + 3)(z + 1)$

6. $(x - 5)(x - 10)$ **7.** $(3 - x)(1 + 3x)$

8. $(3x - 1)(x - 15)$ **9.** $(s + 1)(s + 2)(s - 2)$

10. $(y + 4)(y^2 - 4y + 16)$

---

1. All real numbers **3.** All nonnegative real numbers

5. All real numbers except $x = 2$

7. All real numbers except $x = 0$ and $x = 4$

9. All real numbers greater than or equal to -1

11. $\dfrac{5}{2x} = \dfrac{5(3x)}{6x^2}$ **13.** $\dfrac{x + 1}{x} = \dfrac{(x + 1)(x - 2)}{x(x - 2)}, x \neq 2$

15. $\dfrac{3x}{x - 3} = \dfrac{3x(x + 2)}{x^2 - x - 6}, x \neq -2$ **17.** $\dfrac{3x}{2}, x \neq 0$

19. $\dfrac{x}{2(x + 1)}$ **21.** $-\dfrac{1}{2}, x \neq 5$ **23.** $\dfrac{x(x + 3)}{x - 2}, x \neq -2$

25. $\dfrac{y - 4}{y + 6}, y \neq 3$ **27.** $-(x^2 + 1), x \neq 2$ **29.** $z - 2$

31. $\dfrac{1}{5(x - 2)}, x \neq 1$ **33.** $-\dfrac{x(x + 7)}{x + 1}, x \neq 9$

35. $\dfrac{r + 1}{r}, r \neq 1$ **37.** $\dfrac{t - 3}{(t + 3)(t - 2)}, t \neq -2$

39. $\dfrac{x - 1}{x(x + 1)^2}, x \neq -2$ **41.** $\dfrac{3}{2}, x \neq -y$

43. $x(x + 1), x \neq -1, 0$ **45.** $\dfrac{x + 5}{x - 1}$ **47.** $\dfrac{6x + 13}{x + 3}$

49. $-\dfrac{2}{x - 2}$ **51.** $\dfrac{x - 4}{(x + 2)(x - 2)(x - 1)}$

53. $\dfrac{2 - x}{x^2 + 1}, x \neq 0$ **55.** $\dfrac{1}{2}, x \neq 2$ **57.** $\dfrac{1}{x}, x \neq -1$

59. $\dfrac{-2(x + 1)}{x^2(x + 2)^2}$ **61.** $\dfrac{2x - 1}{2x}$ **63.** $\dfrac{1}{35} \approx 0.029$

65. $\dfrac{194}{21,847} \approx 0.009$ **67.** $\dfrac{1}{50} = 0.02$

69. (a) 12.65% (b) $\dfrac{288(NM - P)}{N(12P + NM)}$ **71.** $\dfrac{7x + 4}{4(2x + 1)}$

73. $\dfrac{2x^2 + x - [(\pi x^2)/4]}{2x^2 + x} = \dfrac{(8 - \pi)x + 4}{4(2x + 1)}$

75.

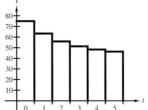

Review Exercises *(page 75)*

1. (a) Natural: $\{11\}$
   (b) Integer: $\{-14, 11\}$
   (c) Rational: $\left\{-14, -\frac{8}{9}, 0.4, \frac{5}{2}, 11\right\}$
   (d) Irrational: $\{\sqrt{6}\}$

3. $-4 < -3$

5. All real numbers less than or equal to 7

7. $x \geq 0$ **9.** -14 **11.** $|-12| > -|12|$
13. 3   **15.** $|x - 7| \geq 4$ or $|7 - x| \geq 4$
17. $2x^2, -3x, 4$ **19.** (a) 2   (b) 0
21. Distributive Property   **23.** -1 **25.** $\frac{2}{3}$
27. 5   **29.** 0.10   **31.** -18 **33.** $8x, x \neq 0$
35. 2.74×10^6 **37.** 483,300,000 miles
39. (a) 11,414.125   (b) 18,380.160

41.

Number of years	Balance
5	$2697.70
10	$3638.79
15	$4908.19
20	$6620.41
25	$8929.94

43. $16^{1/2} = 4$ **45.** 12   **47.** $2x^2$
49. $2 + \sqrt{3}$ **51.** $-3\sqrt{x}$ **53.** $\sqrt{10}$ **55.** $\sqrt{5}$
57. 11.269   **59.** $-6x + 26$ **61.** $x^2 - 8x - 14$
63. $x^2 - x - 2$ **65.** $x^3 - x^2 + 2x - 2$
67. $4(x + 3)(x - 3)$ **69.** $3x(x - 2)(x + 1)$
71. $(x - 4)(x^2 - 2) = (x - 4)(x + \sqrt{2})(x - \sqrt{2})$
73. All real numbers except $x = 3$ **75.** $3x$
77. $\dfrac{x - 2}{2}, x \neq -2$ **79.** $x + 2, x \neq 0$
81. $\dfrac{x - 1}{x - 3}, x \neq -1, \frac{1}{2}$ **83.** $\dfrac{x(3x - 4)}{(x - 1)(x - 2)}$

85. $\dfrac{-1}{xy(x + y)}, x \neq y$ **87.** $\dfrac{1}{10}$ **89.** $\dfrac{x + 4}{2x + 4}$

Chapter Test *(page 78)*

1. -12 **2.** 2   **3.** $-64x^6$ **4.** $-4\sqrt{x}$
5. $5^2 = 25$ **6.** $2\sqrt{2}$ **7.** $2|x|\sqrt{3x}$ **8.** $\dfrac{5 + \sqrt{7}}{9}$
9. $9x^2 + 30x + 25$ **10.** $-5x^2 + 29x$
11. $5(x - 4)(x + 4)$ **12.** $(x^2 - 3)(x - 6)$
13. $\dfrac{x - 4}{3}, x \neq -4$ **14.** $\dfrac{x + 4}{3x + 5}, x \neq -3, \dfrac{5}{3}$
15. $\dfrac{x(4x - 13)}{(x - 3)(x - 4)}$ **16.** $\dfrac{-x - 26}{(x - 2)(x + 5)}$
17.

Number of years	Balance
5	$4469.54
10	$6658.92
15	$9920.76
20	$14,780.41
25	$22,020.53

18. $\dfrac{y + x}{y - x}$ or $\dfrac{x + y}{y - x}$ **19.** $\dfrac{10,000}{500,000} = \dfrac{1}{50} = 0.02$
20. $\dfrac{2}{42} = \dfrac{1}{21}$, or approximately 0.048

CHAPTER 1

Section 1.1 *(page 86)*

Warm Up *(page 86)*

1. $-3x - 10$ **2.** $5x - 12$ **3.** x **4.** $x + 26$
5. $\dfrac{8x}{15}$ **6.** $\dfrac{3x}{4}$ **7.** $-\dfrac{1}{x(x + 1)}$
8. $\dfrac{5}{x}$ **9.** $\dfrac{7x - 8}{x(x - 2)}$ **10.** $-\dfrac{2}{x^2 - 1}$

1. Identity   **3.** Conditional equation

5. Conditional equation

7. (a) No
(b) No
(c) Yes
(d) No

9. (a) Yes
(b) Yes
(c) No
(d) No

11. (a) Yes
(b) No
(c) No
(d) No

13. (a) No
(b) No
(c) No
(d) Yes

15. 5 **17.** -4 **19.** 3

21. 9 **23.** -26 **25.** -4 **27.** $-\frac{6}{5}$ **29.** 9

31. No solution **33.** 10 **35.** 4 **37.** 3 **39.** 5

41. No solution **43.** $\frac{11}{6}$ **45.** No solution **47.** 0

49. All real numbers **51.** No solution

53. $x \approx 138.889$ **55.** $x \approx 62.372$ **57.** $x \approx -2.386$

59. (a) 6.46
(b) 6.41

61. (a) 1.06
(b) 1.06

63. (a) 56.09
(b) 56.13

65. Sometime during 1996 ($t = 16.8$)

67. 61.2 inches **69.** 23 years old **71.** 0.46

Section 1.2 *(page 97)*

Warm Up *(page 97)*

1. 14 **2.** 4 **3.** -3 **4.** 4 **5.** -2
6. 1 **7.** $\frac{2}{5}$ **8.** $\frac{10}{3}$ **9.** 6 **10.** $-\frac{11}{5}$

1. $x + (x + 1) = 2x + 1$ **3.** $50t$ **5.** $0.2x$ **7.** $6x$

9. $1200 + 25x$ **11.** 262, 263 **13.** 37, 185

15. $-5, -4$ **17.** Coworker's check: $300
Your check: $345

19. Coworker's check: $348.65 **21.** 1700% **23.** 860%
Your check: $296.35

25. (a) 1980: 6.6 years
(b) 1990: 7.8 years

27. Guaranteed investment: 540
Money fund: 333
Diversified stock: 311
Aggressive stock: 303
Company stock: 296

29. 0.63 **31.** 15 feet × 22.5 feet **33.** 97

35. $22,316.98 **37.** $1016.77 **39.** 38%

41. 3 hours **43.** $\frac{1}{3}$ hour

45. Family 1 (42 miles per hour): 3.8 hours
Family 2 (50 miles per hour): 3.2 hours

47. 1.29 seconds **49.** 62.5 feet **51.** $563,952

53. $4500 at 10.5% **55.** 11.43%
$7500 at 13%

57. 8824 units per month **59.** 48 feet

61. 32.1 gallons **63.** 12.29 miles per hour

65. $h = \dfrac{2A}{b}$ **67.** $l = \dfrac{V}{wh}$ **69.** $C = \dfrac{S}{1 + R}$

71. $r = \dfrac{A - P}{Pt}$ **73.** $b = \dfrac{2A - ah}{h}$

75. $n = \dfrac{L + d - a}{d}$ **77.** $h = \dfrac{A}{2\pi r}$

Section 1.3 *(page 110)*

Warm Up *(page 110)*

1. $\dfrac{\sqrt{14}}{10}$ **2.** $4\sqrt{2}$ **3.** 14 **4.** $\dfrac{\sqrt{10}}{4}$

5. $x(3x + 7)$ **6.** $(2x - 5)(2x + 5)$

7. $-(x - 7)(x - 15)$ **8.** $(x - 2)(x + 9)$

9. $(5x - 1)(2x + 3)$ **10.** $(6x - 1)(x - 12)$

1. $2x^2 + 5x - 3 = 0$ **3.** $x^2 - 25x = 0$

5. $x^2 - 6x + 7 = 0$ **7.** $2x^2 - 2x + 1 = 0$

9. $3x^2 - 60x - 10 = 0$ **11.** $0, -\frac{1}{2}$ **13.** $4, -2$

15. -5 **17.** $3, -\frac{1}{2}$ **19.** $2, -6$ **21.** $-2, -5$

23. ± 4 **25.** $\pm\sqrt{7} \approx \pm 2.65$ **27.** $\pm 2\sqrt{3} \approx \pm 3.46$

29. $12 + 3\sqrt{2} \approx 16.24$ **31.** $-2 + 2\sqrt{3} \approx 1.46$
$12 - 3\sqrt{2} \approx 7.76$ $\quad -2 - 2\sqrt{3} \approx -5.46$

33. ± 5 **35.** $\pm\dfrac{\sqrt{115}}{5} \approx \pm 2.14$ **37.** ± 8 **39.** 1

41. $\pm\frac{3}{4}$ **43.** $\frac{3}{2}$ **45.** $6, -12$ **47.** $\frac{3}{2}, -\frac{1}{2}$

49. $5, -\frac{10}{3}$ **51.** $9, 3$ **53.** $\frac{1}{5}, 1$ **55.** $-1, -5$

57. $-\frac{1}{2}$ **59.** 34 feet × 48 feet

61. Base: $2\sqrt{2}$ feet **63.** $\dfrac{9\sqrt{13}}{4} \approx 8.11$ seconds
Height: $2\sqrt{2}$ feet

65. 1.43 seconds **67.** 3.54 centimeters **69.** 886 miles

71. 1414 feet **73.** 50,000 units **75.** 2005 ($t \approx 15.4$)

77. 1987 ($t \approx 18.74$). The model was a good representation through 1990.

79. ≈ 0.4490

Section 1.4   *(page 120)*

Warm Up   *(page 120)*

1. $3\sqrt{17}$ **2.** $2\sqrt{3}$ **3.** $4\sqrt{6}$ **4.** $3\sqrt{73}$
5. $2, -1$ **6.** $\frac{3}{2}, -3$ **7.** $5, -1$ **8.** $\frac{1}{2}, -7$
9. $3, 2$ **10.** $4, -1$

---

1. One real solution   **3.** Two real solutions
5. No real solutions   **7.** Two real solutions
9. $\frac{1}{2}, -1$ **11.** $\frac{1}{4}, -\frac{3}{4}$ **13.** $1 \pm \sqrt{3}$
15. $-7 \pm \sqrt{5}$ **17.** $-4 \pm 2\sqrt{5}$ **19.** $\frac{2}{3} \pm \frac{\sqrt{7}}{3}$
21. $-\frac{1}{3} \pm \frac{\sqrt{11}}{6}$ **23.** $-\frac{1}{2} \pm \sqrt{2}$ **25.** $\frac{2}{7}$
27. $2 \pm \frac{\sqrt{6}}{2}$ **29.** $6 \pm \sqrt{11}$ **31.** $x \approx 0.976, -0.643$
33. $x \approx 0.561, 0.126$ **35.** $x \approx 1.687, -0.488$
37. -11 **39.** $\pm\sqrt{10}$ **41.** $\frac{-3 \pm \sqrt{5}}{2}$ **43.** $-2, 4$
45. ± 2 **47.** $50, 50$ **49.** 7, 8 or $-8, -7$
51. 200 units   **53.** 653 units   **55.** 9 seats per row
57. 14 inches $\times$ 14 inches   **59.** Moon: 14.9 seconds
　　　　　　　　　　　　　　　　　　　　　　Earth: 2.6 seconds
61. 259 miles, 541 miles   **63.** (a) 1985 ($t = 5.61$)
　　　　　　　　　　　　　　　　　　　　　　(b) 1995 ($t \approx 15.9$)
65. 1991   **67.** 550 miles per hour, 600 miles per hour
69. 3761 units or 146,239 units

Mid-Chapter Quiz   *(page 124)*

1. $x = 2$ **2.** $x = 6$ **3.** $x = -2$ **4.** No solution
5. 180.244   **6.** 431.398   **7.** $36x + 1400$ **8.** $0.25x$
9. 8000 units   **10.** $\frac{2}{3}, -5$ **11.** $x = \pm\sqrt{5}; x \approx \pm 2.24$
12. $x = -3 \pm \sqrt{17}; x \approx -7.12, 1.12$

13. $x = \dfrac{-7 \pm \sqrt{73}}{6}$ **14.** $x = -1 \pm \sqrt{6}$
15. $x \approx 1.568, -0.068$ **16.** Two real solutions
17. No real solutions   **18.** No real solutions
19. ≈ 3.95 seconds   **20.** 6 inches $\times$ 6 inches

Section 1.5   *(page 133)*

Warm Up   *(page 133)*

1. 11   **2.** $20, -3$ **3.** $5, -45$ **4.** $0, -\frac{1}{5}$
5. $\frac{2}{3}, -2$ **6.** $\frac{11}{6}, -\frac{5}{2}$ **7.** $1, -5$ **8.** $\frac{3}{2}, -\frac{5}{2}$
9. $\dfrac{3 \pm \sqrt{5}}{2}$ **10.** $2 + \sqrt{2}$

---

1. $0, \pm\dfrac{3\sqrt{2}}{2}$ **3.** $3, -1, 0$ **5.** ± 3 **7.** $-3, 0$
9. $3, 1, -1$ **11.** ± 1 **13.** $\pm 3, \pm 1$ **15.** ± 2
17. $\pm\frac{1}{2}, \pm 4$ **19.** $1, -2$ **21.** 50   **23.** 26
25. -16 **27.** $\frac{1}{4}$ **29.** $6, 5$ **31.** $2, -5$ **33.** 0
35. $-59, 69$ **37.** 1   **39.** $\pm\sqrt{69}$ **41.** $\dfrac{-3 \pm \sqrt{21}}{6}$
43. $4, -5$ **45.** $2, -\dfrac{3}{2}$ **47.** -1 **49.** $1, -3$
51. $1, -3$ **53.** $3, -2$ **55.** $\sqrt{3}, -3$ **57.** $10, -1$
59. $x \approx \pm 1.038$ **61.** $x \approx \pm 16.756$ **63.** 34
65. 7%   **67.** 19.2%   **69.** 26,250 passengers
71. 61 years old   **73.** 2,566,025 units   **75.** ≈ 12.12 feet

Section 1.6   *(page 144)*

Warm Up   *(page 144)*

1. $-\frac{1}{2}$ **2.** $-\frac{1}{6}$ **3.** -3 **4.** -6
5. $x \geq 0$ **6.** $-3 < z < 10$ **7.** $P \leq 2$
8. $W \geq 200$ **9.** $2, 7$ **10.** $0, 1$

---

1. (a) Yes (b) No (c) Yes (d) No   **3.** c   **4.** h
5. f   **6.** e   **7.** g   **8.** a   **9.** b   **10.** d

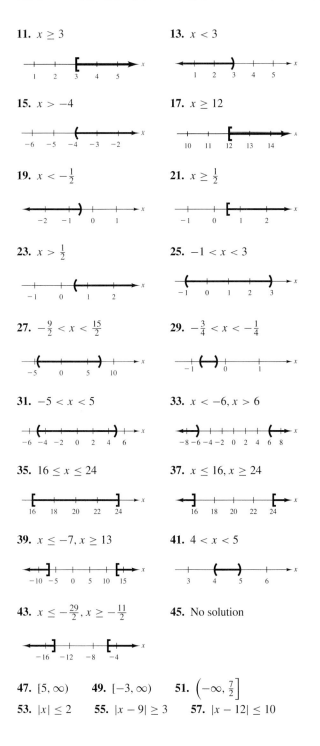

11. $x \geq 3$

13. $x < 3$

15. $x > -4$

17. $x \geq 12$

19. $x < -\frac{1}{2}$

21. $x \geq \frac{1}{2}$

23. $x > \frac{1}{2}$

25. $-1 < x < 3$

27. $-\frac{9}{2} < x < \frac{15}{2}$

29. $-\frac{3}{4} < x < -\frac{1}{4}$

31. $-5 < x < 5$

33. $x < -6, x > 6$

35. $16 \leq x \leq 24$

37. $x \leq 16, x \geq 24$

39. $x \leq -7, x \geq 13$

41. $4 < x < 5$

43. $x \leq -\frac{29}{2}, x \geq -\frac{11}{2}$

45. No solution

47. $[5, \infty)$ **49.** $[-3, \infty)$ **51.** $\left(-\infty, \frac{7}{2}\right]$

53. $|x| \leq 2$ **55.** $|x - 9| \geq 3$ **57.** $|x - 12| \leq 10$

59. $|x + 3| > 5$ **61.** $m > 400$ miles

63. $r > 12.5\%$ **65.** 24 weeks

67. (a)

x	10	20	30	40	50
R	\$1159.50	\$2319.00	\$3478.50	\$4638.00	\$5797.50
C	\$1700.00	\$2650.00	\$3600.00	\$4550.00	\$5500.00

(b) $x \geq 36$ units

69. $m < 24{,}062.5$ **71.** 1996 ($t \approx 16.83$)

73. Overcharged \$0.12 or undercharged \$0.12

75. $[65.8, 71.2]$

Math Matters *(page 147)*

Cube	Ratio of $\dfrac{\text{surface area}}{\text{weight}}$
1	6
2	3
3	2
4	1.5

Section 1.7 *(page 155)*

Warm Up *(page 155)*

1. $y < -6$ **2.** $z > -\frac{9}{2}$ **3.** $-3 \leq x < 1$

4. $x \leq -5$ **5.** $-3 < x$ **6.** $5 < x < 7$

7. $-\frac{7}{2} \leq x \leq \frac{7}{2}$ **8.** $x < 2, x > 4$

9. $x < -6, x > -2$ **10.** $-2 \leq x \leq 6$

1. $-3 \leq x \leq 3$ **3.** $x < -2, x > 2$

5. $-7 < x < 3$

7. $x \le -5, x \ge 1$

9. $-3 < x < 2$

11. $x < -1, x > 1$

13. $-3 < x < 1$

15. $x < 0, 0 < x < \frac{3}{2}$

17. $-2 \le x \le 0, x \ge 2$

19. $x < -1, 0 < x < 1$

21. $x < -1, x > 4$

23. $5 < x < 15$

25. $-5 < x < -\frac{3}{2}, x > -1$

27. $-\frac{3}{4} < x < 3, x \ge 6$

29. $[-2, 2]$ **31.** $(-\infty, 3], [4, \infty)$ **33.** $[-4, 3]$
35. All real numbers   **37.** $-3.51 < x < 3.51$
39. $-0.13 < x < 25.13$ **41.** $2.26 < x < 2.39$
43. $4 < t < 6$ **45.** 13.8 meters $\le L \le 36.2$ meters
47. (a) $90,000 \le x < 100,000$
    (b) $\$30 \le p \le \$32$
49. $r > \sqrt{1.2} - 1 \approx 9.5\%$ **51.** 1997   **53.** 1997

Review Exercises   *(page 160)*

1. Conditional equation
3. (a) No   (b) Yes   (c) Yes   (d) No
5. $-\frac{1}{2}$ **7.** -10 **9.** $-\frac{2}{3}$ **11.** 377.778 **13.** 12
15. May: $\$121,833.51$ **17.** 29.5 feet $\times 59$ feet
    June: $\$103,558.49$
19. $\$399$ **21.** 2 hours   **23.** 2.9 quarts   **25.** $-\frac{1}{2}, \frac{4}{3}$

27. $3, 8$ **29.** $\pm\sqrt{11}, \pm3.32$
31. $-4 + 3\sqrt{2} \approx 0.24$ **33.** 15 feet $\times$ 27 feet
    $-4 - 3\sqrt{2} \approx -8.24$
35. $200,000$ units or $300,000$ units   **37.** Two
39. $6 \pm \sqrt{6}$ **41.** $\dfrac{-19 \pm \sqrt{165}}{2}$ **43.** $-3 \pm 2\sqrt{3}$
45. $1.866, -0.283$ **47.** Moon: ≈ 6.09 seconds
    Earth: 2.5 seconds
49. $0, -1, 4$ **51.** $-3, \sqrt[3]{5}$ **53.** $\frac{25}{4}$
55. $\pm4\sqrt{2}$ **57.** $-3, \frac{7}{5}$ **59.** $2 \pm \sqrt{19}$
61. $\$600$ **63.** $143,203$ units
65. $-3 \le x \le -\frac{1}{4}$ **67.** $-3 < x \le -\frac{4}{3}$

69. $x < -\frac{3}{2}, x > \frac{9}{2}$ **71.** $x > -\frac{5}{2}$

73. $-\frac{9}{2} \le x \le \frac{9}{2}$ **75.** $\$0.56$
77. $-6 \le x \le -2$ **79.** $-11 \le x < -8$

81. $x < -5$ or $x > -1$ **83.** $-2.41 < x < 1.01$

85. $x < 0.50$ or $x > 0.56$
87. 2.5 seconds $< t < 3.75$ seconds
89. $r > 9.5\%$ **91.** $\$37.75 < p < \$62.25$

Chapter Test   *(page 164)*

1. $\frac{17}{23}$ **2.** 175 **3.** April: $\$175,364.00$
    May: $\$140,291.20$
4. $-\frac{5}{3}, \frac{1}{2}$ **5.** $4, -\frac{3}{2}$ **6.** $\pm\sqrt{15}$ **7.** $\dfrac{-13 \pm \sqrt{69}}{2}$
8. $\dfrac{11 \pm \sqrt{145}}{6}$ **9.** $1.038, -0.466$ **10.** $2, -\frac{10}{3}$

11. 4   **12.** $-1, 1, -3, 3$ **13.** $-6, 6$

14. Selling either 341,421 units or 58,579 units will produce a revenue of $2,000,000.

15. $-\frac{11}{3} \le x \le 3$ **16.** $-3 - \sqrt{5} \le x \le -3 + \sqrt{5}$

17. $(-11, -7)$ **18.** $x \le -2, 0 \le x \le 2$

19. $10,839.2 \le x \le 129,160.8$ **20.** $r > 9.6\%$
For a profit of at least $500,000, the company must sell more than 10,839 units but less than 129,161 units.

CHAPTER 2

Section 2.1 *(page 174)*

Warm Up *(page 174)*

1. 5   **2.** $3\sqrt{2}$ **3.** 1   **4.** -2
5. $3(\sqrt{2} + \sqrt{5})$ **6.** $2(\sqrt{3} + \sqrt{11})$ **7.** $-3, 11$
8. 9, 1   **9.** 11   **10.** 4

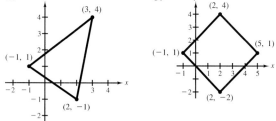

1.   **3.**

5. $(-3, -4)$ is shifted to $(-1, 1)$ **7.** 8   **9.** 5
$(1, -3)$ is shifted to $(3, 2)$
$(-2, -1)$ is shifted to $(0, 4)$

11. Answers vary. $(3, 4), (1, 4)$

13. $a = 4, b = 3, c = 5$ **15.** $a = 10, b = 3, c = \sqrt{109}$

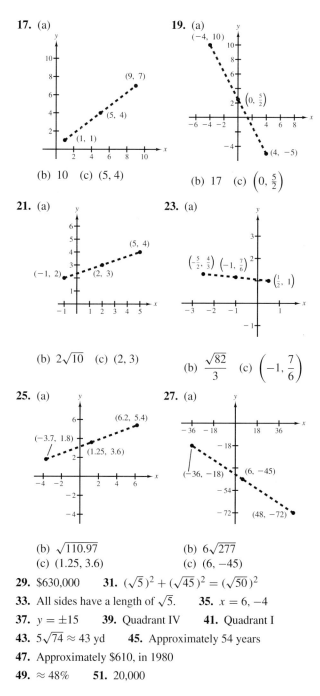

17. (a)

(b) 10   (c) (5, 4)

19. (a)

(b) 17   (c) $\left(0, \frac{5}{2}\right)$

21. (a)

(b) $2\sqrt{10}$ (c) (2, 3)

23. (a)

(b) $\frac{\sqrt{82}}{3}$ (c) $\left(-1, \frac{7}{6}\right)$

25. (a)

(b) $\sqrt{110.97}$
(c) (1.25, 3.6)

27. (a)

(b) $6\sqrt{277}$
(c) (6, −45)

29. $630,000   **31.** $(\sqrt{5})^2 + (\sqrt{45})^2 = (\sqrt{50})^2$

33. All sides have a length of $\sqrt{5}$.   **35.** $x = 6, -4$

37. $y = \pm 15$ **39.** Quadrant IV   **41.** Quadrant I

43. $5\sqrt{74} \approx 43$ yd   **45.** Approximately 54 years

47. Approximately $610, in 1980

49. $\approx 48\%$ **51.** 20,000

Section 2.2 *(page 184)*

Warm Up *(page 184)*

1. $y = \dfrac{3x - 2}{5}$ **2.** $y = -\dfrac{(x - 5)(x + 1)}{2}$

3. 2   **4.** 1, −5   **5.** 0, ±3   **6.** ±2

7. $y = x^3 + 4x$ **8.** $x^2 + y^2 = 4$

9. $y = 4x^2 + 8$ **10.** $y^2 = 3x^2 + 4$

1. (a) Yes (b) Yes   **3.** (a) No (b) Yes

5. (a) Yes (b) Yes   **7.** $(5, 0), (0, -5)$

9. $(-2, 0), (1, 0), (0, -2)$ **11.** $(0, 0), (-2, 0)$

13. $(1, 0), \left(0, \frac{1}{2}\right)$ **15.** y-axis symmetry

17. x-axis symmetry   **19.** Origin symmetry

21. Origin symmetry

23.
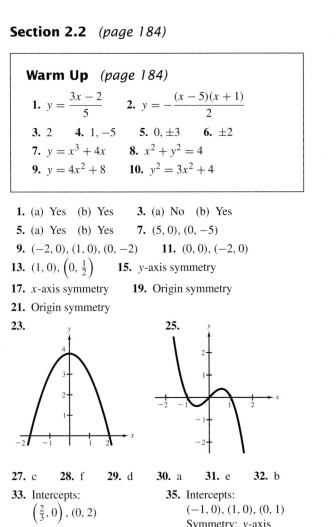

25.

27. c   **28.** f   **29.** d   **30.** a   **31.** e   **32.** b

33. Intercepts:
$\left(\frac{2}{3}, 0\right), (0, 2)$

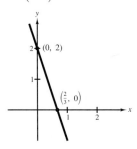

35. Intercepts:
$(-1, 0), (1, 0), (0, 1)$
Symmetry: y-axis

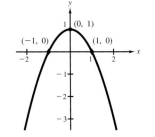

37. Intercepts:
$(3, 0), (1, 0), (0, 3)$

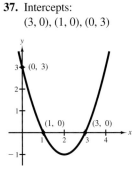

39. Intercepts:
$(-\sqrt[3]{2}, 0), (0, 2)$

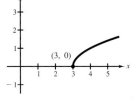

41. Intercepts: $(0, 0), (2, 0)$

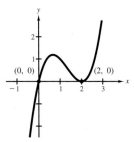

43. Intercept: $(3, 0)$

45. Intercept: $(0, 0)$
Symmetry: Origin

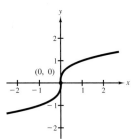

47. Intercepts: $(2, 0), (0, 2)$

49. Intercepts:
$(-1, 0), (0, 1), (0, -1)$
Symmetry: x-axis

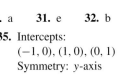
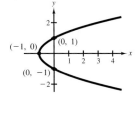

51. Intercepts: $(-2, 0), (2, 0)$
$(0, 2), (0, -2)$
Symmetry:
x-axis, y-axis, origin

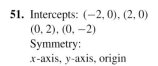
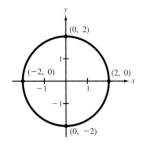

53. $x^2 + y^2 = 9$ **55.** $(x-2)^2 + (y+1)^2 = 16$

57. $(x+1)^2 + (y-2)^2 = 5$ **59.** $(x-3)^2 + (y-4)^2 = 25$

61. $(x-1)^2 + (y+3)^2 = 4$ **63.** $(x-1)^2 + (y+3)^2 = 0$

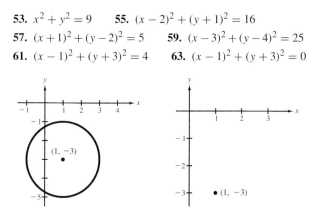

65. $\left(x - \frac{1}{2}\right)^2 + \left(y - \frac{1}{2}\right)^2 = 2$

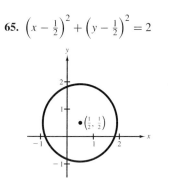

67. $\left(x + \frac{1}{2}\right)^2 + \left(y + \frac{5}{4}\right)^2 = \frac{9}{4}$

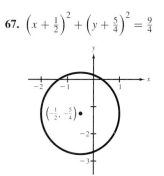

69. 2   **71.** 4

73. $(2, -3), 4, x^2 + y^2 - 4x + 6y - 3 = 0$

75. (a)

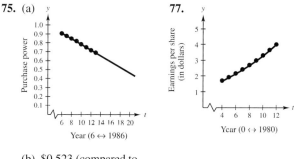

Year ($6 \leftrightarrow 1986$)

(b) \$0.523 (compared to 1982 dollars)

77.

Year ($0 \leftrightarrow 1980$)

Section 2.3 *(page 193)*

Warm Up *(page 193)*

1. $y = 4 - 3x$ **2.** $y = x$ **3.** $y = \frac{2}{3}(1 - x)$

4. $y = \frac{2}{5}(2x + 1)$ **5.** $y = \frac{1}{4}(5 - 3x)$

6. $y = \frac{2}{3}(-x + 3)$ **7.** $y = 4 - x^2$

8. $y = \frac{2}{3}(x^2 - 1)$ **9.** $y = \pm\sqrt{4 - x^2}$

10. $y = \pm\sqrt{x^2 - 9}$

1. d   **2.** g   **3.** a   **4.** f   **5.** i

6. b   **7.** j   **8.** c   **9.** e   **10.** h

11.   **13.**

15.   **17.**

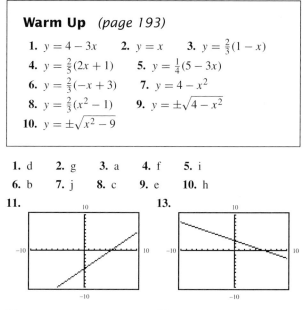
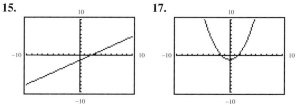

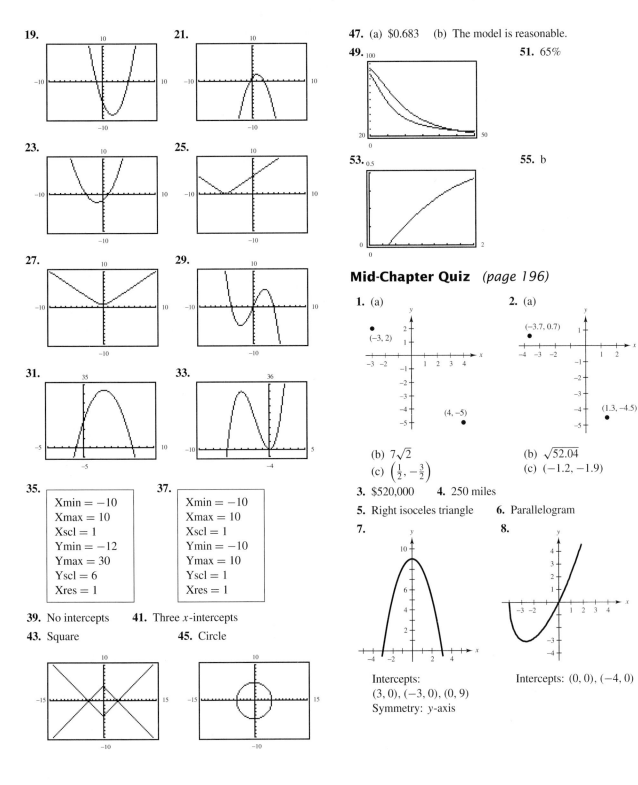

19.

21.

23.

25.

27.

29.

31.

33.

35.

Xmin = −10
Xmax = 10
Xscl = 1
Ymin = −12
Ymax = 30
Yscl = 6
Xres = 1

37.

Xmin = −10
Xmax = 10
Xscl = 1
Ymin = −10
Ymax = 10
Yscl = 1
Xres = 1

39. No intercepts   **41.** Three x-intercepts

43. Square   **45.** Circle

47. (a) $0.683   (b) The model is reasonable.

49.

51. 65%

53.

55. b

Mid-Chapter Quiz *(page 196)*

1. (a)

(−3, 2)

(4, −5)

(b) $7\sqrt{2}$
(c) $\left(\frac{1}{2}, -\frac{3}{2}\right)$

2. (a)

(−3.7, 0.7)

(1.3, −4.5)

(b) $\sqrt{52.04}$
(c) $(-1.2, -1.9)$

3. $520,000   **4.** 250 miles

5. Right isoceles triangle   **6.** Parallelogram

7.

Intercepts:
$(3, 0), (-3, 0), (0, 9)$
Symmetry: y-axis

8.

Intercepts: $(0, 0), (-4, 0)$

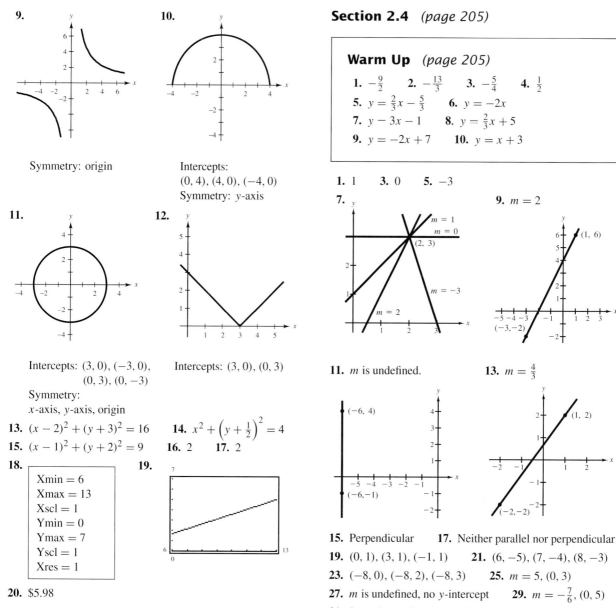

9.

Symmetry: origin

10.

Intercepts:
$(0, 4), (4, 0), (-4, 0)$
Symmetry: y-axis

11.

Intercepts: $(3, 0), (-3, 0)$,
$(0, 3), (0, -3)$
Symmetry:
x-axis, y-axis, origin

13. $(x - 2)^2 + (y + 3)^2 = 16$

15. $(x - 1)^2 + (y + 2)^2 = 9$

12.

Intercepts: $(3, 0), (0, 3)$

14. $x^2 + \left(y + \frac{1}{2}\right)^2 = 4$

16. 2   **17.** 2

18.

Xmin = 6
Xmax = 13
Xscl = 1
Ymin = 0
Ymax = 7
Yscl = 1
Xres = 1

20. $5.98

19.

Section 2.4 *(page 205)*

Warm Up *(page 205)*

1. $-\frac{9}{2}$ **2.** $-\frac{13}{3}$ **3.** $-\frac{5}{4}$ **4.** $\frac{1}{2}$

5. $y = \frac{2}{3}x - \frac{5}{3}$ **6.** $y = -2x$

7. $y - 3x - 1$ **8.** $y = \frac{2}{3}x + 5$

9. $y = -2x + 7$ **10.** $y = x + 3$

1. 1   **3.** 0   **5.** -3

7.

9. $m = 2$

11. m is undefined.

13. $m = \frac{4}{3}$

15. Perpendicular   **17.** Neither parallel nor perpendicular
19. $(0, 1), (3, 1), (-1, 1)$ **21.** $(6, -5), (7, -4), (8, -3)$
23. $(-8, 0), (-8, 2), (-8, 3)$ **25.** $m = 5, (0, 3)$
27. m is undefined, no y-intercept   **29.** $m = -\frac{7}{6}, (0, 5)$
31. $3x + 5y - 10 = 0$ **33.** $x + 2y - 3 = 0$
35. $x + 8 = 0$ **37.** $2x - 5y + 1 = 0$

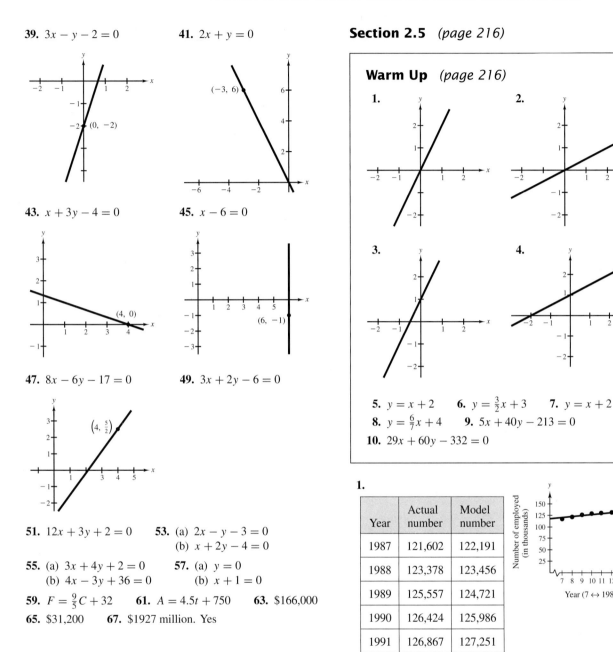

39. $3x - y - 2 = 0$ **41.** $2x + y = 0$

(−3, 6)

(0, −2)

43. $x + 3y - 4 = 0$ **45.** $x - 6 = 0$

(4, 0)

(6, −1)

47. $8x - 6y - 17 = 0$ **49.** $3x + 2y - 6 = 0$

$\left(4, \frac{5}{2}\right)$

51. $12x + 3y + 2 = 0$ **53.** (a) $2x - y - 3 = 0$
                                  (b) $x + 2y - 4 = 0$

55. (a) $3x + 4y + 2 = 0$ **57.** (a) $y = 0$
     (b) $4x - 3y + 36 = 0$ (b) $x + 1 = 0$

59. $F = \frac{9}{5}C + 32$ **61.** $A = 4.5t + 750$ **63.** \$166,000

65. \$31,200   **67.** \$1927 million. Yes

Section 2.5 *(page 216)*

Warm Up *(page 216)*

1.

2.

3.

4.

5. $y = x + 2$ **6.** $y = \frac{3}{2}x + 3$ **7.** $y = x + 2$
8. $y = \frac{6}{7}x + 4$ **9.** $5x + 40y - 213 = 0$
10. $29x + 60y - 332 = 0$

1.

Year	Actual number	Model number
1987	121,602	122,191
1988	123,378	123,456
1989	125,557	124,721
1990	126,424	125,986
1991	126,867	127,251
1992	128,548	128,516
1993	129,525	129,781

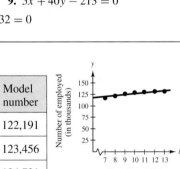

Number of employed (in thousands)

Year (7 ↔ 1987)

3. $y = \frac{12}{5}x$ **5.** $y = 205x$ **7.** $y = \frac{7}{48}x$

9. $I = 0.075P$ **11.** (a) $y = 0.0368x$ (b) $3128

13. (a) $C = \frac{33}{13}I$

(b)

Inches	5.00	10.00	20.00	25.00	30.00
Centimeters	12.69	25.38	50.77	63.46	76.15

15. $V = 125t + 2540$ **17.** $V = -2000t + 20,400$

19. $V = 12,500t + 154,000$ **21.** $V = 875 - 175t$

23. $S = 0.85L$ **25.** $W = 0.75x + 11.50$

27. (a) $h = 7000 - 20t$ (b) 2:13:50 P.M.

29. b; slope $= -10$; the amount owed *decreases* by $10 per week.

31. a; slope $= 0.25$; the amount received *increases* by $0.25 per mile driven.

33. (a) $N = 1200 + 50t$ (b) 2200 students

(c)

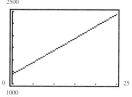

35. Yes   **37.** No

39. $y = 0.4x + 3$ (Answers may vary.)

41. $y = -0.5x + 3$ (Answers may vary.)

43. (a) $y = 57t - 214$

(b) $584,000; $641,000; $698,000

(c) Answers vary.

45. (a) $y = 1.1t + 124.4$

(b) 230 feet

(c) Answers vary.

47. (a) $C = 16.75t + 36,500$ (b) $R = 27t$

(c) $P = 10.25t - 36,500$ (d) 3561 hours

Review Exercises   *(page 223)*

1. 13   **3.** (a) 4 and 3, hypotenuse $= 5$ (b) 5

5. (a)

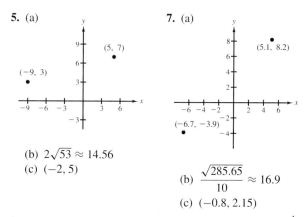

(b) $2\sqrt{53} \approx 14.56$

(c) $(-2, 5)$

7. (a)

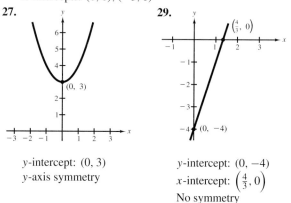

(b) $\dfrac{\sqrt{285.65}}{10} \approx 16.9$

(c) $(-0.8, 2.15)$

9. 725,000   **11.** Slope of line through $(1, 1)$ and $(8, 2)$ is $\frac{1}{7}$.
Slope of line through $(8, 2)$ and $(9, 5)$ is 3.
Slope of line through $(9, 5)$ and $(2, 4)$ is $\frac{1}{7}$.
Slope of line through $(2, 4)$ and $(1, 1)$ is 3.
Consequently, opposite sides are parallel.

13. -10 or 30   **15.** 3 or -7 **17.** (a) Yes   (b) Yes

19. y-intercept: $(0, -6)$
x-intercepts: $(-3, 0), (2, 0)$

21. y-intercept: $(0, 0)$ **23.** Origin   **25.** None
x-intercepts: $(0, 0), (-3, 0)$

27.

y-intercept: $(0, 3)$
y-axis symmetry

29.

y-intercept: $(0, -4)$
x-intercept: $\left(\frac{4}{3}, 0\right)$
No symmetry

31.

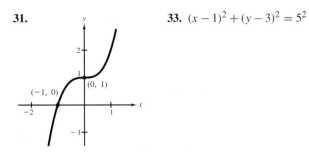

y-intercept: $(0, 1)$
x-intercept: $(-1, 0)$
No symmetry

33. $(x - 1)^2 + (y - 3)^2 = 5^2$

35. $(x - 1)^2 + \left(y + \frac{1}{2}\right)^2 = \frac{25}{4}$

37. $(x - 3)^2 + (y + 2)^2 = 4^2$

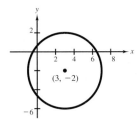

39. b   **40.** a   **41.** e   **42.** d   **43.** f   **44.** c

45.

47.

49. 3   **51.** 2   **53.** Circle

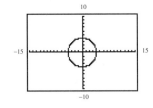

55. $\frac{11}{4}$ **57.** Parallel   **59.** Neither

61. Slope: $\frac{5}{4}$; y-intercept: $\left(0, \frac{11}{4}\right)$

63.

$8x - y - 17 = 0$

65.

$x - 3 = 0$

67.

$3x - 2y - 10 = 0$

69.

$2x + 3y - 6 = 0$

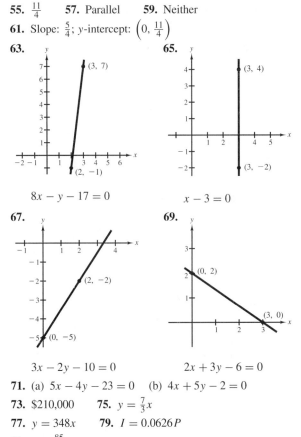

71. (a) $5x - 4y - 23 = 0$ (b) $4x + 5y - 2 = 0$

73. \$210,000   **75.** $y = \frac{7}{3}x$

77. $y = 348x$ **79.** $I = 0.0626P$

81. $y = \frac{85}{3}x$

Ounces	6	8	12	16	24
Grams	170	226.7	340	453.3	680

83. $V = 70 + 3.75t$; \$103.75   **85.** $r = 1.25x - 25$; \$1725

87. $y = 623x + 26{,}194$; 27,440

Chapter Test *(page 227)*

1. Distance: $4\sqrt{5}$ **2.** Distance: 7.81
Midpoint: $(1, 0)$ Midpoint: $(0.44, 4.34)$

3. x-intercepts: $(-5, 0), (3, 0)$ **4.** x-intercept: $(2, 0)$
y-intercept: $(0, -15)$ y-intercepts: none

5. Symmetric with respect to both axes and the origin

6. Symmetric with respect to the origin

7. $y - 4 = 0$ **8.** $3x - 4y - 24 = 0$

9. $4x + 3y - 11 = 0$ **10.** $x - 2 = 0$

11. $4x - 3y - 12 = 0$ **12.** (a)   **13.** (b)   **14.** (c)

15. (d)   **16.** $(x - 2)^2 + (y - 1)^2 = 9$ **17.** \$220,000

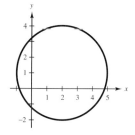

18. $W = 0.55x + 8.75$ **19.** $V = 25{,}000 - 4000x$

20. $S = L - 0.30L = 0.70L$

Cumulative Test: Chapters P–2
(page 228)

1. $-6\sqrt{x}$ **2.** $7^2 = 49$ **3.** $2x^2\sqrt{6x}$ **4.** $\dfrac{3 + \sqrt{5}}{4}$

5. $-4, 4$ **6.** $\dfrac{18}{7}$ **7.** $\pm 1, \pm 4$ **8.** $-1, 4$

9. 1   **10.** $x = \dfrac{-1 \pm \sqrt{41}}{4}$ **11.** \$11,663.24

12. $\frac{1}{15}$, or approximately 0.0667; $\frac{14}{15} \approx 0.9333$, probability of working properly

13. $-\frac{7}{2} \leq x \leq \frac{17}{2}$ **14.** $-6 \leq x \leq -2$

15. $7x + 4y - 13 = 0$ **16.** $2x - 3y - 7 = 0$

17. $y + 3 = 0$ **18.** $(x - 3)^2 + (y + 2)^2 = 16$

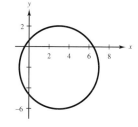

19. \$250,000

20. $x = 79{,}623.23$ or $x = 376.66$. Hence, revenue will equal cost when either 79,623 units or 377 units are sold.

CHAPTER 3
Section 3.1   *(page 238)*

Warm Up *(page 238)*

1. -73 **2.** 13   **3.** $2(x + 2)$ **4.** $-8(x - 2)$

5. $y = \frac{7}{5} - \frac{2}{5}x$ **6.** $y = \pm x$ **7.** $x \leq -2, x \geq 2$

8. $-3 \leq x \leq 3$ **9.** All real numbers

10. $x \leq 1, x \geq 2$

1. (a) -6 (b) 34   (c) $6 - 4t$ (d) $2 - 4c$

3. (a) -1 (b) $\dfrac{1}{15}$ (c) $\dfrac{1}{t^2 - 2t}$ (d) $\dfrac{1}{t^2 - 1}$

5. (a) -1 (b) -9 (c) $2x - 5$ (d) $-\frac{5}{2}$

7. (a) 0   (b) 3   (c) $x^2 + 2x$ (d) -0.75

9. (a) 1   (b) -7 (c) $3 - 2|x|$ (d) 2.5

11. (a) $\dfrac{1}{7}$ (b) $-\dfrac{1}{9}$ (c) Undefined   (d) $\dfrac{1}{y^2 + 6y}$

13. (a) 1   (b) -1 (c) 1   (d) $\dfrac{|x - 1|}{x - 1}$

15. (a) -1 (b) 2   (c) 4   (d) 6

17. 5   **19.** ± 3 **21.** $\frac{10}{7}$

23. All real numbers   **25.** All real numbers except $t = 0$

27. $y \geq 10$ **29.** $-1 \leq x \leq 1$

31. All real numbers except $x = 0, -2$ **33.** Not a function

35. Function   **37.** Function   **39.** Not a function

41. Function   **43.** Function   **45.** Function

47. Not a function. The relationship assigns two elements of B to the element c in A.

49. Not a function. The relationship defines a function from B to A.

51. This is a function from A to B, because each element in A is matched with an element in B.

53. This is not a function, because one element of A is not matched with an element of B.

55. $(-2, 4), (-1, 1), (0, 0), (1, 1), (2, 4)$

57. $(-2, 0), (-1, 1), (0, \sqrt{2}), (1, \sqrt{3}), (2, 2)$

59. (a) $V = 4x(6 - x)^2$ (b) Domain: $0 < x < 6$

61. $h = \sqrt{d^2 - 2000^2}$,   Domain: $d \geq 2000$

63. (a) $A = 2000(1.015)^{4t}$ (b) Domain: $t \geq 0$

65. (a) $C = 12.30x + 98,000$ (b) $R = 17.98x$
    (c) $P = 5.68x - 98,000$

67. $C = 6000 + 0.95x$ (b) $\overline{C} = \dfrac{6000}{x} + 0.95$

69. (a) $R = 12.00n - 0.05n^2, n \geq 80$
    (b)

n	90	100	110	120	130	140	150
R	\$675	\$700	\$715	\$720	\$715	\$700	\$675

    (c) Maximum profit of \$720 occurs when a group of 120 charter a bus.

Section 3.2   *(page 250)*

Warm Up   *(page 250)*

1. 2   **2.** 0   **3.** $-\dfrac{3}{x}$ **4.** $x^2 + 3$ **5.** $0, \pm 4$

6. $\frac{1}{2}, 1$ **7.** All real numbers except $x = 4$

8. All real numbers except $x = 4, 5$

9. $t \leq \frac{5}{3}$ **10.** All real numbers

1. Range: $[0, \infty)$ **3.** Range: $[0, \infty)$ **5.** Range: $[0, 5]$

7. Range: $(-\infty, \infty)$ **9.** Function   **11.** Not a function

13. Function   **15.** Increasing on $(-\infty, \infty)$

17. Increasing on $(-\infty, 0)$, $(2, \infty)$, decreasing on $(0, 2)$

19. Increasing on $(-1, 0)$, $(1, \infty)$, decreasing on $(-\infty, -1)$, $(0, 1)$

21. Increasing on $(-2, \infty)$, decreasing on $(-3, -2)$

23. Even   **25.** Odd   **27.** Neither even nor odd

29. Even   **31.** Neither even nor odd

33. Odd

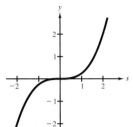

35. Neither even nor odd

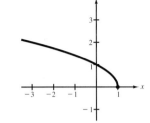

37. Neither even nor odd

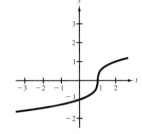

39. Neither even nor odd

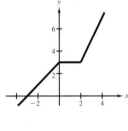

41.

43.

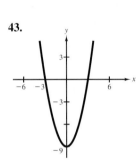

45.

47.

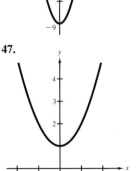

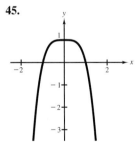

49.

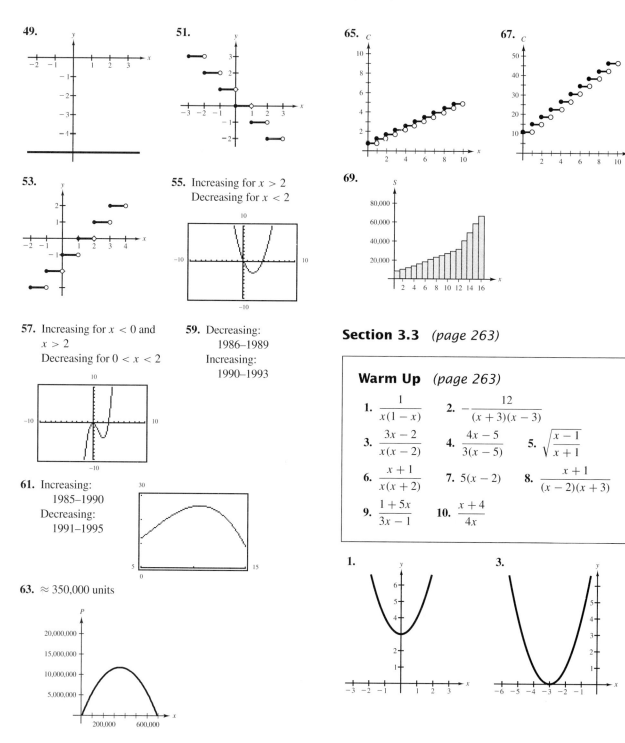

51.

65.

67.

53.

55. Increasing for $x > 2$
Decreasing for $x < 2$

69.

57. Increasing for $x < 0$ and
$x > 2$
Decreasing for $0 < x < 2$

59. Decreasing:
1986–1989
Increasing:
1990–1993

Section 3.3 *(page 263)*

61. Increasing:
1985–1990
Decreasing:
1991–1995

Warm Up *(page 263)*

1. $\dfrac{1}{x(1 - x)}$ **2.** $-\dfrac{12}{(x + 3)(x - 3)}$

3. $\dfrac{3x - 2}{x(x - 2)}$ **4.** $\dfrac{4x - 5}{3(x - 5)}$ **5.** $\sqrt{\dfrac{x - 1}{x + 1}}$

6. $\dfrac{x + 1}{x(x + 2)}$ **7.** $5(x - 2)$ **8.** $\dfrac{x + 1}{(x - 2)(x + 3)}$

9. $\dfrac{1 + 5x}{3x - 1}$ **10.** $\dfrac{x + 4}{4x}$

63. $\approx 350{,}000$ units

1.

3.

5.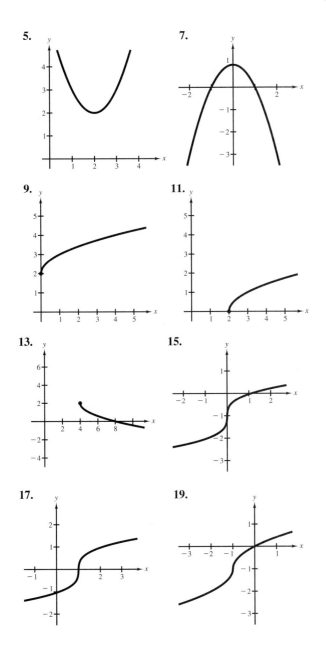

7.

9.

11.

13.

15.

17.

19.

21. (a) Vertical shift of 2 units

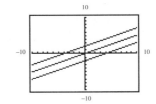

(b) Horizontal shift of 2 units

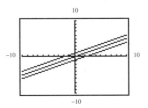

(c) Slope of the function changes

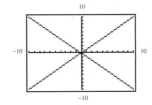

23. (a) $g(x) = (x - 1)^2 + 1$
(b) $g(x) = -(x + 1)^2$

25. $2x; 2; x^2 - 1; \dfrac{x + 1}{x - 1}, x \neq 1$

27. $x^2 - x + 1; x^2 + x - 1; x^2 - x^3; \dfrac{x^2}{1 - x}, x \neq 1$

29. $x^2 + 5 + \sqrt{1 - x}; x^2 + 5 - \sqrt{1 - x};$
$(x^2 + 5)\sqrt{1 - x}; \dfrac{x^2 + 5}{\sqrt{1 - x}}, x < 1$

31. $\dfrac{x + 1}{x^2}; \dfrac{x - 1}{x^2}; \dfrac{1}{x^3}; x, x \neq 0$ **33.** 9

35. $4t^2 - 2t + 5$ **37.** 0   **39.** 26   **41.** 5   **43.** $\frac{3}{5}$

45. (a) $(x - 1)^2$ **47.** (a) $20 - 3x$
(b) x^4 (b) $9x + 20$

49. (a) $\sqrt{x^2 + 4}$ **51.** (a) $x - \frac{8}{3}$ **53.** (a) $\sqrt[4]{x}$
(b) $x + 4, x \geq 4$ (b) $x - 8$ (b) $\sqrt[4]{x}$

55. (a) $|x + 6|$ **57.** (a) 3   **59.** (a) 0
(b) $|x| + 6$ (b) 0   (b) 4

61. (a) $x \geq 0$ (b) All real numbers   (c) All real numbers

63. (a) All real numbers except $x = \pm 1$
   (b) All real numbers
   (c) All real numbers except $x = -2, 0$

65. $P = 30.98 + 0.13t$. Increasing

67. (a) $N(T(t)) = 90t^2 + 590$ (b) ≈ 3.18 hours

69.

Year	P	E	P/E
1985	$ 7.21	$0.55	13.1
1986	$10.42	$0.62	16.8
1987	$12.74	$0.72	17.7
1988	$11.35	$0.86	13.2
1989	$14.41	$0.98	14.7
1990	$15.51	$1.10	14.1
1991	$16.76	$1.18	14.2
1992	$22.10	$1.30	17.0
1993	$26.13	$1.46	17.9

71. $(A \circ r)(t) = \pi(0.6t)^2 = 0.36\pi t^2$
   $A \circ r$ represents the area of the circle at time t.

73. $(C \circ x)(t) = 2800t + 375$
   The cost of production at t hours.

Section 3.4   *(page 274)*

Warm Up   *(page 274)*

1. All real numbers   **2.** $[-1, \infty)$
3. All real numbers except $x = 0, 2$
4. All real numbers except $x = -\frac{5}{3}$ **5.** x **6.** x
7. x **8.** x **9.** $x = \frac{3}{2}y + 3$ **10.** $x = \frac{y^3}{2} + 2$

1.
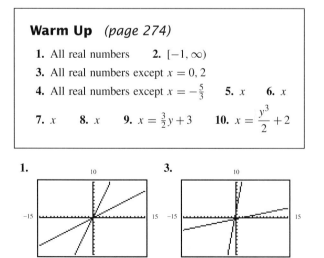

3.

5.
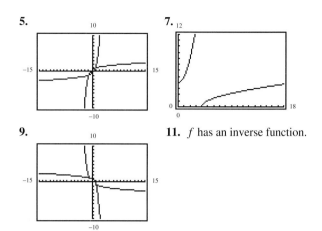

7.

9.

11. f has an inverse function.

13. f doesn't have an inverse function.

15. g has an inverse function.

17. h doesn't have an inverse function.

19. f doesn't have an inverse function.

21. $f^{-1}(x) = \dfrac{x + 3}{2}$ **23.** $f^{-1}(x) = \sqrt[5]{x}$

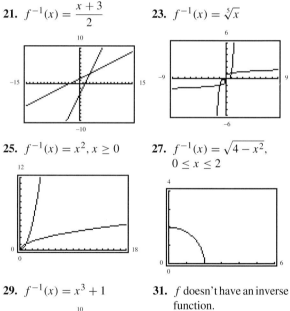

25. $f^{-1}(x) = x^2, x \geq 0$ **27.** $f^{-1}(x) = \sqrt{4 - x^2},$
   $0 \leq x \leq 2$

29. $f^{-1}(x) = x^3 + 1$ **31.** f doesn't have an inverse
   function.

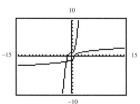

33. $g^{-1}(x) = 8x$ **35.** f doesn't have an inverse function.

37. $f^{-1}(x) = \sqrt{x} - 3, \, x \geq 0$ **39.** $h^{-1}(x) = \dfrac{1}{x}$

41. $f^{-1}(x) = \dfrac{x^2 - 3}{2}, \, x \geq 0$ **43.** g doesn't have an inverse function.

45. $f^{-1}(x) = -\sqrt{25 - x}, \, x \leq 25$

47.

x	0	1	2	3	4
$f^{-1}(x)$	-2	0	1	2	4

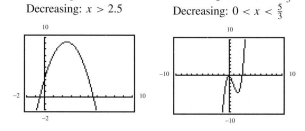

49. 32   **51.** 600   **53.** $(g^{-1} \circ f^{-1})(x) = \dfrac{x+1}{2}$

55. $(f \circ g)^{-1}(x) = \dfrac{x+1}{2}$ **57.** $t = \dfrac{10.88 - y}{0.3}, \, 1986\text{--}87$

59. $t = \sqrt{\dfrac{y - 2615}{30.895}}, \, 1989$ **61.** \$0.59

Mid-Chapter Quiz   *(page 277)*

1. 16   **2.** 4   **3.** 1   **4.** $2a^2 - a + 1$

5. $-2, 0, 2$ **6.** $\dfrac{4}{3}$ **7.** $P = 5.49x - 85{,}000$

8. All real numbers except 1 and 0   **9.** $x \leq 5$

10. Increasing: $x < 2.5$
Decreasing: $x > 2.5$

11. Increasing: $x < 0, \, x > \dfrac{5}{3}$
Decreasing: $0 < x < \dfrac{5}{3}$

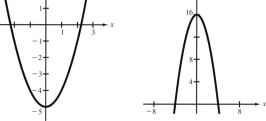

12. Increasing: $x > 0$
Decreasing: $x < 0$

13. Increasing: $x > -1$

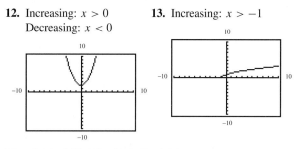

14. $g(x)$ is shifted 2 units to the right.

15. $g(x)$ is shifted 1 unit to the left.

16. $(f \circ g)(x) = (x - 1)^2$ **17.** $(f \circ g)(x) = \dfrac{1}{x^3}$

18. The graphs are reflections of each other in the line $y = x$.

19. The graphs are reflections of each other in the line $y = x$.

20. $t = \dfrac{\sqrt{y} - 96.5}{3.75}, \, 1988$

Section 3.5   *(page 285)*

Warm Up   *(page 285)*

1. $\frac{1}{2}, -6$ **2.** $-\frac{3}{5}, 3$ **3.** $\frac{3}{2}, -1$ **4.** -10

5. $3 \pm \sqrt{5}$ **6.** $-2 \pm \sqrt{3}$ **7.** $4 \pm \dfrac{\sqrt{14}}{2}$

8. $-5 \pm \dfrac{\sqrt{3}}{3}$ **9.** $-\dfrac{3}{2} \pm \dfrac{\sqrt{5}}{2}$ **10.** $-\dfrac{3}{2} \pm \dfrac{\sqrt{21}}{2}$

1. g   **2.** e   **3.** c   **4.** f   **5.** b   **6.** a

7. h   **8.** d   **9.** $f(x) = (x - 2)^2$

11. $f(x) = -(x + 2)^2 + 4$ **13.** $f(x) = -2(x + 3)^2 + 3$

15. Intercepts:
$(\pm\sqrt{5}, 0), (0, -5)$
Vertex: $(0, -5)$

17. Intercepts:
$(\pm 4, 0), (0, 16)$
Vertex: $(0, 16)$

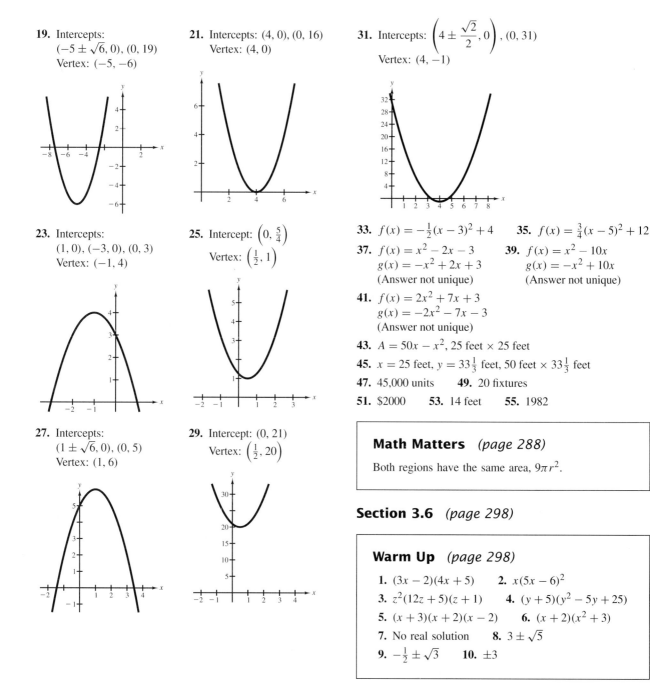

19. Intercepts:
$(-5 \pm \sqrt{6}, 0), (0, 19)$
Vertex: $(-5, -6)$

21. Intercepts: $(4, 0), (0, 16)$
Vertex: $(4, 0)$

31. Intercepts: $\left(4 \pm \dfrac{\sqrt{2}}{2}, 0\right), (0, 31)$
Vertex: $(4, -1)$

23. Intercepts:
$(1, 0), (-3, 0), (0, 3)$
Vertex: $(-1, 4)$

25. Intercept: $\left(0, \dfrac{5}{4}\right)$
Vertex: $\left(\dfrac{1}{2}, 1\right)$

33. $f(x) = -\dfrac{1}{2}(x - 3)^2 + 4$ **35.** $f(x) = \dfrac{3}{4}(x - 5)^2 + 12$

37. $f(x) = x^2 - 2x - 3$
$g(x) = -x^2 + 2x + 3$
(Answer not unique)

39. $f(x) = x^2 - 10x$
$g(x) = -x^2 + 10x$
(Answer not unique)

41. $f(x) = 2x^2 + 7x + 3$
$g(x) = -2x^2 - 7x - 3$
(Answer not unique)

43. $A = 50x - x^2$, 25 feet × 25 feet

45. $x = 25$ feet, $y = 33\dfrac{1}{3}$ feet, 50 feet × $33\dfrac{1}{3}$ feet

47. 45,000 units   **49.** 20 fixtures

51. $2000   **53.** 14 feet   **55.** 1982

27. Intercepts:
$(1 \pm \sqrt{6}, 0), (0, 5)$
Vertex: $(1, 6)$

29. Intercept: $(0, 21)$
Vertex: $\left(\dfrac{1}{2}, 20\right)$

Math Matters *(page 288)*

Both regions have the same area, $9\pi r^2$.

Section 3.6 *(page 298)*

Warm Up *(page 298)*

1. $(3x - 2)(4x + 5)$ **2.** $x(5x - 6)^2$

3. $z^2(12z + 5)(z + 1)$ **4.** $(y + 5)(y^2 - 5y + 25)$

5. $(x + 3)(x + 2)(x - 2)$ **6.** $(x + 2)(x^2 + 3)$

7. No real solution   **8.** $3 \pm \sqrt{5}$

9. $-\dfrac{1}{2} \pm \sqrt{3}$ **10.** ± 3

---

1. e   **2.** c   **3.** g   **4.** d   **5.** f   **6.** h

7. a   **8.** b   **9.** Rises to the left
Rises to the right

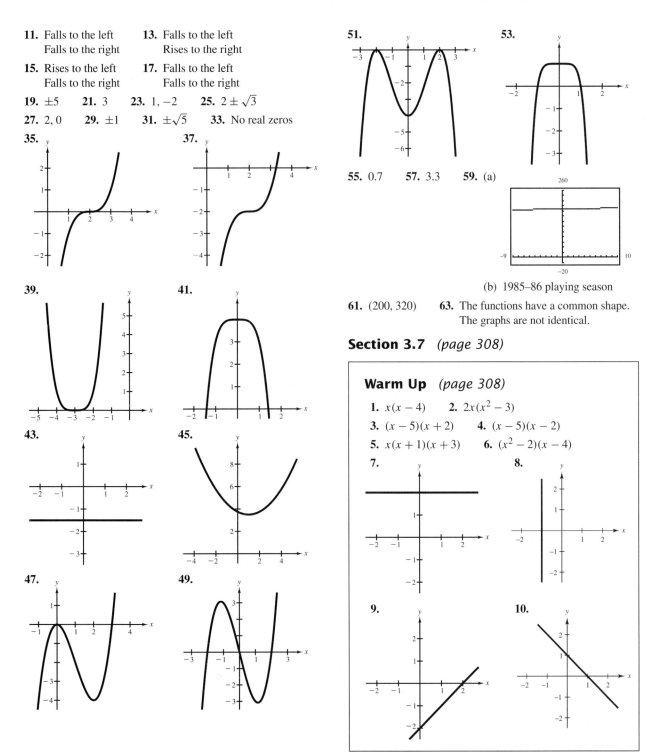

11. Falls to the left   **13.** Falls to the left
Falls to the right   Rises to the right

15. Rises to the left   **17.** Falls to the left
Falls to the right   Falls to the right

19. ± 5 **21.** 3   **23.** 1, -2 **25.** $2 \pm \sqrt{3}$

27. 2, 0   **29.** ± 1 **31.** $\pm \sqrt{5}$ **33.** No real zeros

35.

37.

39.

41.

43.

45.

47.

49.

51.

53.

55. 0.7   **57.** 3.3   **59.** (a)

(b) 1985–86 playing season

61. (200, 320)   **63.** The functions have a common shape.
The graphs are not identical.

Section 3.7   (page 308)

Warm Up   (page 308)

1. $x(x - 4)$ **2.** $2x(x^2 - 3)$

3. $(x - 5)(x + 2)$ **4.** $(x - 5)(x - 2)$

5. $x(x + 1)(x + 3)$ **6.** $(x^2 - 2)(x - 4)$

7.

8.

9.

10.

1. Domain: all $x \neq 0$ **3.** Domain: all $x \neq 2$
5. Domain: all reals   **7.** Domain: all reals
9. f   **10.** e   **11.** a   **12.** b   **13.** c   **14.** d
15. $g(x)$ shifts up 1 unit.   **17.** $g(x)$ is a reflection in the x-axis.

19. $g(x)$ shifts down 2 units.   **21.** $g(x)$ is a reflection in the x-axis.

23. $g(x)$ shifts up 1 unit.   **25.** $g(x)$ is a reflection in the x-axis.

27.   **29.**

31.   **33.**

35.   **37.**

39.   **41.**

43.   **45.**

47.   **49.**

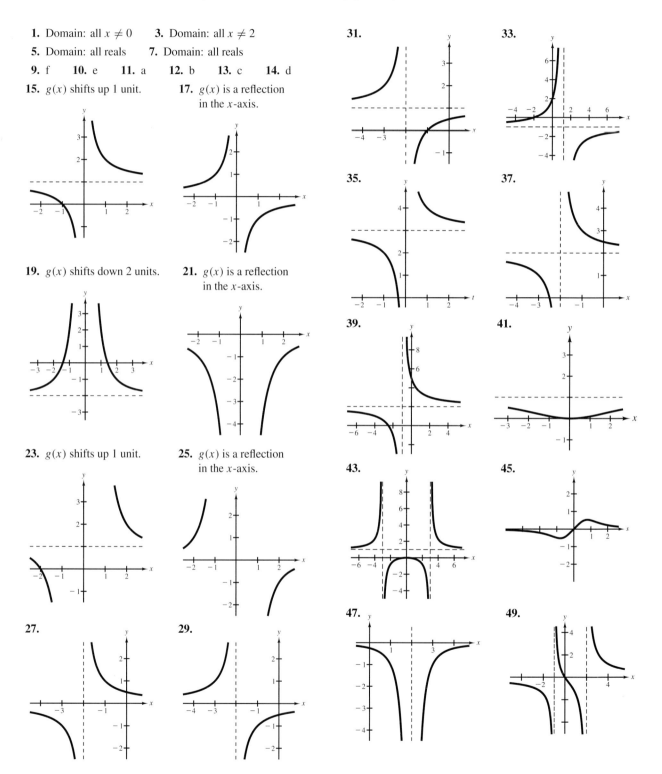

51. (a) $176 million
   (b) $528 million
   (c) $1584 million
   (d) No

53. (a) 167, 250, 400
   (b) 750

55. (a)

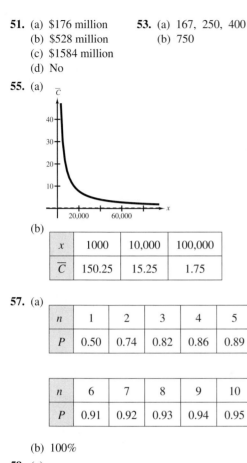

(b)

x	1000	10,000	100,000
$\overline{C}$	150.25	15.25	1.75

57. (a)

n	1	2	3	4	5
P	0.50	0.74	0.82	0.86	0.89

n	6	7	8	9	10
P	0.91	0.92	0.93	0.94	0.95

(b) 100%

59. (a)

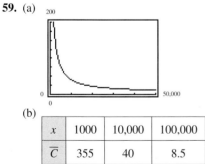

(b)

x	1000	10,000	100,000
$\overline{C}$	355	40	8.5

61. The sale of long-playing albums dropped after cassette tapes and compact discs were introduced.

63. (a) ≈ 1.56 hours

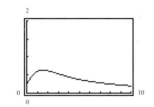

(b) Horizontal asymptote: $y = 0$

Review Exercises   *(page 314)*

1. (a) -2 (b) -11 (c) $3m - 5$ (d) $3x - 2$

3. (a) 5   (b) 8   (c) $5\frac{1}{2}$ (d) $x^2 + 5$

5. $-\frac{7}{2}$ **7.** All real numbers   **9.** $x \geq -9$

11. $t \geq 3$ and $t \neq 5$ **13.** y is a function of x.

15. Function. Every element in A is assigned to an element in B.

17. (a) $V = x(20 - 2x)(20 - 2x) = 4x(10 - x)^2$
   (b) Domain: $0 < x < 10$
   (c)

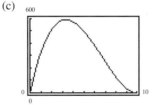

19. (a) Domain: all real numbers   (b) Decreasing: $x < 0$
      Range: $y \geq 1$ Increasing: $x > 0$
   (c) Even

21. (a) Domain: all real numbers
      Range: all real numbers
   (b) Decreasing: $0 < x < \frac{8}{3}$
      Increasing: $x < 0$ or $x > \frac{8}{3}$
   (c) Neither

23.   **25.**

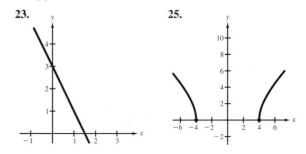

27.

29.

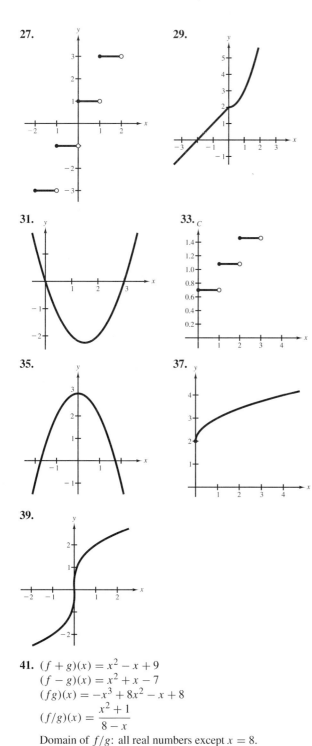

31.

33.

35.

37.

39.

41. $(f + g)(x) = x^2 - x + 9$
$(f - g)(x) = x^2 + x - 7$
$(fg)(x) = -x^3 + 8x^2 - x + 8$
$(f/g)(x) = \dfrac{x^2 + 1}{8 - x}$
Domain of f/g: all real numbers except $x = 8$.

43. $(f + g)(x) = x^2 - x + 2$
$(f - g)(x) = -x^2 + 7x + 2$
$(fg)(x) = 3x^3 - 10x^2 - 8x$
$(f/g)(x) = \dfrac{3x + 2}{x^2 - 4x}$
Domain of f/g: all real numbers except $x = 0$ and $x = 4$.

45. -1 **47.** $-\frac{1}{2}$ **49.** (a) $2x^2 - 1$
                                    (b) $4x^2 - 20x + 27$

51. (a) $\dfrac{1}{3x + x^2}$ (b) $\dfrac{3}{x} + \dfrac{1}{x^2}$

53. $(C \circ x)(t) = 4500t + 800$
$(C \circ x)(t)$ is the daily cost in terms of hours t.

55. $f(g(x)) = \sqrt[3]{(x^3 + 3) - 3}$
$\qquad = x$
$g(f(x)) = (\sqrt[3]{x - 3})^3 + 3$
$\qquad = x$

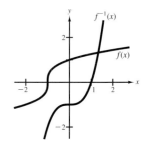

57. $f(x)$ has an inverse.
$f^{-1}(x) = x^3 - 1$

59. f doesn't have an inverse.

61. f doesn't have an inverse.

63.

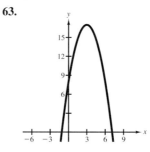

y-intercept: $(0, 8)$
x-intercepts: $(3 \pm \sqrt{17}, 0)$
Vertex: $(3, 17)$

65.

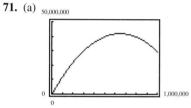

y-intercept: $\left(0, -\frac{4}{3}\right)$

x-intercepts: $\left(-\frac{5 \pm \sqrt{41}}{2}, 0\right)$

Vertex: $\left(-\frac{5}{2}, -\frac{41}{12}\right)$

67. $f(x) = \frac{1}{3}(x^2 - 4x + 13)$ **69.** 40,000 units

71. (a)

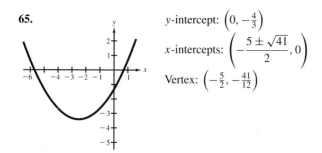

(b) 650,000 units

(c) Write the equation of the parabola in standard form. The vertex is the point of maximum profit.

73. Falls to the left **75.** Rises to the left
Falls to the right   Rises to the right

77. $\pm 2, \pm \sqrt{2}$ **79.** $6, \pm \sqrt{3}$ **81.** 0.4

83. Domain: all real numbers
Horizontal asymptote: $y = 2$
Vertical asymptote: none

85. Domain: all real numbers except $x = 3$
Vertical asymptote: $x = 3$
Horizontal asymptote: $y = 0$ (x-axis)

87. Vertical asymptote: **89.** Vertical asymptote:
    $x = 1$ y-axis
Horizontal asymptote: Horizontal asymptote:
    $y = 0$ x-axis

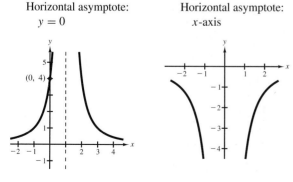

91. (a)

t	5	10	25
N (in thousands)	304	453.3	702.2

(b) 1200 thousand

93. (a) $\approx \$31,666.67$ (b) \$142,500 (c) \$9,405,000
(d) No. The model is not defined for $p = 100\%$.

Chapter Test *(page 318)*

1. $(-\infty, \infty)$ **2.** $[7, \infty)$ **3.** $[2, 7) \cup (7, \infty)$

4. $(-\infty, -2) \cup (-2, 2) \cup (2, \infty)$ **5.** T **6.** F **7.** F

8. (a) Domain: $(-\infty, \infty)$ (b) Increasing: $(0, \infty)$
    Range: $[2, \infty)$ Decreasing: $(-\infty, 0)$
(c) Even

9. (a) Domain: $(-\infty, -2] \cup [2, \infty)$
    Range: $[0, \infty)$
(b) Decreasing: $(-\infty, 2)$
    Increasing: $(2, \infty)$
(c) Even

10.

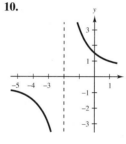

11.

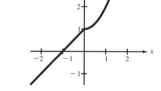

12.

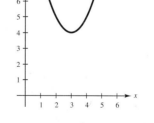

13. $f^{-1} = \sqrt[3]{x + 5}$

14. $(f + g)(x) = x^2 + 2x + 1$ **15.** $(f - g)(x) = x^2 - 2x + 3$

16. $(fg)(x) = 2x^3 - x^2 + 4x - 2$

17. $\left(\dfrac{f}{g}\right)(x) = \dfrac{x^2 + 2}{2x - 1}$ **18.** $f \circ g(x) = 4x^2 - 4x + 3$

19. $g \circ f(x) = 2x^2 + 3$ **20.** 2500 units

CHAPTER 4

Section 4.1 *(page 328)*

Warm Up *(page 328)*

1. $x^3 - x^2 + 2x + 3$ **2.** $2x^3 + 4x^2 - 6x - 4$

3. $x^4 - 2x^3 + 4x^2 - 2x - 7$

4. $2x^4 + 12x^3 - 3x^2 - 18x - 5$

5. $(x - 3)(x - 1)$ **6.** $2x(2x - 3)(x - 1)$

7. $x^3 - 7x^2 + 12x$ **8.** $x^2 + 5x - 6$

9. $x^3 + x^2 - 7x - 3$ **10.** $x^4 - 3x^3 - 5x^2 + 9x - 2$

---

1. $2x + 4$ **3.** $x^2 - 3x + 1$ **5.** $x^3 + 3x^2 - 1$

7. $7 - \dfrac{11}{x + 2}$ **9.** $3x + 5 - \dfrac{2x - 3}{2x^2 + 1}$

11. $x^2 - 3x + 2$ **13.** $x^2 + 2x + 4 + \dfrac{2x - 11}{x^2 - 2x + 3}$

15. $3x^2 - 2x + 5$ **17.** $4x^2 - 9$

19. $-x^2 + 10x - 25$ **21.** $5x^2 + 14x + 56 + \dfrac{232}{x - 4}$

23. $10x^3 + 10x^2 + 60x + 360 + \dfrac{1360}{x - 6}$

25. $x^2 - 8x + 64$ **27.** $-3x^3 - 6x^2 - 12x - 24 - \dfrac{48}{x - 2}$

29. $-x^2 + 3x - 6 + \dfrac{11}{x + 1}$ **31.** $4x^2 + 14x - 30$

33. $(x - 2)(x + 3)(x - 1)$ **35.** $(2x - 1)(x - 5)(x - 2)$

37. $(x + 2)(x + \sqrt{3})(x - \sqrt{3})$

39. $(x - 4)(x^2 + 3x - 2) + 3, \ f(4) = 3$

41. $(x - \sqrt{2})[x^2 + (3 + \sqrt{2})x + 3\sqrt{2}] - 8, \ f(\sqrt{2}) = -8$

43. (a) 1   **45.** (a) 97   **47.** (a) 72
     (b) 4   (b) $-\frac{5}{3}$ (b) 0
     (c) 4   (c) 17   (c) 37.648
     (d) 1954   (d) -199 (d) 30

49. $b, 3, \dfrac{-1 \pm \sqrt{17}}{2}$ **51.** $a, -1, -2 \pm \sqrt{2}$

53. $2x^2 - x - 1$ **55.** $x^2 + 2x - 3$ **57.** $x^2 + 3x$

59. $x^2 + 10x + 24$ **61.** \$192,116

Section 4.2 *(page 337)*

Warm Up *(page 337)*

1. $f(x) = 3x^3 - 8x^2 - 5x + 6$

2. $f(x) = 4x^4 - 3x^3 - 16x^2 + 12x$

3. $x^4 - 3x^3 + 5 + \dfrac{3}{x + 3}$

4. $3x^3 + 15x^2 - 9 - \dfrac{2}{x + (2/3)}$

5. $\frac{1}{2}, -3 \pm \sqrt{5}$ **6.** $10, -\frac{2}{3}, -\frac{3}{2}$ **7.** $-\frac{3}{4}, 2 \pm \sqrt{2}$

8. $\frac{2}{5}, -\frac{7}{2}, -2$ **9.** $\pm\sqrt{2}, \pm 1$ **10.** $\pm 2, \pm\sqrt{3}$

---

1. One negative zero   **3.** No real zeros

5. One positive zero   **7.** Three or one positive zeros

9. $x = 0$ is the only real zero.   **11.** Possible: $\pm 1, \pm 2, \pm 4$
                                   Actual: $-1, \pm 2$

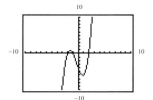

13. Possible: $\pm 1, \pm 3, \pm\frac{1}{2}, \pm\frac{3}{2}, \pm\frac{1}{4}, \pm\frac{3}{4}$
     Actual: $-\frac{1}{4}, 1, 3$

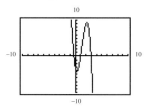

15. Possible: $\pm\frac{1}{4}, \pm\frac{1}{2}, \pm 1, \pm 2, \pm 4$ **17.** (a) Upper bound
     Actual: $\pm\frac{1}{2}, \pm 2$ (b) Lower bound
                                         (c) Neither

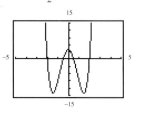

19. (a) Neither   **21.** $1, 2, 3$ **23.** $1, -1, 4$
   (b) Lower bound
   (c) Upper bound

25. $-1, -10$ **27.** $1, 2$ **29.** $\frac{1}{2}, -1$ **31.** $1, -\frac{1}{2}$

33. $-\frac{3}{4}$ **35.** $\pm 1, \pm\sqrt{2}$ **37.** $-1, 2$

39. $0, -1, -3, 4$ **41.** $-2, 4, -\frac{1}{2}$ **43.** $0, 3, 4, \pm\sqrt{2}$

45. (a) $\pm 1, \pm 3, \pm\frac{1}{2} \pm\frac{3}{2}, \pm\frac{1}{4}, \pm\frac{3}{4}, \pm\frac{1}{8}, \pm\frac{3}{8}, \pm\frac{1}{16}, \pm\frac{3}{16},$
     $\pm\frac{1}{32}, \pm\frac{3}{32}$

  (b)

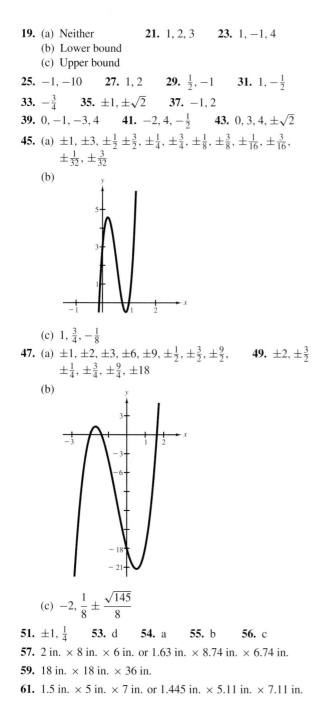

  (c) $1, \frac{3}{4}, -\frac{1}{8}$

47. (a) $\pm 1, \pm 2, \pm 3, \pm 6, \pm 9, \pm\frac{1}{2}, \pm\frac{3}{2}, \pm\frac{9}{2},$ **49.** $\pm 2, \pm\frac{3}{2}$
    $\pm\frac{1}{4}, \pm\frac{3}{4}, \pm\frac{9}{4}, \pm 18$

  (b)

  (c) $-2, \frac{1}{8} \pm \dfrac{\sqrt{145}}{8}$

51. $\pm 1, \frac{1}{4}$ **53.** d   **54.** a   **55.** b   **56.** c

57. 2 in. $\times$ 8 in. $\times$ 6 in. or 1.63 in. $\times$ 8.74 in. $\times$ 6.74 in.

59. 18 in. $\times$ 18 in. $\times$ 36 in.

61. 1.5 in. $\times$ 5 in. $\times$ 7 in. or 1.445 in. $\times$ 5.11 in. $\times$ 7.11 in.

Section 4.3 *(page 345)*

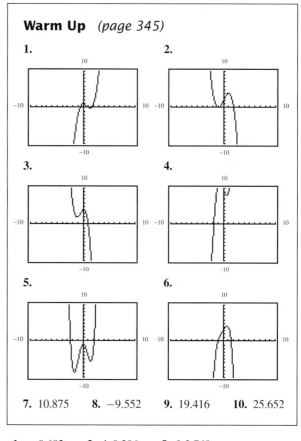

Warm Up *(page 345)*

1.

2.

3.

4.

5.

6.

7. 10.875   **8.** -9.552 **9.** 19.416   **10.** 25.652

1. e, 0.682   **3.** d, 0.206   **5.** f, 2.769

7. $-1.164, 1.453$ **9.** $-4.596, -1.042, 5.638$

11. 0.529   **13.** 3.533   **15.** $-1.453, 1.164$ **17.** 0.246

19. $-2.177, 1.563$ **21.** 1991   **23.** 4.5 hours

25. \$38.4 or \$46.1 (in tens of thousands of dollars)

Mid-Chapter Quiz *(page 347)*

1. $(x + 2)(x - 3)(2x + 1)$ **2.** $(x + 4)(x - 5)(3x - 2)$

3. 13   **4.** -9.8 **5.** $2x^2 + 5x - 12$

6. $P = 2{,}534{,}375$, $x \approx 33.76$ **7.** $\sqrt{5}, -\sqrt{5}, -\frac{7}{2}$

8. $\pm 3, \pm\frac{1}{2}$ **9.** $1, -\frac{4}{3}$ **10.** $\frac{3}{2}$ **11.** $\pm 1, \pm 2$

12. $1, \dfrac{1 \pm \sqrt{13}}{6}$ **13.** $1, -2$ **14.** $1, \dfrac{1 \pm \sqrt{11}}{5}$

15. Possible: 0 or 2 positive, exactly 1 negative
Actual: 2 positive, 1 negative

16. Possible: 0 or 2 positive, 0 or 2 negative
Actual: 2 positive, 2 negative

17. $160 = x(12 - 2x)(14 - 2x)$

18. 2 in. × 8 in. × 10 in. or
$\left(\dfrac{11 - \sqrt{41}}{2}\right)$ in. × $\left(1 + \sqrt{41}\right)$ in. × $\left(3 + \sqrt{41}\right)$ in.

19. 2 in. × 8 in. × 10 in. or 2.298 in. × 7.404 in. × 9.404 in.

20. $\dfrac{11 + \sqrt{41}}{2} \approx 8.702$ results in $l = -5.404$ and
$w = -3.404$

Section 4.4 (page 356)

> **Warm Up** (page 356)
>
> **1.** $2\sqrt{3}$ **2.** $10\sqrt{5}$ **3.** $\sqrt{5}$ **4.** $-6\sqrt{3}$
>
> **5.** 12 **6.** 48 **7.** $\dfrac{\sqrt{3}}{3}$ **8.** $\sqrt{2}$
>
> **9.** $-\dfrac{1}{2} \pm \dfrac{\sqrt{5}}{2}$ **10.** $-1 \pm \sqrt{2}$

1. $i, -1, -i, 1, i, -1, -i, 1,$ **3.** $a = -10, b = 6$
$i, -1, -i, 1, i, -1, -i, 1$

5. $a = 6, b = 5$ **7.** $4 + 3i, 4 - 3i$

9. $2 - 3\sqrt{3}\,i, 2 + 3\sqrt{3}\,i$ **11.** $5\sqrt{3}\,i, -5\sqrt{3}\,i$

13. $-1 - 6i, -1 + 6i$ **15.** $-5i, 5i$ **17.** 8, 8

19. $11 - i$ **21.** 4 **23.** $3 - 3\sqrt{2}\,i$ **25.** $\dfrac{1}{6} + \dfrac{7}{6}i$

27. $-2\sqrt{3}$ **29.** -10 **31.** $5 + i$ **33.** 41

35. $12 + 30i$ **37.** $-40 + 16i$ **39.** 24

41. $-9 + 40i$ **43.** $\dfrac{16}{41} + \dfrac{20}{41}i$ **45.** $\dfrac{3}{5} + \dfrac{4}{5}i$

47. $-7 - 6i$ **49.** $\dfrac{1}{8}i$ **51.** $-\dfrac{5}{4} - \dfrac{5}{4}i$ **53.** $\dfrac{35}{29} + \dfrac{595}{29}i$

55. -10 **57.** $1 \pm i$ **59.** $-2 \pm \dfrac{1}{2}i$ **61.** $-\dfrac{3}{2}, -\dfrac{5}{2}$

63. $\dfrac{1}{8} \pm \dfrac{\sqrt{11}}{8}i$ **65.**

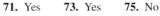

67.

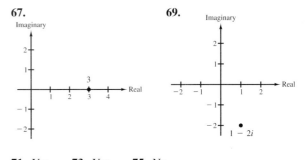

69.

71. Yes **73.** Yes **75.** No

Section 4.5 (page 364)

> **Warm Up** (page 364)
>
> **1.** $4 - \sqrt{29}\,i, 4 + \sqrt{29}\,i$ **2.** $-5 - 12i, -5 + 12i$
>
> **3.** $-1 + 4\sqrt{2}\,i, -1 - 4\sqrt{2}\,i$ **4.** $6 + \dfrac{1}{2}i, 6 - \dfrac{1}{2}i$
>
> **5.** $-13 + 9i$ **6.** $12 + 16i$ **7.** $26 + 22i$
>
> **8.** 29 **9.** i **10.** $-9 + 46i$

1. $\pm 5i, (x + 5i)(x - 5i)$

3. $2 \pm \sqrt{3}, (x - 2 - \sqrt{3})(x - 2 + \sqrt{3})$

5. $\pm 3, \pm 3i, (x + 3)(x - 3)(x + 3i)(x - 3i)$

7. $1 \pm i, (z - 1 + i)(z - 1 - i)$

9. $2, 2 \pm i, (x - 2)(x - 2 + i)(x - 2 - i)$

11. $-5, 4 \pm 3i, (t + 5)(t - 4 + 3i)(t - 4 - 3i)$

13. $-10, -7 \pm 5i, (x + 10)(x + 7 - 5i)(x + 7 + 5i)$

15. $-\dfrac{3}{4}, 1 \pm \dfrac{1}{2}i, (4x + 3)(2x - 2 + i)(2x - 2 - i)$

17. $-2, 1 \pm \sqrt{2}\,i, (x + 2)(x - 1 + \sqrt{2}\,i)(x - 1 - \sqrt{2}\,i)$

19. $-\dfrac{1}{5}, 1 \pm \sqrt{5}\,i, (5x + 1)(x - 1 + \sqrt{5}\,i)(x - 1 - \sqrt{5}\,i)$

21. $2, \pm 2i, (x - 2)^2(x + 2i)(x - 2i)$

23. $\pm i, \pm 3i, (x + i)(x - i)(x + 3i)(x - 3i)$

25. $-2, -\dfrac{1}{2}, \pm i, (x + 2)(2x + 1)(x + i)(x - i)$

27. $x^3 - x^2 + 25x - 25$ **29.** $x^3 - 10x^2 + 33x - 34$

31. $x^4 + 37x^2 + 36$ **33.** $x^4 + 8x^3 + 9x^2 - 10x + 100$

35. $16x^4 + 36x^3 + 16x^2 + x - 30$

37. (a) $(x^2 + 9)(x^2 - 3)$
(b) $(x^2 + 9)(x + \sqrt{3})(x - \sqrt{3})$
(c) $(x + 3i)(x - 3i)(x + \sqrt{3})(x - \sqrt{3})$

39. (a) $(x^2 - 2x - 2)(x^2 - 2x + 3)$
(b) $(x - 1 + \sqrt{3})(x - 1 - \sqrt{3})(x^2 - 2x + 3)$
(c) $(x - 1 + \sqrt{3})(x - 1 - \sqrt{3})$
$(x - 1 + \sqrt{2}i)(x - 1 - \sqrt{2}i)$

41. $\pm 5i, -\frac{3}{2}$ **43.** $\pm 2i, 1, -\frac{1}{2}$ **45.** $-3 \pm i, \frac{1}{4}$

47. $1, 2, -3 \pm \sqrt{2}i$ **49.** $\frac{3}{4}, \frac{1}{2} \pm \frac{\sqrt{5}}{2}i$

51. Solving the equation for $P = 9,000,000$ and graphing shows no real solutions.

53. No, the conjugate pair statement specifies polynomials with *real* coefficients. $f(x)$ has imaginary coefficients.

55. No, $(x + \sqrt{3})$ and $(x - \sqrt{3})$ are not rational factors.

Review Exercises   *(page 368)*

1. $3x + 2$ **3.** $2x^2 + x - 21$

5. $(x - 5)(x - 2)(x + 3)$ **7.** (a) 2   (b) -98

9. $x^2 + 11x + 24$ **11.** Possible positive zeros: 2 or 0
Possible negative zeros: 1
From the graph:
one negative real solution

13. $\pm 1, \pm 3, \pm 5, \pm 15, \pm\frac{1}{2}, \pm\frac{3}{2}, \pm\frac{5}{2}, \pm\frac{15}{2}, \pm\frac{1}{4}, \pm\frac{3}{4}, \pm\frac{5}{4}, \pm\frac{15}{4}$
From the graph $x \approx 2.357$

15. $x^2 + 11x + 28$ **17.** $3, -1, -4$ **19.** $\pm 2, \pm\sqrt{5}$

21. $1, -2, \pm\frac{\sqrt{3}}{3}$ **23.** d   **24.** a   **25.** b

26. c   **27.** $-1.321, -0.283, 1.604$

29. $21.2 or $50.9 (in tens of thousands of dollars)

31. $5 - 3i, 5 + 3i$ **33.** $2\sqrt{6}i, -2\sqrt{6}i$ **35.** $9 - 4i$

37. $(15 - 10\sqrt{2}) + (10\sqrt{5} + 3\sqrt{10})i$ **39.** 89

41. $-10 - 8i$ **43.** $-7 + 24i$ **45.** $\frac{4}{5} + \frac{3}{5}i$

47. $-3 - 4i$ **49.** 10   **51.** $\frac{1 \pm \sqrt{26}i}{3}$

53. $\frac{-11 \pm \sqrt{73}}{8}$ **55.**

57. $(x - 5)(x + 5)(x - 5i)(x + 5i)$

59. $(t + 5)(t - \sqrt{3}i)(t + \sqrt{3}i)$

61. $(x - 2)(2x - 3i)(2x + 3i)$

63. $x^3 - 3x^2 + 16x - 48$

65. (a) $(x^2 + 8)(x^2 - 3)$
(b) $(x^2 + 8)(x - \sqrt{3})(x + \sqrt{3})$
(c) $(x + 2\sqrt{2}i)(x - 2\sqrt{2}i)(x - \sqrt{3})(x + \sqrt{3})$

67. $-3i, 3i, 2$ **69.** $-1 \pm 3i, -1, -4$

Chapter Test   *(page 370)*

1. $(2x - 3)(3x - 1)(2x + 5)$ **2.** $x^2 + 7x + 12$

3. $\pm 1, \pm 2, \pm 3, \pm 4, \pm 6, \pm 12$
$\pm\frac{1}{5}, \pm\frac{2}{5}, \pm\frac{3}{5}, \pm\frac{4}{5}, \pm\frac{6}{5}, \pm\frac{12}{5}$

4. Possible positive zeros: 4, 2, 0
Possible negative zero: exactly 1

5. $16 - 3i$ **6.** $7 - 9i$ **7.** $27 - 4\sqrt{3}i$

8. $23 - 14i$ **9.** $-5 - 12i$ **10.** $21 + 20i$

11. i **12.** $-2 - 5i$ **13.** $\{0, \pm 2, \pm 1\}$

14. $\{-4, -1, 1, 2\}$ **15.** $x = \dfrac{-5 \pm \sqrt{3}i}{2}$

16. $x = \dfrac{5 \pm 3\sqrt{7}i}{4}$ **17.** $\{\sqrt{5}i, -\sqrt{5}i, -2\}$

18. $x^4 - 7x^3 + 19x^2 - 63x + 90$ **19.** $-2 - \sqrt{3}i$

20. $x = 20$; $200,000

CHAPTER 5

Section 5.1   *(page 380)*

Warm Up   *(page 380)*

1. 5^x **2.** 3^{2x} **3.** 4^{3x} **4.** 10^x

5. 4^{2x} **6.** 4^{10x} **7.** $\left(\frac{3}{2}\right)^x$ **8.** 4^{3x}

9. 2^{-x} **10.** $16^{x/4}$

1. 946.852   **3.** 747.258   **5.** 5.256   **7.** 472,369.379

9. 673.639   **11.** 7.389   **13.** 0.472   **15.** g   **16.** e

17. b   **18.** h   **19.** d   **20.** a   **21.** f   **22.** c

23.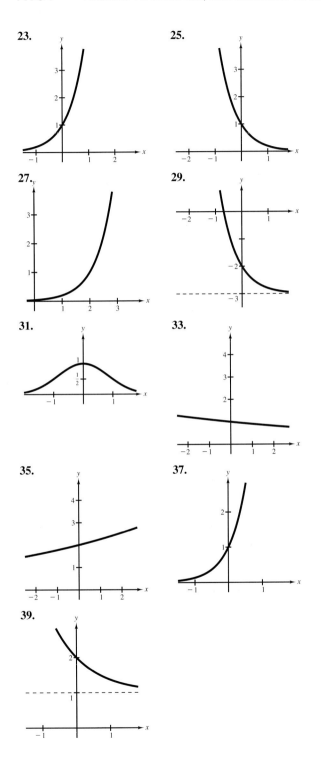

25.

27.

29.

31.

33.

35.

37.

39.

41.

n	1	2	4
A	\$7764.62	\$8017.84	\$8155.09

n	12	365	Continuous Compounding
A	\$8250.97	\$8298.66	\$8300.29

43.

n	1	2	4
A	\$24,115.73	\$25,714.29	\$26,602.23

n	12	365	Continuous Compounding
A	\$27,231.38	\$27,547.07	\$27,557.94

45.

t	1	10	20
P	\$91,393.12	\$40,656.97	\$16,529.89

t	30	40	50
P	\$6720.55	\$2732.37	\$1110.90

47.

t	1	10	20
P	\$90,521.24	\$36,940.70	\$13,646.15

t	30	40	50
P	\$5040.98	\$1862.17	\$687.90

49. \$222,822.57

51. (a) \$472.70   (c)
   (b) \$298.29
   (d) No. Graph shows
   only the demand curve,
   not the cost curve.

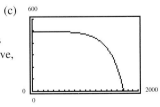

53. (a) 100 (b) 300 (c) 900

55. (a)

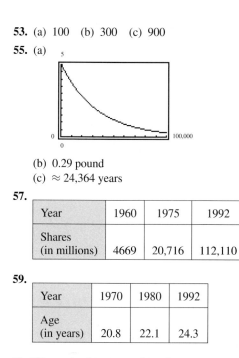

(b) 0.29 pound
(c) $\approx 24{,}364$ years

57.

Year	1960	1975	1992
Shares (in millions)	4669	20,716	112,110

59.

Year	1970	1980	1992
Age (in years)	20.8	22.1	24.3

61. Women tend to marry about 2 years younger than men do. The average ages of both have been rising.

Section 5.2 *(page 391)*

Warm Up *(page 391)*

1. 3 **2.** 0 **3.** -1 **4.** 1 **5.** 7.389
6. 0.368 **7.** Graph is shifted 2 units to the left.
8. Graph is reflected about the x-axis.
9. Graph is shifted down 1 unit.
10. Graph is reflected about the y-axis.

1. 4 **3.** -2 **5.** $\frac{1}{2}$ **7.** 0 **9.** -2
11. 3 **13.** -2 **15.** 2 **17.** $\log_5 125 = 3$
19. $\log_{81} 3 = \frac{1}{4}$ **21.** $\log_6 \frac{1}{36} = -2$
23. $\ln 20.0855 \approx 3$ **25.** $\ln 4 = x$ **27.** 2.538

29. 2.161 **31.** 2.913 **33.** 0.549

35.   **37.**

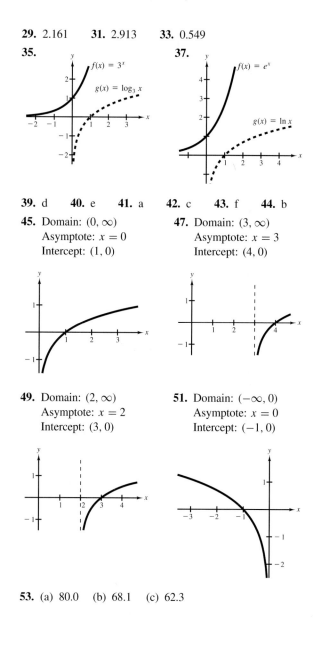

39. d **40.** e **41.** a **42.** c **43.** f **44.** b
45. Domain: $(0, \infty)$ **47.** Domain: $(3, \infty)$
Asymptote: $x = 0$ Asymptote: $x = 3$
Intercept: $(1, 0)$ Intercept: $(4, 0)$

49. Domain: $(2, \infty)$ **51.** Domain: $(-\infty, 0)$
Asymptote: $x = 2$ Asymptote: $x = 0$
Intercept: $(3, 0)$ Intercept: $(-1, 0)$

53. (a) 80.0 (b) 68.1 (c) 62.3

55. (a)

K	1	2	4	6	8	10	12
t	0	7.3	14.6	18.9	21.9	24.2	26.2

(b)

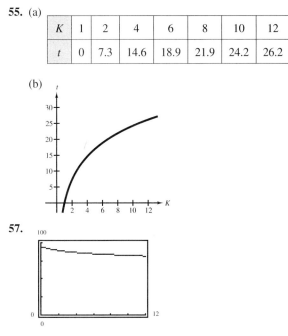

57.

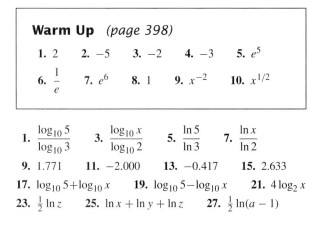

The domain $0 \le t \le 12$ covers the period of 12 months or 1 year. The range $0 \le f(t) \le 100$ covers the possible scores.

59. 29.7 years   **61.** 1980: 29.7 years
2000: 35.0 years
Change of 5.3 years

63. 20 years   **65.** $280,178.40

Section 5.3 *(page 398)*

Warm Up *(page 398)*

1. 2   **2.** -5 **3.** -2 **4.** -3 **5.** e^5

6. $\dfrac{1}{e}$ **7.** e^6 **8.** 1   **9.** x^{-2} **10.** $x^{1/2}$

---

1. $\dfrac{\log_{10} 5}{\log_{10} 3}$ **3.** $\dfrac{\log_{10} x}{\log_{10} 2}$ **5.** $\dfrac{\ln 5}{\ln 3}$ **7.** $\dfrac{\ln x}{\ln 2}$

9. 1.771   **11.** -2.000 **13.** -0.417 **15.** 2.633

17. $\log_{10} 5 + \log_{10} x$ **19.** $\log_{10} 5 - \log_{10} x$ **21.** $4 \log_2 x$

23. $\frac{1}{2} \ln z$ **25.** $\ln x + \ln y + \ln z$ **27.** $\frac{1}{2} \ln(a-1)$

29. $2 \ln(z-1)$ **31.** $2 \log_b x - 3 \log_b y$

33. $\frac{1}{3}(\ln x - \ln y)$ **35.** $\frac{1}{2} \ln y - 5 \ln z$ **37.** $\ln 2x$

39. $\log_{10} \dfrac{z}{y}$ **41.** $\log_{10}(x+4)^2$ **43.** $\ln \dfrac{x}{(x+1)^3}$

45. $\ln(5x)^{1/3}$ **47.** $\ln\left(\dfrac{x-2}{x+2}\right)$ **49.** $\ln \dfrac{3^2}{\sqrt{x^2+1}}$

51. $\ln\left[\dfrac{x}{(x+2)(x-2)}\right]$ **53.** 0.9208   **55.** 0.2084

57. 1.6542   **59.** 0.1781   **61.** 1.8957   **63.** -0.7124

65. 0.9136   **67.** 2   **69.** 2.4   **71.** 4.5   **73.** $\frac{3}{2}$

75. $\frac{1}{2} + \frac{1}{2} \log_7 10$ **77.** $-3 - \log_5 2$ **79.** $6 + \ln 5$

81. $y = \sqrt[4]{x}$ **83.** $y = 2.5x^{-1/4}$

85. They have the same graph.
$$\log_a \dfrac{u}{v} = \log_a u - \log_a v$$

87. Graph the two functions in the same window. Functions and graphs vary.

89. $y = 0.07x^{2/3}$

Mid-Chapter Quiz *(page 402)*

1. **2.**

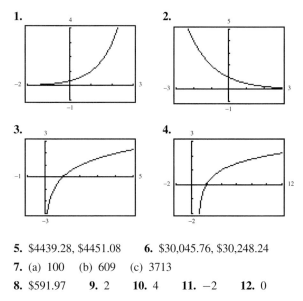

3. **4.**

5. $4439.28, $4451.08   **6.** $30,045.76, $30,248.24

7. (a) 100   (b) 609   (c) 3713

8. $591.97   **9.** 2   **10.** 4   **11.** -2 **12.** 0

13.

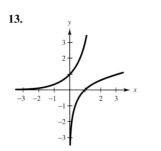

Graphs are reflections about the line $y = x$.

14. $\frac{1}{3}(\log x + \log y - \log z)$ **15.** $\ln(x^2 + 3) - 3\ln x$

16. $\ln\left(\frac{xy}{3}\right)$ **17.** $\log_{10}\left(\frac{1}{64x^3}\right)$ **18.** $\frac{3}{2}$ **19.** $\frac{6}{5}$

20. $y^3 = x$ or $\log_y x = 3$ or $y = \sqrt[3]{x}$

Section 5.4 *(page 410)*

Warm Up *(page 410)*

1. $\dfrac{\ln 3}{\ln 2}$ **2.** $1 + \dfrac{2}{\ln 4}$ **3.** $\dfrac{e}{2}$ **4.** $2e$

5. $2 \pm i$ **6.** $\frac{1}{2}, 1$ **7.** x **8.** $2x$

9. $2x$ **10.** $-x^2$

---

1. 2   **3.** -2 **5.** 3   **7.** 64   **9.** $\frac{1}{10}$

11. x^2 **13.** $5x + 2$ **15.** x^2 **17.** $\ln 20 \approx 2.996$

19. $\ln 18 \approx 2.890$ **21.** $\ln \frac{39}{2} \approx 2.970$

23. $\ln 15 \approx 2.708$ **25.** $\ln \frac{31}{3} \approx 2.335$ **27.** 0

29. $\frac{1}{3}\ln 12 \approx 0.828$ **31.** $-\ln 28 \approx -3.332$

33. $\ln \frac{5}{3} \approx 0.511$ **35.** $\dfrac{1}{0.06}\ln 3 \approx 18.310$

37. $\ln 5 \approx 1.609$ **39.** $\frac{1}{2}\left(-1 + \ln \frac{962}{25}\right) \approx 1.325$

41. $\ln 6 \approx 1.792$ **43.** $e^5 \approx 148.413$

45. $e^{-10} \approx 0.0000454$ **47.** $\frac{1}{2}e \approx 1.359$

49. $e^{7/2} \approx 33.115$ **51.** $\frac{1}{4}$ **53.** $\frac{1}{2}e^{-1} \approx 0.184$

55. $\pm\sqrt{\dfrac{e^5}{2}} \approx \pm 8.614$ **57.** $2e^5 \approx 296.826$

59. $1 + \sqrt{1 + e} \approx 2.928$, $1 - \sqrt{1 + e} \approx -0.928$

61. 8.15 years   **63.** 16.28 years

65. (a) 1426 units
   (b) 1498 units
   (c)

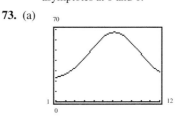

67. (a) 29.3 years   (b) 39.8 years

69.

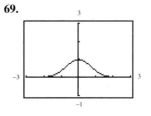

71. (a) 69.71 inches
   (b) The model is an increasing function with horizontal asymptotes at 0 and 1.

73. (a)

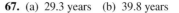

   (b) January, February, March, December
   (c) The model is bell-shaped and peaks.

75. ≈ 35.649

Section 5.5 (page 420)

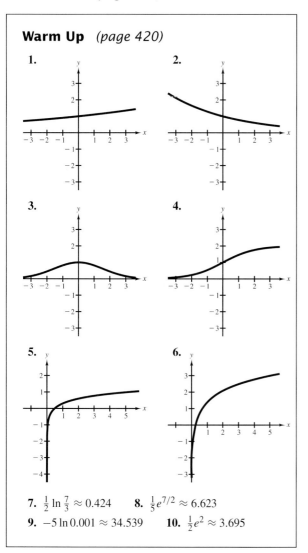

Warm Up (page 420)

1.

2.

3.

4.

5.

6.

7. $\frac{1}{2}\ln\frac{7}{3} \approx 0.424$ 8. $\frac{1}{5}e^{7/2} \approx 6.623$

9. $-5\ln 0.001 \approx 34.539$ 10. $\frac{1}{2}e^2 \approx 3.695$

	Initial Investment	Annual % Rate	Time to Double	Amount After 10 Years
1.	$1000	12%	5.78 yr	$3320.12
3.	$750	8.94%	7.75 yr	$1833.67
5.	$500	9.5%	7.30 yr	$1292.85
7.	$6392.79	11%	6.30 yr	$19,205.00
9.	$5000.00	8%	8.66 yr	$11,127.70

	Half-Life (in years)	Initial Quantity	Amount After 1000 Years
11.	1620	10 g	6.52 g
13.	5730	2.26 g	2 g
15.	24,360	2.16 g	2.1 g

17. Growth   19. Decay   21. $\frac{1}{4}\ln 10 \approx 0.5756$

23. $\frac{1}{4}\ln\frac{1}{4} \approx -0.3466$ 25. 2014

27. $k \approx 0.0447$.
Population in 2010: 61,122

29. 3.15 hours

31. 95.8%   33. (a) -0.05 (b) ≈ 36 days

35.

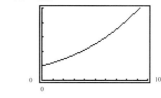

37. 32,200 miles

The number of golfers began to increase at a higher rate in 1985.

39.

64.5 inches

41. (a)

(b) 7.8 months   (c) 1252

43. (a) $k \approx 0.1656$ (b) 203 animals

45. Videocassettes became very popular. Consequently, many companies made them. Supply went up; price came down. (Answers will vary.)

47. (a) 7.91   (b) 7.68   (c) 7.60   (d) 7.30

49. (a) 20   (b) 70   (c) 95   (d) 120

51. 74.9% decrease   53. 7:30 A.M.

Review Exercises *(page 427)*

1. d   **2.** f   **3.** a   **4.** b

5. c   **6.** e   **7.** h   **8.** g

9.   **11.**

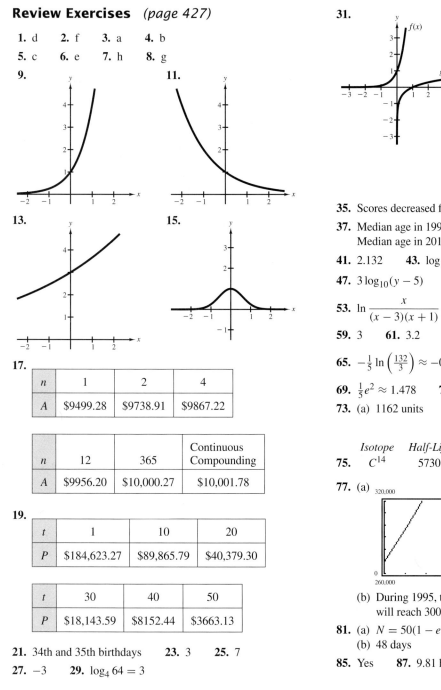

13.   **15.**

17.

n	1	2	4
A	$9499.28	$9738.91	$9867.22

n	12	365	Continuous Compounding
A	$9956.20	$10,000.27	$10,001.78

19.

t	1	10	20
P	$184,623.27	$89,865.79	$40,379.30

t	30	40	50
P	$18,143.59	$8152.44	$3663.13

21. 34th and 35th birthdays   **23.** 3   **25.** 7

27. -3 **29.** $\log_4 64 = 3$

31.

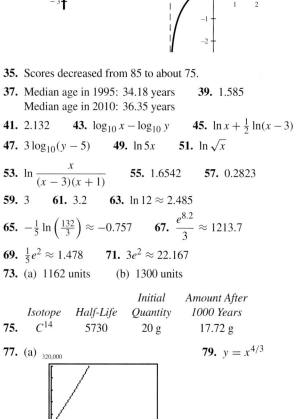

33. Domain: $(-2, \infty)$
Vertical asymptote: $x = -2$
x-intercept: $(-1, 0)$

35. Scores decreased from 85 to about 75.

37. Median age in 1995: 34.18 years   **39.** 1.585
Median age in 2010: 36.35 years

41. 2.132   **43.** $\log_{10} x - \log_{10} y$ **45.** $\ln x + \frac{1}{2} \ln(x - 3)$

47. $3 \log_{10}(y - 5)$ **49.** $\ln 5x$ **51.** $\ln \sqrt{x}$

53. $\ln \dfrac{x}{(x - 3)(x + 1)}$ **55.** 1.6542   **57.** 0.2823

59. 3   **61.** 3.2   **63.** $\ln 12 \approx 2.485$

65. $-\frac{1}{5} \ln\left(\frac{132}{3}\right) \approx -0.757$ **67.** $\dfrac{e^{8.2}}{3} \approx 1213.7$

69. $\frac{1}{5} e^2 \approx 1.478$ **71.** $3e^2 \approx 22.167$

73. (a) 1162 units   (b) 1300 units

Isotope	Half-Life	Initial Quantity	Amount After 1000 Years
75. C^{14}	5730	20 g	17.72 g

77. (a)

79. $y = x^{4/3}$

(b) During 1995, the population will reach 300,000.

81. (a) $N = 50(1 - e^{-0.04838t})$ **83.** (a) $k = 0.3615$
(b) 48 days   (b) 302 deer

85. Yes   **87.** 9.81 hours

Chapter Test (page 431)

1. $y = 2^x$

2. $y = 3^{-x}$

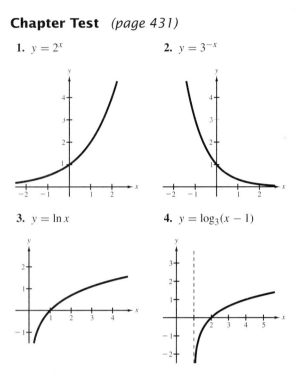

3. $y = \ln x$

4. $y = \log_3(x - 1)$

5. The trust fund will be greater than $100,000 when 18.93 years have elapsed.

6. $2 \ln x + 3 \ln y - \ln z$

7. $\log_{10} 3 + \log_{10} x + \log_{10} y + 2 \log_{10} z$

8. $\ln x + \dfrac{1}{3} \ln(x - 2)$ **9.** $\ln \dfrac{yz^2}{x^3}$ **10.** $\log_{10} \sqrt[3]{x^2 y^2}$

11. $\dfrac{\ln 21}{4} \approx 0.761$ **12.** $x = \ln 6 \approx 1.792$
$x = \ln 2 \approx 0.693$

13. $\frac{1}{4} e^3 \approx 5.021$ **14.** $e^6 - 2 \approx 401.429$ **15.** 87

16. 82.5 **17.** 76.5 **18.** 2005 **19.** 3.4 hours

20. 2.98 grams

3. $6x^3 + 10x^2 - 9x - 15$ **4.** $\dfrac{2x^2 - 3}{3x + 5}, x \neq -\dfrac{5}{3}$

5. $18x^2 + 60x + 47$ **6.** $6x^2 - 4$

7.

8.

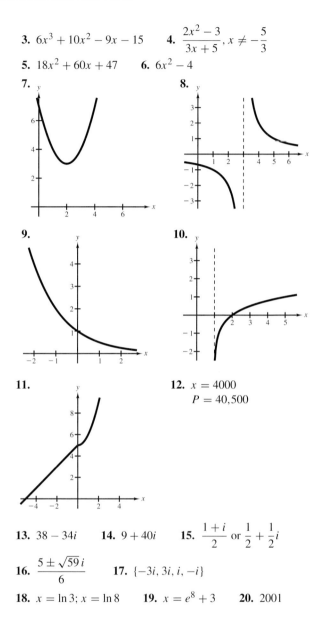

9.

10.

11.

12. $x = 4000$
$P = 40,500$

13. $38 - 34i$ **14.** $9 + 40i$ **15.** $\dfrac{1+i}{2}$ or $\dfrac{1}{2} + \dfrac{1}{2}i$

16. $\dfrac{5 \pm \sqrt{59}\, i}{6}$ **17.** $\{-3i, 3i, i, -i\}$

18. $x = \ln 3; x = \ln 8$ **19.** $x = e^8 + 3$ **20.** 2001

Cumulative Test: Chapters 3–5
(page 432)

1. $2x^2 + 3x + 2$ **2.** $2x^2 - 3x - 8$

CHAPTER 6

Section 6.1 *(page 477)*

Warm Up *(page 477)*

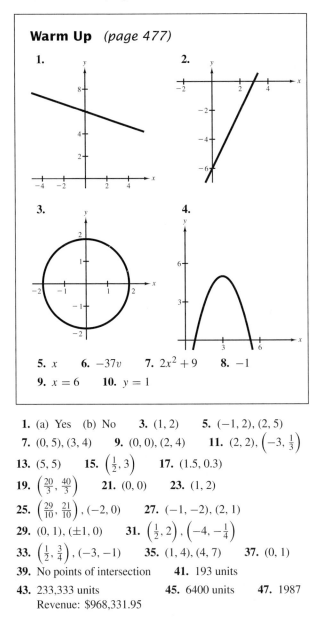

5. x **6.** $-37v$ **7.** $2x^2 + 9$ **8.** -1

9. $x = 6$ **10.** $y = 1$

1. (a) Yes   (b) No   **3.** $(1, 2)$ **5.** $(-1, 2), (2, 5)$

7. $(0, 5), (3, 4)$ **9.** $(0, 0), (2, 4)$ **11.** $(2, 2), \left(-3, \frac{1}{3}\right)$

13. $(5, 5)$ **15.** $\left(\frac{1}{2}, 3\right)$ **17.** $(1.5, 0.3)$

19. $\left(\frac{20}{3}, \frac{40}{3}\right)$ **21.** $(0, 0)$ **23.** $(1, 2)$

25. $\left(\frac{29}{10}, \frac{21}{10}\right), (-2, 0)$ **27.** $(-1, -2), (2, 1)$

29. $(0, 1), (\pm 1, 0)$ **31.** $\left(\frac{1}{2}, 2\right), \left(-4, -\frac{1}{4}\right)$

33. $\left(\frac{1}{2}, \frac{3}{4}\right), (-3, -1)$ **35.** $(1, 4), (4, 7)$ **37.** $(0, 1)$

39. No points of intersection   **41.** 193 units

43. 233,333 units   **45.** 6400 units   **47.** 1987
Revenue: $968,331.95

49. Yes, in 2027

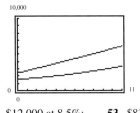

51. $12,000 at 8.5%;   **53.** $8333   **55.** 1992
$13,000 at 8%

Section 6.2 *(page 452)*

Warm Up *(page 452)*

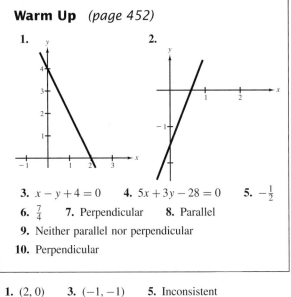

3. $x - y + 4 = 0$ **4.** $5x + 3y - 28 = 0$ **5.** $-\frac{1}{2}$

6. $\frac{7}{4}$ **7.** Perpendicular   **8.** Parallel

9. Neither parallel nor perpendicular

10. Perpendicular

1. $(2, 0)$ **3.** $(-1, -1)$ **5.** Inconsistent

7. $(2a, 3a - 3)$ **9.** $\left(-\frac{1}{3}, -\frac{2}{3}\right)$ **11.** $\left(\frac{5}{2}, \frac{3}{4}\right)$

13. $(3, 4)$ **15.** $(4, -1)$ **17.** $(40, 40)$

19. Inconsistent   **21.** $\left(\frac{18}{5}, \frac{3}{5}\right)$ **23.** $\left(\frac{19}{7}, -\frac{2}{7}\right)$

25. $\left(a, \frac{5}{6}a - \frac{1}{2}\right)$ **27.** $\left(\frac{90}{31}, -\frac{67}{31}\right)$ **29.** $\left(-\frac{6}{35}, \frac{43}{35}\right)$

31. No, lines intersect at $(79, 400, 398)$.

33. 550 miles per hour and 50 miles per hour

35. $6\frac{2}{3}$ gallons at 20%   **37.** $8000 at 12%
$3\frac{1}{3}$ gallons at 50% $4000 at 10.5%

39. 375 adult tickets   **41.** $x = 309,091$, $p = 25.09$
125 children's tickets

43. $x = 2,000,000$, $p = 100$

45. (a) $x = 5$ corresponds to 1985
$N = 0.2x + 4.4$
$T = 0.2x + 3.9$
(b) No

47. $y = 0.97x + 2.1$ **49.** $y = 0.318x + 4.061$

51. $y = -0.243x + 4.757$ **53.** (a) $y = 2.35x + 75.98$
(b) 90.1

Section 6.3 *(page 464)*

> ### Warm Up *(page 464)*
>
> **1.** $(15, 10)$ **2.** $\left(-2, -\frac{8}{3}\right)$ **3.** $(28, 4)$
> **4.** $(4, 3)$ **5.** Not a solution
> **6.** Not a solution   **7.** Solution
> **8.** Solution   **9.** $5a + 2$ **10.** $a + 13$

1. $(1, 2, 3)$ **3.** $(2, -3, -2)$ **5.** $(-4, 8, 5)$

7. Inconsistent   **9.** $\left(1, -\frac{3}{2}, \frac{1}{2}\right)$

11. $(-3a + 10, 5a - 7, a)$ **13.** $\left(-4a + 13, -\frac{15}{2}a + \frac{45}{2}, a\right)$

15. $(-a, 2a - 1, a)$ **17.** $\left(-\frac{3}{2}a + \frac{1}{2}, -\frac{2}{3}a + 1, a\right)$

19. $(1, 1, 1, 1)$ **21.** Inconsistent   **23.** $(0, 0, 0)$

25. $\left(-\frac{3}{5}a, \frac{4}{5}a, a\right)$ **27.** $y = 2x^2 + 3x - 4$

29. $y = x^2 - 4x + 3$ **31.** $x^2 + y^2 - 4x = 0$

33. $x^2 + y^2 - 6x - 8y = 0$

35. $y = \frac{1}{2}x^2 - \frac{1}{2}x$, correct for a polygon with six sides

37. $4000 was invested at 5%
$5000 was invested at 6%
$7000 was invested at 7%

39. $75,000 is borrowed at 10%
$400,000 is borrowed at 9%
$300,000 is borrowed at 8%

41. 25 pounds of grade A paper   **43.** 10 gallons of spray X
15 pounds of grade B paper   5 gallons of spray Y
10 pounds of grade C paper   12 gallons of spray Z

45. One possible solution:
$$C = -\frac{a}{2} + 250,000$$
$$M = \frac{a}{2} + 125,000$$
$$B = -a + 125,000$$
$$G = a$$

47. $y = 0.079x^2 + 0.63x + 2.943$

49. $y = -0.207x^2 - 0.887x + 5.083$

51. $y = 0.013x^2 + 0.264x + 0.874$; 3.36 million

Mid-Chapter Quiz *(page 469)*

1. Answers vary.   **2.** Answers vary.   **3.** $(9, 7)$
$3x + y = 11$ $x + 2y = 11$
$x - 2y = -1$ $2x - y = 7$

4. $(0.927, 2.854)$, $(-1.727, -2.454)$ **5.** 1342 units

6. 500,000 units   **7.** $(2, -1)$ **8.** $\left(1, 1\frac{1}{2}\right)$

9. $18,333$ units   **10.** $17,428.571$ **11.** $(1, -2, 3)$

12. $(-2, 4, 3)$ **13.** $(1.\overline{6}, 2.\overline{8}, -4.\overline{1})$

14. Answers vary. $(1, -1, 3)$, $(0, -2, 0)$, $(3, 1, 9)$

15. Answers vary. $(0, 5, 0)$, $(4, 7, 2)$, $(2, 6, 1)$

16. The system is inconsistent, no solution.   **17.** Infinitely many

18. $3\frac{1}{3}$ gallons of 25% and $6\frac{2}{3}$ gallons of 40%

19. True   **20.** False

Section 6.4 *(page 477)*

> ### Warm Up *(page 477)*
>
> **1.** Line   **2.** Line   **3.** Parabola   **4.** Parabola
> **5.** Circle   **6.** Ellipse   **7.** $(1, 1)$ **8.** $(2, 0)$
> **9.** $(2, 1)$, $\left(-\frac{5}{2}, -\frac{5}{4}\right)$ **10.** $(2, 3)$, $(3, 2)$

1. e   **2.** f   **3.** d   **4.** b   **5.** a   **6.** c

7.

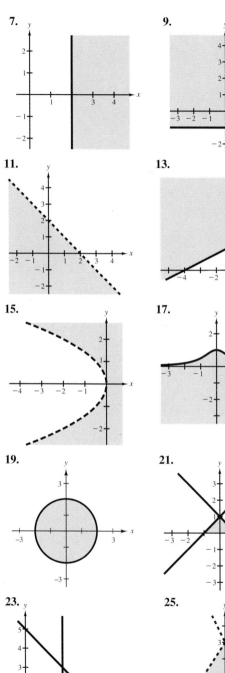

9.

11.

13.

15.

17.

19.

21.

23.

25.

27.

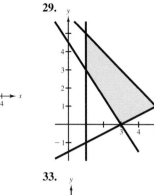

29.

31.

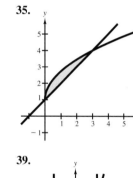

33.

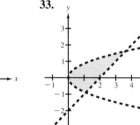

35.

37.

39.

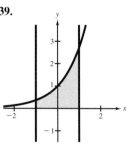

41. $2 \le x \le 5, 1 \le y \le 7$

43. $y \le \frac{3}{2}x, y \le -x + 5, y \ge 0$

45. $x^2 + y^2 \le 16, x \ge 0, y \ge 0$

47. $x + \frac{3}{2}y \le 12, \frac{4}{3}x + \frac{3}{2}y \le 15, x \ge 0, y \ge 0$

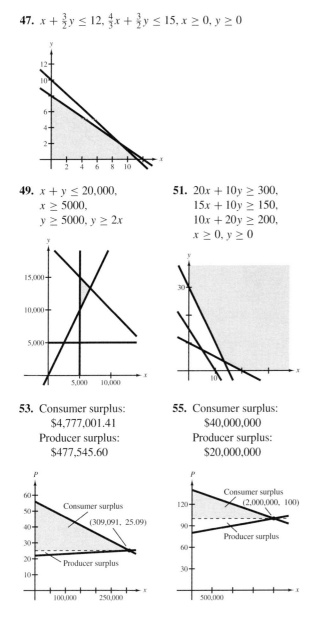

49. $x + y \le 20{,}000,$
$x \ge 5000,$
$y \ge 5000, y \ge 2x$

51. $20x + 10y \ge 300,$
$15x + 10y \ge 150,$
$10x + 20y \ge 200,$
$x \ge 0, y \ge 0$

53. Consumer surplus:
$4,777,001.41
Producer surplus:
$477,545.60

55. Consumer surplus:
$40,000,000
Producer surplus:
$20,000,000

57. (a)

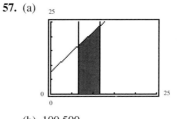

(b) 100,500

Section 6.5 (page 487)

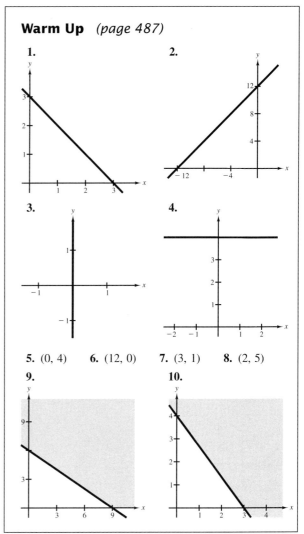

Warm Up (page 487)

5. (0, 4)   **6.** (12, 0)   **7.** (3, 1)   **8.** (2, 5)

1. Minimum value at (0, 0): 0
Maximum value at (0, 6): 30

3. Minimum value at (0, 0): 0
Maximum value at (6, 0): 60

5. Minimum value at (0, 0): 0
Maximum value at (3, 4): 17

7. Minimum value at (0, 0): 0
Maximum value at (4, 0): 20

9. Minimum value at (0, 0): 0
Maximum value at (60, 20): 740

11. Minimum value at (0, 0): 0
Maximum value at any point on line segment connecting
(60, 20) and (30, 45): 2100

13. Minimum value at (0, 0): 0
Maximum value at (5, 0): 30

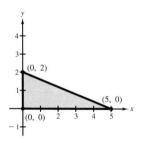

15. Minimum value at (0, 0): 0
Maximum value at (5, 0): 45

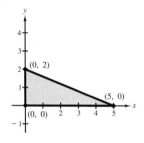

17. Minimum value at (5, 3): 35
No maximum value

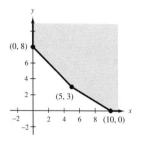

19. Minimum value at (10, 0): 20
No maximum value

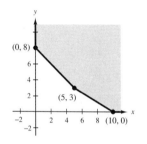

21. Minimum value at (24, 8): 104
Maximum value at (40, 0): 160

23. Minimum value at (36, 0): 36
Maximum value at (24, 8): 56

25. Maximum value at (3, 6): 12

27. Maximum value at (0, 10): 10

29. Maximum value at (0, 5): 25

31. Maximum value at (4, 4): 36

33. 200 units of the $250 model
50 units of the $400 model

35. 3 bags of brand X
6 bags of brand Y
Minimum cost: $195

37. 750 units of model A
1000 units of model B

39. 8 audits
8 tax returns

41. Television: $540,000
Newspaper: $60,000

43. Type A: $62,500
Type B: $187,500

45. z is maximum at any point on the line $5x + 2y = 10$ between the points $(2, 0)$ and $\left(\frac{20}{19}, \frac{45}{19} \right)$.

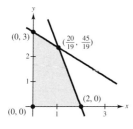

47. The constraint $x \leq 10$ is extraneous. The maximum value of z occurs at $(0, 7)$.

49. The constraint $2x + y \leq 4$ is extraneous. The maximum value of z occurs at $(0, 1)$.

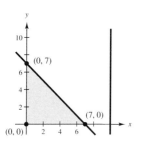

51. Model A: 929, model B: 77

Review Exercises *(page 493)*

1. $(-2, 4)$ **3.** $(8, -10)$ **5.** $(8, 6)$ and $(0, 10)$

7. $(1, 9)$ and $(1.5, 8.75)$ **9.** 4762 units

11. $14,000 at 7%
$20,000 at 7.5%   **13.** $(3, -5)$

15. Infinitely many solutions   **17.** Inconsistent: no solution

19. (a) $0.1a + 0.5b = 0.25(12)$
$a + b = 12$ **21.** $x = 71,429$ units
$p = 22.71$

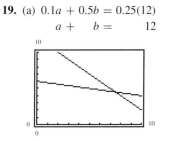

(b) 7.5 gallons of 10% solution
4.5 gallons of 50% solution

23. $(2, -1, 3)$ **25.** $\left(\frac{1}{5}a + \frac{8}{5}, -\frac{6}{5}a + \frac{42}{5}, a \right)$

27. $\left(\frac{2}{7}a + \frac{16}{7}, \frac{3}{7}a - \frac{11}{7}, a \right)$ **29.** $y = 2x^2 + x - 6$

31. $x^2 + y^2 - 4x + 2y - 4 = 0$ **33.** $2000 at 6%
$10,000 at 7%
$3000 at 8%

35. Certificates of deposit: $200,000
Municipal bonds: $100,000
Blue chip stock: $190,000
Growth stock: $10,000

37. $y = 0.114x^2 + 0.8x + 1.59$

39.

41.

43.

45. $x + 7y \geq 15$
$x - y \geq -1$
$3x + y \leq 25$

47. $x + y \geq 16,000$
$20x + 30y \geq 400,000$
$x \geq 9,000$
$y \geq 4,000$

49. August

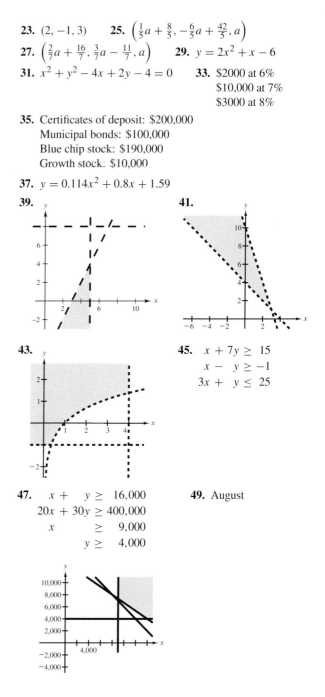

51. Consumer surplus: $4,000,000
Producer surplus: $6,000,000

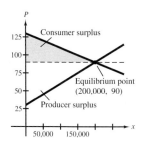

53. Minimum value at $(0, 0)$: 0
Maximum value at $(3, 3)$: 81

55. Minimum value at $(0, 300)$: 18,000
Maximum value at any point on line segment connecting
$(0, 500)$ and $(600, 0)$: 30,000

57. Minimum value at $(0, 0)$: 0
Maximum value at $(5, 3)$: 66

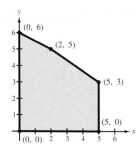

59. Minimum value at $(3, 0)$: 12
Maximum value at $(5, 4)$: 40

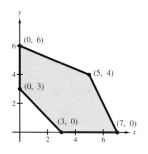

61. 50 of $300 model
50 of $500 model

63. 369 of model A
538 of model B

65. $\frac{1}{3}$ of A
$\frac{1}{3}$ of B

67. 7 audits
10 tax returns

Chapter Test *(page 498)*

1. $(2, 4)$ **2.** $(-2, 5), (0, 3)$ **3.** $\left(\frac{63}{13}, -\frac{17}{13}\right)$

4. $(1.68, 2.38), (-4.18, -26.89)$ **5.** $(3.35, -1.32)$

6. $(2, -1, 3)$ **7.** $(2, -3, 4)$ **8.** No solution

9. $(50,000, 30)$ **10.** No solution

11. Infinitely many solutions   **12.** $y = 2x^2 + 3x + 5$

13. $17,000 at 8%; $18,000 at 8.5%

14. **15.** **16.** **17.**

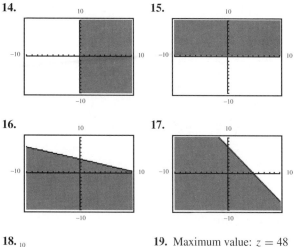

18.

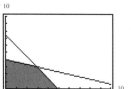

19. Maximum value: $z = 48$
Minimum value: $z = 25$

20. Model A: 297, model B: 570

CHAPTER 7

Section 7.1 *(page 510)*

Warm Up *(page 510)*

1. -3 **2.** 30 **3.** 6 **4.** $-\frac{1}{9}$

5. Solution **6.** Not a solution **7.** $(5, 2)$

8. $\left(\frac{12}{5}, -3\right)$ **9.** $(40, 14, 2)$ **10.** $\left(\frac{15}{2}, 4, 1\right)$

---

1. 3×2 **3.** 5×1 **5.** 4×2

7. Reduced row-echelon form **9.** Not in row-echelon form

11. (a) $\begin{bmatrix} 1 & 2 & 3 \\ 0 & -5 & -10 \\ 3 & 1 & -1 \end{bmatrix}$ (b) $\begin{bmatrix} 1 & 2 & 3 \\ 0 & -5 & -10 \\ 0 & -5 & -10 \end{bmatrix}$

(c) $\begin{bmatrix} 1 & 2 & 3 \\ 0 & -5 & -10 \\ 0 & 0 & 0 \end{bmatrix}$ (d) $\begin{bmatrix} 1 & 2 & 3 \\ 0 & 1 & 2 \\ 0 & 0 & 0 \end{bmatrix}$

(e) $\begin{bmatrix} 1 & 0 & -1 \\ 0 & 1 & 2 \\ 0 & 0 & 0 \end{bmatrix}$

13. $\begin{bmatrix} 1 & 1 & 0 & 5 \\ 0 & 1 & 2 & 0 \\ 0 & 0 & 1 & -1 \end{bmatrix}$ **15.** $\begin{bmatrix} 1 & -1 & -1 & 1 \\ 0 & 1 & 6 & 3 \\ 0 & 0 & 0 & 0 \end{bmatrix}$

17. $\begin{bmatrix} 1 & 0 & 0 \\ 0 & 1 & 0 \\ 0 & 0 & 1 \end{bmatrix}$ **19.** $\begin{bmatrix} 1 & 2 & 0 & 0 \\ 0 & 0 & 1 & 0 \\ 0 & 0 & 0 & 1 \\ 0 & 0 & 0 & 0 \end{bmatrix}$

21. $4x + 3y = 8, x - 2y = 3$

23. $x + 2z = -10, 3y - z = 5, 4x + 2y = 3$

25. $x - 2y = 4$ **27.** $x + 2y - 2z = -14$
$y = -3$ $y + z = 9$
$(-2, -3)$ $z = -3$
$(-31, 12, -3)$

29. $(7, -5)$ **31.** $(-4, -8, 2)$

33. $\begin{bmatrix} 4 & -5 & \vdots & -2 \\ -1 & 8 & \vdots & 10 \end{bmatrix}$

35. $\begin{bmatrix} 1 & 10 & -3 & \vdots & 2 \\ 5 & -3 & 4 & \vdots & 0 \\ 2 & 4 & 0 & \vdots & 6 \end{bmatrix}$

37. $(3, 2)$ **39.** $(4, -2)$ **41.** $\left(\frac{1}{2}, -\frac{3}{4}\right)$

43. Inconsistent **45.** $\left(-\frac{3}{2}a + \frac{3}{2}, \frac{1}{3}a + \frac{1}{3}, a\right)$

47. $(2a + 1, 3a + 2, a)$ **49.** $(5a + 4, -3a + 2, a)$

51. $(0, 2 - 4a, a)$ **53.** $\left(\frac{141}{7}, -\frac{10}{71}, \frac{784}{71}, -\frac{312}{71}\right)$

55. $(0, 0)$ **57.** $(-2a, a, a)$

59. \$100,000 at 9% **61.** $y = x^2 + 2x + 4$
\$250,000 at 10%
\$150,000 at 12%

63. $y = 2x^3 + x + 20$ **65.** $y = 29.91 + 1.58t$; \$42,550

Section 7.2 *(page 524)*

Warm Up *(page 524)*

1. -5 **2.** -7

3. Not in reduced row-echelon form

4. Not in reduced row-echelon form

5. $\begin{bmatrix} -5 & 10 & \vdots & 12 \\ 7 & -3 & \vdots & 0 \\ -1 & 7 & \vdots & 25 \end{bmatrix}$

6. $\begin{bmatrix} 10 & 15 & -9 & \vdots & 42 \\ 6 & -5 & 0 & \vdots & 0 \end{bmatrix}$

7. $(0, 2)$ **8.** $(2 + a, 3 - a, a)$

9. $(1 - 2a, a, -1)$ **10.** $(2, -1, -1)$

---

1. $x = -4, y = 22$ **3.** $x = 2, y = 3$

5. (a) $\begin{bmatrix} 3 & -2 \\ 1 & 7 \end{bmatrix}$ (b) $\begin{bmatrix} -1 & 0 \\ 3 & -9 \end{bmatrix}$

(c) $\begin{bmatrix} 3 & -3 \\ 6 & -3 \end{bmatrix}$ (d) $\begin{bmatrix} -1 & -1 \\ 8 & -19 \end{bmatrix}$

7. (a) $\begin{bmatrix} 7 & 3 \\ 1 & 9 \\ -2 & 15 \end{bmatrix}$ (b) $\begin{bmatrix} 5 & -5 \\ 3 & -1 \\ -4 & -5 \end{bmatrix}$

(c) $\begin{bmatrix} 18 & -3 \\ 6 & 12 \\ -9 & 15 \end{bmatrix}$ (d) $\begin{bmatrix} 16 & -11 \\ 8 & 2 \\ -11 & -5 \end{bmatrix}$

9. (a) $\begin{bmatrix} 3 & 3 & -2 & 1 & 1 \\ -2 & 5 & 7 & -6 & -8 \end{bmatrix}$

(b) $\begin{bmatrix} 1 & 1 & 0 & -1 & 1 \\ 4 & -3 & -11 & 6 & 6 \end{bmatrix}$

(c) $\begin{bmatrix} 6 & 6 & -3 & 0 & 3 \\ 3 & 3 & -6 & 0 & -3 \end{bmatrix}$

(d) $\begin{bmatrix} 4 & 4 & -1 & -2 & 3 \\ 9 & -5 & -24 & 12 & 11 \end{bmatrix}$

11. (a) $\begin{bmatrix} 0 & 15 \\ 6 & 12 \end{bmatrix}$ **13.** (a) $\begin{bmatrix} 0 & -10 \\ 10 & 0 \end{bmatrix}$

(b) $\begin{bmatrix} -2 & 2 \\ 31 & 14 \end{bmatrix}$ (b) $\begin{bmatrix} 0 & -10 \\ 10 & 0 \end{bmatrix}$

(c) $\begin{bmatrix} 9 & 6 \\ 12 & 12 \end{bmatrix}$ (c) $\begin{bmatrix} 8 & -6 \\ 6 & 8 \end{bmatrix}$

15. (a) $\begin{bmatrix} 6 & -21 & 15 \\ 8 & -23 & 19 \\ 4 & 7 & 5 \end{bmatrix}$ (b) $\begin{bmatrix} 9 & 0 & 13 \\ 7 & -2 & 21 \\ 1 & 4 & -19 \end{bmatrix}$

(c) $\begin{bmatrix} 20 & 7 & -8 \\ 24 & 7 & -2 \\ 2 & -5 & 30 \end{bmatrix}$

17. Not possible   **19.** $\begin{bmatrix} -1 & 19 \\ 4 & -27 \\ 0 & 14 \end{bmatrix}$

21. $\begin{bmatrix} 1 & 0 & 0 \\ 0 & 1 & 0 \\ 0 & 0 & \frac{7}{2} \end{bmatrix}$ **23.** $\begin{bmatrix} 60 & 72 \\ -20 & -24 \\ 10 & 12 \\ 60 & 72 \end{bmatrix}$

25. $\begin{bmatrix} -6 & -9 \\ -1 & 0 \\ 17 & -10 \end{bmatrix}$ **27.** $\begin{bmatrix} 3 & 3 \\ -\frac{1}{2} & 0 \\ -\frac{13}{2} & \frac{11}{2} \end{bmatrix}$

29. $A = \begin{bmatrix} -1 & 1 \\ -2 & 1 \end{bmatrix}$, $X = \begin{bmatrix} x \\ y \end{bmatrix}$, $B = \begin{bmatrix} 4 \\ 0 \end{bmatrix}$, $x = 4$, $y = 8$

31. $(1, 1)$

33. $A = \begin{bmatrix} 1 & -2 & 3 \\ -1 & 3 & -1 \\ 2 & -5 & 5 \end{bmatrix}$, $X = \begin{bmatrix} x \\ y \\ z \end{bmatrix}$, $B = \begin{bmatrix} 9 \\ -6 \\ 17 \end{bmatrix}$

$x = 1$, $y = -1$, $z = 2$

35. $AC = BC = \begin{bmatrix} 12 & -6 & 9 \\ 16 & -8 & 12 \\ 4 & -2 & 3 \end{bmatrix}$

37. $\begin{bmatrix} 110 & 99 & 77 & 33 \\ 44 & 22 & 66 & 66 \end{bmatrix}$

39. $BA = \begin{bmatrix} \$1250 & \$1331.25 & \$981.25 \end{bmatrix}$

41. [\$497,500, \$494,500]

Each entry represents the price of the inventory at each warehouse.

43. $\begin{bmatrix} -2 & 3 \\ 1 & -1 \end{bmatrix}$ **45.** $\begin{bmatrix} 0.40 & 0.15 & 0.15 \\ 0.28 & 0.53 & 0.17 \\ 0.32 & 0.32 & 0.68 \end{bmatrix}$

Matrix is unique.

47. Player A: \$4000
Player B: \$3250
Player C: \$4250

Section 7.3   *(page 534)*

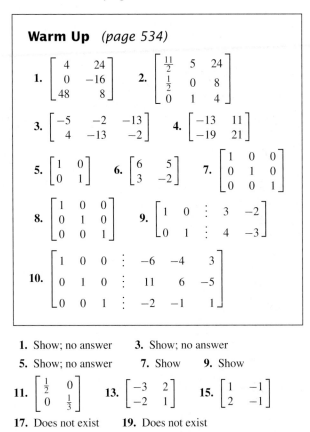

Warm Up   *(page 534)*

1. $\begin{bmatrix} 4 & 24 \\ 0 & -16 \\ 48 & 8 \end{bmatrix}$ **2.** $\begin{bmatrix} \frac{11}{2} & 5 & 24 \\ \frac{1}{2} & 0 & 8 \\ 0 & 1 & 4 \end{bmatrix}$

3. $\begin{bmatrix} -5 & -2 & -13 \\ 4 & -13 & -2 \end{bmatrix}$ **4.** $\begin{bmatrix} -13 & 11 \\ -19 & 21 \end{bmatrix}$

5. $\begin{bmatrix} 1 & 0 \\ 0 & 1 \end{bmatrix}$ **6.** $\begin{bmatrix} 6 & 5 \\ 3 & -2 \end{bmatrix}$ **7.** $\begin{bmatrix} 1 & 0 & 0 \\ 0 & 1 & 0 \\ 0 & 0 & 1 \end{bmatrix}$

8. $\begin{bmatrix} 1 & 0 & 0 \\ 0 & 1 & 0 \\ 0 & 0 & 1 \end{bmatrix}$ **9.** $\left[\begin{array}{cc:cc} 1 & 0 & 3 & -2 \\ 0 & 1 & 4 & -3 \end{array}\right]$

10. $\left[\begin{array}{ccc:ccc} 1 & 0 & 0 & -6 & -4 & 3 \\ 0 & 1 & 0 & 11 & 6 & -5 \\ 0 & 0 & 1 & -2 & -1 & 1 \end{array}\right]$

1. Show; no answer   **3.** Show; no answer

5. Show; no answer   **7.** Show   **9.** Show

11. $\begin{bmatrix} \frac{1}{2} & 0 \\ 0 & \frac{1}{3} \end{bmatrix}$ **13.** $\begin{bmatrix} -3 & 2 \\ -2 & 1 \end{bmatrix}$ **15.** $\begin{bmatrix} 1 & -1 \\ 2 & -1 \end{bmatrix}$

17. Does not exist   **19.** Does not exist

21. $\begin{bmatrix} 1 & 1 & -1 \\ -3 & 2 & -1 \\ 3 & -3 & 2 \end{bmatrix}$ **23.** $\begin{bmatrix} -175 & 37 & -13 \\ 95 & -20 & 7 \\ 14 & -3 & 1 \end{bmatrix}$

25. $\dfrac{1}{2}\begin{bmatrix} -3 & 3 & 2 \\ 9 & -7 & -6 \\ -2 & 2 & 2 \end{bmatrix}$ **27.** $\dfrac{5}{11}\begin{bmatrix} 0 & -4 & 2 \\ -22 & 11 & 11 \\ 22 & -6 & -8 \end{bmatrix}$

29. $\begin{bmatrix} 1 & 0 & 0 \\ -0.75 & 0.25 & 0 \\ 0.35 & -0.25 & 0.2 \end{bmatrix}$ **31.** Does not exist

33. $\begin{bmatrix} -\frac{1}{8} & 0 & 0 & 0 \\ 0 & 1 & 0 & 0 \\ 0 & 0 & \frac{1}{4} & 0 \\ 0 & 0 & 0 & -\frac{1}{5} \end{bmatrix}$

35. $\begin{bmatrix} -24 & 7 & 1 & -2 \\ -10 & 3 & 0 & -1 \\ -29 & 7 & 3 & -2 \\ 12 & -3 & -1 & 1 \end{bmatrix}$

37. $(4, 8)$ **39.** $(10, 30)$ **41.** $(-2, 3)$

43. $(2, 0)$ **45.** $\left(-5, -\frac{81}{2}, 48\right)$ **47.** $(2, 1, 0, 0)$

49. (a) $\begin{bmatrix} 0.2571 & -0.0286 \\ -0.0286 & 0.0044 \end{bmatrix}$ **51.** $\begin{aligned} 2x + y + 3z &= 16 \\ 4x \quad\;\; - 2z &= -2 \\ 3y + 2z &= 1 \end{aligned}$
$21.89 + 1.05t = y$
(b) $\$38,690$

53. $\$22,500$ invested in AAA bonds
$\$5000$ invested in A bonds
$\$10,000$ invested in B bonds

55. $\$20,000$ invested in AAA bonds
$\$6000$ invested in A bonds
$\$12,000$ invested in B bonds

57. 15 computer chips
10 resistors
10 transistors

59. 70 computer chips
30 resistors
10 transistors

61. 75 muffins
200 cookies
100 brownies

Mid-Chapter Quiz *(page 538)*

1. Matrix with 3 rows and 4 columns

2. Matrix with 1 row and 3 columns

3. $\begin{aligned} 3x + 2y &= -2 \\ 5x - y &= 19 \end{aligned}$ **4.** $\begin{aligned} x \quad\;\;\; + 3z &= -5 \\ x + 2y - z &= 3 \\ 3x \quad\;\; + 4z &= 0 \end{aligned}$

5. $(2.769, -5.154)$ **6.** $(4, -2, -3)$

7. $\begin{bmatrix} 2 & -10 \\ 15 & 11 \end{bmatrix}$ **8.** $\begin{bmatrix} -5 & 2 & -13 \\ 5 & 6 & 11 \end{bmatrix}$

9. $\begin{bmatrix} 1 & 2 \\ -3 & 2 \end{bmatrix}$ **10.** $\begin{bmatrix} -6 & -2 \\ 3 & -5 \end{bmatrix}$ **11.** $\begin{bmatrix} 0.4 & 0.2 \\ -0.3 & 0.1 \end{bmatrix}$

12. Not possible **13.** $\begin{bmatrix} 2 & 2 \\ -3 & 5 \end{bmatrix}$ **14.** $\begin{bmatrix} \frac{3}{2} & -2 \\ 3 & \frac{11}{2} \end{bmatrix}$

15. $A = \begin{bmatrix} 1 & -3 \\ -2 & 1 \end{bmatrix}, X = \begin{bmatrix} x \\ y \end{bmatrix}, B = \begin{bmatrix} 10 \\ -10 \end{bmatrix},$
$x = 4, y = -2$

16. $A = \begin{bmatrix} 2 & -1 & 1 \\ 3 & 0 & -1 \\ 0 & 4 & 3 \end{bmatrix}, X = \begin{bmatrix} x \\ y \\ z \end{bmatrix}, B = \begin{bmatrix} 3 \\ 15 \\ -1 \end{bmatrix}$
$x = 4, y = 2, z = -3$

17. $\$34.40$ **18.** $\$72.00$ **19.** $\$101.10$

20. $\begin{bmatrix} 17.8 & 16.6 \\ 37.8 & 34.2 \\ 52.1 & 49.0 \end{bmatrix}$

Total labor for each model listed by plant.

Section 7.4 *(page 546)*

Warm Up *(page 546)*

1. $\begin{bmatrix} 3 & 5 \\ 4 & 0 \end{bmatrix}$ **2.** $\begin{bmatrix} -2 & 8 \\ 2 & -4 \end{bmatrix}$

3. $\begin{bmatrix} 9 & -12 & 6 \\ 3 & 0 & -3 \\ 0 & 3 & -6 \end{bmatrix}$ **4.** $\begin{bmatrix} 0 & 8 & 12 \\ -4 & 8 & 12 \\ -8 & 4 & -8 \end{bmatrix}$

5. -22 **6.** 35 **7.** -15 **8.** $-\frac{1}{8}$

9. -45 **10.** -16

1. 5 **3.** 5 **5.** 27 **7.** -24

9. 6 **11.** 0 **13.** -0.002 **15.** 0

17. 0 **19.** -9 **21.** -18 **23.** 120

25. (a) $M_{11} = -5, M_{12} = 2, M_{21} = 4, M_{22} = 3$
(b) $C_{11} = -5, C_{12} = -2, C_{21} = -4, C_{22} = 3$

27. (a) $M_{11} = 30$, $M_{12} = 12$, $M_{13} = 11$, $M_{21} = -36$,

$M_{22} = 26$, $M_{23} = 7$, $M_{31} = -4$,

$M_{32} = -42$, $M_{33} = 12$

(b) $C_{11} = 30$, $C_{12} = -12$, $C_{13} = 11$, $C_{21} = 36$,

$C_{22} = 26$, $C_{23} = -7$, $C_{31} = -4$,

$C_{32} = 42$, $C_{33} = 12$

29. -75 **31.** 96   **33.** 170   **35.** -58

37. -30 **39.** -108 **41.** 0   **43.** 412

45. Matrices vary. Product of the main diagonals equals -24.

47. Rows 2 and 4 are identical.

Section 7.5 *(page 555)*

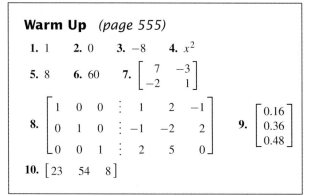

Warm Up *(page 555)*

1. 1   **2.** 0   **3.** -8 **4.** x^2

5. 8   **6.** 60   **7.** $\begin{bmatrix} 7 & -3 \\ -2 & 1 \end{bmatrix}$

8. $\left[\begin{array}{ccc:ccc} 1 & 0 & 0 & 1 & 2 & -1 \\ 0 & 1 & 0 & -1 & -2 & 2 \\ 0 & 0 & 1 & 2 & 5 & 0 \end{array}\right]$ **9.** $\begin{bmatrix} 0.16 \\ 0.36 \\ 0.48 \end{bmatrix}$

10. $\begin{bmatrix} 23 & 54 & 8 \end{bmatrix}$

1. 7   **3.** 14   **5.** $\frac{33}{8}$ **7.** $\frac{5}{2}$ **9.** 28

11. $x = \frac{16}{5}$ or $x = 0$ **13.** 250 square miles

15. Collinear   **17.** Not collinear   **19.** Collinear

21. $3x - 5y = 0$ **23.** $x + 3y - 5 = 0$

25. $2x + 3y - 8 = 0$ **27.** $x = -3$

29. [3 15] [13 5] [0 8] [15 13] [5 0] [19 15] [15 14]

48, 81, 28, 51, 24, 40, 54, 95, 5, 10, 64, 113, 57, 100

31. [20 18 15] [21 2 12] [5 0 9] [14 0 18]

[9 22 5] [18 0 3] [9 20 25]

-52, 10, 27, -49, 3, 34, -49, 13, 27, -94, 22, 54,

1, 1, -7, 0, -12, 9, -121, 41, 55

33. 1, -25, -65, 17, 15, -9, -12, -62, -119,

27, 51, 48, 43, 67, 48, 57, 111, 117

35. -5, -41, -87, 91, 207, 257, 11, -5,

-41, 40, 80, 84, 76, 177, 227

37. HAPPY NEW YEAR   **39.** FILM AT ELEVEN

41. SEND PLANES   **43.** MEET ME TONIGHT RON

Review Exercises *(page 560)*

1. $\begin{bmatrix} 1 & 3 & 0 & 2 \\ 0 & 1 & 1 & 2 \\ 0 & 0 & 1 & 2 \end{bmatrix}$ **3.** $\begin{bmatrix} 1 & 0 & 0 \\ 0 & 1 & 0 \\ 0 & 0 & 1 \end{bmatrix}$

5. $3x - 2y + 4z = 9$ **7.** $x = 3$

$x \qquad + 5z = 11$ $\quad y = -2$

$-2x + y + 3z = -5$

9. $x = 3$ **11.** $x = 10 - 4a$

$\quad y = 2$ $\quad y = a$

$\quad z = -1$ $\quad z = a$

13. No solution   **15.** \$100,000 borrowed at 9%

\$300,000 borrowed at 10%

\$200,000 borrowed at 12%

17. (a) $\begin{bmatrix} 3 & 1 \\ 5 & 5 \end{bmatrix}$ (b) $\begin{bmatrix} -1 & 5 \\ -3 & -9 \end{bmatrix}$

(c) $\begin{bmatrix} 4 & 12 \\ 4 & -8 \end{bmatrix}$ (d) $\begin{bmatrix} -2 & 18 \\ -8 & -29 \end{bmatrix}$

19. (a) $\begin{bmatrix} 3 & 4 & 2 & 1 \\ 3 & -5 & 6 & 0 \end{bmatrix}$ (b) $\begin{bmatrix} -1 & 2 & -6 & 11 \\ -3 & 7 & 0 & 4 \end{bmatrix}$

(c) $\begin{bmatrix} 4 & 12 & -8 & 24 \\ 0 & 4 & 12 & 8 \end{bmatrix}$

(d) $\begin{bmatrix} -2 & 9 & -20 & 39 \\ -9 & 22 & 3 & 14 \end{bmatrix}$

21. (a) $\begin{bmatrix} 11 & 8 \\ 16 & 10 \end{bmatrix}$ (b) $\begin{bmatrix} 0 & 2 \\ 9 & 21 \end{bmatrix}$ (c) $\begin{bmatrix} 7 & 15 \\ 10 & 22 \end{bmatrix}$

23. (a) $\begin{bmatrix} 12 & 8 & -4 \\ -3 & -10 & 5 \\ 8 & 2 & -1 \end{bmatrix}$ (b) $\begin{bmatrix} 2 & 0 & 4 \\ -4 & -1 & 7 \\ 15 & 2 & 0 \end{bmatrix}$

(c) $\begin{bmatrix} 3 & 2 & 4 \\ 4 & -1 & 2 \\ 5 & 2 & 1 \end{bmatrix}$

25. $\begin{bmatrix} 0 & 11 \\ 7 & 21 \\ -2 & 3 \end{bmatrix}$ **27.** $\begin{bmatrix} 1 & 0 & 0 \\ 0 & 1 & 0 \\ 0 & 0 & 1 \end{bmatrix}$ **29.** Not possible

31. $\begin{bmatrix} 5 & -13 \\ -3 & 2 \\ 1 & 0 \end{bmatrix}$ **33.** $\begin{bmatrix} 96 & 84 & 108 & 48 \\ 60 & 36 & 96 & 24 \\ 108 & 72 & 120 & 60 \end{bmatrix}$

35. $ST = \begin{bmatrix} 3000 & 4850 \\ 4600 & 7250 \\ 3000 & 4750 \end{bmatrix}$

This represents the wholesale and retail prices of the inventory at each outlet.

37. BA is the 3×3 identity matrix, and so is AB.

39. $\begin{bmatrix} -5 & 3 \\ 2 & -1 \end{bmatrix}$ **41.** $\begin{bmatrix} -1 & 0 & 0 & 0 \\ 0 & \frac{1}{2} & 0 & 0 \\ 0 & 0 & \frac{1}{4} & 0 \\ 0 & 0 & 0 & \frac{1}{6} \end{bmatrix}$

43. $x = 3$ **45.** $x = 2$ **47.** 24 of model A
$y = 4$ $y = \frac{1}{2}$ 16 of model B
$z = 3$ 7 of model C

49. \$20,000 in blue-chip stock   **51.** 27
\$4000 in common stock
\$8000 in municipal bonds

53. 0   **55.** 44   **57.** 120   **59.** 60   **61.** 36

63. 8   **65.** Not collinear   **67.** $x + 4y - 3 = 0$

69. [8 1] [16 16] [25 0] [1 14] [14 9] [22 5] [18 19]
[1 18] [25 0]
Encoded: 19, 28, 80, 112, 50, 75, 44, 59, 55, 78, 59, 86,
93, 130, 56, 75, 50, 75

71. HAPPY HOLIDAYS

Chapter Test *(page 564)*

1. $\begin{bmatrix} 2 & -1 & \vdots & -3 \\ 1 & 1 & \vdots & 2 \end{bmatrix}$ **2.** $\begin{bmatrix} 3 & 2 & 2 & \vdots & 17 \\ 0 & 2 & 1 & \vdots & -1 \\ 1 & 0 & 1 & \vdots & 8 \end{bmatrix}$

3. $(2, -3)$ **4.** $(-12, -16, -6)$ **5.** $(2, -3, 1)$

6. $\begin{bmatrix} 2 & 4 \\ 7 & 13 \end{bmatrix}$ **7.** $\begin{bmatrix} 14 & -1 & 6 \\ 20 & -2 & 10 \end{bmatrix}$

8. $\begin{bmatrix} 1 \\ 11 \end{bmatrix}$ **9.** $\begin{bmatrix} 7 & 15 \\ 10 & 22 \end{bmatrix}$ **10.** $\begin{bmatrix} 0.8 & 0.2 \\ 0.6 & 0.4 \end{bmatrix}$

11. $\begin{bmatrix} 1 & 0 \\ 0 & 1 \end{bmatrix}$ **12.** $\begin{bmatrix} 0.5 & -0.5 & -0.5 \\ 0.25 & 0.25 & -0.75 \\ -0.5 & 0.5 & 1.5 \end{bmatrix}$

13. 25   **14.** -23 **15.** -20 **16.** $(5, -2, 3)$

17. Matrices vary.

$\begin{bmatrix} 2 & -2 \\ -2 & 2 \end{bmatrix} \times \begin{bmatrix} 3 & 3 \\ 3 & 3 \end{bmatrix}$

18. Yes, collinear   **19.** $\frac{15}{2}$

20. $\begin{bmatrix} 225,000 & 411,000 \end{bmatrix}$

Total expense of stocking each warehouse.

CHAPTER 8

Section 8.1 *(page 574)*

Warm Up *(page 574)*

1. $\frac{4}{5}$ **2.** $\frac{1}{3}$ **3.** $(2n + 1)(2n - 1)$

4. $(2n - 1)(2n - 3)$ **5.** $(n - 1)(n - 2)$

6. $(n + 1)(n + 2)$ **7.** $\frac{1}{3}$

8. 24   **9.** $\frac{13}{24}$ **10.** $\frac{3}{4}$

1. 3, 5, 7, 9, 11   **3.** 2, 4, 8, 16, 32

5. $-2, 4, -8, 16, -32$ **7.** $\frac{1}{2}, \frac{2}{3}, \frac{3}{4}, \frac{4}{5}, \frac{5}{6}$

9. 0, 2, 0, 2, 0   **11.** $\frac{5}{2}, \frac{11}{4}, \frac{23}{8}, \frac{47}{16}, \frac{95}{32}$

13. $3, \frac{9}{2}, \frac{9}{2}, \frac{27}{8}, \frac{81}{40}$ **15.** $5, \frac{19}{4}, \frac{43}{9}, \frac{77}{16}, \frac{121}{25}$

17. $-1, \frac{1}{4}, -\frac{1}{9}, \frac{1}{16}, -\frac{1}{25}$ **19.** 3, 6, 9, 12, 15

21. 0, 1, 2, 3, 4   **23.** 30   **25.** 90

27. $n + 1, 1, 2$ **29.** $\dfrac{1}{2n(2n + 1)}$, undefined, $\dfrac{1}{6}$

31. $a_n = 3n - 2$ **33.** $a_n = n^2 - 1$ **35.** $a_n = \dfrac{n + 1}{n + 2}$

37. $a_n = \dfrac{(-1)^{n+1}}{2^n}$ **39.** $a_n = 1 + \dfrac{1}{n}$ **41.** $a_n = \dfrac{1}{n!}$

43. $a_n = (-1)^{n+1}$ **45.** 48   **47.** 35   **49.** 40

51. 30   **53.** $\frac{9}{5}$ **55.** 14   **57.** 4   **59.** $\displaystyle\sum_{i=1}^{9} \frac{1}{3i}$

61. $\displaystyle\sum_{i=1}^{8} \left[2\left(\frac{i}{8}\right) + 3 \right]$ **63.** $\displaystyle\sum_{i=1}^{6} (-1)^{i+1} 3^i$

65. $\displaystyle\sum_{i=1}^{20} \frac{(-1)^{i+1}}{i^2}$ **67.** $\displaystyle\sum_{i=1}^{5} \frac{2^i - 1}{2^{i+1}}$

69. (a) $A_1 = \$5100.00$, $A_2 = \$5202.00$, $A_3 = \$5306.04$,
$A_4 = \$5412.16$, $A_5 = \$5520.40$, $A_6 = \$5630.81$,
$A_7 = \$5743.43$, $A_8 = \$5858.30$

(b) \$11,040.20   (c) No, the balance is \$24,377.20.

71. (a) $a_1 = 1.064$
(b)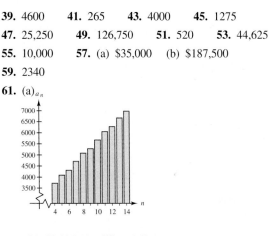
$a_2 = 1.041$
$a_3 = 1.020$
$a_4 = 1.001$
$a_5 = 0.985$
$a_6 = 0.971$
$a_7 = 0.960$
$a_8 = 0.951$
$a_9 = 0.944$

(c) 128.6 million, 121.4 million

73. $a_0 = \$462.30$
$a_1 = \$498.13$
$a_2 = \$538.96$
$a_3 = \$584.79$
$a_4 = \$635.62$
$a_5 = \$691.45$
$a_6 = \$752.28$
$a_7 = \$818.11$

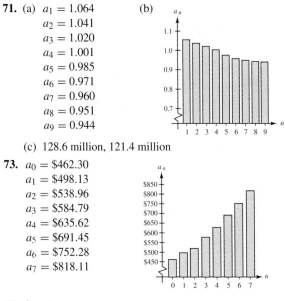

75. $9.75

Section 8.2 *(page 583)*

Warm Up *(page 583)*

1. 36   **2.** 240   **3.** $\frac{11}{2}$ **4.** $\frac{10}{3}$ **5.** 18
6. 4   **7.** 143   **8.** 160   **9.** 430   **10.** 256

1. Arithmetic sequence, $d = 3$

3. Not an arithmetic sequence

5. Arithmetic sequence, $d = -\frac{1}{4}$

7. Not an arithmetic sequence

9. Arithmetic sequence, $d = 0.4$

11. 8, 11, 14, 17, 20; arithmetic sequence, $d = 3$

13. $\frac{1}{2}, \frac{1}{3}, \frac{1}{4}, \frac{1}{5}, \frac{1}{6}$; not an arithmetic sequence.

15. 97, 94, 91, 88, 85; arithmetic sequence, $d = -3$

17. 1, 1, 1, 1, 1; arithmetic sequence, $d = 0$

19. $a_n = 3n - 2$ **21.** $a_n = -8n + 108$

23. $a_n = 3n + 2$ **25.** $a_n = 3n$

27. $a_n = 6n - 4$ **29.** 5, 11, 17, 23, 29

31. 4, 10, 16, 22, 28   **33.** 2, 6, 10, 14, 18

35. $-2, 2, 6, 10, 14$ **37.** 620

39. 4600   **41.** 265   **43.** 4000   **45.** 1275

47. 25,250   **49.** 126,750   **51.** 520   **53.** 44,625

55. 10,000   **57.** (a) \$35,000   (b) \$187,500

59. 2340

61. (a)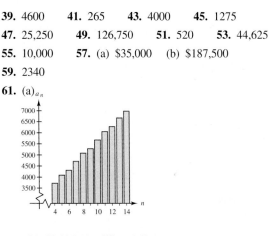

(b) 59,005.10 million dollars

63. 1024 feet   **65.** 126 logs   **67.** \$150,000

Section 8.3 *(page 592)*

Warm Up *(page 592)*

1. $\frac{64}{125}$ **2.** $\frac{9}{16}$ **3.** $\frac{1}{16}$ **4.** $\frac{5}{81}$
5. $6n^3$ **6.** $27n^4$ **7.** $4n^3$ **8.** n^2
9. $\displaystyle\sum_{n=1}^{3} 2(3)^{n-1}$ **10.** $\displaystyle\sum_{n=1}^{3} 3(2)^{n-1}$

1. Geometric sequence, $r = 3$, $a_n = 5(3)^{n-1}$

3. Not a geometric sequence

5. Geometric sequence, $r = -\frac{1}{2}$, $a_n = \left(-\frac{1}{2}\right)^{n-1}$

7. Not a geometric sequence

9. Not a geometric sequence

11. 2, 6, 18, 54, 162, . . .   **13.** $1, \frac{1}{2}, \frac{1}{4}, \frac{1}{8}, \frac{1}{16}, \ldots$

15. $5, -\frac{1}{2}, \frac{1}{20}, -\frac{1}{200}, \frac{1}{2000}, \ldots$ **17.** $1, e, e^2, e^3, e^4$

19. $\left(\frac{1}{2}\right)^7$ **21.** $-\frac{2}{3^{10}}$ **23.** $100e^8$ **25.** $500(1.02)^{39}$

27. 9   **29.** $-\frac{2}{9}$ **31.** 16,368   **33.** 10.67   **35.** 511

37. 29,921.31   **39.** 2   **41.** $\frac{2}{3}$ **43.** $\frac{16}{3}$

45. (a) \$7011.89   (b) \$189.16   (c) \$17,409.45

47. \$13,812.43; \$13,833.34   **49.** \$1746.5 million

51. $3465.71 million **53.** $5,368,709.11 for 29 days
$10,737,418.23 for 30 days
$21,474,836.47 for 31 days

55. 126 square inches

Mid-Chapter Quiz *(page 596)*

1. 2, 5, 8, 11, 14 **2.** $\frac{1}{2}, \frac{2}{3}, \frac{3}{4}, \frac{4}{5}, \frac{5}{6}$ **3.** 210

4. 17,550 **5.** $\frac{2}{n^2}$ **6.** $\frac{1}{n+1}$ **7.** $5n - 3$

8. $3n + 4$ **9.** Arithmetic, common difference is 3

10. Geometric, common ratio is -2

11. Geometric, common ratio is $\frac{1}{2}$

12. Geometric, common ratio is 2 **13.** 25

14. 60 **15.** 650 **16.** 650 **17.** $\frac{3}{2}$ **18.** $\frac{25}{4}$

19. (a) $36,000 (b) $193,500 **20.** $8234.94

Section 8.4 *(page 602)*

Warm Up *(page 602)*

1. $5x^5 + 15x^2$ **2.** $x^3 + 5x^2 - 3x - 15$

3. $x^2 + 8x + 16$ **4.** $4x^2 - 12x + 9$ **5.** $\dfrac{3x^3}{y}$

6. $-32z^5$ **7.** 120 **8.** 336

9. 720 **10.** 20

---

1. 10 **3.** 1 **5.** 15,504 **7.** 4950 **9.** 4950

11. $x^4 + 4x^3 + 6x^2 + 4x + 1$ **13.** $x^3 + 6x^2 + 12x + 8$

15. $x^4 - 8x^3 + 24x^2 - 32x + 16$

17. $x^5 + 5x^4y + 10x^3y^2 + 10x^2y^3 + 5xy^4 + y^5$

19. $a^4 + 8a^3 + 24a^2 + 32a + 16$

21. $64 + 576s + 2160s^2 + 4320s^3 + 4860s^4$
$+ 2916s^5 + 729s^6$

23. $x^5 - 5x^4y + 10x^3y^2 - 10x^2y^3 + 5xy^4 - y^5$

25. $1 - 6x + 12x^2 - 8x^3$

27. $x^8 + 20x^6 + 150x^4 + 500x^2 + 625$

29. $\dfrac{1}{x^5} + \dfrac{5y}{x^4} + \dfrac{10y^2}{x^3} + \dfrac{10y^3}{x^2} + \dfrac{5y^4}{x} + y^5$

31. -4 **33.** $2035 + 828i$ **35.** 1

37. $32t^5 - 80t^4 + 80t^3 - 40t^2 + 10t - 1$

39. $81 - 216z + 216z^2 - 96z^3 + 16z^4$ **41.** $1,732,104x^5$

43. $180x^8y^2$ **45.** $-326,592x^4y^5$ **47.** $70x^8$

49. $\frac{1}{128} + \frac{7}{128} + \frac{21}{128} + \frac{35}{128} + \frac{35}{128} + \frac{21}{128} + \frac{7}{128} + \frac{1}{128} = 1$

51. $\frac{1}{6561} + \frac{16}{6561} + \frac{112}{6561} + \frac{448}{6561} + \frac{1120}{6561}$
$+ \frac{1792}{6561} + \frac{1792}{6561} + \frac{1024}{6561} + \frac{256}{6561} = 1$

53. $0.07776 + 0.2592 + 0.3456 + 0.2304 + 0.0768$
$+ 0.01024 = 1$

55. Fibonacci sequence

57. Rows 8, 9, 10   Rows 6, 7, 8

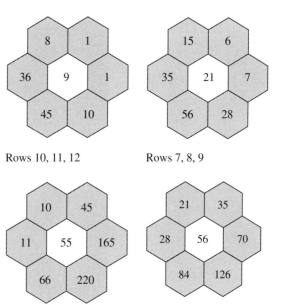

Rows 10, 11, 12   Rows 7, 8, 9

Section 8.5 *(page 611)*

Warm Up *(page 611)*

1. 6656 **2.** 291,600 **3.** 7920 **4.** 13,800

5. 792 **6.** 2300 **7.** $n(n-1)(n-2)(n-3)$

8. $n(n-1)(2n-1)$ **9.** $n!$ **10.** $n!$

---

1. 12

3. 12

1st	2nd	3rd	4th
1	2	3	4
1	2	4	3
1	3	2	4
1	3	4	2
1	4	2	3
1	4	3	2
2	1	3	4
2	1	4	3
2	3	1	4
2	3	4	1
2	4	1	3
2	4	3	1

5. 6,760,000   **7.** $2^6 = 64$

9. (a) 900   (b) 648   (c) 180   **11.** $40^3 = 64,000$

13. (a) 720   (b) 48   **15.** 24   **17.** 336

19. 1,860,480   **21.** 9900   **23.** 120   **25.** 120

27. 11,880   **29.** 4845   **31.** 3,838,380

33. 3,921,225   **35.** 560   **37.** (a) 70   (b) 30

39. 3744   **41.** 5   **43.** 20

45. 420   **47.** 1260   **49.** 2520

51.

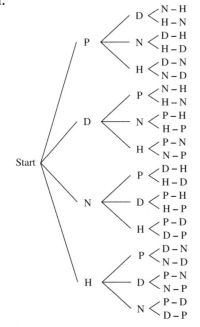

Section 8.6   *(page 623)*

Warm Up   *(page 623)*

1. $\dfrac{9}{16}$ **2.** $\dfrac{8}{15}$ **3.** $\dfrac{1}{6}$ **4.** $\dfrac{1}{80,730}$

5. $\dfrac{1}{804,722,688,000}$ **6.** $\dfrac{1}{24}$ **7.** $\dfrac{1}{12}$

8. $\dfrac{135}{323}$ **9.** 0.366   **10.** 0.997

---

1. $\dfrac{3}{8}$ **3.** $\dfrac{7}{8}$ **5.** $\dfrac{1}{12}$ **7.** $\dfrac{7}{12}$ **9.** $\dfrac{1}{3}$ **11.** $\dfrac{3}{13}$

13. $\dfrac{5}{13}$ **15.** $\dfrac{1}{5}$ **17.** $\dfrac{2}{5}$ **19.** 0.3   **21.** 0.85

23. 0.19   **25.** (a) $\dfrac{5}{12}$ (b) $\dfrac{7}{12}$ (c) $\dfrac{5}{36}$

27. (a) $\dfrac{1}{4}$ (b) $\dfrac{1}{2}$ (c) $\dfrac{9}{100}$ (d) $\dfrac{1}{30}$

29. $\dfrac{3}{28}$ **31.** (a) $\dfrac{1}{169}$ (b) $\dfrac{1}{221}$ **33.** (a) $\dfrac{1}{24}$ (b) $\dfrac{1}{6}$

35. (a) $\dfrac{1}{3}$ (b) $\dfrac{5}{8}$ **37.** $\dfrac{6}{4165}$

39. (a) 0.922   (b) 0.078

41. (a) 0.9702   (b) 0.9998   (c) 0.0002

43. (a) $\dfrac{1}{81}$ (b) $\dfrac{16}{81}$ (c) $\dfrac{65}{81}$

45. (a) $\dfrac{1}{64}$ (b) $\dfrac{1}{32}$ (c) $\dfrac{63}{64}$

47. 0.1024   **49.** 0.004

Review Exercises   *(page 629)*

1. 5, 8, 11, 14, 17   **3.** $\dfrac{2}{3}, \dfrac{4}{5}, \dfrac{6}{7}, \dfrac{8}{9}, \dfrac{10}{11}$

5. $-2, 4, -8, 16, -32$ **7.** 210   **9.** $a_n = 2n + 5$

11. $a_n = \dfrac{1}{n}$ **13.** $a_n = (-1)^n \dfrac{(n+1)}{n+2}$ **15.** 45

17. 100   **19.** $\displaystyle\sum_{k=1}^{12} \dfrac{3}{1+k}$ **21.** $\displaystyle\sum_{n=1}^{13} (-1)^{n-1} 2^n$

23. (a) \$2040.00, \$2080.80, \$2122.42, \$2164.86, \$2208.16, \$2252.32, \$2297.37, \$2343.32

  (b) \$4416.08
  (c) \$21,530.33

25. Arithmetic, $d = 4$ **27.** Arithmetic, $d = -\dfrac{1}{2}$

29. 7, 11, 15, 19, 23; arithmetic, $d = 4$ **31.** $a_n = 5n - 2$

33. 7, 12, 17, 22, 27   **35.** 300   **37.** 6540

39. (a) Company A: $32,000
Company B: $34,000
(b) $169,500
(c) $168,000
(d) For the short term, one should pick Company A, because at the end of 6 years Company A has paid $1500 more in salary.

41. Geometric; $r = -3$ **43.** Geometric; $r = \frac{1}{2}$

45. 2, 6, 18, 54, 162   **47.** 10, -2, $\frac{2}{5}$, $-\frac{2}{25}$, $\frac{2}{125}$

49. $a_{10} = \frac{1}{64}$ **51.** $a_{10} = -\frac{25}{64}$ **53.** 8184

55. 1.5

57. (a) $18,416.57   (b) No; it will be more because the interest is compound.

59. (a)

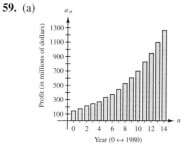

Year (0 ↔ 1980)

(b) $8202.73 million

61. 20   **63.** 1

65. $x^6 + 12x^5 + 60x^4 + 160x^3 + 240x^2 + 192x + 64$

67. $x^{10} + 10x^9 y + 45x^8 y^2 + 120x^7 y^3 + 210x^6 y^4$
$+ 252x^5 y^5 + 210x^4 y^6 + 120x^3 y^7 + 45x^2 y^8$
$+ 10xy^9 + y^{10}$

69. $x^8 - 16x^7 y + 112x^6 y^2 - 448x^5 y^3 + 1120x^4 y^4$
$- 1792x^3 y^5 + 1792x^2 y^6 - 1024xy^7 + 256y^8$

71. $81 - 216y + 216y^2 - 96y^3 + 16y^4$ **73.** $25,344x^7$

75. $\frac{1}{243} + \frac{10}{243} + \frac{40}{243} + \frac{80}{243} + \frac{80}{243} + \frac{32}{243}$

77. 48   **79.** 144   **81.** 1,048,576

83. (a) 90,000   (b) 27,216   (c) 45,000

85. 30   **87.** 720   **89.** 3136

91. (a) 30,240   (b) 1440   (c) 720   **93.** $\frac{5}{13}$ **95.** $\frac{2}{9}$

97. 0.52   **99.** (a) $\frac{1}{120}$ (b) $\frac{1}{6}$ **101.** $\frac{6}{4165}$

103. (a) $\frac{1}{1024}$ (b) $\frac{1}{512}$ (c) $\frac{1023}{1024}$

Chapter Test (page 633)

1. 4, 7, 10, 13, 16   **2.** $-1, 4, -9, 16, -25$

3. 1, 2, 6, 24, 120   **4.** $\frac{1}{2}, \frac{1}{4}, \frac{1}{8}, \frac{1}{16}, \frac{1}{32}$

5. Arithmetic, 5   **6.** Neither   **7.** Geometric, 2

8. Neither   **9.** 2600   **10.** $\approx 29,918.311$

11. $\frac{1}{2}$ **12.** $22,196.40   **13.** $16,469.87

14. $x^7 + 14x^6 + 84x^5 + 280x^4 + 560x^3 + 672x^2 + 448x + 128$

15. $-192,456$ **16.** 120   **17.** $2^{15} = 32,768$

18. 45,000   **19.** $\frac{1}{24}$ **20.** $\frac{1}{32}$

Cumulative Test: Chapters 6–8
(page 634)

1. $x = 8, y = 7$ **2.** $x = 3, y = 2$

3. $x = 2, y = -3, z = 5$ **4.** $x = -2, y = 4, z = -5$

5. $\begin{bmatrix} 1 & 4 & 8 \\ -1 & 4 & 3 \\ 0 & 1 & 5 \end{bmatrix}$ **6.** $\begin{bmatrix} 16 & 9 \\ 8 & 7 \\ 7 & 2 \end{bmatrix}$ **7.** $\begin{bmatrix} 2 & 12 \\ -5 & -3 \\ 1 & 9 \end{bmatrix}$

8. $\begin{bmatrix} \frac{1}{2} & -\frac{1}{2} & 0 \\ \frac{1}{4} & \frac{1}{4} & -1 \\ -\frac{1}{2} & \frac{1}{2} & 1 \end{bmatrix}$ **9.** -39 **10.** -7

11. $x = 18,000$ **12.** $x = 20,000$ at 7.5%
$p = 66$ $y = 25,000$ at 8.5%

13. $BA = \begin{bmatrix} 155,000 & 205,000 & 215,000 \end{bmatrix}$
$155,000 is the value of the inventory in warehouse 1.
$210,000 is the value of the inventory in warehouse 2.
$215,000 is the value of the inventory in warehouse 3.

14. 7, 11, 15, 19, 23   **15.** 3, $\frac{3}{2}, \frac{3}{4}, \frac{3}{8}, \frac{3}{16}$

16. $y^5 - 20y^4 + 160y^3 - 640y^2 + 1280y - 1024$

17. 120   **18.** 18,000   **19.** ≈ 0.985 **20.** 1024

Section A.1 (page A13)

Warm Up (page A13)

1. $9x^2 + 16y^2 = 144$ **2.** $x^2 + 4y^2 = 32$

3. $16x^2 - y^2 = 4$ **4.** $243x^2 + 36y^2 = 9$

5. $C = \pm 2\sqrt{2}$ **6.** $C = \pm\sqrt{13}$

7. $C = \pm 2\sqrt{3}$ **8.** $C = \pm\sqrt{5}$ **9.** 4   **10.** 2

1. (a)   **3.** (d)   **5.** (g)   **7.** (e)

9. Vertex: $(0, 0)$
   Focus: $\left(0, \frac{1}{16}\right)$

11. Vertex: $(0, 0)$
   Focus: $\left(-\frac{3}{2}, 0\right)$

13. Vertex: $(0, 0)$
   Focus: $(0, -2)$

15. Vertex: $(0, 0)$
   Focus: $(2, 0)$

17. $x^2 = -6y$ **19.** $y^2 = -8x$ **21.** $x^2 = 4y$

23. $x^2 = -8y$ **25.** $y^2 = 9x$

27. Center: $(0, 0)$
   Vertices: $(-5, 0), (5, 0)$

29. Center: $(0, 0)$
   Vertices: $(0, \pm 5)$

31. Center: $(0, 0)$
   Vertices: $(\pm 3, 0)$

33. Center: $(0, 0)$
   Vertices: $\left(0, \pm\sqrt{5}\right)$

35. $\dfrac{x^2}{1} + \dfrac{y^2}{4} = 1$ **37.** $\dfrac{x^2}{25} + \dfrac{y^2}{21} = 1$

39. $\dfrac{x^2}{36} + \dfrac{y^2}{11} = 1$ **41.** $\dfrac{x^2}{400/21} + \dfrac{y^2}{25} = 1$

43. Center: $(0, 0)$
   Vertices: $(\pm 1, 0)$

45. Center: $(0, 0)$
   Vertices: $(0, \pm 1)$

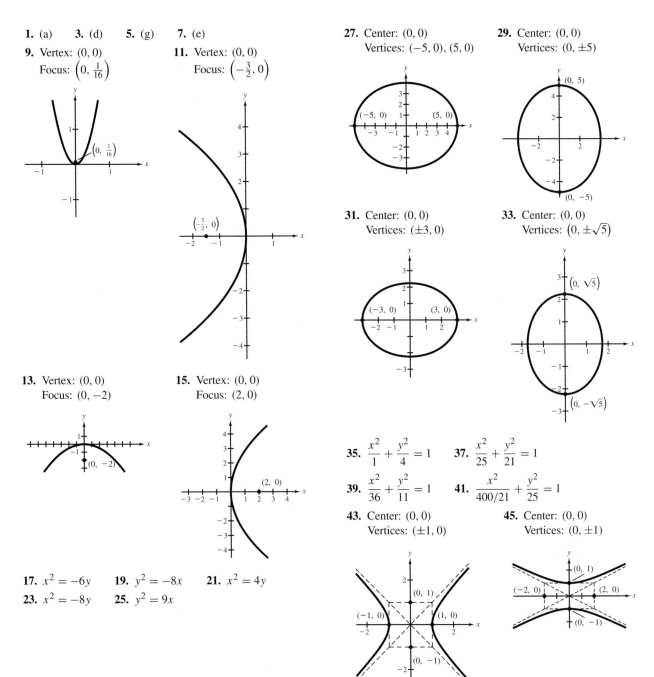

47. Center: $(0, 0)$
Vertices: $(0, \pm 5)$

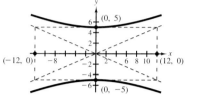

49. Center: $(0, 0)$
Vertices: $\left(\pm\sqrt{3}, 0 \right)$

51. $\dfrac{y^2}{4} - \dfrac{x^2}{12} = 1$

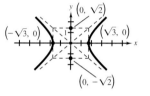

53. $\dfrac{x^2}{1} - \dfrac{y^2}{9} = 1$ **55.** $\dfrac{y^2}{1024/17} - \dfrac{x^2}{64/17} = 1$

57. $\dfrac{y^2}{9} - \dfrac{x^2}{9/4} = 1$ **59.** $x^2 = 12y$

61. Tacks should be placed at $\frac{3}{2}$ feet to the left and right of the
center along the base. The string should be 5 feet long.

63.

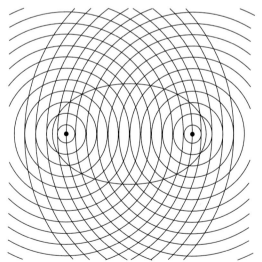

65.

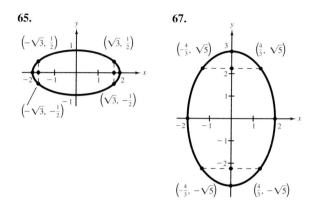

67.

69. $x \approx 110.3$ miles

Section A.2 *(page A23)*

Warm Up *(page A23)*

1. Hyperbola **2.** Ellipse **3.** Parabola

4. Hyperbola **5.** Ellipse **6.** Circle

7. Hyperbola **8.** Parabola **9.** Parabola

10. Ellipse

---

1. Vertex: $(1, -2)$
Focus: $(1, -4)$
Directrix: $y = 0$

3. Vertex: $\left(5, -\frac{1}{2} \right)$
Focus: $\left(\frac{11}{2}, -\frac{1}{2} \right)$
Directrix: $x = \frac{9}{2}$

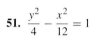

5. Vertex: $(1, 1)$
Focus: $(1, 2)$
Directrix: $y = 0$

7. Vertex: $(8, -1)$
Focus: $(9, -1)$
Directrix: $x = 7$

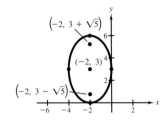

23. Center: $(-2, 3)$
Foci: $\left(-2, 3 - \sqrt{5}\right), \left(-2, 3 + \sqrt{5}\right)$
Vertices: $(-2, 0), (-2, 6)$

25. Center: $(1, -1)$
Foci: $\left(\frac{1}{4}, -1\right), \left(\frac{7}{4}, -1\right)$
Vertices: $\left(-\frac{1}{4}, -1\right), \left(\frac{9}{4}, -1\right)$

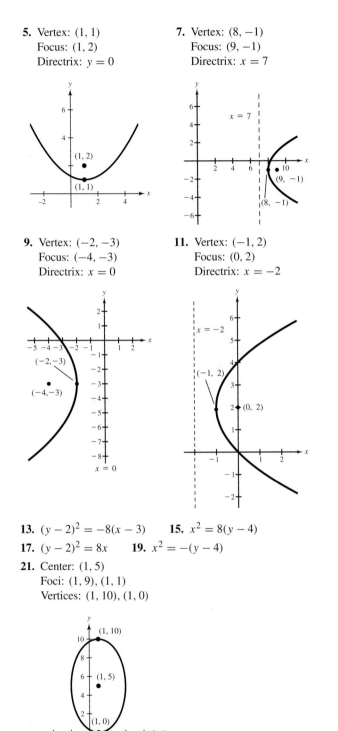

9. Vertex: $(-2, -3)$
Focus: $(-4, -3)$
Directrix: $x = 0$

11. Vertex: $(-1, 2)$
Focus: $(0, 2)$
Directrix: $x = -2$

27. Center: $\left(\frac{1}{2}, -1\right)$
Foci: $\left(\frac{1}{2} - \sqrt{2}, -1\right), \left(\frac{1}{2} + \sqrt{2}, -1\right)$
Vertices: $\left(\frac{1}{2} - \sqrt{5}, -1\right), \left(\frac{1}{2} + \sqrt{5}, -1\right)$

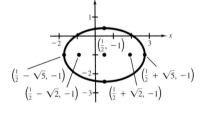

13. $(y - 2)^2 = -8(x - 3)$ **15.** $x^2 = 8(y - 4)$

17. $(y - 2)^2 = 8x$ **19.** $x^2 = -(y - 4)$

21. Center: $(1, 5)$
Foci: $(1, 9), (1, 1)$
Vertices: $(1, 10), (1, 0)$

29. $\dfrac{(x - 2)^2}{4} + \dfrac{(y - 2)^2}{1} = 1$ **31.** $\dfrac{x^2}{48} + \dfrac{(y - 4)^2}{64} = 1$

33. $\dfrac{(x - 3)^2}{9} + \dfrac{(y - 5)^2}{16} = 1$ **35.** $\dfrac{x^2}{16} + \dfrac{(y - 4)^2}{12} = 1$

37. Center: $(1, -2)$
Vertices: $(-1, -2), (3, -2)$
Foci: $\left(1 - \sqrt{5}, -2\right), \left(1 + \sqrt{5}, -2\right)$

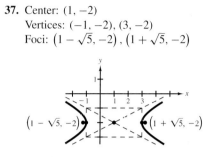

39. Center: $(2, -6)$
Vertices: $(2, -7), (2, -5)$
Foci: $\left(2, -6 - \sqrt{2}\right), \left(2, -6 + \sqrt{2}\right)$

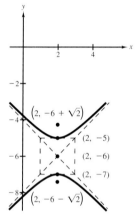

41. Center: $(2, -3)$
Vertices: $(3, -3), (1, -3)$
Foci: $\left(2 - \sqrt{10}, -3\right), \left(2 + \sqrt{10}, -3\right)$

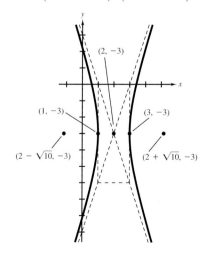

43. Center: $(1, -3)$
Vertices: $\left(1, -3 - \sqrt{2}\right), \left(1, -3 + \sqrt{2}\right)$
Foci: $\left(1, -3 - 2\sqrt{5}\right), \left(1, -3 + 2\sqrt{5}\right)$

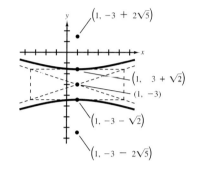

45. Center: $(-1, -3)$
Vertices: $\left(-1 - 3\sqrt{3}, -3\right), \left(-1 + 3\sqrt{3}, -3\right)$
Foci: $\left(-1 - \sqrt{30}, -3\right), \left(-1 + \sqrt{30}, -3\right)$

47. $\dfrac{(x - 4)^2}{4} - \dfrac{y^2}{12} = 1$ **49.** $\dfrac{(y - 5)^2}{16} - \dfrac{(x - 4)^2}{9} = 1$

51. $\dfrac{y^2}{9} - \dfrac{(x - 2)^2}{9/4} = 1$ **53.** $\dfrac{(x - 3)^2}{9} - \dfrac{(y - 2)^2}{4} = 1$

55. Circle   **57.** Hyperbola   **59.** Ellipse

61. Parabola   **63.** (a) 24,748.74 miles per hour
(b) $x^2 = -16,400(y - 4100)^2$

65. $\dfrac{x^2}{25} + \dfrac{y^2}{16} = 1$ **67.** Perihelion: 9.138×10^7 miles
Aphelion: 9.454×10^7 miles

69. $e \approx 5.431 \times 10^{-2}$

71.

$$\frac{(x-h)^2}{a^2} + \frac{(y-k)^2}{b^2} = 1 \qquad \text{Standard form of equation of ellipse}$$

$$\frac{(x-h)^2}{a^2} + \frac{(y-k)^2}{a^2-c^2} = 1 \qquad \text{In an ellipse, } c^2 = a^2 - b^2.$$

$$\frac{(x-h)^2}{a^2} + \frac{(y-k)^2}{a^2/a^2(a^2-c^2)} = 1 \qquad \text{Multiply denominator by } a^2/a^2.$$

$$\frac{(x-h)^2}{a^2} + \frac{(y-k)^2}{a^2(1-c^2/a^2)} = 1 \qquad \text{Distribute } 1/a^2 \text{ into } (a^2-c^2).$$

$$\frac{(x-h)^2}{a^2} + \frac{(y-k)^2}{a^2(1-(c/a)^2)} = 1 \qquad \text{Simplify.}$$

$$\frac{(x-h)^2}{a^2} + \frac{(y-k)^2}{a^2(1-e^2)} = 1 \qquad \text{Definition of } e$$

Section B.1 *(page A36)*

1.

Stems	Leaves
7	0 5 5 5 7 7 8 8 8
8	1 1 1 1 2 3 4 5 5 5 5 7 8 9 9 9
9	0 2 8
10	0 0

3.

Stems	Leaves
6	18 68 71 94
7	16 19 25 25 42 57 58 76 84 88
8	13 28 35 41 46 59 61 62 66 81 83 89 91 92 92
9	05 06 15 18 25 25 26 28 41 44 83 90 92
10	04 10 62 95
11	51 78 86
12	23
13	
14	
15	
16	26

5.

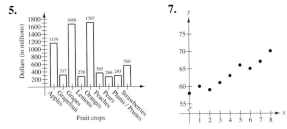

7.

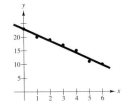

9. $y = 57.49 + 1.43x$; 71.8

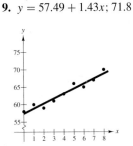

11.

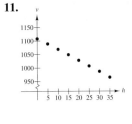

13. $v = 1117.3 + 4.1h$; 1006.6

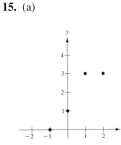

15. (a)

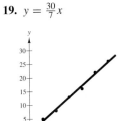

(b) $y = 1.1x + 1.2$; 0.7
(c) $y = \frac{11}{10}x + \frac{6}{5}$; 0.7

17. $y = \frac{16}{35}x + \frac{39}{35}$

19. $y = \frac{30}{7}x$

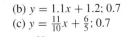

21. $y = -2.179x + 22.964$

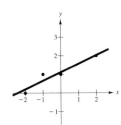

23. $y = 2.378x + 23.546$

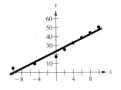

25. (a) and (c) $S = 384.1 + 21.2x$; $479.50
   $r = 0.996$

Section B.2 *(page A45)*

1. Mean: 8.86; median: 8; mode: 7

3. Mean: 10.29; median: 8; mode: 7

5. Mean: 9; median: 8; mode: 7

7. The mean is sensitive to extreme values.

9. Mean: $67.14; median: $65.35

11. Mean: 3.07; median: 3; mode: 3

13. One possibility: {4, 4, 10}

15. Mean: 76.6; median: 82; mode: 42
   The median gives the most representative description.

17. $\bar{x} = 6, v = 10, \sigma = 3.16$ **19.** $\bar{x} = 2, v = \frac{4}{3}, \sigma = 1.15$

21. $\bar{x} = 4, v = 4, \sigma = 2$ **23.** $\bar{x} = 47, v = 226, \sigma = 15.03$

25. 3.42   **27.** 101.55   **29.** 1.65

31. $\bar{x} = 12$ and $|x_i - 12| = 8$ for all x_i.

33. It will increase the mean by 5, but the standard deviation
   will not change.

35. With $\bar{x} = 235$ and $\sigma = 28$:
   At least 75% of the scores in [179, 291].
   At least 88.9% of the scores in [151, 319].
   With $\bar{x} = 235$ and $\sigma = 16$:
   At least 75% of the scores in [203, 267].
   At least 88.9% of the scores in [187, 283].

INDEX OF APPLICATIONS

A133

Chemistry and Physics Applications

Construction Applications

Consumer Applications

U.S. Demographics Applications

INDEX

A

Absolute value, 6
  equations, 129
  inequality, 141
  properties of, 7
  of a real number, 6
Addition, 12, 90
  of complex numbers, 349
  constant, 138
  of fractions, 15, 16
  of functions, 258
  of inequalities, 138
  of matrices, 515
  of polynomials, 45
  of rational expressions, 65
Additive identity
  for a complex number, 349
  for a matrix, 518
Additive Identity Property, 13
Additive inverse for a complex number, 349
Additive Inverse Property, 13
Adjoining, 530
Algebra,
  Basic Rules of, 12
  Fundamental Theorem of, 358
Algebraic expression, 11
  domain of, 62
  equivalent, 62
Algebraic function, 372
Approximately equal to, 3
Area, formulas for, 95
Area of a triangle, 549
Arithmetic
  combinations of functions, 258
  Fundamental Theorem of, 16
  mean, 40
  sequence, 577
    common difference of, 577
    nth term of, 578
Associative Property of Addition, 13
  for complex numbers, 349
  for matrices, 517

Associative Property of Multiplication, 13
  for complex numbers, 350
  for matrices, 517, 521
Asymptote
  of a hyperbola, A11
  vertical and horizontal, 301, 302
Augmented matrix, 500
Average, 40, A39
Axis
  coordinate, 256
  imaginary, 353
  of a parabola, 279, A3
  real, 353
  of symmetry, 279

B

Back-substitution, 435, 456, 505
Bar graph, A31
Base, 23
  of an exponential function, 372
  of a logarithmic function, 384
Basic graphing steps for the *TI-81* graphing calculator, 190
Basic Rules of Algebra, 13
Bell-shaped curve, 417
Bimodal, A39
Binomial, 45, 597
  coefficient, 597
  cube of, 48
  expansion, 600
  square of, 48
  theorem, 597
Bounded, 353
Bounded interval, 136
Bounds for real zeros of a polynomial, 335
Branch of a hyperbola, A9
Break-even point, 439

C

Cancellation Law, 17, 63

Cartesian plane, 166
Center
  of a circle, 181
  of an ellipse, A6
  of a hyperbola, A9
Certain event, 67, 68, 615
Change of base formula, 394
Chebychev's Theorem, A43
Circle
  center of, 181
  general form of equation of, 182
  radius of, 181
  standard form of equation of, 181
Closed interval, 136
Coded row matrices, 552
Coefficient, 11
  matrix, 501
  of a polynomial, 45
Cofactor
  expansion by, 542
  of a matrix, 541
Collinear points, test for, 550
Column
  matrix, 500
  subscript, 500
Combination of n elements taken r at a time, 609, 610
Common formulas
  for area and perimeter, 95
  for volume, 95
Common ratio, 586
Commutative Property of Addition, 13
  for complex numbers, 349
  for matrices, 517
Commutative Property of Multiplication, 13
  for complex numbers, 350
  for matrices, 517
Complement of an event, 621
Complete the square, 114, 182
Complex
  conjugates, 351
  fractions, 66
  number, definition of, 348
    addition of, 349

A137

INSTRUCTOR'S ANSWERS

Answers to all even-numbered exercises are given in the Instructor's Annotated Edition only on pages A145–A201. Answers to Discovery and Technology boxes (pages A189–A191) and Group Activities (pages A192–A196) are also available only in the Instructor's Annotated Edition. For answers to Warm Ups, odd-numbered exercises, Mid-Chapter Quizzes, Chapter Tests, Cumulative Tests, and Math Matters features, see the student answer section (Answers to Odd-Numbered Exercises, Quizzes, and Tests) on pages A71–A132.

CHAPTER P

Section P.1 *(page 9)*

2. (a) Natural: { }
(b) Integers: $\{-7, 0\}$
(c) Rational: $\left\{-7, -\frac{7}{3}, 0, 3.12, \frac{5}{4}\right\}$
(d) Irrational: $\left\{\sqrt{5}\right\}$

4. (a) Natural: $\left\{\frac{8}{2}, 9\right\}$ *[Note:* $\frac{8}{2} = 4$*]*
(b) Integers: $\left\{\frac{8}{2}, -4, 9\right\}$
(c) Rational: $\left\{\frac{8}{2}, -\frac{8}{3}, -4, 9, 14.2\right\}$
(d) Irrational: $\left\{\sqrt{10}\right\}$

6. (a) Natural: $\left\{25, \sqrt{9}\right\}$ *[Note:* $\sqrt{9} = 3$*]*
(b) Integers: $\left\{25, -17, \sqrt{9}\right\}$
(c) Rational: $\left\{25, -17, \frac{12}{5}, \sqrt{9}\right\}$
(d) Irrational: $\left\{\sqrt{8}, -\sqrt{8}\right\}$

8. $-3.5 < 1$

10. $1 < \frac{16}{3}$

12. $-\frac{8}{7} < -\frac{3}{7}$

14. $x \geq -2$ denotes all real numbers greater than or equal to -2.

16. $x > 3$ denotes all real numbers greater than 3.

18. $x < 2$ denotes all real numbers less than 2.

20. $0 \leq x \leq 5$ denotes all real numbers greater than or equal to 0 and less than or equal to 5.

22. $0 < x \leq 6$ denotes all real numbers greater than 0 and less than or equal to 6.

24. $5 < y \leq 12$ **26.** $Y \leq 45$ **28.** $p \leq \$1.45$

30. 0   **32.** $4 - \pi$ **34.** -6 **36.** -1

38. $|-3| > -|-3|$ **40.** $-|-6| < |-6|$

42. $-(-2) > -2$ **44.** $\frac{5}{2}$ **46.** $\frac{5}{2}$

48. $\frac{3}{2}$ **50.** 51   **52.** 14.99

54. $|x - (-10)| \geq 6 \Rightarrow |x + 10| \geq 6$

56. $|z - 0| < 8 \Rightarrow |z| < 8$ **58.** $|y - a| \leq 2$

60. $\frac{381}{220}, 1.73\overline{20}, \sqrt{3}, \frac{26}{15}, \sqrt{10} - \sqrt{2}$

62. $0.\overline{3}$ **64.** $0.\overline{54}$

	$|a - b|$	$0.05b$	Passed Budget Variance Test
66.	$656.52	$5635.00	No
68.	$395.32	$375.00	No
70.	$38.15	$128.75	Yes

	Median Income, y	Household Income, x	$|y - x|$
72.	$35,191	$75,400	$40,209
74.	$48,401	$18,760	$29,641
76.	$37,916	$26,370	$11,546

Section P.2 *(page 20)*

2. $-5, 3x$ **4.** $3x^2, -8x, -11$ **6.** $3x^4, 3x^3$

8. (a) 30   (b) -12 **10.** (a) -10 (b) 0

12. (a) 0   (b) -1

14. (a) $\frac{1}{2}$ (b) Undefined. You cannot divide by zero.

16. Commutative (addition)   **18.** Inverse (addition)

20. Inverse (multiplication)   **22.** Inverse (multiplication)

24. Identity (multiplication)   **26.** Identity (multiplication)

28. Associative (addition)

30. $\frac{1}{7}(7 \cdot 12) = \left(\frac{1}{7} \cdot 7\right)12$ Associative (multiplication)

$\qquad\qquad = 1 \cdot 12$ Inverse (multiplication)

$\qquad\qquad = 12$ Identity (multiplication)

32. $2 \cdot 3 \cdot 5$ **34.** $2^3 \cdot 7$ **36.** 9   **38.** -9

40. 40   **42.** -2 **44.** $\frac{1}{2}$ **46.** $\frac{59}{66}$ **48.** 3

50. $\frac{5}{16}$ **52.** $\frac{11}{12}$ **54.** $-\frac{14}{5}$ **56.** -2.57 **58.** 1.08

60. -13.67 **62.** -0.45 **64.** 10.2   **66.** 4.93

68. 1372.86   **70.** 41.14   **72.** 17.5%

74. The Treasury Department had 21.2% of the total budget.
Treasury: 293 billion dollars
Independent Agencies: 75 billion dollars
Veterans Affairs: 35 billion dollars
Other: 303 billion dollars
Health and Human Services: 258 billion dollars
Agriculture: 57 billion dollars
Defense: 315 billion dollars
Labor: 47 billion dollars

76. $2(-4 + 2)$

78. Scientific:

Graphing: $(-)$ 6 x^2 $-$ (7 + ($-$ 2

) $\wedge$ 3) ENTER

80. ≈ 5237

Section P.3   (page 30)

2. 81   **4.** 8   **6.** 256   **8.** 81   **10.** -16

12. $-\frac{3}{10}$ **14.** $\frac{16}{9}$ **16.** $-512{,}000$ **18.** $\frac{1}{8}$

20. $\frac{1}{4}$ **22.** 1   **24.** 18   **26.** $\frac{7}{16}$ **28.** -135

30. $-125z^3$ **32.** $16x^7$ **34.** $16x^6$ **36.** $-3z^7$

38. $\frac{5y^4}{2}$ **40.** $\frac{1}{r^2}$ **42.** $\frac{5184}{y^7}$ **44.** $1, x \neq 0$

46. $\frac{1}{(z+2)^4}$ **48.** $32y^2$ **50.** $\frac{z^2}{16}$ **52.** $\frac{125x^9}{y^{12}}$

54. 2^{4m} **56.** $\frac{1}{x}$ **58.** 1.394×10^8 square miles

60. 5×10^9 **62.** 1.3837×10^{-2} inch

64. 1×10^{-7} meter   **66.** 350,000,000 air sacs

68. 950,000 insect species   **70.** 0.0009 meter

72. $0.00000006673 \ c^3/\text{gram-sec}^2$

74. (a) 16,405.154   (b) 3818.251

76. (a) 414   (b) 0.015

78. (a) 0.763   (b) 1.422×10^{-5}

80.

Number of Years	5	10	15
Balance	$6734.28	$9070.09	$12,216.10

Number of Years	20	25
Balance	$16,453.31	$22,160.23

82. (a) $6325.12   (b) $6371.97

84. Yes. If the only change between two investments is the principal, the results will yield the same change. In this case, the one investment is twice the other, therefore twice as much money will be obtained after 10 years.

86. 5.0%   **88.** $[1 + 3(2)]^{-2}$

90. $a^0 = 1$ is derived from $1 = a^m/a^m = a^{m-m} = a^0$. If $a = 0$, then we would have $a^m/a^m = 0^m/0^m = 0/0$, which is undefined because of division by zero.

Section P.4   (page 41)

2. $64^{1/3} = 4$ **4.** $-\sqrt{144} = -12$

6. $614.125^{1/3} = 8.5$ **8.** $\sqrt[5]{-243} = -3$

10. $81^{3/4} = 27$ **12.** $\sqrt[4]{16^5} = 32$ **14.** 7   **16.** 4

18. $\frac{3}{2}$ **20.** 0   **22.** 1   **24.** 562   **26.** 3   **28.** 64

30. $\frac{1}{1000}$ **32.** $\frac{2}{3}$ **34.** $-\frac{3}{5}$ **36.** $\frac{2\sqrt[3]{2}}{3}$ **38.** $3x^2$

40. $2x\sqrt[5]{3}$ **42.** $\frac{\sqrt{10}}{2}$ **44.** $\frac{\sqrt[3]{5x}}{x}$ **46.** $\frac{\sqrt{14}+2}{2}$

48. $\frac{2\sqrt{10}+5}{3}$ **50.** $13\sqrt{x+1}$ **52.** $7\sqrt{3}$

54. $13\sqrt{5}$ **56.** $\sqrt{10}$ **58.** $3^{5/3}$ **60.** $2^{1/2}$

62. $1, x \neq 0$ **64.** $|x|$ **66.** $\sqrt{x}$ **68.** $3x^2$

70. 3.557   **72.** 75.686   **74.** 0.516   **76.** -0.134

78. $\sqrt{3} - \sqrt{2} < \sqrt{3-2}$ **80.** $5 = \sqrt{3^2 + 4^2}$

82. $\sqrt{\dfrac{3}{11}} = \dfrac{\sqrt{3}}{\sqrt{11}}$ **84.** $20\sqrt{2}$ feet $\times$ $20\sqrt{2}$ feet

86. $\approx 14.87\%$ **88.** ≈ 1.360 seconds

90. ≈ 523 vibrations per second

92. $\left(2/\sqrt{5}\right)^2 = 4/5$. This is not the same as rationalizing the denominator, which is $2/\sqrt{5} \cdot \sqrt{5}/\sqrt{5} = \left(2\sqrt{5}\right)/5$.

Section P.5 *(page 51)*

2. Degree: 4   **4.** Degree: 0
Leading Coefficient: -3 Leading Coefficient: 3

6. Degree: 1   **8.** Not a polynomial
Leading Coefficient: 2

10. Polynomial, $\frac{1}{2}x^2 + x - \frac{3}{2}$, degree 2

12. Not a polynomial

14. (a) 12   (b) 7   (c) 2   (d) -3

16. (a) -10 (b) -1 (c) 4   (d) 5

18. (a) 0   (b) 0   (c) 0   (d) -6

20. $x^2 + 2x$ **22.** $-2x^2 - 4$ **24.** $2x^4 - 13x - 34$

26. $y^3 - y^2 - 3y + 7$ **28.** $4y^4 + 2y^3 - 3y^2$

30. $4x^4 - 12x$ **32.** $-4x^4 + 4x$ **34.** $x^2 + 5x - 50$

36. $28x^2 - 29x + 6$ **38.** $9x^2 - 12x + 4$

40. $64x^2 - 80x + 25$ **42.** $x^2 + 2x - 2xy - 2y + y^2 + 1$

44. $4x^2 - 9$ **46.** $4x^2 - 9y^2$ **48.** $x^2 + 2xy + y^2 - 1$

50. $9a^6 - 16b^4$ **52.** $x^3 - 6x^2 + 12x - 8$

54. $27x^3 + 54x^2y + 36xy^2 + 8y^3$

56. $25 - x$ **58.** $x^4 - x^3 + 5x^2 - 9x - 36$

60. $x^3 - 8$ **62.** $x^4 - 13x^2 + 4$ **64.** $2x^2 + 8x + 6$

66. ≈ 6595.94 **68.** $1200r^3 + 3600r^2 + 3600r + 1200$

70. Volume $= 4x^3 - 96x^2 + 560x$
$x = 1$ inch, $V = 468$ cubic inches
$x = 2$ inches, $V = 768$ cubic inches
$x = 3$ inches, $V = 924$ cubic inches

72. $11x + 29$ **74.** $40x + 240$

76. $x = -1$, $Y_1 = -8$
    $x = 0$, $Y_1 = 2$
    $x = 1$, $Y_1 = 8$
    $x = 2$, $Y_1 = 28$

Section P.6 *(page 59)*

2. $5(y - 6)$ **4.** $2x\left(2x^2 - 3x + 6\right)$

6. $(3x - 4)(x + 2)$ **8.** $\left(x + \frac{1}{2}\right)\left(x - \frac{1}{2}\right)$

10. $(7 + 3y)(7 - 3y)$

12. $[5 + (z + 5)]\,[5 - (z + 5)] = -z(z + 10)$

14. $(x + 5)^2$ **16.** $(3x - 2)^2$ **18.** $\left(z + \frac{1}{2}\right)^2$

20. $(x + 2)(x + 3)$ **22.** $(t - 3)(t + 2)$

24. $(z - 8)(z + 3)$ **26.** $(x - 6)(x - 7)$

28. $(2x + 1)(x - 1)$ **30.** $(3x + 1)(4x + 1)$

32. $(5u - 2)(u + 3)$ **34.** $(x - 3)\left(x^2 + 3x + 9\right)$

36. $(z + 5)\left(z^2 - 5z + 25\right)$ **38.** $(3x + 2)\left(9x^2 - 6x + 4\right)$

40. $\left(x^2 - 5\right)(x + 5)$ **42.** $\left(5x^2 + 3\right)(x - 2)$

44. $\left(x^3 + 1\right)\left(x^2 + 2\right)$ **46.** $6x^2(2x - 1)$

48. $12(x + 2)(x - 2)$ **50.** $6(x + 3)(x - 3)$

52. $(8 - x)(2 + x)$ **54.** $(3x - 1)^2$

56. $y(2y + 3)(y - 5)$ **58.** $(5x + 3)(x + 2)$

60. $(5 - x)\left(1 + x^2\right)$ **62.** $(u + 2)\left(3 - u^2\right)$

64. $(t + 6)(t - 8)$ **66.** $(x + 2)(x + 4)(x - 2)(x - 4)$

68. $5(x + 2)\left(x^2 - 2x + 4\right)$ **70.** $(3 - 4x)(23 - 60x)$

72. $(3x + 1)^3$

74.

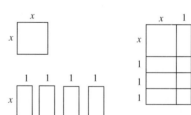

76.

78. c   **80.** d   **82.** $(3x + 2)$ feet

Section P.7 *(page 70)*

2. All real numbers   **4.** All real numbers greater than 0

6. All real numbers except $x = -\frac{1}{2}$

8. All real numbers except $x = -3$ and $x = 3$

10. All real numbers greater than -1

12. $x + 1, x \neq -1$ **14.** $y - 1, y \neq 1$

16. $z + 1, z \neq -1$ **18.** $\dfrac{3}{10y^3}$ **20.** $\dfrac{9x}{2}, x \neq -1$

22. $-(x + 5), x \neq 5$ **24.** $\dfrac{x - 2}{x + 1}, x \neq -10$

26. $\dfrac{1}{x + 10}, x \neq -1$ **28.** $\dfrac{1}{x + 1}, x \neq \pm 3$

30. $\dfrac{y(y - 3)}{y^2 - y + 1}, y \neq -1$ **32.** $-\dfrac{x + 13}{5x^2}, x \neq 3$

34. $\dfrac{x - 3}{(x + 2)^2}, x \neq -5$ **36.** $-\dfrac{8}{5}, x \neq -3, 4$

38. $\dfrac{2\left(y^2 + 2y + 4\right)}{y^2(y - 3)}, y \neq 0, 2$

40. $\dfrac{\left(x^2 + 1\right)\left(x^2 + x + 1\right)}{(x + 1)^2}, x \neq 1$ **42.** $\dfrac{x + 2}{x - 2}, x \neq 3$

44. $\dfrac{x + 1}{x - 1}, x \neq 0$ **46.** $\dfrac{x}{x + 3}, x \neq -3$ **48.** $\dfrac{8 - 5x}{x - 1}$

50. $\dfrac{2x + 5}{x - 5}$ **52.** $\dfrac{1}{(x + 2)(x - 1)}$ **54.** $\dfrac{4x + 1}{x^2 - 1}$

56. $\dfrac{4x^2 - 12x}{x^2 - 16}, x \neq 0$ **58.** $\dfrac{1}{2y + 1}, y \neq 0, -\dfrac{5}{4}$

60. $\dfrac{1}{(x + 1)(x + 5)}$ **62.** $\dfrac{3x - 1}{3}, x \neq 0$

64. $\frac{2}{38} \approx 0.053$ **66.** $\frac{1}{1000} = 0.001$

68. $\dfrac{13}{52} = \dfrac{1}{4} = 0.25$ **70.** (a) 12%   (b) $\dfrac{288(NM - P)}{N(12P + NM)}$

72. $\dfrac{3x + 2}{2(2x + 1)}$ **74.** $\dfrac{7x + 4}{4(2x + 1)}$

[*Note:* This is the same probability as in Exercise 71.]

76.

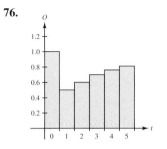

Review Exercises *(page 75)*

2. (a) Natural: { }
(b) Integer: $\{-22, 0\}$
(c) Rational: $\left\{-22, -\frac{10}{3}, 0, 5.2, \frac{3}{7}\right\}$
(d) Irrational: $\left\{\sqrt{15}\right\}$

4. $\frac{1}{2} > \frac{1}{3}$ **6.** All real numbers greater than or equal to 1

8. $2 < x \leq 5$ **10.** 6   **12.** $|9| = |-9|$

14. 10   **16.** $|x - 25| \leq 10$ **18.** $3x^3, -9x$

20. (a) -8 (b) -33

22. Associative Property (addition)   **24.** 4   **26.** $\frac{7}{6}$

28. 256   **30.** -1.5 **32.** $-\frac{1}{3}$ **34.** $-108x^3$

36. 3.048×10^{-1} meter per foot   **38.** 0.00274

40. (a) 67,429.958   (b) 0.713

42.

Number of Years	5	10	15
Balance	$13,140.67	$17,267.71	$22,690.92

Number of Years	20	25
Balance	$29,817.37	$39,182.01

44. $\sqrt[4]{16} = 2$ **46.** 5   **48.** $\dfrac{x\sqrt[3]{2}}{3}$ **50.** $-2\left(3 + \sqrt{10}\right)$

52. $14\sqrt{2}$ **54.** 16   **56.** $\sqrt{|x|}$ **58.** 3.733

60. $-9x^2 - 9x + 6$ **62.** $5x^2 - x$ **64.** $x^2 - 25$

66. $4x^2 + 4x + 1$ **68.** $(x - 5)(x + 1)$

70. $x(x - 4)(x + 4)$ **72.** $(x - 5)\left(x^2 + 5x + 25\right)$

74. All real numbers greater than or equal to 0

76. $x + 2, x \neq -2$ **78.** $x - 5, x \neq -3$

80. $x(x + 3), x \neq 1$ **82.** $\dfrac{4x}{x - 4}, x \neq -2, 0$

84. $-\dfrac{x + 10}{x^2 - 4}$ **86.** $\dfrac{y - x}{y + x}$

88. $\dfrac{10}{2450} \approx 0.004$ **90.** $\dfrac{7x + 16}{8x + 16}$

CHAPTER 1

Section 1.1 *(page 86)*

2. Identity   **4.** Conditional equation

6. Conditional equation

8. (a) No   (b) No   (c) No   (d) Yes

10. (a) No   (b) Yes   (c) No   (d) No

12. (a) Yes   (b) No   (c) No   (d) No

14. (a) No   (b) No   (c) Yes   (d) No

16. -11 **18.** 2   **20.** -4 **22.** 3

24. No solution   **26.** -10 **28.** 6   **30.** 50

32. $-\frac{5}{8}$ **34.** $\frac{1}{2}$ **36.** 0   **38.** $\frac{9}{7}$ **40.** $\frac{7}{4}$

42. $x = 4$ appears to be a solution, but after checking, it is found to be an extraneous solution, so there is no solution.

44. $-\frac{13}{3}$ **46.** 0   **48.** $\frac{1}{5}$ **50.** No solution

52. No solution   **54.** ≈ 1.326 **56.** ≈ 19.993

58. ≈ 1.706 **60.** (a) ≈ 13.93 (b) ≈ 14.38; Yes

62. (a) ≈ 1.63 (b) ≈ 1.63; No

64. (a) ≈ 0.58 (b) ≈ 0.57; Yes

66. 1993   **68.** Yes   **70.** 35 years old   **72.** 0.13

Section 1.2 *(page 97)*

2. $n(25 - n) = 25n - n^2$ **4.** $\dfrac{200}{r}$ **6.** 0.82

8. $10h$ **10.** $3.59x$ **12.** 267, 268, 269   **14.** 19, 95

16. 3, 4   **18.** January: $71,590; February: $85,908

20. January: $\approx$ $87,498.89; February: $\approx$ $69,999.11   **22.** 1950%

24. $\approx 33,233\%$ **26.** Walter Payton: $\approx$ $325,005; Joe Montana: $\approx$ $2,411,894.61

28. Fruit: 341
Potato chips: 256
Cookies: 192
Cake: 119
Corn chips: 55

30. $\frac{128}{414} \approx 0.31$

32. 0.57 feet $\times$ 0.93 feet   **34.** 187 or greater

36. $\approx$ $20,828.10   **38.** $\approx$ $866.21   **40.** 40.04%

42. 4.0625 hours   **44.** 2.25 hours

46. ≈ 46.3 miles per hour   **48.** $\approx 2.93 \times 10^{14}$ miles

50. ≈ 57.14 feet tall   **52.** $518,925

54. $10,000 at 11%; $15,000 at 12.5%

56. First three quarters: 11.5%; last quarter: 10%

58. ≈ 8065 units   **60.** ≈ 0.26 foot   **62.** ≈ 0.48 gallon

64. ≈ 10.81 miles per hour   **66.** $l = \dfrac{P - 2w}{2}$

68. $h = \dfrac{V}{\pi r^2}$ **70.** $L = \dfrac{S}{1 - R}$ **72.** $P = A\left(1 + \dfrac{r}{n}\right)^{-nt}$

74. $\theta = \dfrac{360A}{\pi r^2}$ **76.** $r = \dfrac{S - a}{S - L}$ **78.** $r = \sqrt{\dfrac{A}{4\pi}}$

Section 1.3 *(page 110)*

2. $4x^2 - 2x - 9 = 0$ **4.** $10x^2 - 90 = 0$

6. $-3x^2 - 42x - 135 = 0$ **8.** $2x^2 - x + 5 = 0$

10. $x^2 + 3x - 10 = 0$ **12.** $\pm\frac{1}{3}$ **14.** 1, 9   **16.** $-\frac{7}{4}$

18. $-\frac{3}{2}, 11$ **20.** $-7, 3$ **22.** 2, 6   **24.** ± 12

26. $\pm 3\sqrt{3} \approx \pm 5.20$ **28.** $\pm\frac{5}{3} \approx \pm 1.67$

30. $-13 + \sqrt{21} \approx -8.42$ **32.** $-5 + 2\sqrt{5} \approx -0.53$
$-13 - \sqrt{21} \approx -17.58$ $-5 - 2\sqrt{5} \approx -9.47$

34. $\pm\dfrac{5\sqrt{15}}{3} \approx \pm 6.45$ **36.** $\pm\dfrac{5}{2} = \pm 2.50$

38. $\pm\dfrac{4\sqrt{14}}{7}$ **40.** 1, 5   **42.** $0, -3$ **44.** 7

46. $5 \pm 2\sqrt{2}$ **48.** $\frac{10}{3}, -\frac{8}{3}$ **50.** $16, \frac{9}{4}$

52. $\frac{3}{4}, \frac{5}{2}$ **54.** $-\frac{1}{3}, -1$ **56.** $-1, 5$ **58.** $-\frac{1}{3}, 1$

60. 14 feet $\times$ 24 feet   **62.** $h = b_1 = x = 80$ feet
$b_2 = \frac{3}{2}(80) = 120$ feet

64. ≈ 3.54 seconds   **66.** 2.5 seconds   **68.** ≈ 1.15 feet

70. ≈ 2121.32 feet   **72.** 1993 ($t \approx 10.3$)

74. 30,000 units   **76.** 1998 ($t \approx 8.3$)   **78.** ≈ 42 cards

Section 1.3 *(page 120)*

2. Two real solutions   **4.** No real solutions

6. One real solution   **8.** No real solutions

10. $1, -\frac{1}{2}$ **12.** $\frac{3}{5}, \frac{1}{5}$ **14.** $5 \pm \sqrt{3}$ **16.** $-3 \pm \sqrt{13}$

18. $\frac{1}{2} \pm \frac{\sqrt{5}}{2}$ **20.** $\frac{5}{4} \pm \frac{\sqrt{3}}{4}$ **22.** $-\frac{3}{2} \pm \frac{\sqrt{13}}{2}$

24. $\frac{5}{4} \pm \frac{\sqrt{5}}{2}$ **26.** $-\frac{4}{3}$ **28.** $-\frac{8}{5} \pm \frac{\sqrt{3}}{5}$

30. $-7 \pm \sqrt{13}$ **32.** $-0.178, -0.649$

34. $2.137, 18.063$ **36.** $1.400, -0.150$ **38.** $x = 5$

40. No real solution   **42.** $-4, 1$ **44.** $3, -1$

46. $x = \frac{11}{8}$ **48.** $8, 9$ or $-9, -8$

50. $20, 22$ or $-22, -20$ **52.** 100 units

54. 48 units   **56.** 35 feet $\times$ 20 feet or 15 feet $\times$ 46.7 feet

58. 6 inches $\times$ 6 inches $\times$ 3 inches

60. First sack. When the second sack is thrown, the first sack only needs 2.54 seconds to hit the ground.

62. $30\sqrt{6} \approx 73.48$ feet   **64.** $1995 \, (t \approx 15.8)$

66. $1996 \, (t \approx 16.3)$ **68.** Eastbound: ≈ 578 miles per hour
Southbound: ≈ 478 miles per hour

70. 5279 or $94,721$ units

Section 1.4 *(page 133)*

2. $0, \pm\frac{5}{2}$ **4.** $0, \frac{3}{2}, 6$ **6.** ± 2 **8.** $0, \frac{4}{3}$

10. -2 **12.** ± 2 **14.** $\pm 2, \pm 5$ **16.** $\pm\sqrt{3}, \pm 1$

18. $\pm\frac{\sqrt{7}}{6}$ **20.** $-1, -\sqrt[3]{2}$ **22.** $\frac{9}{16}$ **24.** -4

26. $\frac{124}{3}$ **28.** $\frac{9}{4}$ **30.** $\frac{3}{2} \left(-\frac{5}{2} \text{ is extraneous}\right)$

32. $-2, 3$ **34.** 4 (12 is extraneous)   **36.** $5, -11$

38. ± 5 **40.** $-5, 6, \frac{1 \pm \sqrt{57}}{2}$ **42.** $\frac{1}{3} \pm \frac{\sqrt{31}}{3}$

44. $-12, 2$ **46.** $-1, \frac{3}{4}$ **48.** 2 **50.** $-1, 1$

52. $-1, 5$ **54.** $-3, \frac{5}{3}$ **56.** $-6, -3, 3$

58. $-2, 3, \frac{-1 - \sqrt{17}}{2}$ **60.** $\approx -1.143; \approx 0.968$

62. No real solutions   **64.** $\$900$

66. (a) $\approx 4.61\%$ (b) $\approx 6.31\%$ **68.** $\approx 18.75\%$

70. (a)

t	0	5	10	15	20
F	335	839	1236	1608	1969

(b) $1990 \, (t \approx 20.43)$

72. $1997 \, (t \approx 97.74)$ **74.** 0.26 mile or 1 mile

Section 1.6 *(page 144)*

2. (a) No   (b) No   (c) Yes   (d) No

4. h   **6.** e   **8.** a   **10.** d

12. $x \leq 4$ **14.** $x > \frac{3}{2}$

16. $x < -\frac{5}{2}$ **18.** $x \leq 5$

20. $x < -2$ **22.** $x \geq \frac{1}{3}$

24. $x \leq 1$ **26.** $-2 < x \leq 5$

28. $-3 \leq x < 7$ **30.** $-3 < x < 3$

32. $-3 < x < 3$ **34.** $x < -2$ or $x > 2$

36. $1 < x < 13$ **38.** $x < -28$ or $x > 0$

40. $-2 < x < 3$ **42.** $0 < x < 3$

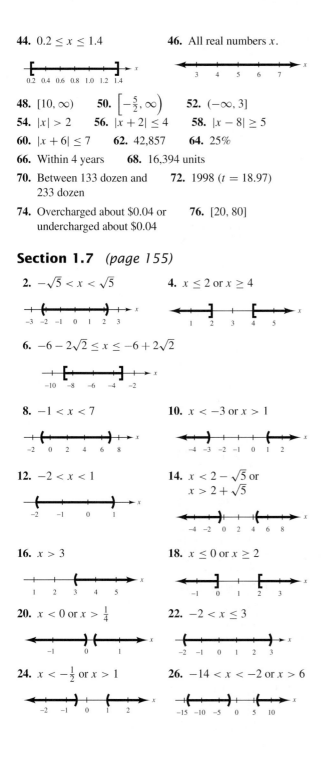

44. $0.2 \leq x \leq 1.4$

46. All real numbers x.

48. $[10, \infty)$ **50.** $\left[-\frac{5}{2}, \infty\right)$ **52.** $(-\infty, 3]$

54. $|x| > 2$ **56.** $|x + 2| \leq 4$ **58.** $|x - 8| \geq 5$

60. $|x + 6| \leq 7$ **62.** 42,857 **64.** 25%

66. Within 4 years **68.** 16,394 units

70. Between 133 dozen and 233 dozen **72.** 1998 ($t = 18.97$)

74. Overcharged about $0.04 or undercharged about $0.04 **76.** $[20, 80]$

Section 1.7 *(page 155)*

2. $-\sqrt{5} < x < \sqrt{5}$

4. $x \leq 2$ or $x \geq 4$

6. $-6 - 2\sqrt{2} \leq x \leq -6 + 2\sqrt{2}$

8. $-1 < x < 7$

10. $x < -3$ or $x > 1$

12. $-2 < x < 1$

14. $x < 2 - \sqrt{5}$ or $x > 2 + \sqrt{5}$

16. $x > 3$

18. $x \leq 0$ or $x \geq 2$

20. $x < 0$ or $x > \frac{1}{4}$

22. $-2 < x \leq 3$

24. $x < -\frac{1}{2}$ or $x > 1$

26. $-14 < x < -2$ or $x > 6$

28. $x < -3$ or $x > 0$

30. $(-\infty, -2], [2, \infty)$ **32.** $[-4, 4]$ **34.** $(-\infty, \infty)$

36. The domain of x is empty. **38.** $-1.13 < x < 1.13$

40. $-4.42 < x < 0.42$ **42.** $1.19 < x < 1.30$

44. $0 \leq t \leq 4 - 2\sqrt{2}$ or $4 + 2\sqrt{2} < t < 8$

46. 6.48 feet $\leq L \leq$ 18.52 feet

48. (a) $40,000 \leq x \leq 50,000$ (b) $\$50 \leq p \leq \$55$

50. 6.27% **52.** 1996 ($t \geq 36.55$) **54.** 1996 ($t \geq 36.4$)

Review Exercises *(page 160)*

2. Identity **4.** (a) No (b) No (c) No (d) No

6. $-\frac{5}{3}$ **8.** 2 **10.** -8 **12.** ≈ 0.078 **14.** $25,970

16. Review/Computer Program: $0.20(320) = 64$ students
Group Study: $0.60(320) = 192$ students
Study Alone: $0.78(320) \approx 250$ students
Review Video Tapes: $0.35(320) = 112$ students
Review Homework: $0.82(320) \approx 262$ students
Study with Tutor: $0.52(320) \approx 166$ students

18. 20 feet × 25 feet **20.** $\approx 24.85\%$ **22.** $\approx \$870,224$

24. 2.625 gallons of 20% **26.** $-3, -12$ **28.** $-\frac{5}{2}, 3$

30. $\pm\frac{5}{4} = \pm 1.25$ **32.** $1 + \sqrt{5} \approx 3.24; 1 - \sqrt{5} \approx -1.24$

34. ≈ 19.36 seconds **36.** ≈ 1414.21 feet

38. No real solutions **40.** $\dfrac{-8 \pm 2\sqrt{31}}{5}$

42. $\dfrac{-3 \pm \sqrt{23}}{2}$ **44.** $\dfrac{11 \pm \sqrt{201}}{20}$

46. ≈ 8.544 or ≈ 0.162 **48.** 17 inches × 17 inches

50. $\pm 1, \pm 2$ **52.** $-3, 1$ **54.** 2 (9 is extraneous)

56. $-2, 7, \dfrac{5 \pm \sqrt{17}}{2}$ **58.** $\pm 2, -4$

60. ± 1 **62.** $\approx 21.18\%$

64. $x < 11$ **66.** $-\frac{13}{2} < x < \frac{11}{2}$

68. $-12 < x < -8$ **70.** $[10, \infty)$

72. $(-\infty, 6]$ or $[9, \infty)$　**74.** $x \geq 36$ units

76. $-1 < x < 3$　**78.** $x < -3$ or $0 < x < 3$

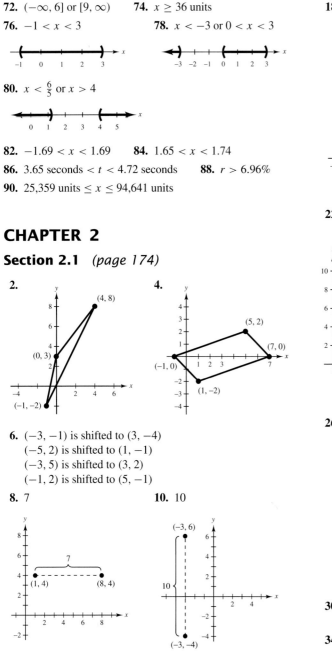

80. $x < \frac{6}{5}$ or $x > 4$

82. $-1.69 < x < 1.69$　**84.** $1.65 < x < 1.74$

86. 3.65 seconds $< t <$ 4.72 seconds　**88.** $r > 6.96\%$

90. 25,359 units $\leq x \leq$ 94,641 units

CHAPTER 2

Section 2.1 *(page 174)*

2.　**4.**

6. $(-3, -1)$ is shifted to $(3, -4)$
$(-5, 2)$ is shifted to $(1, -1)$
$(-3, 5)$ is shifted to $(3, 2)$
$(-1, 2)$ is shifted to $(5, -1)$

8. 7　**10.** 10

12. Answers will vary.　**14.** 13　**16.** $\sqrt{65}$
$(-5, 2)$ and $(-5, 5)$

18. (a)

(b) 13　(c) $\left(\frac{7}{2}, 6\right)$

20. (a)

(b) 15　(c) $\left(-\frac{5}{2}, 2\right)$

22. (a)

(b) $8\sqrt{2}$　(c) $(6, 6)$

24. (a)

(b) $\dfrac{\sqrt{2}}{6}$

(c) $\left(-\frac{1}{4}, -\frac{5}{12}\right)$

26. (a)

(b) ≈ 23.6
(c) $(-5.6, 8.6)$

28. (a)

(b) ≈ 9.428
(c) $(3.6785, 7.2055)$

30. (1993, \$4,925,000)　**32.** Two of the sides are equal
to $\sqrt{29}$.

34. One pair of opposite sides are $\sqrt{45}$ each, and the other pair
of opposite sides are $\sqrt{10}$ each.

36. $x = -20, 4$　**38.** $y = -4, 12$　**40.** III　**42.** II

44. ≈ 180.28 miles　**46.** ≈ 75 years

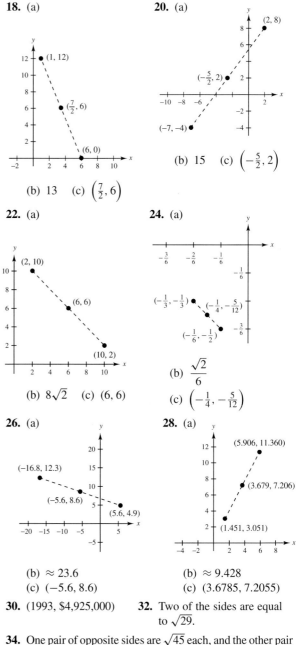

48. ≈ $125 in 1976   **50.** 63%   **52.** 22,000

Section 2.2 *(page 184)*

2. (a) Yes (b) No   **4.** (a) No   (b) Yes

6. (a) No   (b) Yes   **8.** (1, 0), (3, 0), (0, 3)

10. (2, 0), (−2, 0), (0, 4)   **12.** No intercepts

14. (0, 0)   **16.** *x*-axis symmetry   **18.** *y*-axis symmetry

20. Origin symmetry   **22.** *y*-axis symmetry

24.   **26.**

28. f   **30.** a   **32.** b

34. Intercepts:
$\left(\frac{3}{2}, 0\right)$, (0, −3)

36. Intercepts:
(1, 0), (−1, 0), (0, −1)
Symmetry: *y*-axis

38. Intercepts: (0, 0), (−4, 0)   **40.** Intercepts:
(0, −1), (1, 0)

42. Intercepts: (0, 4)
Symmetry: *y*-axis

44. Intercepts: (0, 1), (1, 0)

46. Intercepts: (−1, 0), (0, 1)   **48.** Intercepts:
(−4, 0), (0, 4), (4, 0)
Symmetry: *y*-axis

50. Intercepts: (−4, 0), (0, −2), (0, 2); symmetry: *x*-axis

52. Intercepts: (−4, 0), (0, −4), (0, 4), (4, 0)
Symmetry: *x*-axis, *y*-axis, origin

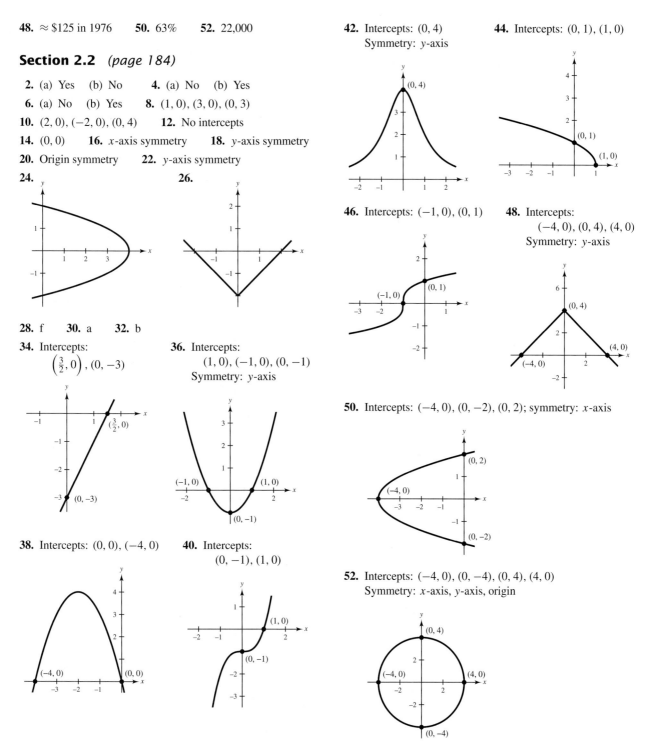

54. $x^2 + y^2 = 25$ **56.** $x^2 + \left(y - \frac{1}{3}\right)^2 = \frac{1}{9}$

58. $(x - 3)^2 + (y + 2)^2 = 25$ **60.** $x^2 + y^2 = 17$

62. $(x - 1)^2 + (y + 3)^2 = 25$ **64.** $x^2 + (y - 1)^2 = \frac{4}{3}$

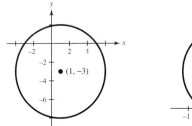

66. $\left(x - \frac{1}{2}\right)^2 + \left(y + \frac{1}{4}\right)^2 = \frac{9}{16}$

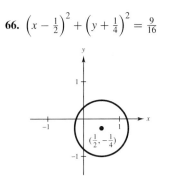

68. $(x - 2)^2 + (y + 1)^2 = 2$

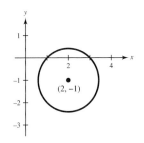

70. $-\frac{1}{8}$ **72.** $-\frac{4}{5}$

74. $\left(-\frac{1}{2}, -1\right), \sqrt{5}, 4x^2 + 4y^2 + 4x + 8y + 15 = 0$

76. (a)

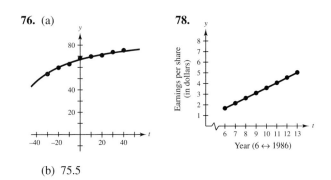

(b) 75.5

78.

Section 2.3 *(page 193)*

2. g   **4.** f   **6.** b   **8.** c   **10.** h

12.

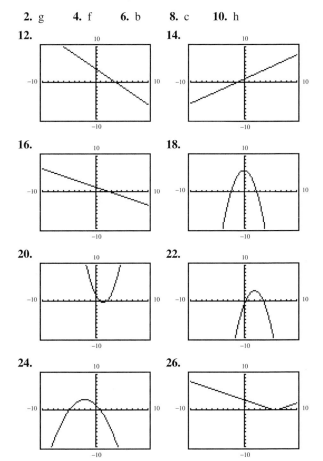

14.

16.

18.

20.

22.

24.

26.

28. 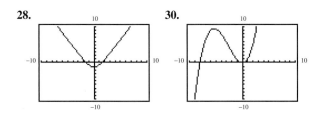   **30.**

32. No. To see both x-intercepts, the graph should be translated to the left. Selecting Xmin $= -4$ and Xmax $= 8$ would be more appropriate.

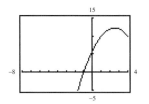

34. Yes. The three x-intercepts are easily seen.

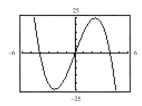

36. Range

| Xmin $= -20$ |
| Xmax $= 8$ |
| Xscl $= 4$ |
| Ymin $= -60$ |
| Ymax $= 10$ |
| Yscl $= 10$ |
| Xres $= 1$ |

38. Range

| Xmin $= -6$ |
| Xmax $= 3$ |
| Xscl $= 1$ |
| Ymin $= -4$ |
| Ymax $= 10$ |
| Yscl $= 1$ |
| Xres $= 1$ |

40. Two x-intercepts

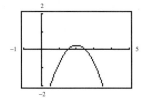

42. One x-intercept

44. Parallelogram   **46.** Isosceles triangle

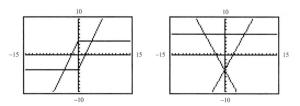

48. (a) ≈ 75.7 years; ≈ 76.8 years
   (b) Yes. Answers will vary. New factors may arise that could either increase or decrease the life expectancy.

50. At any given age between 20 and 45, the percent of unmarried females is greater than the percent of unmarried males. At any given age between 45 and 50, the percent of unmarried males is greater than the percent of unmarried females. Around the age of 45, the two percents are about the same.

52. $\approx 46.7\%$ **54.** $\approx \$0.33$

56. (a) ≈ 0.2092 (b) 0.2407 (c) 0.2501

Section 2.4 *(page 205)*

2. 2   **4.** -1 **6.** $\frac{3}{2}$

8.

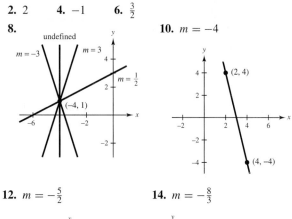

10. $m = -4$

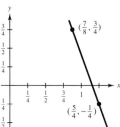

12. $m = -\frac{5}{2}$

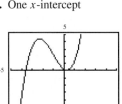

14. $m = -\frac{8}{3}$

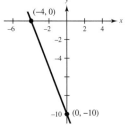

16. Parallel   **18.** Perpendicular

20. $(-4, 0)$, $(-4, 3)$, $(-4, 5)$ **22.** $(0, 4)$, $(9, -5)$, $(11, -7)$

24. $(-4, -1)$, $(-2, -1)$, $(0, -1)$ **26.** $m = -\frac{2}{3}$, $(0, 3)$

28. $m = 0$, $\left(0, -\frac{5}{3}\right)$ **30.** $m = 1$, $(0, -10)$

32. $7x - 8y - 4 = 0$ **34.** $y - 4 = 0$

36. $x + 3y - 4 = 0$ **38.** $3x + 10y + 18 = 0$

40. $x + y - 10 = 0$ **42.** $4x - y = 0$

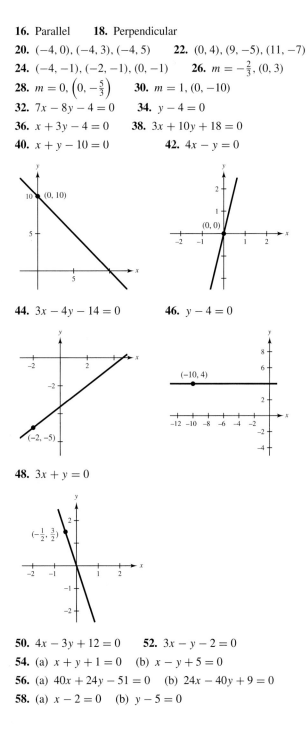

44. $3x - 4y - 14 = 0$ **46.** $y - 4 = 0$

48. $3x + y = 0$

50. $4x - 3y + 12 = 0$ **52.** $3x - y - 2 = 0$

54. (a) $x + y + 1 = 0$ (b) $x - y + 5 = 0$

56. (a) $40x + 24y - 51 = 0$ (b) $24x - 40y + 9 = 0$

58. (a) $x - 2 = 0$ (b) $y - 5 = 0$

60.

°C	−17.8	−10	10	20	32.2	177
°F	0	14	50	68	90	350.6

The Fahrenheit reading is the same as the Celsius reading at $-40°$.

62.

A	$750.00	$754.50	$759.00	$763.50
t	0	1	2	3

A	$768.00	$772.50	$777.00
t	4	5	6

64. 3092 students   **66.** $89.2 million   **68.** 302; yes

Section 2.5   *(page 216)*

2.

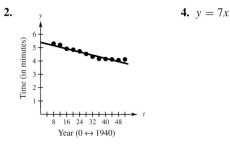

Year (0 ↔ 1940)

The model is a "good fit" for the actual data.

4. $y = 7x$

6. $y = \frac{290}{3}x$ **8.** $y = \frac{23}{150}x$ **10.** $I = 0.0675P$

12. (a) $y = 0.07x$ (b) $\approx \$37.84$

14. (a) $y = \frac{53}{14}x$

(b)

Gallons	5	10	20	25	30
Liters	18.93	37.86	75.71	94.64	113.57

16. $y = 4.50t + 156$, $0 \le t \le 5$

18. $y = -2800t + 45,000$, $0 \le t \le 5$

20. $y = 5600t + 245,000$, $0 \le t \le 5$

22. $V = -2300t + 25,000$, $0 \le t \le 10$

24. $S = 0.75L$ **26.** $W = 2500 + 0.07S$

28. (a) $d = -\frac{28}{29}t + 84$

(b) When $d = 0$, $t = 87$ minutes $\Rightarrow$ 5:57 P.M.

30. c; the amount received   **32.** d; the value decreased
increases by \$1.50 for   \$100 per year.
each unit per hour.

34. (a) $r = \frac{4}{3}x - 50$ (b) $\approx$ \$1683.33

(c)

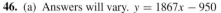
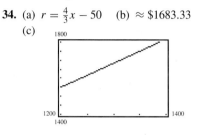

36. No   **38.** Yes

40. Answers will vary.   **42.** Answers will vary.
$y = -1.18x + 7.76$ $y = 0.33x + 1.96$

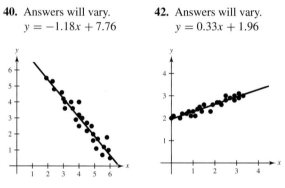

44. Answers will vary. The model may be accurate for a few
years following the first three, but may not be useful after
several years.

46. (a) Answers will vary. $y = 1867x - 950$

(b)

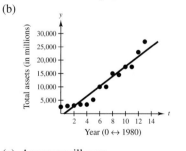

(c) Answers will vary.

48. (a) $x = -\frac{1}{15}p + \frac{226}{3}$ (b) 45 units

Review Exercises *(page 223)*

2. 4   **4.** (a) 5 and 2; hypotenuse $= \sqrt{29}$ (b) $\sqrt{29}$

6. (a)   **8.** (a)

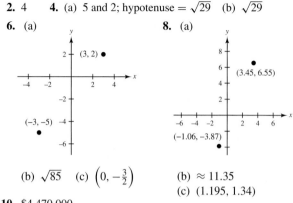

(b) $\sqrt{85}$ (c) $\left(0, -\frac{3}{2}\right)$ (b) ≈ 11.35
(c) $(1.195, 1.34)$

10. \$4,470,000

12. Using the distance formula, the lengths of the sides are
$\sqrt{185}$, $\sqrt{185}$, and $\sqrt{370}$. To prove the triangle is a right
triangle, we show that $\left(\sqrt{185}\right)^2 + \left(\sqrt{185}\right)^2 = \left(\sqrt{370}\right)^2$.

14. -35 or 5   **16.** -4 or 6   **18.** (a) No   (b) Yes

20. $(3, 0), (-3, 0), (0, -9)$ **22.** $(0, 2), (-4, 0)$

24. x-axis, y-axis, origin symmetries   **26.** x-axis symmetry

28. $(5, 0), (-5, 0), (0, 5), (0, -5)$
x-axis, y-axis, origin symmetries

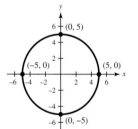

30. $(9, 0)$ **32.** $(3, 0), (0, 3)$

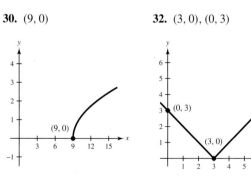

34. $(x + 1)^2 + y^2 = 5^2$

36. $\left(x - \frac{1}{2}\right)^2 + \left(y + \frac{1}{3}\right)^2 = \left(\frac{3}{2}\right)^2$

38. $(x + 1)^2 + (y - 2)^2 = 16$

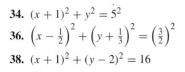

40. a **42.** d **44.** c

46. 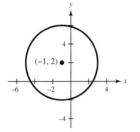 **48.**

50. $\frac{3}{7}$ **52.** 2

54. Square **56.** $\frac{7}{3}$

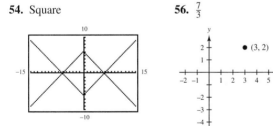

58. Neither **60.** Neither

62. Slope: 0; y-intercept: $\left(0, \frac{2}{3}\right)$

64. $y = -2$

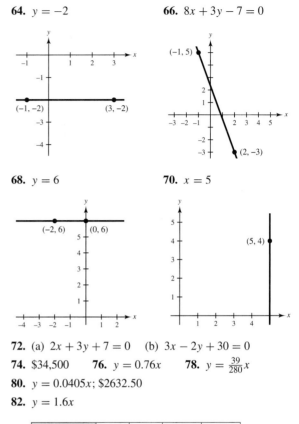

66. $8x + 3y - 7 = 0$

68. $y = 6$

70. $x = 5$

72. (a) $2x + 3y + 7 = 0$ (b) $3x - 2y + 30 = 0$

74. \$34,500 **76.** $y = 0.76x$ **78.** $y = \frac{39}{280}x$

80. $y = 0.0405x$; \$2632.50

82. $y = 1.6x$

Miles	2	5	10	12
Kilometers	3.20	8.00	16.00	19.20

84. $V = -2750t + 30,000, 0 \le t \le 10$

86. $W = 0.60x + 9.50$

88. (a) $C = 15.25t + 24,500$ (b) $R = 35t$
(c) $P = 19.75t - 24,500$ (d) $t \approx 1240.51$ hours

CHAPTER 3

Section 3.1 *(page 238)*

2. (a) $f(4) = \dfrac{1}{(4) + 1}$ (b) $f(0) = \dfrac{1}{(0) + 1}$

(c) $f(4x) = \dfrac{1}{(4x) + 1}$ (d) $f(x + 1) = \dfrac{1}{(x + 1) + 1}$

4. (a) $f(3) = \sqrt{25 - (3)^2}$

(b) $f(5) = \sqrt{25 - (5)^2}$

(c) $f(x + 5) = \sqrt{25 - (x + 5)^2}$

(d) $f(2x) = \sqrt{25 - (2x)^2}$

6. (a) 7   (b) 0   (c) $7 - 3s$ (d) $1 - 3s$

8. (a) 36π (b) 0   (c) $\dfrac{9\pi}{2}$ (d) $\dfrac{32\pi r^3}{3}$

10. (a) 2   (b) 5   (c) $\sqrt{x} + 2$ (d) $\sqrt{x + 16} + 2$

12. (a) $\dfrac{11}{4}$ (b) Undefined   (c) $\dfrac{2x^2 + 3}{x^2}$ (d) $\dfrac{2x^2 + 3}{x^2}$

14. (a) 6   (b) 6   (c) $x^2 + 4$ (d) $|x + 2| + 4$

16. (a) 6   (b) 2   (c) 3   (d) 10   **18.** $\frac{4}{3}$

20. $0, 1, -1$ **22.** $-\frac{3}{2}$ **24.** All real numbers x

26. All real numbers except $y = -5$

28. All real numbers t

30. All real numbers except $x = 0$ and $x = 2$

32. All real numbers $s \geq 1$ except $s = 4$

34. Not a function   **36.** Not a function

38. Not a function   **40.** Function   **42.** Function

44. Because $(1, -2)$ and $(1, 1)$ are both in $\{(0, -1), (2, 2), (1, -2), (3, 0), (1, 1)\}$, this set does not represent a function from A to B.

46. Because 2 is an element of A is not matched with an element of B, $\{(0, 2), (3, 0), (1, 1)\}$ does not represent a function from A to B.

48. Because each element of A is matched with exactly one element of B, $\{(a, 1), (b, 2), (c, 3)\}$ represents a function from A to B.

50. Because each element of A is matched with exactly one element of B, $\{(c, 0), (b, 0), (a, 3)\}$ represents a function from A to B.

52. Because each element of A is matched with exactly one element of B, $\{(a, 1), (b, 2), (c, 4)\}$ represents a function from A to B.

54. Because $(a, 1)$ and $(a, 2)$ are both in $\{(a, 1), (a, 2), (b, 2), (c, 4)\}$, this set does not represent a function from A to B.

56. $\left\{\left(-2, -\frac{4}{5}\right), (-1, -1), (0, 0), (1, 1), \left(2, \frac{4}{5}\right)\right\}$

58. $\{(-2, 1), (-1, 0), (0, 1), (1, 2), (2, 3)\}$

60. (a) $V = 108x^2 - 4x^3$ (b) Domain: $0 < x < 27$

62. $22{,}512; 27{,}610$ **64.** $5.18; 7.78$

66. (a) $C = 110{,}000 + 15.95x$ (b) $R = 23.79x$

(c) $P = 7.84x - 110{,}000$

68. (a) $C = 35{,}000 + 1.15x$ (b) $\overline{C} = \dfrac{35{,}000}{x} + 1.15$

70. (a) $R = 1700x - 10x^2$

(b)

x	25	50	75	100	125
$R(x)$	\$36,250	\$60,000	\$71,250	\$70,000	\$56,250

(c) Answers will vary. The revenue is maximum when there are 85 occupied apartments.

Section 3.2   *(page 250)*

2. $(-\infty, 4]$ **4.** $[0, \infty)$ **6.** $\{-1, 1\}$ **8.** $[0, \infty)$

10. Function   **12.** Not a function   **14.** Not a function

16. Decreasing on $(-\infty, 1)$ **18.** Decreasing on $(-\infty, -2)$
Increasing on $(1, \infty)$ Increasing on $(2, \infty)$

20. Decreasing on $(-\infty, 0)$ **22.** Decreasing on $(-\infty, -1)$
Increasing on $(0, \infty)$ Constant on $(-1, 1)$
   Increasing on $(1, \infty)$

24. Neither even nor odd   **26.** Odd   **28.** Even

30. Odd   **32.** Even

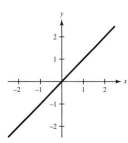

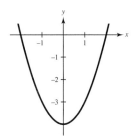

34. Even   **36.** Neither even nor odd

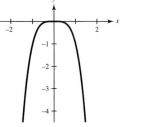

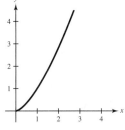

38. Neither even nor odd

40. Neither even nor odd

58. Increasing: $(-1, 1)$
Decreasing:
$(\infty, -1), (1, \infty)$

60. The deposits were decreasing in 1980, 1981, 1982, 1983, 1991, 1992, and 1993 and were increasing in 1984, 1985, 1986, 1987, 1988, 1989, and 1990.

42.

44.

62. Approximately 350,000 units

64.

46.

48.

66.

68.

50.

52.

Section 3.3 *(page 263)*

2.

4.

54.

56. Increasing: $(-\infty, 3)$
Decreasing: $(3, \infty)$

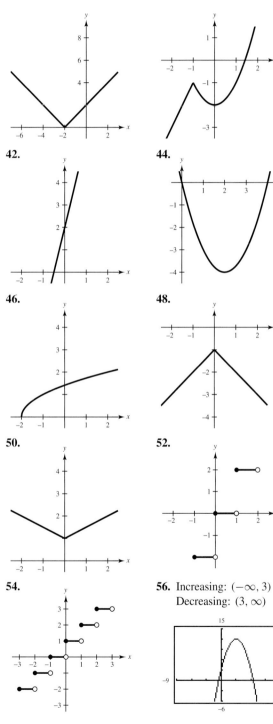
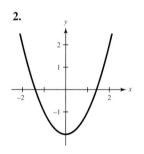
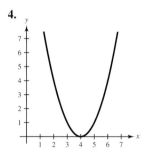
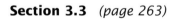

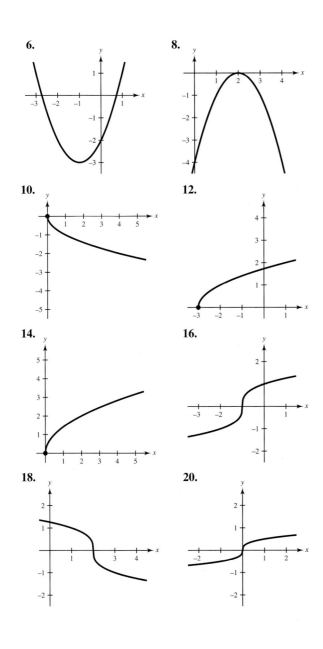

6.

8.

10.

12.

14.

16.

18.

20.

22. (a) Vertical shift of two units

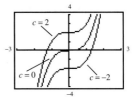

(b) Horizontal shift of two units

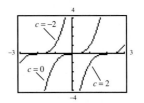

(c) Vertical shift of two units and translated to the right two units

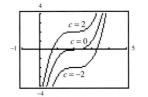

24. (a) $y = -x^3$ (b) $y = (x+1)^3 - 1$

26. $x - 4$
$3x - 6$
$-2x^2 + 7x - 5$
$\dfrac{2x - 5}{1 - x}, x \neq 1$

28. $2x$
$2x - 10$
$10x - 25$
$\dfrac{2}{5}x - 1, -\infty < x < \infty$

30. $\sqrt{x^2 - 4} + \dfrac{x^2}{x^2 + 1}$
$\sqrt{x^2 - 4} - \dfrac{x^2}{x^2 + 1}$
$\dfrac{x^2\sqrt{x^2 - 4}}{x^2 + 1}$
$\dfrac{(x^2 + 1)\sqrt{x^2 - 4}}{x^2}, |x| \geq 2$

32. $\dfrac{x^4 + x^3 + x}{x + 1}$
$\dfrac{x^4 - x^3 + x}{x + 1}$
$\dfrac{x^4}{x + 1}$
$\dfrac{1}{x^2(x + 1)}, x \neq 0, -1$

34. 11 **36.** $t^2 - t - 3$ **38.** -370

40. $-\frac{1}{4}$ **42.** -1 **44.** 52

46. (a) $6x + 3$ (b) $9x$ **48.** (a) $\dfrac{1}{x^3}$ (b) x^9

50. (a) x (b) x **52.** (a) x^{16} (b) x^{16}

54. (a) $4x - 9$ (b) $4x - 9$ **56.** (a) x^4 (b) x^4

58. (a) -1 (b) 0 **60.** (a) 2 (b) 2

62. (a) All real numbers except $x = 0$
    (b) All real numbers
    (c) All real numbers except $x = -3$

64. (a) All real numbers
    (b) All real numbers
    (c) All real numbers

66. $R_1 + R_2 = 733.9 - 7.22t - 0.8t^2$, $t = 0, 1, 2, 3, 4, 5$
    Total sales have been decreasing.

68. (a) $N(T(t)) = 100t^2 + 275$
    (b) ≈ 2.18 hours

70.

Year	P	E	P/E
1985	\$13.73	\$1.32	10.4
1986	\$22.20	\$1.51	14.7
1987	\$25.79	\$1.91	13.5
1988	\$23.01	\$1.77	13.0
1989	\$21.34	\$1.27	16.8
1990	\$15.51	\$0.94	16.5
1991	\$14.01	\$0.87	16.1
1992	\$16.31	\$0.62	26.3
1993	\$15.41	\$0.48	32.1

72. $(C \circ x)(t) = 3000t + 750$
    $C \circ x$ represents the cost of producing x units for t hours.

74. Domain of $(f/g)(x)$: All real numbers $0 \le x < 3$
    Domain of $(g/f)(x)$: All real numbers $0 < x \le 3$
    The two domains differ because of division by zero.

Section 3.4   (page 274)

2.   **4.**

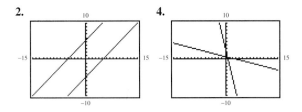

6.   **8.**

10.   **12.** f doesn't have
    an inverse.

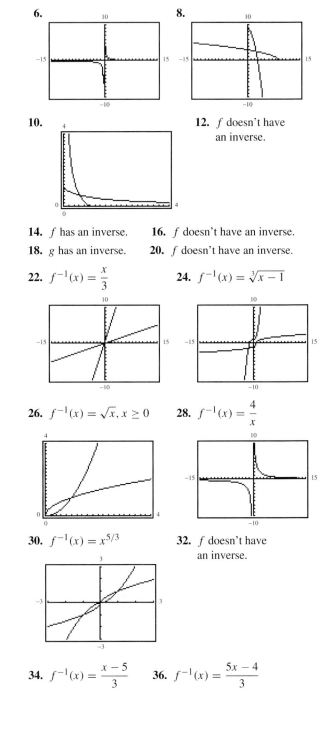

14. f has an inverse.   **16.** f doesn't have an inverse.

18. g has an inverse.   **20.** f doesn't have an inverse.

22. $f^{-1}(x) = \dfrac{x}{3}$ **24.** $f^{-1}(x) = \sqrt[3]{x - 1}$

26. $f^{-1}(x) = \sqrt{x}, x \ge 0$ **28.** $f^{-1}(x) = \dfrac{4}{x}$

30. $f^{-1}(x) = x^{5/3}$ **32.** f doesn't have
    an inverse.

34. $f^{-1}(x) = \dfrac{x - 5}{3}$ **36.** $f^{-1}(x) = \dfrac{5x - 4}{3}$

38. g doesn't have an inverse.

40. $f^{-1}(x) = 2 - x, x \geq 0$ **42.** $f^{-1}(x) = x^2 + 2, x \geq 0$

44. f doesn't have an inverse.

46. $f^{-1}(x) = -\sqrt{x - 36}, x \geq 36$

48.

x	0	2	4	6
$f^{-1}(x)$	6	2	0	-2

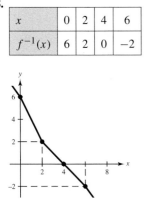

50. 0   **52.** $-\sqrt[9]{4}$ **54.** $\left(f^{-1} \circ g^{-1}\right)(x) = \dfrac{x-3}{2}$

56. $(g \circ f)^{-1}(x) = \dfrac{x-3}{2}$

58. After graphing $f(x) = x^2, x \geq 0$, and $f'(x) = \sqrt{x}$, it is observed that $f(x)$ and $f'(x)$ are reflections of each other about the line $y = x$. Because of this reflection, interchanging the rates of x and y seems to be reasonable.

60. $t = \dfrac{\sqrt{y} - 92.311}{3.98}$, 1990

Section 3.5 *(page 285)*

2. e   **4.** f   **6.** a   **8.** d   **10.** $y = -x^2 + 4$

12. $y = (x + 2)^2 - 2$ **14.** $y = 2(x - 3)^2$

16. Intercepts: $(-2\sqrt{2}, 0), (0, -4), (2\sqrt{2}, 0)$
Vertex: $(0, -4)$

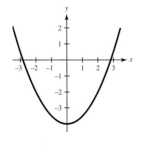

18. Intercepts: $(-5, 0), (0, 25), (5, 0)$
Vertex: $(0, 25)$

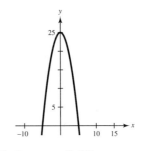

20. Intercept: $(0, 39)$
Vertex: $(6, 3)$

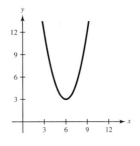

22. Intercepts: $(-1, 0), (0, 1)$
Vertex: $(-1, 0)$

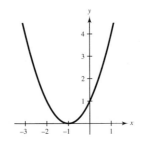

24. Intercepts: $(-4 - \sqrt{5}, 0), (-4 + \sqrt{5}, 0), (0, 11)$
Vertex: $(-4, -5)$

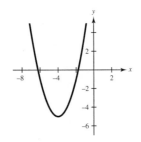

26. Intercepts: $\left(-\frac{3}{2} - \sqrt{2}, 0\right)$, $\left(-\frac{3}{2} + \sqrt{2}, 0\right)$, $\left(0, \frac{1}{4}\right)$
Vertex: $\left(-\frac{3}{2}, -2\right)$

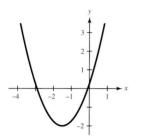

28. Intercepts: $(-2 - \sqrt{5}, 0)$, $(0, 1)$, $(-2 + \sqrt{5}, 0)$
Vertex: $(-2, 5)$

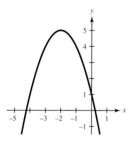

30. Intercept: $(0, 1)$; Vertex: $\left(\frac{1}{4}, \frac{7}{8}\right)$

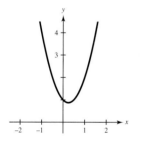

32. Intercepts: $(-2 - \sqrt{6}, 0)$, $(0, -1)$, $(-2 + \sqrt{6}, 0)$
Vertex: $(-2, -3)$

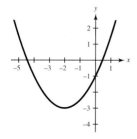

34. $y = -\frac{1}{4}(x - 2)^2 + 3$ **36.** $y = 2(x + 2)^2 - 2$

38. $f(x) = 2x^2 + x - 10$ **40.** $f(x) = x^2 - 12x + 32$
$g(x) = -2x^2 - x + 10$ $g(x) = -x^2 + 12x - 32$
(Answer is not unique)   (Answer is not unique)

42. $f(x) = x^2 - 25$
$g(x) = -x^2 + 25$
(Answer is not unique)

44. $A = (x - 100)^2 + 10,000$; 100 feet $\times$ 100 feet

46. 50 meters $\times$ 31.83 meters   **48.** 250,000 units

50. ≈ 111 units   **52.** 350,000 units

54. $x \approx 22,076$ feet   **56.** End of 1991

Section 3.6 *(page 298)*

2. c   **4.** d   **6.** h   **8.** b

10. Falls to the left   **12.** Rises to the left
Rises to the right   Falls to the right

14. Falls to the left   **16.** Rises to the left
Falls to the right   Rises to the right

18. Rises to the left   **20.** ± 7 **22.** -5
Falls to the right

24. $\dfrac{-5 \pm \sqrt{37}}{2}$ **26.** $1 \pm \sqrt{2}$ **28.** $-4, 0, 5$

30. $0, \pm\sqrt{2}$ **32.** $0, \pm\sqrt{3}$ **34.** $4, \pm 5$

36.

38.

40.

42.

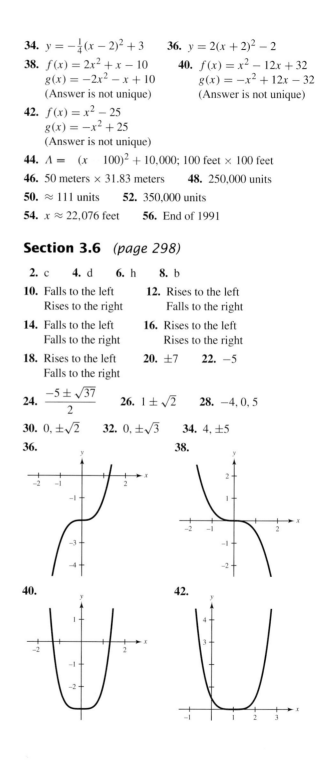

44.

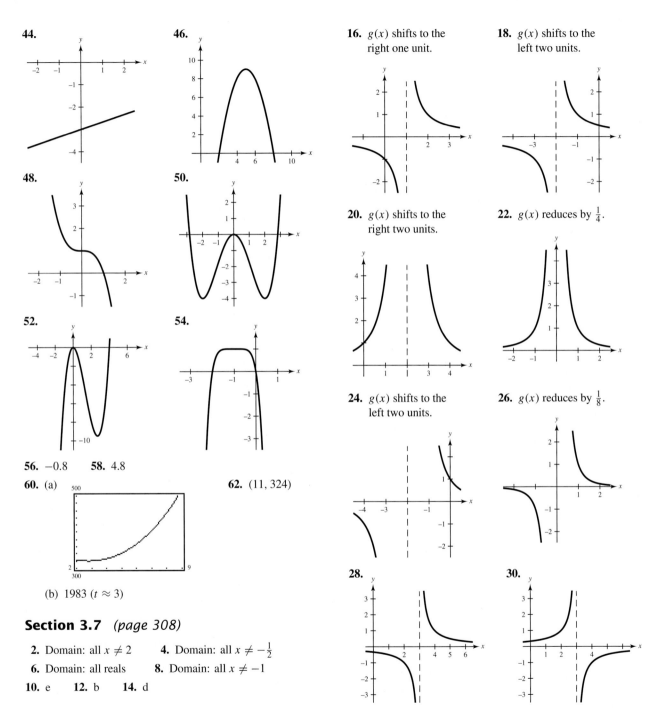

46.

48.

50.

52.

54.

56. -0.8 **58.** 4.8

60. (a)

(b) 1983 $(t \approx 3)$

62. $(11, 324)$

Section 3.7 *(page 308)*

2. Domain: all $x \neq 2$ **4.** Domain: all $x \neq -\frac{1}{2}$

6. Domain: all reals   **8.** Domain: all $x \neq -1$

10. e   **12.** b   **14.** d

16. $g(x)$ shifts to the right one unit.

18. $g(x)$ shifts to the left two units.

20. $g(x)$ shifts to the right two units.

22. $g(x)$ reduces by $\frac{1}{4}$.

24. $g(x)$ shifts to the left two units.

26. $g(x)$ reduces by $\frac{1}{8}$.

28.

30.

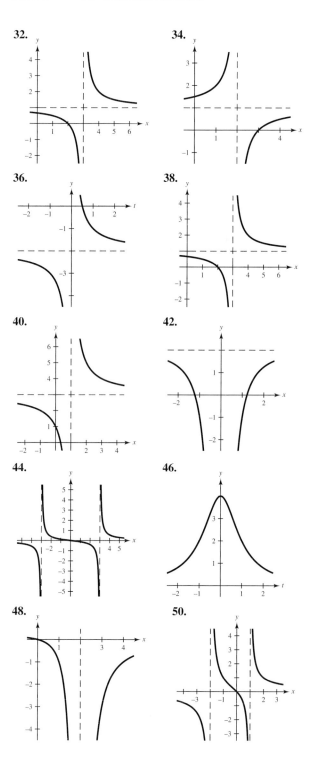

32.

34.

36.

38.

40.

42.

44.

46.

48.

50.

52. (a) ≈ \$14,117.65   (b) \$80,000   (c) \$720,000

    (d) No. The model would generate a zero in the denominator.

54. (a) ≈ 122 elk, ≈ 185 elk, ≈ 309 elk   (b) ≈ 667 elk

56. (a)

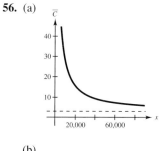

  (b)

x	1000	10,000	100,000
$\overline{C}$	\$253	\$28	\$5.50

Eventually, the average cost per unit will approach the horizontal asymptote of \$3.

58. (a)

n	1	2	3	4	5
P	0.60	0.78	0.85	0.89	0.91

n	6	7	8	9	20
P	0.92	0.93	0.94	0.95	0.95

  (b) 100%

60. (a)

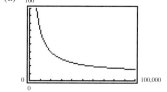

  (b)

x	1000	10,000	100,000
$\overline{C}$	\$610	\$70	\$16

Eventually, the average recycling cost will approach the horizontal asymptote of \$10.

62. Use a reference source to find the time for the 1996 Olympics. This model does not have a horizontal asymptote. For this data the model should have a horizontal asymptote.

64. (a) ≈ 1.9 minutes

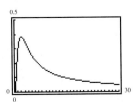

(b) The function has a horizontal asymptote of $y = 0$, which means that eventually the concentration of the medication is untraceable.

Review Exercises *(page 314)*

2. (a) 1   (b) 0   (c) -1 (d) $\sqrt{x+11} - 3$

4. (a) 1   (b) 5   (c) 6   (d) 26   **6.** $0, \pm 2$

8. $t \neq \pm 3$ **10.** All real numbers   **12.** $-3 \le x \le 3$

14. y is not a function of x.

16. Not a function. -9 is not an element of B.

18. (a) $B = 5000 \left(1 + \dfrac{0.0625}{4}\right)^{4t}$ (b) $t \ge 0$

20. (a) Domain: $x \le -3, x \ge 3$
    Range: $y \ge 0$
  (b) Increasing: $x > 3$
    Decreasing: $x < -3$
  (c) Even

22. (a) Domain: All real numbers
    Range: $y \ge 0$
  (b) Increasing: $x > 2$
    Decreasing: $x < 2$
  (c) Neither

24.   **26.**

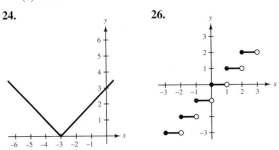

28.

30.

32.

34.

36.

38.

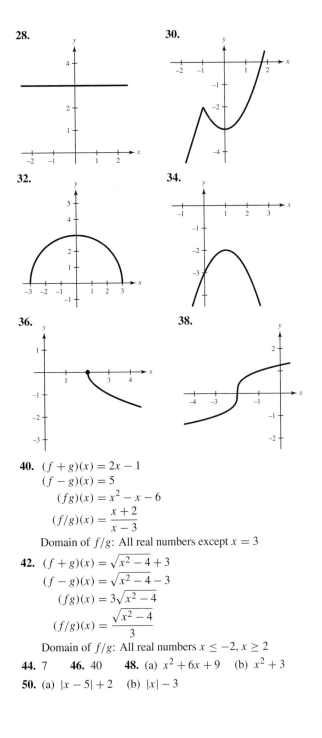

40. $(f + g)(x) = 2x - 1$
    $(f - g)(x) = 5$
    $(fg)(x) = x^2 - x - 6$
    $(f/g)(x) = \dfrac{x + 2}{x - 3}$
  Domain of f/g: All real numbers except $x = 3$

42. $(f + g)(x) = \sqrt{x^2 - 4} + 3$
    $(f - g)(x) = \sqrt{x^2 - 4} - 3$
    $(fg)(x) = 3\sqrt{x^2 - 4}$
    $(f/g)(x) = \dfrac{\sqrt{x^2 - 4}}{3}$
  Domain of f/g: All real numbers $x \le -2, x \ge 2$

44. 7   **46.** 40   **48.** (a) $x^2 + 6x + 9$ (b) $x^2 + 3$

50. (a) $|x - 5| + 2$ (b) $|x| - 3$

52. $R = 594.12 - 2.78t - 0.3t^2, t = 0, 1, 2, 3, 4$

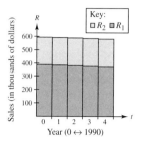

Total sales are decreasing.

54. $f(g(x)) = 3\left(\dfrac{x-5}{3}\right) + 5 = x$

$g(f(x)) = \dfrac{(3x+5) - 5}{3} = x$

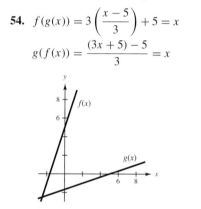

56. f doesn't have an inverse.

58. f has an inverse.

$f^{-1}(x) = \dfrac{1}{x}$

60. f has an inverse.

$f^{-1}(x) = \dfrac{3x+5}{2}$

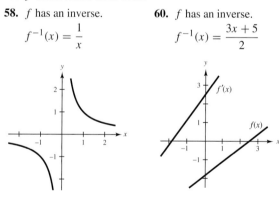

62. y-intercept: $(0, 9)$
Vertex: $(2, 5)$

64. y-intercept: $(0, 11)$

x-intercepts: $\left(\dfrac{6 \pm \sqrt{3}}{3}, 0\right)$

Vertex: $(2, -1)$

66. $f(x) = (x-1)^2 - 4$

68. $A(x) = -(x - 50)^2 + 2500$
50 feet $\times$ 50 feet

70. (a)

(b) ≈ 363 to 364 units

(c) Calculate the vertex from the equation:

$\left(\dfrac{4000}{11}, \dfrac{1,540,000}{121}\right).$

The cost is minimum when $x = \dfrac{4000}{11} \approx 364$ units.

72. Falls to the left
Rises to the right

74. Rises to the left
Falls to the right

76. ± 5 **78.** $0, 2, 5$ **80.** ≈ -2.3

82. Domain: All real numbers except $x = -3$
Vertical asymptote: $x = -3$
Horizontal asymptote: $y = 0$ (x-axis)

84. Domain: All real numbers except $x = \pm 2$
Vertical asymptotes: $x = -2, x = 2$
Horizontal asymptote: $y = 1$

86.

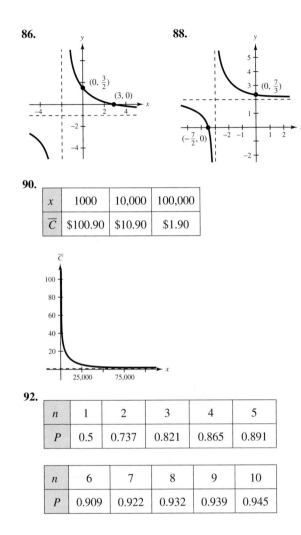

88.

90.

x	1000	10,000	100,000
$\overline{C}$	\$100.90	\$10.90	\$1.90

92.

n	1	2	3	4	5
P	0.5	0.737	0.821	0.865	0.891

n	6	7	8	9	10
P	0.909	0.922	0.932	0.939	0.945

94. (a)

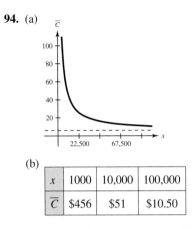

(b)

x	1000	10,000	100,000
$\overline{C}$	\$456	\$51	\$10.50

Eventually, the average recycling cost will approach the horizontal asymptote of \$6.

CHAPTER 4

Section 4.1 *(page 328)*

2. $5x + 3$ **4.** $2x^2 - 4x + 3$ **6.** $x + 4$

8. $4 - \dfrac{9}{2x + 1}$ **10.** $x - \dfrac{x + 9}{x^2 + 1}$ **12.** $x^2 + x + 4$

14. $x^2 + \dfrac{x^2 + 7}{x^3 - 1}$ **16.** $5x^2 + 3x - 2$ **18.** $9x^2 - 16$

20. $3x^2 + 2x + 12$ **22.** $5x^2 - 10x + 26 - \dfrac{44}{x + 2}$

24. $x^4 - 16x^3 + 48x^2 - 144x + 312 - \dfrac{856}{x + 3}$

26. $5x^2 - 15x + 45 - \dfrac{135}{x + 3}$

28. $-3x^3 + 6x^2 - 12x + 24 - \dfrac{48}{x + 2}$

30. $-x^3 - 6x^2 - 36x - 36 - \dfrac{216}{x - 6}$

32. $3x^2 + \dfrac{1}{2}x + \dfrac{3}{4} + \dfrac{49}{8x - 12}$

34. $(x + 4)(x + 2)(x - 6)$ **36.** $(3x - 2)(4x - 3)(4x - 1)$

38. $(x - \sqrt{2})(x + \sqrt{2})(x + 2)$

40. $f(x) = \left(x + \dfrac{2}{3}\right)\left(5x^3 - 2x + 7\right)$

$f\left(-\dfrac{2}{3}\right) = 0$

42. $f(x) = (x - 1 + \sqrt{3})[4x^2 - (2 + 4\sqrt{3})x - (2 + 2\sqrt{3})]$
$f(1 - \sqrt{3}) = 0$

44. (a) 14   (b) 3122   (c) 434   (d) 2

46. (a) -2.5 (b) 20   (c) 65.5   (d) 5668

48. (a) 200   (b) $-\frac{2504}{125}$ (c) -68.125 (d) -52

50. d, $-2, \dfrac{3 \pm \sqrt{5}}{2}$ **52.** c, $3, 1 \pm \sqrt{5}$

54. $x^2 - 7x - 8$ **56.** $2x^2 + 5x + 2$

58. $x^2 + 9x - 1$ **60.** $x^2 + 50x + 400$ square feet

62. (a)

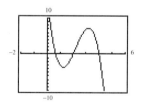

(b) Yes. Another nonnegative real solution is ≈ 28.1078, representing an advertising cost of \$281,078.

Section 4.2   *(page 337)*

2. One negative zero   **4.** Two or zero positive zeros

6. Three or one positive zeros

8. Three or one negative zeros

10. Two or zero positive zeros

12. Possible: $\pm\frac{1}{3}, \pm\frac{2}{3}, \pm1, \pm\frac{4}{3}, \pm2, \pm\frac{8}{3}, \pm4, \pm\frac{16}{3}, \pm8, \pm16$
Actual: $\frac{2}{3}, 2, 4$

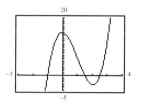

14. Possible: $\pm\frac{1}{4}, \pm\frac{1}{2}, \pm\frac{3}{4}, \pm1, \pm\frac{5}{4}, \pm\frac{3}{2}, \pm\frac{5}{2}, \pm3, \pm\frac{15}{4}, \pm5,$
$\pm\frac{15}{2}, \pm15$
Actual: $-1, \frac{3}{2}, \frac{5}{2}$

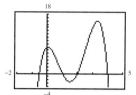

16. Possible: $\pm\frac{1}{2}, \pm1, \pm2, \pm4, \pm8$
Actual: $-\frac{1}{2}, 1, 2, 4$

18. (a) Neither   (b) Upper bound   (c) Neither

20. (a) Neither   (b) Upper bound   (c) Lower bound

22. $-2, -1, 3$ **24.** $1, 2, 6$ **26.** -2 **28.** 3

30. $\frac{1}{3}, 3$ **32.** $\pm\frac{3}{2}, \frac{1}{3}$ **34.** $\frac{2}{3}$ **36.** $\pm\sqrt{3}$

38. $-5, 6$ **40.** $-6, \frac{1}{2}, 1$ **42.** $-2, 0, 1$ **44.** $\pm3, \frac{1}{3}, \frac{3}{2}$

46. (a) $\pm\frac{1}{6}, \pm\frac{1}{3}, \pm\frac{1}{2}, \pm\frac{2}{3}, \pm1, \pm\frac{4}{3}, \pm2, \pm\frac{8}{3}, \pm4, \pm8$

(b)

(c) $1, \dfrac{-5 \pm \sqrt{217}}{12}$

48. (a) $\pm\frac{1}{2}, \pm1, \pm2, \pm\frac{5}{2}, \pm5, \pm10$

(b)

(c) $\dfrac{5}{2}, \dfrac{-5 \pm \sqrt{17}}{2}$

50. $-3, \frac{1}{2}, 4$ **52.** $-2, -\frac{1}{3}, \frac{1}{2}$ **54.** a   **56.** c

58. 2 inches × 6 inches × $\frac{3}{2}$ inches or
3.77 inches × 7.77 inches × 0.61 inch

60. 20 inches × 20 inches × 40 inches

62. (a) 30 and ≈ 38.91

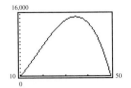

(b) An alternative method is to solve
$-x^3 + 54x^2 - 140x - 3000 = 14{,}000$. Graphing $y = -x^3 + 54x^2 - 140x - 17{,}400$ allows us to see where
$y = 0$. ($x \approx -14.91$, $x = 30$, $x \approx 38.91$). According
to Descartes's Rule of Signs, there may be two positive
real zeros and one negative real zero. Recommend that
the company charge \$38.91.

Section 4.3 *(page 345)*

2. c, 0.755   **4.** a, 0.370   **6.** b, $-1.532, -0.347, 1.879$

8. -0.453 **10.** 0.900, 1.100, 1.900   **12.** 3.540

14. 4.968   **16.** 3.554   **18.** $-1.090, 2.32$

20. $-0.417, 0.337$ **22.** 1994

24. ≈ 4000 units   **26.** ≈ \$320,000

Section 4.4 *(page 356)*

2. (a) 1   (b) i (c) -1 (d) $-i$

4. $a = 13, b = 4$ **6.** $a = 0, b = -\frac{5}{2}$

8. $3 + 4i, 3 - 4i$ **10.** $1 + 2\sqrt{2}\,i, 1 - 2\sqrt{2}\,i$

12. 45, 45   **14.** $-4 + 2i, -4 - 2i$ **16.** $i, -i$

18. $-9, -9$ **20.** $8 + 4i$ **22.** $-3 - 11i$ **24.** 4

26. $-4.2 + 7.5i$ **28.** $-5\sqrt{2}$ **30.** $-375\sqrt{3}\,i$

32. $6 - 22i$ **34.** 85   **36.** $32 - 72i$ **38.** 8

40. $(21 + 5\sqrt{2}) + (7\sqrt{5} - 3\sqrt{10})i$ **42.** $-46 - 9i$

44. $\frac{3}{2} + \frac{3}{2}i$ **46.** $\frac{22}{5} + \frac{9}{5}i$ **48.** $10 - 4i$

50. $-\dfrac{9}{1681} + \dfrac{40}{1681}i$ **52.** $\dfrac{60}{13} - \dfrac{25}{13}i$ **54.** i

56. $-8i$ **58.** $-3 \pm i$ **60.** $\dfrac{1}{3} \pm 2i$ **62.** $\dfrac{1}{3} \pm \dfrac{\sqrt{34}}{3}i$

64. $-\dfrac{3}{5} \pm \dfrac{\sqrt{6}}{5}i$ **66.**

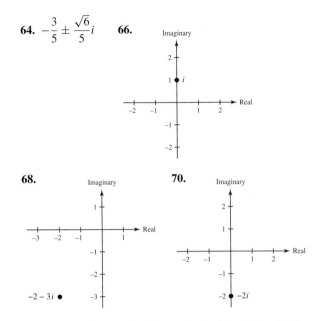

68.

70.

72. The complex number 2 is not in the Mandelbrot Set because, for $c = 2$, the corresponding Mandelbrot sequence
is 2, 6, 38, 1446, 2090, 918, $\approx 4.3719 \times 10^{12}, \ldots$, which
is unbounded.

74. The complex number $-i$ is in the Mandelbrot Set because,
for $c = -i$, the corresponding Mandelbrot sequence is $-i$,
$-1 - i, i, -1 - i, i, -1 - i$, which is bounded.

76. The complex number -1 is in the Mandelbrot Set because,
for $c = -1$, the corresponding Mandelbrot sequence is -1,
$0, -1, 0, -1, 0$, which is bounded.

Section 4.5 *(page 364)*

2. $\dfrac{1 \pm \sqrt{223}\,i}{2}, \left(x - \dfrac{1 - \sqrt{223}\,i}{2}\right)\left(x - \dfrac{1 + \sqrt{223}\,i}{2}\right)$

4. $-5 \pm \sqrt{2}, (x + 5 + \sqrt{2})(x + 5 - \sqrt{2})$

6. $\pm 5, \pm 5i, (y + 5)(y - 5)(y + 5i)(y - 5i)$

8. $1 \pm i, (x - 1)(x - 1 + i)(x - 1 - i)$

10. $3 \pm 2i, (x + 4)(x - 3 + 2i)(x - 3 - 2i)$

12. $-5 \pm 2i, (x + 1)(x + 5 + 2i)(x + 5 - 2i)$

14. $1 \pm 2i, (2s - 1)(s - 1 + 2i)(s - 1 - 2i)$

16. $\frac{1}{3} \pm \frac{2}{3}i, (x - 1)(3x - 1 + 2i)(3x - 1 - 2i)$

18. $-2 \pm \sqrt{3}\,i, (x + 5)(x - 2 + \sqrt{3}\,i)(x + 2 - \sqrt{3}\,i)$

20. $1 \pm \sqrt{3}\,i$, $(3x + 2)(x - 1 + \sqrt{3}\,i)(x - 1 - \sqrt{3}\,i)$

22. -3, $(x + 3)^2(x + i)(x - i)$

24. $\pm 2i$, $\pm 5i$, $(x + 2i)(x - 2i)(x + 5i)(x - 5i)$

26. $1 \pm \sqrt{3}\,i$, $(x - 2)^3(x - 1 + \sqrt{3}\,i)(x - 1 - \sqrt{3}\,i)$

28. $x^3 - 4x^2 + 9x - 36$

30. $x^3 + 4x^2 - 31x - 174$

32. $x^5 - 6x^4 + 28x^3 - 104x^2 + 192x - 128$

34. $3x^4 - 17x^3 + 25x^2 + 23x - 22$

36. $x^5 - 6x^4 + 10x^3 - 8x^2$

38. (a) $(x^2 - 6)(x^2 - 2x + 3)$
   (b) $(x + \sqrt{6})(x - \sqrt{6})(x^2 - 2x + 3)$
   (c) $(x + \sqrt{6})(x - \sqrt{6})(x - 1 - \sqrt{2}\,i)(x - 1 + \sqrt{2}\,i)$

40. (a) $(x^2 + 4)(x^2 - 3x - 5)$
   (b) $(x^2 + 4)\left(x - \dfrac{3 + \sqrt{29}}{2}\right)\left(x - \dfrac{3 - \sqrt{29}}{2}\right)$
   (c) $(x + 2i)(x - 2i)\left(x - \dfrac{3 + \sqrt{29}}{2}\right)\left(x - \dfrac{3 - \sqrt{29}}{2}\right)$

42. -1, $\pm 3i$ **44.** $-3, 5 \pm 2i$ **46.** $-\frac{2}{3}, 1 \pm \sqrt{3}\,i$

48. $-2, -1 \pm 3i$ **50.** $3, \dfrac{-2 \pm \sqrt{2}\,i}{5}$

52. There is no price p that would yield $50 million. This is shown in the graph by the fact that no real zeros exist.

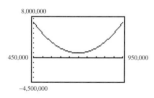

54. No. The conjugate pair statement specifies polynomials with *real* coefficients. $f(x)$ has imaginary coefficients.

56. Counterexample: $x^2 + 1 = (x + i)(x - i)$

Review Exercises *(page 368)*

2. $3x + 5$ **4.** $x^4 - 3x^3 + 7x^2 - 27x + 55 - \dfrac{104}{x + 2}$

6. $(3x - 1)(x + 3)(x + 5)$ **8.** (a) 6.5 (b) 5

10. $x^2 + 5x + 4$ **12.** Possible positive zeros: 3 or 1
Possible negative zeros: 2 or 0

14. From the graph, we see that $h(x)$ has one positive real root and no negative real roots.
$\pm 1, \pm 2, \pm 4, \pm 8, \pm\frac{1}{3}, \pm\frac{2}{3}, \pm\frac{4}{3}, \pm\frac{8}{3}$
Roots: $x \approx -2.611, 0.795$

16. \$256,155

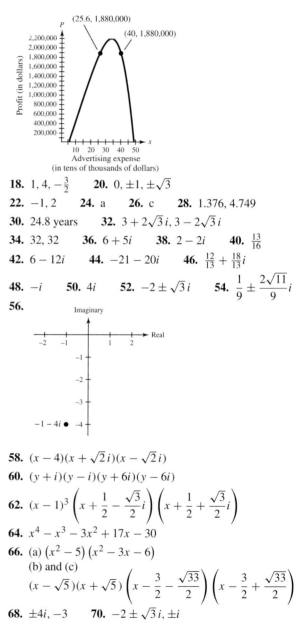

18. $1, 4, -\frac{3}{2}$ **20.** $0, \pm 1, \pm\sqrt{3}$

22. $-1, 2$ **24.** a   **26.** c   **28.** 1.376, 4.749

30. 24.8 years   **32.** $3 + 2\sqrt{3}\,i, 3 - 2\sqrt{3}\,i$

34. 32, 32   **36.** $6 + 5i$ **38.** $2 - 2i$ **40.** $\frac{13}{16}$

42. $6 - 12i$ **44.** $-21 - 20i$ **46.** $\frac{12}{13} + \frac{18}{13}i$

48. $-i$ **50.** $4i$ **52.** $-2 \pm \sqrt{3}\,i$ **54.** $\frac{1}{9} \pm \frac{2\sqrt{11}}{9}i$

56.

58. $(x - 4)(x + \sqrt{2}\,i)(x - \sqrt{2}\,i)$

60. $(y + i)(y - i)(y + 6i)(y - 6i)$

62. $(x - 1)^3\left(x + \dfrac{1}{2} - \dfrac{\sqrt{3}}{2}i\right)\left(x + \dfrac{1}{2} + \dfrac{\sqrt{3}}{2}i\right)$

64. $x^4 - x^3 - 3x^2 + 17x - 30$

66. (a) $(x^2 - 5)(x^2 - 3x - 6)$
   (b) and (c)
   $(x - \sqrt{5})(x + \sqrt{5})\left(x - \dfrac{3}{2} - \dfrac{\sqrt{33}}{2}\right)\left(x - \dfrac{3}{2} + \dfrac{\sqrt{33}}{2}\right)$

68. $\pm 4i, -3$ **70.** $-2 \pm \sqrt{3}\,i, \pm i$

CHAPTER 5

Section 5.1 *(page 380)*

2. ≈ 7.352 **4.** ≈ 1767.767 **6.** ≈ 16.380

8. ≈ 0.006 **10.** ≈ 0.413 **12.** ≈ 1.649

14. ≈ 24.533 **16.** e   **18.** h   **20.** a   **22.** c

24. **26.**

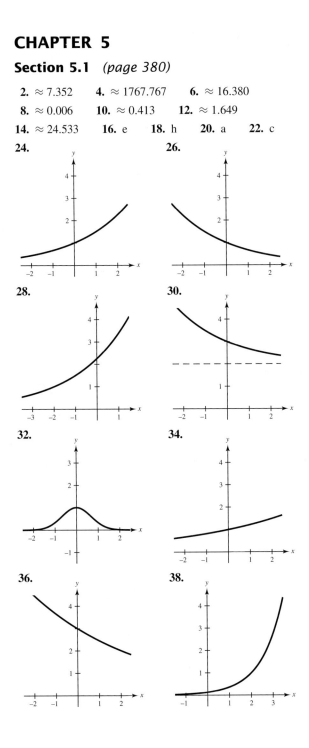

28. **30.**

32. **34.**

36. **38.**

40.

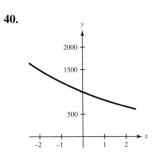

42.

n	1	2	4
A	$2593.74	$2653.30	$2685.06

n	12	365	Continuous compounding
A	$2707.04	$2717.91	$2718.28

44.

n	1	2	4
A	$45,259.26	$49,561.44	$51,977.87

n	12	365	Continuous compounding
A	$53,700.66	$54,568.24	$54,598.15

46.

t	1	10	20
P	$88,692.04	$30,119.42	$9071.80

t	30	40	50
P	$2732.37	$822.97	$247.88

48.

t	1	10	20
P	$93,240.01	$49,661.86	$24,663.01

t	30	40	50
P	$12,248.11	$6082.64	$3020.75

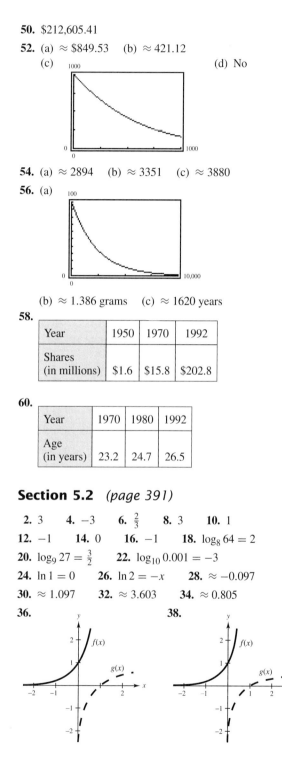

50. $212,605.41

52. (a) $\approx$ $849.53 (b) $\approx$ 421.12

(c) (d) No

54. (a) $\approx$ 2894 (b) $\approx$ 3351 (c) $\approx$ 3880

56. (a)

(b) $\approx$ 1.386 grams (c) $\approx$ 1620 years

58.

Year	1950	1970	1992
Shares (in millions)	$1.6	$15.8	$202.8

60.

Year	1970	1980	1992
Age (in years)	23.2	24.7	26.5

Section 5.2 *(page 391)*

2. 3 **4.** -3 **6.** $\frac{2}{3}$ **8.** 3 **10.** 1

12. -1 **14.** 0 **16.** -1 **18.** $\log_8 64 = 2$

20. $\log_9 27 = \frac{3}{2}$ **22.** $\log_{10} 0.001 = -3$

24. $\ln 1 = 0$ **26.** $\ln 2 = -x$ **28.** ≈ -0.097

30. ≈ 1.097 **32.** ≈ 3.603 **34.** ≈ 0.805

36. **38.**

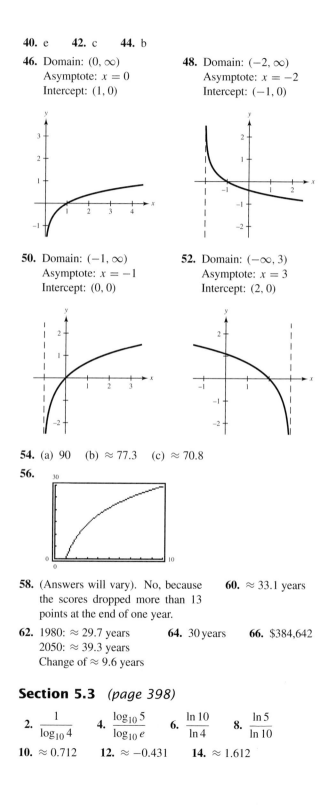

40. e **42.** c **44.** b

46. Domain: $(0, \infty)$
Asymptote: $x = 0$
Intercept: $(1, 0)$

48. Domain: $(-2, \infty)$
Asymptote: $x = -2$
Intercept: $(-1, 0)$

50. Domain: $(-1, \infty)$
Asymptote: $x = -1$
Intercept: $(0, 0)$

52. Domain: $(-\infty, 3)$
Asymptote: $x = 3$
Intercept: $(2, 0)$

54. (a) 90 (b) $\approx$ 77.3 (c) $\approx$ 70.8

56.

58. (Answers will vary). No, because the scores dropped more than 13 points at the end of one year.

60. $\approx$ 33.1 years

62. 1980: $\approx$ 29.7 years
2050: $\approx$ 39.3 years
Change of $\approx$ 9.6 years

64. 30 years **66.** $384,642

Section 5.3 *(page 398)*

2. $\dfrac{1}{\log_{10} 4}$ **4.** $\dfrac{\log_{10} 5}{\log_{10} e}$ **6.** $\dfrac{\ln 10}{\ln 4}$ **8.** $\dfrac{\ln 5}{\ln 10}$

10. $\approx$ 0.712 **12.** ≈ -0.431 **14.** $\approx$ 1.612

16. ≈ 3.823 **18.** $1 + \log_{10} z$ **20.** $\log_{10} y - \log_{10} 2$

22. $-3 \log_2 z$ **24.** $\frac{1}{3} \ln t$ **26.** $\ln x + \ln y - \ln z$

28. $\ln(x + 1) + \ln(x - 1) - 3 \ln x$ **30.** $\ln x - \frac{3}{2} \ln y$

32. $4 \log_b x - 2 \log_b z$ **34.** $\ln x - \frac{1}{2} \ln (x^2 + 1)$

36. $\ln x + \frac{1}{2} \ln (x + 2)$ **38.** $\ln yz$ **40.** $\log_{10} \dfrac{8}{t}$

42. $\log_{10} \dfrac{1}{16x^4}$ **44.** $\ln 64z^5$ **46.** $\ln(z - 2)^{3/2}$

48. $\ln \dfrac{x^3 y^3}{z^4}$ **50.** $\ln t^6$ **52.** $\ln \left[x^3 (x + 1)(x - 1)^2 \right]$

54. 1.3917   **56.** 0.2625   **58.** 1.4854

60. 0.7730   **62.** 0.7396   **64.** 2.5646

66. 0   **68.** $\frac{1}{3}$ **70.** -3 **72.** $\frac{3}{4}$

74. $-1 - \log_5 3$ **76.** $4 + 4 \log_2 3$ **78.** $\log_{10} 3 - 2$

80. $\ln 6 - 2$ **82.** $y = x^{2/3}$ **84.** $Y - 0.5x^{2.5}$

86. They have the same graph. $\log_a(uv) = \log_a u + \log_a v$

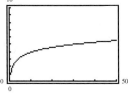

88. Answers will vary. Choose a value for y and graph $\dfrac{\log_a x}{\log_a y}$ and $\log_a x - \log_a y$. Notice the graphs are different. To demonstrate correctly, graph $\log_a \dfrac{x}{y}$ and $\log_a x - \log_a y$, choosing a value for y.

90. $y = 300.55x^{-0.14}$

Section 5.4 *(page 410)*

2. 5   **4.** $\frac{2}{3}$ **6.** 4   **8.** 5   **10.** 1   **12.** $2x - 1$

14. $2x - 1$ **16.** $x^3 - 8$ **18.** $\ln 5 \approx 1.609$

20. $\ln 580 \approx 6.363$ **22.** $\ln 22.75 \approx 3.125$

24. $\ln 32 \approx 3.466$ **26.** $\ln 148.75 \approx 5.002$

28. $\ln \left(\dfrac{25}{3} \right) \approx 2.120$ **30.** $\dfrac{\ln 50}{2} \approx 1.956$

32. $-\dfrac{\ln 14}{2} \approx -1.320$ **34.** $-\dfrac{\ln 14}{5} \approx -0.528$

36. $\dfrac{\ln \left(\frac{8}{3} \right)}{0.07} \approx 14.012$ **38.** $\ln 2 \approx 0.693$ or $\ln 3 \approx 1.099$

40. $-2 \ln 0.625 \approx 0.940$ **42.** $-5 \ln \left(\dfrac{4}{19} \right) \approx 7.791$

44. $e^2 \approx 7.389$ **46.** $e^{-2} \approx 0.135$ **48.** $\dfrac{e^{2.5}}{4} \approx 3.046$

50. $\dfrac{e^{10/3}}{5} \approx 5.606$ **52.** $e^{1/3} - 1 \approx 0.396$

54. $\dfrac{e^6}{3} \approx 134.476$ **56.** $\pm \sqrt{e} \approx \pm 1.649$

58. $-1 \pm \sqrt{1 + e^3}$ (using the positive value for x, $x \approx 3.592$)

60. $e^2 - 2 \approx 5.389$ **62.** ≈ 5.78 years

64. ≈ 14.65 years

66. (a) ≈ 303 units   (b) ≈ 528 units

(c)

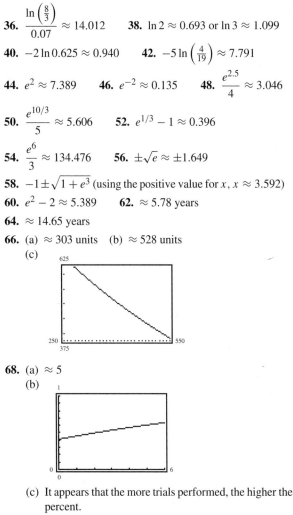

68. (a) ≈ 5

(b)

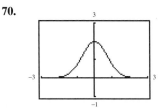

(c) It appears that the more trials performed, the higher the percent.

70.

72. (a) 64.51 inches

(b) The percent of females under height x increases as the height increases until 100% of the females are less than 72 inches tall.

74. May, June, July, August, and September

76. ≈ 26 months

Section 5.5 *(page 420)*

2. Initial Investment: $20,000
Annual % Rate: $10\frac{1}{2}\%$
Time to Double: 6.60 years
Amount After 10 Years: $57,153.02

4. Initial Investment: $10,000
Annual % Rate: 13.86%
Time to Double: 5 years
Amount After 10 Years: $40,000.00

6. Initial Investment: $2000
Annual % Rate: 4.05%
Time to Double: 17.10 years
Amount After 10 Years: $3000

8. Initial Investment: $8986.58
Annual % Rate: 8%
Time to Double: 8.66 years
Amount After 10 Years: $20,000

10. Initial Investment: $250
Annual % Rate: 8.75%
Time to Double: 7.92 years
Amount After 10 Years: $600

12. Isotope: Ra^{226}
Half-Life (years): 1620
Initial Quantity: 2.30 grams
Amount After 1000 Years: 1.5 grams

14. Isotope: C^{14}
Half-Life (years): 5730
Initial Quantity: 3 grams
Amount After 1000 Years: 2.66 grams

16. Isotope: Pu^{230} **18.** Decay
Half-Life (years): 24,360
Initial Quantity: 0.53 gram
Amount After 1000 years: 0.52 gram

20. Growth   **22.** ≈ 0.6212 **24.** ≈ -0.5365

26. 1993   **28.** $\approx 157,232$ people

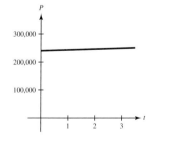

30. ≈ 61.16 hours   **32.** ≈ 160.850 years

34. (a) $S(t) = 100\left(1 - e^{-0.1625t}\right)$ (b) $\approx 55,625$ units

36. (a)

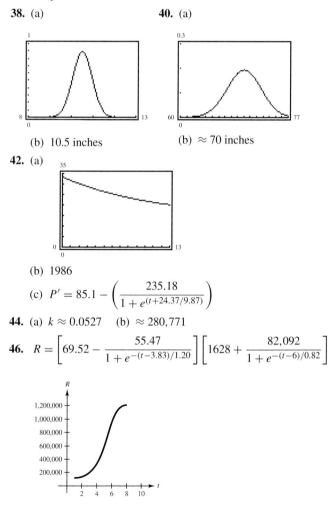

(b) Once a book is published, sales of books and complimentary copies affect the number of books sold each year.

38. (a)   **40.** (a)

(b) 10.5 inches   (b) ≈ 70 inches

42. (a)

(b) 1986

(c) $P' = 85.1 - \left(\dfrac{235.18}{1 + e^{(t+24.37/9.87)}}\right)$

44. (a) $k \approx 0.0527$ (b) $\approx 280,771$

46. $R = \left[69.52 - \dfrac{55.47}{1 + e^{-(t-3.83)/1.20}}\right]\left[1628 + \dfrac{82,092}{1 + e^{-(t-6)/0.82}}\right]$

The revenue was increasing.

48. (a) $\approx 398,107,171$ (b) 5,011,872

50. (a) 30   (b) 85   (c) 90   (d) 115

52. $\approx 93.69\%$ **54.** Yes

Review Exercises   *(page 427)*

2. f   **4.** b   **6.** e   **8.** g

10.   **12.**

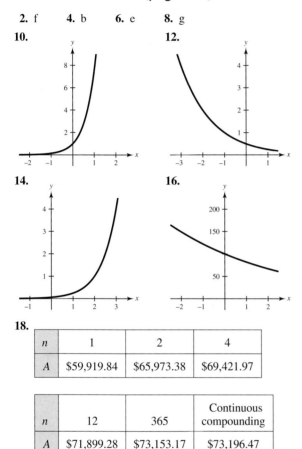

14.   **16.**

18.

n	1	2	4
A	\$59,919.84	\$65,973.38	\$69,421.97

n	12	365	Continuous compounding
A	\$71,899.28	\$73,153.17	\$73,196.47

20.

t	1	10	20
P	\$180,967.48	\$73,575.89	\$27,067.06

t	30	40	50
P	\$9957.41	\$3663.13	\$1347.59

22. (a) 2003   (b) 2017   **24.** $\frac{1}{2}$

26. -1 **28.** 0   **30.** $\log_{25} 125 = \frac{3}{2}$

32.   **34.** Domain: $(-\infty, 3)$
                     Vertical asymptote: $x = 3$
                     x-intercept: $(2, 0)$

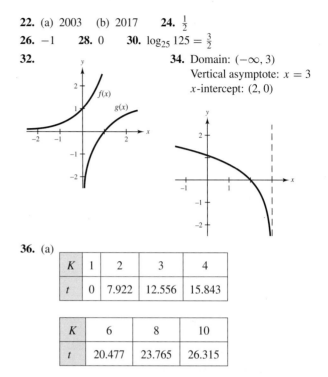

36. (a)

K	1	2	3	4
t	0	7.922	12.556	15.843

K	6	8	10
t	20.477	23.765	26.315

(b) As K increases, t increases logarithmically.

38. The projected change in median age in the United States from 1995 to 2010 is 2.17. The population is gradually getting older.

40. ≈ -2.322 **42.** ≈ -1.159

44. $\ln x + 2\ln y - \ln z$ **46.** $\ln x - \frac{2}{3}\ln y$

48. $\ln 2 + \ln x + \ln y + \ln z$ **50.** $\ln yz^2$

52. $\ln\left(\dfrac{x^2 y^3}{z^5}\right)$ **54.** $\ln\left(\dfrac{x^2(x+2)}{(x+4)^3}\right)$ **56.** 0.525

58. 1.7479   **60.** -3 **62.** $\dfrac{3}{5}$ **64.** $\dfrac{\ln 25}{3} \approx 1.073$

66. $\ln 5 \approx 1.609$ **68.** $\dfrac{e^{7.5}}{4} \approx 452.011$

70. $\pm\dfrac{e^{10.5}}{2} \approx \pm 18,157.751$ **72.** $e^4 - 1 \approx 53.598$

74. (a) ≈ 113 units   (b) ≈ 275 units   **76.** 2.4693 grams

78. $\approx 152,203$ people   **80.** ≈ 4.5388 hours

82. 300

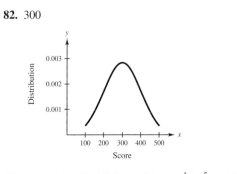

84. (a) $\approx 251{,}188{,}643.2$
(a) $\approx 7{,}079{,}457.844$
(c) $1{,}258{,}925{,}412$

86. (a) $\frac{1}{2}\ln\frac{5}{4}\approx 0.11157$
(b) In 1995: $\approx \$384{,}858$;
In 1996: $\approx \$369{,}231$

CHAPTER 6

Section 6.1 *(page 441)*

2. (a) No   (b) Yes   **4.** $(-2, 3)$ **6.** $(-1, -1), (2, 8)$

8. $(2, 2), (-2, -2), (0, 0)$ **10.** $(-1, 0), (1, 0)$

12. $(0, 3), (2, -1), (1, 1)$ **14.** $(-3, 2)$ **16.** $\left(\frac{4}{3}, \frac{4}{3}\right)$

18. $(1, 1)$ **20.** $(8, 8)$ **22.** $(0, 0)$ **24.** $(-3, 7), (2, 2)$

26. $(3, 4), (5, 0)$ **28.** $(0, -1), (2, 1), (-1 - 5)$

30. $(-1 + \sqrt{6}, -3 + 2\sqrt{6}), (-1 - \sqrt{6}, -3 - 2\sqrt{6})$

32. $(5, 2), (1, 0)$ **34.** $(0, -1), (1, -2)$

36. $(5, 2), (2, -1)$ **38.** $(0, 0), (1, 1)$

40. $\left(\pm\sqrt{\dfrac{-1 + \sqrt{33}}{2}}, \dfrac{-1 + \sqrt{33}}{2}\right)$

42. ≈ 3133 units
Revenue: $\$10{,}307.57$

44. $\approx 294{,}118$ units
Revenue: $\$73{,}529.50$

46. 400 units

48. According to this model, the population of San Diego will be greater than Philadelphia's during the years 2014 to 2032.

50. 1995

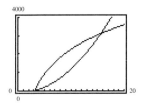

52. $\$6000$ at 7.75%   **54.** $\$500{,}000$
$\$12{,}000$ at 8.25%

Section 6.2 *(page 452)*

2. $(-1, 1)$ **4.** $(3, 4)$ **6.** Inconsistent

8. $\left(\frac{12}{7}, -\frac{23}{14}\right)$ **10.** $(3, 1)$ **12.** $\left(3, \frac{7}{5}\right)$ **14.** $(5, 1)$

16. $\left(\frac{5}{6}, \frac{5}{6}\right)$ **18.** $\left(\frac{12}{7}, \frac{18}{7}\right)$ **20.** Inconsistent

22. $(a, 4 - 4a)$ where a is any real number

24. $(7, 1)$ **26.** $\left(\frac{2}{7}, -\frac{4}{5}\right)$ **28.** $(8, 7)$ **30.** $\left(\frac{13}{3}, -2\right)$

32. No. Lines intersect at $(240, 250)$

34. 600 miles per hour, 550 miles per hour

36. $333\frac{1}{3}$ at 80%   **38.** $\$14{,}000$ at 5.75%
$166\frac{2}{3}$ at 86% $\$18{,}000$ at 6.25%

40. 152 pairs at $\$66.95$ **42.** $x = 900{,}000$ units
88 pairs at $\$84.95$ $p = \$51.00$

44. $x = 250{,}000$ units
$p = \$350.00$

46. (a)

(b) 1990

48. $y = 0.22x + 1.9$ **50.** $y = -\frac{102}{175}x + \frac{566}{105}$

52. $y = -\frac{61}{180}x + \frac{199}{56}$ **54.** (a) $y = 0.12x + 6.48$
(b) 8.52

Section 6.3 *(page 464)*

2. $(0, 4, -2)$ **4.** $(5, -2, 0)$ **6.** $\left(\frac{1}{2}, -\frac{3}{2}, 1\right)$

8. No solution   **10.** $\left(\frac{3}{10}, \frac{2}{5}, 0\right)$

12. $\left(-\frac{1}{2}a + \frac{5}{2}, 4a - 1, a\right)$ **14.** $(-a + 3, a + 1, a)$

16. $(-5a + 3, -a - 5, a)$ **18.** $\left(-\frac{3}{8}a - \frac{1}{4}, -\frac{3}{4}a + \frac{5}{2}, a\right)$

20. $(1, 0, 3, 2)$ **22.** No solution

24. $(0, 0, 0)$ **26.** $\left(\frac{3}{4}a, -2a, a\right)$

28. $y = -x^2 + 2x + 5$ **30.** $y = -2x^2 + 5x - 1$

32. $x^2 + y^2 - 6y = 0$ **34.** $x^2 + y^2 - 3x - 2y = 0$

36. $y = \frac{1}{2}x^2 + \frac{1}{2}x + 1$ **38.** $4000 at 5%
$8000 at 7%
$9500 at 8%

40. $625,000 at 8% **42.** 7500 units at $20
$50,000 at 9% 15,000 units at $10
$125,000 at 10% 10,000 units at $5

44. (a) $8\frac{1}{3}$ liters of 20%; $1\frac{2}{3}$ liters of 50%

(b) $6\frac{1}{4}$ liters of 10%; $3\frac{3}{4}$ liters of 50%

(c) 7 liters of 20%; 1 liter of 10%

46. One possible solution: **48.** $y = \frac{9}{28}x^2 + \frac{173}{100}x + \frac{33}{14}$
$C = -\frac{1}{2}a + 406{,}250$

$M = \frac{1}{2}a - 31{,}250$

$B = 125{,}000 - a$

$G = a$

50. $y = -\frac{5}{56}x^2 - \frac{607}{1400}x + \frac{939}{350}$

52. $y = 0.271x^2 + 0.177x + 27.329$; $38,163

Section 6.4 (page 477)

2. f **4.** b **6.** c

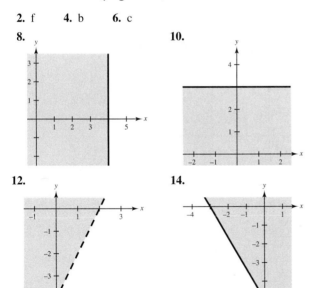

8.

10.

12.

14.

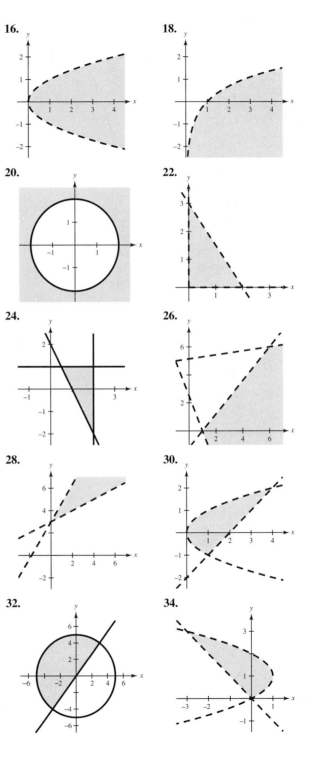

16.

18.

20.

22.

24.

26.

28.

30.

32.

34.

36.

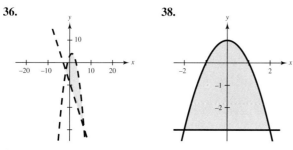

38.

40.

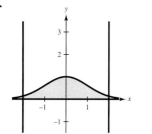

42. $4x - y \geq 0, 4x - y \leq 16, 0 \leq y \leq 4$

44. $y \leq x + 1, y \leq -x + 1, y \geq 0$

46. $x^2 + y^2 \leq 16, x \geq 0, y \geq x$

48. $x \geq 2y, 8x + 12y \leq 200, x \geq 4, y \geq 2$

50. (a) $x + y \geq 15,000, x \geq 8000, y \geq 4000,$
   $15x + 25y \geq 275,000$

(b)

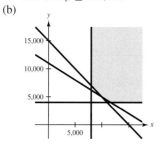

52. (a) $20x + 10y \geq 280, 15x + 10y \geq 160,$
   $10x + 20y \geq 180, x \geq 0, y \geq 0$

(b)

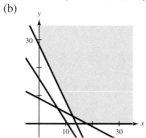

54. Consumer surplus: $\approx \$4,050,000$
   Producer surplus: $\approx \$16,200,000$

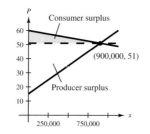

56. Consumer surplus: $\$6,250,000$
   Producer surplus: $\$15,625,000$

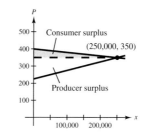

58. (a)

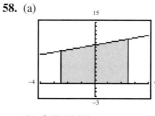

(b) $\$47.6$ billion

Section 6.5 *(page 487)*

2. Minimum at: $(0, 0)$
   Value: 0
   Maximum at: $(0, 4)$
   Value: 32

4. Minimum at: $(0, 0)$
   Value: 0
   Maximum at: $(2, 0)$
   Value: 14

6. Minimum at: $(0, 2)$
   Value: 6
   Maximum at: $(5, 3)$
   Value: 29

8. Minimum at: $(3, 0)$
   Value: 3
   Maximum at: $(0, 4)$
   Value: 24

10. Minimum at: $(0, 600)$
   Value: 21,000
   Maximum at: $(900, 0)$
   Value: 45,000

12. Minimum at: (675, 0) or (0, 600)
Value: 10,800
Maximum at: (0, 800) or (900, 0)
Value: 14,000

14. Minimum at: (0, 0)
Value: 0
Maximum at: (0, 8)
Value: 64

16. Minimum at: (0, 0)
Value: 0
Maximum at: (4, 0)
Value: 28

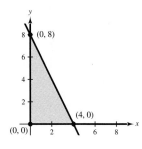

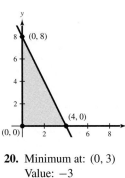

18. Minimum at: (0, 0)
Value: 0
Maximum at: (4, 1)
Value: 21

20. Minimum at: (0, 3)
Value: −3
Maximum at: (5, 0)
Value: 10

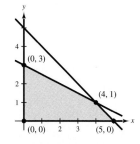

22. Minimum at: (0, 0) or (0, 20)
Value: 0
Maximum at: (12, 0)
Value: 12

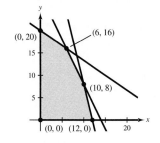

24. Minimum at: (0, 0) or (12, 0)
Value: 0
Maximum at: (0, 20)
Value: 20

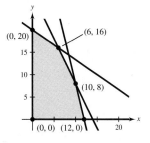

26. Maximum at: (5, 0)
Value: 25

28. Maximum at: (3, 6) or (5, 0)
Value: 15

30. Maximum at: (4, 4)
Value: 24

32. Maximum at: (7, 0)
Value: 28

34. Crop A: 60 acres
Crop B: 90 acres

36. $\frac{1}{6}A$ and $\frac{5}{6}B$

38. Model A: 1000 units
Model B: 500 units

40. Audits: 8
Tax returns: 8

42. First drink: $1666\frac{2}{3}$ liters
Second drink: $833\frac{1}{3}$ liters

44. Type A: $225,000
Type B: $225,000

46. The constraints do not form a closed set of points. Therefore, z is unbounded.

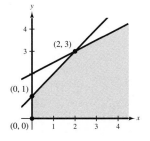

48. The solution region is empty.

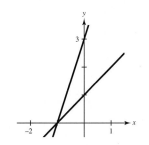

50. z is maximum at any point on the line segment connecting the vertices $(0, 2)$ and $\left(\frac{4}{3}, \frac{4}{3}\right)$.

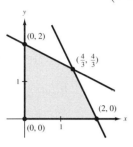

Review Exercises *(page 493)*

2. $(5, 2)$ **4.** $(3, 4)$ **6.** $(0, -3), (3, 0), (-2, -15)$

8. $(-3.99996, 0.01832), (1.328, 3.773)$

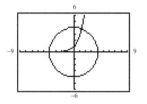

10. $1995(t \approx 5.084)$

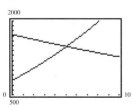

12. $\$500,000$ **14.** $\left(\frac{2}{3}, \frac{7}{5}\right)$ **16.** $(8, 9)$ **18.** $(10, 14)$

20. 400 of the $\$9.95$ tapes   **22.** $x = 250,000$ units
250 of the $\$14.95$ tapes $p = \$95.00$

24. $\left(\frac{38}{17}, \frac{40}{17}, -\frac{63}{17}\right)$ **26.** $(2, 3, -4)$

28. $\left(\frac{274}{93}, -\frac{125}{93}, \frac{359}{93}, -\frac{322}{93}\right)$ **30.** $y = 3x^2 + 11x - 20$

32. $x^2 + y^2 - 2x + 4y - 20 = 0$

34. 20,000 units at $\$20$ **36.** $y = 1.01x + 1.54$
40,000 units at $\$15$
50,000 units at $\$10$

38.

40.

42.

44. $x - y \geq -4$
$2x + y \leq 22$
$x - y \leq 2$
$2x + y \geq 7$

46. $x \geq 2y$
$200x + 300y \leq 4000$
$x \geq 4$
$y \geq 2$

48. Within the first year

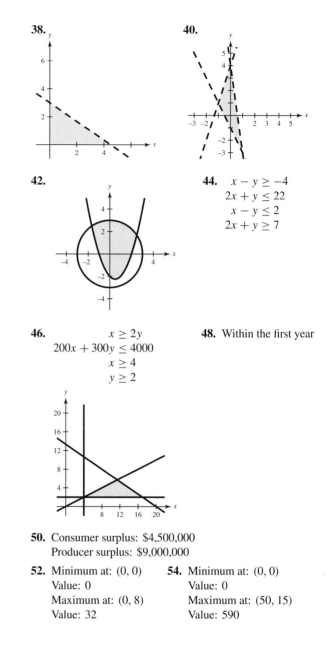

50. Consumer surplus: $\$4,500,000$
Producer surplus: $\$9,000,000$

52. Minimum at: $(0, 0)$ **54.** Minimum at: $(0, 0)$
Value: 0   Value: 0
Maximum at: $(0, 8)$ Maximum at: $(50, 15)$
Value: 32   Value: 590

56. Minimum at: $(0, 0)$
Value: 0
Maximum at: $(4, 3)$
Value: 48

58. Minimum at: $(0, 0)$
Value: 0
Maximum at: $(4, 5)$
Value: 47

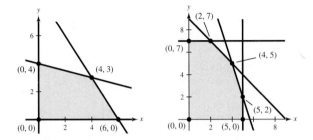

60. 280 of the \$1000 camcorders
None of the \$800 camcorders

62. ≈ 667 units of the basic model
≈ 333 units of the deluxe model

64. Three bags of Brand X
Two bags of Brand Y

66. Seven audits
Ten tax returns

CHAPTER 7

Section 7.1 *(page 510)*

2. 1×4 **4.** 5×4 **6.** 3×3

8. Row-echelon form **10.** Row-echelon form

12. (a) $\begin{bmatrix} 7 & 1 \\ 0 & 2 \\ -3 & 4 \\ 1 & 5 \end{bmatrix}$ (b) $\begin{bmatrix} 1 & 5 \\ 0 & 2 \\ -3 & 4 \\ 7 & 1 \end{bmatrix}$ (c) $\begin{bmatrix} 1 & 5 \\ 0 & 2 \\ 0 & 19 \\ 7 & 1 \end{bmatrix}$

(d) $\begin{bmatrix} 1 & 5 \\ 0 & 2 \\ 0 & 19 \\ 0 & -34 \end{bmatrix}$ (e) $\begin{bmatrix} 1 & 5 \\ 0 & 1 \\ 0 & 19 \\ 0 & -34 \end{bmatrix}$ (f) $\begin{bmatrix} 1 & 0 \\ 0 & 1 \\ 0 & 0 \\ 0 & 0 \end{bmatrix}$

14. $\begin{bmatrix} 1 & 2 & -1 & 3 \\ 0 & 1 & -2 & 5 \\ 0 & 0 & 1 & -1 \end{bmatrix}$

16. $\begin{bmatrix} 1 & -3 & 0 & -7 \\ 0 & 1 & 1 & 2 \\ 0 & 0 & 1 & -\frac{13}{2} \\ 0 & 0 & 0 & 0 \end{bmatrix}$

18. $\begin{bmatrix} 1 & 3 & 0 \\ 0 & 0 & 1 \\ 0 & 0 & 0 \end{bmatrix}$ **20.** $\begin{bmatrix} 1 & 0 \\ 0 & 1 \\ 0 & 0 \\ 0 & 0 \end{bmatrix}$

22. $9x - 4y = 0$
$6x + y = -4$

24. $5x + 8y + 2z \qquad = -1$
$-2x + 15y + 5z + w = 9$
$x + 6y - 7z \qquad = -3$

26. $x - y + 2z = 4$
$y - z = 2$
$z = -2$
$(8, 0, -2)$

28. $x + 2y - 2z \qquad = -1$
$y + z + 2w = 9$
$z \qquad = 2$
$w = -3$
$(-23, 13, 2, -3)$

30. $(-2, 4)$ **32.** $(3, -1, 0)$ **34.** $\begin{bmatrix} 8 & 3 & \vdots & 25 \\ 3 & -9 & \vdots & 12 \end{bmatrix}$

36. $\begin{bmatrix} 9 & -3 & 20 & 1 & \vdots & 13 \\ 12 & 0 & -8 & 0 & \vdots & 5 \\ 1 & 2 & 3 & -4 & \vdots & -2 \\ -1 & -1 & 1 & 1 & \vdots & 1 \end{bmatrix}$

38. $(-1, 3)$ **40.** Inconsistent **42.** $(0.2, 0.5)$

44. $(3a + 5, a)$ **46.** $(8, 10, 6)$ **48.** $(4, -3, 2)$

50. $(-3b + 96a + 100, b, 52a + 54, a)$ **52.** $(-5a, a, 3)$

54. $(-2b - 4a + 5, b, 3, a)$ **56.** $(-2a, a)$

58. $(-2a, -a, a, a)$ **60.** \$200,000 at 10%
\$550,000 at 10.5%
\$50,000 at 11.5%

62. $y = -x^2 + 2x + 10$ **64.** $y = 2x^3 - x + 20$

66. $y = \frac{1000}{33} + \frac{2861}{1650}t$, \$44,174.55 when $t = 8$

Section 7.2 *(page 524)*

2. $x = 13, y = 12$ **4.** $x = -4, y = 9$

6. (a) $\begin{bmatrix} -2 & 0 \\ 6 & 3 \end{bmatrix}$ (b) $\begin{bmatrix} 4 & 4 \\ -2 & -1 \end{bmatrix}$

(c) $\begin{bmatrix} 3 & 6 \\ 6 & 3 \end{bmatrix}$ (d) $\begin{bmatrix} 9 & 10 \\ -2 & -1 \end{bmatrix}$

8. (a) $\begin{bmatrix} 4 & -2 & 5 \\ -4 & 0 & 2 \end{bmatrix}$ (b) $\begin{bmatrix} 0 & 4 & -3 \\ 2 & -2 & 6 \end{bmatrix}$

(c) $\begin{bmatrix} 6 & 3 & 3 \\ -3 & -3 & 12 \end{bmatrix}$ (d) $\begin{bmatrix} 2 & 9 & -5 \\ 3 & -5 & 16 \end{bmatrix}$

10. (a) $\begin{bmatrix} -1 \\ 8 \\ 1 \end{bmatrix}$ (b) $\begin{bmatrix} 7 \\ -4 \\ -3 \end{bmatrix}$ (c) $\begin{bmatrix} 9 \\ 6 \\ -3 \end{bmatrix}$ (d) $\begin{bmatrix} 17 \\ -6 \\ -7 \end{bmatrix}$

12. (a) $\begin{bmatrix} -3 & 3 \\ 12 & -12 \end{bmatrix}$ (b) $\begin{bmatrix} 0 & 0 \\ 3 & -15 \end{bmatrix}$ (c) $\begin{bmatrix} 3 & -6 \\ 6 & -15 \end{bmatrix}$

14. (a) $\begin{bmatrix} 4 & 2 \\ -2 & 4 \end{bmatrix}$ (b) $\begin{bmatrix} 4 & 2 \\ -2 & 4 \end{bmatrix}$ (c) $\begin{bmatrix} 0 & -2 \\ 2 & 0 \end{bmatrix}$

16. (a) $[13]$

(b) $\begin{bmatrix} 6 & 4 & 2 & 0 \\ 9 & 6 & 3 & 0 \\ 3 & 2 & 1 & 0 \\ 0 & 0 & 0 & 0 \end{bmatrix}$

(c) Not possible

18. $\begin{bmatrix} 3 & -4 \\ 10 & 16 \\ 26 & 46 \end{bmatrix}$

20. $\begin{bmatrix} 3 & 0 & 0 \\ 0 & -4 & 0 \\ 0 & 0 & -10 \end{bmatrix}$ **22.** $\begin{bmatrix} 0 & 0 & 0 \\ 0 & 0 & 0 \\ 0 & 0 & 0 \end{bmatrix}$

24. Not possible **26.** $\begin{bmatrix} -2 & -\frac{5}{2} \\ 0 & 0 \\ 5 & -\frac{7}{2} \end{bmatrix}$ **28.** $\begin{bmatrix} 2 & -5 \\ -5 & 0 \\ 5 & 6 \end{bmatrix}$

30. $A = \begin{bmatrix} 2 & 3 \\ 1 & 4 \end{bmatrix}$, $X = \begin{bmatrix} x \\ y \end{bmatrix}$, $B = \begin{bmatrix} 5 \\ 10 \end{bmatrix}$,

$x = -2$, $y = 3$

32. $A = \begin{bmatrix} 2 & -4 & 1 \\ -1 & 3 & 1 \\ 1 & 1 & 0 \end{bmatrix}$, $X = \begin{bmatrix} x \\ y \end{bmatrix}$, $B = \begin{bmatrix} 0 \\ 1 \\ 3 \end{bmatrix}$,

$x = 2$, $y = 1$, $z = 0$

34. $A = \begin{bmatrix} 1 & 1 & -3 \\ -1 & 2 & 0 \\ 0 & -1 & 1 \end{bmatrix}$, $X = \begin{bmatrix} x \\ y \end{bmatrix}$, $B = \begin{bmatrix} -1 \\ 1 \\ 0 \end{bmatrix}$,

$x = 2a - 1$, $y = a$, $z = a$

36. $AB = \begin{bmatrix} 0 & 0 \\ 0 & 0 \end{bmatrix}$

Answers will vary. Other examples:

$C = \begin{bmatrix} -2 & -2 \\ 5 & 5 \end{bmatrix}$, $D = \begin{bmatrix} 1 & -1 \\ -1 & 1 \end{bmatrix}$, $CD = \begin{bmatrix} 0 & 0 \\ 0 & 0 \end{bmatrix}$

$E = \begin{bmatrix} 6 & 6 \\ -1 & -1 \end{bmatrix}$, $F = \begin{bmatrix} 1 & -1 \\ -1 & 1 \end{bmatrix}$, $EF = \begin{bmatrix} 0 & 0 \\ 0 & 0 \end{bmatrix}$

38. $\begin{bmatrix} 72 & 48 & 24 \\ 36 & 108 & 72 \end{bmatrix}$

40. (a) $18,300 (b) $21,260

	Wholesale	Retail	
(c)	$15,770	$18,300	1
	$26,500	$29,250	2 } Outlet
	$21,260	$24,150	3

This represents the wholesale and retail prices of the inventory at each outlet.

42. (a) $25 per hour

(b) $50.40 per hour

(c) $\begin{bmatrix} \$17.70 & \$15.00 \\ \$29.40 & \$25.00 \\ \$50.40 & 43.00 \end{bmatrix}$

This represents the labor cost for each boat size at each plant.

44. $\begin{bmatrix} -2 & 1 \\ 5 & -2 \end{bmatrix}$

Matrix is unique.

46. $P^3 = \begin{bmatrix} 0.3 & 0.175 & 0.175 \\ 0.308 & 0.433 & 0.217 \\ 0.392 & 0.392 & 0.608 \end{bmatrix}$

$P^4 = \begin{bmatrix} 0.25 & 0.1875 & 0.1875 \\ 0.3148 & 0.3773 & 0.2477 \\ 0.4352 & 0.4352 & 0.5648 \end{bmatrix}$

$P^5 \approx \begin{bmatrix} 0.225 & 0.1938 & 0.1938 \\ 0.3139 & 0.3451 & 0.2674 \\ 0.4611 & 0.4611 & 0.5389 \end{bmatrix}$

$P^6 \approx \begin{bmatrix} 0.2125 & 0.1969 & 0.1969 \\ 0.3108 & 0.3265 & 0.2798 \\ 0.4767 & 0.4767 & 0.5233 \end{bmatrix}$

$P^7 \approx \begin{bmatrix} 0.2063 & 0.1984 & 0.1984 \\ 0.3077 & 0.3156 & 0.2876 \\ 0.4860 & 0.4860 & 0.5140 \end{bmatrix}$

$P^8 \approx \begin{bmatrix} 0.2031 & 0.1992 & 0.1992 \\ 0.3053 & 0.3092 & 0.2924 \\ 0.4916 & 0.4916 & 0.5084 \end{bmatrix}$

Note: $p_{12} = p_{13}$ and $p_{31} = p_{32}$

As n increases, $P^n \to \begin{bmatrix} 0.2 & 0.2 & 0.2 \\ 0.3 & 0.3 & 0.3 \\ 0.5 & 0.5 & 0.5 \end{bmatrix}$

Section 7.3 *(page 534)*

2. Show **4.** Show **6.** Show

8. Show **10.** Show

12. $\begin{bmatrix} 7 & -2 \\ -3 & 1 \end{bmatrix}$ **14.** $\begin{bmatrix} -19 & -33 \\ -4 & -7 \end{bmatrix}$

16. $\begin{bmatrix} 0 & -1 \\ 1 & 11 \end{bmatrix}$ **18.** $\frac{1}{5}\begin{bmatrix} 4 & -3 \\ -1 & 2 \end{bmatrix}$ **20.** Does not exist

22. $\begin{bmatrix} -13 & 6 & 4 \\ 12 & -5 & -3 \\ -5 & 2 & 1 \end{bmatrix}$ **24.** $\begin{bmatrix} -10 & -4 & 27 \\ 2 & 1 & -5 \\ -13 & -5 & 35 \end{bmatrix}$

26. $\frac{1}{2}\begin{bmatrix} 2 & -2 & 0 \\ 14 & -17 & 2 \\ -16 & 20 & -2 \end{bmatrix}$ **28.** $\frac{1}{30}\begin{bmatrix} 15 & 0 & 0 \\ 0 & 10 & 0 \\ 0 & 0 & 6 \end{bmatrix}$

30. Does not exist. **32.** $\frac{1}{10}\begin{bmatrix} 10 & -15 & -40 & 26 \\ 0 & 5 & 10 & -8 \\ 0 & 0 & -5 & 1 \\ 0 & 0 & 0 & 2 \end{bmatrix}$

34. $\begin{bmatrix} 1 & 0 & 1 & 0 \\ 0 & 1 & 0 & 1 \\ 2 & 0 & 1 & 0 \\ 0 & 1 & 0 & 2 \end{bmatrix}$ **36.** $\begin{bmatrix} 27 & -10 & 4 & -29 \\ -16 & 5 & -2 & 18 \\ -17 & 4 & -2 & 20 \\ -7 & 2 & -1 & 8 \end{bmatrix}$

38. $(-8, -11)$ **40.** $(-7, -7)$ **42.** $\left(-\frac{9}{5}, \frac{6}{5}\right)$

44. $(2, -1)$ **46.** $(-3, -24, 28)$

48. $(-32, -13, -37, 15)$

50. (a) $y = \frac{7}{30}t^2 + \frac{1}{10}t + \frac{1639}{60}$

(b) 31.45 million workers

52.
$$\begin{aligned} x \quad\quad + z &= 7 \\ x + y + z &= 4 \\ 2x - y \quad\quad &= 12 \end{aligned}$$
54. $10,000 at 8%
$10,000 at 6%
$20,000 at 7%

56. $0 at 8%
$14,000 at 6%
$28,000 at 7%
58. 20 computer chips
15 resistors
10 transistors

60. 90 computer chips
60 resistors
20 transistors
62. 150 muffins
100 cookies
200 brownies

Section 7.4 *(page 546)*

2. -8 **4.** -11 **6.** 14 **8.** 0 **10.** 0 **12.** 3

14. -0.022 **16.** -20 **18.** -2 **20.** -5

22. -16 **24.** -48 **26.** (a) $M_{11} = 2$ (b) $C_{11} = 2$
$M_{12} = -3$ $\quad C_{12} = 3$
$M_{21} = 0$ $\quad C_{21} = 0$
$M_{22} = 11$ $\quad C_{22} = 11$

28.
$M_{11} = 36$	$C_{11} = 36$
$M_{12} = -42$	$C_{12} = 42$
$M_{13} = 85$	$C_{13} = 85$
$M_{21} = -82$	$C_{21} = 82$
$M_{22} = -12$	$C_{22} = -12$
$M_{23} = -68$	$C_{23} = 68$
$M_{31} = 24$	$C_{31} = 24$
$M_{32} = -28$	$C_{32} = 28$
$M_{33} = -51$	$C_{33} = -51$

30. 151 **32.** 650 **34.** -1167 **36.** 2

38. -66 **40.** -168 **42.** 0 **44.** -100

46. Matrices will vary. The product of the main diagonals equals 32.

48. Row 2 is all zeros.

Section 7.5 *(page 555)*

2. $\frac{33}{2}$ **4.** $\frac{31}{2}$ **6.** 55 **8.** $\frac{25}{2}$ **10.** $\frac{23}{2}$

12. $x = 3$ or $x = 19$ **14.** 3100 square feet

16. Not collinear **18.** Collinear

20. Not collinear **22.** $x + y = 0$

24. $7x - 6y - 28 = 0$ **26.** $3x - 2y + 6 = 0$ **28.** $x = 3$

30. [8 5] [12 16] [0 9] [19 0] [15 14] [0 20] [8 5] [0 23] [1 25]

$-21, 29, -40, 52, -9, 9, -38, 57, -44, 59, -20, 20,$
$-21, 29, -23, 23, -27, 28$

32. [16 12 5] [1 19 5] [0 19 5] [14 4 0] [13 15 14] [5 25 0]

$43, 6, 9, -38, -45, -13, -42, -47, -14, 44, 16, 10,$
$49, 9, 12, -55, -65, -20$

34. $16, 35, 42, 0, -26, -65, -16, -74, -137, -3, -44,$
$-101, 28, 31, 1, 95, 215, 265$

36. $58, 122, 139, 1, -37, -95, 40, 67, 55, 23, 17, -19, 47,$
$88, 88, 65, 140, 164$

38. TO BE OR NOT TO BE **40.** CLASS IS CANCELLED

42. RETURN AT DAWN **44.** CANCEL ORDERS SUE

Review Exercises *(page 560)*

2. $\begin{bmatrix} 1 & 2 & -1 & 0 \\ 0 & 1 & 1 & 4 \\ 0 & 0 & 1 & \frac{35}{13} \end{bmatrix}$ **4.** $\begin{bmatrix} 1 & 0 & 0 & 0 \\ 0 & 1 & 0 & 0 \\ 0 & 0 & 1 & 0 \\ 0 & 0 & 0 & 1 \end{bmatrix}$

6. $\begin{aligned} x + \quad\ 2z &= 4 \\ y + 3z &= 9 \\ 2x - 3y - \ z &= -6 \end{aligned}$ **8.** $(-2, 5)$ **10.** $(-1, 3, -2)$

12. $(20 + 32a, -4 - 10a, a)$ **14.** No solution

16. \$400,000 at 9.5%
\$200,000 at 10.5%
\$100,000 at 11%

18. (a) $\begin{bmatrix} 3 & 0 & 3 \\ 2 & -1 & 11 \\ 3 & 0 & 0 \end{bmatrix}$ (b) $\begin{bmatrix} -1 & 0 & 1 \\ -4 & 7 & -1 \\ 1 & -4 & 6 \end{bmatrix}$

(c) $\begin{bmatrix} 4 & 0 & 8 \\ -4 & 12 & 20 \\ 8 & -8 & 12 \end{bmatrix}$ (d) $\begin{bmatrix} -2 & 0 & 5 \\ -13 & 24 & 2 \\ 5 & -14 & 21 \end{bmatrix}$

20. (a) $\begin{bmatrix} 2 \\ 2 \\ 8 \end{bmatrix}$ (b) $\begin{bmatrix} 4 \\ -6 \\ -2 \end{bmatrix}$ (c) $\begin{bmatrix} 12 \\ -8 \\ 12 \end{bmatrix}$ (d) $\begin{bmatrix} 15 \\ -20 \\ -3 \end{bmatrix}$

22. (a) $\begin{bmatrix} -8 & 3 \\ 6 & -5 \end{bmatrix}$ (b) $\begin{bmatrix} -2 & 3 \\ 0 & -11 \end{bmatrix}$

(c) $\begin{bmatrix} 8 & -21 \\ 28 & 13 \end{bmatrix}$

24. (a) $\begin{bmatrix} 16 & 15 & 6 \\ -1 & -2 & 4 \\ 9 & 10 & 6 \end{bmatrix}$ **26.** Not possible

(b) $\begin{bmatrix} 5 & 9 & 7 \\ 3 & 1 & 6 \\ 5 & 9 & 14 \end{bmatrix}$

(c) $\begin{bmatrix} 5 & 4 & 3 \\ -1 & 3 & 3 \\ 3 & 2 & 4 \end{bmatrix}$

28. $\begin{bmatrix} 6 & 0 & -3 \\ 4 & 0 & -2 \\ 8 & 0 & -4 \\ 12 & 0 & -6 \end{bmatrix}$ **30.** $\begin{bmatrix} 0 & 0 & 2 & 2 \\ 3 & 4 & 6 & 7 \\ 4 & 2 & 2 & 2 \end{bmatrix}$

$\begin{bmatrix} 2 & -1 \\ 3 & 5 \\ 3 & 21 \end{bmatrix}$ **34.** $\begin{bmatrix} 27 & 72 & 63 & 18 \\ 45 & 90 & 81 & 81 \\ 54 & 63 & 72 & 90 \end{bmatrix}$

36. (a) \$31.80   (b) \$66.00

(c) $\begin{bmatrix} & A & B \\ & 20.40 & 17.60 \\ & 31.80 & 27.40 \\ & 76.50 & 66.00 \end{bmatrix} \begin{matrix} \\ \text{Basic} \\ \text{Light} \\ \text{Ultra-light} \end{matrix}$

This represents the labor cost for each model of racing bicycle at each plant.

38. BA is the 4×4 identity matrix and so is AB.

40. $\frac{1}{10} \begin{bmatrix} -3 & 1 \\ 4 & 2 \end{bmatrix}$ **42.** $\frac{1}{4} \begin{bmatrix} 2 & -2 & -2 \\ 1 & 1 & -3 \\ -2 & 2 & 6 \end{bmatrix}$

44. $\begin{aligned} x &= -2 \\ y &= 3 \end{aligned}$ **46.** $\begin{aligned} x &= -2 \\ y &= 4 \\ z &= 5 \end{aligned}$ **48.** 180 of model A
120 of model B
40 of model C

50. \$10,000 at 8%   **52.** 12   **54.** -21 **56.** 26
\$8000 at 6%
\$16,000 at 7%

58. -16 **60.** -12 **62.** Not a square matrix

64. 42   **66.** Not collinear   **68.** $x + 6y - 10 = 0$

70. $[13\ \ 5\ \ 5]\ [20\ \ 0\ \ 13]\ [5\ \ 0\ \ 9]\ [14\ \ 0\ \ 9]\ [20\ \ 0\ \ 12]$
$[15\ \ 21\ \ 9]\ [19\ \ 0\ \ 0]$
$23, 41, 36, 7, -12, -51, -4, -26, -53, -5, -48, -105,$
$8, -8, -44, 69, 141, 156, 19, 38, 38$

72. KEEP IT SUPER SIMPLE

CHAPTER 8

Section 8.1 *(page 574)*

2. $1, 5, 9, 13, 17$ **4.** $\frac{1}{2}, \frac{1}{4}, \frac{1}{8}, \frac{1}{16}, \frac{1}{32}$

6. $-\frac{1}{2}, \frac{1}{4}, -\frac{1}{8}, \frac{1}{16}, -\frac{1}{32}$ **8.** $2, \frac{3}{2}, \frac{4}{3}, \frac{5}{4}, \frac{6}{5}$

10. $0, 1, 0, \frac{1}{2}, 0$ **12.** $\frac{3}{4}, \frac{9}{16}, \frac{27}{64}, \frac{81}{256}, \frac{243}{1024}$

14. $1, 1, 2, 6, 24$ **16.** $0, \frac{1}{2}, \frac{8}{11}, \frac{5}{6}, \frac{8}{9}$

18. $-\frac{1}{2}, \frac{2}{3}, -\frac{3}{4}, \frac{4}{5}, -\frac{5}{6}$ **20.** $1, \frac{1}{2^{3/2}}, \frac{1}{3^{3/2}}, \frac{1}{8}, \frac{1}{5^{3/2}}$

22. $2, \frac{14}{9}, \frac{28}{19}, \frac{16}{11}, \frac{74}{51}$ **24.** 210   **26.** 600

28. $(n + 2)(n + 1), 2, 6$ **30.** $(2n + 2)(2n + 1), 2, 12$

32. $a_n = 4n - 1$ **34.** $a_n = \dfrac{1}{n^2}$ **36.** $a_n = \dfrac{n + 1}{2n - 1}$

38. $a_n = \dfrac{2^{n-1}}{3^n}$ **40.** $a_n = 1 + \dfrac{2^n - 1}{2^n}$

42. $a_n = (-1)^{n+1}(2n)$ **44.** $a_n = \dfrac{2^{n-1}}{(n-1)!}$

46. 57 **48.** 42 **50.** 20 **52.** 165

54. $\frac{47}{60}$ **56.** 14 **58.** 10

60. $\displaystyle\sum_{i=1}^{15} \dfrac{5}{1+i}$ **62.** $\displaystyle\sum_{k=1}^{6} \left[1 - \left(\dfrac{k}{6}\right)^2\right]$

64. $\displaystyle\sum_{n=0}^{7} \left(-\dfrac{1}{2}\right)^n$ **66.** $\displaystyle\sum_{k=1}^{10} \dfrac{1}{k(k+2)}$ **68.** $\displaystyle\sum_{k=1}^{6} \dfrac{k!}{2^k}$

70. (a) $A_1 = \$101.00$ (b) $A_{60} = \$8248.64$
$A_2 = \$203.01$ (c) $A_{240} \doteq \$99,914.79$
$A_3 = \$306.04$
$A_4 = \$410.10$
$A_5 = \$515.20$
$A_6 = \$621.35$

72. \$35,516.4 million **74.** \$62,230.6 million

Section 8.2 (*page 583*)

2. Arithmetic sequence, $d = -2$

4. Arithmetic sequence, $d = -\frac{1}{2}$

6. Arithmetic sequence, $d = 4$

8. Not an arithmetic sequence

10. Not an arithmetic sequence

12. 2, 8, 24, 64, 60; not an arithmetic sequence

14. 1, 5, 9, 13, 17; arithmetic sequence, $d = 4$

16. 1, 2, 4, 8, 16; not an arithmetic sequence

18. $-1, 1, -1, 1, -1$; not an arithmetic sequence

20. $a_n = 4n + 11$ **22.** $a_n = \frac{2}{3}(1 - n)$

24. $a_n = 5n - 9$ **26.** $a_n = 11.5n - 27.5$

28. $a_n = 10n - 7$ **30.** $5, \frac{17}{4}, \frac{7}{2}, \frac{11}{4}, 2$

32. 6, 18, 30, 42, 54 **34.** 8, 13, 18, 23, 28

36. 1, 6, 11, 16, 21 **38.** 1850 **40.** 23

42. 20.5 **44.** 16,100 **46.** 10,100 **48.** 26,425

50. 218,625 **52.** 2725 **54.** −898.375

56. 1220 **58.** (a) \$35,550 (b) \$187,050

60. 2379 **62.** \$2179.89 million **64.** 490 meters

66. 470 bricks **68.** \$340,000

Section 8.3 (*page 592*)

2. Geometric sequence, $r = 4$, $a_n = 3(4)^{n-1}$

4. Geometric sequence, $r = -2$, $a_n = (-2)^{n-1}$

6. Geometric sequence, $r = 0.2$, $a_n = 5(0.2)^{n-1}$

8. Geometric sequence, $r = -\frac{2}{3}$, $a_n = 9\left(-\frac{2}{3}\right)^{n-1}$

10. Not a geometric sequence **12.** 6, 12, 24, 48, 96, . . .

14. $1, \frac{1}{3}, \frac{1}{9}, \frac{1}{27}, \frac{1}{81}, \ldots$ **16.** $1, -\frac{1}{4}, \frac{1}{16}, -\frac{1}{64}, \frac{1}{256}, \ldots$

18. $2, 2e^{0.1}, 2e^{0.2}, 2e^{0.3}, 2e^{0.4}, \ldots$ **20.** $5\left(\frac{3}{2}\right)^7$

22. $1\left(-\frac{2}{3}\right)^7$ **24.** $8(\sqrt{5})^8 = 5000$ **26.** $1000(1.005)^{59}$

28. $\dfrac{3}{\frac{1}{4}} = 12$ **30.** $12\left(\dfrac{2}{3}\right)^6$ **32.** 16,380

34. $\dfrac{147,620}{19,683} \approx 7.50$ **36.** 171 **38.** ≈ 592.647

40. 6 **42.** $\frac{6}{5}$ **44.** $\frac{10}{9}$ **46.** $\approx \$3414.47$

48. $\approx \$400,289.64$; $\approx \$401,930.21$

50. $\approx \$4,488.91$ million **52.** $\approx \$6967.08$ million

54. The problem with this gambling strategy is that you may lose all your money before you have a chance to win. You would need

$$\sum_{i=1}^{12} 2i = \$8190$$

to cover the possibility of losing 12 times in a row.

56. ≈ 129.7228 square inches

Section 8.4 (*page 602*)

2. 28 **4.** 1 **6.** 792 **8.** 210 **10.** 210

12. $x^6 + 6x^5 + 15x^4 + 20x^3 + 15x^2 + 6x + 1$

14. $x^4 + 12x^3 + 54x^2 + 108x + 81$

16. $x^5 - 10x^4 + 40x^3 - 80x^2 + 80x - 32$

18. $x^6 + 6x^5 y + 15x^4 y^2 + 20x^3 y^3 + 15x^2 y^4 + 6xy^5 + y^6$

20. $s^5 + 15s^4 + 90s^3 + 270s^2 + 405s + 243$

22. $625 + 1000y + 600y^2 + 160y^3 + 16y^4$

24. $32x^5 - 80x^4 y + 80x^3 y^2 - 40x^2 y^3 + 10xy^4 - y^5$

26. $\dfrac{x^3}{8} - \dfrac{9x^2 y}{4} + \dfrac{27xy^2}{2} - 27y^3$

28. $x^{12} + 6x^{10}y^2 + 15x^8y^4 + 20x^6y^6 + 15x^4y^8$
$\quad + 6x^2y^{10} + y^{12}$

30. $\dfrac{1}{x^6} + \dfrac{12y}{x^5} + \dfrac{60y^2}{x^4} + \dfrac{160y^3}{x^3} + \dfrac{240y^4}{x^2} + \dfrac{192y^5}{x} + 64y^6$

32. $-38 - 41i$ **34.** $-10 + 198i$ **36.** $184 - 440\sqrt{3}\,i$

38. $x^5 - 20x^4 + 160x^3 - 640x^2 + 1280x - 1024$

40. $243y^5 + 810y^4 + 1080y^3 + 720y^2 + 240y + 32$

42. $3{,}247{,}695x^8$ **44.** $720x^2y^8$

46. $8{,}660{,}520x^3y^8$ **48.** $729x^{10}$

50. $\dfrac{1}{4^{10}} + \dfrac{30}{4^{10}} + \dfrac{405}{4^{10}} + \dfrac{3240}{4^{10}} + \dfrac{17{,}010}{4^{10}} + \dfrac{61{,}236}{4^{10}}$
$\quad + \dfrac{153{,}090}{4^{10}} + \dfrac{262{,}440}{4^{10}} + \dfrac{295{,}245}{4^{10}}$
$\quad + \dfrac{196{,}830}{4^{10}} + \dfrac{59{,}049}{4^{10}}$

52. $5.31 \times 10^{-7} + 1.49 \times 10^{-5} + 1.91 \times 10^{-4} + 1.49$
$\quad \times 10^{-3} + 7.80 \times 10^{-3} + 2.91 \times 10^{-2} + 7.92$
$\quad \times 10^{-2} + 1.58 \times 10^{-1} + 2.31 \times 10^{-1} + 2.40$
$\quad \times 10^{-1} + 1.68 \times 10^{-1} + 7.12 \times 10^{-2} + 1.38 \times 10^{-2}$

54. $1.84 \times 10^{-3} + 2.05 \times 10^{-2} + 9.51 \times 10^{-2} + 2.35$
$\quad \times 10^{-1} + 32.8 \times 10^{-1} + 2.44 \times 10^{-1} + 7.54 \times 10^{-2}$

56. The determinants are 1, 3, 6, 10, 15, . . . , which constitute one of the diagonals of Pascal's Triangle.

Section 8.5   *(page 611)*

2. 24   **4.** 30   **6.** 5,760,000   **8.** 1024

10. (a) 9000   (b) 4536   (c) 1800   **12.** 125,000

14. (a) 120   (b) 12   (c) 12   **16.** 120   **18.** 380

20. 100   **22.** 90   **24.** 840   **26.** 720   **28.** 24

30. 66   **32.** 536,878,650   **34.** 24,040,016

36. (a) 126   (b) 108   (c) 486

38. (a) 70   (b) 16   **40.** 40   **42.** 9

44. 35   **46.** 56   **48.** 7560   **50.** 34,650

52.

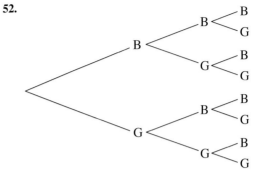

There are eight possible arrangements.

Section 8.6   *(page 623)*

2. $\frac{1}{2}$ **4.** $\frac{1}{2}$ **6.** $\frac{11}{12}$ **8.** $\frac{1}{9}$ **10.** $\frac{19}{36}$

12. $\frac{10}{13}$ **14.** $\frac{6}{13}$ **16.** $\frac{1}{15}$ **18.** $\frac{11}{15}$ **20.** 0.64

22. 0.16   **24.** Moore: 0.25; Jenkins: 0.25; Taylor: 0.50

26. (a) $\frac{112}{209}$ (b) $\frac{97}{209}$ (c) $\frac{574}{627}$

28. (a) $\frac{1}{4}$ (b) $\frac{1}{2}$ (c) $\frac{841}{1600}$ (d) $\frac{1}{40}$ **30.** $\frac{21}{1292}$

32. (a) $\frac{1}{16}$ (b) $\frac{3}{52}$ **34.** (a) $\frac{1}{120}$ (b) $\frac{1}{24}$

36. (a) $\frac{3}{8}$ (b) $\frac{19}{30}$ **38.** $\frac{1}{64{,}974}$

40. (a) $\frac{14}{55}$ (b) $\frac{12}{55}$ (c) $\frac{54}{55}$

42. (a) 0.81   (b) 0.01   (c) 0.99

44. (a) $\frac{1}{1024}$ (b) $\frac{243}{1024}$ (c) $\frac{781}{1024}$

46. (a) $\frac{1}{16}$ (b) $\frac{1}{8}$ (c) $\frac{15}{16}$

48. 0.0256   **50.** 0.474552

Review Exercises   *(page 629)*

2. $-1, 1, 3, 5, 7$ **4.** $4, \frac{9}{2}, \frac{14}{3}, \frac{19}{4}, \frac{24}{5}$

6. $-\frac{1}{2}, \frac{4}{5}, -\frac{13}{14}, \frac{40}{41}, -\frac{121}{122}$ **8.** 462

10. $a_n = 1 + \dfrac{n}{n+1}$ **12.** $a_n = (-1)^{n+1}3^n$

14. $a_n = (-2)^n$ **16.** 42   **18.** 31   **20.** $\displaystyle\sum_{k=1}^{60} \frac{1}{k}$

22. $\displaystyle\sum_{k=1}^{7} 2k^2$ **24.** (a) $A_1 = \$20,100.00$
$A_2 = \$20,200.50$
$A_3 \approx \$20,301.50$
$A_4 \approx \$20,403.01$
$A_5 \approx \$20,505.03$
$A_6 \approx \$20,607.55$
$A_7 \approx \$20,710.59$
$A_8 \approx \$20,814.14$
$A_9 \approx \$20,918.21$
$A_{10} \approx \$20,022.80$
(b) $A_{120} \approx \$36,387.93$

26. Not arithmetic (geometric, $r = 3$)

28. Arithmetic, $d = 0.8$

30. $-3, 2, 7, 12, 17$; arithmetic, $d = 5$ **32.** $31 - 2n$

34. $-1, 1, 3, 5, 7$ **36.** 600   **38.** 5090

40. 1070 seats   **42.** Not geometric (arithmetic, $d = 6$)

44. Geometric, $r = -\frac{1}{3}$ **46.** $1, \frac{3}{4}, \frac{9}{16}, \frac{27}{64}, \frac{81}{256}$

48. $2, 2e, 2e^2, 2e^3, 2e^4$ **50.** $a_5 = \frac{1}{100}$

52. $a_{50} = 10(1.05)^{49}$ **54.** ≈ 5.9988 **56.** $\frac{20}{9}$

58. $\approx \$20,587.34, \approx \$20,604.40$

60. $\approx \$10,439.99$ million   **62.** 252   **64.** 125,970

66. $x^7 - 7x^6 + 21x^5 - 35x^4 + 35x^3 - 21x^2 + 7x - 1$

68. $256x^8 + 3072x^7 + 16,128x^6 + 48,384x^5 + 90,720x^4$
$+ \ 108,864x^3 + 81,648x^2 + 34,992x + 6561$

70. $x^{10} + 20x^8 + 160x^6 + 640x^4 + 1280x^2 + 1024$

72. $243x^5 + 1620x^4y + 4320x^3y^2 + 5760x^2y^3$
$+ \ 3840xy^4 + 1024y^5$

74. $-6,300,000x^4y^5$

76. $\approx 1.02 \times 10^{-7} + 4096 \times 10^{-6} + 7.3728 \times 10^{-5}$
$+ \ 7.86432 \times 10^{-4} + 5.505024 \times 10^{-3}$
$+ \ 2.6424115 \times 10^{-2} + 8.8080384 \times 10^{-2}$
$+ \ 2.01326592 \times 10^{-1} + 3.01989888 \times 10^{-1}$
$+ \ 2.68435456 \times 10^{-1} + 1.07374182 \times 10^{-1}$

78. 60   **80.** 4096

82. (a) 9000   (b) 4536   (c) 4500

84. 120   **86.** 336   **88.** 15,504

90. (a) 6188   (b) 136   (c) 15,504   **92.** $\frac{3}{13}$

94. $\frac{1}{6}$ **96.** 0.25   **98.** (a) $\frac{1}{11}$ (b) $\frac{8}{33}$ (c) $\frac{1}{3}$

100. (a) $\frac{11}{30}$ (b) $\frac{91}{144}$ **102.** (a) $\frac{1}{32}$ (b) $\frac{1}{16}$ (c) $\frac{31}{32}$

104. (a) $\approx 77\%$ (b) $\approx 23\%$

DISCOVERY

Section P.3 *(page 24)*

1000, 100, 10, 1, 0.1, 0.01.
The decimal point is moved one place to the left each time. If $10^1 = 10$, then $10^0 = 1$ and $10^{-1} = 0.1$.

Section 1.4 *(page 116)*

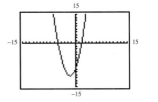

The graph crosses the x-axis two times and there are two solutions in Example 2.
Tracing: $x \approx 1.854$ and $x \approx -4.854$
Decimal approximations: $x \approx 1.854$ and $x \approx -4.854$
The solutions of a quadratic equation are the same as the x-values of its graph at the x-axis.
Yes

Section 2.4 *(page 202)*

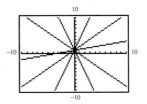

The line falls from left to right. The y-intercept of each graph is 1.

Section 2.4 *(page 203)*

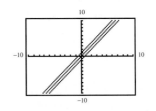

The lines are parallel.

Section 3.1   *(page 234)*

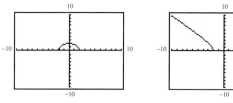

Domain: $-2 \leq x \leq 2$ Domain: $x \leq -2$ and $x \geq 2$
                                    Yes, $x = -2$ and $x = 2$.

Section 3.3   *(page 257)*

$f(x)$ is narrower than $h(x)$.
$g(x)$ is wider than $h(x)$.
Answers will vary.

Section 4.2   *(page 335)*

The x-value at the point of intersection of the x-axis is the real zero of the function.
Four times, four real zeros

Section 4.4   *(page 350)*

$i, -1, -i, 1, i, -1, -i, 1$
Divide the power by 4. A remainder of 1 is i, a remainder of 2 is -1, a remainder of 3 is $-i$, and a remainder of 0 is 1.

Section 5.1   *(page 377)*

(1) \$5466.09   (2) \$5466.35   (3) \$5466.36   (4) \$5466.38
Arguments will vary.

Section 6.1   *(page 435)*

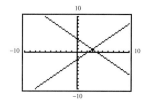

$x \approx 2.98$, $y \approx 1.02$
The coordinates are approximately the same. The coordinates are approximate because of the number of pixels on the screen and the setting of the viewing window. The coordinates of the point become $(3, 1)$.

Section 7.3   *(page 532)*

ERROR. Because $ad - bc = 0$, the matrix is not invertible.

Section 8.6   *(page 621)*

$P(2 \text{ out of } 23) = 0.5073$
$P(2 \text{ out of } 50) = 0.9704$

TECHNOLOGY

Section P.2   *(page 12)*
No answer needed.

Section 1.2   *(page 94)*
\$649

Section 1.4   *(page 117)*
No answer needed.

Section 1.4   *(page 118)*
No answer needed.

Section 1.4   *(page 119)*
1.9 seconds

Section 1.7   *(page 151)*
No answer needed.

Section 2.4   *(page 199)*
No answer needed.

Section 2.4   *(page 204)*

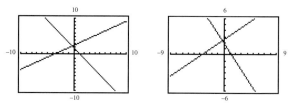

No

Section 2.5   *(page 214)*
No answer needed.

Section 3.2 *(page 246)*

No answer needed.

Section 3.3 *(page 255)*

(a) $g(x)$ is shifted four units to the right.
   $h(x)$ is shifted four units to the right and three units up.

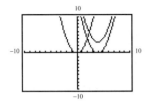

(b) $g(x)$ is shifted one unit to the left.
   $h(x)$ is shifted one unit to the left and two units down.

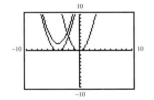

(c) $g(x)$ is shifted four units to the left.
   $h(x)$ is shifted four units to the left and two units up.

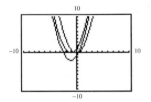

Section 3.3 *(page 256)*

No answer needed.

Section 3.4 *(page 271)*

No answer needed.

Section 3.6 *(page 295)*

No answer needed.

Section 4.1 *(page 324)*

Yes

Section 4.3 *(page 342)*

No answer needed.

Section 4.5 *(page 359)*

Only the real zeros appear on the graph.

Section 5.1 *(page 374)*

No answer needed.

Section 5.4 *(page 404)*

No answer needed.

Section 6.2 *(page 449)*

No answer needed.

Section 6.4 *(page 474)*

(a)

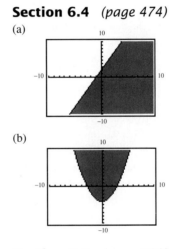

(b)

Section 7.1 *(page 502)*

No answer needed.

Section 7.2 *(page 517)*

No answer needed.

Section 7.4 *(page 540)*

An ERROR message appears.

Section 8.1 *(page 567)*

No answer needed.

Section 8.3 *(page 587)*

No answer needed.

Section 8.5 *(page 609)*

$_8P_5 = 6720$

GROUP ACTIVITIES

Section P.1 *(page 8)*

Equivalent results are given by $-|a|$ and $-|-a|$.

Section P.2 *(page 19)*

(a) $0.167 + 0.667 + 0.167 = 1.001$
(b) $0.167 + 0.333 + 0.333 + 0.167 = 1.000$

For set (b), there seems to be no rounding error when the decimals are added because two fractions are rounded up and two are not, thus "canceling out" the roundoff error.

Section P.3 *(page 29)*

Combination (d) would produce the highest balance ($6651.82). For a fixed interest rate, the balance increases as the number of compounding periods increases.

Section P.4 *(page 40)*

(a) $\bar{x} = 50, \sigma = 2.61$ (b) $\bar{x} = 50, \sigma = 15.81$
(c) $\bar{x} = 10, \sigma = 7.35$ (d) $\bar{x} = 10, \sigma = 1.29$

Parts (a) and (b) have the same mean but different standard deviations. The same is true of parts (c) and (d). In each case, the set of data whose values are spread farther away from the mean has the larger standard deviation.

Section P.5 *(page 50)*

$$(u + v)(u - v) = u^2 - uv + uv - v^2 = u^2 - v^2$$
$$(u + v)^2 = (u + v)(u + v)$$
$$= u^2 + uv + uv + v^2 = u^2 + 2uv + v^2$$
$$(u - v)^2 = (u - v)(u - v)$$
$$= u^2 - uv - uv + v^2 = u^2 - 2uv + v^2$$
$$(u + v)^3 = (u + v)(u + v)^2$$
$$= (u + v)\left(u^2 + 2uv + v^2\right)$$
$$= u^3 + 2u^2v + uv^2 + u^2v + 2uv^2 + v^3$$
$$= u^3 + 3u^2v + 3uv^2 + v^3$$
$$(u - v)^3 = (u - v)(u - v)^2$$
$$= (u - v)\left(u^2 - 2uv + v^2\right)$$
$$= u^3 - 2u^2v + uv^2 - u^2v + 2uv^2 - v^3$$
$$= u^3 - 3u^2v + 3uv^2 - v^3$$

Verbal descriptions may vary. One example is "The product of the sum and difference of two terms is the difference of each term squared."

Section P.6 *(page 58)*

Box 1: $(a - b)a^2$
Box 2: $(a - b)ab$
Box 3: $(a - b)b^2$

The sum of the volumes of boxes 1, 2, and 3 equals the volume of the large cube minus the volume of the small cube, which is the difference of two cubes.

Section P.7 *(page 69)*

Prob(engraved side up) $= \frac{1}{6}$
Prob(blank side up) $= \frac{5}{6}$

The proportion of experiments that resulted in the engraved side facing up should be close to $\frac{1}{6}$.

Section 1.1 *(page 85)*

Answers will vary.

Section 1.2 *(page 96)*

(a) Red herring: how long it takes the diver to ascend
(b) Red herrings: the amount of markup, and the difference between the price and the competitor's price

Section 1.3 *(page 109)*

Examples include (6, 8, 10), (5, 12, 13), and (9, 12, 15).

Section 1.4 *(page 119)*

(a) a shorter time because the force due to gravity is greater
(b) $t \approx 1.9$ seconds

Section 1.5 *(page 132)*

The error occurs in line 3, where values are plugged into the Quadratic Formula directly from line 2 rather than from the equation in standard form. So $a = 3, b = -7$, and $c = -4$, and the only allowable solution is

$$x = \frac{7 + \sqrt{97}}{6}.$$

Section 1.6 *(page 143)*

Let a, b, c, and d be real numbers.

1. *Transitive Property:* If a is less than b and b is less than c, then a is less than c.
2. *Addition of Inequalities:* If a is less than b and c is less than d, then the sum of a and c is less than the sum of b and d.
3. *Addition of a Constant:* If a is less than b, then the sum of a and c is less than the sum of b and c.
4. *Multiplying by a Constant:*
   (i) For $c \geq 0$, if a is less than b, then the product of a and c is less than the product of b and c.
   (ii) For $c \leq 0$, if a is less than b, then the product of a and c is greater than the product of b and c.

Section 1.7 *(page 154)*

For planning purposes, it would be beneficial for a business owner to solve $p < 0$ in order to find the levels of production that would *not* lead to profit.

Section 2.1 *(page 173)*

The graph on the right is misleading: the scale on the vertical axis makes it appear that the rise in profits is dramatic, but the total increase is only 2.4 units.

Section 2.2 *(page 183)*

(a) the y-intercept gives the original purchase price of the car.
(b) the x-intercept gives the price per hour at which no families will demand child care services.
Examples in which it makes sense for there to be no x-intercept will vary.

Section 2.3 *(page 192)*

Each equation must first be solved for y. Additionally, equation (c) must be split into two equations. The standard viewing window shows all of the x-intercepts for equations (b) and (c).

Section 2.4 *(page 204)*

(a) Slope $= \frac{8}{15} = 0.5\overline{3}$
(b) Every 15 days, the high temperature increases by $8°F$ (or the high temperature rises $0.5\overline{3}\,°F$ every day).
(c) Equation: $y = \frac{8}{15}x + 60$
   Day 30: $y = 76°$, day 60: $y = 92°$, day 105: $y = 116°$, day 180: $y = 156°$

No, not all of these predictions are reasonable, because they increase continuously, whereas we know that temperatures in most places very cyclically with the seasons.

Section 2.5 *(page 215)*

Answers will vary.

Section 3.1 *(page 237)*

Examples will vary.

Section 3.2 *(page 249)*

Examples will vary.

Section 3.3 *(page 262)*

(a) $f(x) = (3 - x^3) + 4 = 7 - x^3$
(b) $f(x) = 3 - (x + 2)^3$
(c) $f(x) = [3 - (x - 1)^3] - 2 = 1 - (x - 1)^3$
(d) $f(x) = -3 - x^3$

Section 3.4 *(page 273)*

1. Student confused inverse function notation with an exponent of -1. Correct solution:

$$x = \sqrt{2y - 5}$$
$$x^2 = 2y - 5, \, x \geq 0$$
$$x^2 + 5 = 2y, \, x \geq 0$$
$$f^{-1}(x) = \frac{x^2 + 5}{2}, \, x \geq 0$$

2. Student erroneously combined several techniques to find an inverse. Correct solution:

$$x = \frac{3}{5}y + \frac{1}{3}$$
$$x - \frac{1}{3} = \frac{2}{5}y$$
$$\frac{5}{3}x - \frac{5}{9} = y = f^{-1}(x)$$

Section 3.5 *(page 284)*

(a) x-intercepts: none
    y-intercept: 4
    vertex: $(1.50, 1.75)$

(b) x-intercepts:
    5.63 and 11.37
    y-intercept: 64
    vertex: $(8.50, -8.25)$

(c) x-intercepts: 0 and -0.55
    y-intercept: 0
    vertex: $(-0.28, -0.58)$

(d) x-intercepts:
    1.54 and 8.46
    y-intercept: -6.5
    vertex: $(5, 6)$

(e) x-intercepts:
    -0.04 and -82.29
    y-intercept: 0.19
    vertex: $(-41.17, -89.00)$

(f) x-intercepts: none
    y-intercept: 2
    vertex: $(0.18, 1.55)$

Section 3.6 *(page 297)*

(a) $x^3 - 3x^2 - 10x + 24$
(b) $x^3 - \frac{37}{12}x^2 - \frac{1}{4}x + \frac{3}{2}$
(c) $x^4 + 12x^3 - 54x^2 + 108x + 81$
(d) $x^4 + 2x^3 - 13x^2 - 14x + 24$

Section 3.7 *(page 307)*

The per capita land area is

$$\frac{2316.013}{2.314x + 179.759}$$

and it drops below 8 acres between the years 2007 and 2008.

Section 4.1 *(page 327)*

x	$-10{,}000$	-1000	-100
$f(x)$	-9997.0006	-997.006	-97.06
y	-9997	-997	-97

x	100	1000	$10{,}000$
$f(x)$	103.06	1003.006	$10{,}003.0006$
y	103	1003	$10{,}003$

$g(x) = \dfrac{2x^3 - 5}{x^2} = 2x - \dfrac{5}{x^2}$; slant asymptote: $y = 2x$

x	$-10{,}000$	-1000	-100
$g(x)$	$-20{,}000$	-2000.000005	-200.0005
y	$-20{,}000$	-2000	-200

x	100	1000	$10{,}000$
$g(x)$	199.9995	1999.999995	$20{,}000$
y	200	2000	$20{,}000$

Section 4.2 *(page 336)*

(a) Yes; $x^2 - 2 = 0$ (b) Yes (c) No; yes   (d) No

Section 4.3 *(page 344)*

Examples will vary.

Section 4.4 *(page 355)*

1. Student incorrectly simplified $(-2i)(2i)$ in the denominator. Correct solution:

$$\frac{5}{3 - 2i} \cdot \frac{3 + 2i}{3 + 2i} = \frac{15 + 10i}{9 - 4i^2} = \frac{15 + 10i}{9 + 4} = \frac{15}{13} + \frac{10}{13}i$$

2. Student incorrectly simplified $i\sqrt{-4}$ and $\sqrt{-4}\sqrt{-3}$. Correct solution:

$$\left(\sqrt{-4} + 3\right)\left(i - \sqrt{-3}\right)$$
$$- i\sqrt{-4} - \sqrt{-4}\sqrt{-3} + 3i - 3\sqrt{-3}$$
$$= -2 + 2\sqrt{3} + 3i - 3i\sqrt{3}$$
$$= -2 + 2\sqrt{3} + 3i\left(1 - \sqrt{3}\right)$$

Section 4.5 *(page 363)*

Examples will vary.

Section 5.1 *(page 379)*

(a) Exponential growth, $f(n) = \left(\frac{1}{2}\right)^n$
(b) Linear growth, $f(n) = 4n$
(c) Linear growth, $f(n) = \frac{2}{3}n$
(d) Exponential growth, $f(n) = 5^n$
Other examples of linear and exponential growth will vary.

Section 5.2 *(page 390)*

$h(x)$ is $f(x)$ shifted upward five units.
$j(x)$ is $f(x)$ shifted to the left five units.
$k(x)$ is $f(x)$ shifted to the right three units.
$l(x)$ is $f(x)$ shifted downward four units.
The constant c is responsible for distorting the graph as either a vertical stretch or a vertical shrink.

Section 5.3 *(page 397)*

$\ln y = \frac{3}{2}\ln x$
$\ln y = \ln x^{3/2}$
$\quad y = x^{3/2}$

Section 5.4 *(page 409)*

x	$\frac{1}{2}$	1	2
e^x	1.6487	2.7183	7.3891
$\ln(e^x)$	0.5	1.0	2.0
$\ln x$	−0.6931	0	0.6931
$e^{\ln x}$	0.5	1.0	2.0

x	10	25	50
e^x	22,026.4658	7.2005×10^{10}	5.1847×10^{21}
$\ln(e^x)$	10	25	50
$\ln x$	2.3026	3.2189	3.9120
$e^{\ln x}$	10	25	50

From the table, you can conclude that $\ln(e^x) = x$ and $e^{\ln x} = x$.

Section 5.5 *(page 419)*

Examples will vary.

Section 6.1 *(page 440)*

Agency B cost equation: $y = 100 + 0.25x$

Point of intersection: $(167.08, 141.77)$. This is the point at which the trucks from the two agencies will cost the same. If you plan to drive the truck for less than 167 miles, it is less expensive to choose agency A. If you plan to drive for more than 167 miles, choose agency B.

Section 6.2 *(page 451)*

(a) Infinitely many solutions: $k = \frac{3}{4}$

No solutions: not possible. For all values of k except $k = \frac{3}{4}$, the lines will intersect at the same point, $(-2, 0)$.

(b) Infinitely many solutions: $k = 3$

No solutions: any value of $k \neq 3$. Because the lines have the same slope, a different value of k gives the second line a different y-intercept, making them parallel lines with no intersection.

Section 6.3 *(page 463)*

A quadratic model is appropriate: $y = 2.5x^2 - 15x + 71.5$. Average temperature in both December and February is $51.5°$F.

Section 6.4 *(page 476)*

(a)　　　　　　　　　(b)

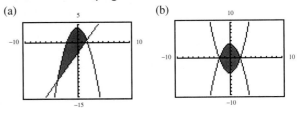

Section 6.5 *(page 486)*

(a) Unbounded region
(b) No region defined by the intersections

Section 7.1 *(page 509)*

Student did not apply row operations to the augmented portion of the matrix. Correct solution:

$$\begin{bmatrix} 1 & 1 & \vdots & 4 \\ 2 & 3 & \vdots & 5 \end{bmatrix} \xrightarrow[-2R_1 + R_2 \to]{} \begin{bmatrix} 1 & 1 & \vdots & 4 \\ 0 & 1 & \vdots & -3 \end{bmatrix}$$

$$\xrightarrow[-1R_2 + R_1 \to]{} \begin{bmatrix} 1 & 0 & \vdots & 7 \\ 0 & 1 & \vdots & 3 \end{bmatrix}$$

Section 7.2 *(page 523)*

(a) Undefined　　　　　　(b) Undefined
(c) Defined; order is 3×2　　(d) Undefined

Examples will vary. One possibility for $AB \neq BA$ is

$$A = \begin{bmatrix} 1 & 3 \\ 2 & 4 \end{bmatrix}; \quad B = \begin{bmatrix} 5 & 1 \\ 3 & 2 \end{bmatrix}.$$

One possibility for $AB = BA$ is

$$A = \begin{bmatrix} 1 & 0 \\ 1 & 2 \end{bmatrix}; \quad B = \begin{bmatrix} 0 & 0 \\ 2 & 2 \end{bmatrix}.$$

Section 7.3 *(page 533)*

$$A^{-1} \approx \begin{bmatrix} -0.1538 & 0.0769 \\ 0.2692 & 0.1154 \end{bmatrix}$$

$$B^{-1} = \begin{bmatrix} 11 & -6 & 2 \\ 3 & -2 & 1 \\ 1 & -1 & 1 \end{bmatrix}$$

$$C^{-1} = \begin{bmatrix} 0.375 & -0.25 & 0.125 \\ -0.125 & 0.75 & -0.375 \\ -0.25 & 0.5 & 0.25 \end{bmatrix}$$

Section 7.4 (page 545)

Answers will vary.

Section 7.5 (page 554)

13	15	14	4	1	25	0	9	19	0
M	O	N	D	A	Y		I	S	

3	15	14	6	9	18	13	5	4	0	0
C	O	N	F	I	R	M	E	D		

Section 8.1 (page 573)

(a) 820   (b) 1,743,392,200   (c) 374.65

Section 8.2 (page 582)

(a) $-7, -4, -1, 2, 5, 8, 11$;
   $a_{n+1} = a_n + 3$

(b) $17, 23, 29, 35, 41, 47, 53, 59, 65, 71$;
   $a_{n+1} = a_n + 6$

(c) Not possible

(d) $4, 7.5, 11, 14.5, 18, 21.5, 25, 28.5,$
   $32, 35.5, 39$; $a_{n+1} = a_n + 3.5$

(e) Not possible

Section 8.3 (page 591)

Sequences will vary. One possibility (when starting with an 8-foot piece of string) is: $96, 48, 24, 12, 6, 3, 1\frac{1}{2}, \frac{3}{4}, \frac{3}{8}, \frac{3}{16}, \frac{3}{32}$ (and the sequence can go no further).

Formula: $a_n = 96 \left(\frac{1}{2}\right)^{n-1}$.

You could theoretically make an infinite number of cuts.

Section 8.4 (page 601)

$A_n = 2^n$, $A_{10} = 2^{10} = 1024$

Section 8.5 (page 610)

$10 \cdot 8 \cdot {}_{13}C_2 = 6240$ menus

Section 8.6 (page 622)

Answers will vary.

Section A.1 (page A13)

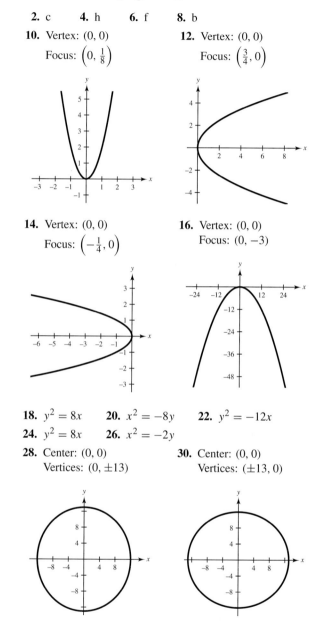

2. c   **4.** h   **6.** f   **8.** b

10. Vertex: $(0, 0)$
   Focus: $\left(0, \frac{1}{8}\right)$

12. Vertex: $(0, 0)$
   Focus: $\left(\frac{3}{4}, 0\right)$

14. Vertex: $(0, 0)$
   Focus: $\left(-\frac{1}{4}, 0\right)$

16. Vertex: $(0, 0)$
   Focus: $(0, -3)$

18. $y^2 = 8x$ **20.** $x^2 = -8y$ **22.** $y^2 = -12x$

24. $y^2 = 8x$ **26.** $x^2 = -2y$

28. Center: $(0, 0)$
   Vertices: $(0, \pm 13)$

30. Center: $(0, 0)$
   Vertices: $(\pm 13, 0)$

32. Center: $(0, 0)$
Vertices: $(0, \pm 8)$

34. Center: $(0, 0)$
Vertices: $(\pm 2, 0)$

62. $3\sqrt{19} \approx 13$ feet

64. $y^2 = \dfrac{b^4}{a^2} \rightarrow 2y = \dfrac{2b^2}{a}$

66.

68.

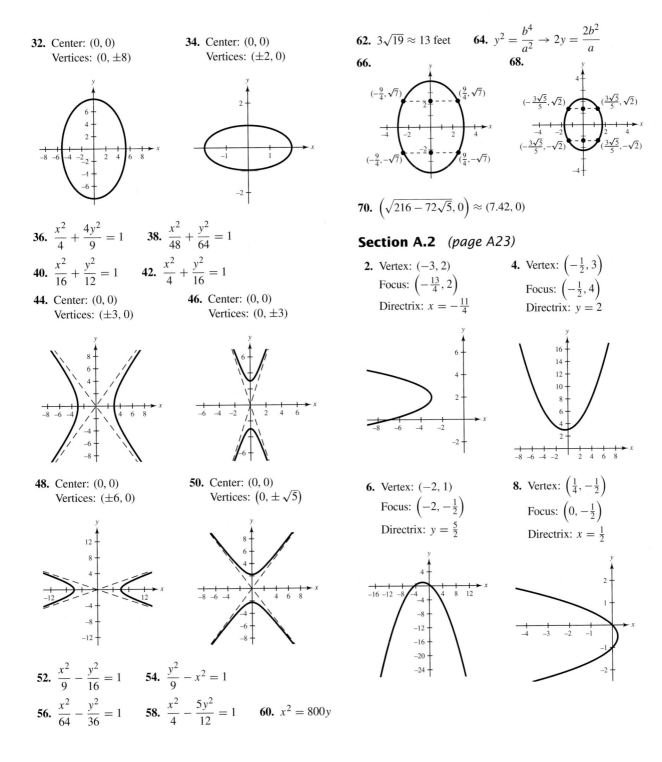

70. $\left(\sqrt{216 - 72\sqrt{5}}, 0\right) \approx (7.42, 0)$

36. $\dfrac{x^2}{4} + \dfrac{4y^2}{9} = 1$ **38.** $\dfrac{x^2}{48} + \dfrac{y^2}{64} = 1$

40. $\dfrac{x^2}{16} + \dfrac{y^2}{12} = 1$ **42.** $\dfrac{x^2}{4} + \dfrac{y^2}{16} = 1$

44. Center: $(0, 0)$
Vertices: $(\pm 3, 0)$

46. Center: $(0, 0)$
Vertices: $(0, \pm 3)$

Section A.2   *(page A23)*

2. Vertex: $(-3, 2)$
Focus: $\left(-\dfrac{13}{4}, 2\right)$
Directrix: $x = -\dfrac{11}{4}$

4. Vertex: $\left(-\dfrac{1}{2}, 3\right)$
Focus: $\left(-\dfrac{1}{2}, 4\right)$
Directrix: $y = 2$

48. Center: $(0, 0)$
Vertices: $(\pm 6, 0)$

50. Center: $(0, 0)$
Vertices: $\left(0, \pm \sqrt{5}\right)$

6. Vertex: $(-2, 1)$
Focus: $\left(-2, -\dfrac{1}{2}\right)$
Directrix: $y = \dfrac{5}{2}$

8. Vertex: $\left(\dfrac{1}{4}, -\dfrac{1}{2}\right)$
Focus: $\left(0, -\dfrac{1}{2}\right)$
Directrix: $x = \dfrac{1}{2}$

52. $\dfrac{x^2}{9} - \dfrac{y^2}{16} = 1$ **54.** $\dfrac{y^2}{9} - x^2 = 1$

56. $\dfrac{x^2}{64} - \dfrac{y^2}{36} = 1$ **58.** $\dfrac{x^2}{4} - \dfrac{5y^2}{12} = 1$ **60.** $x^2 = 800y$

10. Vertex: $(1, -1)$
Focus: $(1, -3)$
Directrix: $y = 1$

12. Vertex: $(-1, 0)$
Focus: $(0, 0)$
Directrix: $x = -2$

26. The graph of this equation is the point $(2, 1)$.

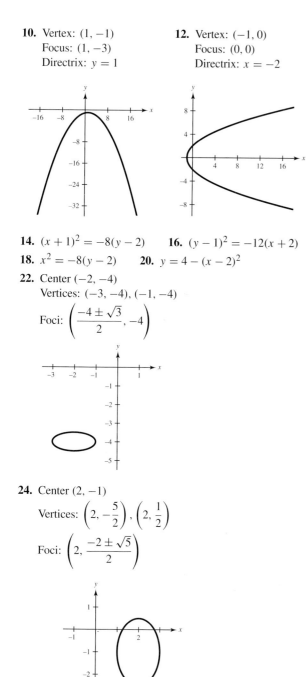

14. $(x + 1)^2 = -8(y - 2)$ **16.** $(y - 1)^2 = -12(x + 2)$

18. $x^2 = -8(y - 2)$ **20.** $y = 4 - (x - 2)^2$

22. Center $(-2, -4)$
Vertices: $(-3, -4), (-1, -4)$
Foci: $\left(\dfrac{-4 \pm \sqrt{3}}{2}, -4 \right)$

24. Center $(2, -1)$
Vertices: $\left(2, -\dfrac{5}{2} \right), \left(2, \dfrac{1}{2} \right)$
Foci: $\left(2, \dfrac{-2 \pm \sqrt{5}}{2} \right)$

28. Center $\left(-\dfrac{2}{3}, 2 \right)$
Vertices: $\left(-\dfrac{2}{3}, 1 \right), \left(-\dfrac{2}{3}, 3 \right)$
Foci: $\left(-\dfrac{2}{3}, \dfrac{4 \pm \sqrt{3}}{2} \right)$

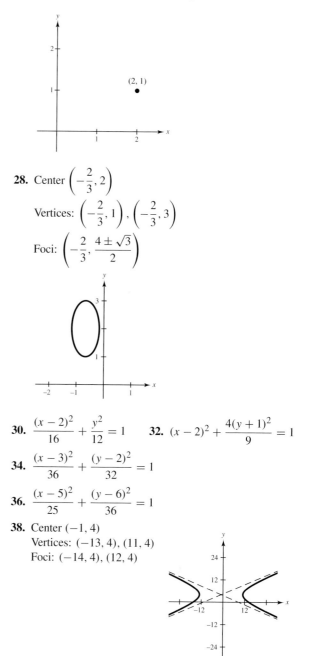

30. $\dfrac{(x - 2)^2}{16} + \dfrac{y^2}{12} = 1$ **32.** $(x - 2)^2 + \dfrac{4(y + 1)^2}{9} = 1$

34. $\dfrac{(x - 3)^2}{36} + \dfrac{(y - 2)^2}{32} = 1$

36. $\dfrac{(x - 5)^2}{25} + \dfrac{(y - 6)^2}{36} = 1$

38. Center $(-1, 4)$
Vertices: $(-13, 4), (11, 4)$
Foci: $(-14, 4), (12, 4)$

40. Center $(-3, 1)$,

Vertices: $\left(-3, \dfrac{1}{2}\right), \left(-3, \dfrac{3}{2}\right)$

Foci: $\left(-3, 1 \pm \dfrac{\sqrt{13}}{6}\right)$

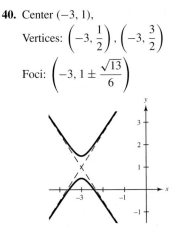

42. Center $(0, 2)$

Vertices: $(-6, 2), (6, 2)$

Foci: $\left(\pm 2\sqrt{10}, 2\right)$

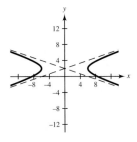

44. The graph of this equation is two lines intersecting at $(1, -2)$.

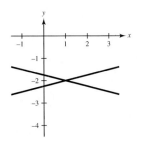

46. Center $(-3, 5)$

Vertices: $\left(-\dfrac{10}{3}, 5\right), \left(-\dfrac{8}{3}, 5\right)$

Foci: $\left(-3 \pm \dfrac{\sqrt{10}}{3}, 5\right)$

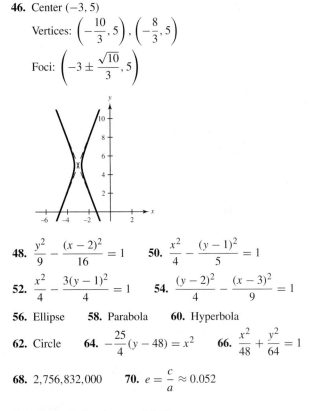

48. $\dfrac{y^2}{9} - \dfrac{(x-2)^2}{16} = 1$ **50.** $\dfrac{x^2}{4} - \dfrac{(y-1)^2}{5} = 1$

52. $\dfrac{x^2}{4} - \dfrac{3(y-1)^2}{4} = 1$ **54.** $\dfrac{(y-2)^2}{4} - \dfrac{(x-3)^2}{9} = 1$

56. Ellipse   **58.** Parabola   **60.** Hyperbola

62. Circle   **64.** $-\dfrac{25}{4}(y-48) = x^2$ **66.** $\dfrac{x^2}{48} + \dfrac{y^2}{64} = 1$

68. $2{,}756{,}832{,}000$ **70.** $e = \dfrac{c}{a} \approx 0.052$

Section B.1   *(page A36)*

2.

Leaves (Exam # 2)	Stems	Leaves (Exam #1)
9	5	
8 8 8 8 7 7 6 6 4 3	6	
8 8 8 7 7 6 4 3 3 1 1 0 0	7	0 5 5 5 7 7 8 8 8
7 4 1 0 0	8	1 1 1 1 2 3 4 5 5 5 5
		7 8 9 9 9
0	9	0 2 8
	10	0 0

By comparing the leaves you see that the scores on Exam 1 were higher.

4.

Stems	Leaves
4	1.2 2.5
5	3.7 5.2
6	6.0 6.7 8.2 8.6 8.6 9.6
7	1.3 5.9 6.5 6.5 9.7
8	4.8 5.6 9.4
9	2.3 2.9
10	3.5 6.3 7.9
11	0.0 1.5 4.9 5.9
12	0.0 4.9
13	
14	2.8

6.

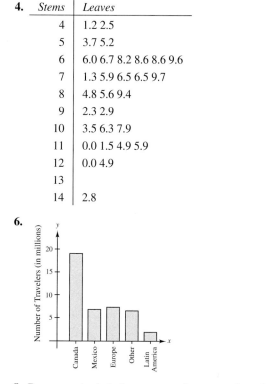

8. Because y tends to increase as x increases, the points are positively correlated.

10. No, the model is not accurate for large values of x.

12. Because v tends to decrease as h increases, the points are negatively correlated.

14. Because the model yields $v = 834$ feet per second when $h = 70$, we see that the model is inaccurate for large values of h.

16. (a)

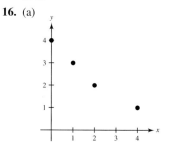

(b) $y = -x + 4$ fits the data well, and the sum of the squared differences is $0^2 + 0^2 + 0^2 + 1^2 = 1$.

(c) The least squares regression line is $y = -\frac{26}{35}x + \frac{19}{5}$. The sum of the squared differences is ≈ 0.17143.

18. $y = 0.65x + 2.6$

20. $y = -1.3143x + 11.2667$

22. $y = 2.2286x + 43.0857$

24. $y = -6.0512x + 148.0349$

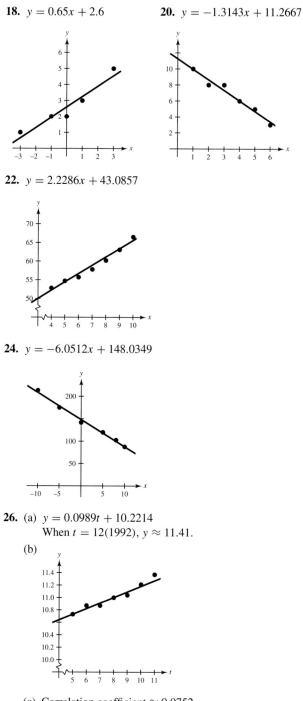

26. (a) $y = 0.0989t + 10.2214$
When $t = 12(1992)$, $y \approx 11.41$.

(b)

(c) Correlation coefficient ≈ 0.9752

Section B.2   *(page A45)*

2. Mean: 33.86; median: 33; mode: 32

4. Mean: 32.43; median: 33; mode: 32

6. Mean: 34; median: 33; mode: 32

8. (a) Mean: 14.86; median: 14: mode: 13
  Each is increased by 6.
  (b) Each will increase by k.

10. Mean: 320; median: 320; mode: 320

12. (a) Average number of hits $= 1$
  (b) Batting average $= 0.250$

14. One possibility: $\{4, 4, 6, 7.5, 8.5\}$

16. Median and mode give most representative descriptions.

18. $\bar{x} = 7, v = 22, \sigma \approx 4.69$ **20.** $\bar{x} = 2, v = 0, \sigma = 0$

22. $\bar{x} = 3, v = 4, \sigma = 2$

24. $\bar{x} = 1.1, v = 0.38, \sigma \approx 0.616$ **26.** 13.53

28. 2.19   **30.** 1.92   **32.** All numbers must be equal.

34. $C < D < A < B$

36. $\bar{x} \approx 40.7177, v \approx 46.74, \sigma \approx 6.837$
  86.7% lie within two standard deviations.

Formulas from Geometry

Triangle

$$h = a \sin \theta$$

$$\text{Area} = \frac{1}{2}bh$$

Law of Cosines:

$$c^2 = a^2 + b^2 - 2ab \cos \theta$$

Right Triangle

Pythagorean Theorem:

$$c^2 = a^2 + b^2$$

Equilateral Triangle

$$h = \frac{\sqrt{3}s}{2}$$

$$\text{Area} = \frac{\sqrt{3}s^2}{4}$$

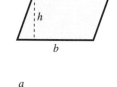

Parallelogram

$$\text{Area} = bh$$

Trapezoid

$$\text{Area} = \frac{h}{2}(a + b)$$

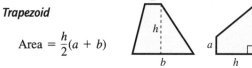

Circle

$$\text{Area} = \pi r^2$$
$$\text{Circumference} = 2\pi r$$

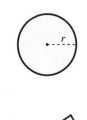

Sector of Circle

$$\text{Area} = \frac{\theta r^2}{2}$$

$$s = r\theta$$

(θ in radians)

Circular Ring

(p = average radius,
w = width of ring)

$$\text{Area} = \pi(R^2 - r^2)$$
$$= 2\pi pw$$

Ellipse

$$\text{Area} = \pi ab$$

$$\text{Circumference} \approx 2\pi \sqrt{\frac{a^2 + b^2}{2}}$$

Cone

(A = area of base)

$$\text{Volume} = \frac{Ah}{3}$$

Right Circular Cone

$$\text{Volume} = \frac{\pi r^2 h}{3}$$

$$\text{Lateral Surface Area} = \pi r\sqrt{r^2 + h^2}$$

Frustrum of Right Circular Cone

$$\text{Volume} = \frac{\pi(r^2 + rR + R^2)h}{3}$$

$$\text{Lateral Surface Area} = \pi s(R + r)$$

Right Circular Cylinder

$$\text{Volume} = \pi r^2 h$$
$$\text{Lateral Surface Area} = 2\pi rh$$

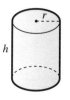

Sphere

$$\text{Volume} = \frac{4}{3}\pi r^3$$

$$\text{Surface Area} = 4\pi r^2$$